The Archaeology, Ethnohistory, and Environment of the Marismas Nacionales

The Archaeology, Ethnohistory, and Environment of the Marismas Nacionales

The Prehistoric Pacific Littoral of Sinaloa and Nayarit, Mexico

Edited by
Michael S. Foster

The University of Utah Press
Salt Lake City

The Defiance House Man colophon is a registered trademark of The University of Utah Press. It is based on a 4-ft-tall Ancient Puebloan pictograph (late PIII) near Glen Canyon, Utah.

LIBRARY OF CONGRESS CATALOGING-IN-PUBLICATION DATA

Names: Foster, Michael S. (Michael Stewart), 1947– editor.
Title: The archaeology, ethnohistory, and environment of the Marismas Nacionales : the prehistoric Pacific littoral of Sinaloa and Nayarit, Mexico / edited by Michael S. Foster.
Description: Salt Lake City : The University of Utah Press, [2017] | Includes bibliographical references and index.
Identifiers: LCCN 2017019090| ISBN 9781607815617 (cloth : alk. paper) | ISBN 9781607815624 (ebook)
Subjects: LCSH: Sinaloa (Mexico : State)—Antiquities. | Nayarit (Mexico)—Antiquities. | Mexico—Antiquities. | Indians of Mexico—Mexico—Sinaloa (State)—Antiquities. | Indians of Mexico—Mexico—Nayarit—Antiquities. | Ethnohistory—Mexico—Sinaloa (State) | Ethnohistory—Mexico—Nayarit.
Classification: LCC F1219.1.S56 A73 2017 | DDC 972/.32—dc23
LC record available at https://lccn.loc.gov/2017019090

Printed and bound in the United States of America.

To those who have gone before.

Contents

Figures

Tables

Publisher's Note

Detailed descriptions of the burial remains of 248 individuals discussed in this work can be found in a free digital supplement available on the book's homepage at the publisher's website: www.UofUpress.com.

Acknowledgments

I begin with thanks to Dr. Stuart D. Scott for the opportunity to have participated in the Marismas Nacionales project. I served as teaching assistant and field director during the 1973 and 1974 field seasons. My master's thesis was on the excavation of Mound C at Venadillo, and data from the project were the subject of many of my graduate seminar papers. The two summers in the Marismas, as well as my early endeavors in Mexican archaeology, were times of great adventure and great archaeology, and despite the intervening 40 years, much of which was spent in the often mind-numbing and petty world of cultural resource management, those times are well remembered.

The impetus for this volume was the result of the assistance of Dr. Ben Nelson and the generosity of an anonymous patron who, through Ben, provided financial support to revive the write-up of the Marismas Nacionales project. This funding allowed the authors of this volume to gather and begin anew their efforts to present the results of the project; it also provided support for the production of graphics for the volume and writing time. Without this support, the preparation and publication of this volume might never have been undertaken and completed.

The chapters in this volume, although generally authored by a single individual, were for the most part the result of a collaborative effort among the authors. My organizational and editorial oversight of this volume and its contents was greatly enhanced by an array of assistance and editorial collaboration with Stuart and Pat Scott. Additionally, a number of individuals provided a range of critical and valued assistance. Very appreciative thanks are given to Dr. J. Andrew Darling (Southwest Heritage Research) for taking time to review most of the chapters presented here. His archaeological and editorial insights added substantively to the chapters he reviewed, and his general support and encouragement are much appreciated. I also thank Chris North for his review of the sites description chapter, a tedious undertaking at best. We thank Dr. James Watson, Arizona State Museum, for his review of the volume at the request of the University of Utah Press and for his helpful comments. I would also like to thank Rebecca Rauch, acquisitions editor for the University of Utah Press, for her interest in the project and support for this volume's publication. Thanks are also owed to Elisabeth A. Graves for her adept copyediting of this volume's manuscript.

Lynn Simon served as the cartographer for the volume. His CAD skills are responsible for most of the maps and profiles in this volume, and he provided assistance well beyond that for which he was compensated. Dr. Arleyn Simon was of considerable assistance. She was frequently pestered with requests to track down digital copies of dissertations, theses, and articles, all of which she provided without hesitation. Additionally, a number of other colleagues, American and Mexican, provided copies of articles and reports or shared data. Their help and professional courtesies are greatly appreciated. In particular, thanks go to Dr. Michael A. Ohnersorgen, Arqglo. Mauricio Garduño Ambriz (Instituto Nacional de Antropología e Historia Centro Nayarit), and Dr. Karen Adams.

Additional funding aiding the publication of this volume was provided by Ann E. Foster, Dr. Michael S. Foster, Patricia K. Scott, Dr. Stuart D. Scott, and Dr. George W. Gill. Also providing funds in support of the publication of this volume was Dr. Arleyn Simon. Her generosity, enthusiasm, and moral support for this project in general are much appreciated.

As I am recently retired and this may well be my final professional publication, I take this opportunity to thank my wife, Ann, for all she has done for me throughout my career and in my efforts to see this volume to an end. Additionally, she hosted the get-together at the beginning of this endeavor, making sure that the authors had nothing to do but concentrate on the subject at hand. Her unwavering support, patience (especially patience), and generosity are deeply appreciated. This volume would not have been completed without her selflessness and friendship, and any success I have had in my career as an archaeologist is in no small part due to her.

—Michael S. Foster

In the eight-year course of fieldwork and research in the Marismas Nacionales many institutions were involved through their project participants, as both students and professional researchers. The project, however, originated at the State University of New York (SUNY) at Buffalo. This volume therefore owes its existence, in considerable measure, to the various departmental and central university administrators who helped to launch the enterprise and subsequently helped in the managerial aspects of the multiyear program.

I am especially happy to express my appreciation to James Blackhurst, director of the Division for Summer Sessions at SUNY/Buffalo. When I first became interested in the idea of the Marismas as a research venue, his enthusiasm took hold. Through his consistent support and his view of the inseparability of teaching and research, he remained the primary administrative mainstay throughout the project's history.

Other SUNY administrators helped me not only to undertake field explorations but also to report the studies in professional publication. Among them are Robert H. Rossberg, vice president for academic affairs; Ross D. MacKinnon, dean of the Faculty of Social Sciences; Albert L. Michaels, director of the Council on International Studies; Mrs. Shirley Ahrens, assistant director of Summer Sessions; and several successive Anthropology Department chairmen: William W. Stein, Marvin A. Opler, and Erwin H. Johnson. For an extended loan of Chametla and Culiacán ceramic type collections, I am indebted as well to the Lowie Museum, University of California at Berkeley.

The list of American advocates and colleagues has its Mexican counterpart. Administratively, the Marismas project could never have been undertaken without the sanction of Mexico's National Institute of Anthropology and History. My initial conversations with Ignacio Bernal and José Luis Lorenzo led to the granting of Federal Concession No. 4/67 in 1967 by the Division for Prehispanic Monuments. The concession was renewed in subsequent years under the divisional directorships of Ignacio Marquina and Eduardo Matos Moctezuma. For many stimulating conversations and shared ideas on Mexican archaeology, I thank Hector Gálvez, Jose Arturo Oliveros, and Otto Schöndube Baumbach. I owe special gratitude as well to Hector Alonzo E., chief of photo interpretation for President Echeverría's Comisión de Estudios del Territorio Nacional. Hector kindly placed a helicopter at our disposal for the aerial survey of several remote locations, but even more importantly he provided the project with accurate prepublication draft versions of the then-new Comisión de Estudios del Territorio Nacional topographic and land-use maps of the coastal plain. I am warmly appreciative, also, to Alfonzo Grave Tirado and Victor Joel Santos for their helpful encouragement by making available data from the most recent coastal investigations in Sinaloa and Nayarit.

It is a pleasure to acknowledge my indebtedness also to friends and colleagues for their facts, suggestions, ideas, and explanations that thereby broadened my outlook on west Mexico. Among many such contacts were Betty Bell, Charles Cazeau, Charles Di Peso, Lawrence Feldman, Peter Furst, J. Charles Kelley, Isabel Kelly, Thomas A. Lee, Jr., Clement Meighan, Joseph Mountjoy, William Mulloy, Ben Nelson, Arturo Oliveros, Otto Schöndube, and Phil Weigand.

My fellow researchers and students who tackled the theme and endured the physical challenges of subtropical marshland prehistory will recognize something of their labors in this final product. And I extend a special word of thanks to the people of Escuinapa and Teacapán who helped with the logistics of foreign residence, guided us through the wetlands domain, labored alongside of us, and generally made these Norteamericanos feel welcome.

This by no means exhausts the list of those whose interest and regard for the Marismas Nacionales was important to the study. Where appropriate, other names will be found acknowledged by individual authors.

Finally, I wish, with deep gratitude, to extend thanks to Dr. Michael Foster, the volume editor. If it were not for his perseverance and guidance over the last decade plus, this report would not have seen the light of day.

—Stuart D. Scott

Preface

Northwestern Mexico is a vast region that is geographically, culturally, and archaeologically diverse. This province encompasses portions of the Mesa del Norte, the Sierra Madre Occidental and its adjacent foothills, and the Pacific coastal plain, the latter of which is the subject of this volume. Despite a long and continuing history of research, much of this area remains an archaeological terra incognita. For decades this region was thought to be a peripheral and culturally simplistic hinterland to the high cultures of central and southern Mexico; both Mexican and American archaeologists shunned the area in favor of ancient cities and the massive pyramids that dominated them.

The first concerted effort to explore the archaeology of the northwest Mexican coast occurred in 1929 when Carl Sauer and Donald Brand undertook an archaeological and geographic reconnaissance of the area from the Culiacán Valley of Sinaloa in the north southward to the Acaponeta Valley in Nayarit, just north of the Río Grande de Santiago. Within this area they identified numerous prehistoric villages and towns, many with civic architecture, and an array of material culture that included numerous and varied polychrome ceramics, some with incised or engraved geometric and stylized designs and motifs, polished monochromes, and a variety of utilitarian ware. Other artifacts included pipes, spindle whorls, ground stone, and obsidian blades. Sauer and Brand noted that across the area they surveyed, a certain level of commonality existed in the material culture, although there was clear local variability within the artifact assemblages they recorded. The work of Sauer and Brand accomplished several important things. They demonstrated that the area was intensively occupied and that an array of complex and variable material culture was present and widespread, they laid a foundation for future work in the region, and they defined the Aztatlán culture. The latter, their proposal of a panregional social and economic cultural phenomenon—the Aztatlán culture—that seemed to bind the cultures of the area together during a given period, may be their most enduring legacy. This concept was to become a focal point of contention regarding cultural development on the west coast of Mexico in terms of not only the relationship of the coastal cultures to one another but their relationship to the cultures of central Mexico and adjacent areas to the northwest. It is a concept that is widely used today, one that remains greatly debated, and one that still eludes, at least to some, a satisfactory cultural definition and contextualization.

Following Sauer and Brand, interest, although still limited, in the northwest Mexican coast grew. Prior to World War II, the pioneering work of Isabel Kelly at Chametla and Culiacán and Gordon Ekholm in the Guasave area provided the first reports on excavations in the area. With World War II, work in the region waned, and it was not until the mid-1950s that interest in the region was renewed when the University of California at Los Angeles began an extended program of archaeological and ethnohistorical studies in west Mexico. In the northwestern sector of the west Mexican coast excavations were carried out at a number of sites, including Peñitas and Coamiles, Nayarit, and most notability Amapa, Nayarit, where Clement Meighan directed excavations at the large ceremonial site on the floodplain of the Río Grande de Santiago.

It was into this background that during the summer of 1967, Dr. Stuart D. Scott of the State University of New York (SUNY) at Buffalo undertook a reconnaissance of southern Sinaloa and northern Nayarit in order to assess the feasibility of conducting archaeological research in the area, which encompassed the vast estuary and lagoon system that marks the boundary between the Mexican states of Sinaloa and Nayarit—the area known as the Marismas Nacionales.

In 1968 Scott began seven consecutive summers of archaeological field investigations in and adjacent to the marshes and swamps that bound the Teacapán Estuary. The work was done within the context of an archaeological field school and included both undergraduate and graduate students. The primary archaeological goals of the research focused on elucidating the occupational history of this sector of the coastal plain; understanding human adaptation to and exploitation of an extensive, resource-rich mangrove swamp-lagoon-estuary system; and understanding the occupational history of the Marismas Nacionales area within a regional context. In addition, a bioanthropological component (physical anthropology in the day, as they say) was part of the research from the beginning as well, as were zooarchaeological studies (including an ethnographic study of local subsistence activities) and ethnohistoric studies. It soon became clear that in the effort to understand the occupational history of the area, it would be necessary to understand the development and growth of the coastal plain itself. Thus, during the 1971 field season a geologic/geomorphological component was added to the project. Archaeological investigations ended in 1974, and a final season of geologic studies was completed in 1975. In addition, although not part of the SUNY at Buffalo investigations, an ethnographic study of the shark-fishing industry centered in Teacapán, Sinaloa (the headquarters of the SUNY project), was completed as a Ph.D. dissertation, and in 1974, that investigator directed an ethnographic field school in Teacapán that coincided with the SUNY field season. Thus, the Marismas Nacionales project evolved from a near-singular archaeological emphasis to the first large-scale multidisciplinary archaeological-geologic project on the northwest coast of Mexico and the first such project in northwest and west Mexico in general; it has remained such to this day.

Summaries and preliminary results of each season's fieldwork were published annually during the project. The project provided data for numerous seminar papers, M.A. theses, and Ph.D. dissertations as well as topics for symposia papers, many of which were subsequently published in symposia proceedings and in a variety of journals or as chapters in a number of edited volumes. However, this volume, completed 40 years after the last Marismas Nacionales field season, represents the first syntheses of the data generated by the project. The primary authors include Dr. Stuart D. Scott, Marismas Nacionales project director; Dr. Michael S. Foster (volume editor); Dr. George W. Gill, bioanthropologist; and Patricia Kay Scott, project ethnohistorian. Henry W. Case coauthored the skeletal biology chapter with Dr. Gill.

The following 13 chapters are drawn from previously unpublished data and from the previously published literature referred to above. These chapters provide details simply alluded to previously. Chapter 1 provides the background and theoretical underpinnings of the Marismas Nacionales project as well as the history of the project. Chapter 2 provides a summary of the physical setting of the Marismas Nacionales, a summary of the geologic and geomorphological studies completed, and a summary of the palynological studies completed. A detailed history of the growth of the coastal plain in the vicinity of the Marismas Nacionales is presented. Chapter 3 presents a summary of past and current research in the Marismas Nacionales and adjacent area as a context for the discussions presented in subsequent chapters. Radiocarbon assays obtained as a result of both archaeological and geomorphic studies are presented, and a tentative cultural chronology is proposed for the Marismas Nacionales in Chapter 4. The age and origin of the enigmatic shell mound Cerro El Calón are discussed in detail.

Chapter 5 summarizes field methods and provides descriptions of the sites tested and surveys conducted, as well as summary results of excavations at sites. Although utilitarian ceramics are discussed, the focus of Chapter 6 is the decorated ceramics recovered as a result of the Marismas Nacionales excavations. Types are illustrated and described, as are their relationship to neighboring areas. An array of ceramic, stone, bone, and some shell artifacts were also recovered; these are described in Chapter 7. The shell lenses and shell middens tested contained not only a variety of mollusk remains but varied and abundant remains of marine, estuarine, and terrestrial animals. In Chapter 8 these are summarized by site, and ethnographic data are reviewed to provide insight into the prehistoric exploitation of the fauna represented.

Chapters 9, 10, and 11 focus on human remains; a number of well-preserved human skeletons and numerous fragmentary remains were recovered. Chapter 9 addresses mortuary practices and variability; paleopathologies and cultural modifications of the skeleton are discussed in Chapter 10; and Chapter 11, on skeletal biology, presents the results of metric and nonmetric analyses of cranial, postcranial, and dental remains.

Chapter 12 summarizes the ethnohistoric period of the northwest Mexican coast with emphasis on the Marismas Nacionales and areas immediately adjacent. Note is also made of possible astronomical alignments associated with Cerro El Calón and other landmarks. Chapter 13 concludes the volume with discussions of the Marismas Nacionales cultural sequence, the relationship of the Marismas sites to neighboring areas and interaction with those areas, mortuary customs, the significance of the bioanthropological studies, and the Marismas Nacionales in the sixteenth century.

Finally, although it took 40 years before the results of the Marismas Nacionales project were synthesized in a single volume and 14 years for this volume to be finished, the completion of this volume at this point in time may have allowed the authors to generate lines of inquiry not considered methodologically or theoretically 40 years ago. Although the volume is couched in a cultural historical rubric, it is not bound by such an approach. It is hoped that the publication of this volume will not only serve to provide baseline information on a little-studied portion of the northwest Mexican coast but serve as a point of inspiration for future work in the Marismas Nacionales, the northwest coast of Mexico, and west and northwest Mexico in general.

The Marismas Nacionales project files are curated at the Center for Archaeology and Society Repository, School of Human Evolution and Social Change, Arizona State University, Tempe, Arizona.

—Michael S. Foster

CHAPTER 1

Introduction and Project History

Stuart D. Scott

In 1967, the State University of New York (SUNY) at Buffalo initiated a nine-year investigation under my direction of the archaeology and natural history of the Marismas Nacionales, a coastal estuarine and mangrove swamp subarea located on the Pacific coast of west Mexico just south of the Tropic of Cancer (Figure 1.1).

As a matter of introduction, the origin of the geographic designation Marismas Nacionales—National Wetlands—may be worth noting. The name represents somewhat of a puzzle in that it is not a phrase derived from any state or federal enactment, nor is it part of any Hispanic or colonial terminology. Rather, it seems to have become a popular sui generis phrase originating as a commonplace term of "legal" language associated with the administration of agricultural policies during the decades of 1930–1960. With the land reform programs that were undertaken by various post-1910 Revolution governments and during the administration of Lázaro Cárdenas (1934–1940), when attention turned to the north coast of Nayarit, a controversy arose. At issue was the question of whether to include, within the agricultural endowments, this micro-region of freshwater lands, beach, lagoon, marshes, estuaries, and tidal flats. The Mexican Constitution clearly states that the waters are the property of the nation, and in recognition of the importance of this coastland to the livestock industry, the demand was to endow them as "national marshes."

Subsequently, scientists with interests in limnology and aquatic ecosystems (see Flores Verdugo 1986) used the term *Marismas Nacionales* for this geographic region on the Pacific coastal plain from the Río San Pedro to the mouth of the Estero Teacapán. The first environmentalists promoting the region's conservation were ornithologists, who then continued the use of the term *Marismas Nacionales* to refer to the beach, tidal flat, and marsh habitats of migratory birds and shorebirds.

The evolution of the name as a toponym might, then, be best thought of in terms described by Blanco y Correa (2011): Initially, during the 1930s, the term *Marismas Nacionales* was commonly associated with agriculture; however, by the 1970s the term was typically associated with a geographic region. More recently, the term has taken on a broad eco-environmental connotation that has resulted in the region's status as a treasured National Biosphere Reserve.

Also known as the Costa de Nayarit (Moriarty 1964, 1965), the Marismas came to my attention through the earlier studies of the physical geography of the region done for the Office of Naval Research, Washington, D.C. (Vann 1972). It was clear from conversations with anyone who had visited the Marismas that here was a rich archaeological zone of relatively unknown precolumbian culture(s) in a relatively unstudied habitat.

The first questions were the obvious ones. When did humans first arrive? Within what local time span did this marginal estuarine culture develop? Since agriculture was practiced in Mexico for thousands of years, were the Marismas people agriculturalists, or were they simply sporadic visitors to the area exploiting the varied and abundant resources of the estuary and ocean? What was the historical situation of this extinct community of people at the time of Spanish contact? What external relations did they have with other precolumbian peoples of Mexico? Could we expect to find the physical remains of the people themselves, and if so, what could the recovery and analysis of these remains tell us about the genetic affinities of populations that occupied the area and the status of their health?

The Marismas Nacionales study began as an archaeological project; therefore, the focus of this volume is on the excavations and their results. The aim, however, is much more. If, as we reasoned, we restricted ourselves to understanding just the cultural succession and the stylistic and chronological issues standard to archaeological theory and practice of the late 1960s, we would be missing the larger perspective of human colonization in this interesting environmental mix of ocean shoreline, estuary, salt marsh, savanna, and mangrove habitats.

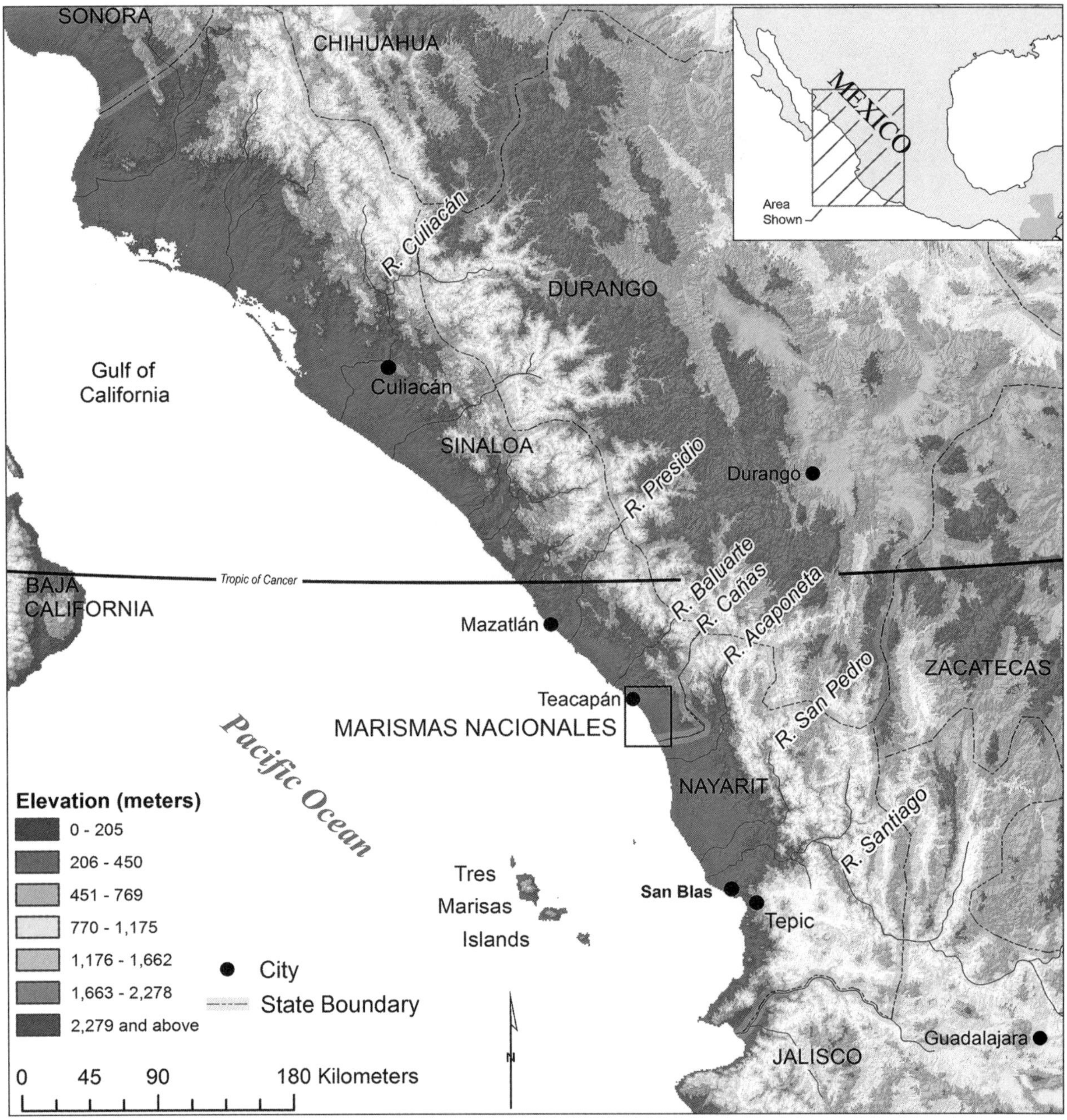

FIGURE 1.1. Northwest Mexican coast showing the location of the Marismas Nacionales.

We wanted to examine the related cultural, biological, and inorganic phenomena, and in order to do so we adopted a general ecological approach (*ecocentric* at the time), more as a research "direction" than as a research design. In hindsight, we must have anticipated in some measure the environmental and ecological approaches of many other archaeological projects worldwide during subsequent decades. The project operated a field school with the gradual addition of researchers from other disciplines who were interested in their own data but whose studies would contribute to a better and broader understanding of the precolumbian use and occupation of the Marismas. In sum, drawing on a diverse range of data including aspects of biology, geography, geology, and history, our purpose was to learn something of the processes of human settlement in this estuarine geographic setting.

An Interpretive Framework—Problems and Expectations

As noted above, several aspects made the Marismas an obvious target for research on human prehistory. First and foremost, historical documents and archaeological field reconnaissance disclosed evidence of high population density in a biogeographic setting that appeared to offer a more productive resource complex than other coastal habitats to the immediate north or south. Second, and perhaps just as important, a long-term study of human colonization and subsistence economies in the resource-rich environment offered by the Marismas was feasible.

The cultural ecological mode (e.g., Sharer and Ashmore 1979:531–532) encompasses a systems view of culture and its interaction with an environmental system including the physical landscape (habitat), the biological environment (biome), and the cultural environment (other human groups). Ecology and systems theory have a long history in archaeological research, as does the closely related concept of adaptation, the latter taken here in its general definition as the process by which a culture adjusts to the particular characteristics of the natural and social environment in which it occurs. Elsewhere in Middle America, particularly in the basins and valleys of the central highlands of Mexico, ecological theory is often at the heart of higher-order theorization about energy, economy, and social change (cultural evolution). In contrast to teleological arguments of the sociopolitical and systemic vs. ecological materialist approach (Blanton et al. 1979; Feinman et al. 1985; Flannery and Marcus 1983; Sanders and Nichols 1988), the adoption of cultural ecology as an orienting framework for the Marismas is, and was, considerably more limited. This ecologically based perspective is not meant to suggest that resource utilization, energy transport, and the use of technology to exploit the estuarine system outweigh the material aspects of human behavior as causal factors in the evolutionary process. Rather, cultural ecology as a conceptual approach serves here as a point of entry for discourse—to provide an objective view of the natural world of the Marismas Nacionales, to examine the record of human occupation there, and to the extent that the data will support them, to consider the implications of human social organization—in this particular environmental setting. This *ecocentric* approach thus stemmed from the simple notion that archaeological knowledge alone could not be expected to answer questions on the material conditions of human life in the Marismas habitat. However, it could reveal a great deal about certain themes of human behavior, social change, and estuarine adaptations. These include:

Theme 1: The Marismas populations as a distinct archaeological presence

Theme 2: Procurement, settlement, and exchange as issues in the sociocultural evolution of Marismas coastal societies

Theme 3: The role of physical and natural science research in amplifying traditional archaeological reconstructions

The Marismas archaeological studies were largely confined to the coastal sector of the Sinaloa/Nayarit coastal borderland—a study region nonetheless quite large, encompassing 140 km^2. The common frame of reference for each presentation in this volume is the Teacapán Estuary and lagoon complex itself, an example of "a semi-enclosed coastal body of water which has a free connection with the open sea and within which sea water is measurably diluted with fresh water derived from land drainage" (Pritchard 1967:3). The Chesapeake, once famously described by H. L. Mencken as an "immense protein factory," is an example of an estuary "where rivers, with all their waterborne materials draining off the altered landscape, meet the sea. But there is more than proximity at work here" (Boicourt 1993:29). Within such a domain, Boicourt contends, "the estuary's unique water motion retains and recycles nutrients essential to living organisms, inducing the richest productivity per square kilometer on the earth's surface" (1993: 29). In reference to the role of humans in estuarine processes, Cronin writes that

> humans have always exhibited a natural affinity for water and these bays and river mouths often present unique advantages. They are semi-enclosed and therefore provide natural harbors; they are effective nutrient traps and therefore are rich in food; they connect the oceans and the inland rivers so they are natural transportation centers [1967:667].

An underlying premise of the Marismas investigations, though perhaps more theoretical than practical, was the complex question of the interrelationship between human populations and the natural world of the Marismas—presumably a sector of phenomenal ecological and resource richness, yet seemingly an inhospitable habitat. Don Matías de la Mota Padilla (1920), drawing on earlier Spanish commentary, speaks of the diversity of pests encountered in the coastal province and likens them to the collected plagues of Egypt. We also might imagine that the excessive seasonal heat, plus periodic devastating floods, would significantly impact human habitability. Yet, although this part of the coastal plain was intolerable for the Spaniards—presumably a force of resolute and adaptable men—the evidence details a high population density, where the general notion of a permanent and productive food supply as a prerequisite to population concentration is borne out by simple observation.

Developing archeological justifications, whether substantive or theoretical, is an essential, ongoing, but problematic debate. Sanders and Nichols offer one conventional view of archaeologists' methodology:

> Our reconstructions tend to be either trivial or highly imaginative with respect to the supporting data. The rather simpleminded which-came-first approach and the preoccupation with taxonomies of cultural behavior rather than with the progressive is an impediment to the testing of evolutionary theories [1988:49].

Explanation, the uppermost level of significance, is truly a goal to aspire to for the maturing of archaeology at large. Taxonomies and so forth are not, nor should they be, ends in themselves. Yet, as we see it, one of the major consequences of studying the Marismas is, unapologetically, the writing of its culture history, in this case with reconstructions augmented by various nonarchaeological disciplines—philosophically a form of archaeological pragmatism (Yorston et al. 1987).

The Marismas, like other coastal lowland regions, lies at the border of and stands in a yet poorly understood relationship to economically and socially complex societies of core Mesoamerica (see Gorenstein and Foster 2000). Even by the "old" standards of diffusionist modeling, historical reconstruction is a legitimate inquiry, not as historical conjecture but done, we might hope, with enough methodological rigor to suggest testable interpretations. Furthermore, such empirical foundations are needed to rectify the underdeveloped state of research summed up by Stark and Voorhies, which regrettably continues to be true almost four decades later: "The human history of the Middle American coastal plains has been inadequately studied and therefore remains largely conjectural" (1978:300).

While each chapter in this volume contributes to our understanding of the land-sea interface, it would be extravagant in the least to suggest that this collective study has any more than broken the surface, particularly in the understanding of human societies in physical interaction with the Marismas ecosystem. Because of the somewhat limited scale of the investigations, the complexity of the natural setting, and the difficulty of extracting data, many detailed answers relevant to the study of human behavioral systems in a coastal habitat still elude us. For example, there is an absence of information on trophic and mortality structures of oysters, shellfish, and shrimp and other critical environmental variables bearing on carrying capacity and population density. Other questions include the human responses to the short-term periodicity in estuarine processes and to larger-scaled environmental changes across the centuries as well as those concerning the contribution of marine products vs. agriculture to human subsistence. For example, to what extent was the energetic efficiency of maritime vs. terrestrial resources a determinant of the cultural vitality of the Marismas populations? In short, critical links are still missing. Nevertheless, the data reported and discussed here contribute significantly to our current understanding of human history and adaptation along the western littoral of Mesoamerica.

Project History

Since the Marismas program grew by accretion, as a composite field school and research venue for physical and natural science as well as archaeological interests, the following provides a year-by-year summary of each season's objectives and personnel. Because, at the time, the Mexican government forbade the exportation of artifacts and samples from the country, each six-week field season took on a sense of urgency out of the need to excavate, map, process, record, photograph, catalog, and in general carry out the many procedural steps that could ordinarily be done with more deliberation after materials are brought home.

1967

In the summer of 1967, I, accompanied by one student, Emerson Pearson, made the initial visit to the proposed west Mexican study area, in part to gain firsthand knowledge of what the area offered archaeologically. Reconnoitering by car and boat as far north as Chametla and south to Mexcaltitlán, we saw a dense distribution of mound sites bordering on or near the edge of the estuary. Pearson (1968) conducted a test excavation on a shell-capped earth mound located along the eastern shoreline of the Panales peninsula. The main objective for this initial season was to examine the range of factors or problems bearing on the possibility of establishing and maintaining a proposed SUNY/west Mexican field school. This preliminary fieldwork was funded by a National Science Foundation Institutional Grant (GU 1864) and a SUNY Research Foundation Grant-in-Aid. A mimeographed summary of the work was distributed in limited numbers under the title of *Archaeological Reconnaissance and Excavations in the Marismas Nacionales, Sinaloa and Nayarit, Mexico. West Mexican Prehistory, Part 1* (Scott 1967; hereafter the Preliminary Report series). These data reports, which were produced annually following the completion of fieldwork, were intended to present capsule summaries of each year's objectives and results, for the benefit, primarily, of incoming personnel in the succeeding field season.

1968

The first field school season incorporated both graduate and undergraduate students: Andrew Anderson, Terry Block, Betty Davis, J. Richard Shenkel, Rosemary Sweetman, and Renata Wolynec from SUNY/Buffalo, and George Gill from the University of Kansas. Graduate students Gill and Shenkel were housed in Teacapán. Gill initiated a multiyear study focusing on the recovery and study of human remains, and Shenkel began a general reconnaissance for data bearing on the types and distribution of shell mounds. The other students were housed in Escuinapa, where they excavated at Juana Gomez, a site that consisted of several mounds and middens adjacent to the ruins of the Templo de Escuinapa, an eighteenth-century Spanish church. During this second year of work, the project was again sponsored by federal and state funds administered through the Research Foundation of SUNY/Buffalo. Also during this year, Elizabeth Wing was the first of several research specialists who became involved in Marismas studies. As curator of zooarchaeology at the Florida State Museum, Dr. Wing began a study of archaeological faunal remains and of the modern estuarine subsistence economy.

Part 2 of the Preliminary Reports (Scott 1968) presented the results of the 1968 season. Anderson, Block, Davis, Wolynec, and I contributed reports related to Juana Gomez; Shenkel added a summary of the Marismas shell mounds, including the discovery of Cerro El Calón; and Gill summarized the first season's data on skeletal remains (burial practices and morphological characteristics). A large portion of this report was devoted to Sweetman's study of the pottery from sites reconnoitered in the foothills of the Sierra Madre Occidental, Juana Gomez, and Panales Test Mound A. Dr. Wing described faunal remains recovered from Juana Gomez, Tecualilla, and Chalpa. Also, Pearson detailed his earlier work on Panales, Nayarit, and set it in the general context of west Mexico in his M.A. thesis entitled "Geocultural Survey of the West Mexican Marismas Nacionales" (1968).

1969

Nine students—Betty Davis, George Gill, William Lyon, Georgia Marzalek, Fred Pasternak, Mark Rucker, J. Richard Shenkel, Michael Steward, and Renata Wolynec—participated in the 1969 field season. Early in the season, excavations at Juana Gomez were terminated, and the Escuinapa crew moved to the site of La Tarjea—a complex of at least seven earthen mounds located about 6 mi southwest of Escuinapa. Meanwhile Gill, with the assistance of Lyon, continued the recovery of skeletal material at Tecualilla and Chalpa, while Shenkel continued his intensive survey of shell mounds, recording their variety, composition, and distribution throughout the central drainage of the Teacapán Estuary. Dr. Elizabeth Wing returned in her capacity as a consultant/visiting researcher and continued to collect faunal samples.

The third Preliminary Report, Part 3 (Scott 1969), included a summary of the work at La Tarjea by Rucker; Shenkel added more information on the shell mounds; and Gill elaborated on the excavation of burial mounds at Chalpa and Tecualilla. In the "Technical Reports" section, Betty Davis authored an essay on ceramics from La Tarjea, Panales Test Mound B, and Tecualilla; Dr. Wing offered a preliminary analysis of the local modern subsistence economy; and Lawrence Feldman provided a brief summary on the identification of archaeomollusks from sites investigated to date.

1970

For the first two years the SUNY program operated without an adequate collections and support facility for processing and instruction. A capital improvement was made during the 1970 field season when laboratory and classroom space with two dormitory rooms was built in Teacapán (Figure 1.2). Students participating in the 1970 field school were Leigh Barker, Constance Foy, George Gill, Tony Goss, Mark Herman, Fred Pasternak, Susan Prince, Jim Stricks, John Volkmann, and Renata Wolynec.

Gill, Prince, and Volkmann continued the excavation of burials at Tecualilla, while Wolynec and other students began work at the site of Arriñitas. Part 4 (Scott 1970) of the Preliminary Reports included a report on the excavations at Arriñitas by Wolynec and one on the 1970 excavations at Tecualilla by Volkmann and Gill. Foy reported on the ceramics from Tecualilla and Arriñitas; and Gill and Prince reported on the skeletal remains recovered that year from Tecualilla, thus expanding Gill's earlier studies of skeletal remains from the site. Gill and Prince also presented a study comparing recorded varieties of human cranial deformation with ceramic figurines from the region. Sweetman, expanding on her 1968 ceramics research, presented her M.A. thesis entitled "Analysis of Pottery from Excavations in Southern Sinaloa and Northern Nayarit, Mexico" (1970), and Wolynec completed her thesis "Obsidian Hydration Dating: A Practical Application" (1970a), in which she applied Meighan and others' (1968) rate for obsidian hydration dating for west Mexican archaeological sites to obsidian from the Marismas.

1971

The 1971 summer season, a geologic component led by Dr. Charles Cazeau from SUNY/Buffalo's Department of Geological Sciences, was added to the Marismas project, building and expanding on earlier work reported by Curray (Curray and Moore 1964; Curray et al. 1969). The initial geologic effort employed power auger borings and the study of recovered cores to investigate beach ridge development to determine stratigraphic relationships and their bearing on the relative ages of the numerous shell mounds.

The student contingent for the 1971 season consisted of Marie Cottrell, Marilyn Hewitt, Lorraine Lambert,

FIGURE 1.2. Marismas Nacionales project headquarters built in 1970.

Curt Peterson, Doug Rautenkranz, Ingrid Rowland, J. Richard Shenkel, Patricia St. George, Steve Taylor, and John Watkins. Excavations were conducted at the site of Rincon de Panal on the island of Panales. Dr. J. Charles Kelley and Dr. Betty Bell visited the project area and consulted on ceramic type identification and lectured on various aspects of west Mexican prehistory. During this season, Dr. Samuel Snedaker of the Department of Aquatic Sciences at the University of Florida initiated a study of estuarine productivity and the Marismas mangrove environment, the latter with specific reference to the El Calón mound. Preliminary Report Part 5 (Scott 1971) included Hewitt, Cottrell, and Cazeau's essay on their geologic investigations in Sinaloa and Nayarit, St. George's report on a glazed rock anomaly, separate contributions by Snedaker and Shenkel on Cerro El Calón, and a summary of the 1971 excavations of Panales by Taylor and Rautenkranz.

1972

Owing to an extended delay in receiving a renewal of the Instituto Nacional de Anthropología e Historia (INAH) excavation permit, the early part of this season was devoted to a combination of continuing geomorphological and archaeological reconnaissance, artifact processing, mapping, and general instruction. Fourteen students made up this season's crew, including Harold Baker, John Cammerata, Jean Clark, Dan Cottrell, Susan Cummings, Bill Fritz, Tony Goss, Rochelle Marrinan, Mike McCoy, Doug Owsley, Ed Pountney, Jim Schutte, Patricia Segmen, and Edward Weil. Collectively, these students represented SUNY/Buffalo, Niagara College, Tulane University, and the University of Wyoming.

Once the INAH permit was received, excavations began at the site of Cristo Rey. Under the direction of Dr. George Gill, now an assistant professor at the University of Wyoming, Owsley, Schutte, and McCoy continued with the analysis of human skeletal remains recovered during 1971 from Rincon de Panal and Tecualilla, as well as those recovered from Cristo Rey. Dr. Cazeau returned for a second season of coastal geologic fieldwork and continued, with Cottrell, research on specific geomorphological aspects of the Marismas Nacionales, including sampling for heavy mineral analysis. Dr. Wing returned to continue her collection and analysis of faunal data. Drs. Bell and Kelley once again contributed their expertise as consultants to the project.

In Part 6 of the Preliminary Reports (Scott 1972), Schutte and Gill reported on their study of human remains from Cristo Rey and Rincon de Panal; Baker, Cammarata, Clark, and Weil summarized the excavations at Cristo Rey; Cottrell contributed an essay about geomorphological aspects of the Marismas; and Cazeau presented a discussion of the occurrence of heavy minerals in both natural sediments and archaeological ceramics. Patricia Kay Segmen (Scott) added notes on the sixteenth- and seventeenth-century accounts regarding the Spanish exploration, conquest, and settlement of the west coast of Mexico in the region of the Marismas. Lesser chapters included a brief description of a modified shell pendant from Chalpa by Shenkel and one by Lawrence Feldman discussing otoliths from archaeological contexts in the Marismas and a summary of fish species reported in the archaeological and ethnohistoric literature from the northwest coast of Mexico. Finally, 1972 was highlighted by the addition of electricity to the SUNY Teacapán field station and the paving of the access road from Escuinapa.

1973

During the 1973 summer season Drs. Parker Calkin and Gordon Connally of the Department of Geological Sciences, SUNY/Buffalo, replaced Cazeau as project geologists, and they were once again accompanied by Dan Cottrell. This season focused on the soil stratigraphy and contemporary beach morphology and processes of peninsular formation. Among the 10 students who participated in the 1973 season were Henry Case, Stephen Cumbaa, Bonnie Fine, Michael Foster, Susan Gessner, John Leedy, Judd Pilosoff, James Schutte, Marla Speas, Ed Weil, and Ronald Yablon. Foster served as field director, and he, Fine, Pilosoff, Yablon, Schutte, and Speas excavated at Venadillo. Dr. Leslie Sirkin of Adelphi University arrived to initiate palynological studies with the aim of developing a biostratigraphic framework for geologic and archaeological research in the Marismas. Dr. Snedaker returned for a brief additional study of the mangrove ecology, and Dr. William Mulloy from the University of Wyoming visited the project. Weil undertook a survey of portions of the Teacapán Peninsula in an effort to evaluate the variability in local site ceramic assemblages, identify local ceramic types, and assess the chronological parameters for identifying ceramic types and their usefulness in developing a statistically based local ceramic sequence.

Preliminary Report Part 7 (Scott 1973) focused on the work completed at Venadillo. Fine, Pilosoff, and Yablon summarized the excavations at the site, and Schutte and Speas reported the Venadillo burial practices. Foster detailed more specifically the excavations in Mound C at Venadillo, and Cumbaa reported an analysis of faunal remains from the mound and observations regarding subsistence practices represented by those remains. Calkin and Cottrell detailed their work on beach formation and the sand ridge system of the Teacapán Peninsula. Segmen included a discussion of the Cañas River as a historical and modern boundary. During the 1973–1974 academic year two master's theses were submitted, one by Dan Cottrell, entitled "The Sedimentary Petrology and Some Geomorphological Aspects of the Marismas Nacionales, West Mexico" (1973), and the other by James Schutte, entitled "Prehistoric Inhabitants of Central West Mexico: Cristo Rey and Rincon de Panal" (1973).

1974

The 1974 season was the eighth and final field season of the SUNY/Buffalo project with archaeology as the focus of investigations. The field crew consisted of Wendy Arbeit, Bob Bohus, Bruce Buck, Henry Case, Bonnie Fine, Meg Forbes, Stephen Hoffman, Mike Matesic, Richard Robrach, and Sue Weigner. Michael Foster once again served as field director. Work focused on the site of Mirador. Elsewhere, mapping of linear shell mounds continued, together with an intensive surface reconnaissance of the distinctive El Calón mound area. Additionally, a limited reconnaissance and testing program occurred at several sites in the vicinity of Playas Novillero and Cuautla on the coast south of the Teacapán Peninsula. Dr. Sirkin and Bonnie Fine continued a study of the vegetation and pollen spectra, while Dr. Connally, accompanied by Steven Hoffman, collected sediment samples for scanning electron microscopy in order to further examine the early formational history of the coastal plain.

By arrangement with the Centro Regional INAH, Guadalajara, some archaeological materials and all human osteological remains had been moved to the storage and laboratory facilities of the private West Mexican Society for Advanced Study at Ajijic, Jalisco. During the summer of 1974, Dr. Gill, Henry Case, and a small team of students continued to conduct osteological and odontometric analyses at the Ajijic facility.

The change in form and direction of the Marismas project from an archaeological to a geology-oriented research program, as initiated by Cazeau in 1971, is reflected in the content of Preliminary Report Part 8 (Connally, ed. 1974). Four reports—"Soil Stratigraphy in the Rio Acaponeta Valley and the Tectonic History of the West Mexican Coastal Plain" (Connally 1974a), "A Soil Stratigraphic Model for the Age and Origin of the West Mexican Coastal Plain" (Connally 1974b), "Provenance, Depositional Environments, and Origin of the Units of the West Mexican Coastal Plain" (Hoffman 1974), and "A Palynologic Model for Reconstructing Vegetation and Environments in the Marismas Nacionales, Sinaloa, Mexico" (Sirkin 1974)—were presented.

Several essays of archaeological interest that might otherwise have appeared in the Preliminary Reports series were instead issued through the publication of *The Archaeology of West Mexico* (Bell 1974). These included a project summary by me, a summary by Shenkel that included a quantitative analysis and population estimates based on his surveys of shell middens in the Marismas, a summary of burial customs by Gill, and a report by Sweetman summarizing her study of ceramics from sites in the area. In 1974, Foster completed his M.A. thesis entitled "Excavations in Mound C, Venadillo, Sinaloa, Mexico" (University of Colorado), and Weil completed his M.A. thesis entitled "Site Location and Local Ceramic Chronology in the Marismas Nacionales, Sinaloa and Nayarit, Mexico" (SUNY/Buffalo).

1975

The principal research agenda during the 1975 field season had shifted to geologically oriented studies almost entirely. In this, the final year of fieldwork, Dr. Connally directed the project. The staff and student roster included those who returned for follow-up data collection as well as several new researchers. Dr. Fred Manley of Georgia State University and Dr. Nicholas Coch of Queen's College, City University of New York, joined the staff to initiate studies of clay mineralogy and beach processes, respectively. The student group, whether new (Patricia O'Brian, Gary Rydout, and Deborah Gilbert) or returning (Patricia Scott, Steve Hoffman), was composed mainly of graduate students who were engaged in their own research or in furthering various project research efforts.

As before, Part 9 (Connally, ed. 1975)—the last of the Preliminary Report series—offered review essays of research in this final year of fieldwork. The volume included articles on the formational history of the coastal plain, including a reexamination of the barrier bar–lagoon complex by scanning electron microscopy (Hoffman 1975a); a study of inferred tectonic events (Connally 1975); and a review of clay petrology/mineralogy (Manley 1975a, 1975b). Sirkin provided a synthesis of palynological research, with supplemental pollen data from El Calón (O'Brien 1975; Sirkin and Gilbert 1975). In her essay, Patricia Scott (1975) presented a translation by June Onesti of Antonio García Cubas's (1890) nineteenth-century document relevant to the coastal geomorphology of the Marismas area. Rydout (1975) offered an interpretation of the nature and function of the enigmatic El Calón mound. It was in this year as well that Steve Hoffman defended his thesis, "Provenance, Depositional Environment and Progressive Diagenetic Changes of Sands from the West Mexican Coastal Plain" (1975b).

Post-Fieldwork

In the years between the end of fieldwork and this publication various professionals and former students continued to analyze data and to publish related articles in journals and books. Gordon Connally published an article entitled "The West Mexican Coastal Plain: Physiographic Definition" in 1977 and a chapter entitled "Soil Stratigraphy and Inferred Tectonic History of the West Mexican Coastal Plain" in 1984. In that same year he and J. T. Sullivan also published an article entitled "Inferred Tectonic History of the West Mexican Coastal Plain."

For a time Michael Foster did research and published many articles related to his work with J. C. Kelley in Durango, and he was awarded a Ph.D. in 1978 from the University of Colorado (Foster 1978). In 1984 Foster authored "Archaeological Investigations in the Marismas Nacionales: Mound C, Venadillo, Sinaloa, Mexico," and in 1985, Foster and Weigand edited *The Archaeology of West and Northwest Mesoamerica*. In 1999 Foster published an essay entitled "The Aztatlán Tradition of West and Northwest Mexico and Casas Grandes: Speculations on the Medio Period Florescence," which references the work in the Marismas. The following year, *Greater Mesoamerica: The Archaeology of West and Northwest Mexico* was published (Foster and Gorenstein 2000)—an edited volume in which Foster and I coauthored "The Prehistory of Mexico's Northwest Coast: A View from the Marismas Nacionales of Sinaloa and Nayarit" (Scott and Foster 2000).

George Gill pursued his interest in the Marismas skeletal remains and dental deformation from his first days with the project. In 1977 he published "Manifestations of a Conceptual Artistic Ideal Within a Prehistoric Culture" in *American Antiquity*. Gill and Henry Case presented "Migration and Population Change in the Prehistoric Marismas Nacionales of Northwestern Mexico" in 1983 at the 18th Mesa Redonda in Taxco, Mexico. In 1985 Gill's chapter entitled "Cultural Implications of Artificially Modified Human Remains from Northwestern Mexico" was published in Foster and Weigand's *The Archaeology of West and Northwest Mesoamerica*.

In 1976, I undertook an extensive study of the excavated ceramics with Betty Bell, and I coauthored with Charles Cazeau an article entitled "Interdisciplinary Approach to Ceramic History: Cultural and Mineralogic Sherd Analysis" (Scott and Cazeau 1976). That same year, although not part of the Marismas project, James R. McGoodwin published an article, "Society, Economy, and Shark-Fishing Crews in Rural Northwest Mexico" (1976), based on his 1973 dissertation and continuing research of the shark-fishing industry centered in Teacapán. In the summer of 1974 McGoodwin had organized an ethnographic field school in Teacapán that coincided with the 1974 Marismas field school.

In 1985 I published "Core Versus Marginal Mesoamerica: A Coastal West Mexican Perspective" in Foster and Weigand's *The Archaeology of West and Northwest Mesoamerica*. My paper "The Marismas Nacionales: Survey and Appraisal of a Mexican Coastal Investigation" was presented at the Seattle Circum-Pacific Prehistory Conference and was published in 1999.

Les Sirkin continued his palynological research and in 1984 published an essay entitled "Late Pleistocene Stratigraphy and Environments of the West Mexican Coastal Plain." In the following year his "Late Quaternary Stratigraphy and Environments of the West Mexican Coastal Plain" appeared in print (Sirkin 1985). In 1977 Sirkin and Connally coauthored "Quaternary Geomorphology and Environments of the West Mexican Coastal Plain," and Sirkin, Connally, and R. Stuckenrath published "The West Mexican Coastal Plain: Stratigraphy and Geologic History" in 1981.

Elizabeth Wing continued at the Florida State Museum and incorporated her Marismas biological research in her 1978 essay entitled "Use of Dogs for Food: An Adaptation to the Coastal Environment." Recently, Wing contributed a chapter entitled "Dog Remains from the Marismas Nacionales" (2013) to a volume entitled *The Archaeology of Mesoamerican Animals*. Bonnie Fine Jacobs expanded her interest in pollen studies, and while associated with the Department of Geosciences at the University of Arizona she published in 1982 an article entitled "Modern Pollen Spectra from Surface Soil Samples, Northern Nayarit, Southern Sinaloa, Mexico." Finally, in 1977, while extending his worldwide mangrove studies, Samuel Snedaker included results from his Teacapán research in an article entitled "Structure of Mangrove Forests in Florida, Puerto Rico, Mexico, and Costa Rica," coauthored with Douglas Pool and Ariel Lugo (Pool et al. 1977).

From this brief project history it can be seen that much has been written about the SUNY/Buffalo Marismas project. However, the information to date has been scattered in master's theses and doctoral dissertations, conference papers, journal articles, and book chapters. Throughout the years following the Marismas fieldwork there was a desire and intent to publish a final descriptive and interpretive monograph of the 1967–1975 research program. This present volume, *The Archaeology, Ethnohistory, and Environment of the Marismas Nacionales: The Prehistoric Pacific Littoral of Sinaloa and Nayarit, Mexico*, fulfills that wish. More importantly, this volume comes at a time when the west coast of Mexico and the region of the Marismas Nacionales are garnering the attention of a new generation of researchers. We are hopeful that this information will contribute to a deeper understanding for those who are now undertaking a new look at the Marismas Nacionales region and the west coast of Mexico in general.

CHAPTER 2

Physical Setting

Michael S. Foster

The West Mexican Coastal Plain extends roughly 200 km in length from Mazatlán, Sinaloa, southward to San Blas, Nayarit, and it was heavily occupied during the Mesoamerican Classic and Postclassic periods as well as at the time of Spanish contact (Figure 2.1). This physiographic region, which ranges from 4 to 40 km wide, forms an interface among the southern Gulf of California, the Pacific Ocean, and the western foothills of the Sierra Madre Occidental (Connally 1977, 1984). Most of the arable land occurs on the beach ridges between Mazatlán and the village of Teacapán or on the alluvial plain between San Blas and the Río San Pedro. The rest of the coastal plain consists of interlacing waterways, salt pans, and mangrove swamps such as the Marismas Nacionales. The area from Mazatlán to Teacapán, including the entire Teacapán Peninsula north of the mouth of the Teacapán Estuary, is characterized by a series of overlapping beach ridges (Figure 2.2). Southward from the mouth of the estuary to the Río San Pedro, the area consists of a narrow sandy barrier that separates the channels and mangrove forests of the Marismas from the ocean.

Much of what follows is based on the works of the late Dr. Charles Cazeau and Dr. Gordon Connally, project geologists; the late Dr. Les Sirkin, project palynologist; and their students and associates. It is hoped that their hard work is accurately represented.

Geologic Framework

The origin of the sands that make up the coastal plain lie in the basaltic, andesitic, and mainly rhyolitic lavas that make up the volcanic mountains of the Sierra Madre Occidental. The volcanic activity that formed the Sierra may have begun as early as 30 million years ago, and although the Sierra remains volcanically active to this day, major volcanic activity ended sometime around 23 million years ago (McDowell and Keizer 1977).

Over time, as they made their way to the ocean, rivers such as the Baluarte, Cañas, Acaponeta, San Pedro, and Grande de Santiago eroded deep canyons into the western slopes of the Sierra. Two rivers in particular supply the Marismas Nacionales, the Río Acaponeta and the Río de las Cañas. The Río Acaponeta enters the Pacific Ocean via the Teacapán Estuary. Its annual discharge averages 1,578 million m^3, of which about 976 million m^3 are diverted southward to Laguna de Agua Brava. The remaining 602 million m^3 combine with the discharge from the Río de las Cañas to form the boundary between the states of Sinaloa and Nayarit, discharging about 945 million m^3 of water through the mouth of the Teacapán Estuary (Vann 1972).

Since the formation of the Sierra, monsoonal rains and consequential flooding have transported millions of cubic meters of alluvial minerals downslope to the coast. These have gradually accumulated at the base of the Sierra. However, what first began as alluvial fans later became prograding beach ridges produced by marine activity and shoreline accretion. Initially, the Gulf of California began as a landlocked trough about 12 million years ago (Ingle 1984). This trough subsequently opened to marine waters about six million years later, when the oceanic crust began spreading northwest and southeast from the East Pacific Rise, separating the tip of the Baja Peninsula from the Trans-Mexican Volcanic Belt. Also at this time, the geologic base or foundation of the West Mexican Coastal Plain was established (Ingle 1984).

The Accretion and Emergence of the West Mexican Coastal Plain

The accretion of the coastal plain has resulted in the formation of approximately 280 subparallel beach ridges with crosscutting floodplains; each ridge rises as much as a meter or more above the intervening swales (Curray et al. 1969). As flooding occurred, sediments were transported across the coastal plain and deposited, forming a broad submarine shelf that extends to the west far beneath the surface of the marine waters of the Gulf of

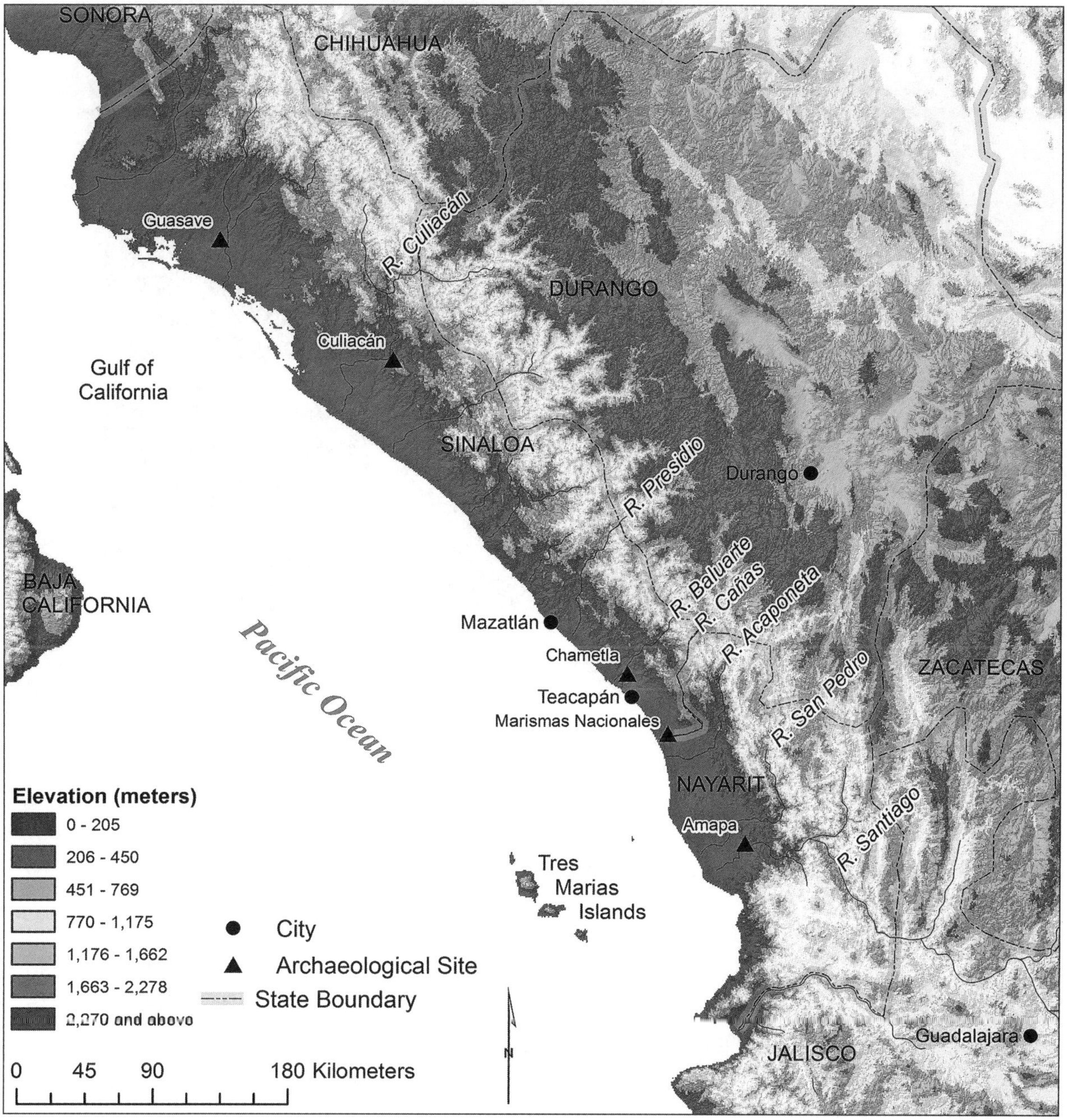

FIGURE 2.1. Map showing the northwest Mexican coastal plain, the Marismas Nacionales, and selected major archaeological sites.

California (Curray and Moore 1964). Over time, wave action redeposited these sediments onto the beaches, where longshore currents transport them laterally in an endless process of coastal straightening. This has resulted in the gracefully curving coastline that exists today (Connally 1984).

Prior to the Marismas project, Curray et al. (1969) proposed a model for the geomorphological evolution of the West Mexican Coastal Plain using aerial photography and ground survey. Based in part on 27 radiocarbon assays, they defined five depositional periods, or units (Table 2.1). Curray and colleagues (1969) also divided the West Mexican Coastal Plain into three broad sectors: the Northern Sector, which includes the Teacapán Peninsula; a narrow Central Sector; and a predominately alluvial Southern Sector.

In 1971, a geologic component was added to the Marismas project (Calkin and Cottrell 1973; Cottrell 1972; Hewitt et al. 1971), and in 1973 a rigorous geomorphic assessment that included widespread reconnaissance accompanied by subsurface geomorphic testing of the West Mexican Coastal Plain was initiated (Connally

FIGURE 2.2. Teacapán Peninsula and surrounding area (© Google Earth 2014).

1974a, 1974b, 1975, 1984; Connally, ed. 1974, 1975; Hoffman 1975a, 1975b; Manley 1975a, 1975b).

This work resulted in a revision of the earlier Curray model (Connally 1984; Sirkin 1985; see Table 2.1). The Connally model differs from the Curray model in three important ways:

1. Two old units were identified within Curray's Period III;
2. Progradation of the beaches took longer than Curray proposed; and
3. Continued, sporadic uplift is inferred based on the identification of soil-forming intervals that punctuate the progradation cycles (Connally 1984).

Based on these observations, Connally proposed three sequential phases for the development of the West Mexican Coastal Plain: the Alluvial phase (pre-Wisconsinan), the Ancestral Peninsula phase (Wisconsinan), and the Prograding Beach phase (6000 BP–present). Sirkin and others (1981) subsequently offered minor revisions of this sequence arguing for bajada deposition along the western slope of the Sierra during the Alluvial phase, development of a barrier-lagoon system on the south-flowing current during Connally's Ancestral Peninsula phase/Barrier-Lagoon phase (which they renamed the Teacapán Peninsula phase), and the deposition of numerous ridge and swale couplets that were laid down in at least four sequences dating from post-eustatic sea level rise during the Prograding Beach phase (Sirkin 1985). These revisions argue that the processes that led to the development of the Teacapán coastal plain and the West Mexican Coastal Plain as a whole extended over a greater period than argued initially by Curray and his associates.

The Alluvial Phase (Pre-Wisconsinan)

The Alluvial phase encompasses the initial event leading to the development of the West Mexican Coastal Plain. The oldest soils on the coastal plain are very dense ultisols that range from as much as 3.8 m thick on the coastal plain to as much as 9 m thick in the upper Río Acaponeta valley. These soils developed on pretransgressive alluvium (Curray's Phase I) derived from the Sierra as alluvial fans coalesced, forming a broad coastal bajada that, at the time, graded into a very low sea level with an as-yet relatively nonexistent coastal plain. The degree of weathering and soil development seen in these soils is taken as evidence of their great age. This event likely occurred after the end of major volcanic activity in the Sierra but long before the shoreline had moved sufficiently west to include the future locations of the El Calón shell mound and the village of Teacapán. At this time, the future estuary and its occupants would have been effectively out to sea (Figure 2.3; Connally 1984).

The Ancestral Peninsula Phase (Wisconsinan)

The Alluvial phase is followed by the Ancestral Peninsula phase. In the original Curray model, the Alluvial phase is followed closely by beach sediment progradation of Periods II–IV. In 1984, however, Connally identified two soils within Curray's Period III that are much

TABLE 2.1. Soil Stratigraphic Models for the Development of the West Mexican Coastal Plain.

Curray Model (Curray et al. 1969)		Connally Model (Connally 1984)			
Period	**Years BP**	**Phase**	**Depositional Event**	**Weathering Event**	**Years BP**
	Present	Prograding Beach			Present
				La Tambora Soil	
Period V					500
			Late Beach		
	500				1,200
				Los Angeles Soil	
Period IV					1,500
			Medial Beach		
	1,500				3,600
				Teacapán Soil	
Period III					5,400
			Early Beach		
	3,600				6,300
				Cañas Soil	
Period II					8,700
	4,500				13,000
"Restricted," Phase III		Ancestral Peninsula		Palmito Soil	
					27,000
			Late Spit		
					34,000
				Mescal Soil	
					70,000(?)
			Early Spit		
					120,000(?)
Period I		Alluvial			
				Oxisols and saprolite	
					?
			Coastal Bajada		

Note: Years BP are presented prior to 1950.

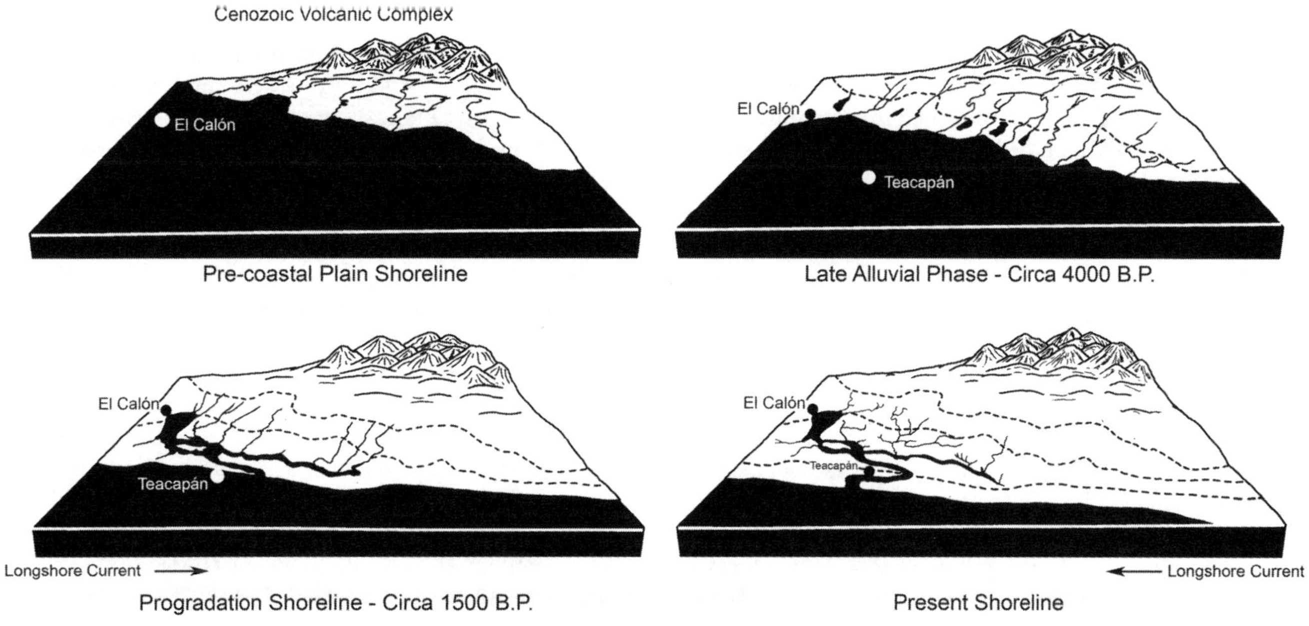

FIGURE 2.3. Schematic view of coastal progradation in the area of the Marismas Nacionales.

older than the inceptisols that make up the bulk of the Period III sediments. The older of the two, named the Mescal Soil, is either a very mature alfisol or possibly an incipient ultisol, and it is much better developed than the younger Palmito Soil. Connally describes the Mescal Soil as being 2.03 m thick, with "an ochric A horizon about 43 cm thick, a clay-rich B_2 horizon 43 to 86 cm thick, and a cambic B_3 horizon that may represent retrograde soil formation under the most recent soil forming conditions" (1984:59).

The younger Palmito Soil is believed to be an immature alfisol or possibly a very mature inceptisol with a mollic epipedon; it is 193 cm deep. Connally describes it as a "mollic A_1 horizon about 36 cm thick and a bleached, hardened (nitric?) A_2 horizon 20 cm thick" (1984:59). The B horizon, like the earlier Mescal Soil, is described as cambic, but it appears to contain little clay.

Both the Mescal and Palmito soils are much more weathered than the Period III and Period II soils reported by Curray et al. (1969; Connally 1974a, 1974b; Hoffman 1975a, 1975b), which they further interpreted for Period III as the consequence of south-flowing currents. This observation was supported a few years later by investigations associated with the Marismas project (Cottrell 1973), although further study also suggested that the Palmito Soil might be storm overwash deposits (Hoffman 1975a). Finally, it was determined that the extent of the Mescal and Palmito soils was restricted to the Northern Sector of the West Mexican Coastal Plain, where they form the ancestral core of the Teacapán Peninsula; but they could not be traced south of the peninsula (Connally 1974a, 1974b, 1984).

The Ancestral Peninsula phase is further defined by the initiation of coastal straightening, with the Mescal and Palmito soils representing embryonic spit developments after the coastal bajada was drowned by rising sea levels (Connally 1984). This initial spit development appears to have started at the base of the highlands near the mouth of the Río Baluarte well to the north of what is now the Teacapán Peninsula before prograding southward. Spit sediments began to form in the north during the early Wisconsinan between as early as 120,000 and 47,600 BP, followed by the appearance of the older Mescal Soil on the Early Spit sediments, which was produced as currents were moving southward (Connally 1984). The Mescal Soil is suggested to have formed between 47,600 and 34,900 BP during a time of low sea level resulting from glaciation, with southward growth, lengthening and widening, occurring between 34,900 and 22,700 BP (Late Spit) during higher interglacial sea levels. Subsequently, the Palmito Soil developed on the newly accreted Late Spit sediments between 22,700 and 14,500 BP, when glacial advance once again resulted in lower sea levels. As the spit prograded, a bay formed, which eventually enclosed and became the Laguna de Agua Grande (Connally 1984).

The Prograding Beach Phase (6000 BP–Present)

During the Prograding Beach phase nearly all the landforms and soils of the West Mexican Coastal Plain were formed (Connally 1984). Connally's (1984) revised model of this period of coastal development generally agrees with that of Curray et al. (1969) (Periods II–V). In addition, the Curray geomorphic map is accepted generally, with modifications of soil boundaries in the Southern Sector of the coastal plain, and the four depositional events are now separated by intervals of nondeposition, weathering, and soil development, although the depositional units are renamed. The four beach units—Initial Beach, Early Beach, Medial Beach, and Late Beach—exhibit distinctive, well-drained soil at the surface and are associated with the Cañas, Teacapán, Los Angeles, and La Tambora soils, respectively (Figure 2.4; Connally 1984).

The oldest of these four soils, the Cañas Soil, developed on the Initial Beach (Period II). It is an inceptisol that is more than 218 cm deep, with an ochric A horizon 50 cm thick and an intense cambic B horizon containing some illuviated clay, and its base is below the water table. The deposition of the Initial Beach is believed to date to 14,500–9,900 years BP, with the development of the Cañas Soil occurring between 9900 and 6900 BP (Connally 1984).

The Teacapán Soil developed on the Early Beach (Period III) deposits. It is a mature inceptisol that ranges from 137 cm deep below swales to 175 cm deep below beach ridges, but it is not as deep as the Cañas Soil. Its maximum thickness occurs at the northwest part of the Teacapán Peninsula where it abuts Ancestral Peninsula sediments, and it is thinnest on the southwest where it is truncated seaward by the Los Angeles Soil. An ochric A horizon is present and is 61–104 cm thick; below it is a well-developed cambic B horizon with iron concentrations at its base. In terms of agriculture, it appears to be the most productive soil on the coastal plain. The deposition of Early Beach deposits is dated between 6900 and 5600 BP, and the development of Teacapán Soil is dated to 5600–4200 BP (Connally 1984).

The Los Angeles Soil developed on Medial Beach (Period IV) deposits. It is a younger and shallower version of the Teacapán Soil and is described as a weakly developed inceptisol that varies from 51 to 86 cm deep, with a 20- to 30-cm-thick ochric A horizon and a cambic B horizon. It was observed that this soil tends to be deeper under beach ridges and to have a thicker A horizon and a thinner profile under swales. It also deepens slightly from its contact with Late Beach deposits. Medial Beach

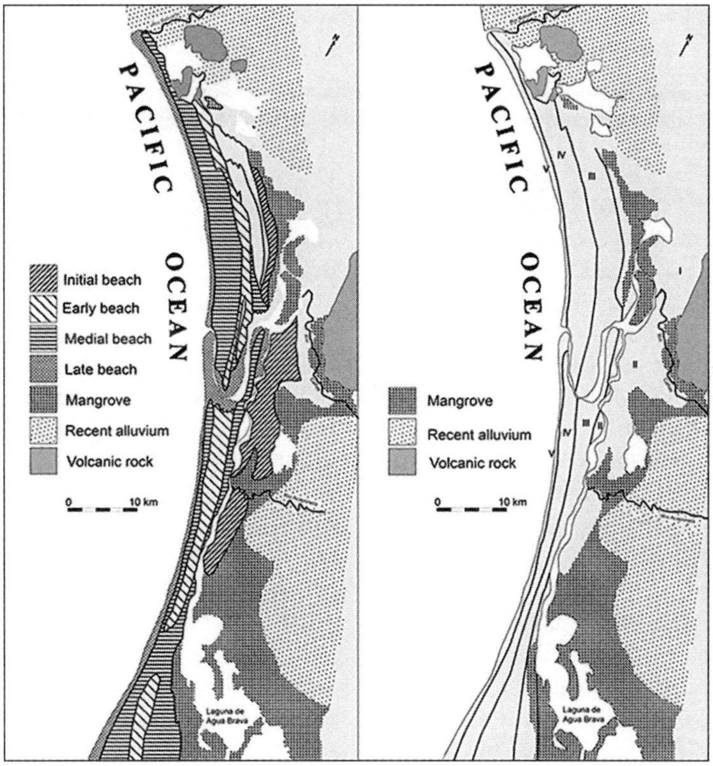

FIGURE 2.4. Soil map of the Northern and Central sectors of the West Mexican Coastal Plain. The Initial, Early, Medial, and Late Beach units are shown on the left (Connally 1984). Curray et al.'s (1969) Periods I–V are shown on the right (adapted from Connally 1984).

deposition is dated from 4200 to 1600 BP, with Los Angeles Soil development occurring between 1600 and 950 BP (Connally 1984). The agricultural potential of the Los Angeles Soil is marginal, with it being most widely used as grazing land for cattle and for copra plantations.

The La Tambora Soil formed on Late Beach (Period V) deposits. It is a weakly developed entisol typified by an ochric A horizon that increases incrementally from 15 cm thick at the modern beach to 30 cm thick at its interface with the Los Angeles Soil. It is further described as a classic A/C soil that lacks a B horizon. Late Beach deposition is dated between 950 and 500 BP, with La Tambora Soil first developing after 500 BP; this development is ongoing (Connally 1984).

The Cañas, Teacapán, Los Angeles, and La Tambora soils developed from similar parent material under similar climatic conditions, in the same topographic situations, and under similar vegetational regimes (Connally 1984; Sirkin and Connally 1977). The soils also developed on similar prograded beach sands—the Initial, Early, Medial, and Late beaches, respectively. The superposition of Late Beach deposits with La Tambora Soil over Medial Beach sediments with Los Angeles Soil has been identified, demonstrating a distinct interval of nondeposition, weathering, and soil formation between Medial and Late Beach deposits. Similar superpositionings of Medial Beach over Early Beach and Early Beach over Initial Beach deposits were not identified, but abrupt and distinctive changes between soils are taken as unambiguous evidence that the soils are of significantly different ages and that these depositional episodes are also separated by distinct intervals of nondeposition, wreathing, and soil formation. Connally (1984) sees this process occurring in distinct episodes "punctuated" by four intervals of soil development over 4,500 years, as opposed to the earlier Curray model (Curray et al. 1969), which suggests that this process occurred gradually over 14,500 years.

Although Connally's 14,500 BP beginning date for this process is tentative, its proposal is tied to the rise of sea level at the time of the Vashol glacier retreat during the Fraser Glaciation of the Pacific Northwest. Furthermore, by 8700 BP, based on core samples from near the El Calón mound, organic sediments had been deposited on the Initial Beach behind the ancestral Teacapán Peninsula. Subsequently, between 8,700 and 6,300 years BP, the Initial Beach was emergent and weathering, and the Cañas Soil began to form. A sea level rise by 6300 BP was sufficient to break through into the protected lagoon (ancestral Laguna de Agua Grande). Submergence from

6300 to 5400 BP allowed the progradation of the Early Beach, and as the beach emerged between 5400 and 3600 BP, the Cañas Soil continued to develop on Initial Beach deposits, and Teacapán Soil began to develop on Early Beach deposits (Connally 1984).

The formation of the Early Beach isolated the lagoon behind the developing Teacapán Peninsula. At this time there is the first evidence of human use of the area, as the initial deposition of the mound at El Calón took place (see Chapters 4 and 5). A subsequent sea level rise resulted in the submergence of the area and its apparent abandonment by humans between ca. 3200 and 1500 BP. It was during this time that the Medial Beach prograded (Connally 1984). After 1,700 years BP, there is evidence of ceramic-producing agriculturalists moving into the area and establishing villages on the coastal plain, although an apparent break in the ceramic sequence at sites such as Venadillo suggests a hiatus in the occupation of the area ca. 1250–1050 BP.

This occupational hiatus of the West Mexican Coastal Plain has been noted elsewhere. Grosscup (1964), based on work at the site of Amapa, Nayarit, to the south of the Marismas, observed evidence for such a break in occupation and proposed that much of the coastal plain had been abandoned. Connally (1984) generally agrees that there is geomorphic evidence for abandonment, but whether a submergence or widespread inundation of portions of the coastal plain had occurred is unclear. Connally proposes that the period between 1700 and 1050 BP was a time of soil emergence, including the development of the Los Angeles Soil and the resumption of the Cañas and Teacapán soil-formation process. As an aside, Connally (1984:65) also suggests that the linear oyster mounds on the Panales Peninsula may have been constructed initially to serve as dikes to combat the rising sea level.

Connally (1984) also assessed the tectonic history of the Teacapán Peninsula, noting that sea level has been rising for the last 14,500 years. He modified the Curray et al. (1969) sea level model for application to the Teacapán Peninsula, suggesting that sea level in the area has risen around 60 m over the aforementioned period. Based on his work and that of his associates, Connally (1984; Connally and Sullivan 1984) indicates that the Teacapán Peninsula has been emergent four times, including presently, which is inferred to mean that the area has risen at a pace equal to the rise of sea level. Again, these emergences interrupted progradation with periods of nondeposition and weathering. It was during these times that the diagnostic soils of the area began to develop on newly deposited sediment units as well as on the older and previously exposed sediment units. Connally (1984) further notes that, as is the case today, the coast has never emerged much above sea level and when submerged, it was never much below sea level. During emergence, the coast must have kept rising through a series of small vertical movements that matched the pace of sea level rise. During submergence, however, the coast continued to rise but clearly did not keep pace with rising sea levels.

In summary, Connally (1984:69) proposes that the coastal plain was submerged from 14,500 to 8700 BP, during which time, he estimates, the Initial Beach prograded, and the coast rose from perhaps 60 m below current sea level to 24 m below current sea level. At 8700 BP the coast emerged, and at 6300 BP, it was again submerged when sea level was at 7.5 m below its current level in response to a major tectonic episode that resulted in a total uplift of the coast of 16.5 m, with the land just managing to stay above sea level. It is during this emergence that Cañas Soil formation began (Connally 1984:69).

For the next 900 years, 6300 to 5400 BP, sea level rose to 5 m below its current level. During this episode, Connally (1984:69–70) proposes a tectonic uplift of 2.5 m, a rate that was slightly outpaced by sea level rise. The Early Beach prograded, followed by a second emergence of the coast at 5400 BP, when sea level was at 5 m below its current level. At 3600 BP, the coast was once again submerged when sea level was at 2 m below current sea level. During this 1,800-year interval of emergence, another major tectonic episode resulted in an uplift of 3 m that managed to keep just ahead of sea level rise. It was during this episode that the formation of the Teacapán Soil was initiated (Connally 1984:70).

Between 3600 and 1500 BP, a period of submergence, tectonic activity resulted in a rise of the coast of near a meter or more. The Medial Beach prograded, and at about 1500 BP, the coast once again emerged, and the Los Angeles Soil began to form. This emergence was short lived. By 1200 BP the coast was again submerged, or the water table was very, very high (Connally 1984:70). Although this is considered a period of submergence, it is probably more appropriate to think of the coastal plain as being awash, more swampy or partially inundated, with some elevated areas remaining relatively dry and accessible. About 500 BP, there was the latest tectonic episode, which resulted in an uplift of the coast of approximately 2 m to its current elevation. It was with this emergence that the Tambora Soil began to form.

A possible explanation for the tectonic activity that resulted in the uplift of the West Mexican Coastal Plain has been identified. A regional Bouger gravity map and a residual anomaly map for the area have resulted in the identification of a 200-km-long positive linear gravity anomaly at the inland edge of the coastal plain. To the southeast this is truncated by the Trans-Mexican Volcanic Belt, and to the northwest it appears to extend offshore in the area of Mazatlán (Connally 1984; Sullivan 1980). The anomaly is believed to have resulted from an upthrust of denser rock along a linear fault zone. Evidence has been identified in the form of northwest-striking vertical faults along the strike of the anomaly at various places on the West Mexican Coastal Plain (Connally 1984; Connally and Sullivan 1984). Because it has the same strike and position, Sullivan (1980) correlates the West Mexican Coastal Plain linear gravity anomaly with Fracture Zone V of the Gulf of California, which is projected northwest to a transpeninsular fault in Baja California identified by Moore and Buffington (1968). Connally (1984) suggests that the tectonic activity manifested by the emergence during postglacial sea level rise of the Teacapán Peninsula was the result of transpressional deformation associated with lateral movement along a major fault that extends from Baja California to the Trans-Mexican Volcanic Belt. Little did the future inhabitants of the region know that such widespread conditions deep within the Earth's crust could dramatically affect the prevailing living conditions along the shoreline of the Mexican Gulf Coast.

Climate

The Marismas coastal plain is situated just south of the Tropic of Cancer. Climatically, the area is a transitional zone from the arid subtropical climate of central and southern Sonora to the wet and dry tropical climate typified by the jungles of coastal southwestern Mexico (Figures 2.5–2.6). This is most obviously manifested in the disparity in rainfall from the northern end of the coastal plain to its southern end (Figure 2.7). At Mazatlán in the north, average annual precipitation is 700 mm, while at Acaponeta in the south, it is 1,200 mm. In general, the climate is dry and pleasant from December to late spring, with temperatures increasing daily to highs between 35°C and 45°C (95°F–113°F) in late June. June can be oppressively hot as the dry season ends and humidity builds prior to the onset of the monsoon season (e.g., Jacobs 1982).

In the south, the rains generally begin by late June, and by mid-July the wet season is established across the entire coastal plain. During July and August precipitation, usually in the form of thunderstorms, occurs late at night. However, from September through November, prolonged, less torrential precipitation is more typical from Teacapán southward. To the north, Mazatlán is usually dry and pleasant by early October. At Teacapán, annual precipitation ranges from 800 to 900 mm, although, as seen in July 1975, when more than 800 mm of rain fell in a 10-day span, extended periods of heavy rainfall can leave much of the coastal plain inundated.

In addition to the monsoonal cycle, the climate of the coastal plain can be affected during the summertime by the migration of the equatorial (intertropical) low-pressure trough. During the summer, a local high-pressure center is formed as the air over the land is heated. This causes an expanding air mass to rise up the slopes of the Sierra, resulting in precipitation at higher altitudes and almost continuous, and welcoming, sea breezes during the day as oceanic air moves inland to replace the rising air. This pattern occurs between mid-March and mid-September. During the winter, a seasonal low-pressure ridge forms over the mountains as the air over the Sierra rapidly cools.

Another major contributor to the region's climate is the Pacific Ocean, which is dominated by three major surface currents that affect the climate of the West Mexican Coastal Plain. The cold California Current transports subarctic waters down the west coast of the Baja Peninsula, while the North Equatorial Current brings warm tropical waters northward. The two currents meet and turn westward, under the influence of the Northeast Trade Winds, just off the tip of Baja California. From December through June, these currents are joined by a current that flows out of the Gulf of California carrying snowmelt from the Rocky Mountains via the Colorado River. With all of the currents moving westward and with little new water added from the land at this time of year, the only source to replace surface flows is an upwelling of cool Pacific water from below. This cool water keeps the ocean breezes from warming and absorbing enough moisture for onshore precipitation. The resulting dry season usually lasts to late June, and in mid-July oceanic upwelling migrates northward up the west coast of Baja California; this usually marks the beginning of the monsoon (e.g., Hubbs and Roden 1964; Vivó Escoto 1964).

From July through October, the surface current of the Gulf of California reverses; warm water is carried northward, replacing that which is lost through evaporation and initiating the monsoon. In the Mazatlán area, the monsoon season extends from July through September, while farther south, it extends from July through November. Seventy to 80 percent of the area's annual precipitation occurs during the summer monsoons.

However, because of the coastal plain's intermediate position between two major climatic zones, climatic conditions in the area can be highly variable. Slight shifts in upper air circulation can bring drought from the north or heavy precipitation from the south in any given year. The region is also subject to tropical cyclonic storms whose torrential rains, winds up to 180 km per hour, and high tides can cause extensive erosion of the coastal plain and widespread, often severe damage to vegetation and property.

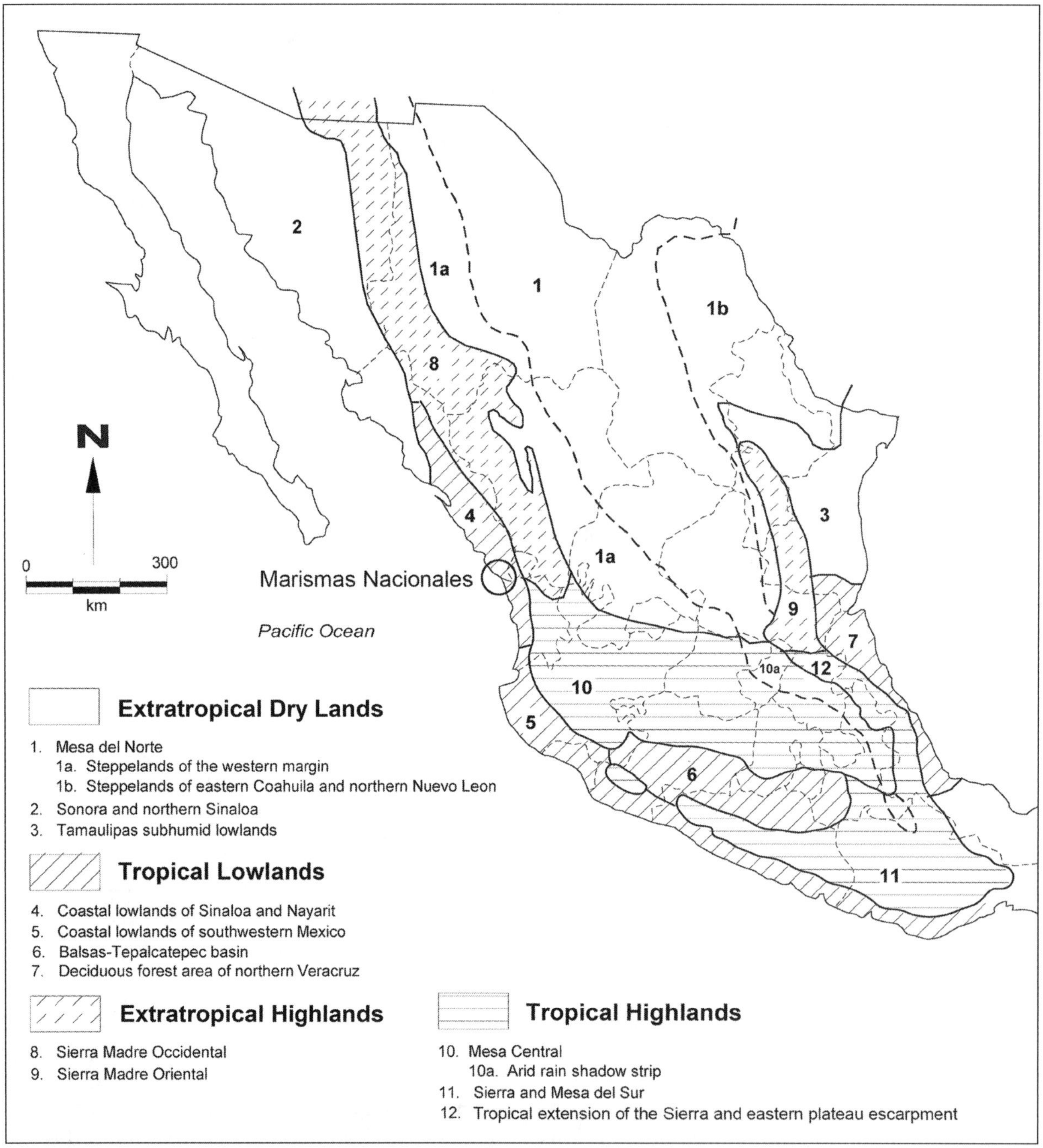

FIGURE 2.5. Natural regions of northwestern Mexico (after West 1964).

VEGETATION

Savanna and Thorny Woodland Scrub

A tropical savanna has developed on the West Mexican Coastal Plain in response to the distinct alternating wet and dry seasons of the area. As a result, the savanna is parched brown during the dry season and lush green during the wet season. During the dry season, the open grassland areas—the *monte* (Wagner 1964:252)—appear lifeless and dry. Today, much of this area has been converted to agricultural use. The savanna supports tall grasses, including saw and pampas grass, interspersed with thorn forest shrub vegetation and scattered trees and palms. The overstory reaches 4 to 10 m and forms a closed canopy, more or less (Wagner 1964:252). Some cacti grow within this grassland area, and there are varieties of spiny leguminous shrubs that are sometimes choked with vines.

During the dry season, only cacti, palms, and the mangroves along the edge of the estuary remain green. However, with the onset of the rains, the area rapidly

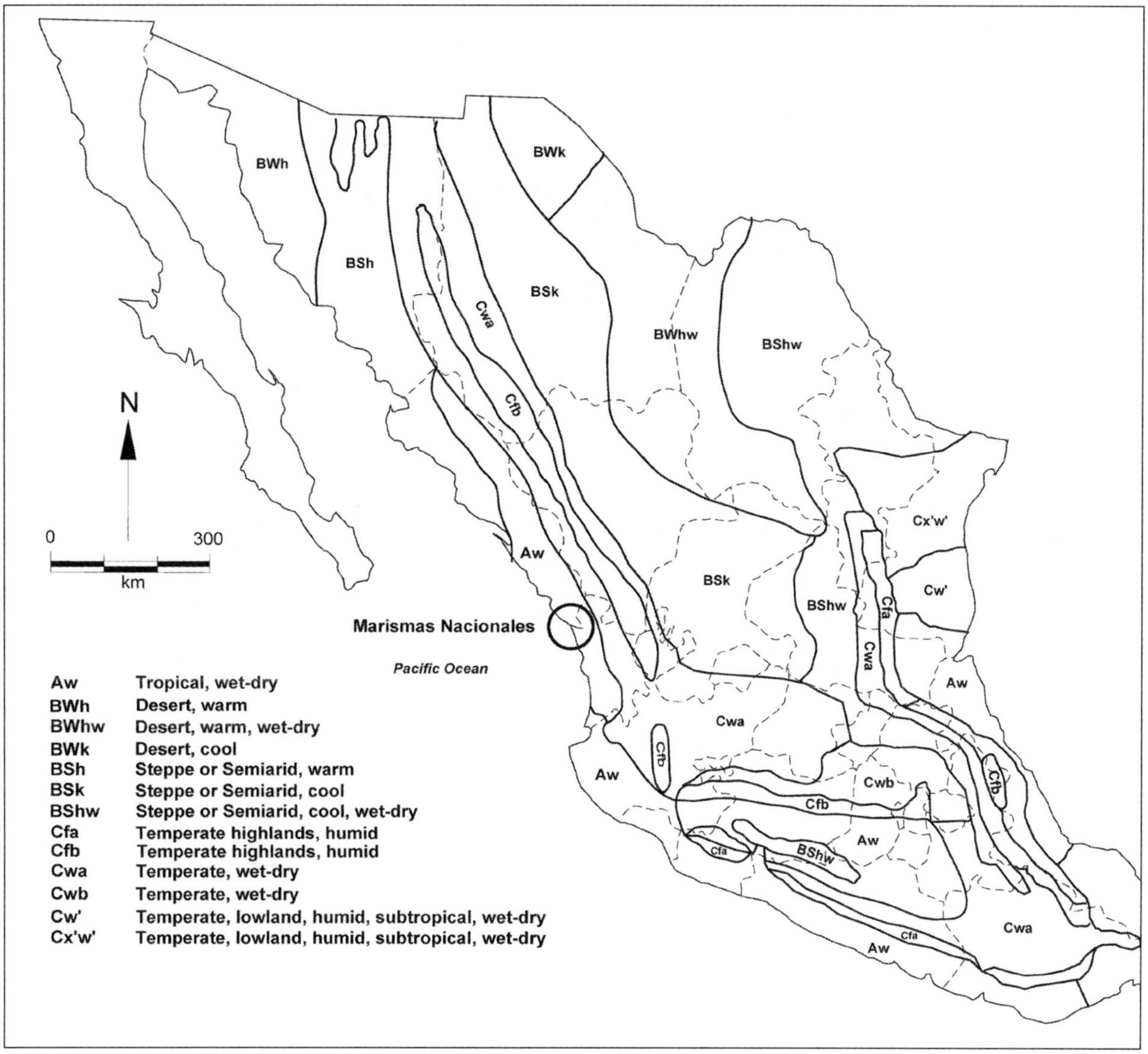

FIGURE 2.6. Climates of northern and central Mexico (after Vivó Escoto 1964).

transforms into a lush subtropical habitat. Scrub vegetation and trees quickly sprout leaves, and the grasses revive and grow (Figure 2.8). Hardy and McDiarmid state that "this formation is not a climatic climax but rather a result of the edaphic conditions of high water-table and poor drainage" (1969:57). They (1969:57) further note, citing the Teacapán area specifically, that although some plants utilize the subsurface water and remain leaved and green year-round, the high water table prevents the coastal plain forest from attaining the stature that typifies the dry forests farther inland.

The foothills of the Sierra are generally covered by a dense thorn forest intermixed with patches of tropical, deciduous growth. The thorn forest, the mangrove forest, and to a lesser extent the tropical deciduous forest (short tree forest) are the primary vegetation complexes on the coastal plain, although it is unlikely that these represent natural associations or climax forests, because of their intensive use by humans for timber, charcoal, dye, and tanbark. Nevertheless, these forests are well adapted to the extended dry season. Many of the trees and other vegetation produce tiny leaves that curl up and drop off when they dry out or large leathery leaves that resist desiccation. A variety of thorny trees, acacias, and mimosas are common within these forests. Also part of the vegetational mix are agave, an organ-pipe cactus, prickly pear cactus, ficus, various bromelias, and lantana (Jacobs 1982; Leopold 1950; Shreve 1934, 1937).

The palm trees (coconut, sabal, and palmetto) are a relatively recent introduction to the area. Palm pollen is not present in the stratigraphic record until about 5,000 years ago (Sirkin 1985). After that time it is common, and groves would have been a normal sight on the coastal plain during prehistoric times. It is likely that coconuts

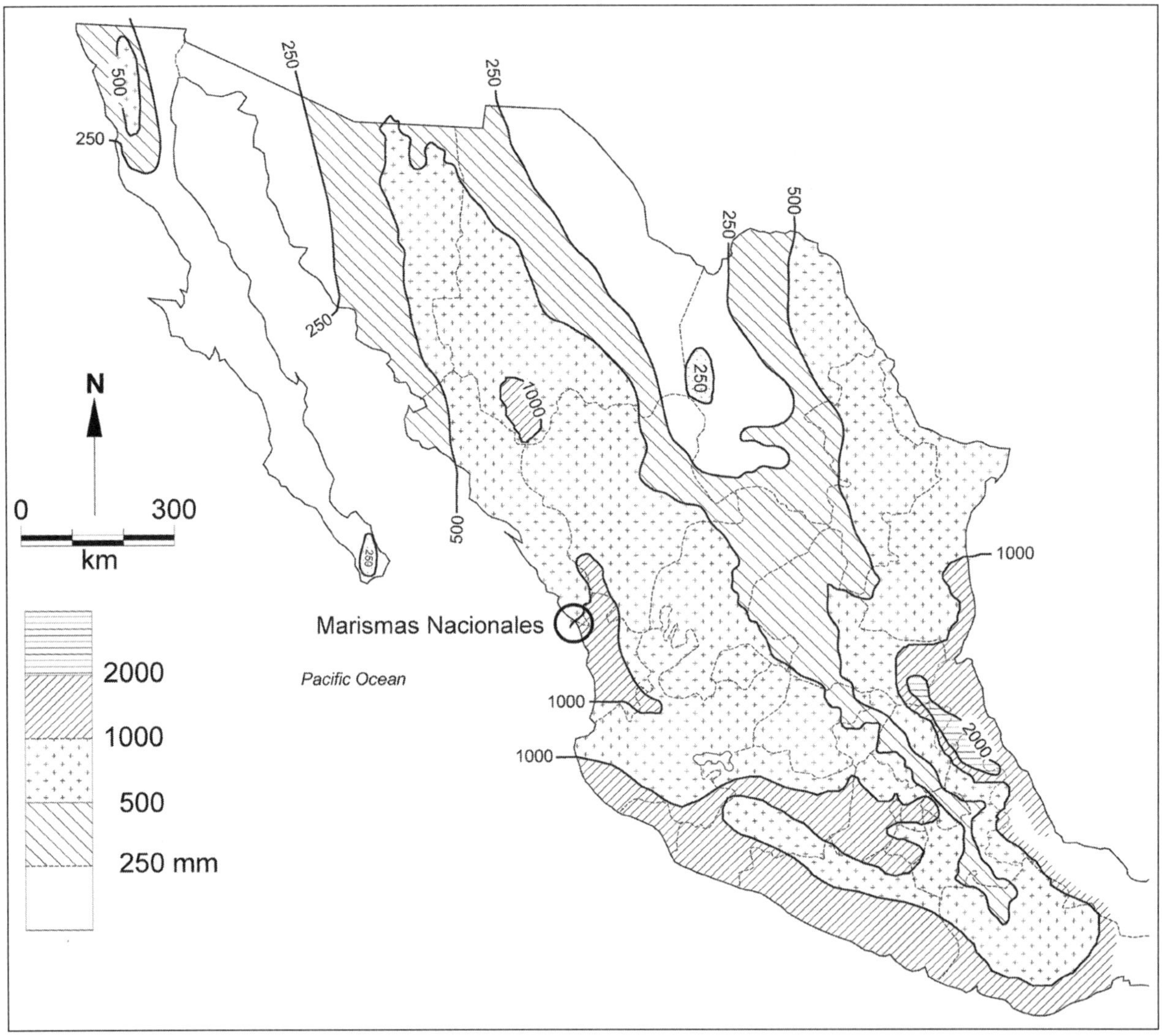

FIGURE 2.7. Mean annual precipitation for northern and central Mexico (after Vivó Escoto 1964).

FIGURE 2.8. Typical savanna and thorny woodland scrub vegetation near Venadillo.

FIGURE 2.9. Red mangrove near El Calón.

floated across the Pacific Ocean, coming to rest in the area, although humans may have also contributed somewhat to their spread as they moved from one area to another along the coast.

Mangrove Forest

The central West Mexican Coastal Plain is bisected by the Teacapán Estuary and a series of lesser estuarine channels associated with the Río Acaponeta and Río de las Cañas. The banks of many of these rivers and estuaries are bounded by a nearly impenetrable jungle of red, white, and black mangrove trunks, roots, and rhizophores (Pool et al. 1977; see also Comisión Nacional para el Conocimiento y Uso de la Biodiversidad 2008). The red mangrove (*Rhizophora mangle*) grows at the border of the estuary and is well adapted to its brackish waters. It is characterized by stilt-like support roots that descend from the branches to facilitate aeration (Figure 2.9). White mangrove (*Laguncularia racemosa*) grow farther inland, beyond the influence of daily tides, and black mangrove (*Avicennia nitida*) and buttonwood (*Conocarpus erecta*) occur even farther inland in brackish to

FIGURE 2.10. Black mangrove forest surrounding El Calón.

freshwater conditions (e.g., Jacobs 1982; Pool et al. 1977; Figure 2.10).

In addition to being well adapted to the brackish waters of the estuary, the red mangrove is also well adapted to the region's extended dry season. With the onset of the dry season, the estuary becomes more saline as less freshwater is available. The red mangrove copes with the increased salinity by passing the saltier water through its trunk and branches and depositing salt on its leaves during transpiration. As the leaves become encrusted with salt, which blocks respiration, they drop and are replaced. With the onset of the rainy season, the salinity of the estuary decreases, and life for the red mangrove is less rigorous. The mangrove forests provide nesting areas and habitats for a variety of birds and terrestrial wildlife, and the wood is used locally for building materials and fuel.

During the rainy season, broad, shallow salt pans are formed adjacent to the mangrove. The vegetation in this part of the mangrove forest is almost exclusively *Sali-*

FIGURE 2.11. Teacapán Estuary at the village of Arriñitas.

cornia (pickleweed). Other plants that inhabit these salt pans include *Acacia*, *Agave*, various *Bromelia*, the lipstick tree (*Bixacea*), *Coccoloba uvifera* (sea grape), and *Lantana* sp. (Jacobs 1982; Leopold 1950; Shreve 1934, 1937).

Although seeming out of place, cacti and other xerophytic vegetation are also found scattered in the marshes. These are almost exclusively distributed on well-drained areas and are most common on archaeological shell and burial mounds. These artificial mounds are frequently the homes of organ-pipe cactus, century plant, Spanish bayonet, prickly pear and ocotillo, the kapok tree (*pachote*), and *indio desnudo* (*Bursera simaruba*—a tree with a dark red bark that peels like birch). Additionally, bromeliads (Bromeliaceae) are frequently found clinging to the upper branches of mangroves and other trees (e.g., Gentry 1942; Leopold 1950; Martin et al. 1998; Shreve 1934, 1937; Vivó Escoto 1964; West 1964).

Tropical Deciduous Forest

A tropical deciduous forest covers the foothills of the Sierra and the relict coastal bajada adjacent to the eastern edge of the coastal plain at elevations from about 30 to about 310 m. The vegetation of this association is more diverse than that of the thorn forest, and its canopy reaches 8 to 15 m in height. Plant species that can be found in this association include species of agave (Agavaceae), bromeliads (Bromeliaceae), various resinous trees and shrubs (Burseraceae), cacti (Cactaceae), dicotyledonous plants such as the sunflower family (Compositae), members of the coca family (Erythroxylaceae), the spurge family (Euphorbiaceae), ficus (Ficeae), legumes (Leguminosae), the knotweed or smartweed family (Polygonaceae), the madder family, bedstraw family or coffee family (Rubiaceae), various flowering plants (Sterculiaceae/Malvaceae), nettles (Urticaceae), and the verbena or vervain family, a family of mainly tropical flowering plants (Verbenaceae). During the dry season, most of these drop their leaves and remain leafless until the rains return. Nevertheless, during this time, some of these trees and plants may bloom. As with the savanna, the only perennial green vegetation is the cacti. Above 300 m to 1,200 m, the slopes of the Sierra are covered with a tropical subdeciduous forest. Oak occurs around 790 m elevation, and above 1,200 m, an oak-pine association eventually transitions to a pine forest (Gentry 1942; Jacobs 1982; Martin et al. 1998; Shreve 1934, 1937).

The Estuary

Estuaries are complex ecosystems. The Teacapán Estuary is a highly dynamic, brackish water, tidal estuary fed by the Río Acaponeta, which empties into the Laguna Agua Grande. Because of tidal flux and rainfall, the estuary is subject to daily and seasonal variation in water levels and, as previously noted, seasonal changes in salinity, chemical composition, and the amount and types of solids transported. These changes can significantly impact the habitat of the marine species that live in and use the estuary. Shrimp and oyster are particularly sensitive to changes in salinity. Other species are less sensitive, and fish such as red snapper, robalo, mullet, and snook can nearly always be found, along with an occasional shark, dolphin, and sea turtle (Wing 1969; Figure 2.11).

FIGURE 2.12. Fallow agricultural fields on the Teacapán Peninsula.

AGRICULTURE

In historical and modern times, agricultural fields (*temporales*) are tilled and planted in late June. New fields and those that have been fallow for a sufficient period of time are prepared by swidden or slash and burn. Beans and maize are the dominant crops, along with some chili and cotton (McGoodwin 1976; Wing 1969). With the onset of the monsoons and summer heat, crops quickly sprout and grow, reaching maturity in the early fall (cotton is harvested in early winter). By late December, the savanna, which also hosts the agricultural lands, becomes dormant as the winter dry season progresses (Figure 2.12). In some areas, agricultural fields can be found along floodplains where the water table is high and streams and rivers are fed by spring snowmelt from the Sierra. These fields can be planted earlier, and thus double-cropping is possible.

Today, in addition to growing traditional crops, the Marismas area has become important in the production of mangos. Mango orchards are numerous, as are groves of now generally unattended coconut palms. The coconut palms were planted in an effort to develop a cash crop in the area. However, as the demand for copra (dried coconut meat) ended at the end of World War II, the area's copra plantations quickly disappeared.

FAUNA

The recent expansion of agriculture, cattle ranching, human settlement, and recreational use of the area has greatly impacted wildlife habitats and the distribution of species. Only lizards, frogs, snakes, rodents, cottontails, possums, and the ubiquitous iguana are encountered on a regular basis, and only occasionally are there signs of jaguar, margay, and jaguarundi. Most snakes that inhabit the area are harmless, although a large boa constrictor or a coral snake can be found, and both black and green iguanas are common (Hardy and McDiarmid 1969). Also found are shrews, armadillos, coyotes, skunks, raccoons, deer, and javelina (Armstrong and Jones 1971; Armstrong et al. 1972; Leopold 1959).

Marine Fauna

The Pacific Ocean adjacent to the West Mexican Coastal Plain and the Teacapán Estuary are fertile with marine species, including fish, shellfish, and marine mammals. *Huacinango* (red snapper), *corbina* (snook), and robalo (rock bass) can be found both offshore and in the estuary. Teacapán is a shark-fishing village (McGoodwin 1973, 1976; Wing 1969), and sharks of all types and sizes are caught offshore, although they will occasionally make their way into the estuary. The Teacapán Estuary is most noted for its shrimp and oysters. After shrimp spawn in the freshwater tributaries to the estuary, the baby shrimp journey to the sea, increasing in size as they go. By the time they approach the mouth of the estuary, they are of sufficient size to be caught by either seining or trapping. Shrimp seem to be most abundant during August and September (Wing 1969:98).

The Teacapán Estuary also provides habitats for oysters, which cling to the roots of the mangroves. Archaeologically and historically, human predation of the estuary's oyster population is a matter of record. However, oysters are sensitive, and populations have been decimated because of hydrologic and environmental changes in the estuary, particularly freshwater flooding. In the mid-1970s, a major effort was undertaken to reintroduce oysters to the estuary; this process can take several years following a die-off (McGoodwin 1973).

Several other marine fauna, including dolphins, whales, other game fish, crocodiles, and caiman, also occur in the estuary or in the adjacent Pacific waters. Pacific green and loggerhead turtles are frequently, and illegally, caught offshore or in the estuary as well. The shores and intertidal zones of the ocean and estuary are also home to a variety of shellfish and crabs.

Birds

A variety of birds take advantage of the climate as well as the abundant food supply and nesting areas found in the Marismas. Nearly 450 species of birds, 80 of which are songbirds, are reported to inhabit the Marismas. Today, the most common species is the American avocet (*Recurvirostra americana*). Man-o'-war birds, egrets, frigate birds, and pelicans are among the larger and more common marine-oriented species present. The roseate spoonbill and the yellow-crowned night heron are also known. The Mexican eagle makes the Marismas its home, as do a variety of other terrestrial birds including orioles, hummingbirds, jays, the yellow-winged cacique, a redheaded woodpecker, and small parrots. Historically, many of these birds were killed for their plumage (e.g., Leopold 1959; Wing 1969).

Insects

Various insects, many of which prey on or greatly annoy humans, are abundant in the Marismas; most notable of these pests are mosquitoes and the tiny gnat locally called the *jejene*. Also of potential irritation and harm to humans are a variety of spiders, scorpions, and wasps, along with ants, cockroaches, beetles, and flies (e.g., Armstrong and Jones 1971).

PALEOENVIRONMENT

The late Dr. Les Sirkin, Adelphi University, served as the Marismas project palynologist. The following discussion is drawn from Sirkin (1974, 1975, 1985) and Sirkin and Gilbert (1975).

Regional Vegetation, Environment, and Palynology

Reconstruction of vegetation and environment is based on the presence of microfloral assemblages, mainly of pollen, that can be favorably compared with the dominant or representative taxa in existing vegetational zones. Based on prior vegetational surveys in western Mexico, the vegetational zones that are found in the Marismas today are identified variously as highland forest (the pine-oak forest of the upper mountain slopes), rain forest (the subdeciduous forest of the lower mountain slopes), and tropical deciduous forest and thorn forest of the piedmont, adjacent alluvial bajada, and well-drained lowlands. The latter three forest types are for the most part indistinguishable in the pollen record because they share many genera in common. Although a few plant species are found to occur in more than one habitat, the remaining lowlands are covered by somewhat more distinctive vegetational zones: the savanna, salt marsh, fresh marsh, mangrove, and field and cropland. As discussed here, these are defined in part by Shreve (1937), Gentry (1942), and Leopold (1950) as well as field observations during the Marismas project by Connally and his associates.

The regional biostratigraphic record for the West Mexican Coastal Plain was developed through a comparison of pollen data from the El Calón cores with the present vegetational zonation. This record was supplemented by additional detailed studies of microflora from a variety of other sedimentary sequences. In the interpretation of the palynological record, changes in prominent associations are believed to reflect changes in the vegetational counterpart. Highland vegetation, for example, is indicated by pine and oak. A substantial representation of this association in a pollen spectrum would suggest an expansion of the pine-oak forest. This, in turn, would be interpreted as evidence of colder climate. A salt marsh or salt pan setting is indicated mainly by chenopod pollen (Chenopodiaceae), generally *Salicornia*, while freshwater conditions are seen in a variety of aquatics, cattail (*Typha*) and alder (*Alnus*), particularly during cooler intervals, and transitional or tidal areas are indicated by sedge (Cyperaceae). While grass (Gramineae) occurs in most vegetational settings, it is used here as an indication of open field or savanna in association with composites (Compositae). Pollen of representative plants for all of the vegetation zones is therefore important to the interpretation of past vegetation and climate, even though certain forms, such as the composites, are often believed to be evidence of human activity. Furthermore, vegetational changes in the estuarine environments, as, for example, in the mangrove succession, have been interpreted as indications of changes in sea level (e.g., Lugo and Snedaker 1974). In this succession, red mangrove (*Rhizophora mangle*) is the dominant form when sea level is low but rising. White mangrove (*Laguncularia racemosa*) is dominant during stable or lowering sea levels, and black mangrove (*Avicennia nitida*) is prominent during the formation of paludal (marshy) basins. Combined with the other pollen data, the mangrove representation can supplement climatic as well as geomorphic inferences.

Pollen in the depositional basins may be derived from the entire watershed or from all of the vegetational zones of the western Sierra via wind, water, and insect vectors. Consequently, even in the modern deposits of

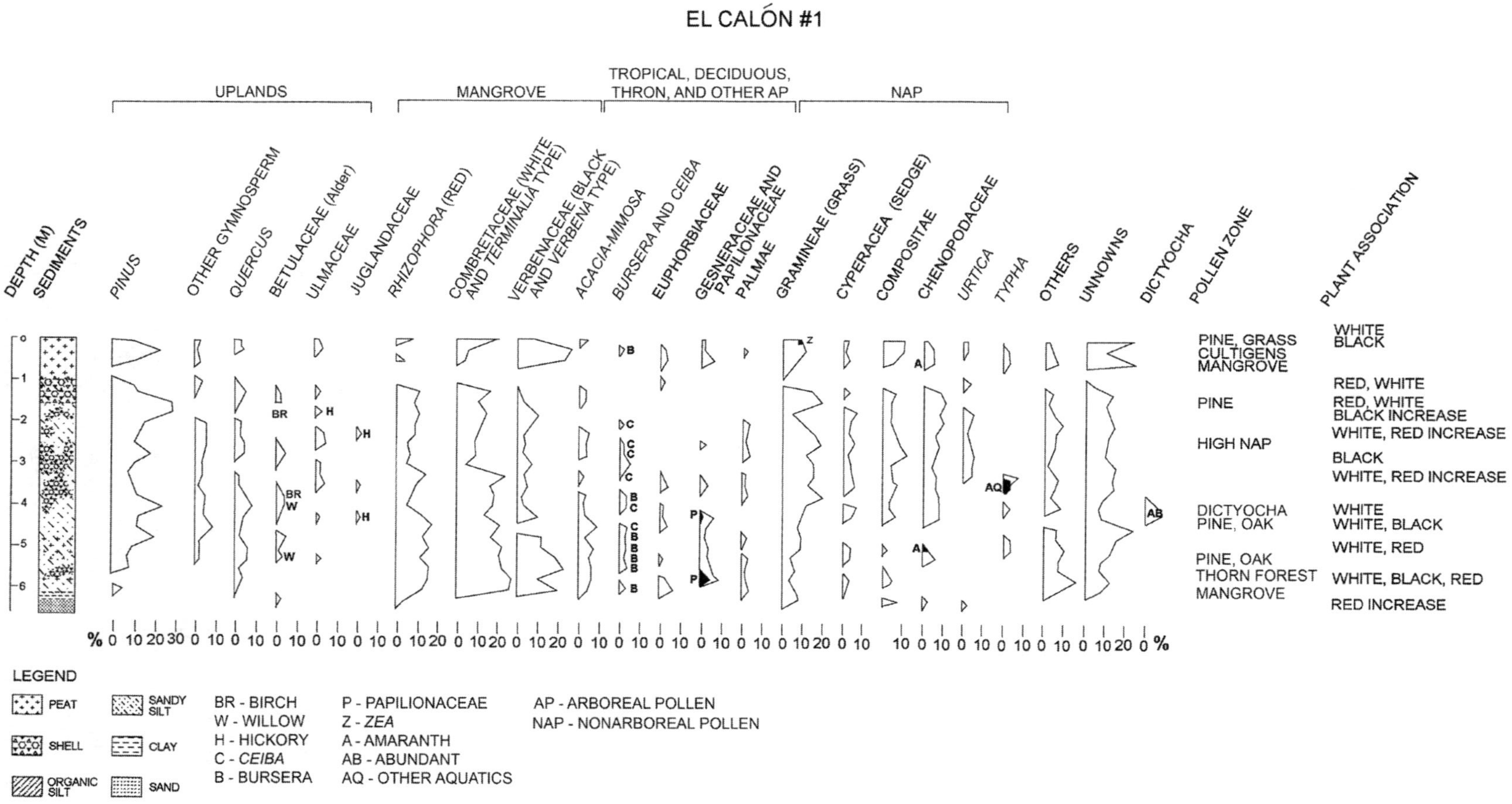

FIGURE 2.13. Pollen diagram, El Calón #1.

the coastal plain, pine and oak from the highlands may be found. While the pine-oak association includes several prominent species of pine and oak, pollen species have not been determined. Alder, juniper, and the Ericaceae also occur in this zone and in the pollen record. Pollen from the minor stands of spruce and holly has not been encountered. Stands of pine may extend downslope to about 1,000 m, while oak may occur somewhat lower, although individuals can be found on the coastal plain.

The rain forest vegetation and related associations from the lower mountain slopes to the bajada have many taxa, mainly families or genera, in common. Prominent in these zones are such families as the Leguminoseae, Euphorbiaceae, Rubiaceae, Convolvulaceae, Papaveraceae, Moraceae, Verbenaceae, Nyctaginaceae, Myrtaceae, and Sapindaceae and species of *Acacia*, *Ficus*, *Bursera*, and *Ceiba*. The thorn forest, in particular, is characterized by species of *Acacia*, *Caesalpina*, and cactus. Some of the species recognized by Shreve (1937) include *Acacia cymbispina*, *Ipomoea arborescens*, *Pachycereus pecten-arboriginum*, *Ceiba acuminate*, *Croton alamosanus*, *Franseria cordifolia*, *Caesalpina platyloba* and *C. gracilis*, *Prosopis chilensis* and *P. juliflora*, and species of *Cassia* and *Opuntia*. More tropical plants, such as *Cedrela* and *Tabebuia*, may also occur in the thorn forest.

Agavaceae, Bromeliaceae, Burseraceae, and Cactaceae and the Compositae might be more abundant in the thorn forest than on the higher slopes. Overall, *Acacia* is most prominent in the rain forest, while other legumes such as *Mimosa* and *Pithecolobium* are common in this forest and in the tropical deciduous forest, along with the Bombacaceae (*Bombax* and *Ceiba*) and Convolvulaceae (*Ipomoea*), among others. It is apparent that many of the same families and genera are distributed through these zones, thus providing a pollen influx with few significant index forms. These similarities would make differentiation of the zones in the pollen record very difficult, particularly when low pollen influx and taxonomic diversity are already seen as problematic.

Marismas Pollen Studies

The palynological studies of the Marismas coastal plain were undertaken in order to develop a biostratigraphic framework for geologic and archaeological research (Sirkin 1974, 1985). Palynological sampling began in the vicinity of the El Calón shell mound, where, at a locus designated El Calón #1 (EC-1), a 6.25- to 6.50-m core of estuarine/lagoonal and mangrove sediments from the Teacapán Estuary was recovered. As a whole, the El Calón cores contain the oldest record of Holocene sediments along the West Mexican Coastal Plain yet identified (Sirkin 1985:8). Eventually included in this study was a 2.0-m core from a small lagoonal swale near the archaeological site of Venadillo, as was a core from Panales. Additionally, soil samples from burial mounds, trenches, middens, ceramic vessel contents, and fill from a shell midden from the site of Venadillo were also processed, as were soil samples collected from the major stratigraphic units identified in earlier geomorphic studies of the area.

The EC-1 core provided significant microfloral data for the establishment of a reference or standard palynological section on which environmental and vegetational reconstructions could be based. In order to supplement this database and obtain samples for radiocarbon dating, additional cores—EC-2 (7.5 m), EC-3 (4.5 m), and EC-4 (6.5 m)—of mangrove, lagoon, and estuarine sediments between the surface and the alluvial substrate were subsequently collected.

Samples from El Calón, Venadillo, and Panales sites were processed by standard laboratory techniques and analyzed for pollen content. The long cores were analyzed at 25-cm intervals, while more detailed studies at 10-cm intervals were applied to the EC-3 and the Panales sections. Data (percent of each taxon at each sampling interval based on the pollen sum at that level) are displayed in the pollen diagrams in Figures 2.13–2.18. The various taxa are grouped in generalized plant associations because many taxa occur in more than one plant association. Also, the mangrove heading may include other taxa in the Verbenaceae and Combretaceae that might occur in the forest associations. Taxa that are only rarely encountered are plotted in separate columns or as "others" with the taxa noted in the legends. Similarly, unknowns are plotted under a separate heading.

The mechanisms of pollen sedimentation in a mangrove regime have not been intensively studied. Nevertheless, it is assumed that depositional parameters are much the same as those of an open estuarine system where runoff and tidal action transport fine-grained and buoyant sediment into deeper-water depositional areas. Pollen not removed in this way would be deposited directly on the ground surface in the mangrove swamp and paludal basins. In general, the pollen influx and relative pollen frequency would be quite low, except possibly in the paludal areas. These would act as enclosed basins, of a temporary nature, because they dry up during the dry season. Since much of the pollen rain would be transported away from the mangrove by water currents or oxidized in place, the pollen influx factor is more accurately one of pollen retention, which is reflected in the very low pollen sums.

Many plants in this region produce relatively little pollen and are usually pollinated by insects and birds. Therefore pollen from these plants would not be common in the pollen record. Low productivity and insect pollination may also account for the low representation of cultigens in the regional record. Furthermore, the croplands are more arid and far enough removed from

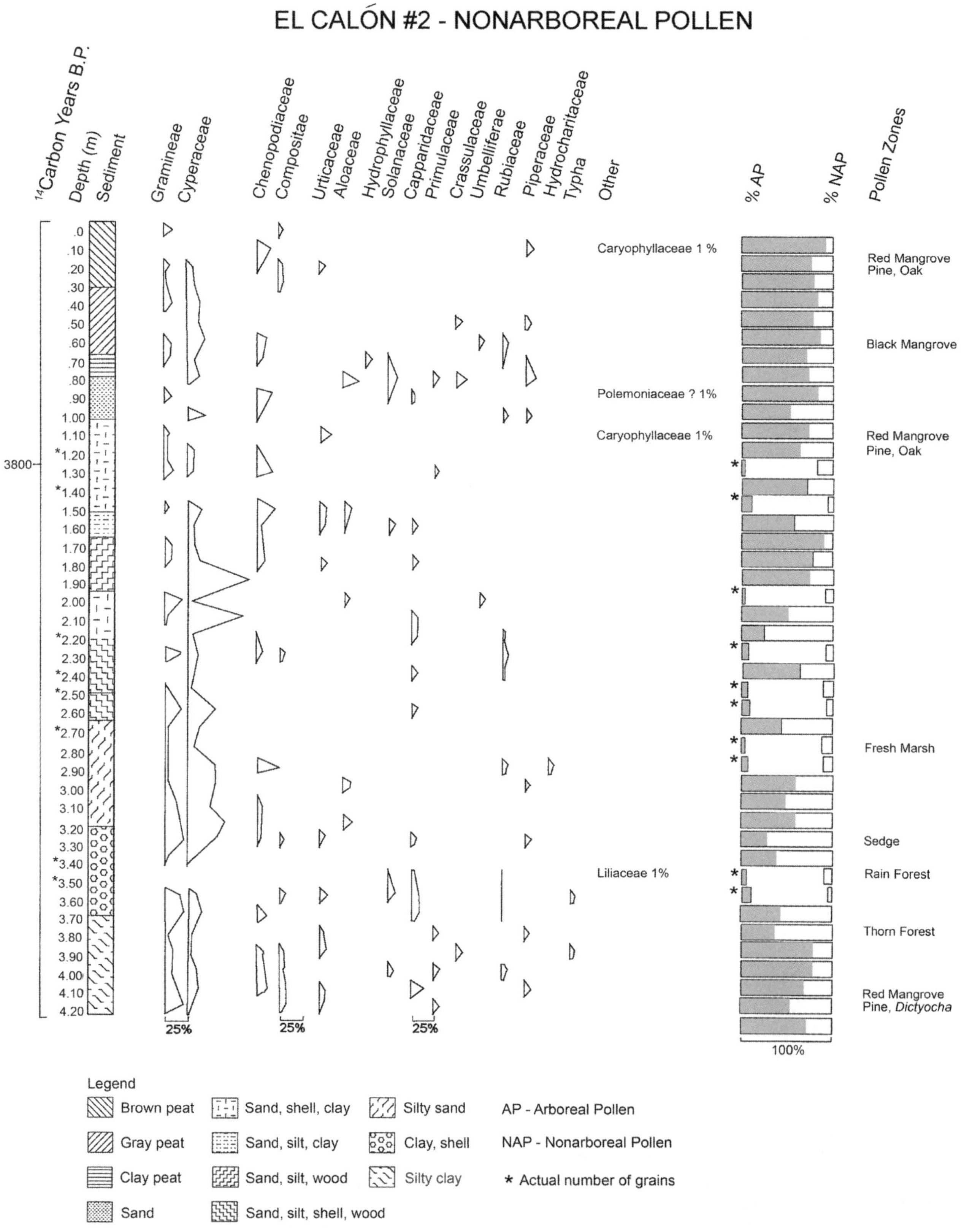

FIGURE 2.14. Nonarboreal pollen diagram, El Calón #2.

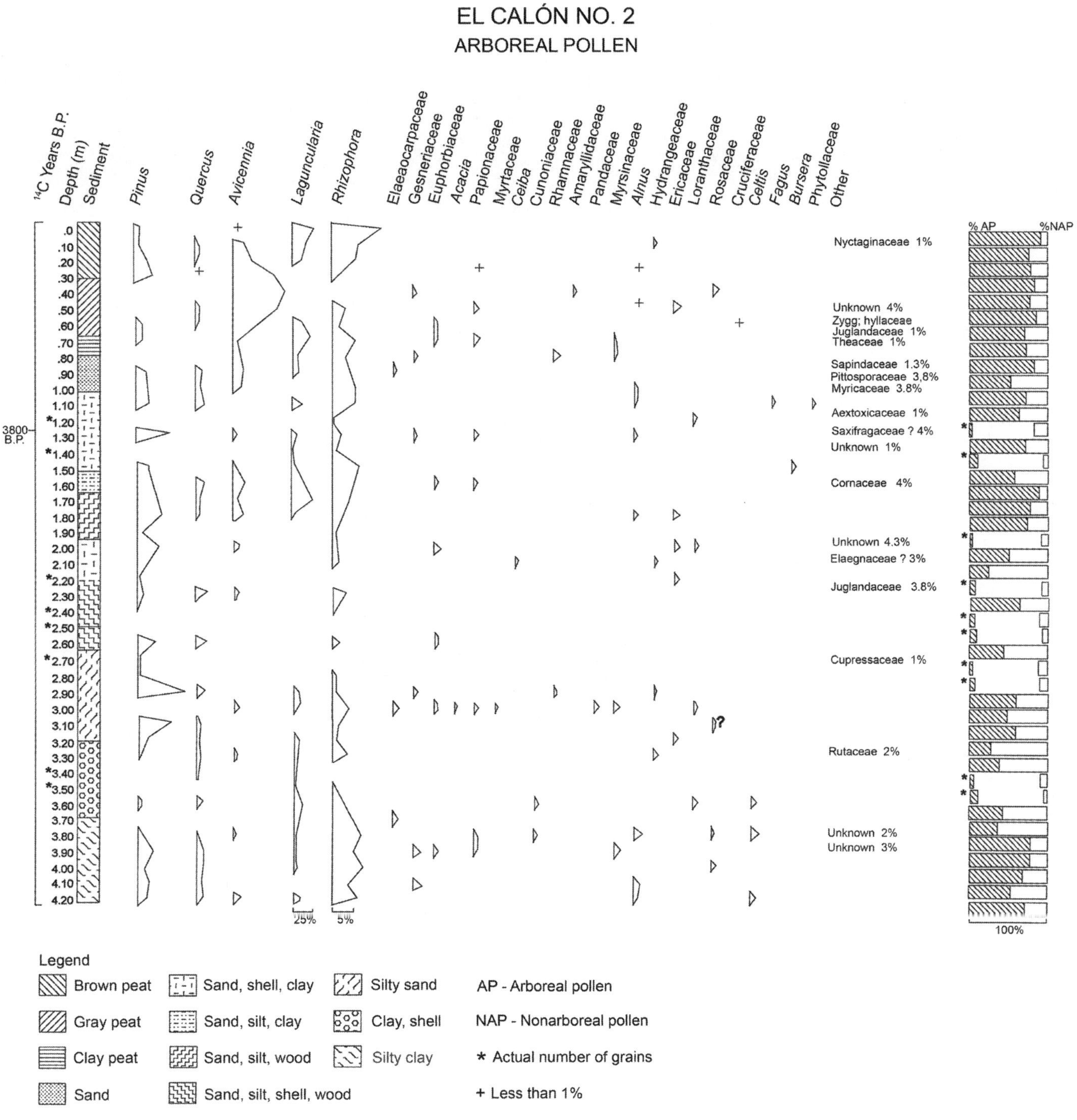

FIGURE 2.15. Arboreal pollen diagram, El Calón #2.

the mangrove area that subaerial oxidation and insect pollination might account for the destruction or lack of pollen from cultigens.

El Calón Sediments and Pollen Stratigraphy

The estuarine region adjacent to the El Calón shell mound was systematically probed in order to locate the longest sedimentary section and to determine the morphology and nature of the substrate. The thickest accumulations of estuarine sediment were found along the western and southwestern margin of the mangrove within 10 m of the shell mound. This greater depth is attributed to a buried channel near the site, which follows the general curve of the mound and then trends westward away from the mound and toward the ocean. Several cores were taken from near El Calón and range

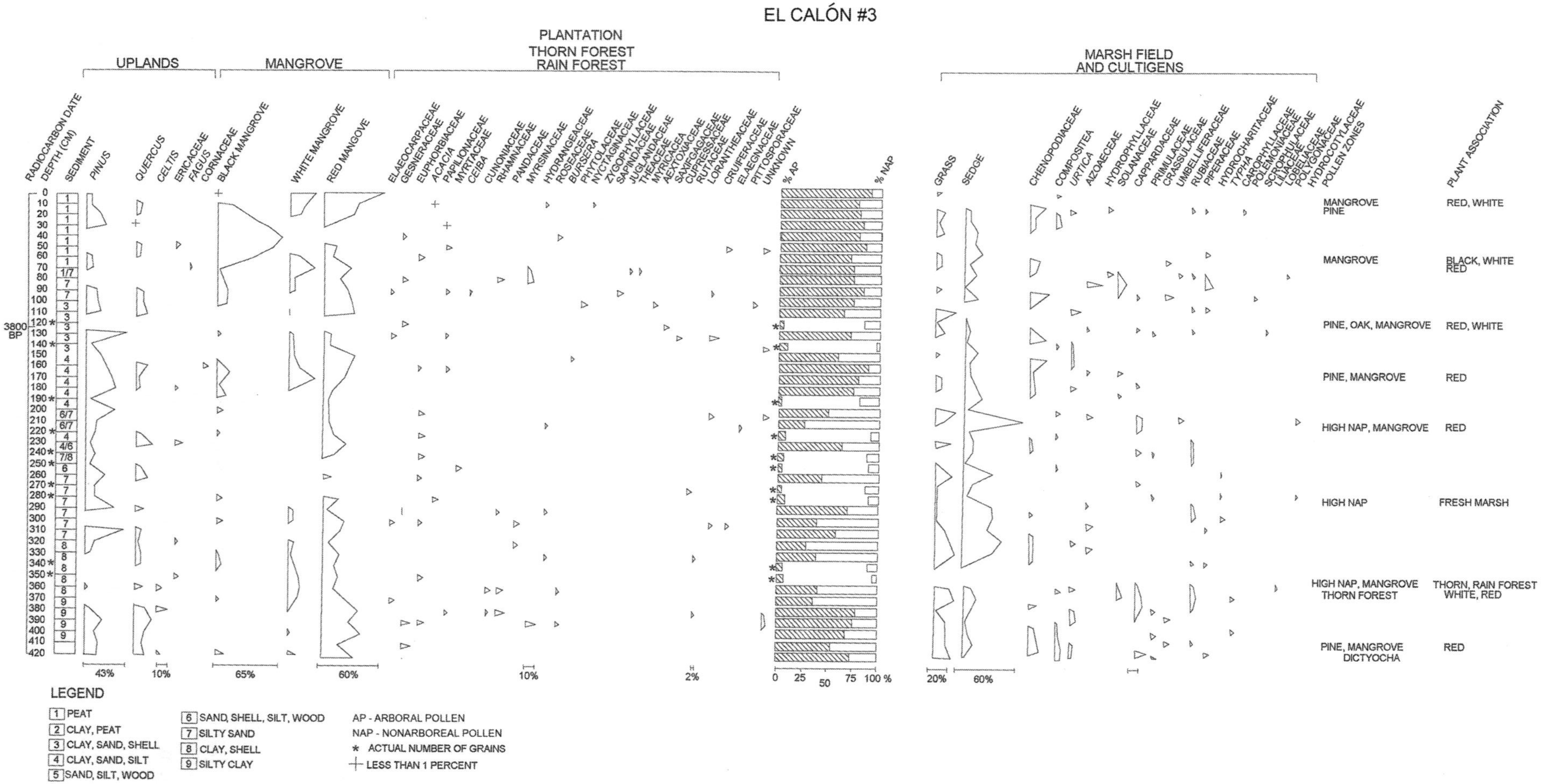

FIGURE 2.16. Pollen diagram, El Calón #3.

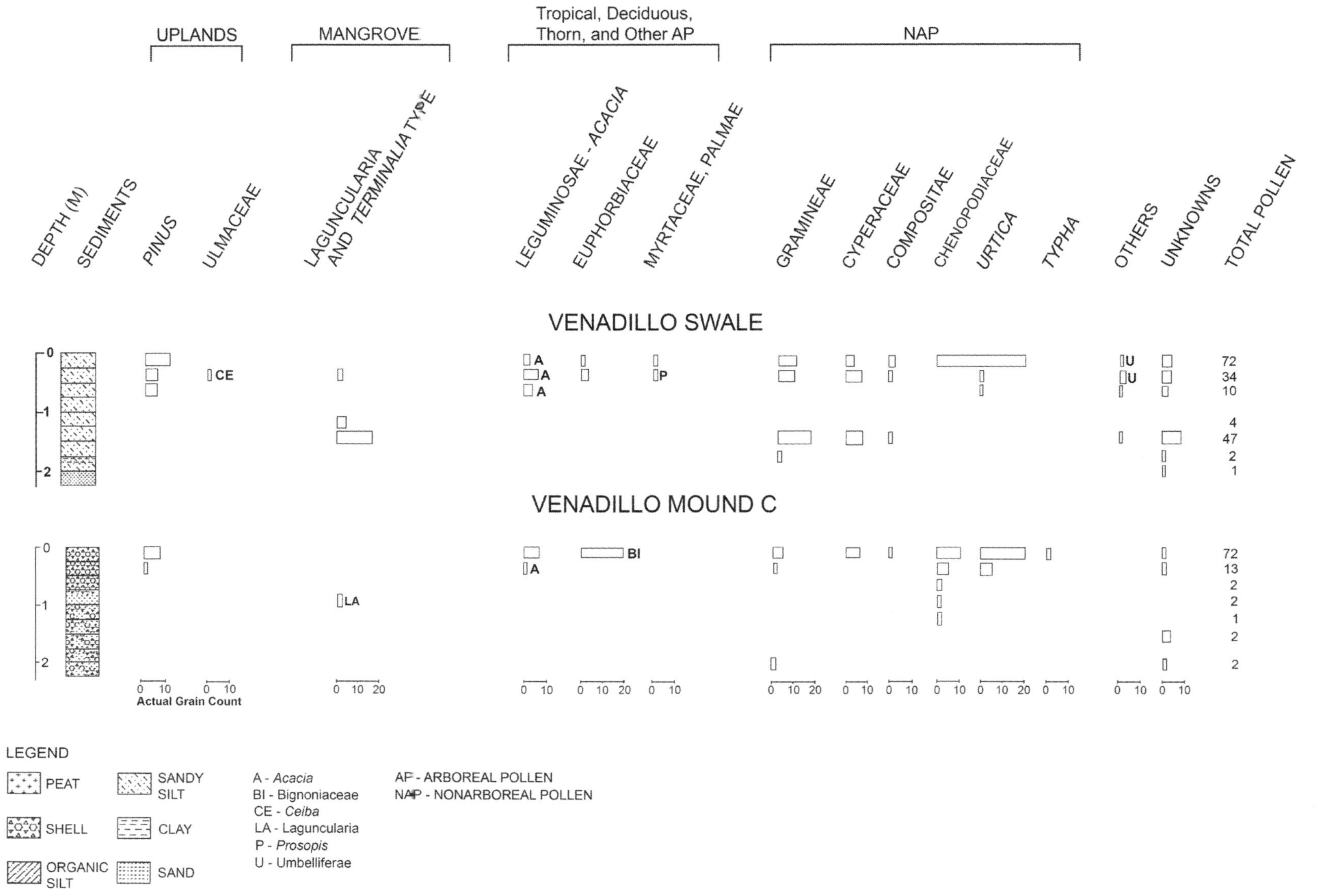

FIGURE 2.17. Pollen diagram, Venadillo.

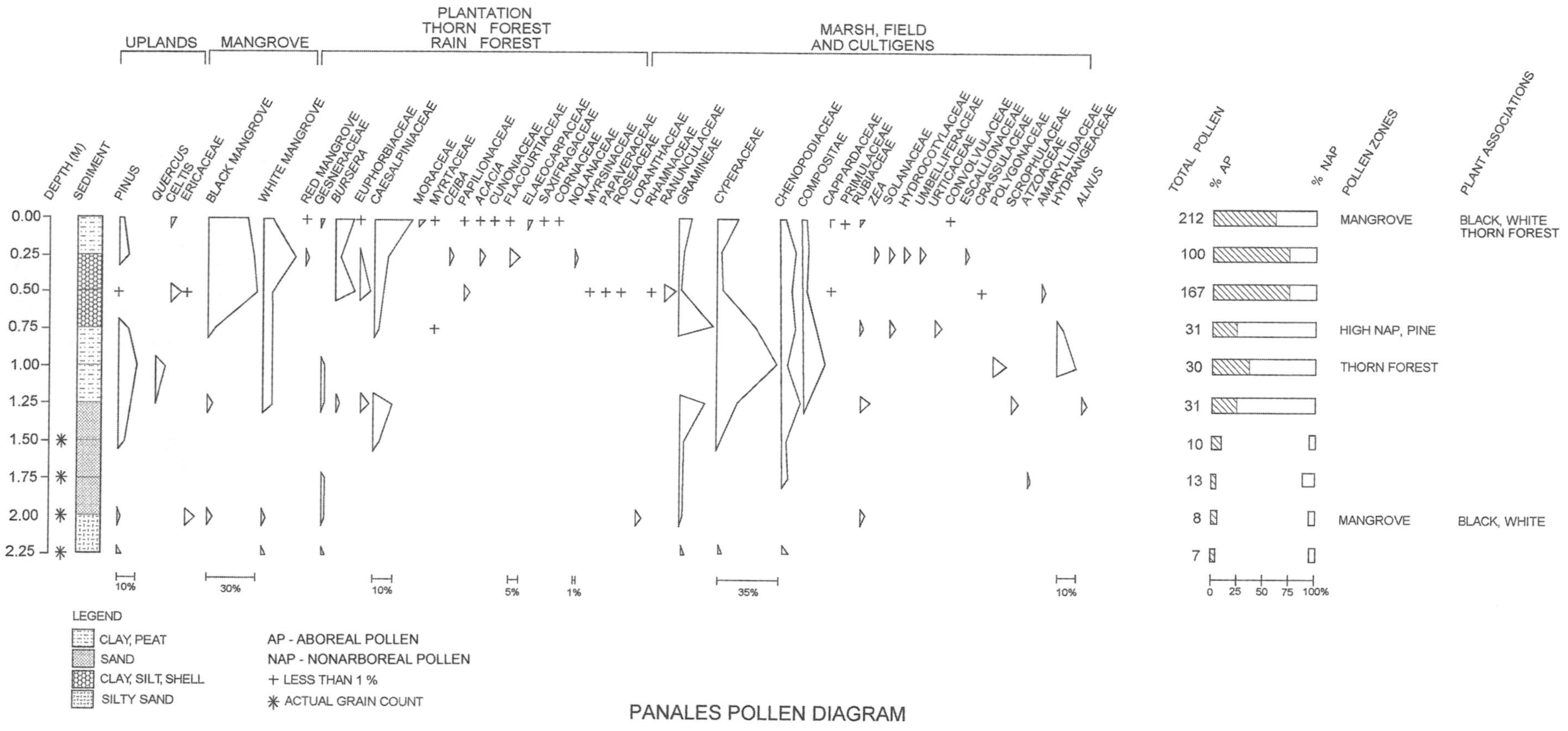

FIGURE 2.18. Pollen diagram, Panales.

in length from 6.0 m to approximately 7.5 m. The longest section was retrieved from the second core, EC-2, while EC-1 and EC-4 bottomed at 6.25 and 6.50 m, respectively. Much of the additional 1.00–1.25 m of EC-2 consists of sandy silt, clay, and sand. Core EC-3 was limited to the upper 4.5 m of the section. All of the cores penetrated the basal clastics, including the sandy alluvium beneath the clays. In EC-2, a thin (about 1-cm) peaty band was found near 6.5 m, but fossils have not as yet been recovered (Sirkin 1985).

The sediments above the basal clay are dominantly lagoonal silts with scattered shell fragments; very finely macerated organic debris; microfossil tests, of which foraminifera are common; and occasional thin (1- to 3-cm) bands of brown peat. Brown peaty sand overlies the clay at 6 m in EC-1. A prominent brown peat was found between 5.00 and 4.75 m in EC-1 and between 5.50 and 5.25 m in EC-2, but in EC-4 this layer may be represented by very thin peat bands in the lagoonal sediments from 5.50 to 5.25 m (Sirkin 1985).

A brown sandy-organic silt lens also occurs in EC-2 at 2.25–2.50 m, and brown fibrous peats are interbedded with and overlay conspicuous shell layers between 1.25 and 1.50 m. These sediments may correlate with the substrate on which the shell mound was built. The upper 0.75–1.00 m of the section consists of peat that is generally brown to gray in color and which becomes more fibrous toward the surface. It is presumed that this peat is derived mainly from the debris and root mass of the mangrove association. The lower 1 m of sand and clay in the EC-2 (see Figures 2.14–2.15) section has not as yet yielded any microfossils (Sirkin 1985).

In the EC-l column, the first pollen occurred at 6.25–6.50 m (see Figure 2.13), where Gramineae (grass) and Compositae (composites) make up 15 out of 22 specimens, along with three of *Rhizophora* (red mangrove) and one each of *Alnus* (alder) and *Urtica*. The very low pollen sum reflects very low pollen influx perhaps due to the oxidation of organics in an arid, terrestrial environment at that time. At 6 m mangrove pollen is abundant and totals 55 percent of the sum, including red mangrove (11 percent), white mangrove (25 percent), and black mangrove (3 percent), as well as additional taxa in the Verbenaceae family (16 percent) and a *Terminalia*-type pollen. Pollen of representative species associated with the lowland forests, such as Caesalpina, Sapotaceae, Bignoniaceae, Proteaceae, Euphorbiaceae, Myrtaceae (cf. Eucalyptus), and Palmae, is common. Upland and marsh species, except for grass (6 percent), are rare at this level (Sirkin 1985).

The pollen assemblage at 5.75 m is quite similar but with slight increases in red and white mangrove. While the Verbenaceae, in general, remain abundant, black mangrove decreases, and both Geraniaceae and Papilionaceae enter the record. Upland and marsh species remain low. At 5.5 m increases are noted in the mangrove species, and thorn forest–rain forest forms such as the Verbenaceae, Terminalia, Bignoniaceae, Mimosa, Caesalpina, and Bursera continue to be prevalent. The overall assemblage between 5.5 and 6.0 m is tentatively designated as a biostratigraphic zone, which is characterized by an assemblage of pollen from both thorn forest and mangrove vegetational associations. The mangrove is represented by three, possibly four, taxa and is initially dominated by white and then red mangrove, although red mangrove is the first to appear in the pollen record (Sirkin 1985).

A marked increase in upland species occurs at 5.25 m. *Pinus* (pine) and *Quercus* (oak) are prominent in this group, along with the first occurrence of the cedar group, *Thuja*, *Taxodium*, *Juniperus*, and *Chamaecyparis* (TTJC), and *Salix* (willow). Lowland species include Verbenaceae, Myrtaceae, Acanthaceae, and Pollemoniaceae. Cactaceae, Chenopodiaceae (chenopods, probably *Salicornia*), and *Typha*, representatives of arid environments and coastal marshes, also appear. Mangroves are represented, in the order of inferred dominance, by white, red, and black (Sirkin 1985).

A continued increase in upland species occurs between 4.25 and 5.00 m, and the TTJC assemblage includes specimens similar to *Taxodium*. Lowland species include *Caesalpina*, Geraniaceae, and *Ceiba* (which first occur at 4.5 m), along with red mangrove, grass, and *Typha*. *Ephedra* first appears at 5.0 m, and black mangrove and other Verbenaceae decrease. At 4.75 m, pine rises to 19 percent; grass, to 9 percent; and alder, to 4 percent. Upland representation continues high at 4.50 m, while mangroves decrease. Marsh and field varieties remain low until 4.25 m, where grass, composite, chenopod, *Typha*, umbellifer, and Cyperaceae (sedge) occur. Red mangrove is also prominent through 4.25 m, where the first evidence of *Dictyocha*, a marine silicoflagellate, appears (Sirkin 1985).

Dictyocha becomes very abundant at 4.00 m and then declines rapidly and disappears at 3.75 m. Upland species increase in this narrow zone, with pine (23 percent), oak (8 percent), TTJC, willow, *Ilex* (holly), *Betula* (birch), and alder occurring at 3.75 m. Red and black mangrove, with white mangrove already at a very low presence, decrease markedly at 4.00 m, but red mangrove increases significantly above this level. Thorn forest species including legumes are also subdued, but marsh and field varieties increase. Grass rises to 19 percent, along with composites, chenopods, *Urtica*, *Typha*, umbellifers, sedge, and Malvaceae. Euphorbiaceae accounts for 3 percent of the sum at 3.50 m, while both upland and lowland forms are poorly represented. Marsh and mangrove forms are prominent here, with red and white mangrove increasing (Sirkin 1985).

Ceiba occurs at 3.25 m, and both *Ceiba* and *Bursera* are in the assemblage of rain and thorn forest species at 3.00 m, along with the marsh forms. Red mangrove again peaks at 3.25 m and then declines in importance along with white mangrove. Pine increases from 13 percent at 3.00 m to 17 percent at 2.75 m in a general rise of upland species. Marsh and field species, such as grass (20 percent), composites (11 percent), chenopods (5 percent), and *Urtica*, also rise. Between 2.0 and 2.5 m lowland forest and marsh species are again well represented, but upland species are only moderately so. The first probable cultigen, *Cucurbita* (squash), appears at 2.25 m, and *Agave* appears at 1.75 m (Sirkin 1985:12). At 1.75 m, most of the groups are poorly represented, although pine peaks at 28 percent of the sum. Pine continues high at 1.50 m, along with grass (18 percent) and chenopods (10 percent). The upland association and the mangroves generally increase through 1.25 m. Pollen from *Cucurbita* and *Agave* occurs at 1.75 m (Sirkin 1985).

The water-saturated marine sediments between 0.5 and 1.0 m are very low in pollen, but they are rich in shell. The upper 1.0 m of the section makes up a separate pollen zone. Black mangrove rises at 0.50 m and peaks at 0.25 m, with 26 percent of the sum. Pine is also very abundant (22 percent), and marsh and field species also rise. Cultigens occur at 0.5 m (squash and melon), and in the surface sample, *Zea* (maize) and the Solanaceae are abundant. The upper zone is also characterized by abundant black mangrove and *Terminalia* pollen, while red and white mangroves are less represented (Sirkin 1985).

The EC-3 (see Figure 2.16) core is a 4.5-m section that exhibits several minor fluctuations in pollen content. Between the base of the core and 3.7 m, red mangrove is dominant and is associated with pine and oak (upland pollens) and sedge, grass, and alder of marsh affinity. Thorn forest species found at 3.8 m include those in the Euphorbiaceae, Cappardaceae, and Rhamnaceae families and *Celtis* (Ulmaceae). Between 2.8 and 3.8 m, possible rain forest species occur, including Rubiaceae, Myrtaceae, Hydrangeaceae, Gesneraceae, and *Ceiba*. Upland vegetation is again represented by pine and oak, and marsh vegetation is indicated by sedge, grass, and chenopods, probably *Salicornia*. A red and white mangrove association also occurs in this section (Sirkin 1985).

Sedge and grass are dominant between 1.9 and 2.7 m, and pine and oak appear only sporadically. Rubiaceae, Euphorbiaceae, and Cappardaceae occur in trace amounts, red mangrove is minor at 1.8 m, and white mangrove is dominant at 1.7 m. All three mangrove species occur in equal amounts at 1.6 m, and red mangrove is again dominant in the 0.7–1.1 m section of the core. Pine, oak, and various thorn forest species are also prominent there. Black mangrove is more abundant than red or white mangrove between 0.2 and 0.6 m. Sedge, grass, and chenopods represent the marsh component; plantation is represented by *Agave*, and cropland is represented by composites. The upper 20 cm of this core has high red mangrove, with minor amounts of white and black mangrove (Sirkin 1985).

Venadillo Sediments and Pollen Stratigraphy

The small lagoonal swale sampled at the archaeological site Venadillo yielded a 2-m core. From a basal 25 cm of sand, the organic content of the sediment increases. From 1.00 to 1.25 m the sample consists of light brown organic silt. Above this the amount of very fine clastic and organic sediment increases, and at 0.5–1.0 m the sediments have a clay binder. The upper 0.5 m of the section is mainly sandy peat grading upward to mangrove peat.

The earthen portion of Mound C at Venadillo, a burial mound capped with a shell midden, consists of as much as 2.10 m of gray sand and shell overlying oxidized barrier beach sands. The upper 0.80 m of this section is mainly oyster shell. A thin (10-cm) darker gray horizon separates the shell cap from the lighter gray sand segment and may represent a buried soil or a zone of organic enrichment (Sirkin 1985).

In the Venadillo swale core (see Figure 2.17), the pollen record corresponds to the upper part of the standard section from El Calón. The lower level of the core has only a sparse representation of grass and fern. Above 1.5 m, the salt marsh and adjacent savanna are indicated by the appearance of grass, sedge, and chenopods at this level. The mangrove succession here is limited to white mangrove, which in its abundance and in the absence of red mangrove attests to the emergent aspect of the land. Above 0.75 m the thorn and deciduous forests are represented by acacias and related trees, euphorbias, and cacti. The presence of pine in the upper part of the core shows a renewed influence of the upland. Evidently, human influence appears in the pollen record at this site, which is immediately adjacent to the site of Venadillo. For example, cropland is suggested by the appearance of *Cucumis* (Cucurbitaceae—cucumbers, melons). A surface sample from the swale demonstrates the pollen record of the salt marsh in the abundance of chenopod, grass (which may also be derived from adjacent fields), sedge, and composite pollen, along with the pine, *Acacia*, *Euphorbia*, and lily pollen from other associations (Sirkin 1985).

The pollen spectrum from Mound C at Venadillo (see Figure 2.17) is sparse but similar to that of the swale. The record provides evidence of salt marsh and white mangrove above 1.0 m and freshwater conditions above 0.5 m. Evidently, freshwater aquatic plants were available nearby or perhaps used by the occupants of the site as seen in the presence of cattail and pondweed. Savanna, field, and cropland are reflected in the presence of grasses, *Urtica*, and composite, while salt marsh is interpreted from chenopod and sedge. Thorn and tropical deciduous forests are represented by *Acacia* and other legumes, and upland, by pine and cedar. An abundance of Bignoniaceae, cf. *Sparattosperma*, is found in the upper part of the section. This form has not as yet been found in reference material and may have been derived from an imported, ornamental plant that may no longer be cultivated locally. The spores in the sample include those of *Lycopodium*, *Sphagnum*, and ferns (Sirkin 1985).

Panales Sediments and Pollen Stratigraphy

The Panales Island, which extends northward into the Teacapán Estuary on the east side of the Novillero barrier beach, is separated from that barrier by a canal. Linear shell mounds constructed mainly of oyster shell are piled to heights of 4.5 m on the west side of Panales. Small bog-like swales have formed between overlapping shell ridges and have filled in with fine sediment and peat. The Panales core, taken from the longest available section and located on the north end of the island, consists of 2.5 m of sediment. The lower 1.5–2.0 m section is composed of organic silt, shell, and sand. Peat with silt and shell forms the upper 0.5–1.0 m of the core (Sirkin 1985).

Pollen is scarce between 1.5 and 2.5 m, with only a few pine, red mangrove, chenopod, and sedge grains occurring in the samples (see Figure 2.18). From 0.75 to 1.50 m field and salt marsh species are dominant, commonly including grass, sedge, chenopod, composite, and Hydrangeaceae. White mangrove, black mangrove, and pine also occur in this interval in small amounts, while *Caesalpina* and *Bursera* represent the forest types. Mangrove, salt marsh, and field associations, as well as thorn forest and deciduous forest representations, increase in the upper 0.75 m of the core, and a slight rise in pine is also noted. These data correspond to similar pollen assemblages observed in the upper levels of the other sites. Regarding mangroves, black mangrove is several times as abundant as white mangrove, with only a trace of red mangrove in the uppermost sediments. In the salt marsh and field associations, first chenopod and then sedge and grass prevail along with a variety of herbs. This diversity of forms may reflect a freshening of the groundwater. A broad variety of forest types are also found in the upper sediments, with the greatest abundances being *Caesalpina* and *Bursera*, with lesser amounts of *Celtis*, Euphorbiaceae, *Acacia*, Rhamnaceae, Papilionaceae, and Moraceae as well as traces of numerous other species. The cropland or cultigen association is seen in the presence of *Zea* and perhaps some of the composites (Sirkin 1985).

Surface Samples

Surface samples provide a good representation of the regional distribution of pollen and of the modern taxa that are deposited and preserved. At El Calón all of the major groups are found, although mangroves are most abundant. Pine and a variety of thorn and rain forest taxa confirm the continued influx of these pollens. Estuarine environments are well represented by mangroves, particularly black and white mangrove, and the chenopods. Grass and composites reflect the nearby savanna and field associations. The presence of cultigens, although sparse in the Venadillo sections and at Panales, shows that agricultural sources can also become part of the regional record in spite of adverse effects on pollen distribution.

Archaeological Samples

Pollen is sparse or absent in the archaeological samples collected. Panales Mound B wall samples, as well as all other burial sediments and fill from vessels, were barren. The various soil samples from beach ridges, the bajada, river terraces, and croplands were also barren. However, Mound B at Venadillo contained the pollen of a possible cultigen as well as orchid, cattail, and sedge, and the shallow trench across the adjacent field, a presumed plaza area between Mounds B and C, contained *Acacia*.

Marismas Pollen Zones and Paleoenvironments

The different species and quantities of pollen in the various sections of the El Calón, Venadillo, and Panales cores and in the soil samples attest to a constantly changing Marismas paleoenvironment. Five major pollen zones can be identified, with fluctuations and changes that indicate changes in paleoclimate, water levels, and water salinity. The basal assemblages (Zone 1) in the El Calón cores contain pollen that indicates a dominance of mangrove vegetation in an early mangrove zone and a succession beginning with red mangrove in the first polliniferous sediments. Rising sea level may be inferred through this zone, from both the mangrove and the sediments, since peat and then estuarine silts cover the blue clay and alluvium. The lack of fossils in the clay may indicate its origin as an underclay or gley in the initial coastal marshes. Thorn and rain forests are also represented in this zone by several indicator taxa.

A pine-oak zone (Zone 2) succeeds the early mangrove zone, based on the dominance of these genera. However, lowland forest taxa, along with cacti, chenopods, and cattail, indicate other aspects of the coastal setting, including the presence of freshwater marsh. The appearance of *Ephedra* reflects the semiarid climate. Thin terrestrial peats in this section also confirm the minor fluctuations in sea level. The subsequent succession of white mangrove suggests more stable coastal conditions.

The *Dictyocha* zone (Zone 3) marks a marine event in the record during which flooding of the estuary occurred, giving access to marine planktonic forms. This transgression may have resulted from a short-term, storm-related breaching of the barrier or perhaps a tectonic adjustment of the coastal region prior to the next episode of coastal progradation. The subsequent pollen record, in which nonarboreal vegetation is prominent, shows increases in the presence of black mangrove, indicating a readjustment and stabilization of sea level.

In general, a zone of high nonarboreal pollen (NAP) (Zone 4), mainly grass, composites, and sedge, follows the *Dictyocha* zone. Increases of red and white mangrove suggest rising sea level. The details of the fluctuations in pollen associations in this zone are best seen in the pollen diagram from the El Calón #3 section (see Figure 2.16). Thorn and rain forest taxa are scattered through the interval, while NAP and mangrove alternate. Furthermore, the prominence of sedge pollen and brown peat horizons is seen as an indication of shifting transitional coastal environments and fresh peat accumulation alternating with mangrove and estuarine conditions.

A pine-mangrove zone (Zone 5) succeeds the NAP-rich interval. Red mangrove alternating with white mangrove indicates rising, perhaps fluctuating, sea level. Above this mangrove dominates, with red and white, then black, and finally a red and white alternation. Again minor sea level changes are the probable cause for the changes seen in the mangrove and marsh associations. These changes may have resulted from minor coastal tectonic movements and subsequent progradation. The increase in pine could be an indication of climatic cooling during this interval. Rain and thorn forest taxa continue to be distributed through this zone.

The Panales and Venadillo swale records are similar to that of the later mangrove zone, particularly in the black and white mangrove alternation. There is a close correlation between the dominant species at Panales and those in the regional standard at El Calón during this mangrove zone, except that the zone is interrupted by a pine-NAP-thorn forest influx. This may reflect the colonization of the linear shell mound and the adjacent open land by thorn forest, the development of freshwater marsh in the depression, and the continued infall of pine. The rise of black mangrove may show the development of paludal conditions nearby, and that of white mangrove may indicate a general emergence of, or increase in, marshland in the estuary with respect to sea level.

The general paleoenvironment, as summarized by Sirkin (1974), involved many fluctuations, including two major mangrove cycles. The first of these two cycles occurred when a red and then a predominately white mangrove environment developed on a sand substrate. The climate was cool, and the sea level was relatively low; but there was an abundant freshwater influx of pine and oak from the adjacent highlands.

Between the lower and upper mangrove sequences, an abundance of marine silicoflagellate in the pollen cores indicates that there was an open marine episode of brief duration. Then, in the second mangrove environment, red mangrove and alder briefly succeeded, followed by white mangrove and then black mangrove. Salt marsh elements such as chenopod, composite, sedge, and grass replaced alder. Then there was the development of savanna and thorn forest, and last, crop and plantation areas appeared along with highland vegetation, including abundant pine.

Radiocarbon Ages and Deposition Rates

Several radiocarbon assays were obtained for sediments from El Calón. For this discussion, the results are presented in uncalibrated ^{14}C years BP per Sirkin's (1985) discussion. The oldest samples that have been dated, using 1950 as the modern base reference date, occur in the 5.50–5.75 m interval at El Calón (EC-1); these are 7,390 ± 165 ^{14}C years BP (S.I. 2722A) and 7,020 ± 85 ^{14}C years BP (S.I. 2722C). A segment of core from EC-1, between 1.00 and 1.25 m, has a radiocarbon age of 4,645 ± 195 ^{14}C years BP (S.I. 2723A), while the age of *Tivela* shell from the archaeological mound at El Calón is 4,050 ± 80 ^{14}C years BP (S.I. 2728a) (4567–4094 cal BP; see Chapter 4) (Sirkin 1985).

Based on the ^{14}C dates cited above, Sirkin approximated the age of the sediments underlying the oldest dated or datable part of the section as well as sediment deposition rates in the area of the El Calón mound. He also calculated the rate of pollen influx between dated levels of the core. Using the radiocarbon ages for EC-1, the segment between 1.25 and 5.75 m, Sirkin suggested an age from about 4,600 to 7,400 ^{14}C years BP, or a span of 2,800 years for a 4.5-m core segment. The deposition rate for this interval was estimated to be about 0.16 cm per year. Thus, the sediment at 6.5 m in the El Calón cores might be 8,650 years old, and that at 7.5 m in EC-2 would be about 9,275 years old, a minimum age for filling of this part of the estuary (Sirkin 1985).

In summary, Sirkin (1985:24) notes that the pollen sequence from the Marismas supports the stratigraphic record of the Prograding Beach phase, the period after 8000 BP. During this period, variations in mangrove dominance are tied to fluctuations in relative sea levels that were caused by the eustatic rise of the sea offset by the tectonic uplift of the coast. The oldest pollen spectra were recovered from lagoonal deposits located behind the Teacapán Peninsula. These date to nearly 8000 BP, and Zones 1 and 2, which contain high percentages of red mangrove pollen, correlate with episodes of lower sea level, the formation of the Cañas Soil, and between 8000 and 6000 BP, a rise in sea level and the deposition of the Early Beach.

White and red mangrove dominate in Zones 3 and 4, which, initially, correlate with the development of the Teacapán Soil after 5650 BP and the subsequent deposition of the Medial Beach between 3000 and 2000 BP. Zone 5 contains evidence of increases in NAP and fluctuations in the mangrove representation. It is during this time that the Los Angeles Soil developed, the Late Beach was deposited, and the La Tambora Soil developed.

Summary

The geomorphic, biological, and paleoenvironmental studies conducted as part of the Marismas project, summarized above, have contributed significantly to the current understanding of the development of the northwest Mexican coastal plain and human occupation of a portion of the northwest littoral of southern Sinaloa and northern Nayarit. The geomorphic studies not only contribute to an understanding of human occupational history of the area; they are significant in their own right as geomorphic and geologic reconstructions. Building on and refining the earlier work of Curray and others (1969), Cazeau, Connally, Sirkin, and their associates defined three phases of development—the Alluvial, Ancestral Peninsula (also known as the Teacapán Peninsula), and Prograding Beach phases. The Alluvial phase (pre-Wisconsin in excess of 100,000 years ago) was a time during which an alluvial bajada formed at the foot of the Sierra Madre Occidental. This was the geomorphic base on which the West Mexican Coastal Plain would develop. During the following Ancestral Peninsula phase an embryonic spit appeared near the mouth of the Río Baluarte, well north of what is now the Teacapán Peninsula. Subsequently, the coastal plain emerged from that point southward with the southerly ocean currents. Connally (1984) dates this process sometime between 47,600 and 120,000 years BP. With the last glacial event and the lowering of sea levels the coastal plain began lengthening and widening. During this time the Laguna de Agua Grande formed when a preexisting large bay was cut off and enclosed as the coastal plain prograded. Subsequently, during the Prograding Beach phase (6,000 years BP to the present), there were four depositional events that led to the formation of the Teacapán Peninsula and coastal plain that is known today—the Initial Beach, Early Beach, Medial Beach, and Late Beach. Each is associated with a distinctive well-drained soil—the Cañas, Teacapán, Los Angeles, and La Tambora, respectively—manifested at the surface.

Additionally, Connally observed that sea level in the area had been rising over the last 14,500 years and that during this period there had been a 60-m rise. His studies concluded that there had been four periods when the Teacapán Peninsula was emergent and that the process of emergence and submergence in the area was the result of both rising sea levels and tectonic uplifting. Repeated emergence and submergence (either complete or partial inundation) of the coastal plain affected the history of human use and occupation. Previous archaeological studies in neighboring areas (e.g., Chametla and Amapa) have also revealed gaps in local archaeological sequences purportedly due to inundation of the coastal plain (e.g., Grosscup 1976). The geomorphic and geologic studies associated with the Marismas project certainly lend credence to the argument that portions of the northwest Mexican coastal plain were subject to inundation and that this significantly impacted local human settlement and mobility.

The geomorphic studies provide an understanding of the platform on which ecological, biological, and cultural components adapted to emerging and submerging landforms. Extant data on the flora and fauna of the area were supplemented by limited biological surveys by Jacobs (1982) and Wing (1969). Wing's (1969) brief ethnographic survey provides significant insight regarding the distribution of modern flora and fauna (terrestrial, estuarine, and marine). The palynological studies also reveal the succession of plant communities and types that appeared as the coastal plain developed. Unfortunately, due to poor preservation, the pollen provides little insight into human exploitation of botanical resources, cultigens, or wild plants. As will be discussed in succeeding chapters, this information, combined with studies of archaeofaunal remains and ethnohistory, contributes to an initial understanding of prehistoric human exploitation of the Marismas Nacionales and of cultural response to change in a dynamic coastal environment.

CHAPTER 3

Archaeological Background

Michael S. Foster

Archaeological investigations of Mexico's northwest coast, encompassing the coastal zones from Jalisco to northern Sinaloa, have been limited compared with those in many other parts of Mesoamerica (Gorenstein and Foster 2000; Figure 3.1). There have been few large-scale systematic surveys of the area, and by today's standards most of the excavation efforts would be considered limited testing projects. Nevertheless, a basic understanding of the temporal and spatial parameters of prehispanic cultural development in the region can be gained from a review of these earlier works. What follows focuses on the northern half of the northwest Mexican coast, from the area of Amapa, Nayarit, northward to northern Sinaloa, followed by a brief summary of the culture history of the southern sector of the northwest Mexican coast where the Marismas Nacionales sites share the greatest cultural affinity.

Sauer and Brand and the Aztatlán Culture

Although the archaeological remains of Mexico's northwest coast are identified in the early literature on the area, it is Sauer and Brand's 1930 survey of Sinaloa and Nayarit that provides the first detailed description and assessment of the area's archaeology and the first recognition of the importance of the region as a center of prehispanic cultural development (see Sauer and Brand 1932). Although Sauer and Brand recognized local cultural evolution on the northwest coast, they also recognized the broad pattern of stylistic elements and commonality of material culture that tied these cultures together. They (1932:31–41) named this the Aztatlán culture.

In December 1929, Sauer and Brand embarked on a three-month archaeogeographic survey of the area between the Culiacán Valley in the north and the Acaponeta Valley, Nayarit, just north of the Río Grande de Santiago in the south. The ceramic wares encountered consisted of well-made and highly polished monochrome and polychrome pottery. Vessel forms varied greatly, as did decoration and design elements, which included dotted circles, human heads with feather headdresses, stylized figures and graphic motifs, geometric and nongeometric patterns, and monochromes and polychromes that were frequently engraved. Additionally, a distinctive utility ware painted with a broad red rim and bands of various widths across the body of the vessel occurred throughout the area investigated. Again, it was the commonality of many of the design patterns on the ceramics across this wide area that led to the definition of the Aztatlán culture.

Other artifact types associated with the Aztatlán culture include figurines, pipes, incised spindle whorls, and cylindrical clay stamps. Urn burials also appear to be an Aztatlán cultural trait as well. Sauer and Brand noted a lack of projectile points on Aztatlán sites, although obsidian blades and debitage were common. However, when projectile points were found, they were usually well made from obsidian. Aztatlán stone carving was not particularly impressive.

Despite the widespread nature of Aztatlán traits and a certain level of homogeneity across the region, Sauer and Brand (1932:37–41) identified variations in local ceramic complexes that resulted in their defining subregions along the west coast. Culiacán Polychrome and a near lack of the red-rimmed and red decorated types found elsewhere defined the Culiacán subregion. A fluted type and a polished black or black-brown type were also apparently restricted to the Culiacán area, as was a "walnut- or mahogany-colored" type with incising on the upper half of the bowl exterior. *Molcajetes* (chili grinders) and tripod vessels with ball feet, many of which also functioned as rattles, were also found.

Chametla Polychrome, Aztatlán Ware, a well-polished blackware, a red-rimmed red-on-buff incised type sometimes decorated with white paint, a local version of the red-rimmed red-on-buff with black painting,

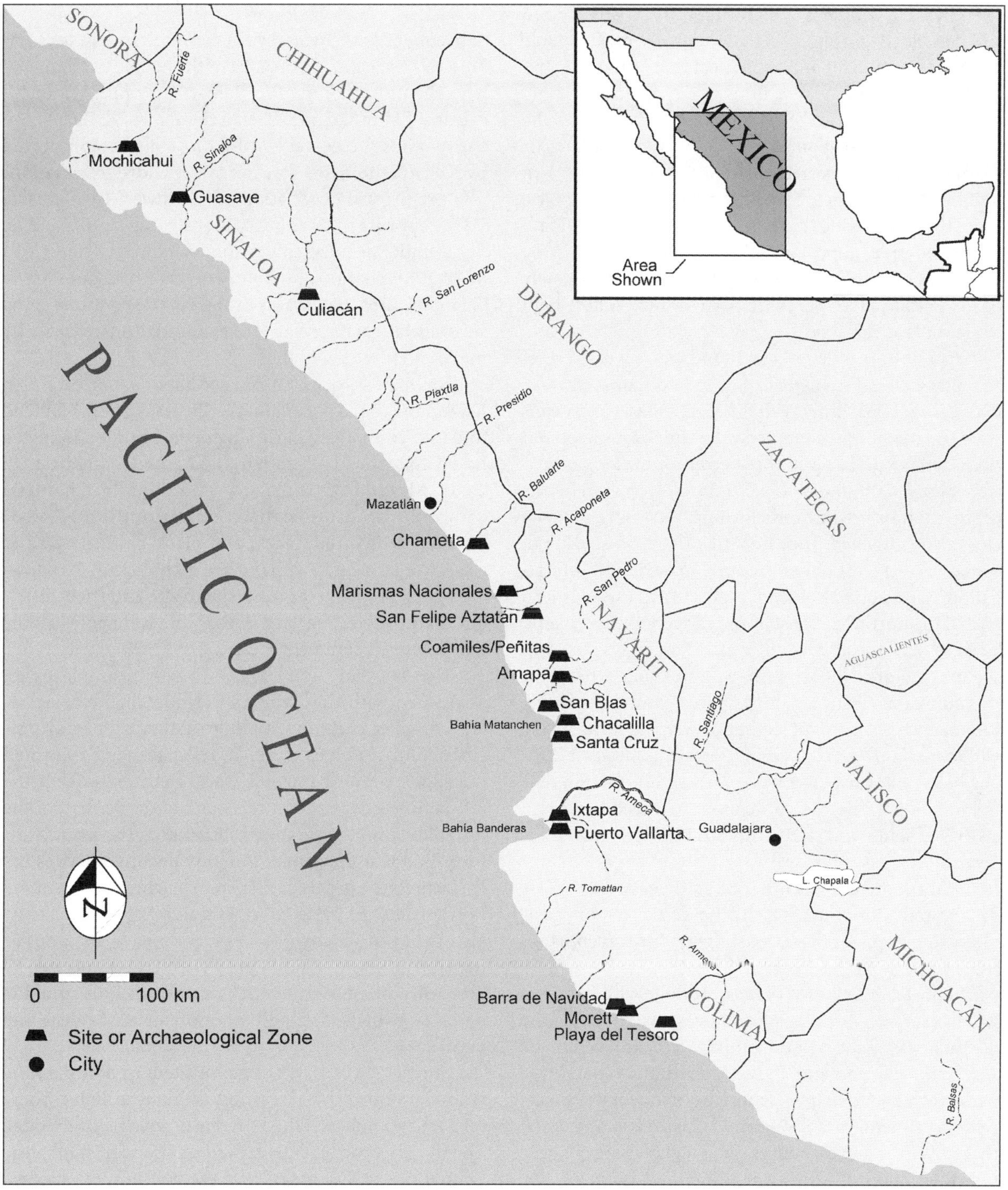

FIGURE 3.1. Selected major archaeological sites and zones of the northwest Mexican coast.

and modeled ceramic figurines defined the Chametla subregion (Sauer and Brand 1932:38–39). Sauer and Brand noted that Mazatlán Polychrome was the only ceramic type that seemed common to both the Chametla and Culiacán areas.

To the north of Culiacán, Sauer and Brand (1932:40–41) defined the Tacuichamona subregion. However, the Tacuichamona subregion appeared less well developed than either Culiacán or Chametla; sites were smaller, and the associated ceramics were crudely decorated. Although the Tacuichamona subregion exhibited some Aztatlán influence, it was not substantial, and the Tacuichamona subregion did not appear to be well integrated into the Aztatlán cultural expression. Kelly (1947:

56) notes that ceramics from the area are mostly incised and punctated and observes that the area is most notable for its unspectacular archaeological remains.

Sites were discovered in a variety of topographic settings, from the lower floodplains of rivers up into the foothills. In the Acaponeta area Sauer and Brand (1932: 22) recorded a truncated earthen mound nearly 11 m high, just over 15 m across at the top, and with a base more than 45 m wide. To the east of Mazatlán Sauer and Brand (1932:26) report a possible ballcourt consisting of a cleared rectangular area measuring approximately 30 m long and 22 m wide, with partially intact chest-high stone walls at each end.

Also of note is Sauer and Brand's (1932:51–56) brief summary of ethnohistoric Spanish documents (see Chapter 12). They observe that maize, squash, and beans were the major crops grown at the time of contact and that in many areas up to three crops of maize per year were raised. Cotton was also a common crop used for the production of textiles and clothing. A variety of fruits were also cultivated. They note that the Spanish did not observe, or at least did not report, any evidence of irrigation. Several types of fowl, especially in the southern part of the northwest coastal zone, apparently were kept, as were bees for honey and wax. Maguey was used in the making of pulque, but they do not state whether it was cultivated. Fish and shellfish were commonly used. Cayman were sources of leather, as were javelina. Along with plumage from birds, shells, pearls, gold, silver, copper bells, and turquoise were used as body ornamentation and jewelry. Native weapons included bows and arrows, shields, and clubs with obsidian inset as blades (*macauitle*).

Spanish documents mention rafts made from mats of cane and gourds. It was also observed that native peoples used large gourds to stash items for transport, as they would swim across rivers.

In the Chametla area houses of adobe with flat roofs were described as being of good construction. To the north, it was noted that the houses were also well made and built of wood and straw. Many houses had large front porches (*ramadas*), where women wove and carried out other domestic duties. In this area, it was noted that above the gables, houses often had painted ceramic figurines, many of which were described as obscene by the Spanish. These towns also had markets where a variety of items could be obtained.

Chiefs were carried about in hammocks. The Spanish also observed differences in dress and in physical appearance between the people in the south and those to the north and that these differences were further marked by linguistic differences. Sauer and Brand note that these groups were clearly of the Mexican cultural tradition.

Chametla, Sinaloa

Subsequent to the work by Sauer and Brand, in 1935 Isabel Kelly excavated in the area of the pueblo of Chametla near the mouth of the Río Baluarte, exploring four localities: El Tamarindo, Tierra del Padre, El Taste, and Cocoyolitos A (see Kelly 1938). Most of what is reported by Kelly came from the Tierra del Padre excavations. The prehistoric occupation of the lower Río Baluarte Valley appears to have been intensive, as manifested by the number and concentration of mounds on terraces of the river channel. In their 1930 reconnaissance of the Chametla area, Sauer and Brand (1932) mention seeing substantial rectangular stone foundations, but Kelly makes no mention of these.

Chametla's occupation spanned a period from AD 300 to 1200, a period that Kelly (1938:34–44) divided into four complexes: Early Chametla, Middle Chametla, Late Chametla II (Aztatlán), and Late Chametla I (El Taste–Mazatlán). The Early and Middle Chametla complexes predate Aztatlán remains in the area. Red-rimmed utility, red, black, and black-banded ceramic wares are associated with the two early complexes. However, their frequencies from site to site and their distribution through time vary. Also found were polychrome and engraved ceramic types.

The Aztatlán and El Taste–Mazatlán complexes are associated with the Aztatlán period. Chametla Aztatlán complex ceramics include Aztatlán Polychrome, Cocoyolitos Polychrome, a red-rimmed decorated ware, and a black-on-buff ware. Kelly notes that the red-rimmed decorated ware was a basic element of the Aztatlán complex ceramic assemblage. The most common El Taste–Mazatlán complex decorated types are El Taste Red-bordered, El Taste Polychrome, Mazatlán Polychrome, El Taste Satin, and El Taste Rough. Kelly also believed Cocoyolitos Polychrome to be a defining trait of the Aztatlán complex. Other associated El Taste–Mazatlán complex artifact types include smoking pipes, large incised spindle whorls, and late El Taste slab figurines.

Thirty-two burials were excavated; 29 of these were secondary burials in large, high-shouldered ollas. These ollas each contained the remains of a single individual. The remains consistently lacked a variety of small postcranial bones. It appears that the bodies were allowed to decompose and then the remaining bones were collected. It is also possible that they were dismembered and then placed in the olla (Kelly 1938:62–64). About a third of the olla burials were associated with funerary offerings, often miniature vessels. The funerary offerings were not abundant and included effigy rattles, some decorated wares, pottery drums, a flute, figurines, an obsidian knife, an obsidian blade, and a clay earplug, to name a few. Kelly (1938:36) suggested that the miniature vessels were specifically made to serve as funerary objects.

The other three burials were extended inhumations, one of which contained the remains of possibly three or four individuals. Subsequent to its discovery, the burial was vandalized, and the bone was damaged so badly that it was not possible to ascertain the actual number of individuals or anything about age or sex. A skull salvaged from the disturbed grave was well preserved, with a broad forehead and a highly vaulted skull. Kelly (1938:62) suggested that this interment might be postcontact and non-Indian.

Subsequent to the excavations at Chametla, J. Charles Kelley, working in the highlands of Durango at the Schroeder site (La Ferreria), a Guadiana Chalchihuites ceremonial site, made a significant contribution to the understanding of west coast archaeology with the discovery of numerous west coast intrusive ceramic types (Foster 1999, 2000; Kelley 1971, 1990; Kelley and Winters 1960). Kelley and Winters (1960), using the contextual data and ceramic associations from the Schroeder site, undertook the first critical assessment of the cultural sequences offered by Kelly for Chametla and Culiacán and Ekholm for Guasave by substituting phase names in place of the general complex designations and assigning temporal boundaries to the phases. Although the Kelley and Winters temporal designations were offered as tentative, subsequent work in the region has generally supported their designations, and they are commonly used today (Foster 1995; Grosscup 1976:248–254). At Chametla, the complex designations were replaced by the Tierra del Padre (AD 300–500), Baluarte (AD 500–700/750), Lolandis (AD 700/750–900), Acaponeta (AD 900–1050/1100), and El Taste (AD 1050/1100–1250) phases.

Guasave, Sinaloa

From 1937 to 1940 Gordon Ekholm conducted two major archaeological investigations in northwest Mexico (see Ekholm 1942). In 1937, he carried out a reconnaissance survey and limited excavations near Guasave as part of the Sonora-Sinaloa Archaeological Project directed by George Vaillant of the American Museum of Natural History, and between 1937 and 1940, Ekholm conducted an extensive reconnaissance survey that extended from Nogales, Sonora, on the U.S.-Mexico border southward to Culiacán, Sinaloa. He recorded 175 sites within this zone as well as surveying farther south near Mazatlán and in the uplands of Jalisco. As a result, he defined the Huatabampo culture and completed excavations at one site on the lower Río Mayo in 1939.

Ekholm also identified an artificial mound on the lower Río Sinaloa about 6 km southwest of the pueblo of Guasave. The mound was about 1 m high and 40 m in diameter. Ekholm described the feature as an unstratified silty sand mound that was of roughly circular form and suggested that periodic silting and changes in the Río Sinaloa drainage channels had obscured other such sites and living surfaces along the river's floodplain. He initially believed the mound to be a trash midden, and he began by excavating a meter-wide trench east-west through the mound.

Burials were immediately encountered, and after several seasons of intermittent excavations, 196 burials and associated artifacts were recovered. Ekholm recovered a variety of funerary offerings including ceramic vessels, shell, metal, and ground-, chipped-, and carved-stone artifacts. A variety of elaborate polychromes, incised and engraved types, polished red-on-buff, red- and plainwares, and grooved wares that came in various forms including bowls, tripods, and jars made up the ceramic vessel assemblage recovered from the Guasave burial mound. Other items included ceramic spindle whorls, smoking pipes, masks, clay plaques, animal and human figurines, whistles, beads, earplugs, and a cylinder stamp. Stone bowls, a pipe, manos, metates, pestles, axes, shell beads, pendants, bracelets, turquoise, copper bells, an earspool(?), a finger ring, a flat plate, and a necklace were recovered as well. There was also evidence of paint cloisonné on gourd fragments. Ekholm (1942:123–124) believed that most of the materials recovered postdated the early complexes at Chametla and that they were associated with the Aztatlán complex.

The richness and elaborate nature of the Guasave artifacts clearly reflect its status as a burial assemblage. Nevertheless, the Guasave materials demonstrate that the Aztatlán phenomenon was more complex and elaborate than described by Sauer and Brand or by Kelly (1938) at Chametla. Also of importance is Ekholm's (1942:125–132) recognition of similarities between the Guasave materials and central Mexican cultures. He (1942:127) listed 44 traits that he believed originated in central Mexican culture, several of which appeared to be directly derived from the Mixteca-Puebla culture. Ekholm believed that the Guasave burial mound reflected a single temporal and cultural component. He suggested that the Guasave materials were the "amalgamation" of three cultural "strains or complexes" (1942:124): one derived from the redware culture (the Huatabampo culture) of northern Sinaloa and southern Sonora, the second from the decorated ceramic-producing culture(s) of Sinaloa, and the third from strong Mixteca-Puebla influence, likely in the form of a migration of people from highland central Mexico.

Ekholm (1942) associated the Guasave burial assemblage with the Aztatlán complex of Kelly (1938, 1945), which, at the time, placed it between ca. AD 750 and 1100. However, as part of their reevaluation of coastal west Mexican archaeological sequences, Kelley and Winters (1960) suggested that the Guasave Aztatlán assemblage dated between ca. AD 1100 and 1300. Additionally, Meighan (1971) reports a radiocarbon date of AD 1220 ± 130 (corrected for the Suess effect). The sample dated was carbonized pitch or gum from a fragment of a coiled basket recovered by Ekholm.

Recently, the Guasave burials have been restudied (Carpenter 1996). Ten additional radiocarbon assays, six of which were from marine shell, one from cloth, one from string, one from a gourd, and one from a carbon sample, were processed (Carpenter 1996:229–232). Carpenter, like Ekholm, recognizes two distinct cultural/temporal components in the Guasave mound, an earlier Huatabampo component and a later Guasave component. Based in part on the radiocarbon assays, Carpenter (1996:232) suggests that the burials in the Guasave mound span a period of 800 years, from ca. AD 600 to 1450, which he (1996:233) separates into two periods, the Huatabampo period, AD 650/750–1050/1100, and the Guasave period, AD 1050/1100–1400/1450. Carpenter (1996:228–229) also cites Hosler (1988, 1992) regarding the technological attributes of the Guasave copper artifacts in his revision of the dating of the Guasave burial assemblage and places the Guasave copper assemblage in her period II, ranging from AD 1200/1300 to contact. Kelley and Winters (1960) proposed a date of AD 1150–1350 for the Guasave Aztatlán complex, which is not in conflict with Carpenter's dates.

Carpenter (1996), however, rejects Ekholm's (1942) suggestion that the "Mexican" or Aztatlán influences (decorative motifs, copper, etc.) seen on Guasave period ceramics and artifacts associated with the burials are the result of a migration of peoples from central Mexico and the argument that Guasave was possibly established to serve as a node in a long-distance Aztatlán mercantile system (Kelley 1986, 2000). Instead, in spite of his own acknowledgment that interaction with neighboring cultures to the south likely influenced burial practices at the Guasave mound, and that the Spanish observed widespread movement of people and items along the west Mexican coastal corridor at contact, he opts for an isolationist, in situ development explanation. Carpenter invokes the concept of prestige exchange to explain the presence of Mexican/Aztatlán iconography and items found in association with the Guasave period burials. However, this offers little in the way of elucidating the cultural processes and interaction that Ekholm, Kelly, and Kelley were seeking to understand on the basis of the archaeological evidence.

More recently, in an exhaustive evaluation of the Aztatlán phenomenon as an ideologically based entity, Mathiowetz (2011:588) argues, based on an extensive analysis of Aztatlán iconographic elements on ceramics and Aztatlán material culture, that the similarities between Aztatlán and central Mexico or Oaxaca pointed out by Ekholm and later J. Charles Kelley are so strong that there was direct interaction in one or both directions by Aztatlán and central Mexican/Oaxacan peoples. During the Postclassic period, the Guasave region did not exist as a backwater taking only what was wanted from the Aztatlán system ideologically or materially.

Culiacán, Sinaloa

In 1935 and 1939, Isabel Kelly conducted excavations and surveys in the vicinity of Culiacán, recording 63 sites (see Kelly 1945). In 1935, she excavated at five sites on the southern edge of the river valley, and in 1939 she surveyed and excavated at four sites in the "playa country" on the coast. Site density was high, with sites ranging from artifact scatters with trash or architectural mounds partially destroyed by agricultural activities to large villages that exhibited complex intrasite distribution of mounds arranged in parallel rows or, more frequently, around a central plaza. Contact-period observations in the Culiacán region indicate that such plazas frequently served as marketplaces.

Kelly recovered an array of material culture. Ceramic types defined include the Aztatlán types; Early Culiacán, Navolato, and Aguaruto polychromes; Aguaruto Incised; and Cerro Izabal and Alamitos Engraved wares. The Red-rimmed and Aztatlán wares common elsewhere along the Sinaloa coast were recovered in limited numbers. Among the other types of artifacts recovered were anthropomorphic ceramic figurines including numerous pancake-head figures with coffee bean eyes; rattles; smoking pipes; incised spindle whorls; cylindrical stamps; horn and bone tools; shell bracelets, beads, and pendants; copper bells, pendants, and wire; and flaked- and ground-stone implements and effigies.

At Las Lomitas, excavations exposed a hard-packed surface with postholes, which was thought to represent a floor of a house with, possibly, a ramada attached as a porch. A few charred macrobotanical remains representing maize and squash were recovered. The 1935 excavations recovered a small amount of marine shell representing food refuse (Kelly 1945:143). However, sites near the seashore excavated during the 1939 season yielded greater quantities of marine shell food refuse, none of which was identified.

Kelly (1945) divided the Culiacán sequence, which she saw as the evolution of a uniform tradition, into four complexes, the Early Culiacán II (Aztatlán complex), Early Culiacán I, Middle Culiacán, and Late Culiacán, all of which postdate the Early and Middle Chametla complexes dating from AD 900 to 1530 (Kelley and Winters 1960). The Kelley and Winters (1960) revision contributed to the identification of four phases that spanned that period: Acaponeta (AD 900–1050/1100), La Divisa (AD 1050/1100–1250), Yebalito (AD 1250–1400), and La Quinta (AD 1400–1530).

Although Aztatlán items were not common at the Culiacán sites investigated, Kelly (1945:118–121) did find what she called a "pure" Aztatlán manifestation in her Early Culiacán I materials, concluding that the Aztatlán expression in the area preceded and was partly contemporaneous with the Early Culiacán I complex. Kelly (1945:Table 17) had made surface collections at the sites she recorded and believed that Sauer and Brand's geographic distribution of the so-called Culiacán culture was supported.

The Culiacán excavations (five sites) also produced 252 burials; 190 were secondary olla burials, two of which contained two individuals (Hulse 1945). The skeletal remains in the ollas either were allowed to deflesh through decomposition or were defleshed prior to interment. The people of the Culiacán area seem to have been more diligent in their collection of small postcranial elements for internment in the burial ollas. Bones from the hands, wrists, feet, and ankles were placed into the cranial vault through the foramen magnum. The radii and ulnae or sometimes fibulae were inserted into the foramen, and the skull was set upon them right side up, forming what Hulse (1945:187) called a "quadrupedal" structure. The long bones, vertebrae, and ribs were then piled neatly beneath and around this structure prior to the sacrum and innominate bones being placed on either side and behind the skull. Most of the ollas contained the remains of a single individual, although there were instances of the burial of multiple individuals in a single olla. The ollas were most frequently buried in trash and very close to the ground surface. There were only three instances where pits were carefully cut into the local hardpan for the placement of a burial olla.

Mortuary offerings were few, and only 36 of the 252 burials were accompanied by grave goods. Plainware ceramics were the most common mortuary offering, with the exception of three cases in which Culiacán Aztatlán types were present. None of the grave goods appeared to have been made for use exclusively as mortuary offerings. Both children and adults had grave goods, but they were most common with adults. Early Culiacán II (Acaponeta phase) burials appear to have been placed in cemeteries rather than in trash deposits as occurred in later times. Both urn burials and inhumations occurred in Early Culiacán II times. Hulse (1945:191) suggested that it was likely the urn burials and inhumations occurred in Early Culiacán I times since they were found in context preceding and subsequent to that period.

Two of the burials recovered are of particular interest in that Hulse (1945:198) believes that they represent contact-period interments. Hulse identified one as unquestionably white, and the second, a probable female, appeared to have a bullet hole in its forehead. Both were buried in urns in the fashion described for the earlier aboriginal urn burials.

More recently, Cabrero G. (1989) reports on the excavation of multiple urn burials and inhumations discovered during excavations associated with the construction of classrooms for a preparatory school on the east side of the town of Culiacán. Five urn burials and four inhumations were excavated. As found by Kelly (1945), four of the five urn burials were not deeply buried, and the urns appear similar in form to those that she describes.

The University of California at Los Angeles Projects

In 1956, the University of California at Los Angeles (UCLA) initiated an extended program of archaeological and ethnohistorical studies in west Mexico (Nicholson and Meighan 1974). Work began with excavations at Peñitas and the excavation of a single test unit at the site of Coamiles. In 1959, excavations were conducted at Amapa. From 1960 through 1962, UCLA conducted Project A, an extensive survey and testing project along the west coast. Portions of the west Mexican coast south of the mouth of the Río Grande de Santiago were surveyed to identify sites for excavation. Subsequently, excavations were carried out at the sites of Barra de Navidad (Long and Wire 1966), Playa del Tesoro on Manzanillo Bay, and Morett (Meighan 1972). Three seasons of work were completed on the coast before the focus of Project A turned to the uplands of Jalisco.

Peñitas, Nayarit

The Peñitas excavations are described in a Ph.D. dissertation by Jacques Bordaz (1964). His work focused on ceramic production. Two loci were examined, Site A, north of the village of Peñitas, which was the focus of the investigation, and Site B, east of Peñitas. The Peñitas sites appear to have been occupied between ca. AD 400 and 1300, although a radiocarbon date of AD 180 is associated with some of the earlier materials (Bell 1971). Bell (1971:700) indicates that an area roughly 100 × 125 m at Peñitas A contained a cobblestone platform mound, two looted burial mounds, two habitation mounds, and three pit kilns. Several trenches were excavated at Peñitas B on the Río San Pedro, about 1.5 km east of Peñitas A. The occupation of Peñitas was divided into three phases, the Tamarindo, Chala, and Mitlán. Based on ceramics, the Tamarindo phase wares appear to be related to shaft tomb materials from Preclassic Nayarit, while Chala and Mitlán phase ceramics, especially the incised wares, exhibit strong affinities to Amapa ceramics. Some of the later Chala and Mitlán phase types also exhibited similarities to or included Chametla types. Chala phase types include Early Chametla Polychrome, Red-rimmed Utility Ware, and Black-banded ware. Mitlán phase ceramics, thought to be early Postclassic in date, exhibit similarities to Lolandis, Acaponeta, and El Taste phase materials at Chametla and to Acaponeta and La Divisa phase materials at Culiacán. The Peñitas materials also exhibited many design element similarities to types from central Mexico, especially Cholula. Von Winning (1956; Bell 1971) reports on a museum collection apparently looted from Peñitas that included copper bells, an alabaster effigy jar with blue and gold painted decoration, and several ceramic conch shells similar to some found at Teotihuacán.

Coamiles, Nayarit

A single test unit was excavated at Coamiles while excavations were being carried out at Peñitas (Bell 1971; Nicholson and Meighan 1974). Bell reports that the small assemblage of ceramic sherds recovered was unlike any at either Amapa or Peñitas. Coamiles is best known as a petroglyph site. The site was revisited in 1980 and only briefly reported (Soustelle 1980). Additional recording of the petroglyphs was undertaken along with the collection of surface artifacts. The ceramics included plainwares, monochromes, bichromes, incised monochrome and painted wares, and polychromes. A number of sherds are illustrated, but there was no attempt to identify the types represented. However, contrary to Bell's (1971) earlier report, Soustelle notes that the ceramics collected appear to be Postclassic period ceramic types related to both Amapa and Peñitas types. The ceramic figurines recovered were similar to those reported from Amapa and included Mazapan types (Grosscup 1961; Meighan 1976a). Other artifacts collected included incised spindle whorls, sherd disks, stone disks (beads?), and obsidian projectile points, blades, and flakes.

Amapa, Nayarit

In 1959, the most extensive excavation of a single site on the west coast occurred at Amapa on the delta plain of the Río Santiago about 15 km from the Pacific coast (Meighan, ed. 1976). The site consisted of as many as 200 mounds and included a concentrated central complex in the familiar Mesoamerican mound-and-plaza arrangement aligned to the cardinal points and mound groupings in patterned arrangements. Also present were well-executed elongated and steep-sided mounds, a ballcourt, mounds with masonry stairways, and cemeteries. The excavations focused on the ceremonial precinct of the site, an area 870 × 700 m. Six clusters of features, Groups A–F, were defined as broad excavation units.

The Amapa cultural sequence is divided into six phases, the Gavilán, Amapa, Tuxpan, Cerritos, Ixcuintla, and Santiago. Grosscup (1964, 1976) dates this sequence from about AD 100 to Spanish contact, with the Classic and early Postclassic period phases generally paralleling those of Chametla. The Tuxpan phase is presented as a hypothetical phase that represents an apparent disconformity between the Amapa and Cerritos phases. It is suggested, as in the case with the occupation of the Marismas, that this apparent hiatus in the occupation of all or part of the northwest Mexican coastal plain was tied to coastal subsidence. The subsequent Cerritos and Ixcuintla phases at Amapa are associated with the Aztatlán tradition.

Meighan (1976b) offered an alternative dating of the Amapa sequence based on a series of obsidian hydration dates. His sequence begins about 200 BC and ends about AD 1400, and although he recognized a hiatus in the sequence, he did not adopt the Tuxpan phase concept. Meighan dated this break between AD 400 and 600, whereas Grosscup dated it between AD 750 and 900.

Artifacts recovered from Amapa included an array of well-made and polished monochrome wares (buff, unslipped), bichromes, incised types (some of which were painted as well), and a series of striking polychromes (Bell 1960; Grosscup 1964, 1976). Vessel forms include plates and shallow bowls, composite silhouette forms, jars, tripods with full-rounded and slab legs, annular-base vessels (including molcajetes), and some double vessels. Grosscup (1976:254–264) noted many specific resemblances between the Amapa ceramic assemblage and the Aztatlán ceramics from Chametla, Guasave, and Culiacán. To date, Grosscup's (1964, 1976) work on the Amapa ceramics remains the most comprehensive study of its kind with regard to the ceramics of the northwest Mexican coast. Figurines (some Mazapan), ceramic stamps, beads, spindle whorls, carved-stone pendants, a variety of ground stone (manos, metates, axes, etc.), and assorted copper items including tweezers, needles, pins, and a fishhook were also recovered (Meighan, ed. 1976; Pendergast 1962a).

Excavations in the Group E area resulted in the exposure of two cemeteries (Meighan, ed. 1976:30–31). Cemetery 1 was located just west of Mound E-7 toward the northern edge of the Group E area. Mound E-1, built during the later portion of the Ixcuintla phase, probably after AD 1200, covered an earlier cemetery (Cemetery 2). The excavations of burials were greatly hindered by the fact that the water table in the area was high and most of the excavation pits kept flooding. This also affected the preservation of the bone, which ranged from partially to completely decomposed. Because of this, little was gleaned regarding burial practices. The few burials from Cemetery 1 recovered above the water table were seated in an upright position with their knees drawn (flexed) up to their chests. The bone material from Cemetery 2 was also poorly preserved.

Forty-five mortuary assemblages (funerary offerings), of which 12 were assigned burial numbers, were recovered. These usually occurred in clusters and included decorated, polished, and plain ceramic vessels; copper bells; spindle whorls; an earspool; obsidian blade fragments; stone beads; copper tweezers; copper slugs; and a Mazapan figurine. Meighan also notes the presence of six tool kits, one of which was for the making of ornaments. Six of the grave lots were assigned to the Santiago phase; eight, to the Ixcuintla phase; and 14, to the Cerritos phase. The remainder was not assigned a phase association (Meighan 1976c:130–135).

Meighan notes some patterns in his discussion of grave lots. The Cerritos phase ceramic offerings included higher numbers of tripod and larger vessels, and ceramic vessels were found to the near exclusion of any other type of offering. Nonceramic offerings were more common in the Ixcuintla phase grave lots. In Cemetery 1, the number of vessels ranged from 1 to 10. Again, the significance of these patterns is not clear based on the sample recovered and the fact that few of the burial lots could be associated with an individual of a known age or sex. Meighan noted some similarities between the burials at Amapa and those recovered at Tizapan el Alto, Jalisco (Meighan and Foote 1968), in terms of burial position and grave goods.

Finally, Meighan suggested that most of the mortuary offerings recovered, because of their quality and their apparent lack of use-wear, were made specifically as burial offerings. Exceptions include the above-cited tool kits, broken cooling vessels, used molcajetes, and several partial ceramic objects that were clearly placed in the grave in their incomplete condition. Meighan suggested that some of these items were the personal property of the deceased. He also pondered whether some of the broken ceramics may have been "killed," especially the *comales* or larger cooking griddles, and whether some of the large sherds recovered may have been used to line the grave.

The State University of New York Marismas Nacionales Project

The Marismas Nacionales project, the subject of this volume, was initiated in the summer of 1967 with a limited general reconnaissance of the Teacapán area to determine the feasibility of a research project there (Pearson 1968; Scott 1967). This was followed by six seasons of archaeological and geomorphic fieldwork that ended in 1973. A final field season of geomorphic work was conducted in 1974, and research continued into 1975 (see Chapter 1).

Recent Investigations

Although archaeological research on the northwest coast of Mexico remains limited, interest and activity have increased with research by archaeologists from the Instituto Nacional de Antropología e Historia (INAH) as well as the United States. In recent years, a considerable amount of work, much in the context of archaeological salvage, has been performed by INAH archaeologists. These include both survey and excavation projects. The following is a brief, although not necessarily comprehensive, summary of some of that work not previously mentioned.

Mochicahui, Sinaloa

At the northern end of the northwest Mexican coastal plain where cultures are associated with the Aztatlán tradition, near the pueblo of Mochicahui about 60 km northwest of Guasave, investigations resulted in the recovery of remains associated with the Aztatlán/Guasave period of northern Sinaloa (Manzanilla López and Talavera Gonzáles 1988). Mochicahui is in the area reported by Ekholm (1939a, 1939b) during his 1937–1939 survey of the Río Fuerte. Ekholm recorded 26 sites, including the site of Mochicahui. In the mid-1980s, during utility excavations by the municipality, 40 burials and associated funerary objects were recovered. The local government requested assistance from INAH for the analysis of both the human remains and artifacts. This, in turn, led to a broader program of archaeological survey of the area and salvage excavations at the locus of Los Bajos, where an additional 15 burials were recovered, six of which had offerings. One of these was an infant, and the remainder was adults between 21 and 35 years in age. Ten were identified as female, and three, as male. The preservation of the skeletons was generally poor. Ten were partially complete, with some evidence of the skull, long bones, and rib cage; the rest were fragmentary. One of the burials was a secondary burial, and two may have been partially removed. All were extended inhumations; two were buried facedown. Eight were buried with the head oriented to the north; five, to the northeast; and one, to the east. One was too fragmentary to determine orientation.

Several burials exhibited indications of heavy dental wear as well as irregularities in tooth enamel and frequent caries. One mandible, from Burial 13, exhibited no teeth and complete alveolar resorption. Some lesions in bone thought to be associated with arthritis were also observed. Although the number of females exceeded that of males, the sample size and the fact that only a small portion of the burial mound was excavated preclude drawing any significant inferences regarding social status and the relative distribution of burials.

Ceramic types identified with the burials and recovered from the excavations at Los Bajos included Mochicahui Plain, Mochicahui Brown, Taxtes Black, Guasave Red, Guasave Red-on-buff, Aztatlán Polychrome, and a type designated Vidriado Moderno. Mochicahui Plain and Guasave Red were the most abundant ceramic types present. Mochicahui Plain appeared in the form of shallow bowls with flaring sides; hemispherical bowls; miniature-necked, *tecomate*, and composite jars; *comales*; and small bowls with a tab on the rim. Manzanilla López and Talavera Gonzáles (1988) state that these types can be included in the group Ekholm (1942:80–81) called heavy red- and plainwares. Taxtes Black is reported as a variant of Guasave Red; it varies from dark brown to black in color and appears to be somewhat heavier and more crudely made than Guasave Red. It was recovered in a variety of forms including composite, effigy, and round vessels with three globular arms/spouts. Flaked and ground stone including ¾-grooved axes and marine shell ornaments, featuring *Glycymeris* shell bracelets, were recovered from the excavations and in association with the burials.

The site is reported as exhibiting both Huatabampo and Aztatlán influences. Manzanilla López and Talavera Gonzáles (1988:61) observe that the peoples of the Mochicahui area were clearly participating in the regional exchange network and cultural sphere known as the Aztatlán tradition. They propose a date range for the site between AD 900 and 1350.

Las Labradas and the Río Piaxtla Area, Sinaloa

The petroglyph site of Las Labradas is located on the coast approximately 50 km north of Mazatlán. Recently, Santos Ramírez and Vicente López (2012) have summarized the efforts undertaken to determine the cultural origin of the rock art. Several sites—La Flor de Océano, La Puntilla, and Lomas del Mar—were tested. La Flor de Océano is located on the Estero de Chicayota southwest of the town of Chicayota. Two mounds were tested. One excavation exposed part of a shell midden. The other exposed an occupational level that included an extensive lithic workshop. Evidence of all stages of lithic reduction was present. Material types were dominated by rhyolite, andesite, and basalt. A variety of lithic tools was present including various types of scrapers and some drills and knives. Some blades, thought to have been imported into the site, were also recovered. The tools are believed to represent a tool kit for the processing of shellfish. A small zoomorphic pendant carved from the apex of a gastropod was found during the excavation in the shell midden. A large amount of ceramics was recovered. Mazatlán Polychrome and El Taste Polychrome dominated, and a local type, Chicayota Red-on-buff, was defined. The presence of these ceramic types indicates a Late Classic–Early Postclassic period (Aztatlán) use of the site.

Additionally, and importantly, an earlier Archaic period component was identified at the site. No ceramics were recovered from a stratum containing a lithic assemblage composed of side scrapers, knives, bifaces/projectile points (leaf-shaped), and cores. The bifaces appear well made (Santos Ramírez and Vicente López 2012:Figures 13–14). The presence of coastal Archaic period sites is not surprising, but the documentation of such sites is a significant advance in the archaeology of the northwest Mexican coast as defined in this volume. Although no radiocarbon date was cited for the La Flor de Océano Archaic period component, the researchers observe that the bifaces they recovered and Pinto-Gypsum materials from the desert Southwest of the United States are similar and therefore date the site to as early as 8000–7600 BC, potentially.

Santos Ramírez and Vicente López (2012) also describe excavations at a second site, La Puntilla. It is located on the extreme southeast edge of Estero de Chicayota. The site consists of archaeological material scattered along the top of a hill on which there are five small shell middens (less than a meter high). The artifact assemblage, both ceramics and lithics, was far more limited than that identified at La Flor de Océano. The ceramics indicated a date of AD 750–1200 for the site.

Another site, Lomas del Mar, is also discussed by Santos Ramírez and Vicente López (2012). The site is located on the coast approximately 19 km north of the Estero de Chicayota. A variety of artifacts are scattered over a 600-m area on a rocky outcrop that is 5 to 6 m in height. The flaked stone includes a number of gray and green obsidian blades. Excavations were conducted in three parts of the site. Marine shell was abundant and indicative of shellfish processing and consumption. As at the other two sites, ceramics indicate an occupation between AD 750 and 1250. Based on the work at the three sites, Santos Ramírez and Vicente López (2012:28–29) suggest the presence of a Late Classic and Early Postclassic regional culture called Chicayota-Aztatlán.

Río Baluarte, Sinaloa, Surveys

The lower Río Baluarte Valley has recently been subject to two extensive survey projects (Grave Tirado and Nava Burgueño 2010, 2012). During the first season, 71 sites were recorded. Despite years of modern agricultural use of the fertile soils of the river valley's floodplain, many sites still have readily identifiable mounds on them. Some of the mounds are large. Cocoyolitos has four sets of mound complexes covering a 40-ha area. The largest of these are on a platform 400 m long × 160 m wide. An array of plain and decorated ceramic artifacts including sherds and figurines, ground stone (manos, metates, grooved axes), flaked stone, obsidian tools and flakes, and marine shell artifacts were recorded. During the second season of survey 38 additional sites were identified, most of which are described as small, sometimes less than 1 ha in size. Grave Tirado and Nava Burgueño (2012) suggest that these may be the remains of one or a small number of dwellings. During the first season of work, an attempt to develop a settlement classification was made. Grave Tirado and Nava Burgueño (2010) define four site types based on the size and complexity of the remains documented. Level 1 sites were defined on the presence of ceremonial architecture and ballcourts; only one such site, RB-1 (Chametla), was identified. Level 2 sites are defined as villages and hamlets with mounds, some of which are likely ceremonial structures. Level 3 sites were defined as small hamlets, which have abundant artifacts associated with them, and Level 4 sites are small, sometimes with shell middens, and are thought to be field houses or fishing/shellfishing camps occupied for short periods. Of the 38 sites recorded during the second field season, 28, or 74 percent, were Level 3 sites, and seven, or 25 percent, were Level 4 sites. Only a single Level 2 site was recorded, and no Level 1 sites were identified. Based on the survey results, the occupation of the Río Baluarte Valley extended from the Chinesco (Chinesco complex—200/100 BC–AD 200/250) through El Taste phases (200 BC–AD 1250/1350). It is of note that Grave Tirado and Nava Burgueño (2012:46) appear to be extending the terminal date of the El Taste phase to AD 1350. It is during this time that the recorded ceramic assemblages are most varied in terms of the types represented. In addition to the survey, the site of Loma de Ramírez was tested, resulting in the recovery of several human burials and a dog burial along with a variety of material culture including Chinesco, Chametla, and Amapa ceramic types, Chametla- and Amapa-style figurines; spindle whorls; whistles; and flaked and ground stone including fully grooved axes, overhanging manos, and metates. Based on the ceramic types present, the site spans a period from AD 0/200 to 1100.

Tierra del Padre Chametla, Sinaloa

Recently, Santos Ramírez (2012a) has summarized work conducted in the vicinity of Chametla subsequent to Isabel Kelly's (1938), including specifically the locus referred to as Tierra del Padre. In 1995, five urn burials were recovered from Tierra del Padre as the result of a salvage effort (Alducin Hidalgo y Terán 1995). The urns ranged from 45 to 80 cm in diameter, and three lacked lids. The largest also contained an obsidian projectile point 15 cm in length. The five urns contained the remains of six individuals; two were children (one very young), and the other four were young adults. Three were identified as females, and one, as male; the sex of the remaining individuals could not be determined. Alducin Hidalgo y Terán (1997) subsequently dated these burials to the Aztatlán period (Late Chametla II or Acaponeta phase—AD 900–1150) at Chametla (Acaponeta phase); however, Santos Ramírez (2012a) indicates that they are actually earlier and date to the Baluarte phase (AD 500–700). Talavera Gonzáles (1997) subsequently completed an osteological analysis of the remains from the urns and concluded that the corpses had been allowed to deflesh before the remains were collected and placed in the urns. The individuals in Urn 1 consisted of a young adult female and a young adult male.

In 2002 nine burial urns were salvaged (Santos Ramírez 2002). Generally, the remains were fragmentary and incomplete. Urns 1 and 3 contained miniature vessels, and Urn 3 contained a broken figurine. The urns ranged from 50 to 70 cm in diameter, with orifices that averaged 20 cm wide; they were approximately 60 cm in height. The urns were undecorated and were assigned to the Baluarte phase. Also recovered during the excavations were biconical and round spindle whorls, pieces of green obsidian, broken manos and metates, white-filleted figurine fragments, a small number of figurine heads with crested headdresses, and modeled figurines with a brown ceramic paste.

At Loma del Panteón, on the eastern edge of Chametla, construction associated with a baseball stadium resulted in the discovery of quantities of archaeological materials (Santos Ramírez 2008a, 2008b). A large 5- to 8-m-high mound (Loma del Panteón) is present, but it has been badly damaged by use as a material source for soil. Another such mound, Loma del Arco, is located approximately 100 m south of Loma del Panteón. A series of 1-×-1-m units were excavated, from which six secondary burials, two in urns and four others, were recovered. The remains were incomplete and in poor condition. Burial 1 was an urn burial. The plate covering the urn was a Red-rimmed Utility Ware, *mano colorado*. Also found with Burial 1 were two small, possibly turquoise, beads with incising on both faces. One also had, in addition to the central perforation, two holes drilled into one surface. The type of burial represented in Burial 2 could not be determined; the remains were disarticulated and fragmentary. It was thought to be a secondary burial. A miniature vessel was present and is thought to be a funerary offering. Burial 3 was a poorly preserved secondary burial. Below the burial was a complete monochrome bowl. No further discussion of the remaining burials is presented (Santos Ramírez 2012a). Additional excavations revealed the remains of a wall; however, work was curtailed by the onset of rains. The remains at Loma del Panteón were assigned to the Tierra del Padre and Baluarte phases (Santos Ramírez 2012a).

Coamiles, Nayarit

Even more recently, INAH archaeologists have been investigating Coamiles (Garduño 2008, 2009, 2012, 2013; Garduño and Vásquez del Mercado 2005), which is located approximately 5 km southeast of the pueblo of Tuxpan on the south side of the Río San Pedro, approximately 5 km southwest of the Peñitas sites, and approximately 11 km north-northeast of the site of Amapa. The site lies approximately 80 km directly south-southeast of the Teacapán Peninsula. As noted above, the site was recorded by UCLA, and one test pit was excavated at the time. Surveys by INAH in the mid- and late 1980s indicated that the area was the location of a substantial precolumbian occupation of considerable duration. More recently, prior to the excavations by Garduño and his colleagues, the Coamiles site was recorded, and the area was surveyed (Duverger 1998; Duverger and Levine 1993). The site consists of at least 40 structures—platform mounds and mounds of various forms and sizes that are isolated structures or parts of clusters of structures around plazas. These included large stepped platform mounds built on hillsides that were part of a civil-ceremonial complex that was clearly distinguishable from the residential portions of the site. The site appears well laid out, and it is estimated to cover more than 120 ha on the southwest portion of the Cerro de Coamiles.

Garduño (2013) identifies four primary architectural complexes or zones. Zone I, the South Complex, is located at the foot of the Cerro de Coamiles and consists of four main clusters of petroglyphs on large monoliths and boulder outcrops. These and two mounds oriented on an east-west axis serve to partially define a large, west-facing square. Garduño suggests that this complex may have served as the main access to the site's ceremonial precinct and settlement as a whole. Zone II corresponds to the major architectural center of the site. Much of the area was leveled and filled to provide open areas on which plazas and mounds could be built. This architectural complex includes the so-called North Acropolis, which incorporates an astronomical observatory (Platform 4) that was used to observe the equinoxes. Zone III is on the southern slope of the Cerro de Coamiles. Little investigation of this area has been completed, but terracing of the area, apparently for habitation, is extensive, and archaeological materials are abundant. Zone IV is on the lower fringe of the hill above the coastal plain. Some mounds and archaeological materials are present, but the area is not well documented. It appears that these mounds may have extended down onto the coastal plain, but they have been destroyed by land leveling associated with modern agriculture.

There is only a limited discussion of material culture other than ceramics. Obsidian workshop areas have been identified and have a variety of ground-stone implements and figurine fragments. Ceramics, both domestic utilitarian and decorated, were the most common artifact type recovered during the excavations. The ceramics appear to be dominated by types initially defined at Amapa (Grosscup 1976), and the ceramic sequence indicates a long occupation of the site from 100 BC though AD 1350. The earliest phase is represented by Chinesco Polychrome. This is followed by phases designated by the Amapa sequence including the Gavilán, Amapa, Cerritos, and Ixcuintla phases (Garduño 2013:87–88). Most notable is the dating. Garduño identifies a 100-year gap between the Amapa and Cerritos phases, which corresponds to the hypothetical Tuxpan phase (see Chapter 4) at Amapa, a time that saw a break in the sequence at Amapa and no apparent occupation in the Marismas. Still, it is not clear whether he is suggesting a break in the sequence at Coamiles as well. Given the elevation at the site, it is unlikely that the area would have been subject to coastal inundation during Late Classic and Early Postclassic times as discussed in Chapters 2 and 4.

San Felipe Aztatán

Garduño and Gámez Eternod (2005) report on salvage investigations at another large and important site, San Felipe Aztatán, located on the south side of the Río Acaponeta, approximately 37 km southeast of Teacapán. A previous record of the site identified the presence of a complex settlement with temple mounds, elite residences, workshops, and storage facilities that covered 100–120 ha. The main feature at the site is Loma de la Cruz, first reported by Sauer and Brand (1932:21–22 [Plate 4], 24). It is believed to be a major ceremonial temple mound that not only served the site of San Felipe Aztatán but perhaps was the principal ceremonial center for the lower Rio Acaponeta (Perez et al. 2000). Up to 92 structures have been recorded at the site, some of which ranged in height from 0.4 to 9.0 m. The site is oriented on near east-west and north-south axes. A similar but somewhat smaller temple mound, known as La Montosa, is reported at the site of La Guasima (Perez et al. 2000:85–91).

Among the material culture recovered were copper axes; a hook, a needle, a bell, and a heavy hanger-like hook; obsidian artifacts including prismatic blades; and plain and decorated ceramic artifacts. Many of the decorated wares were elaborately painted Aztatlán types. Obsidian sources include La Joya in Jalisco; Ixtlán del Rio, Nayarit; and farther afield, the Sierra de Navajas in Hidalgo, based on neutron activation analysis. Identified ceramic types include Tuxpan Engraved, Botadero Incised, Sentispac Red-on-buff, and Mangos Polychrome, as well as a Tohil Plumbate effigy jar that was also recovered.

Subsistence remains recovered included maize and *guaje* (*Leucaena esculenta*); marine mollusks including *Rangia*, *Tivela*, and *Donax*; and the medium-sized freshwater mussel *Unio*. Fish remains recovered consisted of *Cathorops* (sea catfishes), *Lutjanus* (snappers), *Eugerres* and *Chichlasoma* (mojarras), and *Ophisternon* and *Gobiomorus* (guavinas or sleeper gobies). Faunal remains included turkey, dog, white-tailed deer, hares and rabbits, and terrestrial turtle (Manrique 2004).

The occupational history of the site is also interesting. A Chinesco phase (AD 0–200/250) is represented and followed by Amapa's Gavilán, Amapa, Cerritos, and Ixcuintla phases, indicating an occupation from ca. AD 0/200 to 1350. A single radiocarbon assay based on excavations in the trash midden yielded a date range of AD 860–1160. Based on the amount of Aztatlán-related material in the excavated portions of the trash mound, the Cerritos phase occupation of the site appears to have been substantial; it is during that phase that it appears that an extensive architectural expansion of the site occurred.

Marismas Nacionales

Recently, the Marismas Nacionales has once again become the focus of archaeological research, through the efforts of Dr. Luis Alfonso Grave Tirado (2003a, 2003b, 2003c, 2010, 2011, 2012). Grave Tirado's work included testing of the large shell mound at El Calón and survey of the Teacapán Peninsula and parts of the adjacent piedmont, including investigations at the Juana Gomez, a site investigated early on in the Marismas project reported here. Grave Tirado's work at El Calón is discussed in detail in Chapters 4 and 5 of this volume.

During the first survey season the focus was on the Teacapán Peninsula, and 91 sites were recorded. Most of the sites were small shell middens. Four site types were defined: 43 were sites with large accumulations of oyster shell; 31 were sites with large accumulations of clamshell; six sites had middens with two types of shell present; and 11 were artifact scatters with no shell present. Tirado notes that some of the habitation sites appeared to have been occupied for an extended period but surprisingly lack any significant accumulations of shell. He further states that, based on the ceramic types present, sites with clamshell dated to the Baluarte phase, while sites with predominantly oyster shell dated to the later El Taste phase. Sites lacking shell or abundant shell remains are reported to have been occupied during the Baluarte phase, and some were occupied subsequently after a period of abandonment. Of particular note is Grave Tirado's (2012:20) observation that after AD 750 the area appears to have been unoccupied, except for Panales (see Chapter 5).

During the second season, fieldwork continued off the Teacapán Peninsula proper. A block area north-northeast of the peninsula between the villages of Cristo Rey and El Palmito was surveyed, as was a block area in the foothills of Cerro del Muerto, the marsh area surrounding Cerro de las Cabras, as well as part of the hill itself. Thirty-three new sites were recorded: 26 in the area between Cristo Rey and El Palmito, two in the wetlands at the foot of Cerro de las Cabras, one on the Cerro de las Cabras, and four on the foothills of Cerro del Muerto. Grave Tirado notes the presence of modern material on some of the sites. Other sites exhibited structural remains in the form of rock alignments, rock/masonry room and house foundations, and platforms (house and public or ceremonial structures), and one site (MSS-96 El Macho) may have a ballcourt. A variety of ceramic types and flaked- and ground-stone artifacts were recorded.

Grave Tirado places the occupation of the sites recorded during the second season of fieldwork between AD 500 and 1100. During the Aztatlán period, AD 750–1100, he sees an expansion in the number of sites, including those with more civic-ceremonial or ritualistic elements. He also notes the lack of colonial-period remains at the sites he documented while acknowledging the ethnohistoric accounts of substantial populations on the northwest Mexican coast at the time of the Spanish entrada (see Chapter 12).

Chacalilla, Nayarit

In 2003, Ohnersorgen began a research project at Chacalilla, Nayarit, beginning with survey and mapping followed by excavations (see Ohnersorgen 2004, 2007). The site is about 10 km north of San Blas on the small Río Zauta on the Cerro de Chacalilla. Chacalilla is both large and architecturally complex; and it is clearly a primary ceremonial center with a series of formally laid-out civic-ceremonial precincts made up of platform mounds and plazas. A ballcourt is also present. Excavations produced evidence of a predominately Early and Middle Postclassic occupation of the site. Ohnersorgen's research focuses on household and community organization, patterns of craft production, and trade in order to evaluate various models of political and economic interaction with the Aztatlán phenomenon. The 2008 excavations also produced evidence of human sacrifice.

The Southern Sector of the Northwest Coast

The coastal zone south of Amapa appears to have a longer history of occupation than the coastal region north of Amapa. Environmentally, the area differs from the northern segment of the Mexican west coast in that it is more subtropical. Work in the San Blas area and in the Tomatlán and Banderas valleys by Joseph Mountjoy has contributed significantly to our understanding of the prehistoric occupation of the southern sector of the northwest Mexican coast (southern Jalisco, southern Nayarit, and Colima). Mountjoy (2000) divides the region's prehistory into six general traditions, the earliest being the Coastal Archaic Shellfishing tradition, defined by excavations at Matanchén Bay just south of San Blas, Nayarit (Mountjoy 1970a, 1970b; Mountjoy et al. 1972). Radiocarbon dates suggest an occupation between 2200 and 1730 BC.

The following Capacha tradition, associated with the initial expansion of farmers into the coastal zone, includes three related local developments, the Capacha of Colima (Kelly 1974, 1980); El Opeño, Michoacán (Noguera 1942; Oliveros 1974); and a San Blas Capacha (Mountjoy 1970a, 1970b, 2000, 2012). In Colima, the Capacha tradition is dated between 1200 and 800 BC, and in the Tomatlán and Banderas valleys it is dated from as early as the 800s BC to as late as the 300s and 200s BC. There, ceramics include zone incised and punctuated pottery, some of which is painted, and ceramic figurines (including a Capacha-like figurine in the Tomatlán Valley) (Mountjoy 2000). Mountjoy (2000:85) further notes the presence of Capacha-like pottery at Amapa (Meighan, ed. 1976:478, Plate 172h–t) that is quite similar to Middle Preclassic pottery from the Río Balsas drainage (Chadwick 1971).

The Shaft Tomb and Tuxcacuesco traditions follow (Mountjoy 2000). After around 300 BC there was a significant influx of people into the southern coastal area. The Shaft Tomb tradition, dating between 300 BC and AD 500, is a unique mortuary complex centered in coastal and highland Colima, Nayarit, and Jalisco (e.g., Foster 2001; Kan et al. 1989; Townsend 1998). The Tuxcacuesco tradition (Kelly 1949) covers an area west of the Shaft Tomb arc, from the highlands of Jalisco to coastal Colima and Nayarit. On the coast, Tuxcacuesco materials have been identified in the Río Tomatlán Valley (Mountjoy 2000), at San Blas, and at the Morett site (Meighan 1972; Mountjoy 1974, 2000).

Along the southern west Mexican coast, the period after AD 300 to perhaps the early 800s is assigned to a generalized and somewhat amorphous Red-on-buff tradition (Mountjoy 2000:92–95). Mountjoy suggests that this tradition is expressed in the early phases at Amapa, Morett, Playa del Tesoro, parts of the San Blas, Banderas Valley, Tomatlán Valley, and the Río Armeria Valley in Colima. The Red-on-buff tradition is noted for considerable local variation in its ceramic assemblages, which include pottery with red designs on a buff-, brown-, or orange-slipped background as well as solid figurines. This was also a time of significant population expansion and a greater number of sites. Additionally, there is an increase in the architectural complexity of many of the sites, including large platform mounds with sunken courtyards and high masonry walls.

The Aztatlán tradition, AD 800/900–1100, follows the Red-on-buff tradition (Mountjoy 2000). In the Banderas and Tomatlán (Nahuapa I and II) valleys, many sites have distinct Aztatlán components. These sites range from small hamlets to larger villages with one or more small ceremonial mounds to sites with large platform mounds and ballcourts. Aztatlán ceramics include red-on-buff, black-on-buff incised, red-on-buff incised, and red-and-white-on-buff incised types. In the Tomatlán Valley, some of the red-and-white-on-buff ceramic types have codex-style designs, and the elaborate Iguanas Polychrome of Amapa is present. Mazapan figurines are found in Aztatlán contexts in the Banderas Valley.

Typical for the Aztatlán tradition, there is evidence for widespread trade. Obsidian prismatic blades and cores from the highlands of Jalisco (Glascock et al. 1997; Weigand and Spence 1982) and copper bells, plaques, needles, tweezers, buttons, a chisel-like object, and fishhooks are among the items found in Aztatlán contexts along the southern coast (Mountjoy 2000; Mountjoy and Torres 1985). In the San Blas area, Aztatlán materials are scarce.

The Post-Aztatlán tradition represents the terminal cultural tradition defined for the southern west Mexican coast. In San Blas, the Banderas Valley, and the Tomatlán Valley there is widespread evidence of local cultures that had their roots in the earlier Aztatlán tradition. However, the ceramic and metal items produced take on more localized styles. Post-Aztatlán times saw an increase in populations as well as movement of populations into more agriculturally marginal areas and into the mountains.

The Post-Aztatlán tradition is dated sometime beginning after AD 1000 to the early 1100s or later, depending on the area. The end of the Post-Aztatlán tradition occurred after Spanish contact, sometime during the mid- to late AD 1500s to the early 1600s.

Overall, Mountjoy (2000:106) sees the southern coast during Post-Aztatlán times as a place of fragmentation and diversification. Although there was contact between groups living in the foothills and mountain valleys and those living along the coast, the widespread cultural homogeneity of the preceding Aztatlán tradition was no longer.

Summary

Evidence indicates that portions of Mexico's northwest coast, particularly the southern sector (San Blas area), were utilized and possibly occupied by Archaic period hunters and gatherers and perhaps early agricultural groups exploiting shellfish and other marine resources (Mountjoy 1970a, 1970b, 2000; Santos Ramírez and

Vicente López 2012). Evidence of such an occupation farther north, in the area of the Marismas and northward, is limited or ambiguous. In the Marismas, the shell mound at El Calón (see Chapters 4 and 5) has been considered (e.g., Scott and Foster 2000; Shenkel 1969a, 1971a, 1971b) within an early Mesoamerican Preclassic period context; however, as discussed below, it may represent a late Archaic period feature based on radiocarbon dates derived from marine shell.

In the southern sector of the northwest Mexican coast, the Early Preclassic period is manifested in the Capacha tradition (e.g., Kelly 1980; Mountjoy 2000) and subsequent Shaft Tomb tradition (e.g., Foster 2001; Mountjoy 2000). The Capacha tradition has not been identified in the northern portion of the northwest coast, and although there are some stylistic similarities in the material cultures of the Shaft Tomb tradition and the early ceramic cultures on the northern west coast, the Shaft Tomb tradition does not appear to have spread northward into coastal northern Nayarit and Sinaloa; nor have Tuxcacuesco tradition materials and traits been identified.

Although the Shaft Tomb tradition is not represented in the northern section of the northwest Mexican coast, recent investigations have identified the presence of Chinesco Polychrome and Chinesco Black-on-cream, types associated with the Chinesco complex (100 BC–AD 200/250) related to the latter portion of the Shaft Tomb tradition of southern Nayarit. Where found at sites in the northern sector of the northwest Mexican coast, Chinesco ceramic types are generally dated between AD 0 and 200/250 (e.g., Grave Tirado and Nava Burgueño 2010, 2012), although to the south, the Chinesco complex is dated as early as 200/100 BC.

Mountjoy (2000:92) identifies the period from sometime after AD 300 to around 800 as the Red-on-buff tradition. He notes that the definition of this tradition is amorphous and lumps together "all that stuff" that occurs between the Shaft Tomb/Tuxcacuesco tradition and the Aztatlán tradition. The extent and nature of the Red-on-buff tradition in the northern reaches of the northwest coast are poorly understood, although Mountjoy (2000:93) notes its presence at Amapa. In the Marismas, this time is represented by the Late Estero period. No ceramic types associated with the Red-on-buff tradition have been identified, but it is unclear whether this is due to the lack of a complete and systematic analysis of materials recovered from the Marismas or whether it is not represented. Human settlement of the Marismas area was well established by this time, which is consistent with the expansion and increased settlement of Red-on-buff tradition peoples on the coast especially in the south, but Mountjoy (2000:95) also notes ties to the west Mexican highlands and possibly to the highlands of northwestern and north-central Mexico.

The northern sector of the northwest Mexican coast, including the Marismas area, was extensively settled during Aztatlán and post-Aztatlán times, as in the south. Settlements in the Marismas appear to have been limited to hamlets and villages that lack formal civic-ceremonial architecture. Nevertheless, they appear to have been economically and politically tied or allied to the surrounding areas where such sites are found. Although the region was heavily inhabited at contact, archaeologically the Late Conchera period, ca. AD 1350/1400 to the early 1500s, is not known in the Marismas. However, as discussed by Patricia Scott in Chapter 12, the Marismas is mentioned and described in the early colonial documents pertaining to the area, and it was noted as a significant source of oysters, as manifested in the linear mounds of oyster shell that line the Teacapán Estuary today.

Finally, as witnessed by the brief summary above of examples of recent and ongoing work by Mexican archaeologists from INAH, extant knowledge of northwest Mexican coastal archaeology has grown significantly since the end of archaeological fieldwork conducted by the State University of New York at Buffalo in 1974, especially in the last two decades. This work has greatly expanded our knowledge of the distribution of sites over the landscape, site types, site function, and material culture. Because of a lack of basic understanding of the scale and scope of the precolumbian occupation of the region, ongoing work continues to emphasize traditional cultural-historical issues, with some attempts to address broader anthropological questions simultaneously. Nevertheless, much of the region remains unsurveyed systematically, excavations are limited to relatively small parts of sites, and many basic archaeological questions built on assumptions, which in some cases are more than half a century old, have yet to be addressed.

CHAPTER 4

Chronology

Michael S. Foster, with Stuart D. Scott

In order to provide a temporal framework for the discussions that follow, this chapter presents a preliminary cultural chronology for the Marismas Nacionales. Stuart Scott and Betty Bell (Scott 1999; Scott and Foster 2000) initially divided the prehistoric occupation of the Marismas area into two broad periods—the Estero and Conchera periods. The Estero period was dated to the early to middle Mesoamerican Classic period. There was no initial date for the period, and it ended at approximately AD 600. The Estero period was considered to be generally contemporaneous with the Tierra del Padre through Baluarte phases of Chametla (Kelley and Winters 1960), the late Gavilán through Tuxpan phases at Amapa (Early period at Amapa) (Grosscup 1976), and the Tamarindo and Chala phases at Peñitas (Bordaz 1964).

The Conchera period was assigned a beginning date of ca. AD 800 with an end date of AD 1450/1500, the time of the Mesoamerican Postclassic period. It was seen as contemporaneous or overlapping with the Acaponeta and El Taste phases of Chametla (Kelley and Winters 1960), the Acaponeta through La Quinta phases at Culiacán (Kelley and Winters 1960), the Cerritos through Santiago phases at Amapa (Late period at Amapa) (Grosscup 1976), the Guasave phase (Carpenter 1997; Grosscup 1976; Kelley and Winters 1960), and the Mitlán phase of Peñitas (Bordaz 1964).

The Estero and Conchera periods covered the precontact ceramic period from the Mesoamerican Early/Middle Classic period through the Postclassic period, but the Scott and Bell scheme did not include the earliest cultural manifestation identified as a result of the Marismas project, the large shell mound El Calón, which was simply cited as the earliest evidence of a human presence in the area. The dating of the Estero and Conchera periods is based primarily on ceramic cross-dating and the temporal boundaries established for the neighboring phase sequences, particularly those defined by Kelley and Winters (1960) and Grosscup (1976; Table 4.1).

The available chronometric data are made up of a series of radiocarbon assays obtained from charcoal and shell. Additionally, Meighan and others' (1968) obsidian hydration dating for west Mexican archaeological sites was assessed.

Cultural Chronology

This chapter presents a modification of the Scott and Bell scheme in an effort to define more precisely the occupational history of the Marismas region. In addition to the Estero and Conchera periods, a third period—the Costanero period—which precedes the Estero period, is added (Figure 4.1). As will be noted there are also several breaks (see also Chapter 2) in the geomorphic and cultural chronological sequence representing times of rising sea levels, subsidence, or inundation when there was either limited use or a series of hiatuses in human settlement of the Marismas coastal plain area.

Costanero Period

The term *Costanero* is meant to reflect the fact that only a narrow coastal strip was present during this period. Temporally, the period spans the Late Archaic and Preclassic periods and is thought to date from 2500 BC to AD 300. It is likely that any early occupation of the Marismas coastal plain at this time was limited and sporadic and, perhaps, based on seasonal exploitation of marine resources (e.g., Mountjoy 1970a, 1970b, 2000; Voorhies et al. 2002). Little is known regarding this period with the exception of El Calón, which, based on radiocarbon assays, may have been constructed between 2400 and 2200 cal BC, with at least one apparent addition sometime between 1600 and 1200 cal BC. Suffice it to say, El Calón is a Costanero period feature, although nothing is known of the period between 1200 BC and AD 300, the

Table 4.1. Major Coastal West Mexican Ceramic Types Identified by Period at Marismas Nacionales Sites.

Marismas Nacionales Period	Chametla Types[a]	Amapa Types[b]	Culiacán Types[c]	Peñitas Types
Early Estero	Tierra del Padre phase	Gavilán phase		Chala phase
	Early Chametla Polychrome			
	Early Black-banded Engraved			
Late Estero	Baluarte phase	Amapa phase		Chala phase
	Late Black-banded Engraved	Amapa Red-on-cream		
	Black-banded Nonengraved			
	Middle Chametla Polychrome			
	Middle Chametla Polychrome Engraved			
	Scalloped Rim			
	White-on-buff			
	Black-on-buff			
	Red-rim Decorated (Lolandis Red Rim)			
	Aztatlán Ware			
	Polished Buff			
	Buff Exterior, Red Interior			
	Red-rimmed Utility Ware			
	Red-rimmed Utility Ware, *mano colorado*			
	Baluarte Red-on-buff/ Red-on-brown			
	Polished Whiteware			
Early Conchera	Acaponeta phase	Cerritos phase[b]	Acaponeta phase	Chala phase
	Cocoyolitos Polychrome	Cerritos Polychrome	Aguaruto Polychrome	
		Botadero Incised	Aztatlán Ware (Culiacán variety)	
		Botadero Black-on-buff	Culiacán Incised Ware	
		Mangos Engraved	Culiacán Band Incised	
		Sentispac Red-on-buff	Navolato Polychrome	
	Roblitos Polychrome[d]	Tuxpan Red-on-orange		
	Cixto Incised[d]	Tuxpan Black-on-orange		
Middle Conchera	El Taste phase	Ixcuintla phase[c]	La Divisa phase	Mitlán phase
	Mazatlán Polychrome	Botadero Red-on-white	Aguaruto Exterior Incised	Red Incised/ Engraved (Entries 1–4, 7, 8)
	El Taste Polychrome	Cerritos Engraved	Aguaruto Interior Incised	Polychrome (Entries 43–49)
	El Taste Red-bordered	Ixcuintla Polychrome	Aztatlán Ware (Culiacán)	Polychrome (Entries 50–54)
	El Taste Satin	Ixcuintla White-on-orange	Red-rim Decorated (Culiacán)	
	Heavy (Deep) Ribbed	Mangos Polychrome	Early Culiacán Polychrome	
	El Taste Rough	Otates Polychrome	Navolato Polychrome	
		Peñitas Engraved		
		Santiago Engraved		
		Sentispac Buff		
		Sentispac Polychrome		
		Sentispac Black-on-buff		
		Tuxpan Engraved		
Late Conchera		Santiago phase	La Quinta phase	

[a] Kelly 1938.; [b] Grosscup 1976.; [c] Kelly 1945.; [d] Putative Marismas types (see Chapter 6).

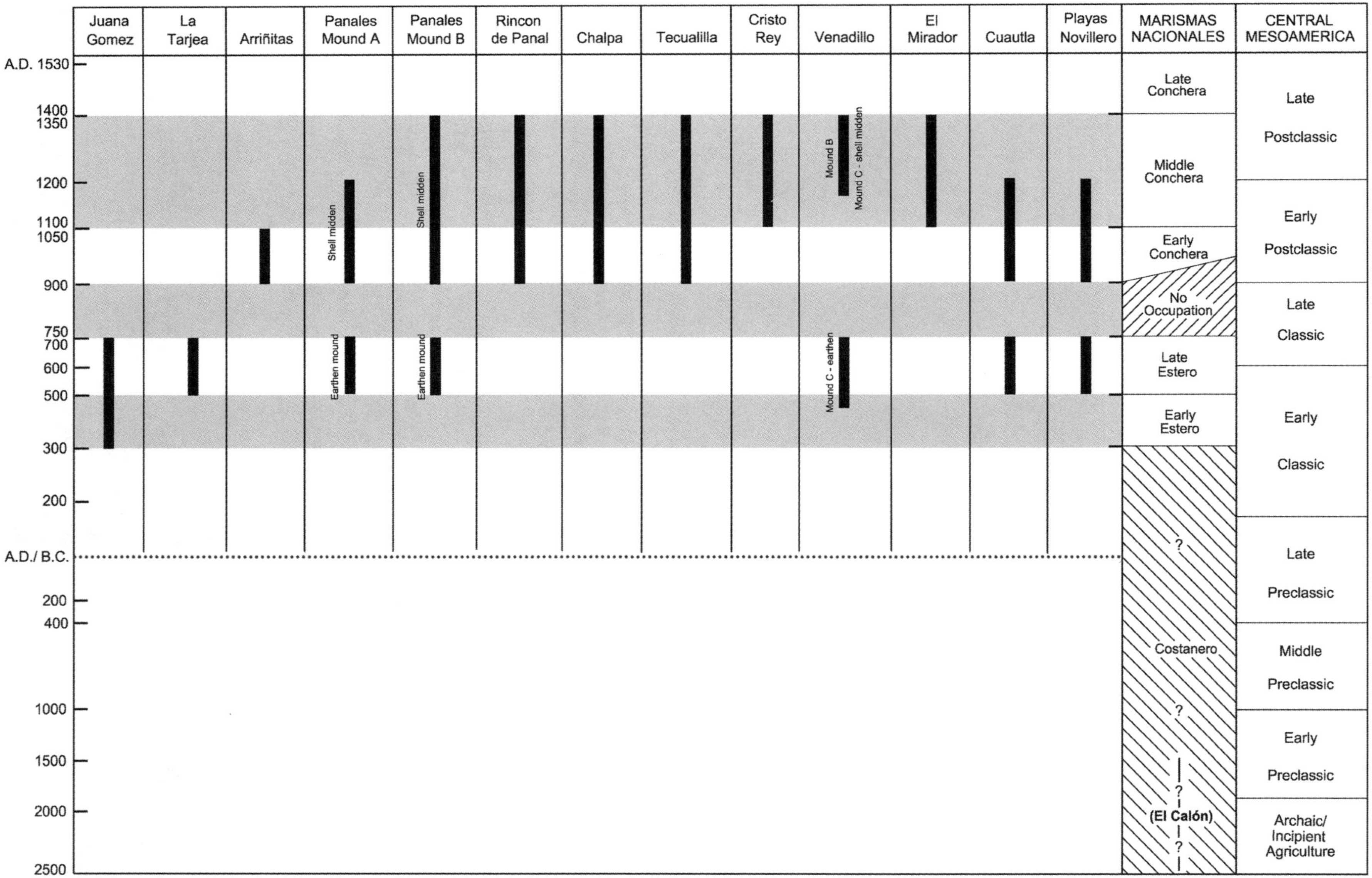

FIGURE 4.1. West Mexican and Marismas Nacionales cultural chronologies.

time after what is believed to have been the last depositional event at El Calón and when a sea level rise likely made the area uninhabitable.

Estero Period

The Estero period represents the initial ceramic-period settlement of the yet emerging coastal plain and developing estuary. Human occupation of the northwest Mexican coastal zone from southern Nayarit to central Sinaloa was widespread during this time; village life was well established, and there is clear evidence of social complexity and formal ceremonialism. The Estero period is suggested to date between ca. AD 300 and 700 and is here divided into two subperiods, Early and Late. The Early Estero, AD 300–500, is contemporaneous with the Tierra del Padre at Chametla and in part with the Gavilán and Amapa phases at Amapa. It is worth noting that Curray and others (1969) report radiocarbon assays in the AD 300s and 400s for some of the *Tivela* middens in the Marismas. The archaeological significance of the Curray dates is unclear, although they suggest further evidence of the Early Estero period. The Late Estero phase is dated AD 500–700 and is contemporaneous with the Baluarte phase at Chametla and Amapa phase at Amapa.

Occupational Hiatus

Following the Late Estero period, between approximately AD 700 and 900/1000, there appears to have been a general abandonment of the Marismas coastal zone (see Chapter 2). This period equates to Grosscup's (1976) hypothesized Tuxpan phase at Amapa. As stated earlier, Grosscup hypothesized the Tuxpan phase because of an unexplained sharp break in the Amapa cultural sequence between the Amapa and Cerritos phases. This break was particularly observable in the Amapa ceramic tradition. Grosscup (1964, 1976) suggested that much of the West Mexican Coastal Plain, at least in the Amapa area and northward, was abandoned during this time. He (1976:255–256) notes that the Lolandis phase deposits at Chametla were limited as well and that this also may be an indication of a restricted occupation or abandonment of the coastal zone in the Chametla area. J. Charles Kelley (1986; Kelley and Winters 1960) has long suggested, based on the presence of the Lolandis Red-rimmed ceramic type—a type analogous to Tuxpan Red-on-orange at Amapa (Grosscup 1976)—and a variety of other introduced traits and items in the uplands of Durango, that coastal inhabitants moved into the western foothills of the Sierra Madre and possibly into the uplands of Durango during Lolandis (Tuxpan) phase times. The geomorphic data from the Marismas project (see Chapter 2), when combined with the archaeological data from Amapa, Chametla, and Durango, suggest that during this time there is a strong likelihood that much of the north-central West Mexican Coastal Plain was submerged or otherwise widely inundated. A radiocarbon assay (AD 835–1155) on oyster shell from the base of one of the linear shell middens along the edge of the estuary suggests that the hiatus may have ended as early as the late AD 900s in some parts of the Marismas and perhaps lasted beyond AD 1000 in others.

Conchera Period

Following the hiatus in the occupation of the Marismas and the reemergence and progradation of the coastal plain, the area appears to have been reoccupied beginning as early as AD 900. New sites were established, and many of the previously occupied sites were reoccupied. The time between AD 900 and the early 1500s is assigned to the Conchera period, the local Postclassic period occupation of the area. This designation reflects the intensive use of shellfish as a supplement to an agriculturally based diet. During this time, the cultures of the region were well integrated into the Mesoamerican cultural tradition, as reflected in the Aztatlán tradition and via their participation in a wide-ranging exchange network—the Aztatlán trading or mercantile system (Kelley 1986, 2000)—and possibly a widespread common worldview (Mathiowetz 2011) that linked the cultures of the northwest Mexican coast with those of west and central Mexico.

The Conchera period is divided into three subperiods. Early Conchera dates between AD 900 and 1100 and is approximately contemporaneous with the Acaponeta phase of Chametla and Culiacán, the Cerritos phase of Amapa, and possibly the Chala phase of Peñitas. The Middle Conchera period dates from AD 1100 to 1400 and is, in part, contemporaneous with the El Taste phase of Chametla, the La Divisa and Yebalito phases of Culiacán, the Ixcuintla phase of Amapa, the Mitlán phase of Peñitas, and the Aztatlán period at Guasave (Guasave phase; Carpenter 1997). The Late Conchera period dates from AD 1400 to Spanish contact, making it contemporaneous with the La Quinta phase of Culiacán and the Santiago phase of Amapa. The Late Conchera period, however, is not well documented in the Marismas.

In summary, the prehistoric occupation of the Marismas Nacionales is divided into three general periods: the Costanero, Estero, and Conchera periods. The Costanero period is seen as representing the initial use or occupation of the area by late Archaic period and early Preclassic period peoples. In the Marismas proper, the 23-m-high shell mound El Calón is the single archaeological manifestation associated with the Costanero period. Based on radiocarbon assays, El Calón is thought to date to between 2500 and 1500 BC. The only other early date in the vicinity of the Marismas comes from a site

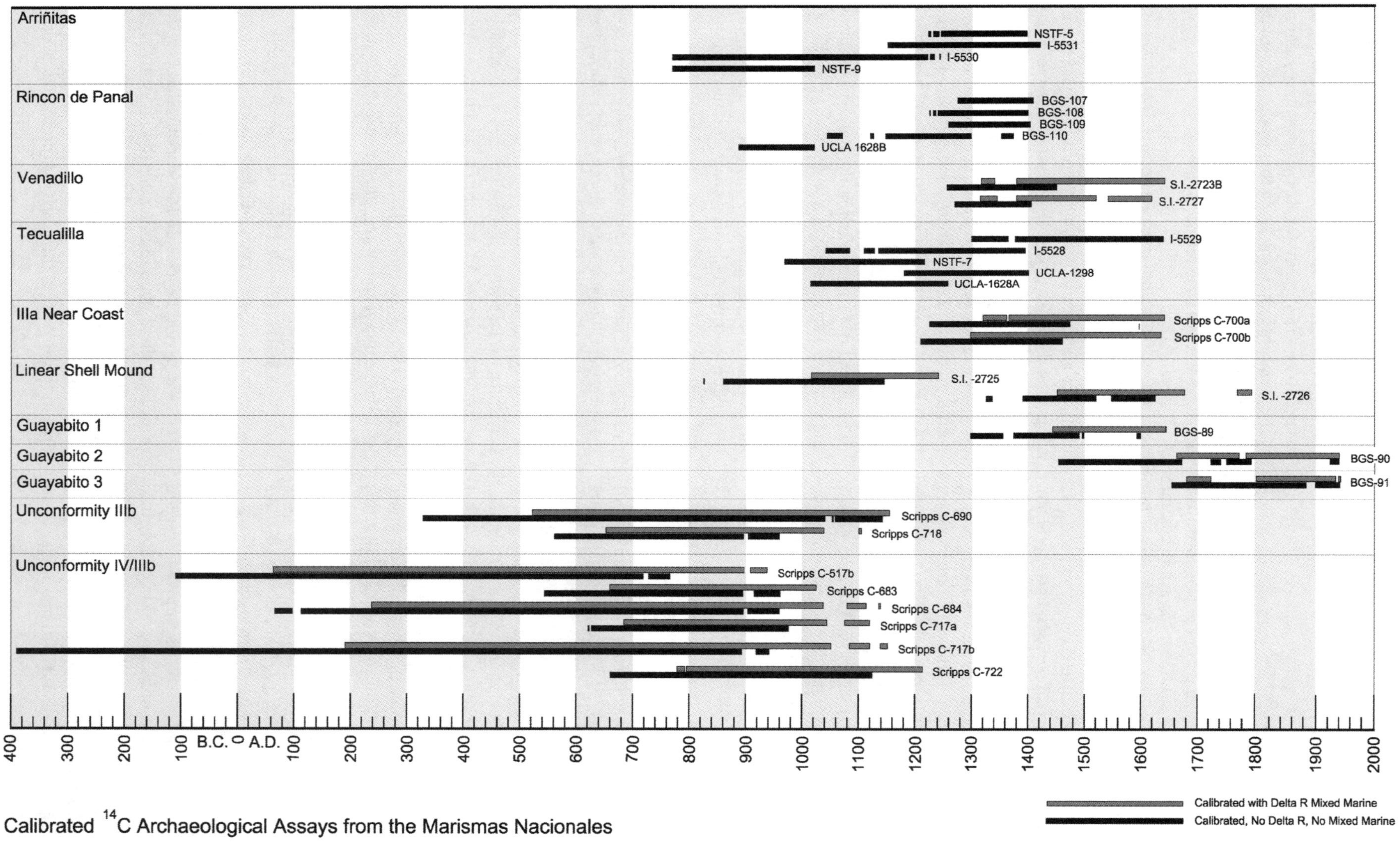

FIGURE 4.2. Archaeological and other calibrated radiocarbon assays from the Marismas Nacionales.

on the Río Acaponeta where a ^{14}C assay on oyster shell yielded a calibrated date between 2894 and 2491 BC. The occupational history of the northern west coast in the vicinity of the Marismas between roughly 1500 BC and AD 300 is not yet known, although, as suggested previously, rising sea levels likely contributed to the paucity in human occupation and consequently archaeological evidence.

The Costanero period is followed by the Estero period, which is divided into two subperiods, the Early Estero period (AD 300–500) and the Late Estero period (AD 500–700), the time of the Early and Middle Mesoamerican Classic period. The Estero period appears to be truncated about AD 700, with the inundation or submergence and apparent abandonment of the coastal plain from possibly the area of Amapa, Nayarit, south of the Marismas to Chametla, Sinaloa, north of the Marismas.

The Mesoamerican Postclassic period is manifested in the local Conchera period, which is suggested to start around the late AD 900s or 1000, with the reemergence and subsequent reoccupation of the coastal plain. Three subperiods, Early, Middle, and Late, constitute the Conchera period. Suggested dates for the Early Conchera period are AD 900 or 1000 to 1100; the Middle Conchera period is dated AD 1100 to 1400; and the Late Conchera period is dated from AD 1400 to contact (AD 1530).

Radiocarbon Dating

Thirty-seven radiocarbon assays (Figure 4.2; Table 4.2) from both archaeological and geomorphic contexts are available as a result of the Marismas project. Of the 37, 15 of the assays are from archaeological features (shell middens) sampled to date the geomorphic features (beach ridges etc.) on which they were situated. These assays are derived from charcoal or marine and brackish water/estuarine mollusk shells. The dating of marine shell is problematic, although generally acceptable if the dates are appropriately calibrated. The dating of oyster shell can also be problematic; the brackish water inhabited by the oysters contaminates the shells with freshwater carbonates that increase the age of radiocarbon assays (Michels 1973:161–162). Inferentially, the same may be true of *Anadara grandis*, another brackish mollusk. It should be noted that great care was taken in the selection of shell samples for radiometric processing. An effort was made to only select shell that appeared to be in good condition and from apparent primary archaeological contexts. The processing of samples by the various radiocarbon laboratories included a thorough cleaning of the specimens before analysis.

The radiocarbon assays reported are from samples processed in the late 1960s and 1970s, using the calibration program CALIB version 6.1 (Stuiver et al. 2005). As discussed, the calibration of marine shell dates requires additional corrections beyond those applied to non–marine shell samples. Marine shell is formed in reservoirs that differ in specific ^{14}C content from the atmosphere, and thus, when dated by the radiocarbon technique, the resulting age determinations may be older than the actual age of the shell (Stuiver et al. 1986:980). In order to correct these ages, the appropriate reservoir age (ΔR age) must be subtracted from the reported ^{14}C date. ΔR ages for various regions were determined using historic shell samples of known ages. Stuiver and others (1986:995) indicate that the ΔR age for the west coast area of Mexico is 185 ± 20 years. Additionally, the standard deviation of the sample age should be corrected, by first squaring it, then adding to the squared standard deviations σ of the ΔR age, and finally taking the square root of the sum. This produces the "total" standard deviation for the reported ^{14}C date minus the ΔR age—the formula is $(^{14}C \text{ age} - \Delta R) \pm \sigma_{total}$.

In order to assess the calibration of marine and brackish water shell dates from the Marismas with the use of the marine reservoir age for the west coast of Mexico, the 185 ± 20 years ΔR age was applied to the calibration of two shell dates obtained from Mound C at the site of Venadillo following Stuiver and others (1986). The results of the calibrations were compared with the ceramic dates of the shell midden capping Mound C (e.g., Grosscup 1976; Kelley and Winters 1960). Based on the identifiable ceramic types recovered, the shell midden capping Mound C is thought to date from AD 1100–1400. The Mound C uncalibrated shell dates are 620 ± 60 BP (AD 1330) and 580 ± 85 BP (AD 1370). The two-sigma calibrated dates without the application of the ΔR correction were AD 1279–1417 (S.I. 2727) and AD 1265–1460 (S.I. 2723B); both are compatible with the ceramic dates. The application of the 185 ± 20 ΔR correction yielded a two-sigma age of AD 1788–1952 and AD 1718–1953, respectively. These dates are unquestionably inconsistent with the archaeology by around 400 years. Based on these results, it is clear that uncritical application of the ΔR correction can be problematic.

As a further test, the Mound C dates were calibrated using the mixed marine northern hemisphere calibration option with a 25 percent marine carbon calibration (e.g., Voorhies et al. 2002) to account for the brackish water environment of the Marismas. Also, in this instance, marine reservoir data from Isla Isabel, Nayarit, an area close to the Marismas, was used. That ΔR correction is 229 ± 50 years. This calibration yielded two-sigma ranges for the two Mound C dates of AD 1322–1631 and AD 1325–1656. The lower ends of these ranges approach the ceramic dates for the Venadillo Mound C shell cap, but overall they still generally appear more recent than suggested by the ceramic dates. Obviously, reducing the

TABLE 4.2. Archaeological and Geomorphic Radiocarbon Assays from the Marismas Nacionales.

Location/Site	Material and Date Collected	Sample Provenience	Lab and Sample Number	^{14}C Age Years BP	Calibrated Age BC/AD (2σ)[a]	Calibrated Age BC/AD (2σ; ΔR 229 ± 50; 25% Mixed Marine)[a]
El Calón 1	*Anadara grandis*, 1973	Surface south face of mound	S.I. 1540	3235 ± 110	1865–1849 BC (0.005) 1773–1259 BC (0.99) 1230–1219 BC (0.004)	1612–1042 BC (1.0)
El Calón 2	*Anadara grandis*, 1973	Surface south face of mound	S.I. 1541	3265 ± 75	1737–1710 BC (0.024) 1659–1408 BC (0.976)	1603–1574 BC (0.015) 1559–1546 BC (0.006) 1543–1192 BC (0.977) 1139–1135 BC (0.002)
El Calón 3	*Anadara grandis*, 1973	Surface south face of mound	S.I. 1542	3845 ± 85	2563–2534 BC (0.019) 2494–2115 BC (0.93) 2099–2038 BC (0.052)	2427–2425 BC (0.001) 2403–2376 BC (0.012) 2357–2354 BC (0.001) 2351–1880 BC (0.986)
El Calón	*Tivela* sp.	Mound	S.I. 2728a	4050 ± 80	2878–2451 BC (0.97) 2445–2438 BC (0.004) 2430–2405 BC (0.009) 2378–2350 BC (0.016)	2617–2610 BC (0.003) 2598–2594 BC (0.001) 2584–2142 BC (0.995)
El Calón	*Anadara grandis*	Mound	ANU-1611 (recalibrated)	3910 ± 60	2568–2520 BC (0.07) 2498–2267 (0.85) 2260–2206 BC (0.08)	2434–2419 BC (0.01) 2407–2022 BC (0.99)
La Concha	Oyster, 1974	Río Acaponeta site	S.I. 2728	4135 ± 80	2894–2560 BC (0.94) 2552–2551 BC (0.0017) 2536–2491 BC (0.057)	2857–2806 BC (0.042) 2760–2281 BC (0.958) 2246–2245 BC (0.0008)
Tecualilla 1d	Charcoal, 1968	Feature A, Test Unit (TU) A, 3 × 3 m, Mound peak, Level 3	Isotopes 1-5529	460 ± 90	AD 1304–1365 (0.12) AD 1384–1642 (0.87)	
Tecualilla	Charcoal, 1968	Feature A, Level 6, 1.8 m deep	UCLA 1628 A	870 ± 80	AD 1023–1273 (1.00)	
Tecualilla	Charcoal, 1967	Mound A, 1.83 m deep	UCLA 1298	700 ± 80	AD 1178–1413 (1.00)	
Tecualilla 1a	Charcoal, 1968	Feature A, NW 1.5 × 1.5 m, 3 × 3 m	NSTF-7	946 ± 60	AD 980–1213 (1.00)	
Tecualilla Ib	Charcoal, 1968	Feature A, TU C; 2.13 m deep	Isotopes 1-5528	755 ± 90	AD 1045–1096 (0.07) AD 1119–1142 (0.03) AD 1147–1401 (0.90)	
Arriñitas 1a	Charcoal, 1970	Mound a, TU A, Level 7, lot 7	NSTF-5	685 ± 56	AD 1228–1232 (0.01) AD 1239–1248 (0.01) AD 1251–1401 (0.98)	

Arriñitas 1d	Charcoal, 1970	Mound A, TU A, Level 7, lot 7	Isotopes 1-5531	700 ± 90	AD 1158–1425 (1.00)	
Arriñitas 1e	Charcoal, 1970	Well Unit, Feature 1, 1.28 m deep (humic acid not removed)	Isotopes 1-5530	1015 ± 110	AD 776–1227 (0.993) AD 1233–1240 (0.005) AD 1248–1251 (0.002)	
Arriñitas 1g	Charcoal, 1970	Mound A, TU A, Level 8, SE quarter	NSTF-9	1108 ± 62	AD 775–1027 (1.0)	
Rincon de Panal	Charcoal, 1967	Feature. A, bottom	UCLA 1628B	985 ± 80	AD 894–1216 (1.0)	
Rincon de Panal	Charcoal, 1971	TU C, Level 4, 1.60 m deep	Brock BGS-107	609 ± 56	AD 1283–1418 (1.0)	
Rincon de Panal	Charcoal, 1971	TU C, Level 3–4, lot 2	Brock BGS-110	775 ± 60	AD 1052–1080 (0.03) AD 1128–1133 (0.003) AD 1152–1305 (0.95) AD 1363–1385 (0.02)	
Rincon de Panal	Charcoal, 1971	TU C, Level 3	Brock BGS-109	640 ± 64	AD 1268–1417 (1.0)	
Rincon de Panal	Charcoal, 1971	TU B, Level 9	Brock BGS-108	683 ± 55	AD 1229–1231 (0.004) AD 1240–1247 (0.009) AD 1251–1403 (0.987)	
Venadillo	Clam, 1974	Mound C, shell cap	S.I. 2723B	580 ± 85	AD 1265–1460 (1.00)	AD 1325–1350 (0.02) AD 1390–1656 (0.98)
Venadillo	Oyster, 1974	Mound C, shell cap	S.I. 2727	620 ± 60	AD 1279–1417 (1.00)	AD 1322–1355 (0.05) AD 1388–1531 (0.84) AD 1553–1631 (0.11)
IIIa Near Coast[a]	*Donax punctatos*	69 cm deep; shell midden in IIIa near coast and lagoon; mainly *Tivela* and *Donax* intertidal sand forms; evidently represented late occupation	Scripps C-700a	600 ± 100	AD 1217–1485 (0.9996) AD 1605–1605 (0.0004)	AD 1307–1373 (0.10) AD 1377–1654 (0.90)
IIIa Near Coast[a]	*Tivela sp.*		Scripps C-700b	620 ± 100	AD 1209–1468 (1.0)	AD 1305–1643 (1.0)
Linear shell mound	Oyster, 1974	Gravel pit, base number 1	S.I. 2725	1050 ± 60	AD 835–836 (0.0025) AD 869–1155 (0.9975)	AD 1024–1251 (1.0)
Linear shell mound	Oyster, 1974	Gravel pit, 1.75 m from top	S.I. 2726	450 ± 55	AD 1329–1340 (0.012) AD 1396–1525 (0.832) AD 1557–1632 (0.156)	AD 1455–1682 (0.97) AD 1779–1799 (0.03)
Guayabito 1	Oyster, 1971	0.91 m above sea level	Brock BGS-89	491 ± 56	AD 1304–1364 (0.17) AD 1384–1500 (0.81) AD 1503–1511 (0.007) AD 1601–1616 (0.16)	AD 1453–1656 (1.0)

TABLE 4.2. (cont'd.) Archaeological and Geomorphic Radiocarbon Assays from the Marismas Nacionales.

Location/Site	Material and Date Collected	Sample Provenience	Lab and Sample Number	^{14}C Age Years BP	Calibrated Age BC/AD (2σ)[a]	Calibrated Age BC/AD (2σ; ΔR 229 ± 50; 25% Mixed Marine)[a]
Guayabito 2	Oyster, 1971	2.13 m above sea level	Brock BGS-90	274 ± 56	AD 1462–1682 (0.85) AD 1737–1757 (0.021) AD 1761–1803 (0.101) AD 1936–1951 (0.028)[b]	AD 1676–1783 (0.39) AD 1796–1952 (0.61)[b]
Guayabito 3	Oyster, 1971	3.66 m above sea level	Brock BGS-91	152 ± 57	AD 1663–1894 (0.82) AD 1904–1953 (0.18)[b]	AD 1689–1736 (0.14) AD 1804–1943 (0.79) AD 1950–1955 (0.07)[b]
Unconformity IIIb	*Tivela* sp.	Surface veneer	Scripps C-718	1300 ± 100	AD 569–903 (0.095) AD 914–969 (0.005)	AD 660–1048 (0.997) AD 1109–1116 (0.003)
Unconformity IIIb (near LV/IIIb)	*Tivela* sp.	67 cm deep in midden	Scripps C-690	1310 ± 195	AD 340–1058 (0.971) AD 1066–1072 (0.001) AD 1075–1155 (0.027)	AD 534–1267 (1.0)
Unconformity IV/IIIb	*Tivela* sp.	12 cm deep in midden	Scripps C-517b	1670 ± 200	102 BC–AD 725 (0.99) AD 738–771 (0.009)	AD 70–901 (0.993) AD 919–945 (0.007)
Unconformity IV/IIIb	*Tivela* sp.	Surface veneer	Scripps C-683	1320 ± 100	AD 545–899 (0.97) AD 918–962 (0.03)	AD 660–1030 (1.0)
Unconformity IV/IIIb	*Tivela* sp.	Surface veneer	Scripps C-684	1500 ± 200	AD 88–103 (0.005) AD 122–902 (0.986) AD 915–968 (0.019)	AD 245–1052 (0.991) AD 1089–1121 (0.008) AD 1143–1146 (0.001)
Unconformity IV/IIIb	*Agoronia testaceo*	Surface veneer	Scripps C-717a	1260 ± 95	AD 624–627 (0.003) AD 631–981 (0.997)	AD 688–1058 (0.956) AD 1081–1124 (0.035) AD 1138–1150 (0.009)
Unconformity IV/IIIb	*Tivela* sp.	Surface veneer	Scripps C-717b	1700 ± 300	386 BC–AD 898 (0.995) AD 920–945 (0.005)	BC 205–AD 1054 (0.993) AD 1085–1122 (0.006) AD 1140–1148 (0.001)
Unconformity IV/IIIb	*Tivela* sp.	1.50 m deep in *Tivela* midden	Scripps C-722	1170 ± 100	AD 661–1024 (1.0)	AD 780–794 (0.011) AD 798–1218 (0.989)

[a] Hughen et al. 2004; Reimer et al. 2004.
[b] Suspect due to impingement on the end of the calibration data set.

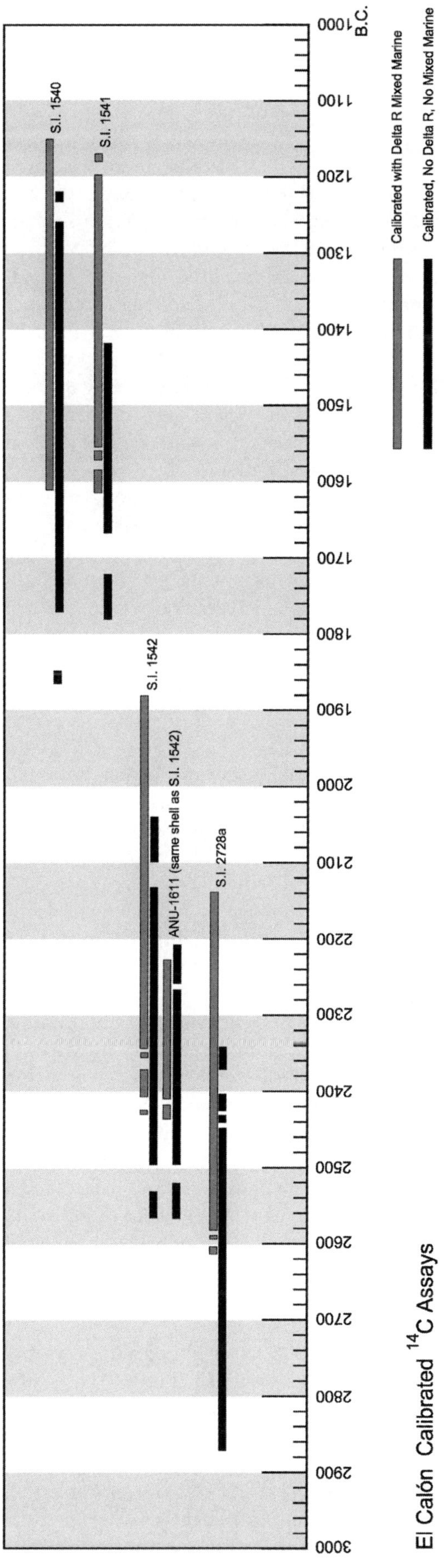

FIGURE 4.3. Calibrated marine shell dates from El Calón.

percentage of marine carbon affecting the calibration would bring the dates more into line with the suggested ceramic dates for the deposit. However, because there was undoubtedly fluctuation in the percentage of marine carbon across both time and space in the brackish water environment of the Marismas, further adjustment of the marine carbon calibrations was not undertaken. Because of the problem arising in the use of the ΔR correction on the Venadillo Mound C dates, the shell dates discussed below are presentations of the calibrated conventional date without the use of the ΔR and mixed marine corrections unless otherwise noted, and all the ^{14}C assays discussed are presented as calibrated two-sigma ranges.

The following discussion presents the results of radiocarbon analyses for each of the archaeological sites from which dates were derived (El Calón, Arriñitas, Rincon de Panal, Tecualilla, and Venadillo). As demonstrated above, not just shell dates in particular but all dates in general can be informative. However, they must be interpreted equally in their archaeological and environmental contexts and calibrated or even corrected accordingly. Whenever possible, the cross-correlation of dates with diagnostic artifacts, particularly pottery, may serve to refine date ranges derived from radiocarbon samples, which otherwise might suggest unreasonably long occupations. Such observations are hardly trivial and are useful for fine-tuning individual site histories. When recognizing that cycles of human occupation and abandonment of the Marismas are as important to understanding the cultural chronology of the region as its environment, these dates provide genuine insights into the natural history of the Marismas to which the human ecology is tied, inextricably, and can serve as an index of local and regional environmental change.

El Calón

To date, the oldest archaeological manifestation identified in the Marismas is the large shell mound El Calón (see Chapter 5), which is the only Marismas site assigned to the Costanero period. The shell mound is near the northwestern edge of the Laguna de Agua Grande (see Figure 1.1) and is constructed primarily of the marine mollusk *A. grandis* (Keen 1971). Our estimate of an early age for El Calón centers on two issues of chronology. First is the apparent antiquity of this structure based on five radiocarbon assays (Figure 4.3; Table 4.2) on shell (*A. grandis* and *Tivela* sp.). The shell samples were collected from excavations in an erosional cut about 15–18 m above the base of the mound and at more than 1.5 m below the surface of the mound in an area below any visible root zone. Four of the assays were processed by the Smithsonian Institution (S.I.), and two of the assays, S.I. 1542 and ANU-1611, are from the same specimen. Because of the surprising antiquity, sample S.I. 1542 was reprocessed and analyzed by the Australian National University (ANU) to verify the results.

The assays fall into two distinct groups, suggesting that at least two different construction or depositional episodes are represented (see Figure 4.3). In this instance the calibrated radiocarbon assays using the ΔR and mixed marine corrections are consistent with archaeological data regarding the Archaic period occupations of adjacent regions as well as the inferences based on El Calón's geomorphic setting and the age of those deposits as discussed in Chapter 2. Based on the visual overlap of the ^{14}C assays, the earliest construction period is suggested to have occurred between 2400 and 2200 cal BC, although it could be as early as 2600 to as late as 1900 cal BC. The more recent construction episode appears to have occurred between 1600 and 1200 cal BC.

As previously suggested, argument for the antiquity of the El Calón mound is further supported by its location. Details of the long-term mechanisms of successive beach ridge formation, sea level change, and transgressive shorelines have been demonstrated amply in the works of Curray and others (1969), Vann (1972), and Connally (1984). Following the conclusions of these researchers regarding the geomorphic context of the coastal plain, we hypothesize that at the time of its construction, El Calón was on or very near the shoreline of the Pacific Ocean. Vann describes the deltaic and alluvial landforms in this section of the coast: "In the past 3,000 to 3,500 years the shoreline has advanced seaward by deposition of a vast sheet-like accumulation of littoral sands" (1972:17). Simply put, sites well to the west of El Calón date to the AD 1300s, and the geologic evidence clearly indicates that the coastal plain between the current shoreline and the location of El Calón required a substantial period to develop.

Today, the El Calón structure is located some 15 km east of the current ocean shoreline and is embedded in a dense forest canopy of red mangrove and black mangrove. The site is notably difficult to access through mangrove forest sediments of soft, organic, anaerobic muds that are typically associated with this ecosystem type. In Snedaker's view, which largely mirrors our own, "the shell mound at Calón appears to be an artifact of an earlier and different environmental regime" (1971:16). He adds, "The working hypothesis is that at the time of construction, the mangrove forest did not sequester the site, the ground level elevation was lower by some measure, and the areas experienced a significantly different hydroperiod" (1971:16).

In addition to the narrative of his discovery of El Calón in 1969, and its original setting and age, Shenkel's research also focuses on the shells themselves. As stated previously, the large ribbed clam, *A. grandis*, was the primary building material, with lesser percentages of *Ostrea corteziensis*, *Anomalocardia subragosa*, *Hexaplex (Muricanthus) nigritus*, *Chione guidea*, *Hexaplex brassica*, *Melongena patula*, and *Anadara tuberculosa*. *A. grandis*, in particular, inhabits intertidal mudflats along the edges of the same mangrove swamps (MacKenzie 2001:1), and it can be harvested at low tide on sandbars (Keen 1971:46), whereas other American west coast *Anadara* species, that is, *A. similis* and *A. tuberculosa*, occur in level mud sediments in mangrove swamps. The numerous secondary component species in the El Calón matrix embrace a wider variety of ancient habitats, namely, mangrove, intertidal mud- and sand flats, possibly rocky reefs, saline bays, and even subtidal sources, leading Shenkel (1969a, 1971a) to posit an ancient pass or open water interface for easier overwater transportation of shells at the time of mound construction.

Such intuitive reasoning about El Calón's environmental characteristics accords with Snedaker's opinion that while some colonization of mangrove may have been one of the paleoenvironmental associations, "unless the site was conveniently accessible by land or water, the transporting of shell through the mangroves would have been impossible without constructing a manmade causeway" (1971:16). "The interiors of mangrove forests," he continues,

> are notably undesirable as sites for human occupation because (1) the microclimate is hot and humid, (2) there is an abundance of mosquitos and other potential pests, (3) movement is difficult due to the soft sediments and vegetative tangle, and, (4) the mangrove forest interior is not productive in terms of man's subsistence [1971:16].

Another potentially confirming, or confounding, factor in the challenge to investigate and interpret El Calón's chronology is the discovery of a grain of melon pollen in a core taken near El Calón (Sirkin 1984; see also Chapter 2). Albeit a tenuous bit of evidence for humans in the area, a stratum containing melon pollen was identified at 2.25 m below the ground surface, and it is thought to date to ca. 3800 BC. Evidence of *Cucurbita* (squash) and *Agave* occurs at 1.75 m below the ground surface, and Sirkin (1974) believed that this was clear evidence of human disruption to the local environment at 3,000 years ago.

In further support of the suggested antiquity of El Calón is a date from an oyster shell recovered from a midden at the foothill site La Concha on the Río Acaponeta. The La Concha assay, with ΔR and mixed marine corrections like those reported above for El Calón, falls between 2760 and 2280 cal BC (S.I. 2728; see Table 4.2). The La Concha assay appears slightly earlier than the earliest El Calón assays but overlaps substantially with them. The El Calón and La Concha assays suggest that Late Archaic period peoples were on the northwest coast of Mexico prior to the development of the Teacapán Peninsula. The presence of Paleoindian and Archaic period peoples in the uplands of northwest Mexico and coastal desert areas to the north is well documented. Massive Archaic period constructions are not unknown in northwest Mexico (e.g., Hard and Roney 1998), thus, it is no surprise that Archaic and early Formative peoples were utilizing the southern Sinaloa and northern Nayarit coastal zone. Additionally, Mountjoy (1970a, 1970b, 2000) has identified an early coastal, late Archaic period use of Matanchén Bay just south of San Blas. He dates the shell middens there from 2200 to 1730 BC. Thus, there is solid evidence of human expansion into and use of the west Mexican coast at an early date (see also Mountjoy and Claassen 2005).

Clearly, determining the age of El Calón poses numerous investigative and interpretive challenges. One of the remarkable attributes of this enormous and exceptional structure is that the majority of mollusks used as constructive material are mostly still articulated. For some, this is evidence that the shell was mined from fossil shell beds and that the age of the mound is likely far more recent than suggested here. This is exemplified in the following translation:

> The recent radiocarbon analysis of one shell from El Calón yielded a date of approximately 4000 years. Is El Calón, then, as old as Scott argues? The shell with which it was built, apparently yes; the date of construction, not necessarily. One probable alternative is that in order to erect El Calón the builders may have used already-dead molluscs harvested from "natural shell beds," which would explain the somewhat irrational presence of still-closed shells, since when molluscs die a natural death, they remain closed [Bióloga Raquel Briseño, personal communication May 2008; Grave Tirado, personal communication 2014].

This contrasting view holds that mollusks remain closed following natural death. Yet another common understanding is that once a bivalve mollusk dies, the adductor muscle or ligament that closes and holds the shell tight starts to decompose, the shell opens, and it does not take long for valves to become separated. In fact, the issue of shell articulation is far more complicated; as Claassen notes, "Articulation of valve pairs is a phenomenon thought to be a common occurrence in natural marine shell deposits but is, in fact, by no means a frequent phenomenon and varies according to the habit of the bivalve" (1998:74). Species-specific die-offs are more common in the marine setting (Claassen 1998), and it is, of course, possible that in the Marismas a sudden, mass die-off of *A. grandis* resulted in the accumulation of articulated shells that were later mined. Personal observations by the senior author of this chapter of fossil shell deposits in the Puerto Peñasco, Sonora, area at the northern end of the Gulf of California indicate that mass die-offs of clams (*Chione* spp.) can produce fossil beds that are composed mostly of articulated shell. It seems reasonable to suppose, however, that the excavation, transport, and dumping from such a deposit could almost certainly lead to the disarticulation of substantial quantities of shell.

Thus, we argue that the *A. grandis* used to create El Calón were collected while the individuals were still alive or very recently dead; and as for the age of El Calón, we contend, based on the radiocarbon estimates and the relation between the mound and past and present shorelines, that this enigmatic site is a Late Archaic and perhaps an Early Preclassic period cultural manifestation, with the initial formation of the mound occurring between 2600/2400 and 2200/1900 cal BC. At least one subsequent additional period of construction, based on radiocarbon assays dating between 1600 and 1200 cal BC, is represented; however, this is a period when much of the coastal plain in the vicinity of El Calón is thought to have been submerged. Nevertheless, questions remain. Why was the mound built where it is? Were there two construction episodes, and if so, was the second episode perhaps related to rising sea levels as people were forced off the coastal plain? What was the mound's purpose, and how long was it used? Finally, what is the significance of the articulated shell?

Arriñitas

Four radiocarbon assays, all from charcoal, were obtained from Arriñitas. Three were from Mound A, Test Unit (TU) A, Levels 7 and 8, and all were from contexts containing Tuxpan Red-on-orange, Tuxpan Engraved, and Roblitos Polychrome ceramics. Tuxpan Red-on-orange and Tuxpan Engraved are associated with the Cerritos phase at Amapa, with Tuxpan Engraved being present throughout the Late period sequence at Amapa, the Cerritos through Santiago phases, or a period from roughly AD 900 to possibly as late as AD 1400. Grosscup (1976:243) notes that Tuxpan Red-on-orange declines in frequency throughout the Cerritos phase. Roblitos Polychrome is a tentative local (Marismas) type (Sweetman 1968:71–72). Because of its association with Tuxpan Red-on-orange and Tuxpan Engraved, Roblitos Polychrome is thought to be associated with the Cerritos phase, AD 900–1100. The Mound A, Level 7, assays are AD 1218–1401 (NSTF-5) and AD 1158–1425 (I-5531), respectively.

The Level 8 assay is AD 775–1027 (NSTF-9) (see Figure 4.2 and Table 4.2).

A fourth assay was obtained as a result of excavations in Feature 1, the Well Unit, at Arriñitas. It yielded a date of AD 776–1251 (I-5530). The ceramics recovered were dominated by Tuxpan Red-on-orange and Tuxpan Engraved; also included were Botadero Incised and Aztatlán Ware. The presence of these ceramics indicates that Feature 1 was contemporaneous with Mound A, or partially so, and while the occupation suggested by the radiocarbon assays from the two mounds is much longer, from the AD 700s to 1400, the ceramics suggest a significantly shorter occupation between AD 900 and 1100 (Grosscup 1976). No ceramic types diagnostic of the subsequent neighboring Ixcuintla or El Taste phases were represented. Thus, the two later radiocarbon assays, ca. AD 1200s–1400s, are problematic, although it is possible that some of the Tuxpan ceramic types persisted into later Ixcuintla phase times. Admittedly, the analysis of the Arriñitas ceramic assemblage as a whole is incomplete, and it is possible that later ceramic types associated with the Ixcuintla and El Taste phase might be present. However, based on limited ceramic and chronometric data, the occupation of Arriñitas falls within the Early Conchera period of the local Marismas sequence between AD 900 and 1000/1100.

Rincon de Panal

Five radiocarbon assays, all from charcoal, resulted from the excavations in a single mound at Rincon de Panal. Three of the assays are from Levels 3 and 3–4 of TU C; these are AD 1268–1417, AD 1052–1385, and AD 1283–1418. The ceramics from Levels 3 and 4 include El Taste Polychrome, El Taste Satin, Ixcuintla Polychrome, Sentispac Buff, and Tuxpan Engraved, types associated with the neighboring El Taste and Ixcuintla phases (AD 1100–1300). Several Peñitas types, including an engraved, a red incised, and polychromes (Bordaz's types 40–54), were also represented (Bordaz 1964). The bulk of the identified decorated types were recovered from Level 4.

A fourth assay, AD 894–1216, from the bottom of Feature A (Level 5) in TU C is of the Acaponeta and early Ixcuintla phase times. The assay is associated with burned daub that may have been part of an unconfirmed structure or redeposited construction debris (Taylor and Rautenkranz 1971). Identified ceramic types from Level 5 were limited but included El Taste Red-bordered, Sentispac Buff, Tuxpan Engraved, and Tuxpan Red-on-buff. Although these ceramic types are associated with the time span represented by the radiocarbon assay, the presence of Tuxpan Engraved and Tuxpan Red-on-buff suggests, as does the stratigraphic position of the radiocarbon assay and ceramics, a slightly earlier time. A fifth assay of AD 1229–1403 with no reported ceramic associations resulted from a sample from Level 9 in TU B. Based on the radiocarbon assays and identified ceramic types present, it appears that Mound A accumulated over a period from ca. AD 900 to 1350 and is associated, therefore, with the local Marismas Early and Middle Conchera periods.

Tecualilla

Five radiocarbon assays, all from charcoal, were obtained from Feature A, the largest and most intensively investigated mound at Tecualilla. Test Unit A was excavated at the highest point on Feature A; it measured 10 × 10 m and was subsequently extended with a 1.5-m-wide × 10-m-long trench to the northwest and a 1.5-m-wide × 14-m-long trench to the northeast. An assay from Level 3 of TU A yielded a date range of AD 1304–1642 (I-5529). Among the ceramics removed from Level 3 were Chametla types—Red-rimmed Decorated and El Taste Red-bordered and Amapa types, Botadero Black-on-orange, Mangos Engraved, Santiago Engraved, Tuxpan Engraved, and Tuxpan Red-on-buff—that date from as early as AD 900 to as late as AD 1400+. Overall, the ceramics from Level 3 suggest a mixed deposit, although the majority of the types suggest a date well earlier than the ^{14}C assay. The presence of a ceramic type associated with the Santiago phase at Amapa suggests that this ^{14}C assay may reflect a late occupation at Tecualilla, and when considered against the distribution of the other ^{14}C assays and its relative position stratigraphically in Feature A, it may best represent the latter portion of the local Middle Conchera period and the Late Conchera period.

A second assay, from Level 6 of Feature A, yielded a date of AD 1023–1273 (UCLA 1628A). While no identifiable ceramic types are reported for this level, it was 1.8 m below the modern ground surface and is clearly earlier in age. A third assay, AD 1178–1413 (UCLA 1298), is also reported from the same level, which is more recent but overlaps UCLA 1628A; however, the contextual relationship between the two is unclear.

A fourth assay is reported from the northwest 5-×-5-m subunit of TU A of Feature A. However, detailed provenience information is unavailable. The assay, AD 980–1213 (NSTF-7), is the earliest reported from Feature A. A fifth assay, AD 1045–1401 (I-5528), was derived from the excavation of TU C in Mound A; it was from Level 7. No identifiable ceramic types are reported for TU C.

Combined, the minimum and maximum ranges for the five Tecualilla calibrated ^{14}C assays range from AD 980 to 1642, while the ceramic assemblage suggests a range in dates for the site from AD 900 to just beyond

AD 1400. Based on the ceramic and the ^{14}C assays, an occupation spanning the local Marismas Early and Middle Conchera periods is suggested for Tecualilla.

Venadillo

Two radiocarbon assays were obtained from the shell midden that capped Mound C at Venadillo. These are AD 1265–1460 (S.I. 2723B) and AD 1279–1417 (S.I. 2727); S.I. 2723B was derived from a clam, taxa unspecified, and S.I. 2727 was taken from oyster. As discussed above, the ΔR and the mixed marine corrections are not applied to either assay. The Mound C shell midden contained ceramic types associated with the El Taste phase (AD 1050/1100–1250) at Chametla (Mazatlán Polychrome, El Taste Polychrome, El Taste Satin, and El Taste Red-bordered) and Ixcuintla phase (AD 1100–1400) ceramics (Ixcuintla White-on-orange, Ixcuintla Polychrome, and Peñitas Engraved) at Amapa; the Chametla types dominated. Ceramically, this suggests a time span from ca. AD 1100 to 1400. Based on the ceramic assemblage, the shell midden capping Mound C was deposited during the Marismas Middle to Late Conchera period. The ^{14}C assays appear generally consistent with the ceramic dates suggested for the shell midden capping Mound C.

Obsidian Hydration Dating

Obsidian hydration was developed as an archaeological dating technique in the 1960s, although its widespread use as a relative or chronometric dating method has been slow to develop, while being extensively used in some areas and practically ignored in others (Eerkens et al. 2008). Nevertheless, west Mexico was one of the focal points of early use of the technique, and it has continued to be an area where the technique has been commonly used (e.g., Meighan 1972, 1976b; Meighan et al. 1968; Mountjoy 1990, 2000). At the time the Marismas project was undertaken, the technique was in its early development. As a result of work stemming from Project A (see Chapter 3), Meighan and others (1968) undertook an analysis of obsidian artifacts from the Morett site (Meighan 1972) in Colima in an effort to develop an obsidian hydration rate that would be applicable to the coast and adjacent uplands of west Mexico including portions of Nayarit, Colima, Jalisco, and Michoacán. Their efforts yielded a hydration rate of 260 ± 15 years per micron. Meighan (1976b) subsequently used this rate for some of the Amapa obsidians in order to date the phase sequence for the site. Meighan's temporal scheme is at significant variance with that proposed by Grosscup (1976) (and favored here). More recently, Mountjoy (1990, 2000) has made extensive use of obsidian hydration dating of sites in southern Nayarit and Jalisco and has found the technique to be of considerable value. However, it is often difficult to evaluate his results because he rarely reports the hydration rate or rates used.

The formation of a hydration band is affected by several important variables including time, the chemical composition of the obsidian, and soil temperature, although water vapor pressure and soil alkalinity may also play a role in some situations (Freter 1993; Friedman and Obradovich 1981; Friedman et al. 1994; Friedman et al. 1997; Hull 2001; Michels and Tsong 1980; Morganstein et al. 1999; Ridings 1996; Skinner and Tremaine 1993; Stevenson et al. 1993; Stevenson et al. 1998; Stevenson et al. 2000). Once a hydration band has formed, it can be measured and used to determine the relative ages of the specimens analyzed. Theoretically, the thicker the band, the older the specimen. If a hydration rate can be determined based on an artifact's association with materials of a known age or radiometric dates, it is possible that a hydration rate can be assigned an estimated chronometric value and used to assess the temporal association of otherwise difficult-to-date contexts (e.g., Eerkens et al. 2008; Meighan 1976b).

Obsidian artifacts—projectile points, blades, and debitage—were a relatively common occurrence at the sites investigated in the Marismas. Wolynec (1970a) undertook an effort to critically assess the application of the Meighan west Mexican hydration rate (260 ± 15 years) to obsidian artifacts from the Marismas. Her sample universe consisted of 215 specimens of green obsidian, all of which were determined to be chemically uniform through neutron activation. Samples from five sites were analyzed: Juana Gomez ($n = 60$), La Tarjea ($n = 9$), Test Mound A ($n = 21$) and Test Mound B ($n = 27$), Tecualilla ($n = 58$), and Chalpa ($n = 31$).

The samples Wolynec analyzed were from excavated contexts and were collected from arbitrarily defined excavation levels. Once determined, the hydration band values were plotted against the depth of the excavation unit from which the specimens were recovered. Wolynec then calculated and plotted the median value for the hydration bands (in microns) by excavation level in order to determine the central chronological tendency of each excavation level. She next calculated and plotted a trend line against the data (median values), with a positive trend indicating superpositioning of the data and a negative line indicating mixing of deposits. Any variances, below or above, between the median points and the trend line are taken as indications of specific points of the mixing of deposits.

Juana Gomez

The 60 obsidian samples measured from Juana Gomez were recovered from Mound 1, one of the largest and the most intensively excavated large feature at the site; thus, as Wolynec (1970a) states, the data are only applicable to Mound 1. The hydration band widths ranged from 1.7 to 6.6 μ. This, based on Meighan's suggested hydration rate, yields a temporal range of AD 234–1508, with an average date of AD 1222 (2.2 μ). Wolynec (1970a:15) notes a weakly positive trend line and interprets this as evidence for some superpositioning.

Based on the analysis of ceramics recovered from Mound 1 (Sweetman 1970, 1974), it appears that Mound 1 formed during the Early through Late Estero period, AD 300–700. Only three of the 60 samples analyzed fall within this temporal span; one sample is earlier, and the remaining 56 are more recent, with most ($n = 54$) dating AD 988–1508.

In an effort to explain the divergence between the hydration dates and the ceramic dates, Wolynec (1970a:27) suggests that Chametla Baluarte phase (Marismas Late Estero period) pottery types persisted longer in the Marismas than in the Chametla areas and that there was a mixing of earlier (Early Estero period) materials with later materials. However, there is little evidence to support the hypothesis that particular Chametla ceramic types present were being made long enough to account for the discrepancy between the hydration dates and ceramic dates.

La Tarjea

Nine hydration determinations were made for La Tarjea. These range in age from AD 754 to 1456 (1.9–4.6 μ), with a mean of AD 1248 (2.7 μ). The origin of the nine flakes analyzed is not specified, but it is assumed here that they were derived from Mounds A and B both. The lack of an adequate sample size, the lack of association between the obsidian artifacts analyzed and other temporally diagnostic artifacts, and the lack of temporally diagnostic artifacts in general preclude any useful interpretation of the La Tarjea obsidian hydration data (Wolynec 1970a:28). The trend line was negative, implying an inverse correlation between age and depth (Wolynec 1970a:17). The few diagnostic ceramic types identified suggest that the La Tarjea occupation occurred between AD 500 and 700 (Late Estero period), although there is some indication of an earlier Early Estero period (AD 300–500) occupation. It appears that all the obsidian hydration dates postdate the ceramic dates, although the presence of a later unidentified occupation of the site is possible.

Panales Test Mounds A and B

Wolynec's (1970a:17–21) discussion of the results for Panales Test Mounds A and B is problematic. Her initial discussion combines the data from the two mounds, and those data are graphically displayed in her Figure 7. Although the features are located in the same general area on the island of Panales, they are spatially discrete, and it is not clear that they are part of a single site or that they share a common formational history, and Wolynec does not explain why the data were combined. Nevertheless, she (1970a:Figures 5–6) also presents graphic plots of individual data sets for Test Mounds A and B both, but with significant conflicts between the individually plotted data points from each of the test mounds and the combined data represented in her Figure 7. In Wolynec's Figure 5, 21 data points are plotted, and in Figure 6, 27 points are plotted; in Wolynec's Figure 7, 57 data points are plotted. The origin of the additional nine points is unclear, and interpretation and verification of these data and Wolynec's accompanying commentary is further exacerbated by apparent inaccuracies in the plots themselves. In addition, the graphs indicate a substantial mixing of deposits. Wolynec (1970a:21) states that the combined hydration data from Test Mounds A and B indicate a span of time from AD 754 (4.6 μ) to 1586 (1.4 μ).

Unfortunately, Wolynec's (1970a:28) discussion of hydration dates for the individual mounds is highly inconsistent with her initial discussion of the combined results. She states that the hydration dates for Test Mound A range from AD 685 to 1612 (although this is difficult to verify visually based on her Figures 5 and 7), a range that exceeds both the earliest and latest dates for the aforementioned combined Panales data set. Wolynec, citing Sweetman (1968:54–55), states that the ceramics recovered from Test Mound A indicate a temporal range from AD 500 to 950. She further indicates that four of the five hydration dates from the shell midden capping the earthen portion range from AD 932 to 1430 and notes "rough" correspondence between the hydration and ceramic dates. Be that as it may, the hydration dates for the earthen portion of Test Mound A are much later than the ceramics indicate and are too few in number to derive any meaningful inferences. The trend line for Panales Mound A is only slightly sloping, indicating no superpositioning, and the plotted data show a great deal of mixing (Wolynec 1970a:28). Wolynec (1970a:29) suggests that the variance between the hydration dates and the ceramic dates may be attributable to variations in soil temperature, the retention of ceramic styles beyond previously suggested times, or the reuse or breakage of obsidian artifacts, resulting in more recent age determinations. Such explanations are potentially reasonable. However, they also suggest that the obsidian hydration technique, at least by 1970s standards, is only useful when other, more reliable methods for direct or indirect chronological control can be used to better situate the data in temporal and archaeological space.

As for Test Mound B, the 27 hydration dates reported range from AD 850 to 1638 (Wolynec 1970a:Figure 6), while the ceramic assemblage suggests a range of

AD 300–1250/1300 (Davis 1969:91–94). Based on the data in Wolynec's Figure 6, nearly all the plots for Mound B date after AD 1300. Like that for Panales Mound A, the trend line for Panales Mound B is only slightly sloping, indicating no superpositioning, and the plotted data show a great deal of mixing (Wolynec 1970a:28). In general, the obsidian hydration dates derived from the Panales Mound A and Mound B samples do little to clarify the temporal boundaries of the use of the mounds.

Tecualilla

As with the Panales mounds, there is some discrepancy between Wolynec's (1970a:21) discussion and the data plotted in her Figure 8. She reports a range of hydration band widths from 1.1 to 6.5 μ (AD 260–1664); however, the data points plotted in her Figure 8 seem to range from 1.1 to 5.5 μ (AD 520–1664). The data discussed were derived from two different mounds, Features A and B, at the site; these are combined into a single data set and are not discussed separately, because of their contemporaneity. Wolynec (1970a:29) notes that the majority of the band widths, 40 of the 58, fall between 3 and 4.5 μ (AD 780–1170). Ceramic dates suggest a range of AD 700–1150 (Davis 1969:94–95). This is the best example of the five sites studied where there is a high correspondence between the ceramic dates and the obsidian hydration dates. Further, it is worth noting that 30 of the 40 dates cited fall between AD 900 and 1150. If in fact these obsidian hydration dates are applicable, they, by their frequency, suggest that the post–AD 900 occupation of Tecualilla was more intense, and perhaps this also marks the end of the proposed hiatus (AD 700–900/1000) in the occupation of the Marismas area. Wolynec (1970a:21) reports a mean hydration date of AD 1066 (3.4 μ) for Tecualilla.

Chalpa

Wolynec's (1970a:221) discussion of the Chalpa hydration dates appears to reflect more accurately the data plotted in her Figure 9. The dates range from AD 910 (4.0 μ) to AD 1508 (1.7 μ), with a reported mean band width of 3.3 μ (AD 1092). There is a weak negative trend line, indicating an inverse correlation of depth and age, and there appears to be significant mixing of the deposits. Limited analysis of the decorated ceramics from Chalpa was completed after Wolynec's study, and the ceramic types identified place the occupation of the site between AD 1100 and 1250, the early portion of the Marismas Middle Conchera period.

In summary, Wolynec (1970a:30–31) believes that the Meighan hydration rate of 260 years per micron may be generally applicable to the Marismas, despite the array of potential environmental and cultural factors that can affect hydration rates and the size of the hydration bands. As she notes, the data from only one of the five sites studied, Tecualilla, produced a relatively strong correlation between the bulk of the hydration dates and the ceramic dates for the features analyzed. The only other possible example where there was some correspondence of the hydration and ceramic dates was the shell midden portion of Panales Test Mound A. Nevertheless, as Wolynec acknowledges and as noted above, the validation of the technique as a chronometric device for dating cultural contexts within the Marismas requires additional work, and it clearly works best in concert with other, more reliable techniques.

Discussion

The prehistoric occupation of the Marismas has been divided into three broad periods. The earliest is the Costanero period, which is arbitrarily defined between ca. 3000 BC and AD 300 and is associated with the Late Archaic period and early Mesoamerican Preclassic period. Other than the existence of the El Calón mound, we know little of this early period of human presence in the Marismas area. Five ^{14}C assays on shell place the El Calón mound within the Costanero period, and the definition of the period is based on the assumption that the mound was constructed or accumulated over a period from as early as 2600 BC to as late as 1200 BC. This argument is further supported by the geomorphic setting of the mound, palynological evidence from sediment cores taken in the vicinity of El Calón that suggests anthropogenic disturbances ca. 3000 BC, and a single ^{14}C assay of 2760–2280 BC from an oyster shell from the site of La Concha on the Río Acaponeta. The La Concha date is taken as additional evidence of the presence of humans in the area at an early date, which is partially contemporaneous with the older El Calón dates. Other than El Calón and La Concha, nothing is known of the nature of human use of the Marismas during the Costanero period or the transition to village life as manifested in the village sites that appear ca. AD 300.

The ceramic-period occupation of the Marismas falls within the span of the Mesoamerican Classic and Postclassic periods, with ethnohistoric reports of occupied villages at the time of Spanish contact. The Mesoamerican Classic period occupation of the Marismas is manifested in the Estero period, ca. AD 300–700. The Estero period is divided into the Early Estero period, ca. AD 300–500, and Late Estero period, ca. AD 500–700. The Estero period is followed by an apparent abandonment or greatly diminished occupation of the Marismas as the area appears to have been inundated through the submergence of the coastal plain and/or rising sea levels. This hiatus is dated ca. AD 700–900/1000. Beginning about AD 900 it appears that the coastal plain had reemerged sufficiently to allow some resettlement. The period between AD 900 and Spanish contact is defined as the Conchera period, which is subdivided into Early (AD 900–1050/1100), Middle (AD 1050/1100–1350/1400), and Late (AD 1350/1400–1530).

The temporal boundaries for the Estero and Conchera periods are defined exclusively on ceramic cross-dating with the Chametla and Amapa cultural sequences as defined by Kelley and Winters (1960) and Grosscup (1976). For the most part, the ^{14}C assays available from archaeological contexts support the ceramic dates; however, most are of no use in refining the temporal boundaries established based on ceramic cross-dating with neighboring areas. The experiment with obsidian hydration dating, although a worthwhile effort, generally proved to be of little value. Only in one case, the site of Tecualilla, did the hydration dates approximate the ceramic dates for the site. Wolynec (1970a) cites a number of probable causes for the suspect dates, including mixing of deposits, artifact reuse, environmental conditions, and an incomplete understanding of the temporal boundaries of the associated ceramic types. Additionally, today we know that establishing a local hydration rate for obsidian from one or more raw material sources is complex and the application of a rate for even one type of obsidian across a broad geographic range is not always valid. Finally, although there was an attempt to assess the context of the obsidian samples used in the study, the lack of stratigraphy in the burial mounds and the shell middens that capped many of them, the use of arbitrary excavation levels, bulk sampling from general excavations rather than from targeted well-controlled contexts, and the lack of general contextual information likely contributed significantly to the results, which were clearly problematic.

The only site within the Marismas proper assigned to the Costanero period is the shell mound El Calón. It is the only site investigated that is dated exclusively by radiocarbon assays and an analysis of the feature's geomorphic setting. Small scattered shell middens present in the area of El Calón are assumed to date to the Costanero period as well and are taken as further evidence of a human presence in the area during the Archaic period.

Early Estero period components or features have been identified at the lowest level of the Juana Gomez mound, while the stratum above the floor of the Juana Gomez mound dates to the Late Estero period. La Tarjea, albeit based on limited data, and the earthen portions of Panales Mound A, Panales Mound B, and Mound C at Venadillo also date to the Late Estero period, and a Late Estero component may be present in the lower levels of Mound B at Venadillo.

The occupation at Arriñitas appears to date to the Early Conchera period. The shell cap at Mound A at Panales dates to the Early and Middle Conchera periods, as does the shell midden capping Panales Mound B. The site of Chalpa has both Early and Middle Conchera period components. The Rincon de Panal mound on Panales seems to have accumulated over several hundred years during the Middle Conchera period. Feature A at Tecualilla falls mostly into the Middle Conchera period, as does the shell midden capping Mound C at Venadillo. Mound B at Venadillo also appears to be associated primarily with the Middle Conchera period. The sites of Cristo Rey and El Mirador are also believed to date to the Middle Conchera period. The limited number of temporally diagnostic ceramics identified from the Cuautla and Playa Novillero sites suggest that Late Estero period and Early Conchera and Middle Conchera period components are present in the area. Most of the lineal shell mounds in the Marismas generally date to the Early, Middle, and Late Conchera periods, although there are early Spanish contact-period components and features present. The earliest of the prehistoric mounds seem to have started accumulating subsequent to the occupational hiatus between the Late Estero period and the Early Conchera period.

Although it is tentative, we have offered up a named local cultural sequence distinct from those of the surrounding area. Three periods—the Costanero, Estero, and Conchera—are defined, with the Estero and Conchera periods being further subdivided into Early and Late and Early, Middle, and Late, respectively. The designations assigned were to reflect the cultural—settlement and subsistence—history and, in part, the geomorphic history of the area as they relate to the occupational history of the area. Recent researchers in the Marismas and adjacent areas, as noted in Chapter 2, have opted to use phase designations from the Amapa sequence (Grosscup 1961, 1964, 1976; Meighan, ed. 1976) and the revised Chametla and Culiacán phase designations offered by Kelley and Winters (1960). These phase designations are assigned based on the ceramic types identified at a site or archaeological component, and they have been used with little if any modification or refinement of their previously established temporal boundaries. The utility of using the previously established phase sequences across the region is clear, especially if archaeological components are being assigned to a given phase based on the prevalence of a specific ceramic type or co-occurring ceramic types. Furthermore, the use of previously established phase designations avoids the need to constantly correlate competing local phase sequences. In retrospect, it may have been more prudent to utilize Chametla phase designations for the earlier Marismas components and sites and Amapa phase designations for the later ones.

In addition to using previously established phase designations, recent researchers have added an early phase designation borrowed from the Shaft Tomb area from southern coastal Nayarit, the Chinesco phase (Chinesco complex—200/100 BC–AD 200/250). This phase is assigned based on the presence of Chinesco Polychrome or Chinesco Black-on-cream ceramic types (e.g., Garduño 2013; Grave Tirado and Nava Burgueño 2010, 2012). To date, these have not been reported in the area discussed in this volume; however, they have been reported in adjacent areas, and their presence clearly suggests that a Preclassic occupation of portions of the northern Nayarit and Sinaloa coasts and adjacent uplands occurred and that there was an early movement of goods and people through the region.

Despite the wide use of previously established phase designations across the region by recent investigators, there appears to have been little effort to refine the temporal boundaries of these phases. As noted in Chapter 3, there is the occasional reference that a phase ends later than suggested by earlier investigators; however, the reason for this is not fully explained. Despite recent and ongoing work, there seems to be little focus on chronology building and refinement. Although cited, recent radiocarbon assays do not appear abundant, and they are generally offered as support of ceramic cross-dates, as is the case here, rather than being tools to reinterpret extant temporal schemes. These comments are not offered as criticism but, rather, simply as a call to more rigorously assess the temporal boundaries of the northwest Mexican coastal phase boundaries. All this said, it is of note that there have been no new data indicating that the temporal boundaries posed by Kelley and Winters and by Grosscup are significantly in error.

Finally, the chronometric data, radiometric and relative, generated by the Marismas project have provided significant insight into the geomorphic and cultural evolution of the region. Geomorphic evidence for the growth of the coastal plain in the Marismas area as well as for periodic inundation has been substantiated by the distribution of prehistoric sites across the coastal plain within the Marismas. The applicability of the Marismas geomorphic and cultural findings, both local and regional, as reported in this volume will be validated, modified, or rejected as additional work is undertaken and completed. Regardless, chronology remains a significant and essential starting point in archaeology, and prior to the Marismas project, little was known of the area's cultural chronology.

CHAPTER 5

Site Descriptions and Excavations

Michael S. Foster and Stuart D. Scott

At first appearance, the subtropical coastal zone that is the Marismas Nacionales, with its mangrove swamps, extended dry season, periodic floods, oppressive monsoon-season humidity, and abundant insects, looks to be a hostile and uninviting environment for human occupation. This intimidating veneer all but masks the amazingly rich ecosystem that is in the truest sense an ecotone, a place where different ecozones meet and, by definition, a place abundant in resources (e.g., Odum 1971:157–159). The Marismas is not unique along the coasts of Mexico and northern Central America. Similar sites and environs, although more tropical in climate, occur on both the Pacific and the Gulf and Caribbean coasts, and it is clear that some of the earliest inhabitants of Middle America utilized these environments, as did their ceramic-period descendants (e.g., Coe 1961; Coe and Flannery 1967; Mountjoy 2002; Stark 1977; Stark, ed. 2001; Stark and Voorhies, eds. 1978; Voorhies 1976; Voorhies et al. 2002).

Archaeologically, the sites in the Marismas (Figure 5.1) appear as earthen or shell mounds or as earthen mounds capped with shell middens. In addition, there are shell middens and linear shell middens that extend for hundreds of meters and are in excess of several meters high. In terms of site types, they appear to be small villages and hamlets consisting of burial mounds, house mounds, and trash middens separated by open spaces that likely functioned as communal plazas, exterior work areas, or perhaps household garden plots. Other than the burial mounds and likely El Calón, no formal communal or ceremonial architecture was identified in the sites investigated by the Marismas project. The residents of these sites exploited an array of estuarine, marine, and terrestrial resources to supplement their diet of corn, beans, and squash.

The sites investigated, with the exception of El Calón, date to the Mesoamerican Classic and Postclassic periods. These temporal assignments are supported by inter- and intraregional ceramic styles, the presence of some well-defined temporal markers such as copper, figurine styles, ceramic smoking pipes, burial practices, a smattering of radiometric dates, and the geomorphic data regarding the development and accretion of the coastal plain (see Chapter 2).

The first excavations of earth mounds produced human skeletal material with associated funerary offerings including whole ceramic vessels and animal burials. Mound fill contained sherds, obsidian, stone tools and debitage, bone and other artifact types, and shell. This material was generally mixed, indicating that the mounds grew by accretion. There was no evidence to indicate that these mounds were constructed or modified in a way that resulted in the formation of sealed, temporally stratified deposits.

The occurrence of slight depressions indicating nearby borrow pits suggests that at first the dead were laid out at or near ground level and the earth mounds were then raised above them. Subsequently the mounds were heightened in now indiscernible construction levels to cover additional burials, and/or burial pits were excavated into the same mounds. Also, the occasional presence of scattered, disarticulated human bone indicates that earlier burials were sometimes disturbed during the subsequent burial of the dead (Gill 1974:88). It also appears that some of the earthen mounds served as house platforms. Several of the mounds investigated yielded evidence of floors, hearths, other architectural features, or the occasional chunk of daub with impressions or remnants of organic matter.

The artifact assemblages recovered were dominated by a variety of pottery sherds and, to a lesser degree, flaked and ground stone. The ceramic assemblage, consisting mostly of sherds and some whole vessels, included elaborately decorated polychrome pottery

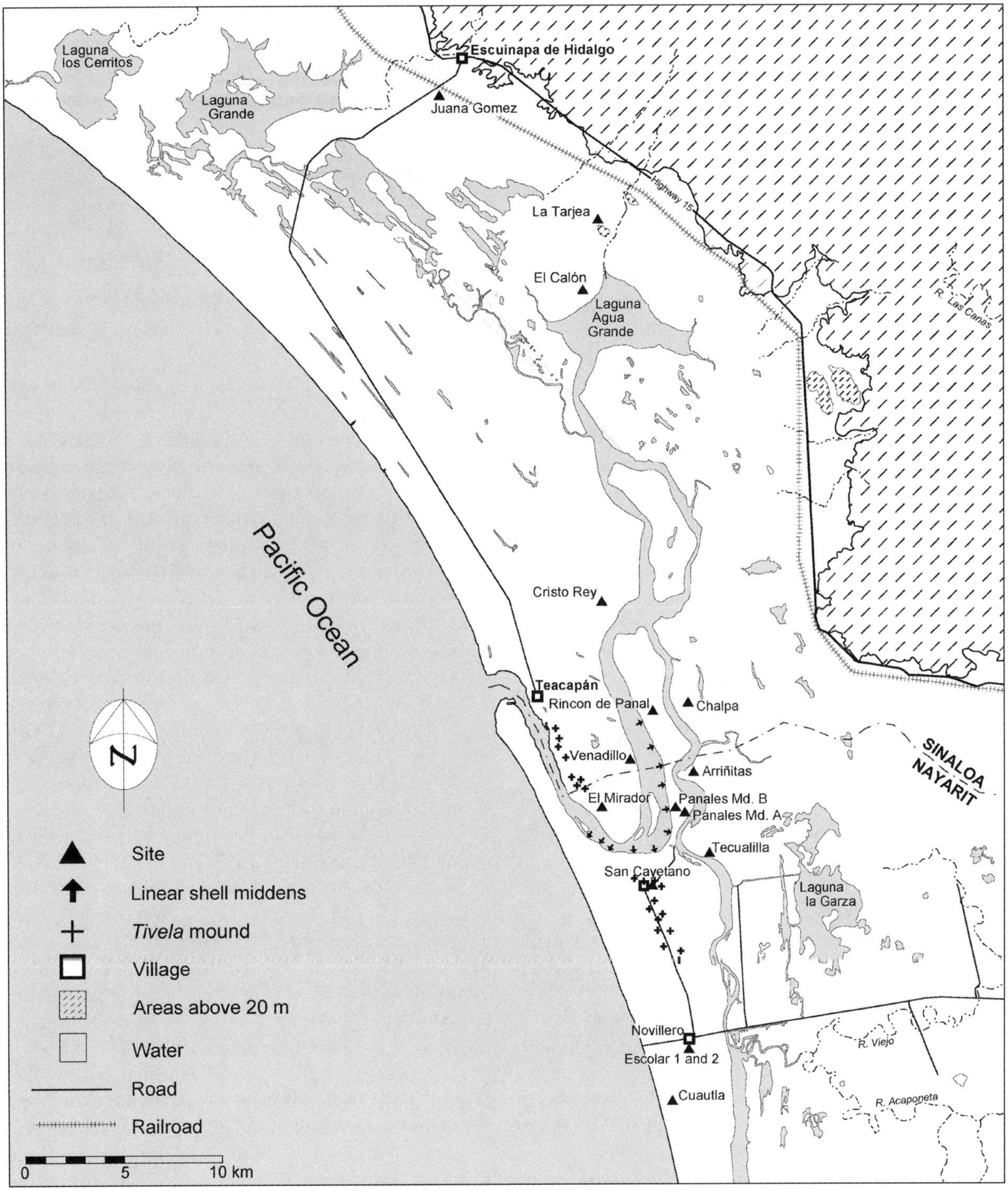

FIGURE 5.1. Archaeological sites investigated and features recorded during the Marismas Nacionales project.

types, bichromes, polished monochromes, and utilitarian plainware vessels. Other ceramic artifact types included plain and decorated spindle whorls, smoking pipes, and some figurines. Flaked-stone artifacts included projectile points, prismatic blades, and debitage. Obsidian blades and debitage were common, along with an occasional obsidian projectile point. Ground-stone items included manos, metates, pestles, and an incised stone spindle whorl. A few pieces of copper and a few bone artifacts were also present. Surprisingly little marine shell jewelry was recovered. Overall, the excavations in the Marismas produced an array of artifactual material that provides insight into the daily life of the precolumbian inhabitants of the area, their worldview, and their place in local and regional economic and political systems.

Field Methods

Sites were assigned a name based on their location or proximity to a local landmark such as a village or named geographic feature. Features were assigned either alpha or numeric designations, and site maps were completed using a plane table and alidade. Excavations at the sites investigated were limited, and in terms of current archaeological vernacular, they can be best thought of as test excavations. Most excavations were completed using combinations of 1-×-1-m or 2-×-2-m test units, which were usually excavated in arbitrary 10-, 20-, or 25-cm levels. These test units were expanded as necessary to expose further features or burials. Note was made of any natural strata encountered. However, stratified deposits were virtually nonexistent within either the excavated burial mounds or the shell midden deposits, except where earth mounds were overlaid with a shell capping.

Excavations were usually terminated when coarse beach sand lacking cultural material was encountered. This stratum was generally present across the area investigated. Screening, using ¼-inch or ⅛-inch mesh, was employed to enhance the recovery of archaeological remains or to systematically sample certain deposits (e.g., middens) for environmental and subsistence-related remains.

Chronometric sampling focused on the recovery of charcoal and shell for radiocarbon analysis. However, surprisingly little charcoal was recovered. Artifacts were collected as they were excavated or from screens, and they were sorted and bagged by material type (e.g., ceramics, flaked stone, ground stone, bone, etc.). Pollen samples were generally collected from columns within excavation units or through augering.

The remains of 246 individuals, most from burial contexts, were recovered during the Marismas project. As stated in the introductory chapter, human osteology was a primary component of the project. Burial recovery procedures were established by Dr. George Gill, and he or one of his students was usually in charge of the excavation and recovery of burials in the field. Burial feature designations were assigned consecutively across sites or, sometimes, assigned consecutively within the feature from which they were recovered. Funerary offerings were cataloged referencing burial feature numbers. The data presented in the burial summary tables presented in this chapter are based on field data and subsequent laboratory assessments of the skeletal remains recovered. More comprehensive descriptions of the recovered burials are presented in Appendix A (see also Chapters 9, 10, and 11).

Artifact Recording and Storage

Field notes were created in duplicate by each excavator, and correspondence, contracts, and additional paperwork related to the survey and excavation were kept by the director. Artifacts were generally photographed or drawn, and a card file was established for datable finds. All Marismas materials, both surface-collected and excavated, remain in various Mexican repositories, including the Museo Nacional de Antropología, Museo Regional de Tepic, and Museo Regional de Guadalajara.

Shell Mound Survey and Testing

Early in the Marismas project, the shell mounds that abound in the area were investigated through survey and limited testing (Shenkel 1969a, 1971a, 1974). Although Shenkel identified hundreds of shell mounds, his study was limited to the central and southern portions of the Marismas. His goals were to develop a typology based on the dominant species represented, identify their temporal associations based on artifacts present, and generate population estimates based on the estimated amount of animal protein represented by the shells in the midden. Four types of shell mound (middens), defined by dominant species present or shape, were identified: oyster, *Tivela*, linear shell mounds, and *Anadara*.

Oyster middens were the most common of the shell mounds, with 557 mounds reported by Shenkel (1971a). Several species of oyster, *Ostrea corteziensis* and *O. palmula*, inhabit the estuary (oysters are brackish water species) in the Marismas, but *O. corteziensis* appears to be the most common species found in the oyster middens. Oyster mounds ranged in size from 3 m in diameter to ovoid mounds measuring 30 × 60 m in size; these were up to 6 m in height. Shenkel (1971a) estimated that oyster middens ranged in volume from 1 to 16,680 m^3, with a mean volume of 308 m^3. He (1971a:91) indicates that his volume estimates are just that and should be seen as minimum estimates since many of the mounds were partially buried and the actual size of each midden could not be determined with great accuracy.

The second midden type identified by Shenkel (1971a) is the *Tivela* midden, so named because *Tivela byronensis* is the most abundant of the species present. *Tivela* is a small- to medium-size smooth clam; three species of *Tivela* occur in the archaeological record of the Marismas (Foster 1974, 1984). *Tivela* inhabit intertidal sand beaches and sand flats to a depth of up to 10 m and nearshore sand and mud from depths of 11 to 26 m. Twenty *Tivela* mounds were recorded by Shenkel, with most being located south of the Teacapán Estuary. Those recorded ranged in size from about 12 m in diameter and 60 cm high to ovoid or linear mounds 30 × 60 m and 3 m high. Calculated volumes ranged from 6 to 500 m^3, with a mean of 157 m^3 (Shenkel 1971a).

Bounding the course of the Teacapán Estuary are

numerous linear shell mounds. Shenkel (1971a) recorded 48 such mounds, the longest of which extends approximately 2.8 km in length and was 100 m wide with peaks of over 10 m. These mounds are composed almost entirely of oyster, and Shenkel (1971a:105) estimated that more than five billion oysters are represented. Volumes of these shell mounds were estimated to range from approximately 9 to 613,000 m^3, with a mean of 24,730 m^3. Radiocarbon dates indicate Early Conchera and late Middle to Late Conchera dates for the linear shell mounds sampled. The later date at a two-sigma range extends well into the early contact period, suggesting that these mounds are at least in part the product of Spanish exploitation of oysters at a commercial level (see Chapter 12).

Shenkel (1971a) also reports the presence of two *Anadara grandis* mounds, one being El Calón (Scott 1999; Scott and Foster 2000; Shenkel 1974). Shenkel estimated the volume of El Calón to be 355,840 m^3, which would have included several billion individuals. More recently, Grave Tirado and Pulido Méndez (2005:194–195) have calculated that El Calón is made up of more than 275 million individual *Anadara*. *Anadara* is a low-salinity lagoon and mangrove (brackish water) and intertidal sand beach species. At least six other species of shellfish were mixed with the *Anadara* that make up El Calón.

Another *Anadara* mound was recorded by Shenkel on the opposite side of the Laguna Agua Grande. Additionally, situated some 50 m or more from El Calón itself is another outlying low mound of disarticulated *Anadara*. Rydout (1975:64) noted two low *Anadara* mounds near El Calón that were no higher than 2 m and 100–150 m in width. These low *Anadara* mounds are similar in size and all superficial respects to another *Anadara* mound observed near the suspected location of the contact-period settlement of Olita (see Chapter 12).

The *Anadara* mound observed in the likely location of Olita was about 25 to 35 m long and no more than a meter high. It was found on the south side of the Río Cañas channel about 200 m from the point where the river today discharges into the Estero Puerta del Río or the Teacapán Estuary channel, east of Isla Panales. The Olita mound is at the edge of an open tidal flat, free of mangroves. Although it is not known how much the local setting has changed since the formation of the mound, such an open tidal flat would have been ideal for solar salt production resulting from the evaporation of trapped pools of seawater.

Subsequent to Shenkel's survey, Weil (1974) conducted a relatively broadscale survey within the Marismas with goals focusing on establishing a typology of local ceramics and refining the cultural chronology of the area. He established 19 tentative local plainware (utilitarian) ceramic types and conducted a comparative analysis of Chametla and Amapa ceramic types found within his sample to the type descriptions of Kelly (1938) and Grosscup (1964). Weil hoped that the analysis of the plainwares would demonstrate that some were temporally sensitive. Unfortunately he was not able to do so, and equally unfortunate is the fact that circumstances did not allow his typology to be tested or further refined. In terms of the decorated wares, he concluded that the application of the established neighboring ceramic sequences and chronology to the Marismas was appropriate.

Cerro El Calón

El Calón represents an anomaly among the archaeological vestiges of Marismas prehistory, for the questions of both its age and its function. Chapter 4 presents a summary of arguments for and against the view of El Calón as the oldest archaeological manifestation yet dated in the Marismas. As discussed in Chapter 4, based on radiocarbon assays, it may have been constructed between 2400 and 2200 cal BC, with at least one addition sometime between 1600 and 1200 cal BC. Beyond the issue of its age, we focus here instead on understanding the nature of El Calón as a remarkably conspicuous feature of the landscape.

What is El Calón? To open the discussion, what it is not is a coastal accumulation of storm-deposited shells, nor was it formed by other natural geomorphological processes. It is a massive, human-engineered mound constructed primarily of the marine mollusk *Anadara grandis*. Located at the far northern edge of the main estuarine channel (see Figure 5.1), El Calón appears to represent an example of platform architecture in the rough form of a truncate pyramid, measuring approximately 80 × 100 m at its ovoid base. Its platformed summit stands approximately 23 m above the dense surrounding mangrove forest, which is dominated by *Rhizophora mangle L.* and *Avicennia nitida Jacq.* (Snedaker 1971:16; Figures 5.2–5.6). From its squared and leveled top the view in all directions, to mountain and sea, is unobstructed.

If El Calón was a purposeful construction, as we believe it was, what motivated the builders to collect the immense amount of marine shell to create this enigmatic component of Marismas Nacionales prehistory? As noted above, Grave Tirado and Pulido Méndez (2005:194–195) estimate that El Calón is made up of more than 275 million mollusk shells, mostly *Anadara grandis*. The genus *Anadara* is an important source of protein in many tropical, subtropical, and warm temperate areas of the world (Broom 1985:1; Keen 1971). Yet the height and form of the mound and the fact that most of the shells were deposited unopened, indicating that they were perhaps collected alive, suggest that the mound is something other than a large shell midden.

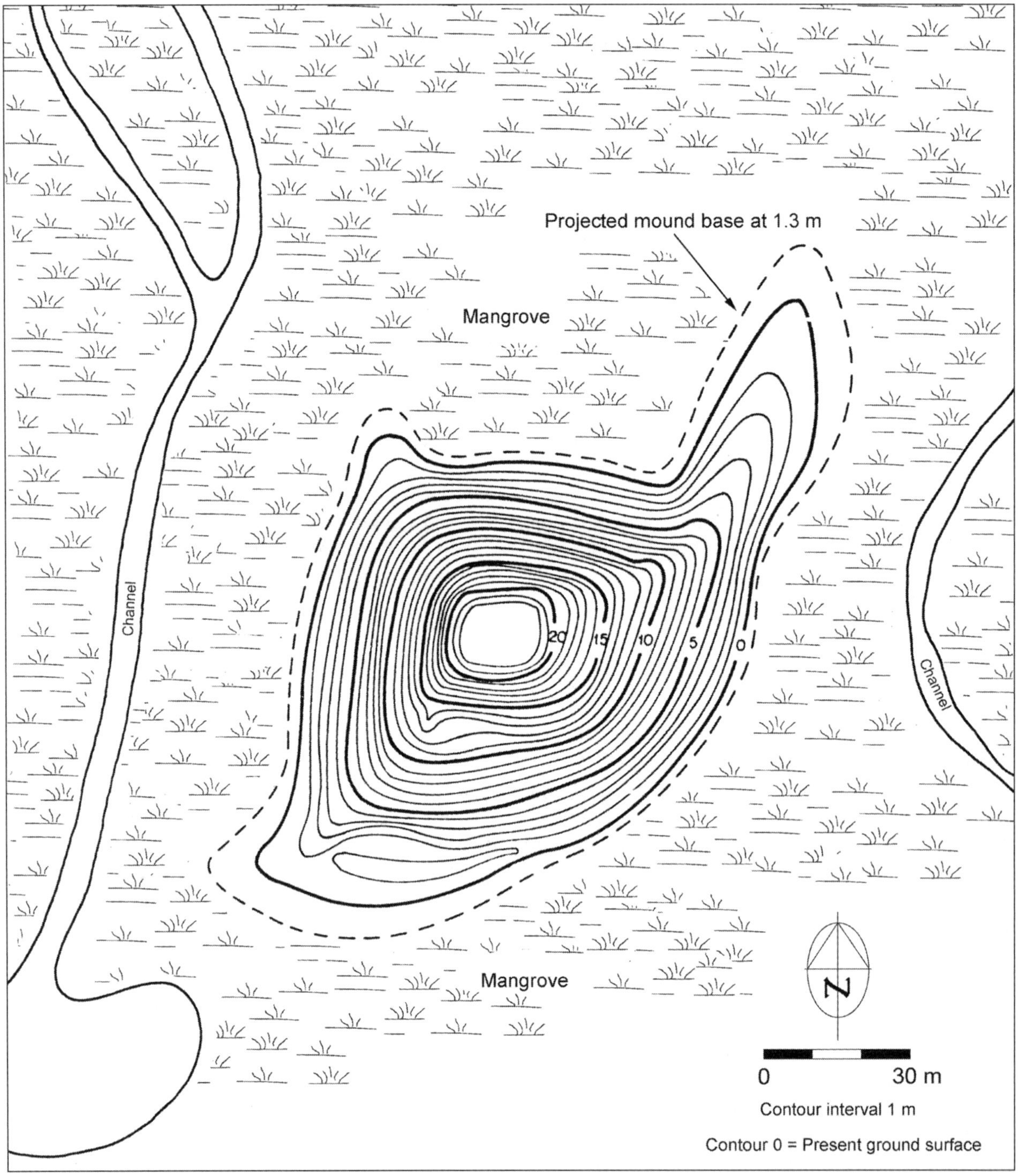

FIGURE 5.2. Topographic map of Cerro El Calón (adapted from Shenkel 1974:63).

Could still-paired shellfish have played some symbolic role as a food offering? Claassen (1998:229–230) cites examples from Europe and North America of the use of unmodified (unopened) shells in mounds of religious or sacred special use. Whatever the true inspiration was for El Calón, this scale of exploitation of closed marine mollusks would appear to have been at the loss of a great bulk of consumable biomass.

Clearly, as it is not midden- or habitation-related, we have in El Calón a structure of the type that if found in Veracruz or Guerrero would be interpreted as a form of religious or corporate architecture. The open question is, How plausible does it seem that El Calón was built as a feature associated with the ideological or symbolic realm of human behavior? The first evidence that might attribute some spiritual or ritual use to this site began to

FIGURE 5.3. Aerial view of Cerro El Calón surrounded by mangrove forest, facing southeast with the Laguna Agua Grande in the background.

emerge soon after the mound first came to our attention. Interestingly, although no systematic excavations of El Calón were attempted during the Marismas project, clay figurine fragments began to appear, initially because of an inadvertent discovery from a workman's casual turning up of shells while resting at the top of the mound.

In all 17 figurine fragments, mostly torso and limb fragments, were recovered. One of the specimens was a nearly complete figurine, and another was a head. These two specimens were found at the southern edge of the top of the mound 8 to 10 cm below the surface; the remaining fragments were found in the erosional scree down the south-facing slope of the mound. No sherds or other artifacts were identified. The figurine fragments were solid and hand-molded, and the paste was heavily granular, giving the figurines the initial appearance of stone. Owing possibly to excessive leaching, or weathering, there is little detail in the modeling beyond gross shape and size. The two specimens with faces have similar bulbous noses, and the head fragment has elongated blobs for ears and engraved slashes for eyes and a mouth (Figure 5.7a–b). Scott (e.g., Scott and Foster 2000:132) has suggested correlations with Preclassic figurine types of central Mesoamerica.

However, these El Calón specimens also bare some resemblance to the figurine styles (Figure 5.7c) from the early part of the Chametla sequence (Kelly 1938:Plate 15i–l, see also 54–57). Regardless of the suggested affinities, the figurines collected during the Marismas project lack significant stylistic attributes to allow specific comparisons or assignment of their temporal or cultural association. Nonetheless, the occurrence of figurines as an exclusive artifact class, we concluded, gave the site a cultural "separateness," adding to its semblance as an isolated case of civic or religious architecture.

In that vein, recent investigations of El Calón have yielded additional insight into the function of the structure (Grave Tirado 2003b, 2003c, 2007, 2010; Grave Tirado and Pulido Méndez 2005). Five units, one on the northwest edge of the mound, one on the southwest edge of the mound, two on the elongated projection of shell at the northeast edge of the mound, and one on top of the mound, were excavated. The results confirmed that the mound was almost entirely constructed of marine shell. Additionally, a very limited amount of small lithic debitage was found in several of the excavation units. Grave Tirado (2010:26) importantly notes that the excavation of the five units yielded only a very small number of artifacts. He suggests, therefore, that El Calón is not a shell midden like those found on the adjacent coastal

FIGURE 5.4. View of Cerro El Calón from its base, facing east-southeast.

plain that contain quantities of marine shell as well as cultural debris and other types of faunal remains. During his (2010:26) first season's efforts, artifacts recovered included six grinding slabs or metates; three manos; 26 green obsidian flakes, of which seven exhibited retouch or other edge damage; and several modern-era ceramic sherds. Grave Tirado (2010:Figure 5), for unspecified reasons, suggests that one of the metates is a fragment of a stela. This is not evident in his illustration of the specimen, of which only about half remains. As illustrated, the left edge of the specimen appears to exhibit some pecking marks; these may represent battering from shaping the implement. Additionally, a mano appears to have been partially shaped before fracturing of the material prevented the specimen from being completed (Grave Tirado 2010:26, Figure 6). Grave Tirado suggests that the specimen was being modified to represent a zoomorphic figure of some type. A small ax (not described) and a pounder (maul/pulverizer?), both made of basalt, were also recovered.

Additional artifacts were recovered during a second season of Grave Tirado's excavation at El Calón. In addition to six small metates, a partial greenstone ax was recovered from the excavation unit atop El Calón. The bit is broken away, although it appears that a rough, unmodified cutting edge that would allow it to function

FIGURE 5.5. Rendering of Cerro El Calón, facing south (computer-generated rendering by Lynn Simon).

FIGURE 5.6. Aerial view of a rendering of Cerro El Calón, facing east (computer-generated rendering by Lynn Simon).

FIGURE 5.7. (a–c) Examples of figurine fragments recovered from Cerro El Calón during the Marismas Nacionales project.

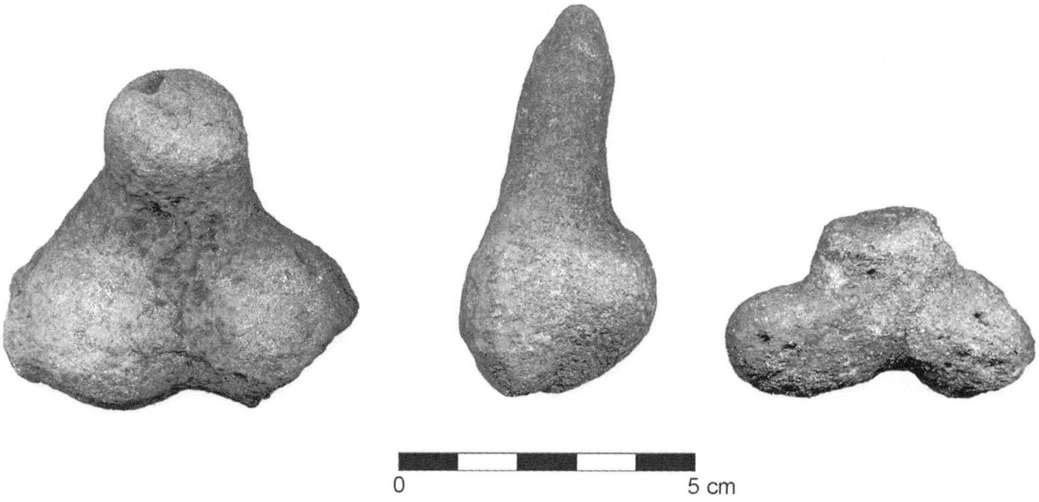

FIGURE 5.8. Three stone phalluses recovered from Cerro El Calón (photograph courtesy of Luis Alfonso Grave Tirado).

is present. It is not clear whether the butt is broken. A rounded edge is apparent in the illustration; however, it appears that the butt may also be broken. It is reported as a ¾-grooved specimen (Grave Tirado 2010:28, Figure 8). Of further significance was the recovery of three small stones that appear to represent phalluses (Figure 5.8). All were recovered from 30–35 cm below the surface in adjacent subunits of the excavation unit atop the mound, and all three appear to be unmodified concretions. Other lithics include 14 green obsidian artifacts; 11 were flakes, and the remaining specimens appear to be scrapers or knives.

A few plainware sherds were recovered, but there continued to be an apparent near absence of prehistoric ceramics in the mound fill. Nevertheless, two significant discoveries were made. Two figurines, one complete and the other fragmentary, were recovered. The complete specimen was exposed at the surface atop El Calón. Grave Tirado (2010:29–30) reports it as an anthropomorphic figure (Figure 5.9). It clearly has a human head with a hat or turban or perhaps hair. The body lacks any anatomical detail, and it appears to be that of a fat torso with short legs. On the back, in the area of the buttocks, are two additional legs or supports that appear to have allowed the figurine to stand upright. Grave Tirado (2010:29) suggests that it is a seated figure. It almost appears to be a human head on the body of a quadruped. The surface of this hollow specimen is eroded, but it appears to have been covered or partially covered with a white slip, and there are vestiges of either red or orange paint on parts of the body. It appears that a necklace was painted on the figure and perhaps clothing too. Based on the painting, Grave Tirado (2010:30) suggests that this figurine is similar to Kelly's Chametla White-filleted, painted figurine, but because of its other attributes he suggests that it is most similar to Kelly's (1938:59, Plate 19) Cocoyolitos hollow figurines. However, other than the surface treatment and the fact that the figurine is hollow, it bears little if any stylistic similarity to either the White-filleted painted or Cocoyolitos hollow figurine types. Kelly (1938:59) assigns both types to her Middle Chametla pottery complex: the Baluarte phase, AD 500–700 (Kelley and Winters 1960). The fragmentary specimen is only a torso; the head, arms, and legs are missing. It too is white-slipped and is thought to be similar to a Chametla White-filleted–type figurine (Grave Tirado 2010:30).

Regardless of the type associations of the figurines, Grave Tirado's (2006, 2010) work at El Calón confirms and expands the association of figurines as apparent votive offerings left on or buried in El Calón. When the figurine fragments recovered during the Marismas project are considered with the figurines and the three stone phalluses reported by Grave Tirado, the ritualistic or spiritual function of El Calón postulated by Scott (1985, 1999; Scott and Foster 2000) and Shenkel (1971a, 1974) becomes more apparent.

In addition to the figurines and phalluses left as offerings, Grave Tirado (2010:30–31) reports the presence of a high number of fish bones and the remains of possibly several turtles to a depth of 3 m below the surface of the top of the mound. As described, this faunal material may represent midden deposits; however, the possibility that the remains may represent food offerings as proposed by Grave Tirado cannot be ruled out. However, he did not indicate whether the faunal remains were either partially or completely articulated, which would have indicated likely offerings. Furthermore, Grave Tirado (2010:33–34) suggests that the location of El Calón and its alignment with Cerro del Muerto (to the east) may be significant. In Chapter 12 Patricia Scott discusses the possible astronomical significance of El Calón in relationship to Cerro del Muerto as well as the significance of Cerro del Muerto to the ethnohistoric populations of the region. Based on the above, Grave Tirado (2010, 2012) argues that El Calón served as a temple where ceremonies were conducted related to fertility, with propitiatory rites to favor good fishing.

There is, then, mounting evidence of El Calón's purpose as a place for supplicatory or expiatory rites. This religious interpretation of the structure as a sacred place is correlated, in general and perhaps specific terms, with

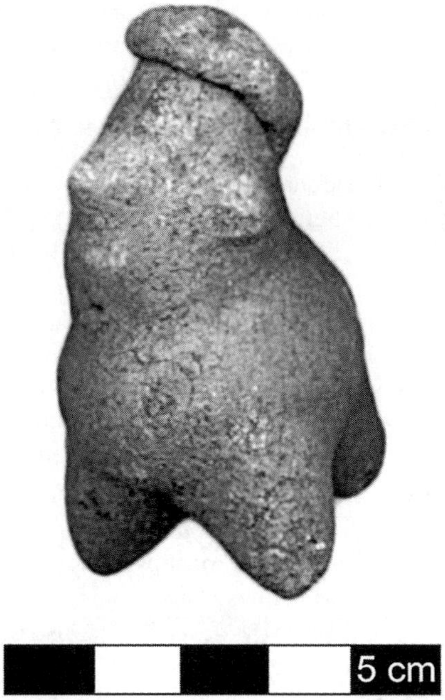

FIGURE 5.9. Figurine recovered from Cerro El Calón (photograph courtesy of Luis Alfonso Grave Tirado).

the cosmology recorded by the Spanish chronicler Father Arias de Saavedra (1975 [1899]). His observations of clay figurines, fragments of such, phallic symbols, and so forth are discussed further in Chapter 12. However, here it can be noted that in the seventeenth century such offerings were deposited in places considered sacred. The offerings were left to supplicate deities, who were thought to be located in special places in relation to sun risings and settings and were believed to have powers over fish, birds, crops, weather, and women.

Despite the evidence, albeit limited, suggesting that El Calón served as a ritualistic feature, numerous interrelated questions remain as to when and how it was constructed, why it was built in this location, and how it was used. First, how old is El Calón? Scott (1999; Scott and Foster 2000) has maintained that the radiocarbon assays for El Calón indicate that the feature dates to sometime between 2500 and 1500 BC. Radiocarbon assays, pollen data, and geomorphological reconstructions of long-term land and waterscape changes have been cited as supporting evidence for this antiquity (see Chapter 2). Scott, within a normative Preclassic to Postclassic central Mexican temporal and cultural paradigm, initially interpreted El Calón as a temple mound. Subsequent work on the west coast and in the northwestern uplands of Mexico has demonstrated that human use of the region dates well into the Archaic period and that likely there was cultural use of the coastal areas and resources during that period. It is possible that El Calón originated during part of an earlier cycle of human activity in the

Marismas Nacionales area similar to the mound at La Venta on the Gulf Coast. If the radiocarbon assays are indicative of the period of construction or accumulation, it thus seems appropriate to interpret the feature as a manifestation of a local or regional coastal Late Archaic or Early Preclassic adaptation similar to that identified to the south in southern Nayarit and Jalisco (e.g., Mountjoy 2000:81–84). If El Calón is the product of this earlier time, it would have been partially submerged by the 1650 BC–AD 450 sea rise and then again about AD 750, when there was another short submergence or period of very high water.

In a very different interpretation, Grave Tirado (2007:35) dates the construction and use of El Calón to AD 750–1100. In a later article, he (2010) suggests that El Calón was built through cooperative labor by Middle Classic period "fishermen" occupying the villages scattered across the Marismas coastal plain. His dating estimate would seem to put El Calón's construction just prior to the area being submerged around AD 750, and its use would have extended into the period when there was a several-hundred-year hiatus of occupation on the coastal plain. Grave Tirado's dating would make El Calón contemporaneous with several of our excavated earthen mounds, which date to just prior to this last coastal submergence and before the noted hiatus of occupation.

Other large mounds are reported in the region (Kelly 1938; Rydout 1975; Sauer and Brand 1932). However, as described, these earthen mounds stand about one-third the height of El Calón. The first, Loma de La Cruz (Sauer and Brand 1932:21–22 [Plate 4], 24), is located on the Río Acaponeta just west of San Felipe, approximately 47 km southeast of El Calón. Sauer and Brand indicate that it was about 12 m high and had a quadrangular base approximately 50 m wide; however, a more recent assessment of the mound indicates that its base is 100 m long × 60 m wide, that it stands 9 m tall, and that its long axis is oriented east-west (Garduño and Gámez Eternod 2005). The second mound is called La Loma, which Sauer and Brand indicate is similar to Loma de La Cruz. It is approximately 10 m high with a flat top and is located approximately 66 km northwest of El Calón, south of Villa Union, Sinaloa, and near the pueblo of Alamo. Both sites were littered with sherds and lithic artifacts. Sauer and Brand (1932) mention other such mounds in the vicinity of Chametla. Kelly (1938) expanded the description of those mounds, noting Loma de Ramírez, nearby mounds between Chametla and Apoderado, and two such mounds at Cocoyolitos. She also mentions large shell mounds near the coast in the Chametla area; however, no description is provided. While these are all big mounds, there are significant differences between the earthen mounds and the El Calón shell mound, with the latter being nearly three times the height of those described by Kelly or Sauer and Brand, and to date very few artifacts have been recovered at El Calón. The question can be asked: Were all or any of these earthen and shell mounds built in response to rising sea levels?

Second, how was El Calón built? Everyone seems to agree that it was built almost entirely of articulated *Anadara* shells. While no one really knows where the shell was procured, Grave Tirado (2010:34–35) contends that the shell was mined from fossil shell beds predating the building of the mound. No matter where the shell came from, how the mound was constructed remains to be answered. Was El Calón deliberately erected in one stage to its present shape and height? Scott (1985, 1999; Scott and Foster 2000) and Shenkel (1971a, 1971b, 1974) have steadfastly argued that El Calón was constructed as a temple mound, based on the traditional central Mesoamerican Preclassic through Postclassic paradigm of pyramidal religious architecture. Scott observes that at least two periods of accumulation are represented in the ^{14}C assays; alternatively, these dates may indicate the extended period over which the mound accumulated. Grave Tirado (2010:26) states that the apparent homogeneity of the mound's shell matrix is the product of a well-planned single effort. He further states that the structure is actually a conical mound built on a formal platform. This seems to be simply a declarative statement, and it is not evident in either Shenkel's (1974:Figure 2) topographic map of El Calón or Grave Tirado's (2010:Figure 2) topographic map of the mound. The two renditions give very similar details, in that both seem to show El Calón as having a truncated rather than a conical form. Additionally, the three-dimensional renderings of Shenkel's map (see Figures 5.6–5.7) suggest that the arm coming off the northeast corner of the mound looks to be a long inclined ramp, which could have been used in the construction of the mound or to access its flattened summit.

Waselkov (1987:115–116) reports that in physically circumscribed areas shell middens can accumulate vertically, often up to 20 m or more in height, and usually in the form of a truncated cone. While no one argues that El Calón is only a midden, could its basal portions have at least in part begun as an extensive shell midden? Further, could such a midden have been augmented later to the height and shape we see today, including what appears to be a ramp to allow easy access to the top? As previously noted, Scott dates El Calón to the Late Archaic or Early Preclassic, while Grave Tirado dates it to the Classic. Forgetting functional aspects, if Grave Tirado's concept about El Calón's one-stage Classic construction is accepted, then all the radiocarbon dates must be ignored or associated with his suggested fossil shell beds. Some might agree, but as shown in Chapter 4, there are valid reasons to accept the ^{14}C dates based on how they were collected and calibrated, along with the relationship of the mound to the geomorphological development of the surrounding landscape. If El Calón accumulated or was constructed in stages, then it is possible that both Scott's and Grave Tirado's datings are accurate and should both be considered when modeling the construction of El Calón.

Third, why was El Calón built where it is? As noted in Chapter 4, the geologic findings suggest that El Calón was built on a beach ridge that was formed between 2600/2400 and 2200/1900 cal BC, with one additional period of beach ridge formation dating between 1600 and 1200 cal BC. Therefore, El Calón was built nearer the ancient shoreline of the Pacific Ocean, and geologists believe that it stands on the first of the two ridges mentioned above. Today the mound is 15 km east of the ocean, surrounded by dense areas of mangroves near the shores of Laguna Agua Grande. It stands approximately 48.5 km due west of Cerro del Muerto and in a visual line toward the east with the highest peak of that large mountain. This alignment was captured in a photo by Luis Alfonso Grave Tirado (2010:Figure 13). As can be seen in Chapter 12, from the top of El Calón the sunrises of the vernal and autumnal equinoxes appear directly over the highest peak of Cerro del Muerto (Appendix B). It is not obvious whether this alignment is coincidental or whether those who began the accumulation of millions of *Anadara* shells that led to the feature of El Calón as it is known today recognized the relationship between its location and Cerro del Muerto regarding equinox sunrises. The feature may have started as a simple midden with no known relationship to Cerro del Muerto or astronomical knowledge of potential equinox or solstice alignments. However, as people settled the region and knowledge of the landscape grew, relationships between geographic features and the location of the El Calón mound may have been discovered. If indeed the El Calón mound started as a midden, perhaps with the discovery of its relationship to sacred geographic features on the landscape the mound itself was turned into a sacred locus and augmented to its present size and shape in one or more stages. It is also possible that a later population recognized the relationship of the existing midden to the equinox sunrises and then expanded its size and shape to meet their ritualistic needs. As evidenced by the discovery of figurines, in whatever manner, El Calón, at least during some part of its history, seems to have become or to have been built as part of a ritual feature of one or more populations of the region. The question remains, How far back in time did such a ritual or belief system extend? The premise that El Calón incorporated some sense of ideological or supernatural power during the Classic and Postclassic, as Grave Tirado suggests, and/or later does not preclude the possibility that this prominent feature was built as a religious structure much earlier.

At this time we are left with the two distinct views: El Calón was built as a Late Archaic or Early Preclassic period temple mound, or it was a Classic period temple mound. If El Calón was in fact a product of early foragers and shell fishermen of the West Mexican Coastal Plain in the area that is now southern Sinaloa and northern Nayarit, or if it was constructed during the Classic period, what was El Calón's history between the time it formed or was built and the occupation of the area as the modern coastal plain emerged? As discussed in Chapter 2, the paleogeomorphic reconstruction of the Sinaloa plain shows that El Calón and especially any low-lying associated features would have been abandoned during periods when the sea rose and inundated the land. We come then to the missing chapter in the human settlement history of the coast. If El Calón dates to the Late Archaic or Early Preclassic, the absence of any archaeological record of continual occupation indicates that the region was not inhabited again until the Classic and Postclassic period earth mound and shell mound builders recolonized the emergent coastal plain. If it was built in the Middle Classic, there was a period after its construction of 200 to 300 years when there seems to have been a break in the occupation. Clearly, El Calón remains an enigmatic archaeological feature awaiting further understanding of its origin, age, and function.

Juana Gomez

The sites of Juana Gomez and La Tarjea (La Atarjea) were selected for excavation during the first two full field seasons of the Marismas project because of their location at the inner rim of the coastal plain and, more practically, for their easy accessibility from temporary project headquarters in Escuinapa de Hidalgo, the *cabecera* of the Escuinapa *municipio*. The modern town straddles the Arroyo Buñigas, and like many other municipalities that developed near water channels, Escuinapa was built over a substantial accumulation of prehistoric cultural debris.

The Juana Gomez site takes its name from an intermittent, meandering stream of the same name that drains the rolling foothills northeast of the town of Escuinapa, Sinaloa (see Figure 5.1). The site itself is situated on an elevated basaltic ridge that extends from the lower mountain slopes, and it consists of nine more or less distinct mounds, plus the stone foundation of a postconquest church (Figure 5.10). The mounds are distributed along the wide and sloping crest of the ridge in two groupings: a cluster of seven mounds at the southwest end and the remaining two, plus the church foundation, in the upper part of the site toward the northeast.

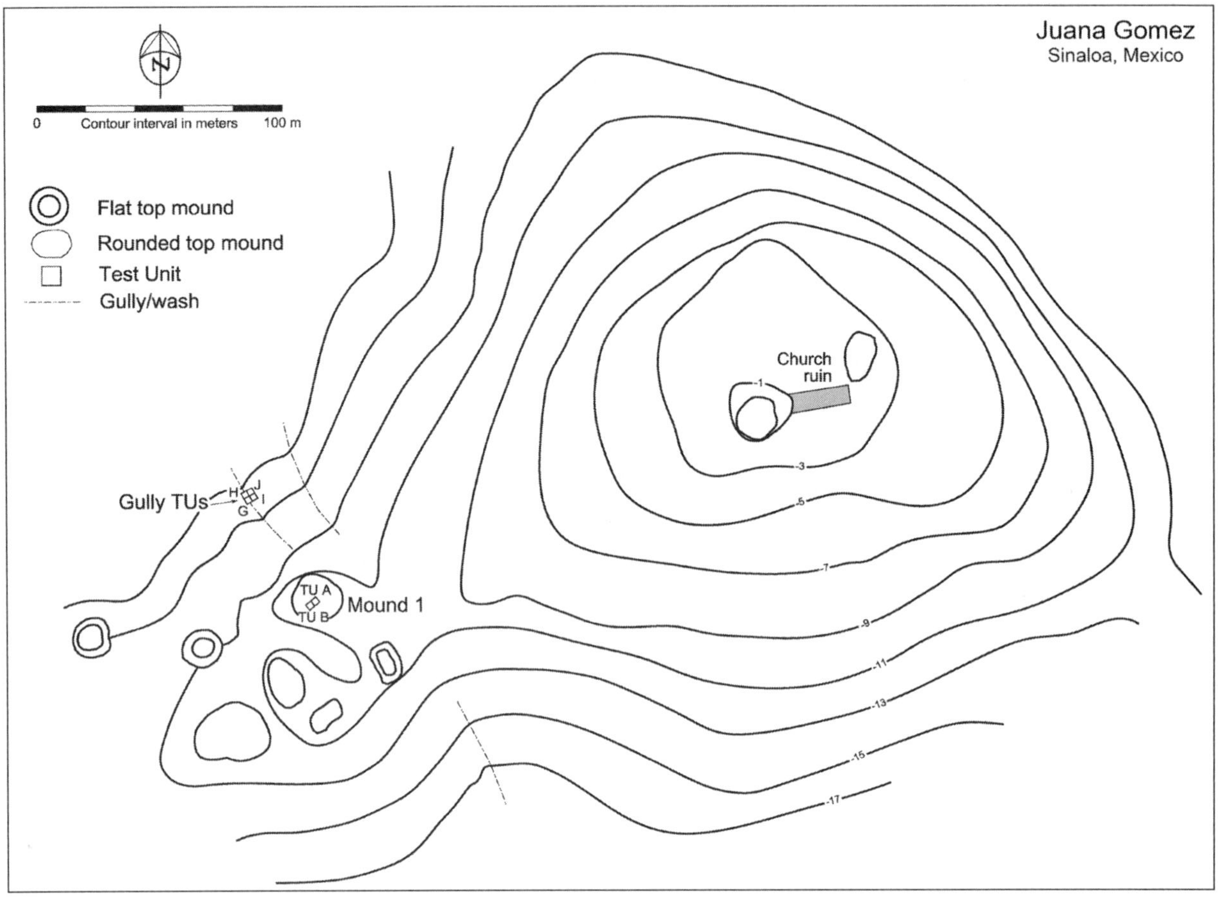

FIGURE 5.10. Site map of Juana Gomez. TU = Test Unit.

In what would appear to be the major part of the site in the southwest, a borrow pit was clearly visible. The mounds themselves range from barely discernible rises to unmistakable protrusions on the surface of the ridge. An abundance of artifacts was scattered across the site, including sherds, spindle whorls, figurine fragments, obsidian tools and fragments, chert flakes, grooved axes, manos, and metates.

What is known archaeologically of Juana Gomez comes primarily from excavations in a single mound, Mound 1, and supplementary excavations on the lower northern flank of the ridge where bone fragments and sherds appeared in an erosional channel feeding into the Juana Gomez arroyo (Wolynec et al. 1968). Two test units, Test Units (TUs) A and B, encompassing an area of 2 × 5 m, were placed in Mound 1 at the southwestern edge of the ridge. The ridge extended from the summit of the mound to its outer edge and was excavated in arbitrary 30-cm levels (Wolynec et al. 1968:Figure 3). The profiles in the test units showed stratification—changes in texture and color from highly calcic pebbly sand to darker noncalcic soils—reflecting the local geomorphic variability at the margin of the mountains as opposed to the more homogeneous nature of the estuarine sediments (Figure 5.11). Similar geomorphic components were identified at La Tarjea during excavations in the summer of 1969 (see below).

The clearest example of a living surface (Feature 1) from any site investigated during the Marismas project was identified in Mound 1; it was exposed in the upper third of TU A. A deposit of stick-impressed daub, probably roofing or wall material, was uncovered at 77 to 90 cm below the mound top. This construction material had collapsed onto and was scattered across a uniform and highly compacted earth floor containing three postholes, each about 4 cm in diameter (Figure 5.12). In addition to some charcoal, seeds, and obsidian, two reconstructible Red-rimmed Utility Ware vessels, apparently broken by the weight of the structural collapse, were on the floor. The function of the feature, whether habitation or storage, is unclear, as is the abandonment process. It is not known whether the feature was abandoned with vessels left on the floor, whether there was a catastrophic collapse of the structure during its use, or whether the structure was deliberately destroyed by its occupants during the abandonment process. Following abandonment, construction fill (see Figure 5.11-A2) was subsequently added to raise the height of the mound.

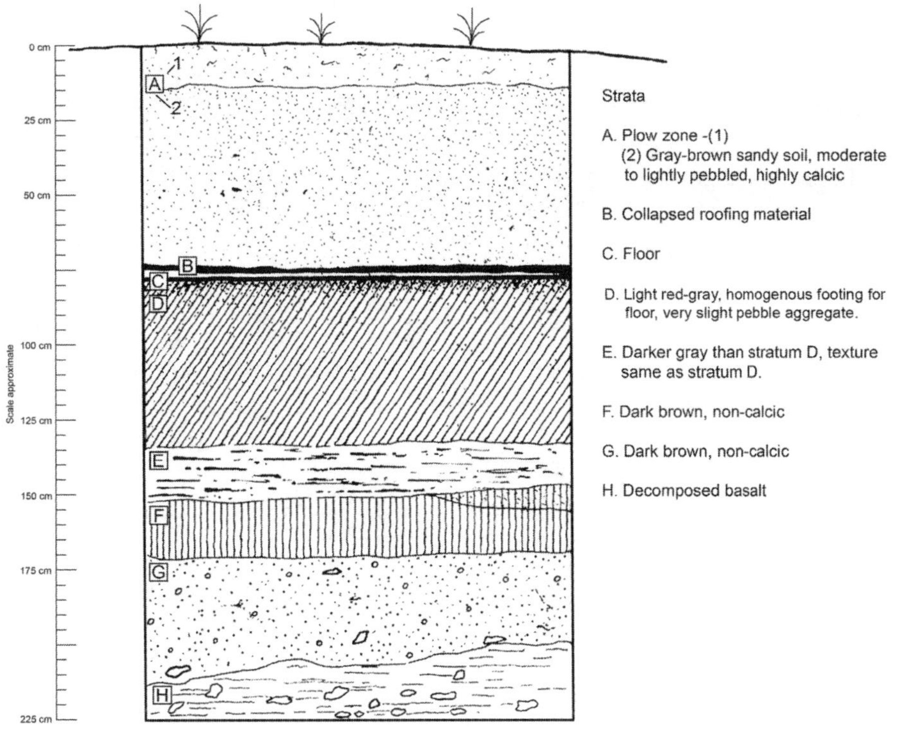

FIGURE 5.11. Profile of Test Unit A, Mound 1, Juana Gomez.

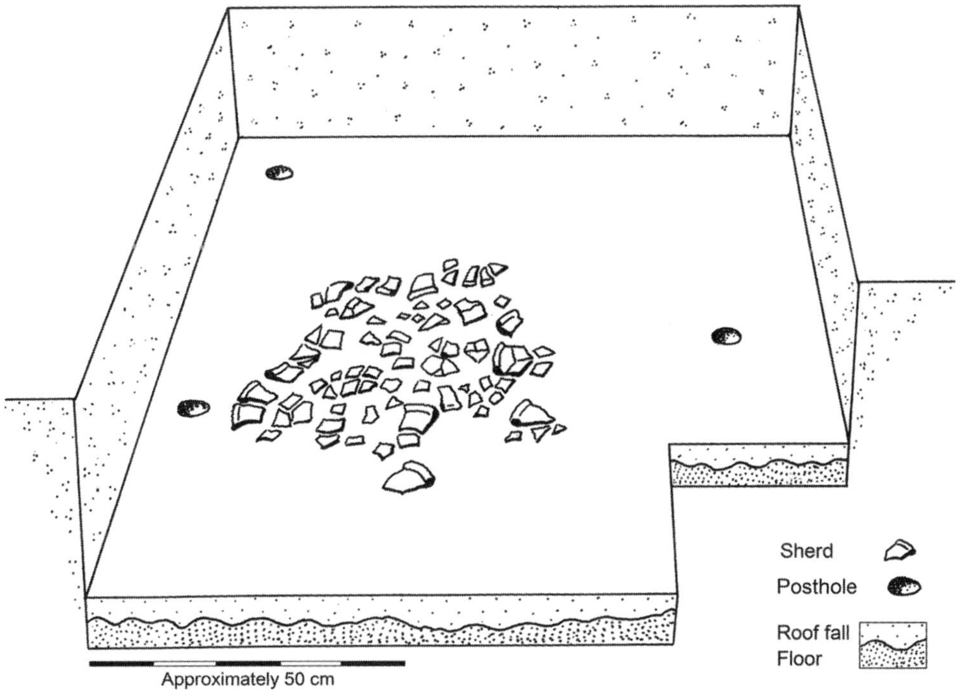

FIGURE 5.12. Schematic illustration of Feature 1, Mound 1, Juana Gomez.

TABLE 5.1. Summary of Recovered Artifacts by Level from Test Unit A, Mound 1, Juana Gomez.

		Level								
Artifact	Surface	1	2	3	4	5	6	7	8	Total
Obsidian blades				1					1	2
Obsidian projectile points							1		1	2
Obsidian debitage	1	16	6	14		7	66	X	X	110+
Faunal bone	1		2	2			X	X	X	
Human tooth								1		1
Whole shells		3	13	4	6		X	X	X	26+
Shell fragments	1	1	13	3		20	X	X	X	38+
Figurines		2			2	1		1		6
Spindle whorls		1			1					2
Miscellaneous clay objects			1			1	6	7	9	24
Worked sherds								1		1
Seeds			5	3	1			1	1	11
Crab claws			1			2	X			3+
Pipe fragments	1								1	2
Stone axes			1							1
Beads								1		1

Note: X = present but not tabulated.

TABLE 5.2. Summary of Recovered Artifacts by Level from Test Unit B, Mound 1, Juana Gomez.

		Level			
Artifact	Surface	1	2	3	Total
Obsidian debitage		6	6	1	13
Unspecified bone	1		4		5
Whole shells		2	5	6	13
Shell fragments		5			5
Figurines			1	1	2
Miscellaneous clay objects		2			2
Worked sherds	1				1
Seeds			1		1
Crab claws				1	1
Metates			1		1

The fill above the floor contained Middle Chametla Polychrome, Middle Chametla Polychrome Engraved, Black-banded Nonengraved, Late Black-banded Engraved, White-on-buff, and Polished Buff, each of which suggests a Late Estero period occupation. A number of other artifact types were also recovered from TUs A and B (Tables 5.1–5.2).

The levels below the floor were marked, at 1.8 m below the surface, by a change in color and a ceramic assemblage that included Early Chametla Polychrome, Early Black-banded Engraved, and Red-rimmed Utility Ware. The Red-rimmed Utility Ware from this level is the Tierra del Padre phase (Chametla) variant with a predominance of orange paste and less red painted decoration (Kelly 1938:31). A stray Amapa Red-on-cream sherd (Amapa phase) was also recovered. In her analysis of the ceramics from Juana Gomez, Sweetman (1970:51) noted that the initial level below the floor contained a clear mix of Chametla types associated with the Baluarte and the earlier Tierra del Padre phases. This mixed collection included an intact unclassified red-on-orange vessel similar to those described at La Tarjea (Sweetman 1970:Plate Va). These levels containing mixed Early Estero and Late Estero period (AD 300–700) materials appear to represent either a transition between the two periods or the simple mixing of two temporal components.

The basal levels of the mound produced pottery assignable only to the Early Estero period. Recovered from these levels were two reconstructible vessels: a small and shallow well-made Red-rimmed Utility Ware bowl and an Early Chametla Polychrome bowl (Sweetman 1970:Plates IVa–b, Vb–c). None of the whole or reconstructible vessels were recovered from funerary contexts.

Four additional test units (the Gulley TUs G, H, I, and J) were excavated on the lower western slope of the Juana Gomez ridge, where sheetwash had redeposited surface artifacts from above and where gully erosion had revealed sherds and bone fragments in the subsurface. The test units were excavated to depths that varied from 54 to 64 cm; the widths of the units varied from 60 to 125 cm. Excavations were terminated when bedrock was encountered.

Two burials, those of a young adult male and a young adult female, were recovered along with a variety

TABLE 5.3. Ceramic Types Identified at Juana Gomez.

Type	Name
Polychrome	Early Chametla Polychrome
	Middle Chametla Polychrome
	Middle Chametla Polychrome Engraved
Engraved/Incised	Early Black-banded Engraved
	Late Black-banded Engraved
	Aztatlán Ware (Botadero Incised)
Other	Black-banded Nonengraved
	White-on-buff
	Polished Buff
	Buff Exterior, Red Interior
	Red-rimmed Utility Ware
	Blackware
	Redware

Source: Sweetman 1968:50.

of sherds, obsidian flakes, shell, and figurine fragments. The male was a bundle burial, and the female was an extended inhumation. No definite funerary offerings were associated with either burial. The sherds were Chametla types including redware and Red-rimmed Utility Ware along with some blackware and Polished Buff. Two badly eroded sherds of a Chametla Polychrome were also recovered. The artifacts were not directly associated with the burials and may represent trash deposits from the upper portion of the site (Block 1968; Sweetman 1974:70). Identified ceramic and artifact types recovered from Juana Gomez are summarized in Tables 5.3 and 5.4.

The church, the foundation of which can be seen in the uppermost part of the Juana Gomez site, seems likely to have been the work of the Spanish colonists at Escuinapa, possibly associated with the Franciscan friars from the convent at Acaponeta, Nayarit (Macias 1996). Within the church ruins were remains of the painted and plastered adobe. Elsewhere at Juana Gomez, a few sherds of San Luis Polychrome, one of the Majolica types of the seventeenth and eighteenth centuries, were found. Historically, the settlement at Chametla has the greater longevity, the original church there having been founded in 1564. A quadricentennial was celebrated at Chametla in 1964. Construction of the church at Rosario was begun in 1757 and was completed in 1759. No date could be found for the church at Escuinapa, but it is undoubtedly one of the many churches of Sinaloa built in the eighteenth century or later. Not only was the church at Escuinapa later and probably smaller than those at Chametla and Rosario, but one might judge from the following quotation (of unknown authorship) that it was artistically less impressive as well: "Templo de Escuinap. Es una simple obra de albanilería cuyo proyectista indudablemente se instruyó en algunas formas arquitectónicas do otros templos, ejecutando en este ealizaciones de muy pobre valor artistico" (in Macias 1996:27). It should be noted here that Macias (1996:27) calls our attention to the discovery of a second small chapel at the site, the foundations of which were partially destroyed in the building of the Nogales-Guadalajara highway.

La Tarjea and Juana Gomez lie near but are not physically part of the marsh and mudflat-shore region of the Marismas that was subject to the dynamic relationship of changing tides and deltaic landforms. Instead, both sites were located at the inner "hard rock" margins of the coastal plain. Predictably, midden material at both sites contained small quantities of fish and mammal bone, shell, and crab claws, but these two communities probably also shared a primary subsistence economy based heavily on farming. Juana Gomez, however, illustrates rather different settlement aspects.

As described, the ridgetop mounds that lie east of and above the Juana Gomez streambed are the most visible and defining aspects of the site. If we take into account the more recent discoveries of archaeological remains from beneath the streets of Escuinapa, then the observed limits of the Juana Gomez occupation should perhaps be expanded. Figure 5.13 shows the Escuinapa city plan in its open setting between the Arroyo Juana

TABLE 5.4. Summary of Recovered Artifacts by Level from Test Units G, H, I, and J, Juana Gomez.

		Level			Burial		
Artifact	Surface	1	2	3	1	2	Total
Obsidian projectile points		1					1
Obsidian debitage	2	29	13		1		45
Faunal bone	8	3					11
Whole shells	1	1	1		1	4	2
Figurines	2	3			2		7
Miscellaneous clay objects		3	7		1	1	12
Worked sherds		3					3

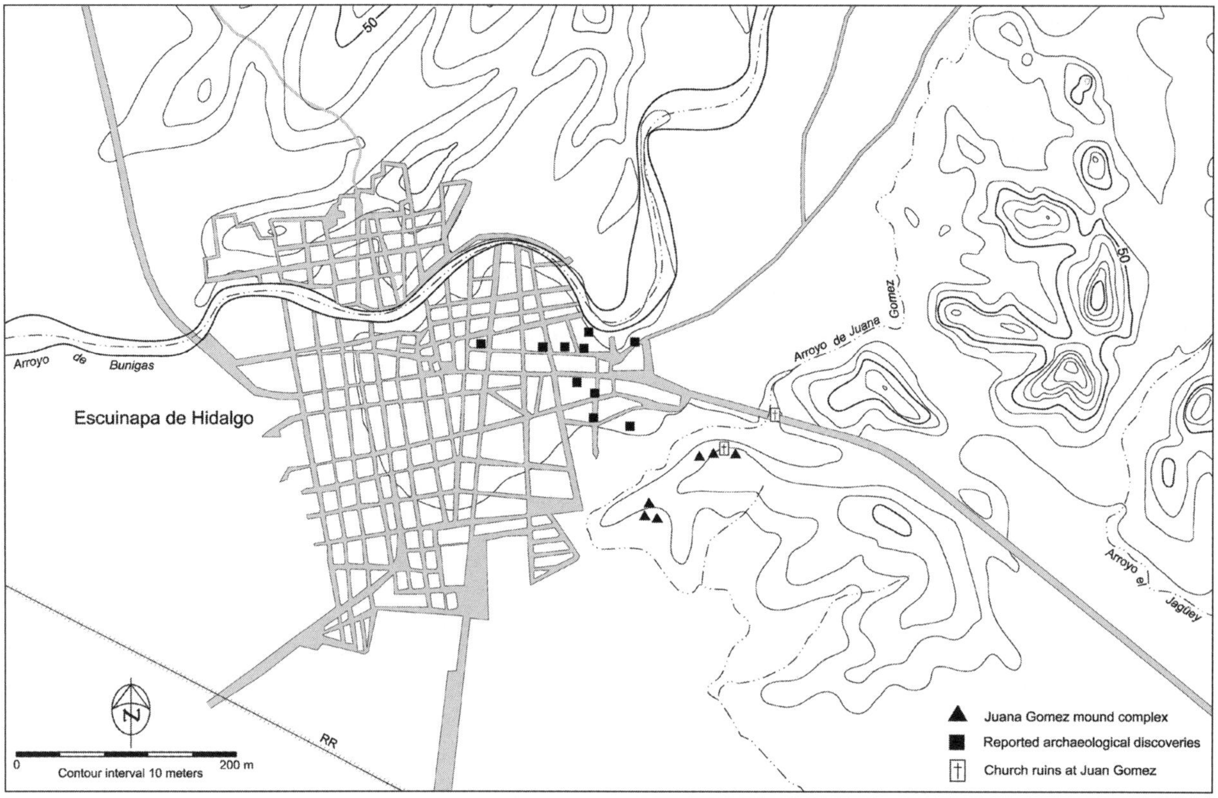

FIGURE 5.13. Map showing reported archaeological discoveries within Escuinapa de Hidalgo and their relationship to the site of Juana Gomez.

Gomez and the larger channel of the Arroyo de Buñigas. As in many other such cases in antiquity, Escuinapa was established and expanded on top of the residual remains of an earlier epoch. Utility lines or water system improvements that require trenching and excavation within the urban area often reveal precolumbian ceramics under the modern city infrastructure. Macias (1996) has documented many of these locations, and he notes that these historical remains are distributed east of the modern city center.

La Tarjea appears to have been a relatively small settlement of several families, like many others situated along the fertile coastal plain at the interface of the Sierra Madre Occidental and the estuary-lagoon complex. In contrast, Juana Gomez represents a much higher relative population figure, based on the assumption that the mounds situated on the higher basaltic erosional remnant were contemporaneous and part of a large settlement that included residential and farming land in the corridor between the Juana Gomez and Buñigas rivers.

Unfortunately, Juana Gomez's location at the edge of Escuinapa has made the site highly vulnerable to the destructive forces of commercial agriculture and modern development. At the time of the State University of New York/Buffalo field investigations in 1968 the site itself and its nine distinct mounds were intact and easily accessible. About 30 years later, in response to a proposed housing development on the land occupied by the site, a brief salvage project revealed that only five of the original mounds were still extant and that there was no longer any vestige of the eighteenth-century chapel (Grave Tirado and Nava Burgueño 2012; Téllez 1997). Over time, the large and open promontory adjacent to Escuinapa was favored for activities destructive to the archaeological resource base. In addition to its appeal for residential development, there have been 40 years of agriculture in the form of commercial mango orchards, road building, use of the ancient mounds for a municipal entertainment park with trails for motorcycles and recreational vehicles, and extensive looting of the site.

Elsewhere in this volume (Chapter 3), the productive effect of new investigations by the Instituto Nacional de Antropología e Historia in this northwestern coastal corridor, especially Grave Tirado's return to Juana Gomez in the 2000s, is discussed. In response to the proposed new residential development on the Juana Gomez site, Grave Tirado carried out salvage fieldwork early in 2008 (see Grave Tirado 2012). The main objective was to define the boundaries of the prehistoric settlement for site conservation. The results from the excavation of 18 test units added significantly to the informational context of the site, addressing such questions as the probable age of occupation, the time of abandonment, and the site's importance in a regional context (Grave Tirado 2012:131).

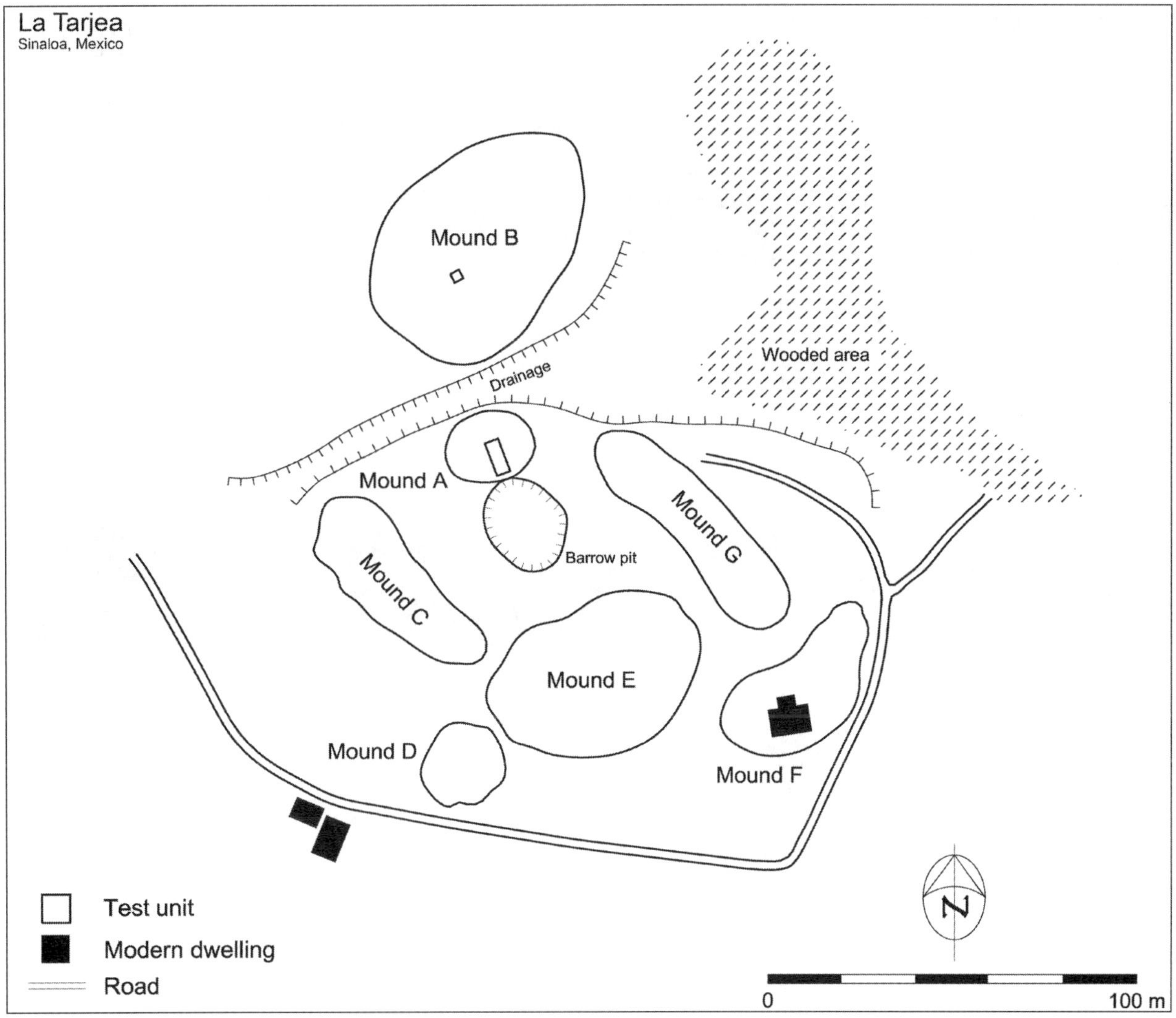

FIGURE 5.14. Site map of La Tarjea.

In general concordance with our earlier study, and based largely on ceramic typology, Grave Tirado concluded that Juana Gomez was a locus of cultural development that spanned our nominative Estero period, dating to the early to middle Mesoamerican Classic period and corresponding with Chametla's Tierra del Padre and Baluarte phases.

Grave Tirado (2007) speculates that the mound complex of Juana Gomez was a principal or capital center having an organizational structure and a residing group in power. Further, he suggests that the functional basis of the site, situated at the edge of the biologically rich wetlands of the Marismas, was a sort of marketplace in which elites controlled or at least supported the production of salt and other resources, including oyster and shrimp, which would have been dried and smoked before being sold at market. In ascribing this higher level of organization to Juana Gomez, Grave Tirado (2007:55, 2012:134–136) believes that these Marismas product resources served as exchange commodities for obsidian and other nonlocal "prestige goods," which could then be redistributed in collective rituals or communal celebrations, possibly at the time of harvest or planting. Beyond the economic links, as Grave Tirado suggests, certain Marismas sites were probably under the political aegis of Juana Gomez. Although there are plausible grounds for Grave Tirado's speculation, the supporting evidence for Juana Gomez's prominence in the prehistoric period is tenuous. The presence, however, of two Spanish churches suggests that priests were administering to the needs of a significant and enduring Juana Gomez population.

La Tarjea

La Tarjea, located about 10 km to the southeast of Escuinapa (see Figure 5.1), is a grouping of seven mostly distinct earthen mounds that are up to 3 m high (Figure 5.14). Two mounds, Mounds A and B, were selected for testing. Four contiguous 2-×-2-m test units were laid out north-south, across the southern half of Mound A,

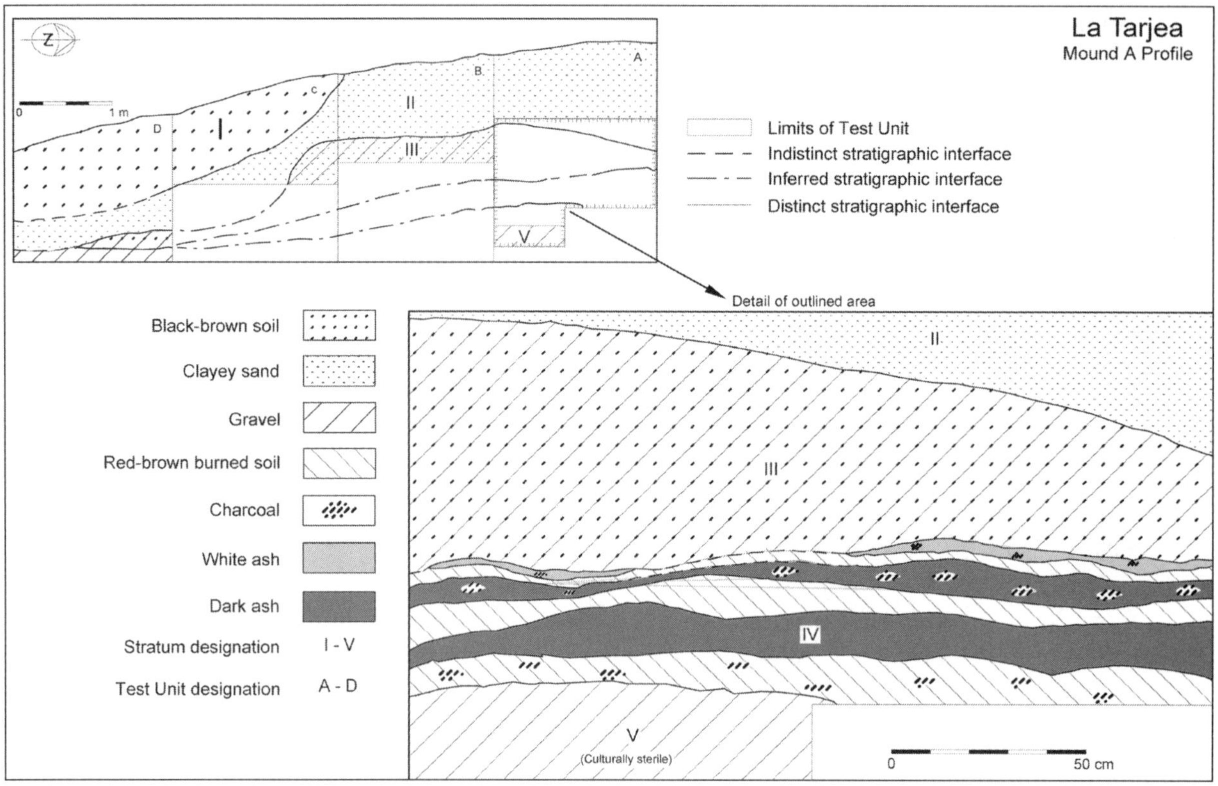

FIGURE 5.15. Profile of Mound A, La Tarjea.

TABLE 5.5. Summary of Burials from La Tarjea.

Mound and Burial Number	Condition	Sex	Age	Burial Type	Funerary Offerings
Mound A					
B 1	Incomplete	Undetermined	Undetermined	Undetermined	None
B 2	Incomplete	Undetermined	Undetermined	Undetermined	None
B 3	Nearly complete	Undetermined	Undetermined	Extended	1 small Middle Chametla Polychrome Engraved jar
B 4	Nearly complete	Undetermined	Undetermined	Semiflexed	1 small Middle Chametla Polychrome Engraved jar
B 5	Nearly complete	Undetermined	Undetermined	Undetermined	1 small, necked jar; 1 ceramic turtle effigy; and fragments of large red *tecomate*
B 6	Incomplete	Undetermined	Undetermined	Flexed	1 necked jar, possible variant of Chametla Red-rim Decorated
B 7	Nearly complete	Undetermined	Undetermined	Semiflexed	None
B 8	Nearly complete	Undetermined	Undetermined	Extended	1 necked jar
B 9	Nearly complete	Undetermined	Undetermined	Semiflexed	1 frog or toad effigy jar
B 10	Nearly complete	Undetermined	Undetermined	Extended	Present but not described
Mound B					
B 1	Nearly complete	Undetermined	Undetermined	Flexed	1 small bowl
B 2	Fragmentary	Undetermined	Undetermined	Undetermined	1 orange-on-red jar
B 3	Nearly complete	Undetermined	Undetermined	Undetermined, possibly seated	1 polychrome jar
B 4	Nearly complete	Undetermined	Child	Extended	1 polychrome jar

a small but relatively high mound near the center of the site. Unlike estuarine mounds formed of lagoon sediments from sand islands and bars with little or no stratification, the profiles in the La Tarjea test units exposed a series of strata representing major additions to the mound. Although derived from local materials, these structural units differed in soil color and textures (gravel or sand or both). Well-developed soil horizons and alluvium reflected the site's proximity to the foothills of the Sierra Madre.

Four culture strata were identified in Mound A (Rucker 1969:7–15; Figure 5.15). Stratum I consisted of a dark, black-brown clayey soil containing a few sherds, obsidian, and fragmentary faunal remains. This layer appeared only on the lower periphery of the mound, although it probably originally extended over most of its entire surface. Stratum II consisted of a compact yellowish clay with very little artifactual material; however, the fragmented remains of 10 human burials were present (Table 5.5). Stratum III, appearing intermittently in the excavation trench, was distinguishable by a dark black-brown soil with a gravelly texture containing only a few sherds and occasional obsidian flakes.

Stratum IV was identified only in one excavation unit that was dug to a deeper level than the other units. It consisted of alternating layers of ash and burned earth, containing quantities of sherds, along with some fish and mammal bone and shell. Rucker (1969) viewed this stratum as evidence of numerous, large and sustained burning events, although he offers no explicit interpretation of their origins.

It is possible that these strata are either midden or redeposited midden materials. These four strata overlaid a culturally sterile layer of clayey sand containing charcoal but no artifacts above the gravel stratum.

To augment the ceramic sample from Mound A, a single 2-×-2-m test unit was excavated at the top of Mound B. This mound also contained clear evidence of stratification. Rucker (1969:15) describes the uppermost stratum, which extended to a depth of about 1.2 m, as a compact red to yellowish brown soil that contained four human burials and a relative abundance of artifacts, including two clay pipe stem fragments, a decorated spindle whorl, an incised sherd, a mano, obsidian flakes, several figurine fragments, and faunal remains. At the lower levels of Mound B, excavators found discontinuous lenses of ash, a burned earth layer, and what appeared to be a compacted living surface or house floor. No postholes, building materials, or other associated features were exposed within the limits of the test unit. The floor level and the dark red-brown soil stratum above it were also devoid of any cultural debris. The remains of four human burials were recovered from Mound B (see Table 5.5).

If, as Rucker suggests, the various mound levels at La Tarjea were functionally distinct, the pottery content, it was hoped, would vary by distinct strata. Of the approximately 2,000 sherds recovered, only about a half dozen were diagnostic decorated types. These were divided between redware and blackware sherds. Although the diagnostic ceramic sample is small, it nevertheless suggests a temporal association for La Tarjea. Near the basal level (Level 6, TU D) of Mound A were Gavilán Polychrome, a Gavilán phase Amapa type, and Early Black-banded Engraved sherds, the latter representing the earliest (Tierra del Padre phase) of the known Black-banded types at Chametla (Kelly 1938; Sweetman 1968). The remaining decorative sherds from Levels 1 and 2 (Davis 1969:89–90) in Mound B were Black-banded Nonengraved (Level 1, TU A) and Late Black-banded Engraved (Level 2, TU A), ceramic types that bridge the Baluarte and Lolandis phases at Chametla. Subsequent study of whole vessels recovered from Burials 3 and 4 resulted in the identification of two Middle Chametla Polychrome vessels as well as two likely Chametla Red-rim Decorated vessels, one of which was from Burial 6. In terms of the local Marismas sequence, the decorated ceramics present at La Tarjea indicate a Late Estero period (ca. AD 500–700) occupation of the site.

Arriñitas

The site of Arriñitas, located approximately 1.5 km northwest of the village from which it takes its name, was tested in 1970 (Wolynec 1970b). Three earth mounds, aligned more or less north-south, were present (Figure 5.16). The northernmost mound was approximately 2 m high and roughly L-shaped. The central, Mound A, was circular and approximately 2 m high. The southern mounds, also roughly circular, were about 50 cm in height and were barely discernible from the surrounding landscape. All three mounds were relatively flat-topped and were covered with a light scatter of pottery.

Excavations at Arriñitas were limited to two 3-×-3-m test units in Mound A, one at the top of the mound and TU E at the mound's base. A third test unit (Well Test Unit) was excavated next to a well on the southern edge of the site. It was thought that the excavation of a test unit adjacent to the well might result in the identification of an earlier component at the site. The excavation of the Well TU exposed a lens of shell containing cultural material about a meter below the ground surface. However, the diagnostic ceramics from both Mound A and the Well TU were Amapa Cerritos and Ixcuintla phase types and Chametla Acaponeta phase types, generally placing the occupation of the site between AD 900 and 1200 (Early Conchera to mid–Middle Conchera period) (Tables 5.6–5.8). Foy (1970) suggests that the site may have been occupied as early as AD 750 based

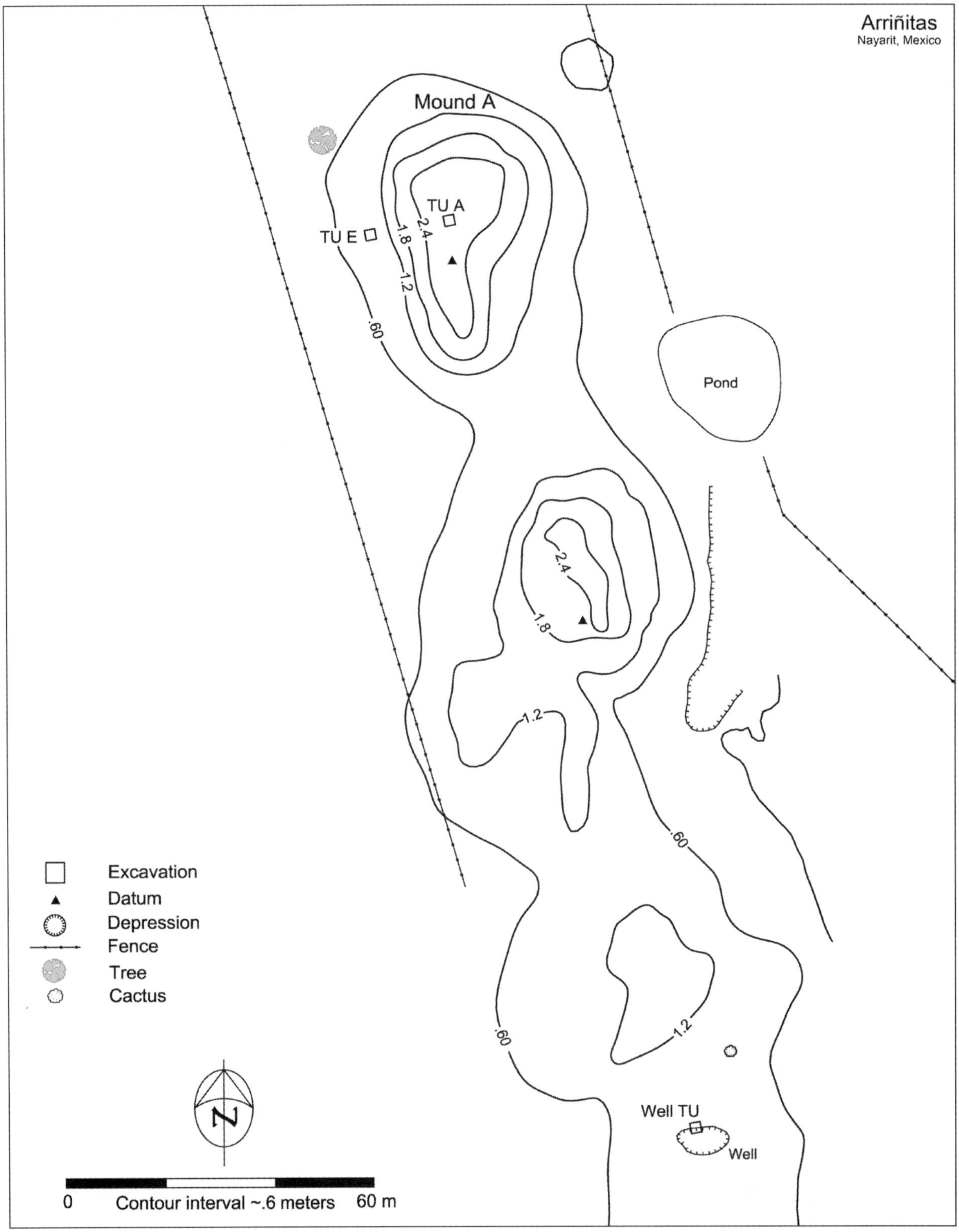

FIGURE 5.16. Site map of Arriñitas. TU = Test Unit.

Table 5.6. Frequencies of Identified Ceramic Types from Test Unit A, Mound A, Arriñitas.

	Chametla Type			Amapa Type						Miscellaneous/Other			
Level	Aztatlán Ware	Baluarte Red-on-buff(?)	Subtotal	Tuxpan Red-on-orange	Tuxpan Engraved	Roblitos Polychrome	Botadero Incised	Amapa Red-on-cream(?)	Subtotal	Miscellaneous Decorated	Subtotal	Cixto Incised[a]	Subtotal
Level 1 (Lot 1)										1	1		
Level 3 (Lot 2)				5					5				
Level 3 and 4 (Feature 1, NE corner)				8					8				
Level 4 (Lot 4)				4					4				
Level 5 (Lot 5)				6	2	2			10				
Level 6 (Lot 6)						3	1		4				
Level 7 (first 1–8 cm)										20	20		
Level 7 (Lot 7)				2	3	3			8			1	1
Level 8 (Lot 8)	1		1	11	6	2			19			1	1
Level 9 (Lot 9)	1	1	2	1			1		2				
Level 10 (30 cm level)				16	3		1	1	21				
Level 11/12[b]													
Subtotal	2	1		53	14	10	3	1		21		2	
Total			3						81		21		2

[a] Cixto Incised is a possible local (Marismas) red-on-orange incised type (Foy 1970:6, 16; Sweetman 1968:63).
[b] Pottery not analyzed.

Table 5.7. Frequencies of Identified Ceramic Types from Test Unit E, Mound A, Arriñitas.

	Chametla Type		Amapa Type					Other	
Level	Aztatlán Ware	Subtotal	Tuxpan Red-on-orange	Tuxpan Engraved	Subtotal	Miscellaneous Decorated	Subtotal	Cixto Incised	Subtotal
Level 1 (Lot 1)						1	1		
Level 3 (Lot 2)									
Level (Lot 3)									
Level 4 (Lot 4)	2	2		4	4				
Level 5 (Lot 5)	1	1							
Level 6 (Lot 6)	2	2	26	9	35			1	1
Level 7 (Lot 7)			5	5	10				
Level 8[a]									
Level 9 (Lot 9)[b]									
Level 10 (Lot 10)						2	2		
Subtotal	5		31	18		3			
Total		5			49		3		1

[a] Ceramics from this level were not analyzed.
[b] All sherds from this level were badly eroded and impossible to type.

Table 5.8. Frequencies of Identified Ceramic Types from the Well Test Unit, Arriñitas.

	Chametla Type		Amapa Type					
Level	Aztatlán Ware	Subtotal	Tuxpan Red-on-orange	Tuxpan Engraved	Roblitos Polychrome	Botadero Incised	Cerritos Polychrome(?)	Subtotal
Overburden (Lot 1)			10			2		12
Overburden/ shell interface	1	1	16	2		1		19
Level 2 (shell layer)			34	10		5		49
Level 2 (Lot 2) (shell layer)				30	5	14	10	59
Shell layer near center of wall	7	7		1	2	4		7
Level 3 (Lot 3) (compact shell layer)	1	1	9	1		3		13
Level 3 (South side to 1.04 m)				10	6	6		22
Level 4				10	6			16
Level 4 (Lot 4)				14		1		15
Level 4 (Lot 5)[a]								
Subtotal	9		69	78	19	36	10	
Total		9						212

[a] Pottery not analyzed.

on the presence of Tuxpan Red-on-orange, a Tuxpan phase type. However, this type also extends into Early Conchera times.

The excavation of TU A in Mound A produced the remains of an isolated hearth or fire pit. Chunks of burned clay were present in Levels 4 and 5. The feature, located in the northwestern corner of the unit, was basin-shaped with a flat bottom, measured 18 cm deep, and was roughly circular, measuring 90 to 100 cm in diameter. It was filled with chunks of burned orange-red and yellow clay like those found in the surrounding levels, and it did not appear to be associated with any habitation feature. The function of this feature is unknown.

The number of artifacts in the first seven levels of TU A was low and included sherds, animal and fish bone, ground and flaked stone, obsidian debitage, and projec-

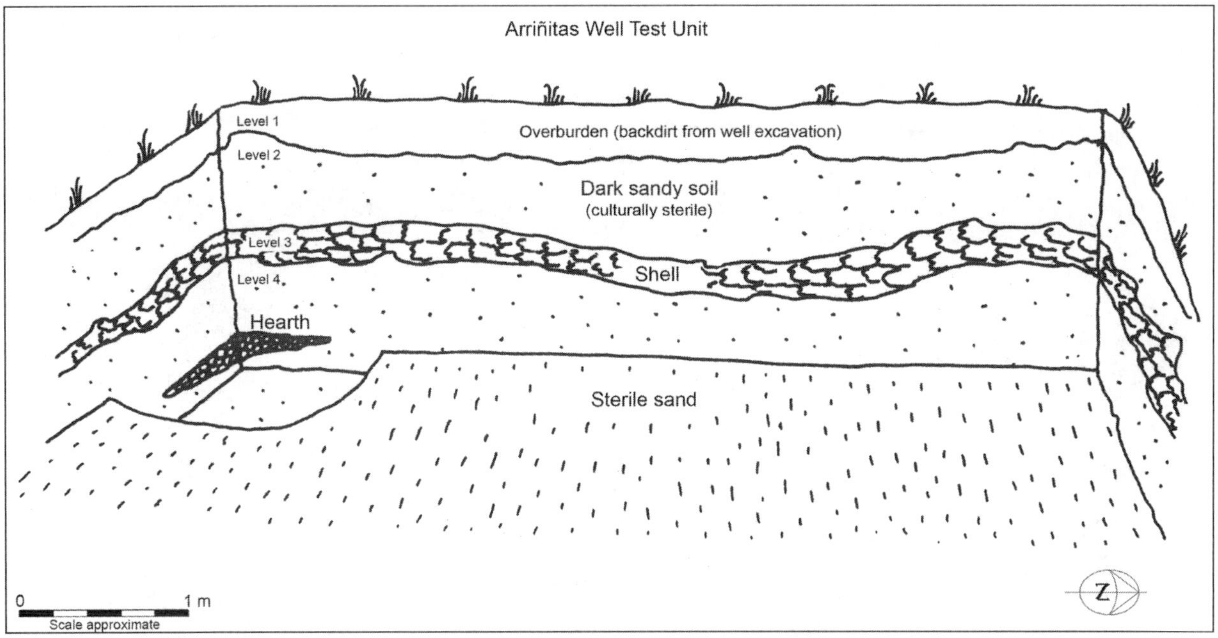

FIGURE 5.17. Schematic illustration of the profile of the Well Test Unit at Arriñitas.

tile points. In Level 8 the frequency of artifacts increased significantly. In addition, an isolated human talus bone was recovered from Level 8, a human incisor and vertebra were recovered in Level 9, and a human molar was recovered from Level 10. These were the only indications of human remains excavated; no human burials were exposed. Artifact density again diminished in Level 10, although Level 11 produced a small copper bell and several pipe bowl fragments. Excavations were terminated at 2.5 m below the surface when an orange-yellow culturally sterile sand layer was reached. The excavation of TU E produced a similar array of material. However, no features were discovered. A small copper bell was recovered from Level 6 along with a few pieces of baked clay (daub?).

Another test unit was excavated at the western edge of the modern well. A cross section approximately 4.9 m long was exposed (Figure 5.17). Levels 1 and 2, which occurred above a shell lens, consisted of fill from the excavation of the well and a culturally sterile sandy layer directly overlying the shell lens. The shell lens, Level 3, was a midden deposit consisting mainly of oyster shell with a mix of artifacts. Artifacts were abundant and included a variety of animal and fish bone; obsidian flakes and blades; polychrome, engraved, and incised sherds; and an assortment of other ceramic artifacts. Level 4 below the shell lens quickly became culturally sterile, with the only artifacts present being those at the interface with the shell lens. Nevertheless, a hearth or roasting pit was present in the southwestern corner of the test unit, near the base of Level 4. The feature is basin-shaped and about a meter in diameter. It consists of a 25-cm-thick layer of compact burned and fire-cracked rock containing a quantity of charcoal. No artifacts or other remains were found in association with the feature.

In terms of the local Marismas sequence, the occupation of Arriñitas, based on diagnostic ceramic types, is assigned to the Early Conchera period (AD 900–1100). Foy (1970) notes that except for what may be several sherds of an aberrant El Taste Polychrome, no ceramic evidence of a Middle Conchera period occupation is evident at Arriñitas. However, dating the occupation of Arriñitas is problematic. As discussed in Chapter 4, four ^{14}C assays, all from charcoal, are available from the site. Two of the ^{14}C assays, one from Level 8 in Mound A dating AD 775–1027 (NSTF-9) and the other from the Well Test Unit, AD 778–1251 (I-5530), overlap with the dates suggested by the diagnostic ceramics. The two other assays, both from Mound A, Level 7, yielded dates of AD 1251–1401 (NSTF-5) and AD 1158–1425 (I-5531). These fall within the mid- to late Middle Conchera period; however, supporting evidence in the form of diagnostic decorated ceramics was not identified at the site.

Panales Mound A

The peninsula of Panales (see Figure 5.1) is the locale for several hundred prehistoric mounds, mostly of shell but also many discrete earth mounds and knolls of mounded earth that underlie larger shell heaps. Figure 5.18, a plan view of mound distribution near the narrow waist of Panales, illustrates the density and apparent formless character of shell mound loci. What at first glance would appear to be a random scatter is actually a linear distribution, with mounds placed on the Medial Beach ridges,

FIGURE 5.18. Map of a section of Panales showing the archaeological features in the area.

possibly to minimize the effect of seasonal flooding or the use of available dry land. The amorphous mounds of earth and shell are broken by mounds of more determinant form. The peninsula, sometimes referred to as an island, is now cut off from the mainland by a modern canal. The canal had been cut in the hope of increasing shrimp, oyster, and clam populations in parts of the estuary by opening a new path for them to move and to reduce travel time for boats within the estuary (Pearson 1968).

Mound A, located on the eastern shoreline of Panales (see Figure 5.1), was one of the first sites tested in 1967 (Pearson 1968). It was oval, with its long axis measuring about 18 m, and it stood about 3 m above the present ground level. An unknown but substantial part of the site had been washed away by tidal erosion.

A trench about 1.5 m wide and nearly 2 m long was excavated into the eastern (estuary) side of the mound (Figure 5.19). The earthen mound was covered by a shell cap that was up to 75 cm thick, which was made up of oyster and clam intermixed with an abundance of artifacts. Below the shell cap, significantly fewer artifacts were identified, but two poorly preserved secondary burials were recovered; these were not analyzed further. Shenkel (1968) subsequently excavated an additional four 3-×-3-m test units into the mound but was unable to add any new information regarding the nature and age of the deposits.

Sweetman (1968, 1974) analyzed the sherd material from Mound A, noting that the ceramics from the earth mound exhibited unmistakable affiliations with Amapa and Baluarte phase types (Middle Chametla Poly-

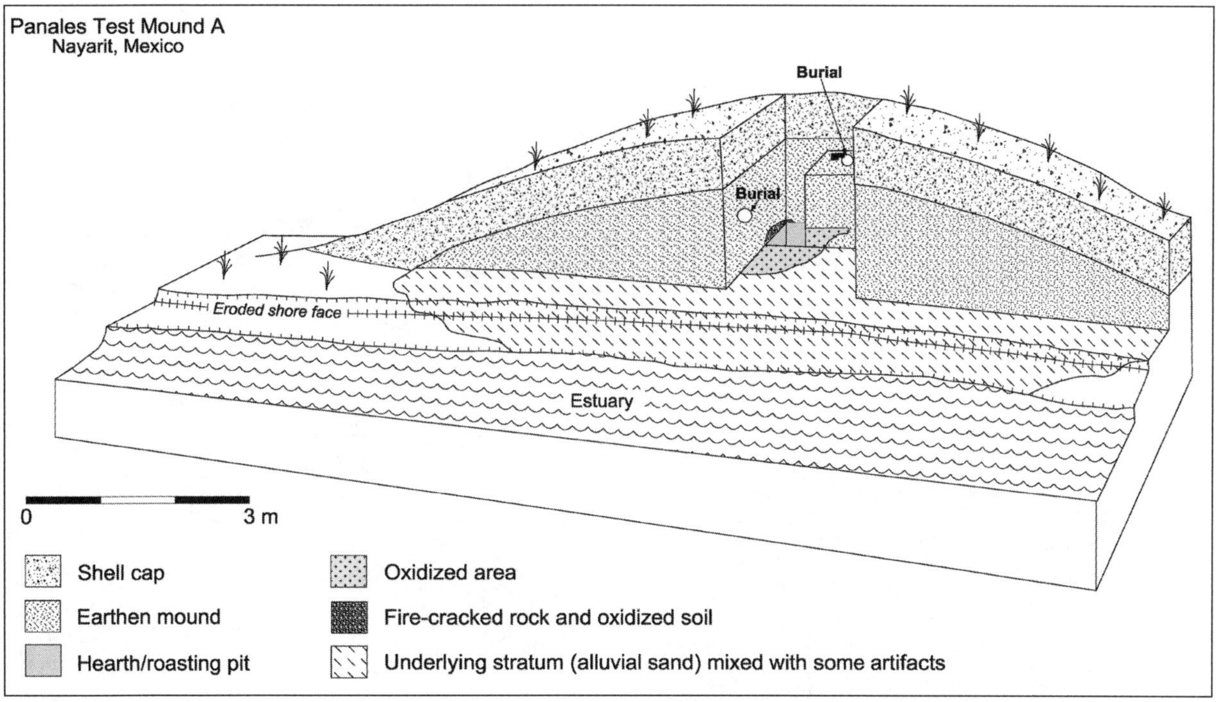

FIGURE 5.19. Illustration of the cross section of Panales Test Mound A, facing west (adapted from Pearson 1968).

TABLE 5.9. Summary of Recovered Artifacts by Level from Panales Test Mound A.

		Level						
Artifact	Surface	Shell Cap	Shell Layer	Level 1	Level 2	Level 3	Burials	Total
Obsidian blades	1		9	1				11
Obsidian projectile points			1			2		3
Obsidian debitage	1	1	3	2	6	10		23
Unspecified bone		1	23	1	1	8		34
Whole shells			2	8		2		12
Shell fragments		3		5	8	12	22	50
Figurines			2		1	1		4
Spindle whorls			1					1
Worked sherds			1		1			2
Miscellaneous clay objects				3	3	3	1	10
Pipe fragments			1		1			2
Stone axes	1							1
Manos		3	1	1				5
Metates		1						1

chrome, Middle Chametla Polychrome Engraved, Late Black-banded Engraved, and Amapa Red-on-cream). Ceramic types recovered from the shell cap, on the other hand, were diagnostic of the Acaponeta, Cerritos, and Ixcuintla phases (Botadero Incised, Botadero Black-on-buff, Tuxpan Red-on-orange, Tuxpan Engraved), as well as local Roblitos Polychrome and Roblitos Red; Amapa types dominated. A number of artifact types were also recovered (Table 5.9). Based on the ceramic cross-dates, the earth mound dates to the local Late Estero period, and the shell cap, to the Early and Middle Conchera periods. This pattern was to be repeated at many of the sites subsequently investigated during the Marismas project.

Panales Mound B

Subsequent to his testing of Test Mound A, in 1969, Shenkel (1969a, 1971a) excavated four test units (cuts) in a large earthen mound (Test Mound B) partially covered with an oyster shell midden at Panales (see Figure 5.1). The mound measured 36 × 21 m and was 5.5 m high

FIGURE 5.20. Panales Test Mound B, facing southeast.

(Figure 5.20). Shenkel's initial excavation, TU 1 (Cut 1), was a 2-×-8-m trench into the southwest face of the mound, approximately in line with the mound's central short axis (Figures 5.21–5.22). TU 2 (Cut 2) was a 2-×-2-m unit excavated at the approximate center of the exposed portion of the earthen mound. TU 3 (Cut 3) was a 2-×-2-m unit excavated in the open agricultural field 20 m west of Test Mound B. TU 4 (Cut 4) was a 1-×-3-m unit excavated at the juncture of the shell midden and earthen mound on the north side of the shell midden. All excavations were in arbitrary 30-cm levels within four general stratigraphic zones as described below and illustrated on some of the excavation profiles.

Shenkel (1971a) begins his discussion of the excavation results by first discussing TU 4, which revealed the most numerous and distinct stratigraphic layers, which he designated Zones I–IV, with Zone I being the uppermost (Figure 5.23). Zone I was a layer of mostly whole oyster shell. Shenkel observed a thin stratum of oyster shell mixed with crushed shell and sand, noting that it appeared much like the weathered surface of the shell midden itself. Although not explained, the inference is that this may have been an earlier depositional episode that was left exposed during a period of no deposition and that the midden deposits above it accumulated afterward. Zone II, exposed in the northern two-thirds of the test unit, was composed of a brown sandy loam. The southern third of the Zone II loam stratum was less distinct and partly intermixed with oyster shell. Zone III was another stratum of predominantly oyster shell; a variety of other marine mollusks were also present but in small numbers. Excavations were terminated in Zone IV, which was a uniformly brown sandy loam that contained occasional flecks of charcoal and some patches of oxidized earth. A single feature was exposed in Zone IV. It was a layer of black "greasy" soil containing burned clay balls that ranged in size from 10 to 20 cm in diameter. The feature was present in both walls of the test unit and measured 30 cm thick and 2.3 m in length. The function of this apparent thermal feature is unclear. No other artifacts of note, other than the burned clay balls, were recovered from the feature.

In TU 1, only three (Zones I, II, and IV) of the four zones identified in TU 4 were present, although there was some variation within them not noted in TU 4 (Figure 5.24). Zone I contained two thinner lenses of the crushed shell and sand noted in TU 4; Zone III was generally thicker; and Zone IV exhibited greater variability in color with multiple substrata. Shenkel (1971a:81) notes that two poorly preserved burials were recovered

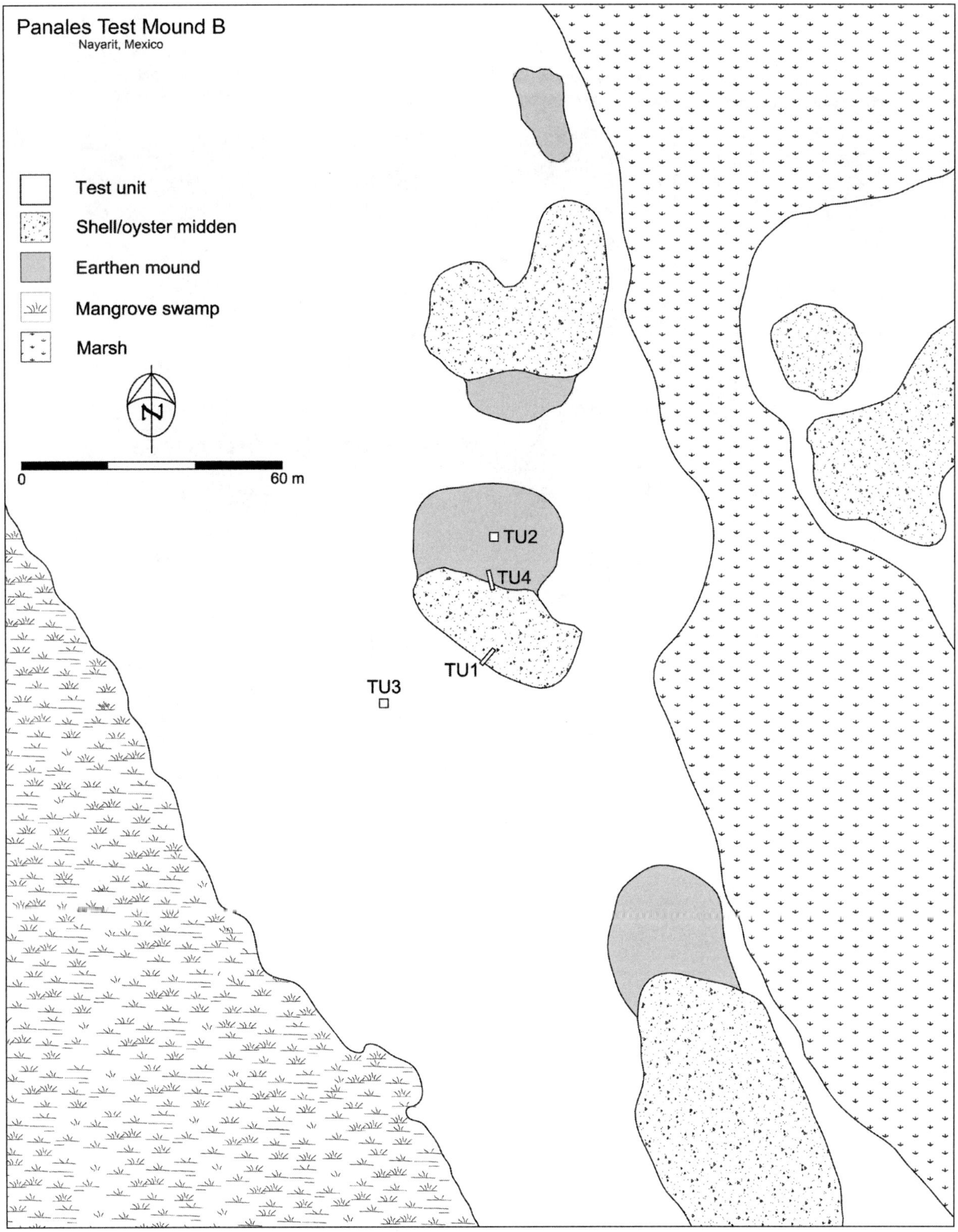

FIGURE 5.21. Map showing Panales Test Mound B, test units excavated, and vicinity. TU = Test Unit.

FIGURE 5.22. Photograph of the excavation of Test Unit 1, Panales Test Mound B, facing northeast.

from the interface of Zones III and IV in an area where Zone III intruded Zone IV. A nearly complete skeleton of a dog was also recovered from Zone III. A variety of artifacts were recovered from all zones, including an overhanging mano fragment from the upper portion of Zone IV. Zone IV also yielded obsidian artifacts including a well-made projectile point and prismatic blades.

TU 2, near the center of the earthen mound north of the shell midden cap, was initiated as a 2-×-2-m unit but was expanded to 2 × 3 m to facilitate the recovery of burials that were discovered there. Three of the zones, Zones II, III, and IV, identified in TU 4 were present in TU 2 (Figure 5.25). The oyster shell lens that made up Zone III, as exposed in TU 2, was not as dense as that exposed in TU 4. Overlaying Zone III was a thin lens of oxidized soil. Shenkel (1969a:41) suggests that this may represent a burning of brush prior to the deposition of Zone II. Zone IV contained scattered pockets of charcoal and of oxidized soil. The remains of at least four human burials were encountered in Zone II. In all, five burials were exposed in Panales Test Mound B (Table 5.10). Burial 1 was located 40 cm below the surface of the mound immediately above Zone III; Burial 2 was exposed at 1.3 m below the surface within Zone II at the end of the shell lens that made up Zone III; Burial 3 was exposed at a depth of 1.4 m; and Burial 4 was at 1.75 m

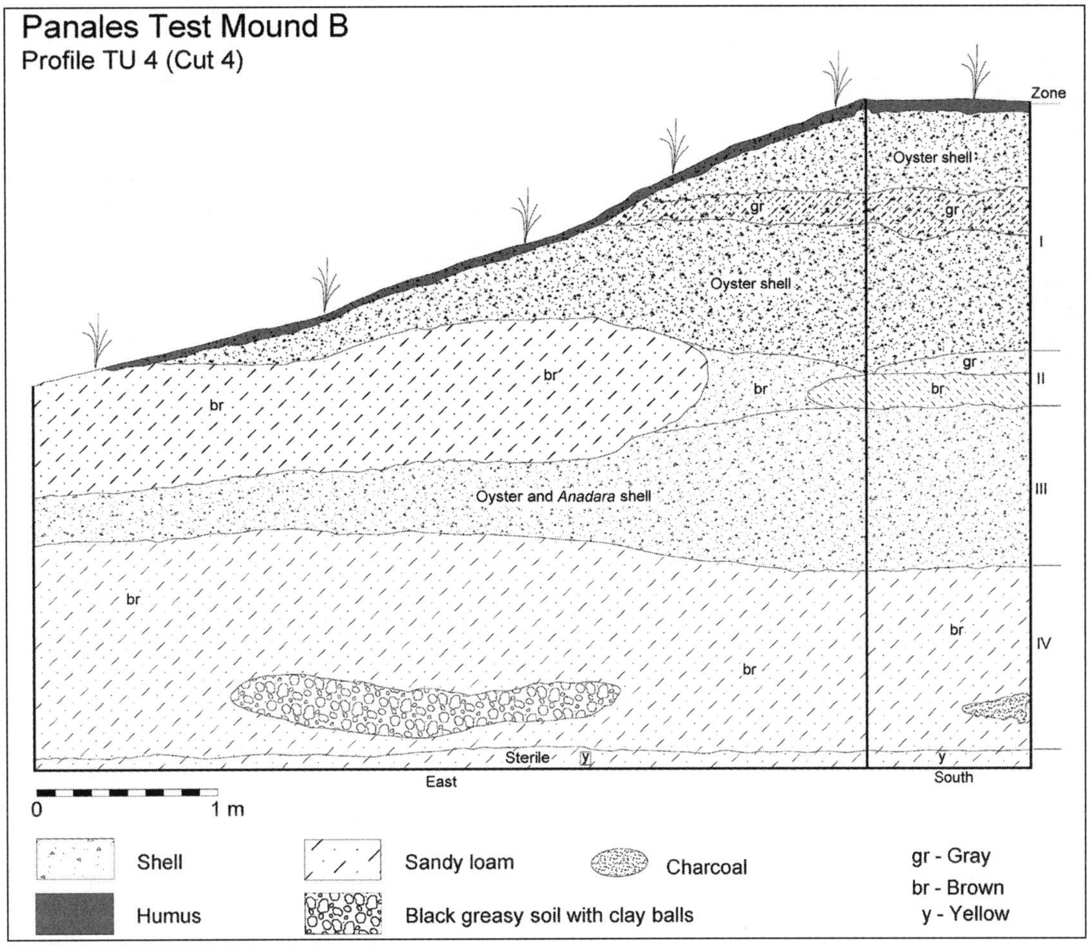

FIGURE 5.23. Profile of Test Unit 4 in Panales Test Mound B.

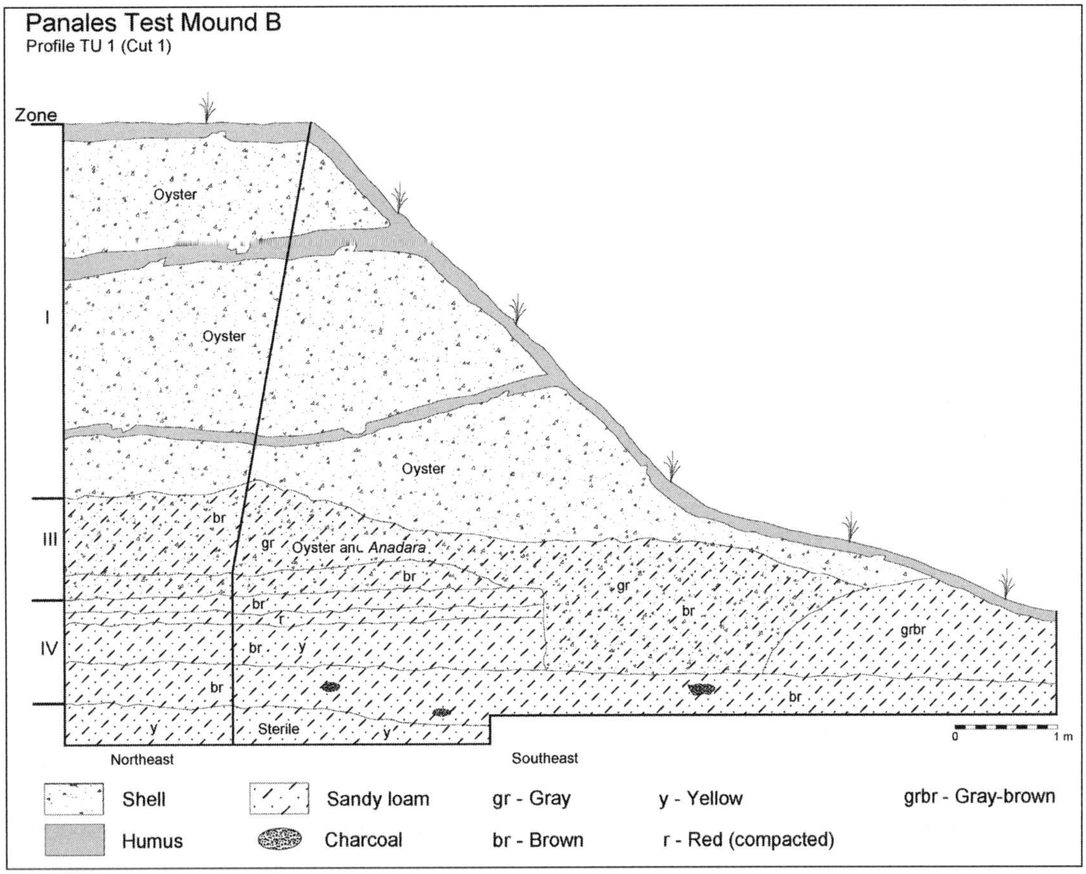

FIGURE 5.24. Profile of Test Unit 1 in Panales Test Mound B.

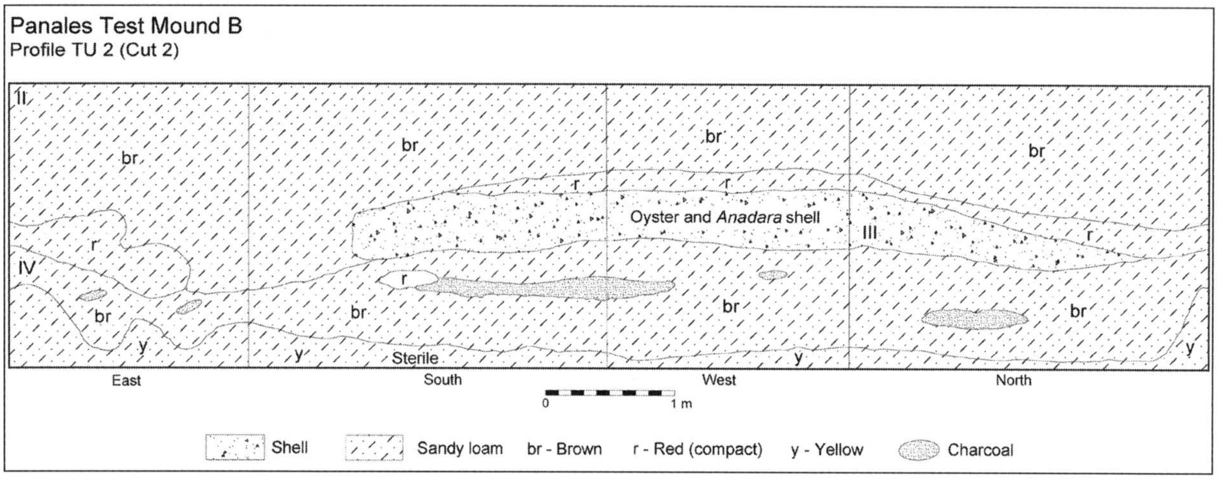

FIGURE 5.25. Profile of Test Unit 2 in Panales Test Mound B.

TABLE 5.10. Summary of Burials from Panales Test Mound B.

Burial Number	Condition	Sex	Age	Burial Type	Funerary Offerings
B 1	Nearly complete	Male	Young adult	Extended	None
B 2	Nearly complete	Probably female	Young adult	Extended	Possibly a broken metate
B 3	Incomplete	Undetermined	Adolescent	Undetermined	2 vessels, 6 small white beads, 16 tooth-shaped beads, 5 spindle whorls
B 4	Incomplete	Undetermined	Child	Undetermined	1 carved red bead, 1 polished bone pendant or bead, 1 tripod vessel
B 5	Incomplete	Undetermined	Young adult	Undetermined	None

below the surface. Both Burials 3 and 4 were in the east wall of TU 2, and because of their poor condition the osteological material from these burials was not recovered for further study.

Zone II also produced the complete remains of a dog burial. In addition to the human and dog burials, Zone II and Zone III produced quantities of sherds, stone artifacts, and disarticulated animal bone.

TU 3 was a 2-×-2-m unit excavated in the cleared field adjacent to and approximately 20 m west of Test Mound B. It was excavated to determine whether any features or other cultural deposits occurred in proximity to Test Mound B. The unit was excavated to a depth of 1 m, and two strata were exposed. The bulk of the unit's fill consisted of a brown sandy loam, which contained only a few sherds and no cultural features. The second stratum was yellow sand encountered at the bottom of the pit; it contained no cultural material or features.

The artifact assemblages and composition of the earth and shell components were similar to those recovered from Mound A, namely, in both cases the earthen mounds contained predominantly earlier Chametla pottery types (Table 5.11). In fact, as seen in the ceramic assemblage recovered from the earthen mound exposed in TU 1, there is a near exclusion of any ceramic type other than Chametla types. By contrast, the ceramic assemblage represented in the massive shell cap, which reached a height of 5 m, was found to contain Late Classic and Postclassic period Chametla and Amapa types in near equal proportions (Davis 1969). The shell midden cap is dated in the local sequence to the Early and Middle Conchera periods, while the earlier earthen mound is mostly Late Estero period.

San Cayetano

Also in summer 1969, Shenkel (1971a:91–100) tested a shell mound where the dominant species present was *Tivela byronensis*, a small smooth-shell clam. The tested mound (Shell Site #5) was located on the outskirts of the small village of San Cayetano. Test Mound A (also illustrated in Shenkel 1971a:Plate 4 as Test Mound #1) was a small circular midden 13.6 m in diameter and 79 cm in height. A trench 1.81 m long and 91 cm wide was excavated in the northwest quadrant of the mound. Three distinct strata overlying sterile sand were exposed (Figure 5.26). Stratum 1, the uppermost layer, consisted of loosely compacted shell that contained some incipient soil development, rootlets, and a small number of sherds. Stratum 2 was similar to Stratum 1 but lacked any sign of soil development and rootlets. A single feature, a pocket of the marine snail *Natica chemnitzii*, was exposed just below the interface of Strata 1 and 2. No artifacts or other materials were associated with the shell, and Shenkel (1971a) does not explain this feature. Perhaps it represents a single dumping episode of shell after the consumption of a large number of the individuals. Stratum 3 is also a shell midden deposit, but it is distinguished from Stratum 2 by the presence of brown sand in the interstices between the shells. The three cultural strata overlaid culturally sterile coarse brown sand.

Linear Shell Mound Testing

As part of his survey, Shenkel (1969a, 1971a) also conducted limited testing at four of the 51 linear shell mounds he had recorded. He notes that the linear shell mounds are almost exclusively composed of *Crassostrea corteziensis* and that they range up to 2.8 km in length and are 5 to 10 m in height. Shenkel provides little specific detail on the test excavations but, rather, summarizes the results. A single map and profile are provided (Figure 5.27). His summary follows:

> The LSM [linear shell mound] profile typically displays a stratified column of layers of loosely packed whole shell with little or no intervening material, alternating with layers of loosely packed, predominantly whole shells mixed with a small amount of gray sand and crushed shell fragments. Toward the base of the column, the layers lose distinctiveness. At the water table the shell is intermixed with dense, highly plastic gray to gray-brown clay. This shell-clay mixture continued in all tests well below the water table and beyond the excavation limits permitted by our equipment. In two of the tests, Test 1 (Figure 8 [Figure 5.27]) and Test 3, a layer of black carbonaceous, clayey sand mixed with oyster occurred immediately above the shell-clay layer at the column base. In all tests and in every instance during the survey, tiny flecks of charcoal, predominately the carbonized remains of twigs, were found scattered all through the mound matrix. They were not associated with pockets of ash or found in noticeable concentrations. Neither in the excavations nor during the survey, was any artifactual material of any kind found in association with the LSM [1971a:104].

Subsequent radiocarbon assays from two of the linear shell mounds (see Table 4.2) indicate that some may have accumulated during the prehistoric occupation of the area (Early and Middle Conchera periods), while others are products of Spanish exploitation of the area during the contact period (see Chapter 12).

Rincon de Panal

The site of Rincon de Panal on Panales (see Figure 5.1) was tested during the summer of 1971. One of the mounds at the site appeared to be T-shaped in form, and because of this unusual shape it was selected for excavation (Taylor and Rautenkranz 1971). Excavations were initiated at the point of junction of the arms of the T. The mound had been vandalized; nevertheless, because of its shape it was anticipated that excavations in the mound would likely produce quantities of cultural remains relevant to the study of the prehistoric occupants of the area. Five 3-×-3-m test units (A, B, C, G, and H) were initiated, and as excavations progressed, three additional units (D, E, and F) were added as extensions to the original five test units. These were generally excavated in 20-cm levels, and a datum was established at the northwest corner of TU B, although most subsequent excavations are reported as centimeters below the surface.

The excavation of TU A focused on cleaning up one of the vandal pits and establishing stratigraphic control of the mound. Based on the description provided (Taylor and Rautenkranz 1971:27), it appears that the zone to a depth of 1.6 m contained oyster and other marine shell, some sherds, and traces of charcoal. Although not specified, the inference is that this zone was disturbed. Taylor and Rautenkranz (1971:27) indicate that the first "uniform" level exposed was Level 9 (1.6–1.8 m). Level 10 (1.8–2.0 m) was defined by the presence of a darker soil. An obsidian blade fragment was recovered, and there was a general increase in the abundance of cultural material. Between 2.0 and 2.2 m, a thin ash lens and a highly compact floor were encountered. Course yellow sand was encountered at 2.2 m, and the unit was excavated to a depth of 2.6 m without the discovery of any cultural material. To check that the yellow sand was culturally sterile, a small test pit in the middle of the unit was excavated to a depth of 3.46 m, and no cultural remains were discovered.

TU B was excavated immediately south of TU A. A plow zone approximately 40 cm thick was present, and only a few pieces of what were referred to as green-glazed rock were recovered. This green-glazed rock was to become somewhat of an anomaly at several sites as the excavation program in the Marismas continued (Appendix A). The excavation of the next 40 cm in TU B also produced little cultural material—a few sherds, some fire-cracked rock, oyster shell, and more green-glazed rock. However, at about 85 cm, pieces of fire-reddened adobe-like material (daub?) were encountered. Just below this layer of burned material was a stratum containing sherds, a decorated spindle whorl, some shell, and a light scattering of ash and charcoal. Below this was another layer of the fire-reddened adobe-like material. Most of the subsequent test units excavated also contained these

TABLE 5.11. Frequencies of Identified Ceramic Types from Panales Test Mound B.

Test Unit and Level	Chametla Type								
	Early Black-banded Engraved	Black-banded Engraved	Black-banded Nonengraved	Late Black-banded Engraved	Early Chametla Polychrome	Middle Chametla Polychrome	Middle Chametla Polychrome Engraved	El Taste Polychrome	El Taste Red-on-gray/ Red-on-cream
Test Unit 1									
Level 1		2	1			1			1
Level 2				2		5			
Level 3									
Test Unit 2									
Level 1			2					2	
Level 2			1			4		1	
Level 3				2		2	2		
Level 4				2		1	1		
Level 5	1		1		2				
Level 6									
Test Unit 4									
Level 1								2	
Level 2[a]				1				3	1
Level 3[a]								2	
Level 4[a]									
Level 5[b]									
Level 6		1[c]				2			
Level 7				1					
Subtotal	1	3	5	8	2	15	3	10	2
Total									

Test Unit and Level	Amapa Type										
	Cocoyolitos Polychrome	Botadero Black-on-orange	Botadero Incised	Roblitos Polychrome	Sentispac Buff	Tuxpan Engraved	Tuxpan Red-on-orange	Otates Polychrome(?)	Amapa Red-on-cream	Amapa White	Subtotal
Test Unit 1											
Level 1											
Level 2											
Level 3											
Test Unit 2											
Level 1			1							1	2
Level 2		1		1			1				3
Level 3			1	11		1					13
Level 4			1				2	1	1		5
Level 5							1		3		4
Level 6											
Test Unit 4											
Level 1											
Level 2[a]	1										1
Level 3[a]	1				1						2
Level 4[a]	1						1				2
Level 5[b]											
Level 6											
Level 7											
Subtotal	3	1	3	12	1	1	5	1	4	1	
Total											32

Table 5.11. (cont'd.) Frequencies of Identified Ceramic Types from Panales Test Mound B.

Test Unit and Level	Chametla Type											Culiacán Type		
	El Taste Red-bordered	Mazatlán Polychrome	El Taste Satin	White-on-buff	Red-on-buff	Lolandis Red Rim	Red-rim Decorated	Red-rimmed Utility Ware	Ribbed	Scalloped Rim	Subtotal	Aguaruto Polychrome	Early Culiacán Polychrome	Subtotal
Test Unit 1														
Level 1							11	3	1	2	22			
Level 2	1	1					1	4			14			
Level 3														
Test Unit 2														
Level 1			1			7		4			16			
Level 2						3					9			
Level 3	2		3	1		1		2			15			
Level 4								2			6			
Level 5						3		8			15			
Level 6								10			10			
Test Unit 4														
Level 1	1	8									10			
Level 2[a]	1										6			
Level 3[a]	1	2			1		1	1			8		2	2
Level 4[a]		1					7	8			16	1		1
Level 5[b]														
Level 6							3	3		1	11			
Level 7								3			4			
Subtotal	6	12	4	1	1	14	23	48	1	3		1	2	
Total											162			3

Test Unit and Level	Peñitas Type			
	Polychrome (Bordaz 43–49)	Polychrome z (Bordaz 50–54)	Red Incised (Bordaz 5, 8, etc.)	Subtotal
Test Unit 1				
Level 1				
Level 2				
Level 3				
Test Unit 2				
Level 1				
Level 2				
Level 3				
Level 4				
Level 5				
Level 6				
Test Unit 4				
Level 1				
Level 2[a]	3			3
Level 3[a]	4	1	1	6
Level 4[a]	5	1		6
Level 5[b]				
Level 6				
Level 7				
Subtotal	12	2	1	
Total				15

[a] Oyster shell layer.
[b] One unidentifiable sherd.
[c] Possibly engraved.

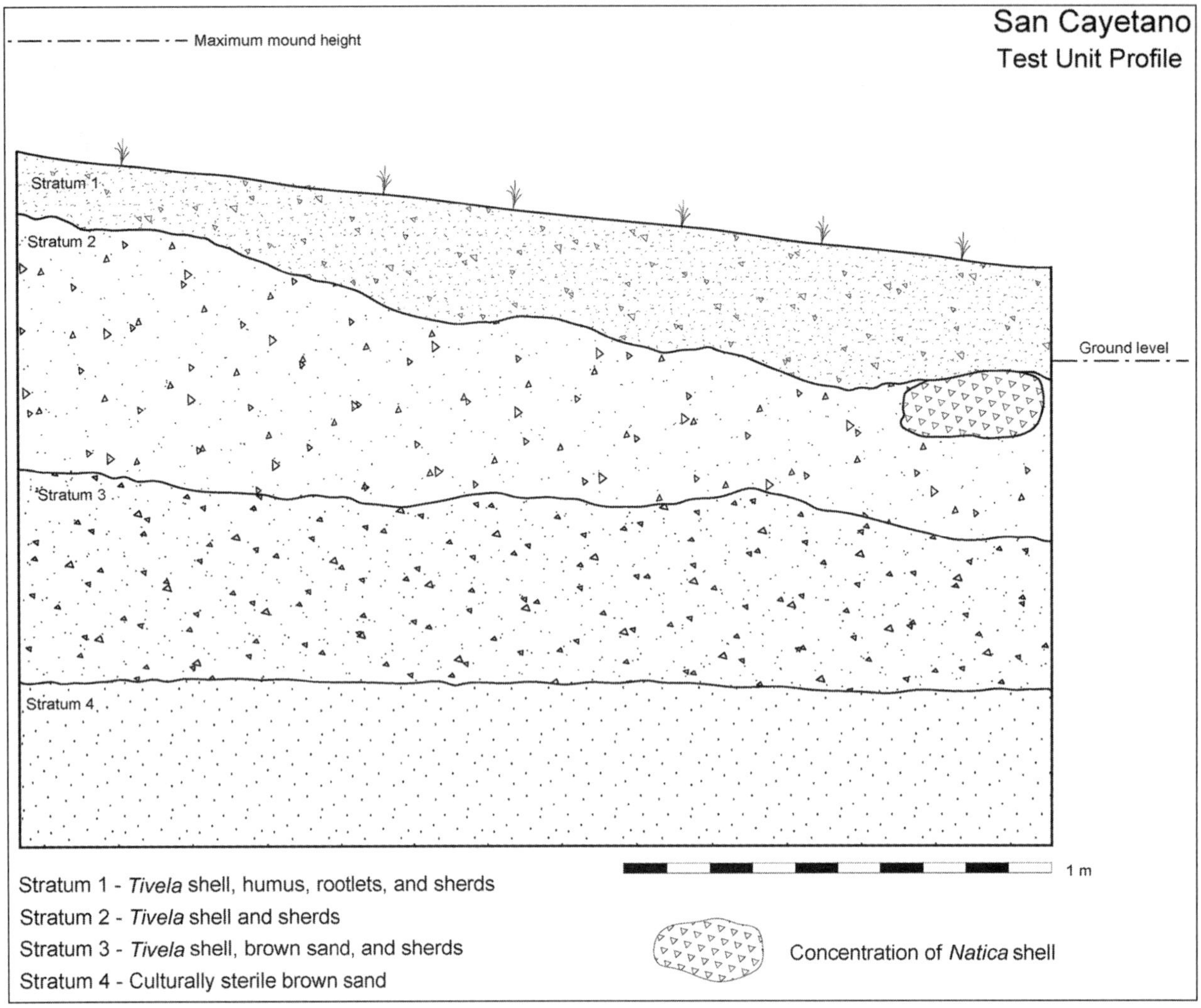

FIGURE 5.26. Profile map of a section of the San Cayetano test unit.

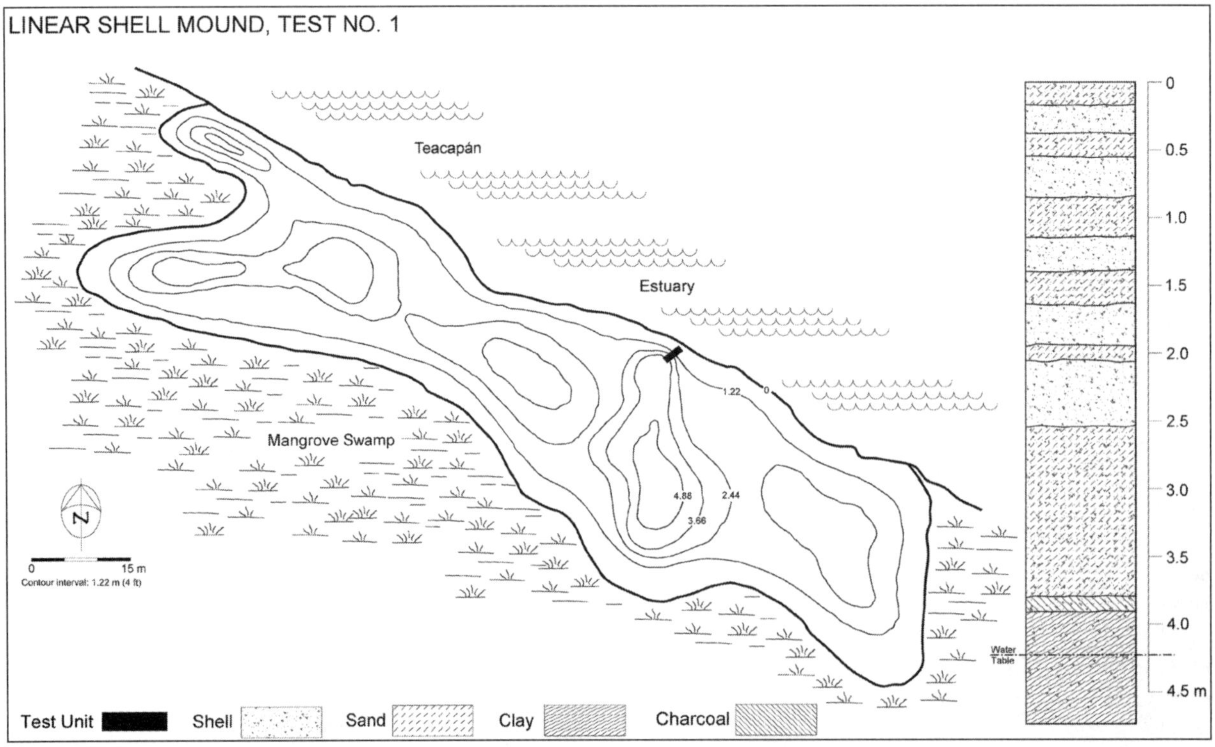

FIGURE 5.27. Map of one of the linear shell middens tested showing the location of the test unit and a profile map of the strata exposed.

layers of material at a similar depth, and in several of the test units these layers formed a single stratum. At 1.6 m, a 20-cm-thick shell midden containing oyster and other marine shell, crab claws, barnacles, a raccoon mandible, and sherds was encountered. At 2.11 m and 2.38 m, two flexed adult human burials (Burials 3 and 4, respectively) were recovered. Burial 3 had a complete Polished Buff, Cocoyolitos-style pipe resting on its chest. Burial 4 exhibited cranial deformation and a hole about 2.5 cm in diameter in the back of the skull.

TU C was excavated north of TU A. The area between TUs A and C had previously been vandalized. The disturbed area was also excavated and designated as TU A to C. The excavations in the extension produced few remains. However, at 2.45 m and 2.77 m below the surface, two human burials, Burials 1 and 2, respectively, were recovered. Both were adults and in flexed positions. Burial 2 was oriented with its head to the east and was accompanied by a broken vessel. Some shell and animal bone were in proximity to the burials was well.

Excavations in TU C exposed a 40-cm-thick plow zone, and at the base of the plow zone, a possible floor was discovered. However, it was inadvertently destroyed during excavations. Artifacts were sparse from 40 to 80 cm below the surface; however, from 80 to 120 cm the number of artifacts increased dramatically. Plain and incised red-on-white sherds were abundant, along with small amounts of obsidian. Similar quantities and varieties of material were found in the next excavated levels. At 1.45 m a shell midden layer containing a variety of marine shell, animal bone, sherds, and some obsidian was encountered. Excavation of the midden deposit continued in portions of Level 5, from 1.6 to 2.0 m; however, the midden did not extend all the way across TU C or TU F (the extension of TU C). The midden contained a concentration of shell with a high number of artifacts, and it is of note that both Levels 4 and 5 produced high quantities of obsidian blade fragments.

At 1.9 m below the surface, near the bottom of Level 5, three oxidized pieces of adobe were exposed. One was fire-blackened on one side. The largest piece measured 10 × 15 × 3 cm. It was believed that these might be structural material, and they were designated as Feature A. Upon excavation additional pieces of the material were exposed. Although the material appeared to be structural daub from a collapsed or burned structure, no evidence of a structure was identified, and it is likely that they represent redeposited debris. The material was concentrated in an area measuring 56 × 43 cm and 10 cm thick.

At 2.0 m below the surface, a yellow sand layer was encountered. Below the midden, two ceramic vessels were discovered that were thought to be associated with Burial 1 excavated in TU A to C. Additional human bone also thought to be from Burial 1 was recovered. A dog burial was exposed in the northeast corner of TU C at a depth of 2.1 m. At second dog burial was discovered about 1.2 m south of the first at a slightly greater depth, and finally, a third dog burial appeared in the northwest corner of TU C, at approximately the same level as the second dog burial.

TU F was excavated as an extension of TU C to fully expose Feature A. Excavation to 1.3 m below the surface was generally unproductive; however, between 1.3 and 2.0 m below the surface artifact density increased significantly. Materials encountered included sherds, shell, animal bone, some charcoal, and a considerable number of obsidian artifacts. Level 5, at 1.7 m below the surface, produced a small ceramic container or miniature vessel. In his field notes, Rautenkranz suggested that it might be an amulet. It was 5 cm long, with a maximum diameter of 2.5 cm, and it tapered slightly from the top to the bottom. Two small suspension holes were present near the rim of the specimen, and striations were visible on its surface near its base. The function of this object is unknown, but it appears to have been suspended from a cord; perhaps it was worn about the neck, or it was a miniature vessel used for ceremonial purposes.

TU G was opened in order to excavate the burial exposed during the excavation of TU C. The artifacts recovered from TU G were limited, and as in the other excavated units, they tended to increase with depth. At 2.2 m below the surface a fourth dog burial was exposed, as was a feature that was thought to be a possible floor. Investigation of the feature, however, failed to confirm its function. Burial 5, a human bundle burial, was exposed at 2.5 m below the surface of the mound. A slight discoloration of the soil around the burial was noted, and it was thought that this perhaps represented a stain from a decomposed body wrapping. The body was oriented with its head to the east. The burial was located on top of the yellow sand layer, and excavations in TU G were terminated with the removal of the burial.

TU H was not part of the contiguous block of test units excavated. It was located at the westernmost end of one of the arms of the T-shaped mound. The first meter that was excavated contained few artifacts. However, at 50 cm below the surface a burned area containing pieces of adobe was found. The adobe was similar to that found in other parts of the mound, and the soil containing the adobe appeared to have been burned. The significance of this remains unresolved. From 1.2 to 1.8 m an extensive trash deposit was encountered. Abundant quantities of animal bone, obsidian, and pottery were present along with a small well-preserved wedge of red ochre. No shell was present.

The disarticulated remains of a dog and another human burial (Burial 6) were recovered at 1.6 m below the surface. Burial 6 was the semiflexed remains of an adult with its head oriented to the northeast. The teeth

TABLE 5.12. Summary of Burials from Rincon de Panal.

Burial Number	Condition	Sex	Age	Burial Type	Funerary Offerings
B 1	Incomplete	Probably male	Young adult	Undetermined	10 spiral shells, 2 ceramic vessels (1 red-on-buff)
B 2	Nearly complete	Male	Mid-adult	Flexed	None
B 3	Incomplete	Male	Young adult	Seated, flexed	1 plain buff Cocoyolitos-style smoking pipe
B 4	Nearly complete	Male	Young adult	Flexed	1 reconstructible ceramic vessel
B 5	Incomplete	Female	Young adult	Bundle	None
B 6	Incomplete	Female	Young adult	Semiflexed	1 crude plainware vessel
B 7	Fragmentary	Undetermined	Adult	Semiflexed	1 reconstructible ceramic vessel

TABLE 5.13. Frequencies of Identified Ceramic Types from Test Unit C, Rincon de Panal.

	Chametla Type						Amapa Type							Peñitas Type				
Level	Black-banded Engraved	El Taste Polychrome	El Taste Red-bordered	El Taste Satin	Mazatlán Polychrome	Subtotal	Ixcuintla Polychrome	Ixcuintla White-on-orange	Santiago Engraved	Sentispac Buff	Tuxpan Engraved	Tuxpan Red-on-orange	Subtotal	Engraved	Red Incised	Polychrome (Bordaz 43–49)	Polychrome (Bordaz 50–54)	Subtotal
1							1			1	1		3	2				2
2											1		1	1				1
3		4	5		4	13		1		2	13		16	12		1(?)		13
4	1		9	2	1	13	4		1	4	19	2	30	36	3	2	1	42
5			3			3				1	3	1	5	3		3		6
Subtotal	1	4	17	2	5		5	1	1	8	37	3		54	3	6	1	
Total						29							55					64

were mutilated (notched), and a crudely made plainware pot was found in association. Another burial, Burial 7, was exposed at a depth of 1.83 m. The burial pit, which was not distinguishable from the surrounding soil, had been excavated into the coarse yellow sand, which was normally culturally sterile. The remains were those of an adult in a semiflexed position with a single associated broken vessel. A few artifacts and a small concentration of shell continued to a depth of 2.2 m, where excavations were terminated. In total, the remains of seven individuals were recovered from Rincon de Panal (Table 5.12).

A sketch in the site's field notes indicates that TUs D and E were eastern extensions of TUs B and C, respectively. No detailed discussion of the investigation of these units, if in fact they were excavated, was provided, although it is possible that the descriptions and results of their excavations were subsumed in the discussion of TUs B and C.

Five radiocarbon assays, all from charcoal, are available from the excavations of the Rincon de Panal mound. Three of the dates, all from Level 3 or Levels 3–4 of TU C, have ceramic types associated with the El Taste and Ixcuintla phases (AD 1100–1300) at Chametla and Amapa. These types include El Taste Polychrome, El Taste Satin, Ixcuintla Polychrome, Sentispac Buff, and Tuxpan Engraved. Several Peñitas types were also represented (Bordaz 1964). In terms of the local Marismas sequence, the ceramics indicate a Middle Conchera period occupation (Tables 5.13–5.14). The radiocarbon assays (AD 1052–1385 [Brock BGS-110], AD 1268–1417 [Brock BGS-109], and AD 1283–1418 [Brock BGS-107]) are consistent with the ceramic dates and suggest, perhaps, a mid- to late Middle Conchera period occupation. These dates also imply that these ceramic types may have persisted past AD 1300.

A fourth date, AD 1229–1403 (Brock BGS-108), from a charcoal sample from Level 9 in TU B has no reported ceramic associations but is generally contemporaneous

Table 5.14. Frequencies of Identified Ceramic Types from Test Unit H, Rincon de Panal.

	Chametla Type											Culiacán Type			Amapa Type								Peñitas Type					
Level	Black-banded Engraved	Chametla Black-banded Engraved	El Taste Polychrome	El Taste Polychrome Red-on-gray	El Taste Red-bordered	El Taste Rough	Mazatlán Polychrome	Red-on-buff[a]	Red-rim Decorated	Ribbed	Subtotal	Aztatlán Ware	Culiacán Banded Incised	Subtotal	Black-on-buff	Cerritos Engraved	Ixcuintla Polychrome	Mangos Engraved	Sentispac Buff	Tuxpan Engraved	Tuxpan Red-on-orange	Subtotal	Engraved	Red Incised (Bordaz 1–5?)	Polychrome (Bordaz 44–49)	Polychrome (Bordaz 44–49)	White-on-red	Subtotal
1[b]																												
2							2				2																	
3					3	2			3		8	3		3			1		3	2	1	7	7	1	1			9
4					12		1		3	2	18				4			2	7	4		17	1	7	5	3		16
5			1		5		1		1		8		1	1					2	1		3	2	3	2			7
6				2	7	2	2	4	4		21		3	3			1		10	5		16	3	3	3	5		14
7																												
8	1				4				2		7														5	4	1	10
9		2									2				2	1	1		5	4		13		1				1
Subtotal	1	2	1	2	31	4	6	4	13	2		3	4		6	1	3	2	27	16	1		13	15	16	12	1	
Total											66			7								56						57

[a] A note on the original tally sheet indicates that one of the Red-on-buff jar neck sherds appears to be very similar to that illustrated in Kelly 1938:41, Figure 16.
[b] No Level 1 data.

with the three dates from TU C. A fifth date, AD 894–1216 (UCLA 1628B), from the bottom of Feature A in TU C, is affiliated with an Early Conchera occupation. The date is associated with burned daub. It was suggested that the daub material may have been part of a structure or perhaps redeposited structural materials (Taylor and Rautenkranz 1971). The date appears to be valid and, like other radiocarbon assays from the mound, suggests that Rincon de Panal accumulated over a period from as early as AD 900 to perhaps as late as AD 1400, the Early and Middle Conchera periods.

Additionally, the ceramic pipe found in direct association with Burial 3 was a Cocoyolitos-style pipe (Kelly 1938:52–53), similar to Porter's (1956) type 3A, "sencillo sin soporte." The presence of this Cocoyolitos-style pipe lends further support to the Middle Conchera date for the site of Rincon de Panal.

A small oyster shell midden located about 0.5 km from the Rincon de Panal excavations was also tested in summer 1971 (Taylor and Rautenkranz 1971). The expressed purpose of this excavation was to recover faunal remains that would provide information on the subsistence activities of the prehistoric occupants of the area. It was also hoped that excavation of such a midden would yield artifacts such as fishhooks, net sinkers, or harpoon points that might have been used in the procurement of fish and other wild animals. A trench, 1.9 m wide, 3.3 m long, and with a maximum depth of 50 cm, was excavated through the center of the mound, resulting in the recovery of faunal and fish remains, obsidian blades and flakes, and an abundance of pottery.

Chalpa

The Chalpa site is named for a small hamlet on the east bank of the Teacapán Estuary's main tidal channel (see Figure 5.1). The addition of Chalpa to the list of excavated sites came about by the reported discovery of burials in the course of well digging. Several earth mounds 30 to 40 m in diameter border the modern village, and all were within 200 m of the modern estuary shoreline. These extremely low-profile mounds, rising almost imperceptibly less than a meter above the surrounding terrain, had become the target of intensive looting. Excavations at Chalpa focused on the recovery of surviving osteological material (Gill 1974:86–88) and associated artifacts, the study of burial practices, and as elsewhere, an examination of mound construction detail and general content. Situated on land under intensive cultivation, the Chalpa mounds had been nearly leveled by plowing, resulting in the exposure of skeletal and artifactual materials within

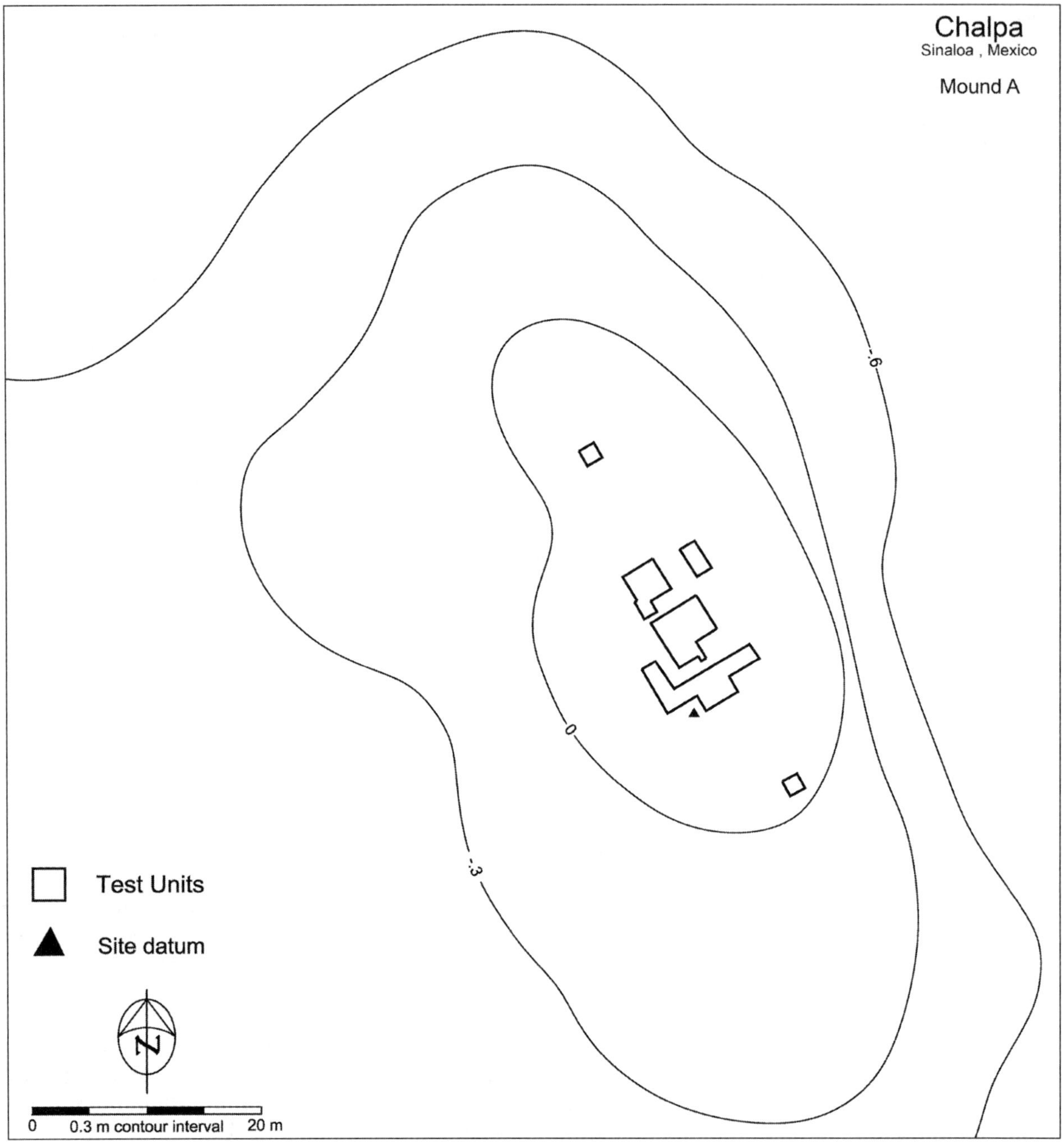

FIGURE 5.28. Map of Mound A at Chalpa showing the test units.

a few centimeters of the surface. Skeletal material, found up to a depth of 2 m, was generally abundant at Chalpa. In the absence of observable stratigraphy, excavations were completed in arbitrary 30-cm levels.

Excavations at Chalpa were initiated in 1968 in Mound A (Feature A), an ovoid, northwest-to-southeast-oriented earth mound measuring approximately 39 m long and 24 m wide (Figure 5.28). A large test unit (no letter designation) was initiated in a heavily vandalized area at the center of the mound in order to salvage any information that might be left. This test unit, excavated to a depth of about 60 cm, where culturally sterile sand was encountered, produced only scattered fragments of human remains.

A 3-×-3-m unit designated TU A was opened immediately to the southeast of the first large test unit. Excavated to a depth of about 1 m, TU A contained seven human burials. A large thick shell lens was encountered in the southeastern two-thirds of TU A. The burials were all interred above the thick shell lens. However, Gill (1971), citing Shenkel (1968), indicates that some burials were found below a small, thin lens of shell but above the larger, thicker shell lens.

Two additional 1.5-×-1.5-m test units were placed

FIGURE 5.29. Distribution of burials in Mound A test units, Chalpa (Gill 1971).

along the northern edge of Mound A. A thin shell lens was exposed in the southern portion of the southernmost unit. It appeared that a shell midden, or several, formed the base of most of the southern portion of Mound A and that the midden was directly deposited on an old beach prior to the construction of the burial mound.

A second season of excavations occurred at Chalpa in 1969. TU A was extended with a 1.5-×-3.0-m unit, and an additional 3-×-3-m test unit was excavated in the northwestern portion of the mound. TU B was opened in the northwest part of the mound as well and produced a single human burial and six animal burials. TU C was placed just northeast of TU B. No burials were recovered from its excavation. By the end of the 1969 season, Mound A had produced 25 individuals from 20 burials, along with nine dog and raccoon burials (Figure 5.29; Table 5.15).

Upon completion of excavations at Mound A, efforts focused on two mounds, C and D, closer to the village of

TABLE 5.15. Summary of Burials from Mounds A, B, C, and D, Chalpa.

Mound and Burial Number	Condition	Sex	Age	Burial Type	Funerary Offerings
Mound A					
B 1	Fragmentary	Male	Young adult	Unknown	None
B 2	Complete	Female	Adolescent	Semiflexed	Dog burial, 2 mollusk shells
B 3	Complete	Female	Old adult	Seated, flexed	1 vessel, 1 smoking pipe
B 4A	Incomplete	Female	Adult	Flexed	1 decorated vessel
B 4B	Fragmentary	Male	Mid-adult	Unknown	None
B 5A	Nearly complete	Female	Mid-adult	Semiflexed on side	1 bowl, 1 spindle whorl
B 5B	Fragmentary	Unknown	Newborn	Unknown	Unknown
B 6	Nearly complete	Male	Adolescent	Seated, flexed	1 vessel, 2 carved shell beads
B 7	Fragmentary	Unknown	Infant	Unknown	None
B 8	Nearly complete	Female	Young adult	Flexed	1 obsidian projectile point
B 9A	Nearly complete	Unknown	Infant	Extended	Unknown
B 9B	Fragmentary	Female	Adult	Unknown	None
B 10	Nearly complete	Female	Young adult	Semiflexed	1 spindle whorl; 1 broken decorated bowl; 1 smoking pipe; 1 small, handled vessel
B 11	Nearly complete	Unknown	Infant	Extended	None
B 12A	Incomplete	Male	Young adult	Flexed on side	1 smoking pipe
B 12B	Fragmentary	Unknown	Infant	Unknown	None
B 13	Nearly complete	Unknown	Infant	Extended	None
B 14	Nearly complete	Probably male	Young adult	Extended	1 animal burial, 1 stone bead, 1 chert flake
B 15	Complete	Female	Old adult	Seated, flexed	1 large smoking pipe, 1 decorated vessel
B 16	Nearly complete	Unknown	Child	Flexed	1 decorated vessel
B 17A	Fragmentary	Male	Adult	Unknown	None
B 17B	Fragmentary	Unknown	Infant	Unknown	None
B 18	Complete	Female	Old adult	Seated, flexed	1 decorated vessel, 1 spindle whorl
B 19	Incomplete	Male	Young adult	Flexed	1 decorated vessel
B 20	Complete	Female	Young adult	Flexed	1 orange plainware vessel, 1 spindle whorl
Mound B					
Salvage	Incomplete	Undetermined	Undetermined	Undetermined	Undetermined
Mound C					
B 1A[a]	Fragmentary	Male	Old adult	Secondary mass inhumation	None
B 1B[a]	Fragmentary	Female	Young adult	Secondary mass inhumation	None
B 1C[a]	Fragmentary	Unknown	Child	Secondary mass inhumation	None
B 1D[a]	Fragmentary	Male	Young adult	Secondary mass inhumation	None
B 1E	Fragmentary	Female	Young adult	Secondary mass inhumation	None
B 1F	Fragmentary	Male	Young adult	Secondary mass inhumation	None
B 1G	Fragmentary	Male(?)	Adult	Secondary mass inhumation	None
B 1H	Fragmentary	Female(?)	Adult	Secondary mass inhumation	None

Table 5.15. (cont'd.) Summary of Burials from Mounds A, B, C, and D, Chalpa.

Mound and Burial Number	Condition	Sex	Age	Burial Type	Funerary Offerings
B 1I	Fragmentary	Unknown	Newborn	Secondary mass inhumation	None
B1J	Fragmentary	Unknown	Child(?)	Secondary mass inhumation	None
B 2	Complete	Female	Young adult	Flexed	2 large ollas, 1 small dish
Mound D					
B 1A	Incomplete	Female	Old adult	Unknown	Unknown
B 1B	Fragmentary	Unknown	Adult	Unknown	Unknown
B 2A	Incomplete	Male	Young adult	Unknown	Unknown
B 2B	Fragmentary	Probably female	Adult	Unknown	Unknown
B 2C	Fragmentary	Unknown	Adolescent	Unknown	Unknown
B 3A	Incomplete	Unknown	Infant	Semiflexed	Many small white beads
B 3B	Fragmentary	Probably male	Adult	Unknown	Unknown
B 4A	Nearly complete	Male	Young adult	Semiflexed	None
B 4B	Fragmentary	Probably male	Old adult	Unknown	Unknown
B 5	Incomplete	Male	Mid-adult	Unknown	Unknown
B 6	Nearly complete	Female	Old adult	Semiflexed	1 orange plainware vessel, 1 cylindrical ceramic stamp
B 7	Nearly complete	Probably male	Old adult	Flexed	None
B 8	Fragmentary	Unknown	Young adult	Unknown	2 plainware vessels[b]
B 9A	Incomplete	Female	Mid-adult	Semiflexed	1 reddish plainware vessel
B 9B	Incomplete	Female	Mid-adult	Unknown	Unknown
B 9C	Fragmentary	Male	Adult	Unknown	1 broken tripod vessel[c]
B 9D	Fragmentary	Female	Adult	Unknown	Unknown
B 9E	Fragmentary	Possibly male	Adult	Unknown	Unknown
B 9F	Fragmentary	Unknown	Child	Unknown	Unknown
B 9G	Fragmentary	Possibly female	Adult	Unknown	Unknown
B 9H	Fragmentary	Unknown	Adult	Unknown	Unknown
B 10A	Nearly complete	Female	Old adult	Flexed	1 smoking pipe, 1 spindle whorl, 1 obsidian projectile point, 1 broken vessel[d]
B 10B	Fragmentary	Unknown	Adult	Unknown	Unknown
B 10C	Fragmentary	Unknown	Infant	Unknown	Unknown
B 11A	Incomplete	Unknown	Fetal[e]	Unknown	1 unmodified stone
B 11B	Fragmentary	Unknown	Fetal	Unknown	Unknown
B 12A	Complete	Male	Young adult	Seated, flexed	2 small dishes
B 12B	Fragmentary	Female	Adult	Unknown	Unknown
B 13A	Complete	Female	Adolescent	Semiflexed	1 white plainware vessel
B 13B	Fragmentary	Unknown	Adolescent	Unknown	Unknown
B 14A	Complete	Unknown	Adolescent	Seated, with legs extended	1 handled vessel, 1 red pendant, 4 tooth-shaped beads
B 14B	Fragmentary	Unknown	Newborn	Unknown	Unknown

Source: Gill 1971.

[a] These four specimens were within the large mass burial and were represented by complete or nearly complete skulls arranged in a circular pattern.

[b] The vessels listed in association with B 8 could be associated with the B 9 group of burials. The stratigraphic position of these vessels and their prior disturbance in that part of the mound preclude their positive assignment to a specific burial.

[c] This broken bowl could also be associated with any of the B 9 individuals, as this part of the mound was highly disturbed.

[d] This burial may be intrusive into an earlier deposit; thus, it is possible that the obsidian projectile point found with the burial is from another context and therefore not a funerary offering associated with B 10A.

[e] The length of the long bones of this specimen indicates that the birth must have been premature. If not an actual stillborn, the premature infant must have died soon after parturition. Its comparable size and its proximity to the B 11B remains suggest that the two individuals represent a multiple birth.

Chalpa. Mound C was approximately 30 m in diameter, and it too had been extensively vandalized. In an effort to determine whether undisturbed burials were present, a 1.5-×-6.0-m test unit was excavated. The mound consisted of approximately 2 m of sandy fill containing sherds and bone fragments. Based on the results of the test excavations and information from local informants, it appeared that the mound rose approximately 2 m above the sterile beach sand on which it was constructed. Two burials, one of which was intrusive into the sterile sand zone, were recovered. Additionally, a small copper ax was found at a depth of 30 cm below the mound's surface. Burial 1 in Mound C at Chalpa is of particular note. It was thought to be a mass burial that contained the remains of at least 10 individuals of various ages. The skeletal remains were disarticulated, suggesting that the bodies had been allowed to decompose before being buried or that the remains represent reinterred (secondary) burials. Many of the bones were in poor condition and highly fragmentary, and a number of elements were missing. The skeletal material had been placed in an oval grave that measured 90 × 107 cm. At the bottom of the grave pit were four skulls placed in a tight ring facing one another with the faces tilted slightly downward; their mandibles were not articulated to the skulls. A mass of postcranial material and mandibles along with some skull fragments covered the skulls, all of it tightly packed into the grave. No grave goods were associated with Burial 1 (Gill 1971:58–59). Burial 2 was a flexed adult female with a large ceramic vessel buried with her.

Excavations subsequently focused on Mound D (Feature D). It measured approximately 40 m in diameter, and there was little apparent evidence of vandalism. A 3-×-3-m test unit was established at the center of the mound and excavated in arbitrary 30-cm levels. The mound, in terms of its construction, was similar to Mounds A and C, and its fill was homogeneous with no discernible cultural strata evident. The test unit was excavated to a depth of 1.5 m. Human bone appeared immediately after excavations began. Much of the bone was scattered, indicating postdepositional disturbance of the feature. Gill (1969a:55) attributes this to prehistoric disturbance of burials during the construction of the mound. In other words, as burials were added, previously interred graves were disturbed. Burials were continuous to the bottom of the test unit. The remains of at least 33 individuals, many represented only by fragmentary elements, were recovered (see Table 5.15). Several other burials were also identified, but because of time limitations they were left unexcavated, and the test unit was refilled.

All three of the tested mounds contained great numbers of the small marine gastropod *Cerithidea montagnei*. These were scattered throughout the fill of the mounds and did not appear to be associated with any particular burials. Their presence had to be the result of human activity; perhaps they represent the residue of some ritualistic behavior. Also found in the three mounds were reddish clumps of burned clay. These were irregular in form, measured 2.5 × 15.0 cm in diameter, and sometimes exhibited black smudge marks on their exterior. Several examples were flat or disk-shaped and measured over 40 cm in diameter (Gill 1969a:53–54, 1971).

No radiocarbon assays are available from the excavations at Chalpa. Diagnostic ceramics recovered from the site indicate an Early through Middle Conchera period (AD 900–1350) age for the site (Table 5.16).

Tecualilla

The site of Tecualilla is located near the southern end of the study area (see Figure 5.1) and was selected for investigation because it appeared to be large and seemed likely to yield significant information on the prehistory of the area. It consisted of a broad, high artificial embankment with low sloping borders that extended about 160 m along the shoreline. Today, the inland boundary of the site lies 80 m from the water. An unknown portion of the site had been eroded by storm surges and the tidal action of the estuary. Toward the northern edge of the site is a prominent earthen mound that rises significantly above the larger mound base on which it sits. It was anticipated that excavations would result in the recovery of a significant number of human burials and associated funerary offerings. Of particular interest is the fact that a number of the burials recovered were located below the modern water table, making their recovery only possible during low tide. The Tecualillans would not have buried their dead within the tide range of the estuarine basin, and this site, like Arriñitas, provides clear evidence of estuarine and coastal plain hydrologic dynamics and the corresponding effect on prehistoric sites.

Tecualilla was first investigated in 1967, during the first full field season for the Marismas project (Scott 1967; Figure 5.30). A 2-×-2-m test unit was initially excavated in what would subsequently be designated Mound A. Measuring 30 × 15 m, the asymmetrical mound dominated the site. From its high point at the feature's northern edge, the mound sloped southward toward the estuary. The first test unit was excavated in arbitrary 15-cm levels, and it was eventually expanded into a 2-×-4-m test unit. The fill of the mound generally consisted of soft and dark sandy soil, although a highly compacted pocket of light-colored sand was also exposed. Scattered through both matrices, dark and light, were sherds from a small olla that was subsequently reconstructed. A small amount of human bone was also scattered throughout the mound fill, as were some oyster shells and a few clam- and *Conus* sp. shells. Two human burials (Table 5.17), both extended, were recovered at depths of 2.1 m and 2.9 m below the surface of the mound, respectively.

Table 5.16. Frequencies of Identified Ceramic Types from Chalpa.

	Chametla Type										Culiacán Type				Amapa Type			Peñitas Type			
Test Unit and Level	El Taste Polychrome	El Taste Red/Cream	El Taste Red-bordered	El Taste Rough	El Taste Satin	Mazatlán Engraved	Mazatlán Polychrome	Red/Buff Rim	Red-rim Decorated	Subtotal	Culiacán Banded Incised	Culiacán Incised	Early Culiacán Polychrome	Subtotal	Black/Buff	Sentispac Buff	Subtotal	Engraved	Red Incised	Polychrome (Bordaz 44–49)	Subtotal
North Extension, Test Unit A																					
Level 1			1			1	1		2	5											
Level 2			1							1											
Level 3			1							1						1	1			1	1
Test Unit B																				1	1
Level 3			2					2		4		1?		1		1	1	2	5		7
Level 4			2							2						1	1		1	1?	2
Level 5																					
Test Unit C																					
Level 1							1			1											
Mound A, Test Unit B																					
Level 1	1						2			3											
Level 2			1				7		1	9								1	1?		2
Level 3	1		1	1	1		2		2	8								1			1
Level 4			2						1	3									1		1
Level 5													1	1					1		1
R 22 B		6	1				3			10	4			4	1		1				
Burial 2																				2	2
Burial 14													1?	1							
Burial 17							10		1	11											
Subtotal	2	6	12	1	1	1	26	2	7		4	1	2		1	3		4	9	5	
Total										58				7			4				18

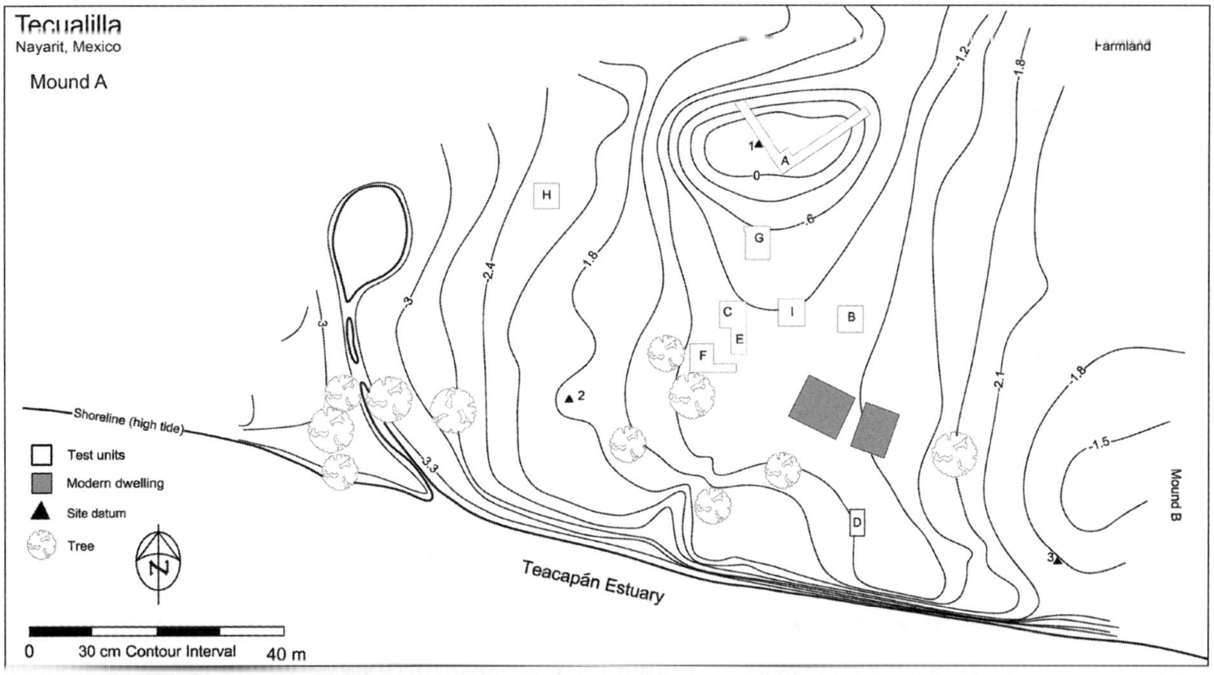

Figure 5.30. Map of Tecualilla showing the location of test units excavated in and near Mound A.

TABLE 5.17. Summary of Burials from Mounds A and B, Tecualilla.

Mound and Burial Number	Condition	Sex	Age	Burial Type	Funerary Offerings
Mound A					
B 1	Complete	Male	Mid-adult	Seated, with legs extended	2 plainware vessels
B 2	Complete	Male	Young adult	Seated, with legs extended	2 plainware vessels, 1 large decorated olla
B 3A	Nearly complete	Male	Young adult	Seated	None
B 3B	Fragmentary	Male	Adult	Secondary inhumation	None
B 4	Incomplete	Male	Young adult	Extended	1 tripod polychrome vessel, 1 plainware bowl
B 5	Nearly complete	Probably female	Old adult	Extended	1 plainware vessel
B 6A	Incomplete	Unknown	Child	Unknown	Unknown
B 6B	Incomplete	Unknown	Child	Unknown	1 carved pendant,[a] 1 plainware vessel
B 7	Complete	Male	Young adult	Extended	1 plainware vessel
B 8	Complete	Male	Old adult	Seated, with legs extended	None
B 9	Complete	Male	Mid-adult	Extended	1 decorated vessel
B 10	Incomplete	Male	Young adult	Extended	9 shell beads, 1 obsidian blade
B 11	Complete	Female	Old adult	Semiflexed	None
B 12A	Complete	Male	Young adult	Unknown	1 decorated vessel[a]
B 12B	Nearly complete	Unknown	Child	Unknown	Unknown
B 12C	Nearly complete	Unknown	Child	Seated, with legs extended	2 plainware vessels,[a] 3 spindle whorls
B 12D	Fragmentary	Unknown	Old adult	Unknown	Unknown
B 13	Incomplete	Female	Mid-adult	Semiflexed	None
B 14A	Incomplete	Male	Young adult	Undetermined	5 shell pectorals
B 14B	Fragmentary	Unknown	Child	Unknown	Unknown
B 15	Nearly complete	Unknown	Child	Semiflexed, on left side	None
B 16	Incomplete	Unknown	Infant	Extended	Animal burial[b]
B 17	Complete	Female	Young adult	Semiflexed	1 plainware vessel,[c] 3 spindle whorls
B 18	Incomplete	Male	Young adult	Secondary, skull only	None
B 19	Nearly complete	Female	Young adult	Extended	None
B 20	Nearly complete	Male	Old adult	Bundle	None
B 21A[d]	Complete	Male	Old adult	Extended, on right side	None
B 21B[d]	Complete	Female	Young adult	Extended, on right side	None
B 21C	Nearly complete	Possibly male	Adolescent	Extended	1 animal burial, 1 projectile point
B 22	Complete	Male	Old adult	Bundle	1 plainware bowl, 2 obsidian blades, 4 bone beads
B 23	Complete	Male	Young adult	Extended	1 plainware bowl, 1 small plainware vessel, 3 beads
B 24	Incomplete	Female	Mid-adult	Seated, legs extended(?)	1 plainware vessel, 1 spindle whorl

TABLE 5.17. (cont'd.) Summary of Burials from Mounds A and B, Tecualilla.

Mound and Burial Number	Condition	Sex	Age	Burial Type	Funerary Offerings
B 25	Complete	Male	Young adult	Seated, with legs extended	1 polychrome vessel, 1 plainware bowl, 2 obsidian blades
B 26A[e]	Fragmentary	Female	Young adult	Unknown	Unknown
B 26B	Fragmentary	Probably female	Adult	Unknown	Unknown
B 26C	Fragmentary	Probably male	Young adult	Unknown	Unknown
B 27A	Fragmentary	Male	Adult	Unknown	Unknown
B 27B	Fragmentary	Probably female	Adult	Unknown	Unknown
B 28	Fragmentary	Unknown	Unknown	Unknown	Unknown
Mound B					
B 1A	Incomplete	Male	Adolescent	Seated	4 triangular stone beads
B 1B	Incomplete	Probably female	Adult	Urn(?)	Urn/vessel fragments
B 1C	Incomplete	Female	Adult	Seated, with legs extended	1 small plainware vessel, 1 spindle whorl
B 1D	Incomplete	Male	Mid-adult	Urn (skull only)	Urn fragments, 1 small vessel within the urn
B 1E	Fragmentary	Male	Adult	Unknown	Unknown
B 1F	Fragmentary	Male	Adult	Unknown	Unknown
B 1G	Fragmentary	Unknown	Newborn	Unknown	Unknown
B 1H	Fragmentary	Unknown	Adult	Urn	Urn fragments
B 2A	Incomplete	Unknown	Adolescent	Bundle	1 decorated bowl
B 2B	Incomplete	Unknown	Child	Unknown	None
B 3	Fragmentary	Unknown	Mid-adult	Unknown	Unknown
B 4A	Incomplete	Male	Mid-adult	Extended, facedown	1 small plainware vessel
B 4B	Fragmentary	Female	Adult	Unknown	Unknown

[a] These items were found in a multiple burial complex and have been assigned to the most probable individual burial; however, it is possible that they are associated with another individual within this group of burials.
[b] The animal burial was situated above the human burial, and its association with the human burial is uncertain.
[c] The plainware vessel was found near the skull of B 18 but also close to the body of B 17. Its association with a specific burial is uncertain.
[d] These two individuals were interred side by side in the same grave.
[e] The skull of this burial was found in the lap of the seated burial, B 25. Other bones of this individual were found near the skull.

A single radiocarbon assay was obtained from a charcoal sample recovered from Level 13 (1.8 m). The assay yielded a date of AD 1178–1413 (UCLA 1298), which suggests a Middle Conchera period association.

Tecualilla was further investigated in 1968 when two mounds, designated Mounds A and B (Features A and B), were tested. The investigation of Mound A included the excavation of two trenches, one oriented northwest-southeast and the other oriented northeast-southwest; they were excavated perpendicular to one another, and both were approximately 1.5 m wide and 12 m long. Excavation of the trenches began at the modern ground level in order to expose a full profile of the mound from its base to its highest point. At the junction of the trenches at the apex of Mound A, a 3-×-3-m test unit (TU A) was initially excavated adjacent to the original 1967 excavation unit. The test unit was dug in arbitrary 30-cm levels, and all fill was screened through ⅛-inch mesh. Mound B, found to be extensively disturbed, was located to the east-southeast of Mound A along the shoreline of the estuary. Richard Shenkel was in charge of the 1968 excavations in Mound A, and George Gill directed the work in Mound B.

Excavation of the trenches failed to expose any features within Mound A, and little, if any, variation was seen in the mound's fill. The mound fill consisted of a highly compact, dry, light gray sandy soil that was difficult to excavate by hand. Shenkel's description of the mound's fill varies considerably from that reported during the initial 1967 testing of the mound (Scott 1967). Shenkel does note that his excavations occurred at the end of the dry season and suggests that the mound may have been more easily excavated after the onset of the rainy season. Because of the difficulty of excavating the

Table 5.18. Frequencies of Identified Ceramic Types from Test Unit A, Tecualilla.

Level	Chametla Type											Culiacán Type						
	Black-on-buff	Red-on-buff	El Taste Polychrome	El Taste Red-bordered	El Taste Red-bordered	El Taste Rough	Mazatlán Polychrome	Aztatlán Ware	Red-rim Decorated	Red-rimmed Utility Ware	Subtotal	Early Culiacán Polychrome	Red-rim Decorated	Aguaruto Exterior Incised	Aguaruto Interior Incised	Navolato Polychrome	Aztatlán Heavy Ware	Subtotal
First 30 cm																		
1	3				4			2		21	30	1				1		2
2		1			16		1	1	11		30			1				1
3	1			9					32		42							
4[b]					10				28		38		3			1		4
5[c]			1						1		2							
6[c]					1				13		14							
7[c]								1	10		11	1					1	2
8[d]					2			2	49		53			3	1			4
9						1			2		3							
Subtotal	4	1	1	9	33	1	1	6	146	21		2	3	4	1	2	1	
Total											223							13

Level	Amapa Type															Peñitas Type							
	Botadero Red-on-white	Botadero Incised	Mangos Engraved	Ixcuintla Polychrome	Ixcuintla Engraved	Cerritos Engraved	Sentispac Polychrome	Sentispac Red-on-buff	Sentispac Buff	Santiago Engraved	Peñitas Engraved	Tuxpan Engraved	Tuxpan Red-on-orange	Tuxpan Red-on-buff	Subtotal	Peñitas Polychrome	Red-on-white-on-orange	Polychrome (Bordaz 50–54)	Red Engraved (Bordaz 1–2)	Unspecified Polychromes	Red-on-white	Red Engraved	Subtotal
First 30 cm							1								1								
1			1		1					2		7	5		16	1		5		15	2		23
2	3						2	1					1		7		2	3		6	1		12
3			1						1	1		1		4	8			6		22[a]		1	29
4[b]				3					3	2	1	1		2	12			1		13			14
5[c]																							
6[c]				1							1	1	4		7					8			8
7[c]						2						2		2	6					11			11
8[d]		1			1						2	9	7		20	1		7	3	4[e]			15
9																							
Subtotal	3	1	2	4	2	2	3	1	4	5	4	21	17	8		2	2	22	3	79	3	1	
Total															77								112

[a] Some of these resemble Iguanas Polychrome.
[b] These tallies are from half of Test Unit A.
[c] These tallies are from the northwest 1.5-x-1.5-m subunit of Test Unit A.
[d] Level 8 produced three sherds identical to what Sauer and Brand (1932) identified at Navolato as Culiacán Polychrome; however, that style is not described by Kelly (1945) (note on original tally sheet by Betty B. Bell).
[e] Bordaz polychromes 44–49 or 67–68.

mound fill and time constraints, TU A was reduced to a 1.5-×-1.5-m test unit midway through the excavation. Additionally, because of the homogeneity observed in the Mound A fill, Shenkel (1968, 1969a; see also Gill 1971: 30–32) postulated that Mound A had been constructed in a single episode.

Excavations of the first 1.8 m (Levels 1–6) of TU A yielded little material. A copper pin was recovered from Level 7, and a copper finger ring, a red bead, and a spindle whorl were recovered from Level 8. The upper portion of Level 9 produced two whole vessels: one large olla covered with carbon and a smaller bottle-shaped vessel. These two vessels had no apparent association with a feature or a burial. At depths between 3 and 4 m, three human burials were exposed. These appeared to be slightly disturbed, and all had associated ceramic vessels as funerary offerings. They were in seated positions with their legs extended, and their upper torsos were upright, with their skulls in their laps. The deepest of the burials was just above coarse, reddish, loose sand that marked the current water table. The recovered sherd assemblage included types identified as Chametla, Culiacán, Amapa, and Peñitas (Table 5.18).

Gill's work at Mound B was limited and focused on the recovery of human remains eroding from the bank of the estuary about 60 m south of the excavations in Mound A. It appeared that most of Mound B had been washed away by the estuary, and much of what remained was covered by modern village habitation features and animal pens. In all, 13 individuals were represented in the recovered remains; all remains were fragmentary or incomplete (see Table 5.17). Eleven of the 13 burials were recovered from a natural sand layer 1.5 m below the surface of the mound. This layer of sand was about 1 m thick and above the water table. Two of the burials, 2A and 2B, were recovered from the mound fill above the primary burial zone.

Based on the 1968 work at Tecualilla, Gill (1971:31) had surmised, despite Shenkel's (1968, 1969a) opposing view, that it was likely numerous human burials were present in the area between Shenkel's excavations in Mound A and those Gill had recovered from Mound B. Thus, Gill returned to Tecualilla for a third season at the site in 1969. Four test units, B, C, D, and E, were excavated in the southward-sloping area below the peak of Mound A. TUs B and C were 3 × 3 m, and TUs D and E were 1.5 × 3.0 m in size; TU E abutted TU C. The excavations in TUs C and E resulted in the recovery of the remains of 16 individuals from at least 11 separate burials (B 4–B 14) (Figure 5.31; see Table 5.17). Fourteen of the burials were found interred within a natural sand layer below the mound fill, which consisted of a homogeneous loamy soil that contained numerous decorated sherds, some thin lenses of oyster shell, and scattered animal bone. The sand layer was about a meter thick and occurred at a depth of 1.5 to 1.9 m below the surface of the mound. The two other burials (B 6A and B 6B), both children, were found at the bottom of the mound fill directly above the sand layer. One of the other burials, B 8, was buried in a seated position within the sand layer, although its back and shoulders extended into the mound fill above. The water table occurred directly below the sand layer at 2.7 to 3.0 m below the surface of the mound. Burial 14 was located within the water table at the time of excavation in a heavy, mineralized matrix. It is likely that the water table was lower at the time of its internment. No burials were recovered from TUs B and D, although quantities of plain and decorated sherds and some animal bone were recovered. Both test units were excavated to a depth of approximately 2.6 m, well into the presumed culturally sterile soil, to verify that no burials or other cultural materials were present at the tested loci (Gill 1969a).

Additional testing was conducted at Tecualilla in 1970. Four additional 3-×-3-m test units were excavated, TUs F, G, H, and I. TUs F, G, and H were excavated in arbitrary 15-cm levels, and all fill from the units was screened with ⅛-inch mesh. TU H was excavated approximately 23 m west of Mound A to determine whether burials occurred in nonmound contexts; none were identified. The other three test units, F, G, and I, were placed on the southern slope of Mound A in general proximity to the test units excavated in 1968 and 1969. No burials were recovered from TU I, the fill of which was excavated expediently in bulk as the field season was ending.

TUs F, G, and I produced an abundance of plain and decorated ceramics (Tables 5.19–5.22); spindle whorls; manos, including overhanging types; a ground and polished ax head; solid ceramic figurine fragments (Mazapan-style); ceramic pipe stem and bowl fragments; and obsidian debitage, blades, and stemmed projectile points (Volkmann and Gill 1970:49–50). TU H yielded the remains of four individuals (B 15–B 18), and TU G contained the remains of 12 individuals (B 19–B 26C) (see Table 5.17).

Shenkel (1968, 1969b) and Gill (1969a, 1969b, 1971; see also Volkmann and Gill 1970) extended much effort in debating the formation processes that led to the mounds at Tecualilla, especially Mound A. Shenkel argued, based on his 1968 excavation of the two trenches of the test unit in Mound A, that the lack of stratigraphy was evidence that the mound was constructed in a single episode for the purposes of interring the burials that were discovered. He further suggested that the southern slope area adjacent to Mound A was perhaps an adjacent midden or possibly a deposit resulting from some alluvial or eolian process. Gill argued that Shenkel had

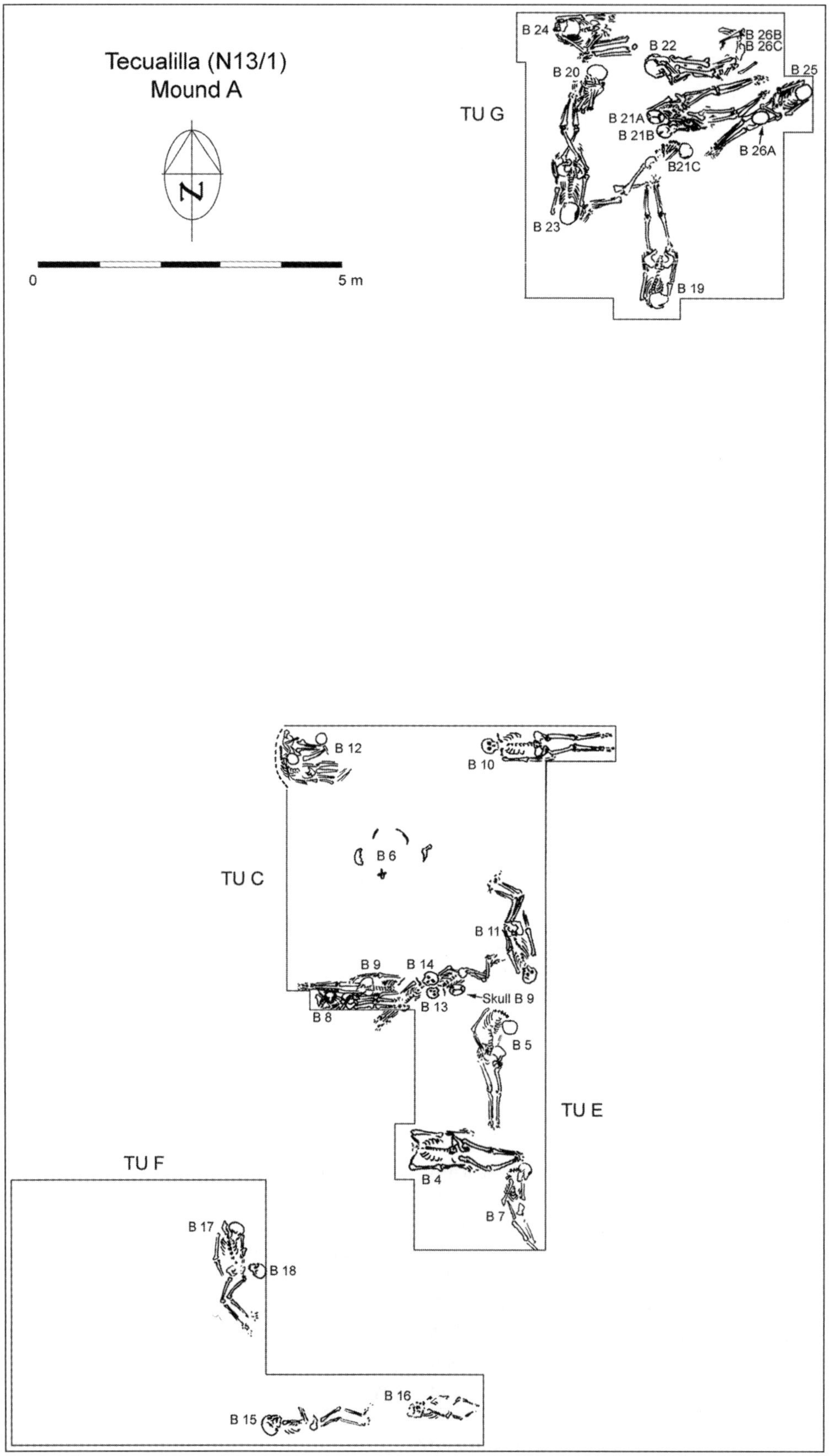

FIGURE 5.31. Distribution of burials in Test Units (TUs) C, G, E, and F, Mound A, Tecualilla (Gill 1971).

TABLE 5.19. Frequencies of Identified Ceramic Types from Test Unit F, Tecualilla.

	Chametla Type										Culiacán Type	
Level	Black-on-buff	Cocoyolitos Polychrome	Middle Chametla Polychrome	El Taste Polychrome	El Taste Red-bordered	El Taste Rough	Mazatlán Polychrome	Lolandis Red Rim	Red-rimmed Utility Ware	Subtotal	Aztatlán Ware	Subtotal
First 15 cm							3	2		5		
Second 15 cm								4		4		
Third 15 cm		1		1	3	1	16	1	4	27		
Fourth 15 cm							4	6	1	11		
4	3	1	1		2		33	3		43		
5					6		26	10		42		
6		2			9		1	11		23	2	2
7	2				24		1	49		76	8	8
8	3				8			31		42	7	7
9					1	1			25	27	7	7
10								14		14	5	5
11								9		9	3	3
Subtotal	8	4	1	1	53	2	84	140	30		32	
Total										323		32

	Amapa Type												Peñitas Type							
Level	Botadero Incised	Botadero Red-on-white	Botadero Black-on-buff	Cerritos Engraved	Cerritos Polychrome	Ixcuintla Polychrome	Mangos Engraved	Sentispac Buff	Sentispac Red-on-buff	Tuxpan Engraved	Tuxpan Red-on-orange	Subtotal	Engraved	Incised	Painted	Red Engraved	Red Incised (Borda 5, 8, etc.)	White-on-red	Possibly Painted (Iguanas Polychrome?)	Subtotal
First 15 cm																				
Second 15 cm																				
Third 15 cm				2				1				3		1						1
Fourth 15 cm																				
4				1				6				7								
5				1				7		3		11					2	1	1	4
6								7				7			7					7
7				2			1	18		12		33			17		3		1	21
8			1		3	3	1	7	4	11		30	9		7	1				17
9	1	1		1		1		3	2	2		11			1	1				2
10		1						1	2		2	6			2					2
11				1					1		1	3	1		1					2
Subtotal	1	2	1	8	3	4	2	50	9	28	3		10	1	35	2	5	1	2	
Total												111								56

TABLE 5.20. Frequencies of Identified Ceramic Types from Test Unit G, Tecualilla.

	Chametla Type										Culiacán Type						
Level	El Taste Polychrome	El Taste Red-bordered	El Taste Red-rimmed	El Taste Rough	Heavy Ribbed	Mazatlán Polychrome	Red-rim Decorated	Red-rimmed Utility Ware	Ribbed Ware	Subtotal	Aguaruto Polychrome	Red-rim Decorated	Aztatlán Heavy Ware	Aztatlán Ware	Early Culiacán Polychrome	Navolato Polychrome	Subtotal
1		4				4	3		1	12				1			1
2		8		3		30	16			57					1		1
3		9		1		11	9			30				1	2		3
4	1	19		3		11	35		1	70			1	5	8		14
5		27					24		1	52			3		5	2	10
6		5					10			15			1	3	5		9
7		7				1	24			32				2	3	1	6
8		5					19			24			1	4			5
9		6			1		34			41	1	3	1	3			8
10		7			2		23	1		33				2			2
11			1			1	23			25				9	1		10
12		1				3	45			49				4			4
13														1			1
14							1			1							
15							3			3							
Subtotal	1	98	1	7	3	61	269	1	3		1	3	7	35	25	3	
Total										444							74

	Amapa Type																Peñitas Type						
Level	Black-on-buff	Botadero Red-on-white	Cerritos Engraved	Cerritos Polychrome	Iguanas Polychrome	Ixcuintla Polychrome	Mangos Engraved	Mangos Polychrome	Mangos Red-on-white	Santiago Engraved	Sentispac Buff	Sentispac Black-on-buff	Sentispac Red-on-buff	Tuxpan Engraved	Tuxpan Red-on-orange	Subtotal	Engraved	Engraved Through White-on-red	Painted	Polychrome	Red Engraved	White-on-red	Subtotal
1	2		1				1							4	1	9	1						1
2	2	2					1				5			5	7	22			2		1		3
3	2			1			1				6		1	10	3	24			1		1		2
4	2						2	1	1		15		1	9	8	39			4		6		10
5				1	1	2					7		1	8	12	32			7				7
6											6		1	6	4	17	1		9		1		11
7										1	4			2	2	9			11		1		12
8						1				2	1			4	9	17	1						1
9					17					4	1			15	17	54				3	1		4
10			2			1				3	2			20	9	37				32		1	33
11								1			1	1	1	11	16	31		3	3	34	2	1	43
12			5			2	2			2				8	15	34			7	25			32
13											1					1				2	1		3
14																							
15																				1	2		3
Subtotal	8	2	8	2	18	6	7	2	1	12	49	1	5	102	103		3	3	44	97	16	2	
Total																326							165

TABLE 5.21. Frequencies of Identified Ceramic Types from Test Unit H, Tecualilla.

	Chametla Type					Culiacán Type				Amapa Type									Peñitas Type	
Level	El Taste Red-bordered	El Taste Rough	Red-rim Decorated	Red-rimmed Utility Ware	Subtotal	Aztatlán Red-rimmed Decorated	Aztatlán Heavy Ware	Aztatlán Ware	Subtotal	Black-on-buff	Botadero Red-on-white	Mangos Engraved	Santiago Engraved	Sentispac Black-on-buff	Sentispac Buff	Tuxpan Engraved	Tuxpan Red-on-orange	Subtotal	Polychrome	Subtotal
2															1			1		
3			5		5										2	2		4		
4	1		23		24		2	1	3											
5			14		14	2			2		2		1			2	3	8	2	2
6	1	1	7		9					1		2					2	5		
7			6	3	9						1			1	1			3	2	2
Subtotal	2	1	55	3		2	2	1		1	3	2	1	1	4	4	5		4	
Total					61				5									21		4

TABLE 5.22. Frequencies of Identified Ceramic Types from Test Unit I, Tecualilla.

	Chametla Type					Culiacán Type			Amapa Type						Peñitas Type				
Level	El Taste Satin	Aztatlán Ware	Red-rim Decorated	Heavy Ribbed(?)	Subtotal	Aguaruto Exterior Incised	Aztatlán Ware	Subtotal	Cerritos Engraved	Santiago Engraved	Sentispac Red-on-buff	Tuxpan Engraved	Tuxpan Red-on-orange	Subtotal	Red Engraved (Bordaz 1–2?)	Painted Polychrome (Bordaz 50–54)	Incised Through White-on-red (Peñitas Engraved?)	Polychrome	Subtotal
First 30 cm												1		1					
4	1				1							1		1					
5		5	9	1	15				1	2	2	2	5	12	1	2	4	6[a]	13
6		2	20		22	1	1	2		1		2	1	4				5	5
7			5		5														
Subtotal	1	7	34	1		1	1		1	3	2	6	6	1	1	2	4	11	1
Total					43			2						18					18

[a] Appears to be related to Iguanas Polychrome.

not considered all the evidence from the excavations in Mounds A and B and that the formation processes were more complex than Shenkel envisioned. Gill suggested, based on the presence of burials in the same geomorphic context in both mounds (the above-referenced layer of natural sand), that additional burials were likely widely distributed across the site. Furthermore, Gill argued that the mounded or elevated area at the northern end, which lacked any evidence of burials or fragmentary human remains, likely represented a second phase of construction that was not associated with the use of Mound A as a burial area. During the 1969 excavations at Tecualilla, Gill sought evidence to further support his argument that at least two construction stages were represented at the site and at Mound A in particular. Based on the 1969 work, Gill (1971) concluded that his argument for

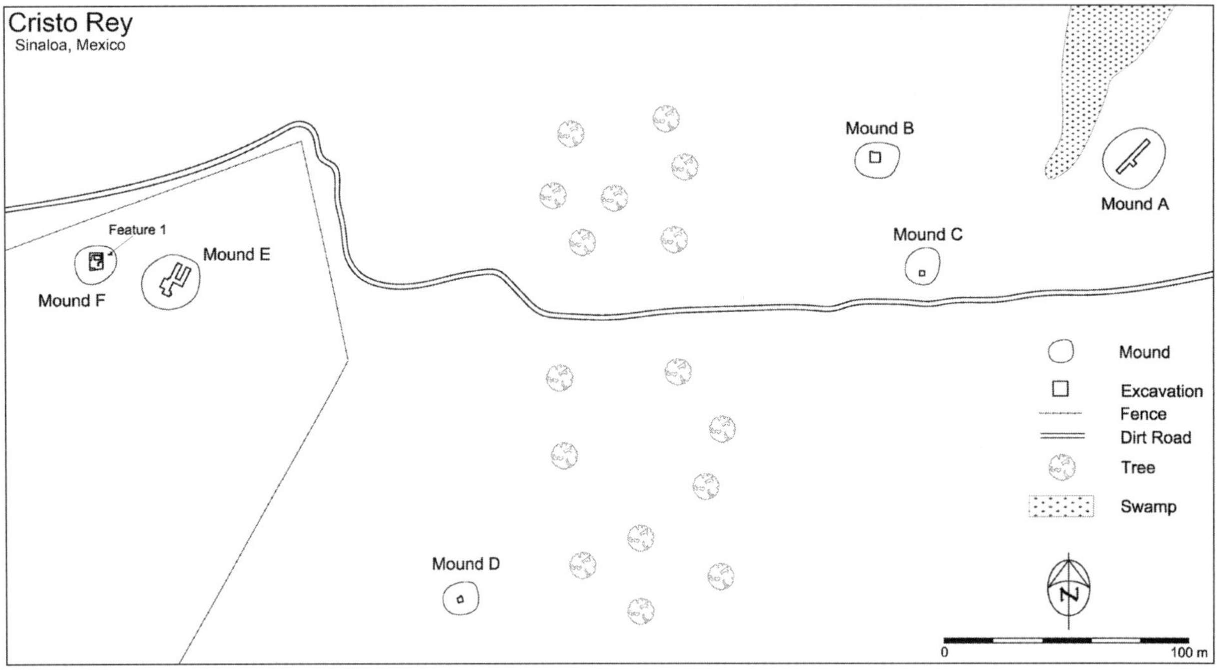

FIGURE 5.32. Site map of Cristo Rey showing locations of test units and Feature 1.

at least two phases of mound construction was generally confirmed, although some of his newly generated data raised other issues, including the possibility that three construction episodes were represented. Although Shenkel (1969b) subsequently revised his one construction sequence thesis, he continued to argue that the earlier (lower) mounds were middens in which burials were interred after the mound had formed through the accumulation of midden deposits. Conversely, Gill (1971) continued to argue that the lower mounds were constructed as burial mounds on which mound fill was subsequently added.

The debate was never resolved regarding the Tecualilla mounds, even as other sites were investigated during the Marismas project. The general lack of stratigraphy or other clear evidence of mound construction often hindered interpretation of how the mounds were built or how they accumulated. Many of the earthen mounds functioned exclusively as burial mounds. Whether these mounds were built and then burials were interred into them or whether they grew by accretion as burials were added to the burial locus is not clear. There is also evidence that some of the earthen mounds likely had habitation or perhaps storage features built on them. Also there are instances where earlier burial mounds became loci of midden deposition. Undoubtedly these mounds served a variety of functions. However, the lack of stratigraphy in the mounds and the lack of systematic excavations that focus on mound construction prevent resolution of the issues raised by Shenkel and Gill.

Five radiocarbon assays, all from charcoal, were obtained from Mound A at Tecualilla (see Chapter 4): AD 980–1213 (NSTF-7), AD 1023–1273 (UCLA 1628A), AD 1045–1401 (I-5528), AD 1178–1413 (UCLA 1298), and AD 1304–1642 (I-5529). The ceramic assemblages suggest a date range from AD 900 to just beyond AD 1400, which is generally consistent with the ^{14}C assays. Thus, based on the ceramics present and the ^{14}C assays, these relative and chronometric dates suggest an occupation spanning the local Marismas Early and Middle Conchera periods. An unanswered question is whether Tecualilla represents a single extended occupation or whether multiple distinct temporal components are present.

Cristo Rey

During the 1972 field season a complex of mounds in cleared and cultivated fields north of the village of Cristo Rey was selected for testing (see Figure 5.1). The Cristo Rey site consisted of at least six distinguishable low earth mounds distributed over an area of about 150 × 500 m (Figure 5.32); each of these mounds was tested (Baker et al. 1972; Schutte and Gill 1972). A series of 2-×-2-m test units were excavated in arbitrary 25-cm levels. It should be noted that burials (Burials 2, 6, 7, 8, 9, 10, and 13) within Mound A were assigned sequential numbers exclusive to that mound. Burials from Mound B, Mound D, and Mound E were assigned sequential numbers starting with 1 and ending with 30 (Table 5.23). In both instances, some assigned numbers were subsequently voided. Excavations were somewhat disrupted when several test units were looted one night, destroying the remaining archaeological contexts.

Table 5.23. Summary of Burials from Mounds A, B, and D, Cristo Rey.

Mound and Burial Number	Condition	Sex	Age	Burial Type	Funerary Offerings
Mound A					
B 2	Fragmentary	Unknown	Adolescent	Flexed	3 spindle whorls, 1 polychrome vessel
B 6	Fragmentary	Male	Young adult	Flexed, tightly	1 whole vessel
B 7	Fragmentary	Male	Mid-adult	Flexed, tightly	1 whole plainware vessel
B 8	Fragmentary	Probably male	Young adult	Bundle	None
B 9	Fragmentary	Indeterminate	Child	Flexed, tightly	1 whole plainware vessel
B 10	Fragmentary	Indeterminate	Child	Flexed, tightly	2 beads, 1 plainware vessel, 2 sherds
B 13	Fragmentary	Unknown	Child	Bundle	None
Mound B					
B 1	Fragmentary	Unknown	Unknown	Urn	None
B 3	Fragmentary	Probably male	Adolescent	Bundle	Green-glazed rock
B 4	Fragmentary	Unknown	Adult	Flexed	1 whole vessel
B 5(11)	Fragmentary	Probably male	Young adult	Unknown	Green-glazed rock
B 12	Fragmentary	Male	Young adult	Flexed, tightly	1 whole plainware vessel
Mound D					
B 14	Fragmentary	Unknown	Young adult	Seated, semiflexed	None
B 15	Fragmentary	Probably female	Young adult	Undetermined	None
Mound E					
B 16	Fragmentary	Female	Young adult	Semiflexed	None
B 17	Fragmentary	Male	Young adult	Semiflexed	None
B 18	Fragmentary	Male	Adult	Semiflexed	None
B 19	Fragmentary	Male	Young adult	Flexed, tightly	2 smoking pipes, 1 vessel, 1 obsidian projectile point
B 20	Fragmentary	Unknown	Child	Unknown	Unknown
B 21	Incomplete	Unknown	Child	Semiflexed	None
B 22	Fragmentary	Female	Young adult	Flexed, tightly	1 vessel, 2 spindle whorls
B 23	Fragmentary	Male	Old adult	Flexed, tightly	1 large studded vessel
B 24	Incomplete	Male	Mid-adult	Flexed, tightly	None
B 25A	Incomplete	Male	Old adult	Flexed, tightly	1 crocodile tooth
B 25B	Fragmentary	Unknown	Child	Unknown	B 25A(?)
B 26	Incomplete	Male	Young adult	Flexed, tightly	None
B 27	Fragmentary	Male	Young adult	Flexed	1 whole vessel
B 28	Fragmentary	Female	Adult	Unknown	1 smoking pipe(?)
B 29	Fragmentary	Female	Young adult	Flexed	None
B 30	Incomplete	Female	Young adult	Flexed, tightly	1 vessel, 1 spindle whorl, 1 smoking pipe, 1 obsidian flake

Source: Schutte 1973.

Mound A

Nine 2-×-2-m test units were laid out along one axis of Mound A, the easternmost mound tested. Not all the test units were excavated, and a tenth unit was established as an extension of TU 1. The excavation of TU 1 produced few artifacts, mainly sherds, in Levels 1 through 4, although fragmentary human bone was present in Levels 3 and 4.

At 1.08 m below the surface in the southwest corner of the unit, a cluster of adobe chunks was exposed. The origin of this material was not determined, although it was suggested that it might be part of an adobe floor.

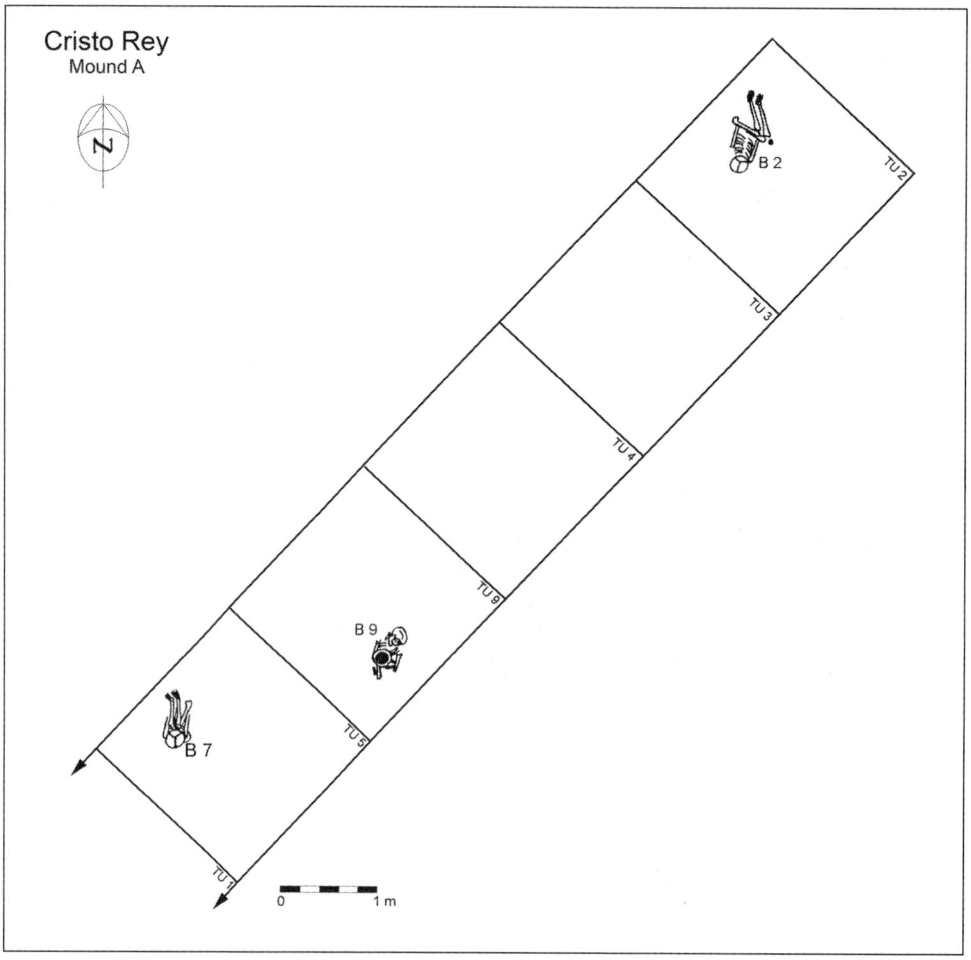

FIGURE 5.33. Distribution of human burials in Mound A, Cristo Rey (Schutte 1973). TU = Test Unit.

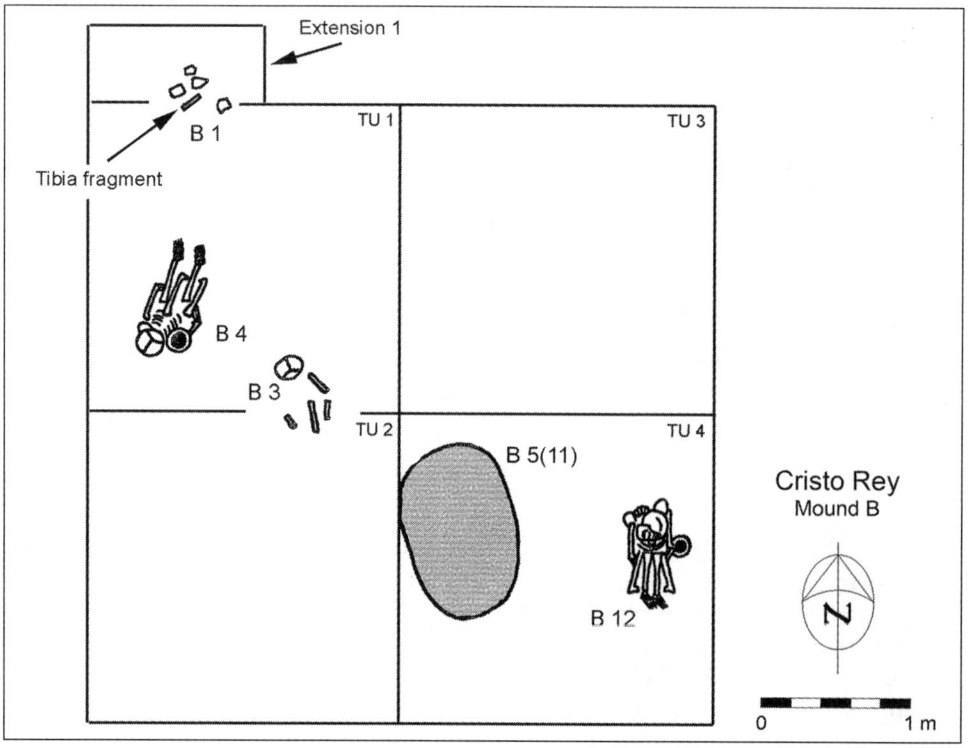

FIGURE 5.34. Distribution of human burials in Mound B, Cristo Rey (Schutte 1973). TU = Test Unit.

However, no other evidence of structural remains was identified. At 1.2 m a human burial (Burial 2) was exposed (Figure 5.33). Associated with this burial were three spindle whorls and a polychrome vessel. Excavation of TU 1 was terminated in sterile soil at 1.75 m. TUs 2, 3, and 4 were excavated to sterile soil without producing any cultural materials.

TU 5 yielded no cultural remains in Level 1. At 32 cm in Level 2, a possible hearth was exposed in the eastern part of the unit. Level 3 produced a large broken vessel, and Level 6 yielded a dog burial. At 1.5 m Burial 7 was exposed. Associated with it was a single plainware vessel. TU 6 was excavated adjacent to the south edge of TU 1. Levels 1 through 4 produced a pipe bowl, a spindle whorl, and some sherds. Level 4 also produced some unidentified bone fragments. Level 5 produced two dog burials, and Level 6 produced another. Excavations ceased in Level 7.

Excavations in TU 9 were not productive, with the exception of the recovery of a single human burial (Burial 9) in Level 5. A whole vessel was found in the chest cavity of the skeleton. Schutte and Gill (1972) suggest that this burial may be similar to one found at Tecualilla (Burial 23), where it appeared that the rib cage of the individual was opened up and a vessel was placed in the left side of the chest cavity. However, the poor condition of Burial 9 precludes positive identification of this practice at Cristo Rey.

TU 10, an extension of TU 1, was the final test unit excavated in Mound A. Levels 1 through 4 yielded no cultural remains. In Level 5, at 1.1 m, a flexed human burial (Burial 13) was exposed. Two other burials (Burials 6 and 8) were recovered from along the west wall of the unit in Levels 7 and 5, respectively. Associated with Burial 6 was a complete decorated vessel. Level 7 produced a second burial (Burial 10) and an associated coarse-paste plainware vessel.

Mound B

The investigation of Mound B, a small earth mound, included the excavation of four 2-×-2-m test units. TU 1 was placed in the center of the mound. Level 1 produced two pieces of obsidian and several pieces of green-glazed rock. Artifacts in Level 2 were somewhat more abundant and included sherds, some oyster shell, obsidian, a fragment of a spindle whorl, and a possible pipe bowl fragment. Also found were two dog burials, one of which may have had a vessel associated with it. Several chunks of adobe (daub?) were also recovered. Level 4 produced a secondary human burial (Burial 4), which extended into TU 2 (Figure 5.34). Associated with this burial were several of the spherical green-glazed rocks. The excavation of TU 1 was terminated in Level 5 when culturally sterile soil was reached.

In order to recover one of the dog burials from Level 2, TU 2 was opened on the south side of TU 1. Level 1 produced a number of sherds, a fragment of a pipe bowl, and a small whole plainware vessel. Level 3 yielded a fragmentary human burial (Burial 1). A portion of TU 2 was extended 50 cm to the north in order to recover the remains. A redware olla was associated with the burial, and one of its femurs was wedged in the vessel, leading Schutte and Gill (1972) to speculate that this might be the remains of an urn burial. At a depth of 40 cm three dog burials were also recovered. Two of the individuals were next to one another, suggesting that they had been buried at the same time; the third was incomplete and not associated with the other two. Level 4 produced Burial 4 (located below Burial 3 in TUs 1 and 2). This burial had a plainware vessel placed at its right shoulder. Excavation of TU 2 continued into Level 5, where additional scattered fragments of green-glazed rock were recovered about 10 cm above a culturally sterile layer.

TU 3 was excavated along the east side of TU 1. Little cultural material was recovered. TU 4 was opened in an effort to recover portions of Burial 5, which was disturbed by overnight looting of the excavations. Level 1 produced an obsidian blade and some fragments of dog bone. Also found were several peculiar pieces of what are described as "hard, dried marsh component," apparently some kind of organic or sedimentary material. One piece is described as having a J-like inscription on it. The origin and function of these pieces of material are unclear; however, they may be pieces of daub from a structure. The remains of another dog burial were recovered from Level 2, as were the rest of the remains of Burial 5. Level 3 produced Burial 12, which had an associated plainware vessel. The excavation of TU 4 was also terminated in Level 5.

Mound C

Mound C was a low-lying earthen mound. Work in this mound was abandoned after the excavation of four levels in a test unit failed to produce any significant amounts of cultural material.

Mound D

A single test unit was opened in Mound D. Artifacts in Levels 1 through 3 were sparse; however, two human burials (Burials 14 and 15) were encountered in Level 4. Burial 14 was semiflexed and seated, and Burial 15 appeared to be a secondary burial of the skull of a young adult (Figure 5.35).

Mound E

Mound E was located near the presumed western edge of the site. Seventeen test units were established, but not all were excavated due to time constraints. Levels 1 and 2 of TU 1 produced a few plainware sherds. Level 3 yielded some fish bones and obsidian blade fragments. A clay

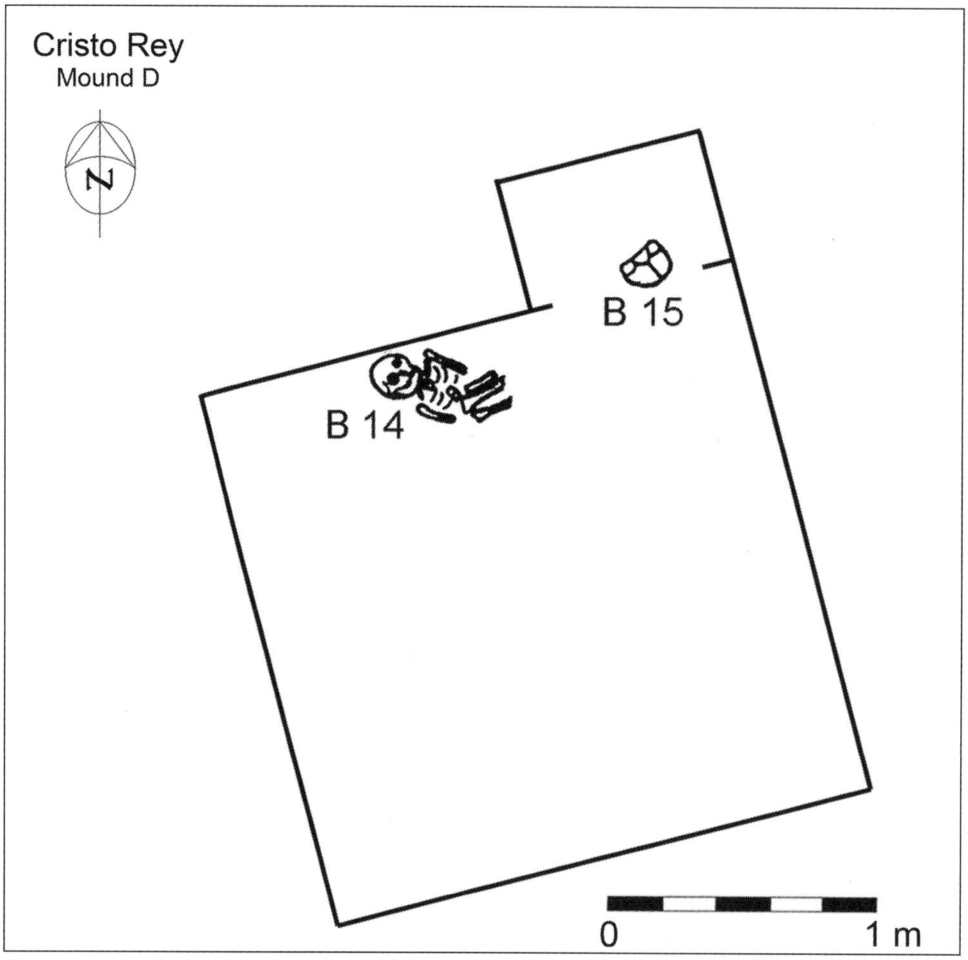

FIGURE 5.35. Distribution of human burials in Mound D, Cristo Rey (Schutte 1973).

object with a small perforation (bead or pendant?) was also recovered along with a concentration of clamshell and chunks of adobe. Level 4 produced two reconstructible decorated vessels and a spindle whorl. Another reconstructible vessel was recovered from Level 5. The rim of the vessel exhibited two small mend or suspension holes. Level 5 terminated in culturally sterile soil.

Levels 1 through 3 of TU 2 produced varying amounts of sherds and some dog bones. Burial 19 was recovered at approximately 95 cm in Level 4 (Figure 5.36). Associated with it were two pipes, an obsidian projectile point, and an overturned vessel near its head. A layer of adobe may have covered the burial pit. Other items found in Level 4 included a broken pot, dog bones, and some sherds. Excavations were terminated in Level 5.

In TU 3, Level 1 produced few artifacts. Level 2 contained Burial 18. No funerary offerings were associated, and Levels 3 through 6 produced only a few sherds. Levels 1 through 3 of TU 4 yielded no cultural material. A few pieces of unidentified bone were exposed in the wall of Level 4. The unit was extended to recover the bones. In addition to the bone, the extension produced a figurine head and a copper hook. Level 5 produced no cultural remains, but Burial 16 was exposed in Level 6. No funerary offerings were associated with this burial. Excavation of TU 4 continued into Level 7, where Burial 24 was recovered. No funerary offerings were found with Burial 24. Excavations were terminated at 1.75 m.

Excavation of TU 5 was nonproductive and was terminated at 1.25 m. Levels 1 and 3 of TU 6 produced a scattering of oyster shell, some pieces of adobe, and a few sherds. A nearly complete shallow bowl was also recovered from the north wall of the test unit in Level 3. Levels 4 and 5 yielded no archaeological remains, and excavations were ended. Levels 1 and 2 of TU 9 produced some scattered pieces of green-glazed rock and adobe. Level 3 yielded Burial 17, which lacked any associated funerary items. TUs 10, 11, and 12 were not excavated.

Levels 1 through 3 of TU 13 did not produce any archaeological materials. At 90 cm in Level 4, Burial 20, the fragmentary and isolated remains of a child, was exposed. Burial 22 was exposed at 1.18 m in Level 5. The test unit was extended in order to recover the entire burial. During the excavation of the extension, another

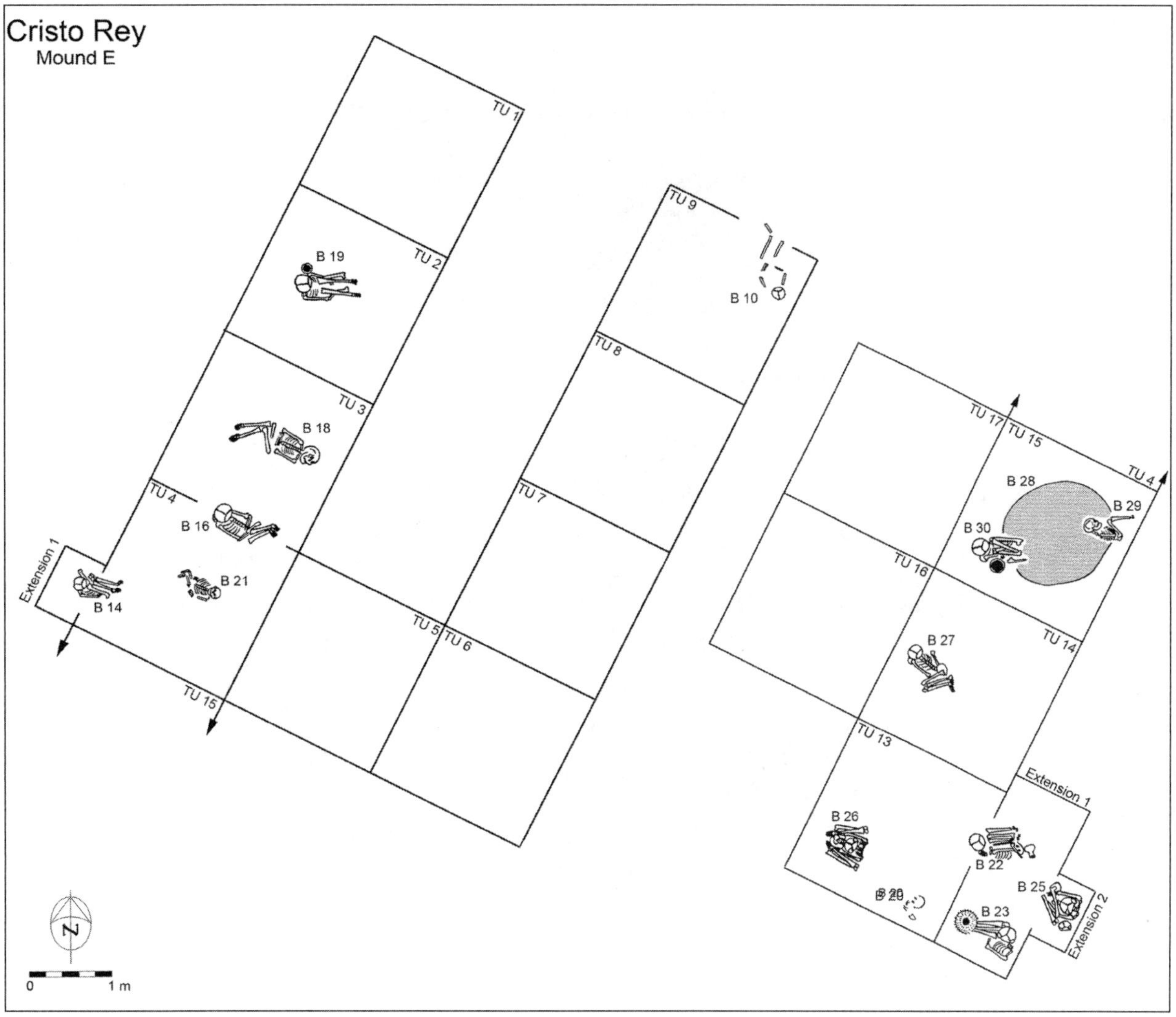

FIGURE 5.36. Distribution of human burials in Mound E, Cristo Rey (Schutte 1973). TU = Test Unit.

burial, Burial 23, was exposed at 1.1 m below the surface. A single vessel with stud-like protrusions spaced evenly across its exterior (datura pot?) (Figure 5.37) was associated with the burial. Burial 22 was recovered after the removal of Burial 23. Associated with Burial 22 were two spindle whorls and a small vessel. Level 6 produced Burial 26, which lacked associated funerary offerings. Excavations of TU 13 were ended at 1.5 m. However, as the extension of TU 13 was excavated to the level of the rest of the unit, Burial 25A was exposed in the east wall of the extension. The extension was enlarged yet again in order to recover the burial. This led to the recovery of Burial 25B. The only object accompanying these two burials was a crocodile tooth found near the head of Burial 25A.

Levels 1 and 2 of TU 14 produced only a few sherds, and Levels 3 and 4 contained no archaeological materials. The only recovery from TU 14 was Burial 27 and an associated vessel in Level 5. The excavation of TU 15 produced generally similar results, with three burials being recovered. Burial 28, badly disturbed by pothunters, was encountered in Level 5. A probable pipe stem was thought to be associated with the burial. Burial 29, lacking any funerary offerings, was recovered from Level 6. However, Burial 30, also from Level 6, had associated with it a pipe, a vessel, a spindle whorl, and an obsidian flake. Excavation was terminated at 1.5 m, and the excavations of TUs 16 and 17 failed to produce any archaeological remains.

Mound F

Mound F was selected for excavation because of its proximity to Mound E. Additionally, the mound was made up of gray-colored soil in the form of a hard-packed layer high in clay content rather than the sandy plowed knolls that made up the other mounds. TU 1 was placed at the center of the mound. Levels 1 and 2 contained few artifacts. At the base of Level 2 an adobe wall was exposed

FIGURE 5.37. Possible datura pot from Burial 23 at Cristo Rey.

along with a large circular darkened area (Feature 1). TU 1 was expanded to expose more of the feature. A second wall measuring 2.4 m in length and 15 to 35 cm high was exposed (Figure 5.38). An orange-colored floor was exposed within the confines of the wall. The exterior area surrounding the walls consisted of a hard-packed gray use surface or floor that extended 1.1 m south. Unfortunately, overnight looting destroyed most of the feature. An attempt was made to delineate further the extent and nature of the feature after it was vandalized, but so little remained that nothing more can be said of its form or function.

No radiocarbon assays are available from Cristo Rey. The presence of El Taste Polychrome and Cocoyolitos pipes indicates a Middle Conchera period (AD 1100–1400) occupation of the site.

Venadillo

The Venadillo site, a complex of low earth mounds, several of which are capped with shell middens, is located near the eastern shoreline of the Teacapán Peninsula (see Figure 5.1). Testing was initiated in Mound A but was abandoned in favor of more intensive excavations of Mounds B and C when excavations in those two mounds became more productive in terms of material culture and human remains (Fine et al. 1973). The excavation of the Mound C shell cap, a presumed midden, was also seen as an opportunity to systematically collect faunal remains associated with subsistence activities. Additionally, a pit (Test Pit A) was opened up in a depressed area between Mounds A and B, and a long trench approximately 180 m in length was excavated across a low area thought perhaps to be a plaza or living area between mounds (Figure 5.39).

Mound B, like Mound A, was an earthen mound consisting of undifferentiated beach sand. It measured approximately 35 × 50 m, with its long axis oriented northwest-southeast. The mound appears to have been a burial mound that grew by accretion as burials were added. Thirty-five burials were recovered from Mound B (Schutte and Speas 1973); subsequent laboratory studies resulted in the identification of 39 individuals

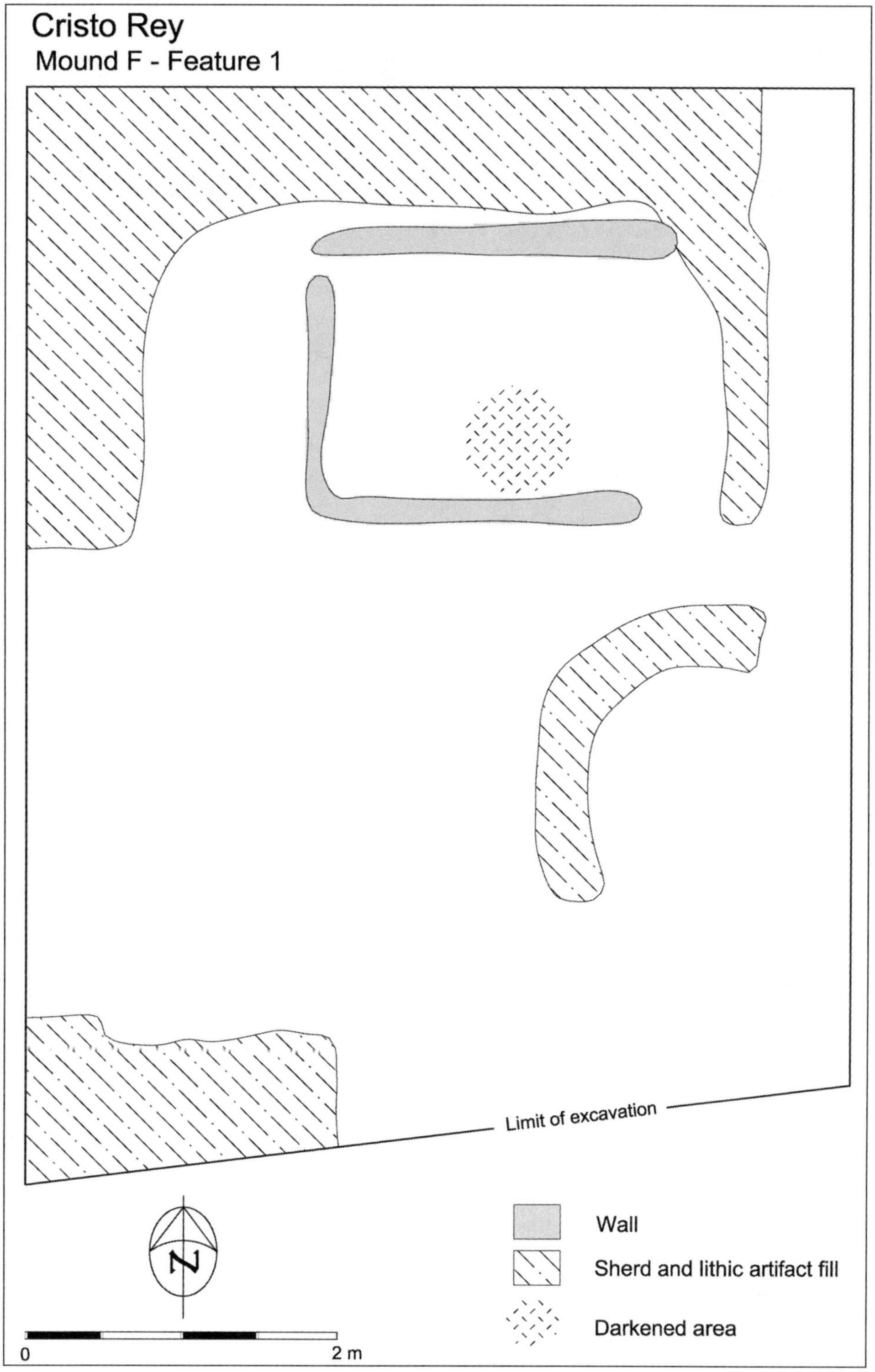

FIGURE 5.38. Plan view of Feature 1, Mound F, Cristo Rey.

(Table 5.24). These were primarily flexed burials, although two were in extended supine positions. The flexed burials ranged from semiflexed to tightly flexed, and the scattered isolated remains of several individuals were also present.

Funerary offerings were the most common artifact class recovered and included ceramic vessels, spindle whorls, obsidian projectile points and blades, figurines, beads, a ceramic ball, and a ¾-grooved ax. Quantities of Chametla, Culiacán, Amapa, and Peñitas ceramic types

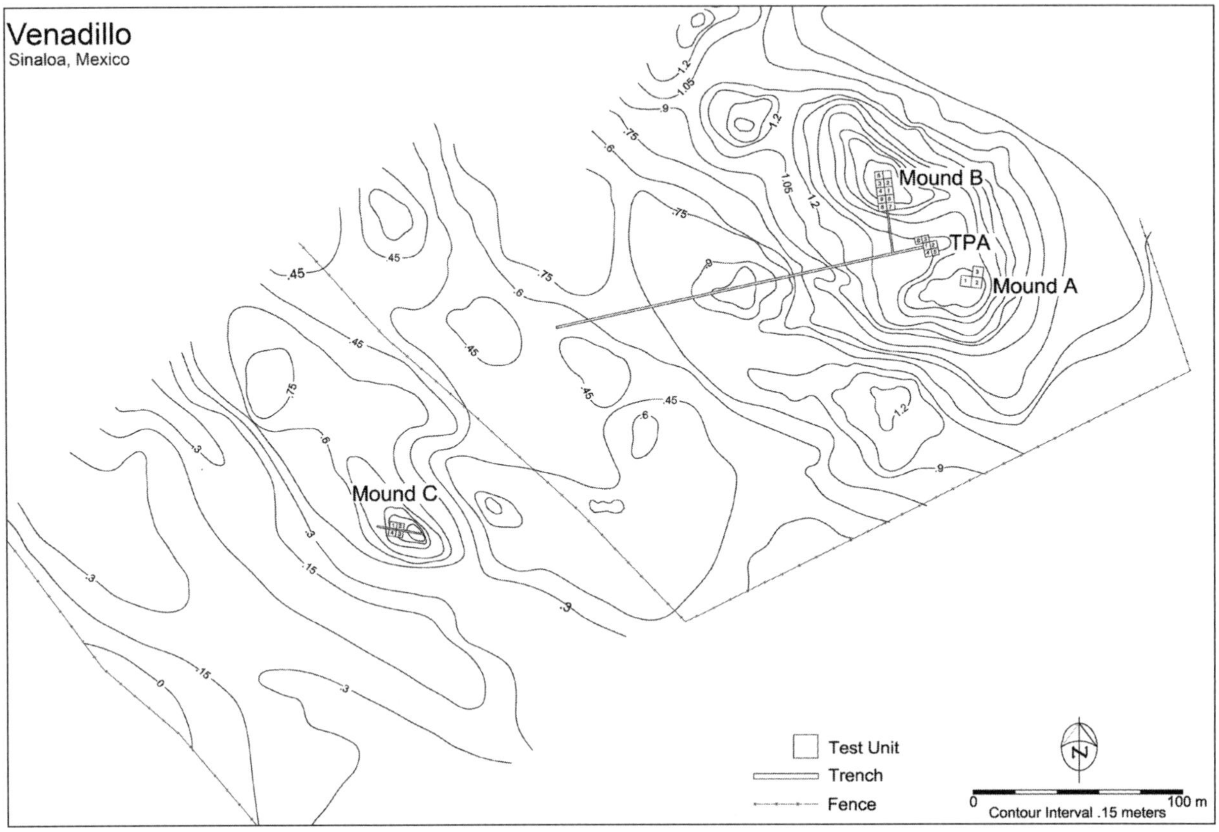

FIGURE 5.39. Site map of Venadillo showing locations of test units and test trenches. TPA = Test Pit A.

TABLE 5.24. Summary of Burials from Mounds B and C, Venadillo.

Mound and Burial Number	Condition	Sex	Age	Burial Type	Funerary Offerings
Mound B					
B 1A	Fragmentary	Undetermined	Child	Undetermined	None
B 1B	Fragmentary	Undetermined	Infant	Undetermined	None
B 2	Nearly complete	Female	Young adult	Flexed	1 orange-slipped jar with flaring neck
B 3	Nearly complete	Male	Old adult	Flexed	1 orange-slipped, unpolished jar with a slightly outcurved neck
B 4A	Incomplete	Male	Mid-adult	Seated, flexed	None
B 4B	Fragmentary	Female	Young adult	Undetermined	None
B 5A	Incomplete	Female	Young adult	Bundle	2 vessels, possibly sherds
B 5B	Incomplete	Undetermined	Infant	Bundle	Same as B 5A?
B 6	Incomplete	Female	Mid-adult	Flexed	3 vessels
B 7	Complete	Female	Young adult	Flexed	1 vessel
B 8A	Nearly complete	Female	Young adult	Flexed, tightly	None
B 8B	Nearly complete	Male	Young adult	Undetermined	None
B 9	Complete	Male	Old adult	Flexed, tightly	1 vessel, sherds
B 10	Complete	Female	Young adult	Flexed, tightly	1 decorated bowl, 1 orange plainware olla, 1 spindle whorl
B 11	Nearly complete	Male	Young adult	Flexed	1 orange plainware vessel, 2 plates

TABLE 5.24. (cont'd.) Summary of Burials from Mounds B and C, Venadillo.

Mound and Burial Number	Condition	Sex	Age	Burial Type	Funerary Offerings
B 12	Incomplete	Male	Old adult	Flexed	3 vessels
B 13	Fragmentary	Undetermined	Child	Flexed	None
B 14	Fragmentary	Undetermined	Old adult	Undetermined	None
B 15	Complete	Male	Old adult	Seated, flexed	Possibly sherds
B 16	Incomplete	Female	Young adult	Flexed	1 vessel, 1 spindle whorl, 1 obsidian projectile point, 1 obsidian blade
B 17A	Complete	Male	Young adult	Flexed, tightly	4 vessels
B 17B	Fragmentary	Undetermined	Infant	Undetermined	None
B 18	Nearly complete	Male	Young adult	Flexed	1 buff jar
B 19A	Complete	Female	Young adult	Semiflexed	1 polished buff jar, half of a spindle whorl, 2 small pinch pots
B 19B	Fragmentary	Male	Mid-adult	Seated	1 polychrome jar
B 20	Fragmentary	Male	Young adult	Flexed, tightly	None (B 20 was associated with Burials 5A and 5B)
B 21	Fragmentary	Undetermined	Adolescent	Undetermined	None
B 22	Fragmentary	Probably female	Adult	Undetermined	None
B 23	Incomplete	Male	Young adult	Skull only	Associated with B 19 and B 20
B 24	Fragmentary	Female	Mid-adult	Flexed	None
B 25	Incomplete	Female	Young adult	Flexed	3 vessels, 1 dog tooth
B 26A	Incomplete	Female	Young adult	Flexed	1 large vessel, 1 spindle whorl, 1 figurine head
B 26B	Incomplete	Male	Mid-adult	Flexed	1 ¾-grooved ax, 3 beads, 1 jar
B 27	Incomplete	Male	Old adult	Flexed	2 vessels
B 28	Nearly complete	Male	Old adult	Flexed, tightly	1 large decorated olla, 2 pinch pots, 1 fragmentary pinch pot
B 29	Fragmentary	Female	Adolescent	Flexed	1 large vessel, 1 reconstructible vessel, 1 spindle whorl, beads
B 30	Fragmentary	Undetermined	Infant	Extended	None
B 31	Fragmentary	Undetermined	Child	Flexed	None
B 32A	Fragmentary	Female	Old adult	Flexed	2 jars
B 32B	Fragmentary	Female	Mid-adult	Undetermined	None
Mound C					
B 1A	Fragmentary	Undetermined	Infant	Undetermined	None
B 1B	Fragmentary	Undetermined	Infant	Undetermined	None
B 1C	Fragmentary	Undetermined	Infant	Undetermined	None
B 2A	Incomplete	Female	Young adult	Urn (with B 2B)	None
B 2B	Fragmentary	Male	Young adult	Urn (with B 2A)	None
B 3	Complete	Undetermined	Adolescent	Flexed	None
B 4A	Fragmentary	Female	Adult	Urn	3 small anthropomorphic beads, 1 dog burial
B 4B	Fragmentary	Undetermined	Infant	Urn	See B 4A
B 5	Fragmentary	Female	Mid-adult	Undetermined	Green-glazed rock
B 6	Nearly complete	Female	Old adult	Flexed	None
B 7A	Incomplete	Male	Young adult	Flexed	None
B 7B	Complete	Female	Mid-adult	Flexed	None
B 8	Incomplete	Undetermined	Adult	Flexed	Ceramic rattle

Source: Schutte and Speas 1973.

Table 5.25. Frequencies of Identified Ceramic Types from Test Unit 2, Mound B, Venadillo.

	Chametla Type										Culiacán Type			Amapa Type			Peñitas Type	
Level	Black-banded Nonengraved	Middle Chametla Polychrome	El Taste Polychrome	El Taste Red-bordered	El Taste Satin (Polished Buff?)	Late Black-banded Engraved	Red-rim Decorated	Red-rimmed Utility Ware	Scalloped Rim	Subtotal	Early Culiacán Polychrome	Navolato Polychrome	Subtotal	Sentispac Buff	Tuxpan Engraved	Subtotal	Polychrome (Bordaz 43–49)	Subtotal
1	3		1		10	1	3	3		21								
2	7		1	1	13	1	1	8	3	35	1	1	2	2	2	4	1	1
3	7	11	1		9	15	13	36	1	93					1	1		
4	2	2			1	1	1	7	1	15								
Subtotal	19	13	3	1	33	18	18	54	5		1	1		2	3		1	
Total										164			2			5		1

(Table 5.25); a pipe stem fragment; lithic debitage; and faunal remains were scattered in the fill of the mound as well.

Excavation in TU 8 in Mound B produced a unique feature. As initially exposed it appeared as a clay- or adobe-lined circular pit or cist within the burial mound. The feature was exposed near the end of the workday, and although it was covered with dirt to conceal it, the feature was vandalized overnight. It was completely cleaned out and mostly destroyed. It may have been a special burial pit. Some human bone fragments and the remains of several vessels were found in the vandal's backdirt, but nothing more can be said of its form or function.

Just to the south of Mound B, Test Pit A was excavated in order to determine whether evidence of a prehistoric living surface was present. Upon the discovery of a hearth and associated food remains, excavations in this area were extended to include five additional 2-×-2-m test units.

The hearth measured approximately 55 cm in diameter and was associated with a scatter of oyster shell. The area appears to have been an oyster cooking and consumption area (Figure 5.40). Associated with this were some of the green-glazed rocks. Also recovered was a well-shaped slab metate (Figure 5.41). Based on the excavations in the Test Pit A area, it appears that 80 to 100 cm of postoccupational fill (water-deposited silts) accumulated at Venadillo subsequent to its prehistoric occupation.

The excavations at Mound C, the westernmost mound excavated at Venadillo, were productive for a variety of reasons (Cumbaa 1973; Foster 1974, 1984; Gill 1984). The mound measured approximately 15 × 25 m, with its long axis oriented northwest-southeast. Four 2-×-2-m test units were excavated in arbitrary 25-cm levels near the center of the mound. Two trenches, one 9.00 m long extending east from the test units and the other 5.75 m long extending west, were also excavated. The excavations revealed five geomorphic and cultural strata within the mound (Figures 5.42–5.45). Strata 1 and 2 are postoccupational layers of soil and silt, with Stratum 1 being a plow zone that averaged about 40 cm thick. Stratum 3 (Levels 1–3, 0–80 cm below ground surface) was a shell midden that capped an earthen burial mound; Stratum 4 (Level 3, 80–90 cm) was thought to represent a noncultural layer marking a hiatus in the occupation of Venadillo (ca. AD 700–900). Stratum 5 (Levels 3–8, 85–190 cm) was a burial mound composed of extinct beach ridge sand.

The shell midden was predominately oyster, but a variety of other mollusks and faunal remains were present. At its deepest, the midden was 80 cm thick and contained an estimated 85 m^3 of material. These materials were mixed with a variety of artifacts, forming an undifferentiated shell cap sitting on what turned out to be a small earthen burial mound. The remains of 13 individuals were recovered from the excavated portion of the mound (see Table 5.24). Both isolated elements and complete burials were recovered. Two urn burials, one with an adult male and adult female and one containing an adult female and an infant, were recovered, along with flexed, semiflexed, and semiseated burials. Among the human burials were several dog burials (Figure 5.46; Foster 1974, 1984; Gill 1984; Schutte and Speas 1973). The fill of the burial mound also contained sherds and lithic debris (Tables 5.26–5.28).

At least two distinct temporal components are represented at the site of Venadillo. Ceramics from Mound B are a mix of Late Estero period and Middle Conchera pe-

FIGURE 5.40. Hearth with green-glazed stones and oyster shell from Test Pit A, Venadillo.

riod types; Late Estero types are dominant. This mixing of earlier and later ceramic types is likely related to the construction of the burial mound. Based on the ceramic types present, predominantly Chametla Baluarte phase types, the earthen portion of Mound C dates to the Late Estero period (ca. AD 450–750). Associated ceramics date the shell midden capping Mound C to the Middle Conchera period (AD 1200–1400).

Two ^{14}C assays were obtained from the shell midden capping Mound C at Venadillo. These are AD 1265–1460 (S.I. 2723B) and AD 1279–1417 (S.I. 2727). S.I. 2723B was derived from a clam, taxa unspecified, and S.I. 2727 was from oyster. As discussed in Chapter 4, the ΔR correction was not applied to these samples, because the results were thought to be inconsistent with what is generally accepted as the likely temporal framework for the phase sequences of the Sinaloa-Nayarit area. If taken at face value, the two Mound C shell midden uncorrected ^{14}C assays support the dating of the Mound C shell midden based on the ceramic types present. The Mound C shell midden is also dated to the Middle Conchera period, making it contemporaneous with Mound B.

El Mirador

El Mirador is located on the southern end of the Teacapán Peninsula (see Figure 5.1). The site was tested during the final season of the Marismas project's archaeological component. The area investigated was in a pasture bounded by palm trees. Fifteen small- to medium-size earthen mounds and shell middens were scattered over an area roughly 300 × 860 m (Figure 5.47). The feature designated Mound D was rather strange in

FIGURE 5.41. Shaped slab metate from Test Pit A, Venadillo.

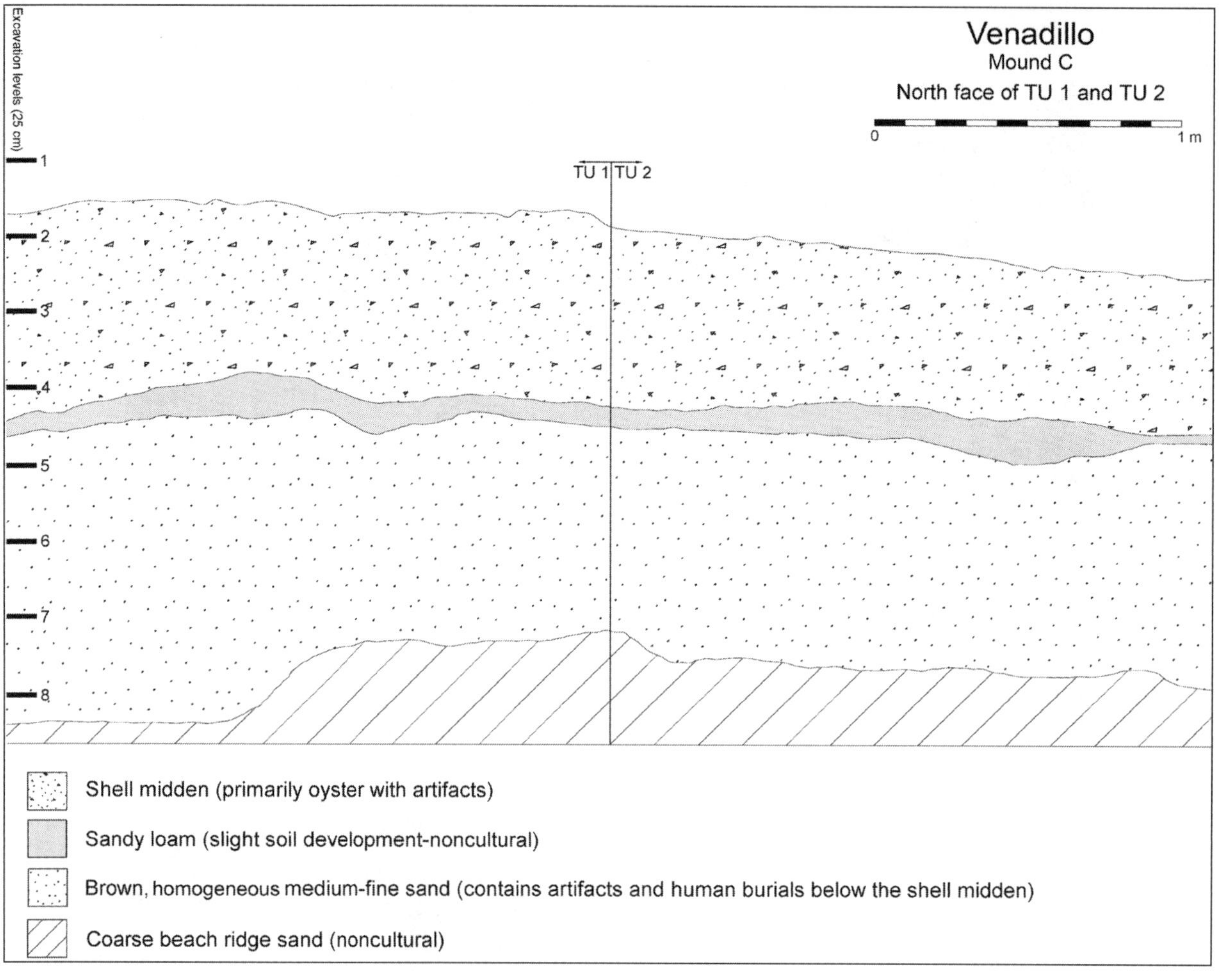

FIGURE 5.42. Profile map of the north faces of Test Units (TUs) 1 and 2, Mound C, Venadillo.

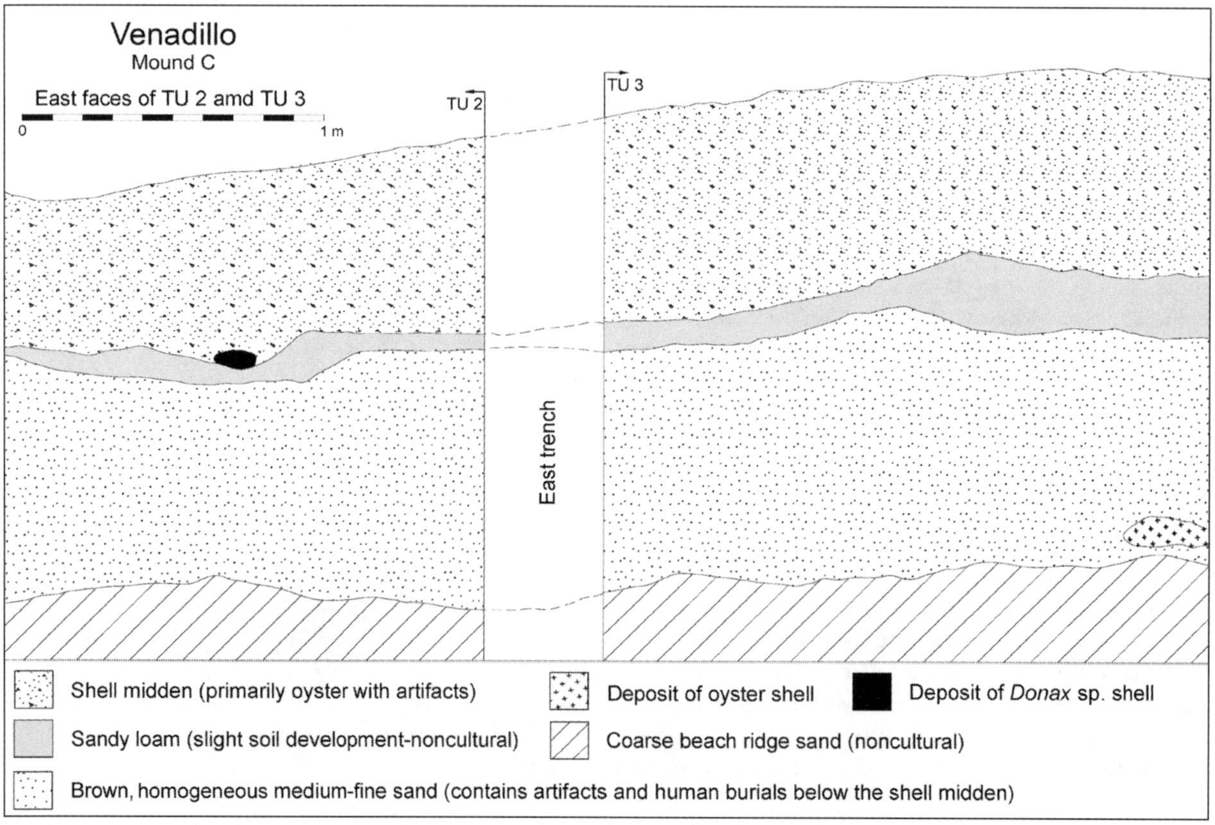

FIGURE 5.43. Profile map of the east faces of Test Units (TUs) 2 and 3, Mound C, Venadillo.

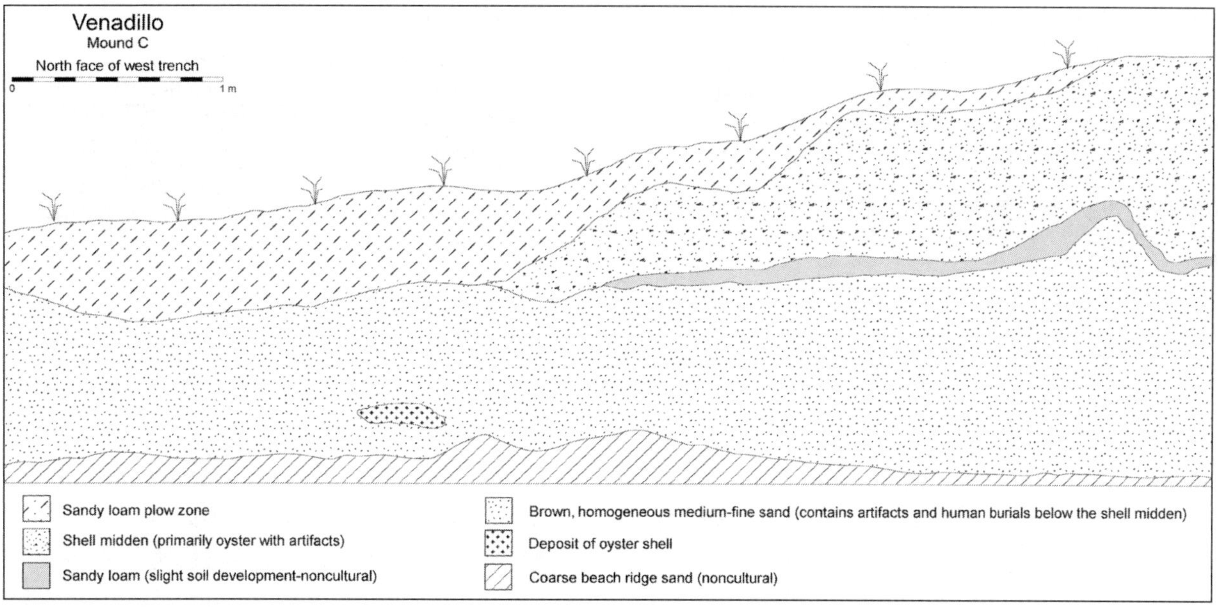

FIGURE 5.44. Profile of the north face of the west trench of Mound C, Venadillo.

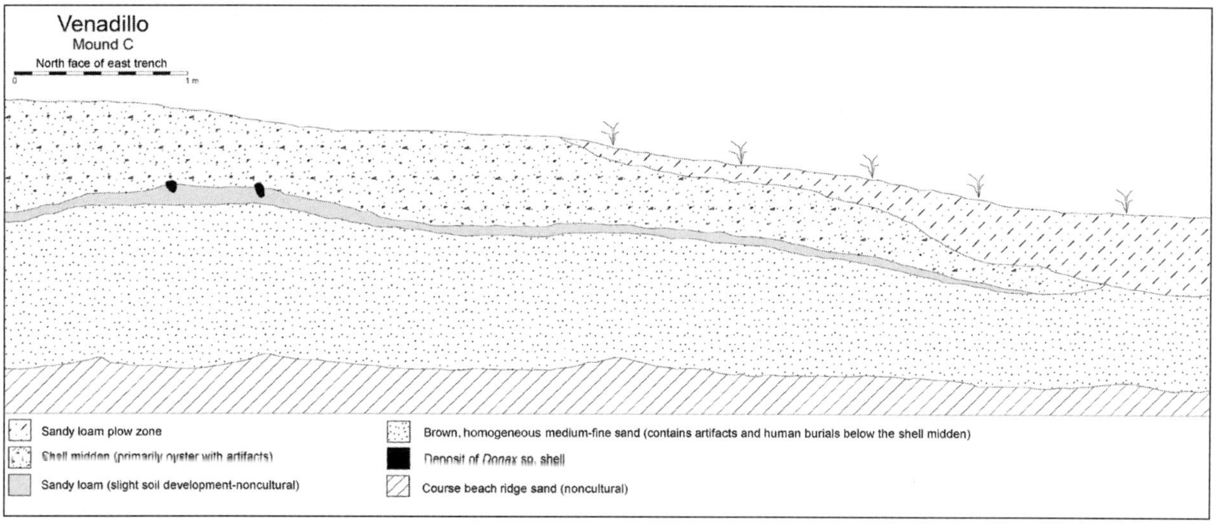

FIGURE 5.45. Profile of the north face of the east trench of Mound C, Venadillo.

form. It was round with a finger extending to the northwest. The northern end is raised, and about a third of it is covered with shell. Two smaller shell heaps are present on the mound. The earthen mound on which the shell is deposited drops to the southeast in a steplike fashion to Mound E. Extending from the southeast edge of the mound was a low platform-like rise that extended to the southeast. It was about 38 m wide and appeared to be artificial. Another shell midden was located on it to the south of Mound E.

A series of 2-×-2-m test units were excavated in five of the mounds. The four test units excavated in Mound A produced a variety of ceramics, some green-glazed rocks, burned bone, obsidian blade fragments and flakes, figure fragments, and some shell. Several beads, a spindle whorl fragment, and several figurine fragments were also recovered. Level 2 produced a partially articulated dog burial. Four human burials were encountered in Level 3. Two were disturbed and incomplete, one was an urn burial, and the fourth was a bundle burial. In all, the remains of nine humans were recovered from Mound A (Table 5.29).

Three test units were excavated in Mound B. TU 1 produced a variety of sherds, burned bone, some shell, several figurine fragments, lithic debris, and chunks of adobe (daub). The presence of these pieces of adobe suggests structural remains. However, no evidence of postholes, foundations, or walls was found. Nevertheless, in Level 4 (75–100 cm) a probable unprepared floor was encountered. The soil was darker, several pieces of a

TABLE 5.26. Frequencies of Identified Ceramic Types from Mound C, Venadillo.

	Decorated Sherd Types from the Earthen Burial Mound														Decorated Sherd Types from the Shell Midden Capping the Earthen Burial Mound											
	Chametla Type														Amapa Type						Peñitas Type					
Level[a]	Middle Chametla Polychrome, Engraved	Chametla Polychrome	Buff Exterior, Red Interior	Late Black-banded Engraved	Late Black-banded Nonengraved	Middle Chametla Polychrome	Polished Buff	Scalloped Rim	Red-rimmed Utility Ware	Red-rimmed Utility Ware, Mano Colorado	Black-on-buff	Early Chametla Polychrome	Early Black-banded Engraved	Subtotal	Mazatlán Polychrome	El Taste Polychrome	El Taste Satin	El Taste Red-bordered	El Taste Polychrome Red-on-cream	Subtotal	Ixcuintla White-on-orange	Ixcuintla Polychrome	Peñitas Engraved	Tuxpan Engraved	Sentispac Buff	Subtotal
Shell Midden																										
Level 1	1	1					23		2	2				29	17	8	5	4		34	7	1	2			10
Level 2			1				34							35	28	19	7	18	3	75	7	8	15			30
Level 3			2				76		7	2		3		90	6	24	15	15	2	62	17	5	10	1		33
Level 4			1				87		10	1				99	3	11	13	15	2	44	1	1	3		1	6
Earthen Burial Mound																										
Level 5	7		5	3	3	1	110	2	54	15	2			202	1	2	12	1	2	18		1	5		8	14
Level 6	5		1	10	12	7	81	1	130	69			2	318	1					1				5	4	9
Level 7	5		4	1	8	1	14		39	70			3	145												
Level 8	13			1			2		2	8			5	31											1	1
Subtotal	31	1	14	15	23	9	427	3	244	167	2	3	10		56	64	52	53	9		32	16	35	6	14	
Total														949						234						103

[a] Arbitrary 25-cm levels.

TABLE 5.27. Frequencies of Utility Wares from Mound C, Venadillo.

	Ware/Type			
Level[a]	El Taste Rough	Red Ware	General Utility Wares	Total
Shell Midden				
Level 1	12	33	331	376
Level 2	11	38	830	879
Level 3	12	33	475	520
Level 4	6	30	542	578
Earthen Burial Mound				
Level 5	6	80	967	1,053
Level 6	3	294	2,092	2,389
Level 7	6	199	1,250	1,455
Level 8	—	28	154	182
Subtotal	56	735	6,641	
Total				7,432

Source: Foster 1974.

[a] Arbitrary 25-cm levels.

Table 5.28. Distribution of Artifacts and Burials from Mound C, Venadillo.

	Shell Midden Level				Earthen Burial Mound Level				
Artifact/Burial	1	2	3	4	5	6	7	8	Total
Obsidian blades	1	2	8	7	18	1			37
Obsidian projectile points				1		7	2		10
Obsidian debitage			1		23	47	35		106
Quartz debitage					3	10	6		19
Granitic debitage							1		1
Pumice polishing stone					1				1
Manos		3	4	2					9
Metates			3	2					5
Pestles						2			2
Figurines			1						1
Figurine fragments		1				5			6
Whistles					1	2			3
Pipe fragments		2	5	1	4	4			16
Spindle whorls	1				2	5	1		9
Possible net sinkers				1					1
Bowl fragments				1					1
Copper					1				1
Infant burials							1		1
Flexed adult burials							2	2	4
Urn burials						1	1		2
Multiple burials							1		1
Dog burials					1	7			8
Total	2	8	22	15	54	91	50	2	244

Figure 5.46. Dog burial from Mound C, Venadillo.

metate were found, and under the metate fragments was a concentration of clamshells and utilitarian plainware sherds. Also present were more adobe chunks, as well as some of the green-glazed rock. This likely represented a kitchen or food preparation area.

TU 2 was opened immediately to the west of TU 1 to explore more fully the kitchen area. Excavations in Level 2 produced a pestle about 13 cm long. Both ends exhibited flake scars. Also, a fragment of a bone rasp was recovered, as were the fragmentary remains of a dog burial. Again, in Level 4 charcoal-stained soil and burned and stained plainware sherds were encountered. Additionally, the poorly preserved remains of a small fire pit or hearth were recorded.

A third test unit was excavated immediately west of TU 2. The ceramics that were recovered were mostly plainwares with some Chametla Polychromes. Quartz and obsidian flakes were also present, along with a possible scraper. A dog skull and bones were also recovered. The kitchen lens present in TUs 1 and 2 did not extend into TU 3. Excavations in all three units were terminated at 1.5 m below the surface when culturally sterile soil was reached.

Mound C had one 2-×-2-m test unit excavated into it. Most notable in terms of recovered remains was a

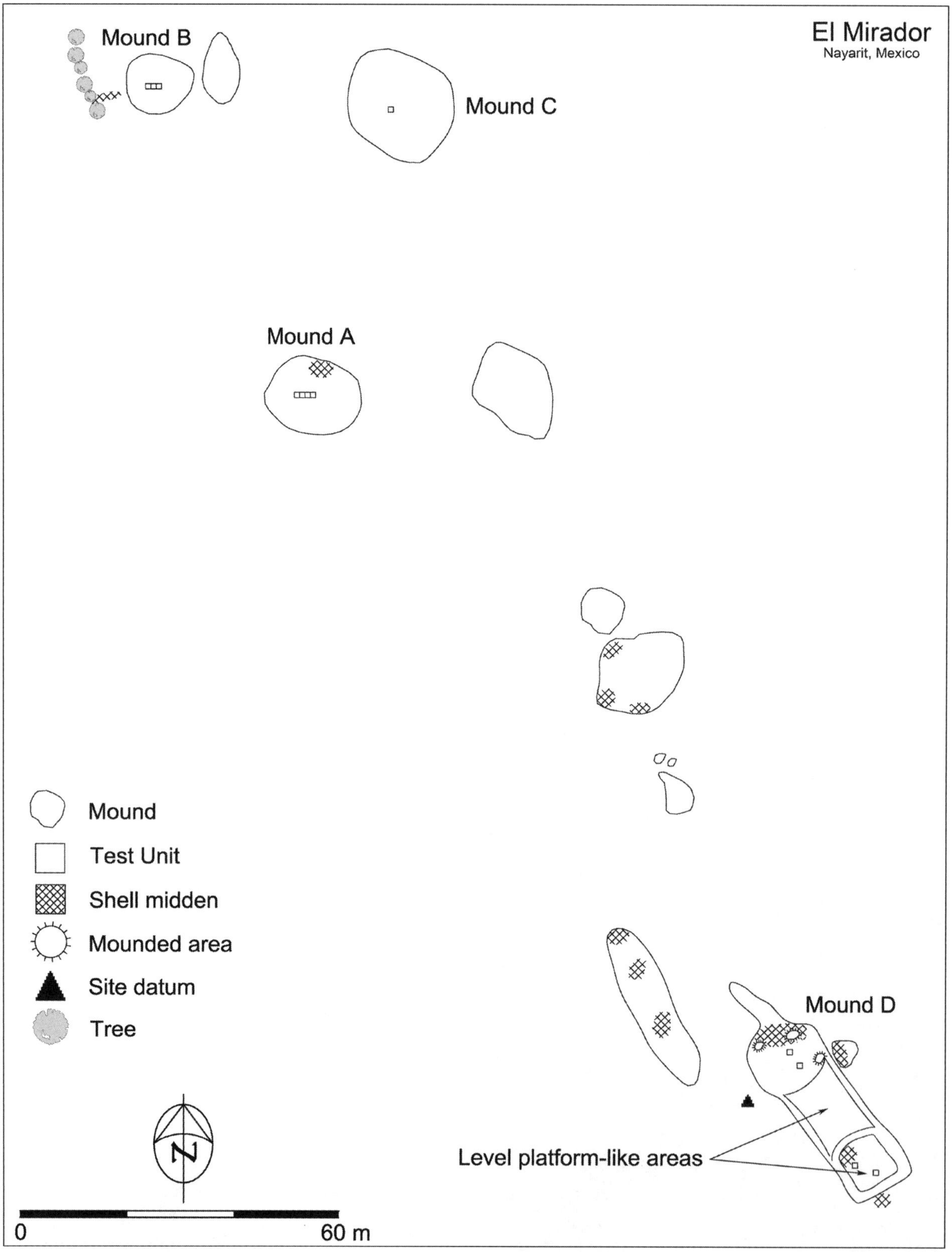

FIGURE 5.47. Site map of El Mirador.

Table 5.29. Summary of Burials from El Mirador.

Burial Number	Condition	Sex	Age	Burial Type	Funerary Offerings
B 1	Fragmentary	Probably female	Child	Urn	None
B 2	Fragmentary	Probably female	Young adult	Urn	None
B 3	Fragmentary	Probably male	Young adult	Undetermined	None
B 4	Fragmentary	Probably male	Young adult	Undetermined	None
B 5A	Fragmentary	Probably female	Young adult	Undetermined	None
B 5B	Fragmentary	Undetermined	Child	Undetermined	None
B 6	Fragmentary	Undetermined	Mid-adult	Flexed, tightly	Possibly a metate fragment
B 7	Fragmentary	Undetermined	Young adult	Flexed	None
B 8	Fragmentary	Undetermined	Young adult	Extended	Possibly 1 piece of obsidian, possibly a cluster of adobe balls

concentration of redware sherds and two well-preserved dog burials, all from Level 2 of TU 1. The bones of one dog were stained dark brown, a characteristic observed on some of the human remains in the region. This is apparently the result of the mineralization of the bone. The other dog was in a semicircular flexed position. Excavations were terminated at 2 m in culturally sterile soil.

Mound D had two 2-×-2-m test units excavated in it. These were excavated on what could be called the first step or terrace below and south of the shell midden at the western end of the mound. TU 1 was excavated at the edge of the shell midden, and TU 2 was excavated at the edge of the second drop-off or step. Excavations in TU 1, Level 2, produced some Mazatlán Polychrome and Chametla Buff and Utility wares. A concentration of adobe in the center of the unit quickly disappeared as excavations continued, as did the layer of shell encountered. Only some small bone fragments, an obsidian flake, a metate fragment with a leg, and a large Sentispac Buff sherd were recovered from Levels 3, 4, and 5. Excavations were terminated at 1.5 m in sterile beach sand.

TU 2 excavations produced sherds, some more of the green-glazed stones, and a poorly preserved dog burial in Level 3. Level 4 produced more adobe chunks, quantities of shell, and burned and blackened sherds. A pipe stem fragment was found midway through Level 5. It was decorated with black and red paint and incised and probably was from a Cocoyolitos-style pipe. Just below the pipe, culturally sterile beach sand was encountered, and the excavations were terminated.

Mound E, a shell midden, had two 2-×-2-m test units excavated at its southern edge, just off the shell midden. A poorly preserved dog burial was exposed in Levels 2 and 3 in TU 1, and a well-preserved dog burial was recovered from lower in Level 3. Level 3 contained less shell and was sandier. The unit was excavated into Level 5, where beach sand was encountered, and the excavation was ended.

TU 2 was placed in the highest area of the earthen portion of the mound in an attempt to determine whether it was natural or cultural in origin. Excavations produced a few sherds and no other cultural material.

Ceramics indicate a Middle Conchera period date for El Mirador. This is further supported by the presence of a probable Cocoyolitos-style pipe stem recovered at the site.

Playas Novillero and Cuautla

During the 1974 season, limited collecting and testing were carried out in the vicinity of Playas Novillero and Cuautla on the coast south of the Teacapán Peninsula (see Figure 5.1). This was the southernmost area investigated during the project. Grab samples of diagnostic sherds were surface-collected from two sites (Site 1 Escolar and Site 2 Escolar) on the southern edge of Playas Novillero (Table 5.30). Shell from these sites had been mined for road construction. Sherds, mainly plainwares, were plentiful, along with some green-glazed rock and obsidian flakes. A local informant indicated that a number of burials had come out of the shell mounds but none from the earth below them. Diagnostic sherds were also collected from three sites (Sites 1, 2, and 4) near Cuautla. Chametla, Amapa, and Peñitas ceramic types were present. The sites consisted of *Tivela* shell middens, and much of the shell had been hauled away for use in road construction.

Cuautla Site 3 was tested in this same area. It consisted of several *Tivela* middens, some of which were a meter or more in height and 8 to 12 m in diameter. One intact mound at Site 3 was chosen for testing. Two 1-×-2-m test units were established approximately 5 m apart. Excavations indicated that the typical pattern of early earthen burial mounds covered by later shell middens also applied to Cuautla. Excavations in TU 1 produced three human burials. Burial 1 was encountered 80 to 95 cm below datum in the middle of the test unit. It was fragmentary, consisting of pieces of mandible, ribs, teeth, cranium, phalanges, and long bone. It was unclear why the remains were so fragmentary. It could have been a primary burial disturbed by subsequent interments,

TABLE 5.30. Frequencies of Identified Ceramic Types from Cuautla and Playas Novillero.

Site, Test Unit, and Level	Chametla Type												Culiacán Type				Amapa Type								Peñitas Type						
	Early Chametla Black-banded Engraved	Late Chametla Black-banded Engraved	El Taste Polychrome	El Taste Red-on-gray	El Taste Red-on-cream	El Taste Red-bordered	El Taste Satin	Mazatlán Polychrome	Red-rim Decorated	Red-rimmed Utility Ware	Scalloped Rim	Subtotal	Aztatlán Heavy Ware	Aztatlán Ware	Navolato Polychrome	Subtotal	Botadero Red-on-white	Ixcuintla Engraved	Ixcuintla Polychrome	Mangos Engraved	Santiago Engraved	Tuxpan Engraved	Tuxpan Red-on-orange	Subtotal	Polychrome (Bordaz 43–49)	Polychrome (Bordaz 50–54)	Red Engraved (Bordaz 1–2?)	Red Incised	Red-on-white	Unidentified[a]	Subtotal
Cuautla																															
Site 1				1		1	1	14				17									1			1				1	3		4
Site 2			1					14				15							1					1					3	2	5
Site 3									2			2										1		1							
Test Unit 1																															
Level 1									1			1								1		1		2							
Level 2								1	4			5	1		1	2		2		1		5		8	1						1
Level 3					1							1										1		1		1					1
Level 4		1										1		1		1		1				3	2	6							
Level 5														1		1	1							1						1	1
Test Unit 2																		1						1							
Level 1																															
Level 2									2			2								3		2		5							
Level 3									2			2								1				1							
Level 4	2											2																			
Level 5	1											2																			
Level 6	1									1		2																			
Site 4								3				3										1		1		2	1				3
Playas Novillero																															
Site 1	2										1	3																			
Site 2										4		4																			
Subtotal	6	1	1	1	1	1	1	32	11	5	1		1	1	1		1	4	1	6	1	14	2		1	3	1	1	6	3	
Total												62				3								29							15

[a] Two sherds like Sentispac Red-rimmed but with stick-polished design.

or it could have been all that remained of a secondary burial. Subsequent osteological analysis of Burial 1 resulted in the identification of a single, unrelated molar, thus Burial 1 was reassigned the designation of Burial 1A, and the molar was assigned the designation Burial 1B. Burial 2 was the fragmentary remains of what may have been a bundle burial. It was located about 1.15 m below datum and not associated with the first. The third, poorly preserved burial, Burial 4, was recovered from TU 1. Burial 4 exhibited C-4 or a variant of C-4 dental modification on its upper four incisors (see Chapter 10).

The excavations in TU 2 of Cuautla Site 3 produced a variety of pottery, an obsidian blade, and some scattered animal bone in the first level. A layer of *Tivela* shells was encountered near the bottom of Level 1 at 20 cm below datum, and it extended into Level 2 to a depth of about 40 cm below datum. Redware, polychrome, and incised sherds were recovered from the shell layer. Shell was still abundant in Level 3. This level yielded a whole plain spindle whorl, some obsidian, and a variety of pottery. In Level 4 the number of artifacts present declined. At 1.01 m below datum a large, well-made obsidian corner-notched projectile point was recovered. Levels 5 and 6 continued to produce sherds, but the number of deco-

Table 5.31. Summary of Burials from Cuautla.

Burial Number	Condition	Sex	Age	Burial Type	Funerary Offerings
B 1A	Fragmentary	Undetermined	Young adult	Undetermined	None
B 1B	Fragmentary	Undetermined	Young adult	Undetermined	Undetermined
B 2	Fragmentary	Undetermined	Young adult	Undetermined	None
B 3A	Incomplete	Female	Mid-adult	Semiflexed	1 stone spindle whorl, 4 dog tooth beads, 2 shell bracelets, 1 cowry shell, 1 red-on-buff bowl
B 3B	Fragmentary	Undetermined	Child	Undetermined	Undetermined
B 4	Incomplete	Male	Young adult	Extended	None

rated wares decreased, giving way to all plainwares. Little else was recovered, and it appeared as though an earlier temporal component had been exposed.

At the bottom of Level 6 and the beginning of Level 7, human bone was encountered. This was one of the more elaborate burials (Burial 3A) recovered during the Marismas excavations. It was the burial of an older female in a semiflexed position. She was on her right side with her right arm bent and her right hand under her head. She appeared to be a rather tall individual. The bone was very poorly preserved. A variety of funerary offerings accompanied the burial. A red-on-buff vessel was near the head, and a white stone, incised spindle whorl was next to the bowl. Two *Glycymeris* (cf. *G. gigantea*) bracelets were present at the middle of the lower left arm. Also present were four drilled dog tooth beads, a large cowry shell, and several large *Oliva* sp. shells. The dog teeth were in the upper chest and neck area, indicating that they had been part of a necklace. The shells were unmodified and were located in the area of the pelvis. Burial 3B was fragmentary, consisting of a partial skull of a child and some associated mandible fragments and teeth. These remains were found in the area of the pelvis. They were gracile and poorly preserved. The remains were initially thought to represent a funerary offering, perhaps a trophy head; however, since the osteological analysis indicates that they are the remnants of a child's head, the idea that B 3B is a funerary offering seems unlikely. In all, the remains of six individuals were recovered from Cuautla Site 3 (Table 5.31).

The diagnostic ceramics recovered from the limited investigations of the Cuautla–Playas Novillero sites are few and represent a mix of types that range from the Late Estero to the mid–Middle Conchera periods.

Summary

The Marismas Nacionales investigations clearly show a substantial precolumbian occupation of the area. The area's abundant biotic resources undoubtedly drew people to this region, and the archaeological evidence indicates that they intensively used the resources available to them. Midden deposits, as well as other features, contain abundant amounts of shellfish remains as well as aquatic and terrestrial animal remains. These same deposits also produced ground-stone implements such as manos and metates that provide indirect evidence of agriculture. Like other precolumbian cultures throughout Mesoamerica and the American Southwest, the ceramic-period Marismas inhabitants subsisted on a diet of corn, beans, squash, and other cultigens supplemented by shellfish, fish, wild game and birds, and wild plant foods. Most of such resources were concentrated on the estuary and adjacent coastal plain, but it is likely that the Sierra Madre foothills were also exploited for food, minerals, and other resources. The variety of subsistence resources and material culture recovered and the presence of burial mounds indicate that sites were occupied year-round and likely for generations. However, it is also clear that some sites were abandoned as portions of the Marismas coastal plain were inundated and that the Marismas coastal plain and some of these sites were reoccupied as sea levels subsided.

Sites were located either adjacent to the estuary or inland on the monte. In either case, they were close to the estuary and to areas suitable for agriculture. Shell middens also occurred along the edge of the estuary and inland as well. The linear shell middens along the edge of the estuary may be mostly the by-product of commercial oyster production during the early colonial period. Other middens, particularly *Tivela* middens, appear to be loci where either household or cooperative labor groups processed quantities of shell. It is likely that most of these represent multiple episodes of shellfish processing; however, some may also represent community-level mass collecting and processing for ceremonial or festival events.

The Classic and Postclassic period sites produced an array of material culture dominated by ceramics. The majority of the ceramics recovered were plain utilitarian wares, and of the decorated ceramics (polished monochrome, bichrome, polychrome, and incised types) most were types first identified at the neighboring archaeological sites of Chametla and Amapa, as well as from other areas of the west Mexican coast. Flaked and ground stone, copper, smoking pipes, spindle whorls, and figurines were among the other artifact types and classes represented. Obsidian was a common material type found in the flaked stone assemblage, and vesicular basalt was a common material type used for the making of ground stone. Several pieces of copper were recovered, and there was a surprising lack of shell jewelry despite the proximity and availability of marine shell in the area.

The most common archaeological features encountered and investigated were shell middens, earthen burial mounds, and human burials interred in the mounds. Some evidence for architectural remains, burned daub for the most part, was encountered. The recovered pieces of adobe are assumed to be parts of wattle-and-daub buildings that housed the occupants of the area, and some of these appear to have been built on the earthen burial mounds once they had ceased receiving burials. The scarcity of architectural remains and the limited extent of excavations at the sites have made any understanding of intrasite settlement organization difficult. The sites themselves appear to have been small villages and hamlets. Although a point of speculation, it appears that populations at the sites may have been limited to a relatively small number, perhaps fewer than 100 people made up of several extended family groups. Additionally, intrasite settlement organization seems to have been lacking. The mounds do not appear to be arranged in any particular fashion, and there was no evidence of houses being arranged around courtyards or plazas. However, this also remains a point of speculation because much of the work focused on the burial mounds. No evidence of ceremonial or communal architecture such as small platform mounds was apparent other than the burial mounds and El Calón.

The occupational history of the ceramic-period sites investigated is summarized in Figure 4.1 (Chapter 4). As noted in our discussion of the Marismas cultural chronology, El Calón, because of the radiocarbon dates, had initially been discussed as a Preclassic period feature using the traditional cultural chronology for central Mexico as a temporal and behavioral reference point (Scott 1985, 1999; Scott and Foster 2000). However, as further discussed, El Calón may be better thought of as a Late Archaic or Early Preclassic period manifestation, which perhaps predates the widespread sedentism and use of ceramics in the general area. The feature may have accumulated or been constructed over an extended period or in multiple episodes. Regardless of its temporal and therefore cultural association, it is clear that the feature became either a focal point of communal ceremonial activity in the area or a sacred place or shrine where people left offerings to engender the favor of their deities.

As with other littoral zones, early sites such as those predating the Early Estero period may lie buried by cycles of mixed fluvial-marine deposits. The history of the alluvial plain was punctuated by periods of submergence, one of which, Medial Beach progradation, accounts for the gap during which no material remains or other signs of human occupation have been found (ca. 4200–1600 BP). Sometime following the accumulation or building of the El Calón shell mound, it became the victim of that long interval of environmental instability, the immediate consequence of which was the rupture of whatever coastal tradition the El Calónists had evolved. For archaeology, it represents a long period of obscurity between what we have termed the Costanero (El Calón) period and the later Estero and Conchera periods.

The Estero period marks the reoccupation (ca. AD 300) of the Marismas coastal plain on the emergent Los Angeles Soil (Connally 1977, 1984; Connally and Sullivan 1984). This reoccupation is typified by the presence of shell middens dominated by the small smooth *Tivela* clam. *Tivela*, unlike oyster, has a strictly marine habitat, which would account for the special subareal distribution of these middens on the seaward San Cayetano Peninsula. These *Tivela* mounds are seen as remnants of the food quest by the earliest Estero period occupants. Elsewhere earthen burial mounds scattered widely through the central estuarine area during the Estero period indicate a substantial population density with a Classic period material culture inventory that included a variety of plainware and decorated pottery, obsidian, molded clay spindle whorls, human figurines, grooved axes, and manos and metates.

What followed was a second occupational break of shorter duration (ca. AD 700–900/1000) that, similar to those noted at Amapa and Chametla, bridged the Late Classic and Early Postclassic periods. As the sea level subsided humans again recolonized the coastal lowlands. As seen at Venadillo and on Panales, evidence of this Postclassic reoccupation sometimes takes the form of oyster middens accumulated on earth mounds that may not, themselves, have been inundated. Elsewhere, Postclassic burials were found in purely earth mound contexts, occasionally above thin buried lenses of oyster shell (Gill 1974:86–87). The cultural sequence of Marismas prehistory spans the Late Archaic or Early Preclassic period into the Postclassic, with an uncertain connection to the Spanish colonial era.

CHAPTER 6

Ceramics

Michael S. Foster

Sherds from ceramic vessels were the most abundant artifacts recovered from excavations at sites investigated during the Marismas Nacionales project. The sherd assemblages were dominated by plainwares, although decorated wares were also generally abundant. In addition to the sherds, a number of complete or nearly complete and reconstructible vessels, mostly from burial contexts, were recovered. The analysis of the ceramic assemblages was limited to macroscopic assessments of surface (exterior and interior) treatment (surface finish/texture: unslipped, slipped, polished, painted, appliqués), paste color, and temper coarseness. The intensity of the analyses of the ceramic assemblages varied from site to site, with the focus being primarily on decorated types and wares (bichromes, polychromes, and vessels with appliqués). The primary focus of the study of decorated ceramics was the development of site-specific ceramic sequences that in turn would be used, through ceramic cross-dating (e.g., Grosscup 1976; Kelley and Winters 1960), to estimate the dates for the various archaeological components investigated. It became clear from the outset of the project that the majority of the decorated types conform to type descriptions for assemblages from neighboring sites and areas including Chametla (Kelly 1938) and Amapa (Grosscup 1976). Additionally, some of the utility wares (plain and monochrome) also resemble or are identical to previously established types.

Utility Wares/Plainwares

The analysis of sherds classified as utility wares or plainwares (including monochromes such as black- and redwares) was uneven, and when it occurred, it was limited to simple counts, although many of the complete or nearly complete vessels were described in detail if they were illustrated. Where tallies are available (Panales Mound B, Test Unit 2, and Venadillo, Mound C), the percentage of utility wares averaged approximately 85 percent of the total recovered ceramic assemblages. In the study of Mound C at Venadillo, I (Foster 1974) sorted utility wares into three groups: El Taste Rough, Red Ware, and a third category (General Utility Wares) that served as a catchall for sherds that did not fall into the previous two groups. Sherds in the El Taste Rough and Red Ware categories conform to the descriptions by Kelly (1938:26, 29), although the Red Ware category did not include any highly polished redwares. Most redwares from Mound C had a matte-like appearance. The General Utility Wares category included sherds of various thickness that were unpolished and made of coarse paste, with surface colors (slipped and unslipped) that ranged from buff to red to gray. These three categories totaled 7,428 sherds, 2,355 (32 percent) of which were recovered from the Middle Conchera period (AD 1100–1250) shell midden and 5,073 of which were recovered from the Late Estero period (AD 500–700) earthen burial mound. El Taste Rough accounted for approximately 2 percent of the nondecorated sherds from the shell midden; Red Ware, for approximately 5 percent; and the General Utility Wares category, for 93 percent. Not surprisingly, El Taste Rough was nearly absent in the nondecorated assemblage from the earthen burial mound. Kelly (1938) notes its association with El Taste phase deposits at Chametla, thus its absence in the earlier Late Estero period is expected. The Red Ware category, at approximately 12 percent, makes up a larger percentage of the nondecorated earthen burial mound assemblage, suggesting that perhaps redwares were more commonly made in earlier times, a trend noted at other Marismas sites (Sweetman 1968, 1970) and at the Chametla sites (Kelly 1938). General Utility Wares account for 88 percent of the assemblage. Two nearly complete plainware vessels, a small jar and a bowl, were recovered in association with a dog burial from Mound C at Venadillo (Level 6, earthen burial mound).

The utility ware and plainware type descriptions provided below are based in part on my work (Foster 1974), that of Sweetman (1968, 1970), and that of Weil (1974). These are followed by a brief, illustrated summary based on the utility ware and plainware whole vessel assemblage, illustrated with examples of those wares.

Utility Ware/Plainware Ceramic Types

El Taste Rough (Kelly 1938:26–27, Plate 10n)

Kelly states that El Taste Rough is the coarsest and crudest ware in the Chametla ceramic series. It has an "excessively" rough paste that does not appear well mixed, and the temper includes large bits of rock and small pebbles. Vessel walls can be up to 1 cm thick, and the surface is rough and is sometimes coarsely rasped. Vessel forms typically included hemispherical bowls and deep bowls with incurved walls. Some have necks, and rarely, bowls are lipped.

El Taste Rough was identified in low frequencies at Rincon de Panal, Chalpa, and Tecualilla. This is likely due in part to the fact that the plainware assemblages at these sites were not intensively analyzed. Sweetman (1968, 1970:39) describes an assemblage of rough wares, equating them with El Taste Rough. She notes that it is a highly variable group and that it ranged from 8 to 15 mm thick. El Taste Rough was somewhat more prevalent at Mound C at Venadillo, where the plainware assemblage was fully tallied (Foster 1974, 1984); nevertheless, it made up less than 1 percent of the total plainware assemblage, which included unspecified redwares and other general utility wares.

Blackware

Sweetman (1968, 1970:23) briefly describes a blackware recovered in relative abundance from Juana Gomez. She appears to equate the Marismas blackware to Kelly's (1938:27–29) blackware. Sweetman reports that the temper ranged from very fine to fine and that the paste color ranged from dark gray to brownish black to black. No cores or slips were noted, and the surfaces were polished. Although not specified by Sweetman, it is assumed that the black finish was achieved by sooting a vessel before polishing it (see Kelly 1938:27). Sherd thickness ranged from 4.5 to 7.5 mm. At Chametla, blackware was most prevalent in the Tierra del Padre and Baluarte phases (Early and Late Estero periods in the Marismas) and declined through the Lolandis and Acaponeta phases. It appears that a similar pattern is also repeated in the Marismas.

Redware

Sweetman (1968, 1970:33) also describes a redware recovered in relative abundance at Juana Gomez. She, like me (Foster 1974), indicates that these fall within the class of redware ceramics described by Kelly (1938:29) at Chametla. Kelly notes that this is a variable group that ranges from coarse, heavy utilitarian vessels (jars and ollas) to well-polished redware jars. The temper of the sample studied by Sweetman ranged from very fine to fine to medium, and the temper sometimes included coarse red inclusions. Although generally buff, the color of the paste was highly variable, and within a given color there was great variation. Jars and ollas were red slipped on the exterior, while bowls were slipped on both surfaces. Vessel thickness ranged from 5.0 to 18.5 mm. Sweetman (1970:33) defined three varieties of redware:

a. Coarse variety. Based on her discussion, this category is apparently made up primarily of various-sized ollas and *tecomates*. The exterior was red slipped, and the slip extends onto the interior rim of the vessel. The interior is either smoothed or left rough. On vessels with constricted orifices, the area just below the interior rim or neck was smoothed; however, because the interior of the vessel was difficult to reach, it was left rough. Occasionally interiors exhibited a thin cream- or orange-colored wash. Vessel thickness ranges from 8.0 to 18.5 mm.
b. Fine variety. This variety is defined by a typically brighter red slip on the exterior or interior or both. Vessel thickness ranged from 5 to 9 mm. Sweetman states that some vessels have "ridged" decoration. It is unclear, but it is assumed here that she is indicating the presence of some modeling or appliqués. She also states that there are sherds that are intermediate between the coarse variety and fine variety.
c. Fine variety with polished red exterior. As indicated, this category is defined by a well-polished exterior. The red slip also extends onto the interior of the vessel rim. Below the red-slipped area, the interior of the vessel is unslipped but polished. Interior surface colors range from buff to cream. Vessel thickness ranges from 8 to 12 mm. This variety was common in the earlier levels at Juana Gomez.

Sweetman also describes what she classifies as Roblitos Red, a type with the exterior covered with a "thick, glossy coat of low value red paint or slip" (1970:39; see 1968). She further states that, apparently based on a color difference, this variety is unlike Santiago Red (Grosscup 1976:234–235). Like Chametla Blackware, redware was most common in the Tierra del Padre and Baluarte phases (Early and Late Estero periods in the Marismas) and declined through the Lolandis and Acaponeta phases (Early Conchera period in the Marismas). Again, this also appears to be the case in the Marismas.

Unnamed Polished Orangeware (Sweetman 1968, 1970:34)

Based on 32 sherds, Sweetman describes a plain, unslipped orangeware that has very fine temper, sometimes with coarse red inclusions. The paste is orange in color, and all sherds have a light gray core, except one that is black. No sooting was noted. Sherd thickness ranged from 5 to 9 mm. Sweetman, in the discussion of this type, mentions seven other orange-colored sherds; however, these had an orange-gray-colored paste and a polished orange slip.

The Weil Plainware Typology

A more detailed analysis and classification of Marismas plainwares was completed by Weil (1974) as part of his survey of sites in the area (Table 6.1). The development of a plainware typology was a stated goal of his work, which also focused on an attempt to determine whether the distribution of plainware types had any significant temporal associations. Despite his effort, Weil (1974:55–56) was unable to identify any temporally diagnostic types definitively. Nevertheless, some trends were noted. Of the 19 diagnostic types, the most common appear through the entire occupational sequence. Type IX and Type XIV were most common early in the sequence when they co-occurred with Early Chametla Polychrome (Early Estero period). Type V and Type XII appeared to be most common in the middle of the sequence (i.e., the Early Conchera/Middle Conchera period), while Type I, Type II, and Type IV were concentrated in the latter portion of the sequence (i.e., the Middle/Late Conchera period). Despite the tentativeness of his results, Weil (1974:56) suggested that there was likely temporal variability in the surface treatment of the nondecorated pottery.

Unfortunately, Weil's typology was developed late in the Marismas project and was not subsequently tested or otherwise employed. Nevertheless, it was a laudable effort and is certainly a starting point for further study of plainware assemblages from the region. Despite the intervening 40 years since his work, no such effort has been made.

Summary of the Utility Wares/Plainwares

The utility wares/plainwares recovered from the Marismas, like those from coastal northwest Mexico as a whole, exhibit a great deal of variability in terms of surface and paste color, technology, finish, and form (Figures 6.1–6.10), and as a whole, this class of ceramic wares is poorly documented and little studied. The utility ware/plainware ceramics recovered ranged from very crudely made, asymmetrical vessels to well-formed and well-finished (polished and slipped) vessels. Forms include bowls, tecomates, ollas, necked jars, composite silhouette jars, and tripods. Utility wares were found in both nonmortuary and mortuary contexts. None of those found in mortuary contexts appears to have been made especially as a mortuary offering. Some appear to have had a relatively long use-life, as reflected by unbroken yet worn rims on ollas and necked jars. The presence of carbonized exteriors indicates that utility wares were used for cooking, while the lack of carbonized exteriors indicates that the vessels functioned as storage containers or as serving vessels. Although, as noted above, Weil (1974) suggested some temporal correlation in the variability seen in the surface finish of undecorated pottery, undoubtedly further study would identify additional temporal, technological, and spatial variability in the plainwares.

Decorated Types

As noted above, the majority of the decorated ceramics recovered from the Marismas appear to correspond to previously established pottery types from neighboring sites in the region, particularly those from the Chametla, Amapa, Culiacán, and Peñitas excavations (e.g., Bordaz 1964; Grosscup 1976; Kelly 1938, 1945). Thus, most decorated ceramic sherds and vessels were sorted and classified based on previously established ceramic typologies. A number of the ceramic vessels recovered are related stylistically to regional types but exhibited sufficient decorative variance to prevent classification as a previously established type. Other decorated vessels and sherds could not be typed, and they likely represent either local idiosyncratic types, previously undefined local or regional types, unidentified variants of established types, or imports from unknown localities.

The initial studies of Marismas ceramic assemblages were those of Sweetman (1968, 1970, 1974), Davis (1969), and Foy (1970). These studies focused on the classification of decorated ceramics in an effort to identify temporally diagnostic types that could be used to date archaeological components at the sites being investigated.

Sweetman's (1968, 1970) work, although limited, represents the most comprehensive analysis of ceramics from the Marismas project. Assemblages studied included surface and excavated materials from sites in the Sierra Madre (not discussed in this report), Juana Gomez, and Panales Test Mound A. Her analysis consisted of a macroscopic examination of temper (size—very, very fine through very coarse), notation of surface treatment (unslipped, slipped), identification of vessel/sherd types, sherd descriptions, and assignment to previously established regional types. Additionally, she completed profile drawings and descriptions of whole or nearly whole vessels recovered from excavations and from a private collection of archaeological material. Sweetman (1968:63, 71–72, 1970:36–38) also established two tentative sherd types that may or may not be of local (Marismas) origin, Cixto Incised and Roblitos Polychrome.

Davis (1969) undertook the analysis of ceramics recovered during the 1969 investigations at the sites of La Tarjea and Panales Test Mound B. A smaller sample from the site of Tecualilla was also analyzed. La Tarjea yielded six whole vessels and several thousand sherds, most of which were plainwares. Only four decorated sherds were identified in the site's analyzed ceramic

Table 6.1. Suggested Typology for Plainware Sherd Types from the Marismas Nacionales.

Type Designation	Description	Variant	Number of Specimens	Wall Thickness (mm)	Comments
I	Unslipped, the paste color ranges from yellow-brown to red-brown; most are yellow-brown. The tempers vary from fine to medium. The interior surface is smooth to medium, and the exterior surface is always smooth.		31	4–11	
		1. Yellow-brown paste	27	4–11	
		2. Dark brown paste	1	9	
		3. Red-brown paste	1	12	
		4. Yellow-brown paste	2	9 and 12	These two rim sherds have painted lips (red-brown); the paint extends slightly below the exterior rim.
II	Unslipped, with the paste colors being the same as Type I. The temper varies from very fine to medium. The interior surfaces are rough to medium, with the exterior surface medium.		88	4–16	
		1. Yellow-brown paste	76	4–16	
		2. Dark brown paste	3	6.0–7.5	
		3. Red-brown paste	3	8.5–11.0	
		4. Yellow-brown paste	6	8–16	Red-brown painted lip with the painting extending slightly below the exterior rim.
III	Unslipped, with the paste color varying within the yellow-brown range.		9	6–15	
IV	The exterior surfaces are slipped, and the interior surface is not slipped. The paste varies from yellow-brown to dark brown to red-brown in color. The temper varies from fine to medium. The interior surfaces are smooth to rough, with the exterior surface always smooth. Sherds with the yellow-brown slip are the most common.		260	3–14	
		1. Yellow-brown slip	171	3–13	
		2. Dark brown slip	21	5–10	
		3. Red-brown slip	57	3–12	
		4. Light gray to pinkish gray slip	6	3–10	
		5. Yellow-brown slip		4–14	This category includes five rim sherds with red-brown painted lips; the painting extends slightly onto both the interior and exterior rims.
V	The exterior surfaces are slipped, and the interior surfaces are unslipped. Paste colors vary within yellow-brown, dark brown, and red-brown ranges. Temper varies from fine to medium. Both the interior and exterior surface finishes vary from medium to rough.		38	4–14	

		1. Yellow-brown slip; medium exterior surface	30	4–14	
		2. Dark brown slip; medium exterior surface	1	7	
		3. Red-brown slip; medium exterior surface	5	5.0–9.5	
		4. Light gray to pinkish gray slip; medium exterior surface	1	9	
		5. Yellow-brown slip; rough exterior surface	1	7	
VI	The exterior surfaces are slipped, and the interior surfaces are unslipped. The exterior surfaces are polished, and the interior surfaces are smooth. Temper varies from fine to medium.		3	8–14	Weil (1974:29) states that this Type IV is considered a separate type because of its polished surface.
VII	The exterior surfaces are slipped, and eight interior surfaces are unslipped. Paste colors vary within the yellow-brown color range. Both the interior and exterior surfaces are always smooth.		9	5.0–9.5	One of the nine sherds is a rim sherd with a red-brown painted lip with the painting extending onto the interior of the lip.
VIII	This type is the same as Type VI but with a medium to rough exterior surface finish.		2	5–8	
IX	Both the interior and exterior surfaces are slipped with the same color. The colors of the paste vary within the yellow-brown, dark brown, and red-brown ranges. Tempers vary from fine to coarse. The interior surface finishes range from medium to smooth.		340	3–14	
		1. Yellow-brown slip	258	3–14	This variety makes up the largest percentage (20%) of this sherd assemblage analyzed by Weil.
		2. Dark brown slip	31	4–11	
		3. Red-brown slip	12	5–10	
		4. Light gray to pinkish gray slip	12	5–10	
		5. Yellow-brown slip	4	6–11	These four rim sherds have a yellow-brown painted lip with the paint extending slightly below the lip onto the interior and/or exterior of the rim.
		6. Yellow-brown slip	3	5–7	These three rim sherds have a red-brown painted lip with the paint extending slightly below the lip onto the interior and exterior of the rim.
		7. Yellow-brown slip	7	5–10	These seven rim sherds have a red-brown painted lip with the paint extending slightly below the lip onto the interior and/or exterior of the rim.
		8. Dark brown slip	2	4.5–5	Two rim sherds with red-brown painted lips.

TABLE 6.1. (cont'd.) Suggested Typology for Plainware Sherd Types from the Marismas Nacionales.

Type Designation	Description	Variant	Number of Specimens	Wall Thickness (mm)	Comments
X	Both the interior and exterior surfaces are slipped in the same color. Paste colors vary within the yellow-brown, dark brown, and red-brown ranges. Temper varies from fine to medium. Both the interior and exterior surface finishes range from medium to rough.		63	3–14	
		1. Yellow-brown slip; medium exterior surface finish	57	3–14	
		2. Dark brown slip; medium exterior surface finish	1	9	
		3. Red-brown slip; medium exterior surface finish	3	7–9	
		4. Light gray to pinkish gray slip	1	9	
		5. Yellow-brown slip	1	9	
XI	Both the interior and exterior surfaces are slipped in the same color. The colors of the pastes vary with the color of the slip, possessing lower chroma values in each occurrence. Both the interior and exterior surfaces are polished.		8	4–7	
		1. Yellow-brown slip	4	4–7	
		2. Light gray to pinkish gray slip	1	7	
		3. Dark brown slip (7.5YR4.5/0)	1	5	Possibly burnt.
		4. Red-brown slip	1	4.5	
XII	Both the interior and exterior surfaces have a primary slip. The color of the paste and the color of the primary slip fall within the yellow-brown range. The color of the secondary slip that is present on both the exterior and interior surfaces defines the varieties. Temper varies from medium to fine. Both the interior and exterior surface finishes are smooth.		22	4–11	
		1. Yellow-brown secondary slip	11	4–11	
		2. Dark brown secondary slip	1	7	
		3. Red-brown secondary slip	7	5–7	
		4. Yellow-brown secondary slip	2	5.0, 7.5	These two rim sherds have a red-brown painted lip with the paint extending slightly onto the exterior of the rim.
		5. Yellow-brown slip; both interior and exterior surfaces polished	1		Weil states that if there had been more than one of these specimens of this variety, it would have been designated as a separate type.

XIII	Both the interior and exterior surfaces have a primary slip. The color of the paste varies within the yellow-brown color range; the primary slip varies within the light gray–pinkish gray range. The color of the secondary slip that is present on both the exterior and interior surfaces defines the varieties.		4	4–8	
		1. Yellow-brown secondary slip	2	4–8	
		2. Red-brown secondary slip	1	4.5	
XIV	Both the interior and exterior surfaces have a primary slip. The color of the paste and the color of the primary slip fall within the yellow-brown range. The color of the secondary slip, which is found only on the exterior surface, defines the varieties. Temper varies from fine to medium. The interior surface finish ranges from medium to smooth, and the exterior surface finish is always smooth.		115	3–13	
		1. Yellow-brown secondary slip	69	3–11	
		2. Dark brown secondary slip	23	3–8	
		3. Red-brown secondary slip	23	3–8	
		4. Yellow-brown secondary slip	4	5–11	These four rim sherds have a yellow-brown painted lip with the paint extending slightly below the lip onto the interior and/or exterior of the rim.
XV	Both the interior and exterior surfaces have a primary slip. The color of the paste and the color of the primary slip fall within the dark brown range. The color of the secondary slip, which is found only on the exterior surface, defines the varieties. Temper varies from fine to medium. The interior surface finish ranges from medium to smooth, and the exterior surface finish is always smooth.		9	4–9.5	
		1. Red-brown secondary slip	7	4–8	
		2. Yellow-brown secondary slip	2	8.0, 9.5	These two rim sherds have a yellow-brown painted lip with the paint extending slightly below the lip onto the interior rim.
XVI	Both the interior and exterior surfaces have a primary slip. The color of the paste and the color of the primary slip fall within the red-brown range. The color of the secondary slip, which is found only on the exterior surface, defines this type. Temper varies from fine to medium. The interior surface finish ranges from medium to smooth, and the exterior surface finish is always smooth.		10	4–9	

TABLE 6.1. (cont'd.) Suggested Typology for Plainware Sherd Types from the Marismas Nacionales.

Type Designation	Description	Variant	Number of Specimens	Wall Thickness (mm)	Comments
XVII	Both the exterior and interior surfaces have a primary slip, the colors of which fall within the yellow-brown, dark brown, and red-brown ranges. With minor exceptions (Variety 4), the secondary slip is only on the interior surface. It falls within the red-brown color range. The temper varies from fine to medium. Both the interior and exterior surface finishes are always smooth.		9	5–10	
		1. Yellow-brown primary slip	5	5–10	
		2. Dark brown primary slip	1	10.5	
		3. Red-brown primary slip	1	5	
		4. Red-brown primary slip; yellow-brown secondary slip	2	4.5, 6.	
XVIII	This type has a primary slip on both the interior and exterior surfaces and a different-color secondary slip on either surface. Temper is fine. Both the interior and exterior surfaces are smooth.		2	6.0, 6.5	Both sherds have a yellow-brown primary slip. One sherd has a yellow-brown interior secondary slip and a dark brown exterior secondary slip. The second sherd has a red-brown interior secondary slip and a yellow-brown exterior secondary slip.
XIX	The primary slip color on the interior surface varies from the primary slip color on the exterior surface. Temper is fine. Both the interior and exterior surfaces are smooth.		4	4.5–8.0	Two of the sherds have a yellow-brown interior primary slip and a red-brown exterior primary slip. One of the sherds has a light gray–pinkish gray interior primary slip and a yellow-brown exterior primary slip. One sherd has a yellow-brown primary slip on both surfaces, but the colors differ from one another.

Source: Weil 1974:27–35.

Figure 6.1. First example of utility ware/plainware ceramic vessels from the Marismas Nacionales.

FIGURE 6.2. Second example of utility ware/plainware ceramic vessels from the Marismas Nacionales.

FIGURE 6.3. Third example of utility ware/plainware ceramic vessels from the Marismas Nacionales.

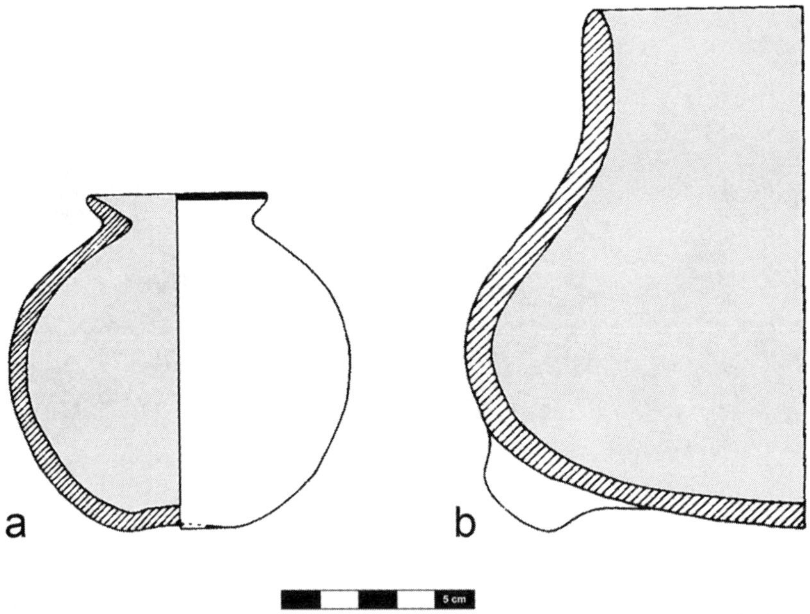

FIGURE 6.4. Utility ware/plainware vessel profiles from La Tarjea. (*a*) Small ovoid jar with low, everted rim and narrow mouth. The rim, exterior neck, and narrow band below it are painted red. Vertical red stripes, pendant from a lower red band, meet at the bottom of the vessel. Type: possibly related to Chametla Red-rim Decorated. Association: Burial 6. (*b*) Small globular jar with very low, everted rim. It is an unslipped buffware with a red painted rim.

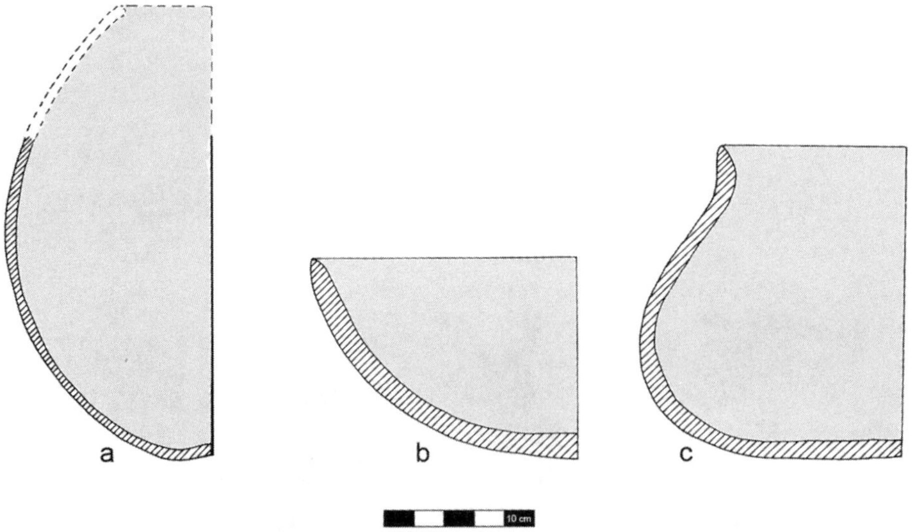

FIGURE 6.5. Utility ware/plainware vessel profiles from Chalpa. (*a*) Partial unslipped, plainware olla. Association: Burial 2. (*b*) Small open plain buffware bowl, unslipped and crudely executed. Association: Burial 18. (*c*) Well-made, slipped plainware globular bowl with concave base and short straight neck. Association: Burial 20.

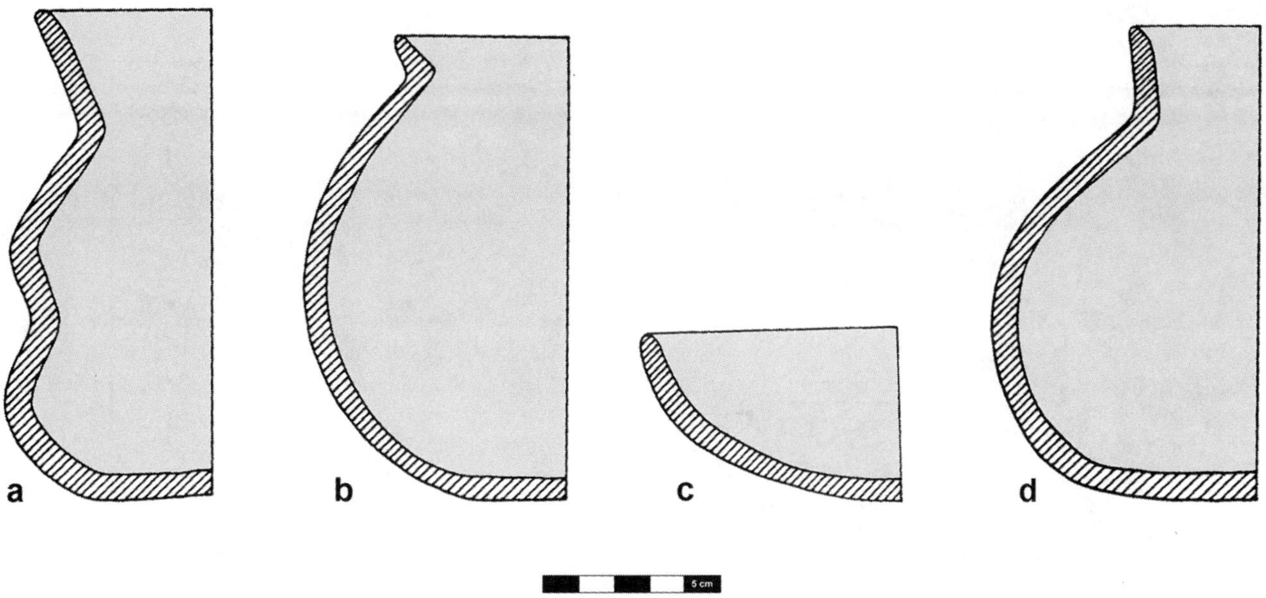

FIGURE 6.6. Utility ware/plainware vessel profiles from Tecualilla. (*a*) Botellon-shaped jar (imitation of double gourd), slipped brown, slightly polished. This form occurs in Tuxpan Brown at Amapa, but Amapa vessels are polished and always have tripod legs or ring bases. Association: Burial 2, Feature A. (*b*) Hemispherical buffware jar with a wide mouth and short rim, unslipped and unpolished. The surface is uneven and not well smoothed. Association: Burial 2, Feature A. (*c*) Small open plainware bowl. Association: Burial 22, Feature A. (*d*) Globular plainware jar, roughly finished and unslipped. Association: Burial 17, Feature A.

assemblage. Ceramics from La Tarjea included redwares, blackwares, Early Black-banded Engraved, Late Black-banded Engraved, and Black-banded Nonengraved. Davis indicates that these all appeared to fall within type descriptions for Chametla (Kelly 1938). Davis also observed that the number of blackware sherds increased in the lower levels of the mound, as did the number of slipped and polished sherds.

The ceramic assemblage from Panales Mound B consisted of several thousand sherds and three whole vessels. Davis (1969:91–94) focused on the decorated material, and the only sorting that was done was the separation of the decorated wares from the undecorated wares; the decorated wares, when possible, were typed. The types present in the earthen portion of the mound (Test Unit/Cut 1) were Chametla types; no Amapa types or other types from areas to the south were identified. Ceramic types identified in Test Unit/Cut 2 included both Chametla and Amapa types, indicating a Late Estero period and Early into the Middle Conchera period age for the mound. Davis's (1969:94–95) analysis of sherd material from Tecualilla was limited to a relatively small number of decorated sherds. Most were Chametla types, with some Amapa types present as well.

The analysis of the ceramic assemblages recovered during the 1970 field season at Tecualilla and Arriñitas fell to Foy (1970), who methodologically followed Sweetman's study as well. In addition to his study of utility ware/plainwares, Weil (1974) also described the decorated ceramic types he recovered during his survey. For the most part, he reports that these conformed to the original descriptions of Chametla and Amapa types, although he comments on any variation observed.

The late Dr. Betty Bell served as consulting project ceramicist and was responsible for the final identification and tallying of decorated ceramics recovered from the sites investigated. The results of her work by site are presented in the decorated ceramics tables presented in Chapter 5. The ceramics from one site, Cristo Rey, were not assessed. At the time, Dr. Bell was one of the foremost experts on prehistoric ceramics from the northwest coast of Mexico. Her (1960) dissertation topic was a stylistic analysis of ceramic vessels recovered from mortuary contexts at Amapa, and she (1958, 1971) aided in the analysis of ceramics recovered during the Peñitas project (Bordaz 1964).

The following provides summaries and some illustrations of the decorated ceramic types recovered during the Marismas project. For each a summary of the original type description is presented followed by a summary of the expression of that type in the Marismas. The reader is directed to the original descriptions for detailed discussion of the types (Bell 1960; Bordaz 1964; Grosscup 1964, 1976; Kelly 1938, 1945). It should be

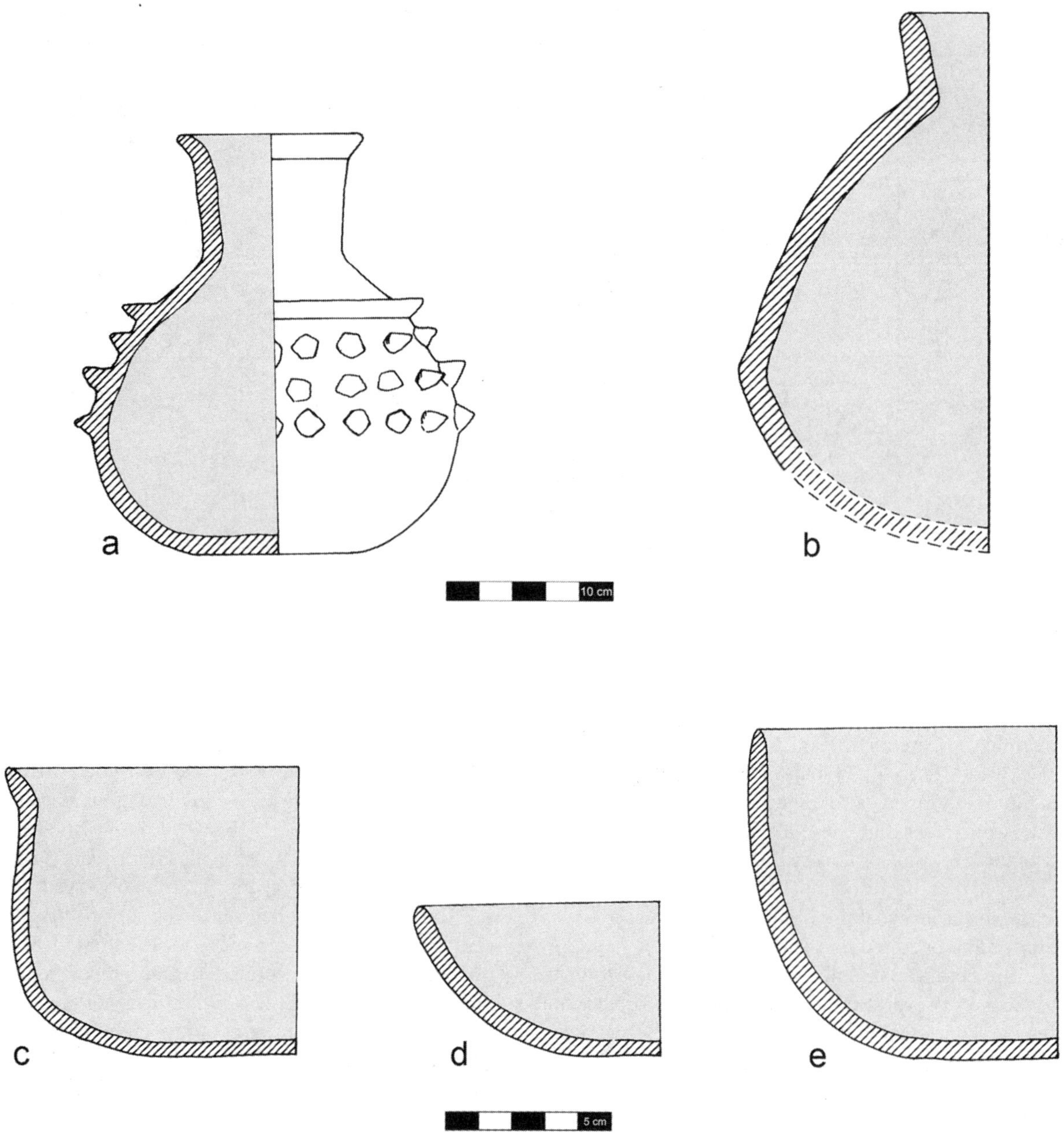

FIGURE 6.7. Utility ware/plainware vessel profiles from Cristo Rey. (*a*) Large, heavy, gray-slipped globular jar with incised linear decoration on flattened shoulder rim and three encircling rows of hobnails. Association: Burial 23. (*b*) Narrow-mouthed unslipped jar with short flaring neck. Association: Burial 4. (*c*) Deep, straight-sided, wide-mouth orangeware bowl with a slightly flaring rim and orange-slipped and unpolished interior and exterior surfaces. Association: Burial 27. (*d*) Small, unslipped buff plainware bowl, somewhat crudely made. Association: Burial 22. (*e*) Large unslipped plainware bowl. Association: Burial 30.

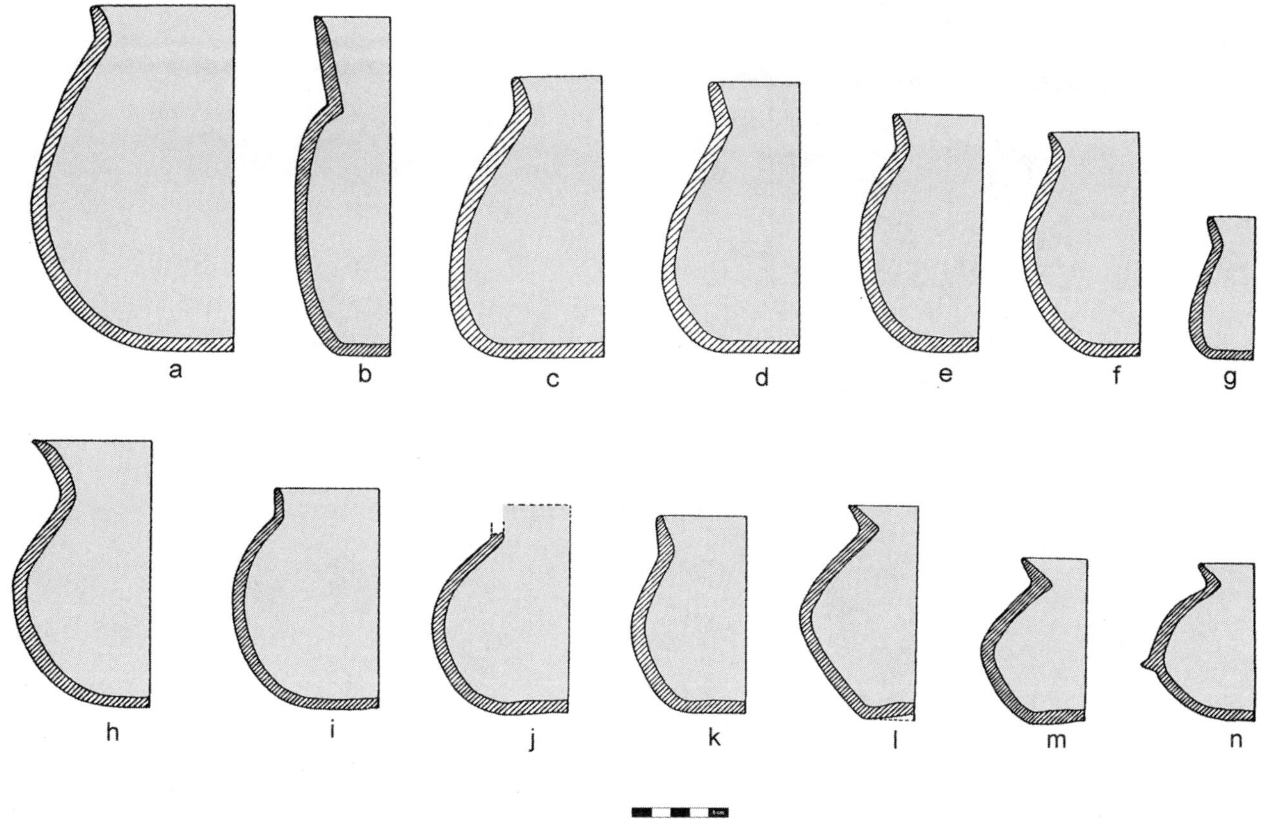

FIGURE 6.8. Examples of utility ware/plainware vessel profiles from Venadillo. (*a*) Jar with a short, slightly flaring rim and reddish orange slip; the exterior surface is well smoothed and covered with root marks (see Figure 6.2, center right). Association: Burial 26B. (*b*) High-necked jar, with a slipped and polished buff exterior that extends over into the interior of the neck. Probably Chametla Polished Buff (Middle Chametla), although Kelly does not describe this jar form. Not quite fine enough for Sentispac Buff. Association: Burial 17A. (*c*) Flattened jar, almost canteen shaped, with a short, straight neck. The surface, which is uneven, is weathered, but it appears to be covered in unpolished buff slip. The bottom is broken out. Association: Burial 6. (*d*) Orange-slipped jar with a flaring neck, fire clouded. It appears rather lumpy, and it is not well smoothed. Association: Burial 2. (*e*) Globular jar with a low straight neck, orange-slipped and unpolished; the surface is uneven and partially covered with a black mineral deposit. Association: Burial 26A. (*f*) Jar that appears to have been orange-buff slipped, but the surface is now much weathered. The surface was smoothed. Association: Burial 25. (*g*) Small orange plainware jar with a low straight neck. It was crudely made and has a rough surface. Association: Dog burial, Late Estero period, earthen portion of Mound C. (*h*) Globular jar with a widely flaring rim, with buff slip outside and inside with red painted rim, once polished but now somewhat worn. Association: Burial 9. (*i*) Buff-colored jar, low, with a slight shoulder and wide mouth and a low, straight rim. It is slipped and highly polished with a mirrorlike finish. Association: Burial 19A. (*j*) Globular jar with an orange slip, smoothed but unpolished. The neck is missing. It may have been made in two sections and luted together. (*k*) Jar with neck slightly outcurved (not flared) and orange slip, unpolished but in good condition. The surface is quite well smoothed. Association: Burial 3. (*l*) Small jar with a low flaring rim. The upper body is somewhat globular, but the lower body is almost conical. It has a dark orange slip on the exterior that extends over the outer edge of the rim. It is unpolished but well smoothed. Association: Burial 32A. (*m*) Small globular jar with a low flaring rim (again possibly manufactured in two parts). It appears to have had a cream-colored slip, the only remnants of which remain around the interior and exterior of the rim. (*n*) Small jar with a low rim. It has a flattened globular shape but with projecting flanges instead of shoulders (see Figure 6.10). The surface is uneven and not well smoothed. It appears to have thick white slip, though some small portions are orange. This appears to be staining or partial painting, for there is no indication of weathering of an orange slip or paint over the white. Association: Burial 32A.

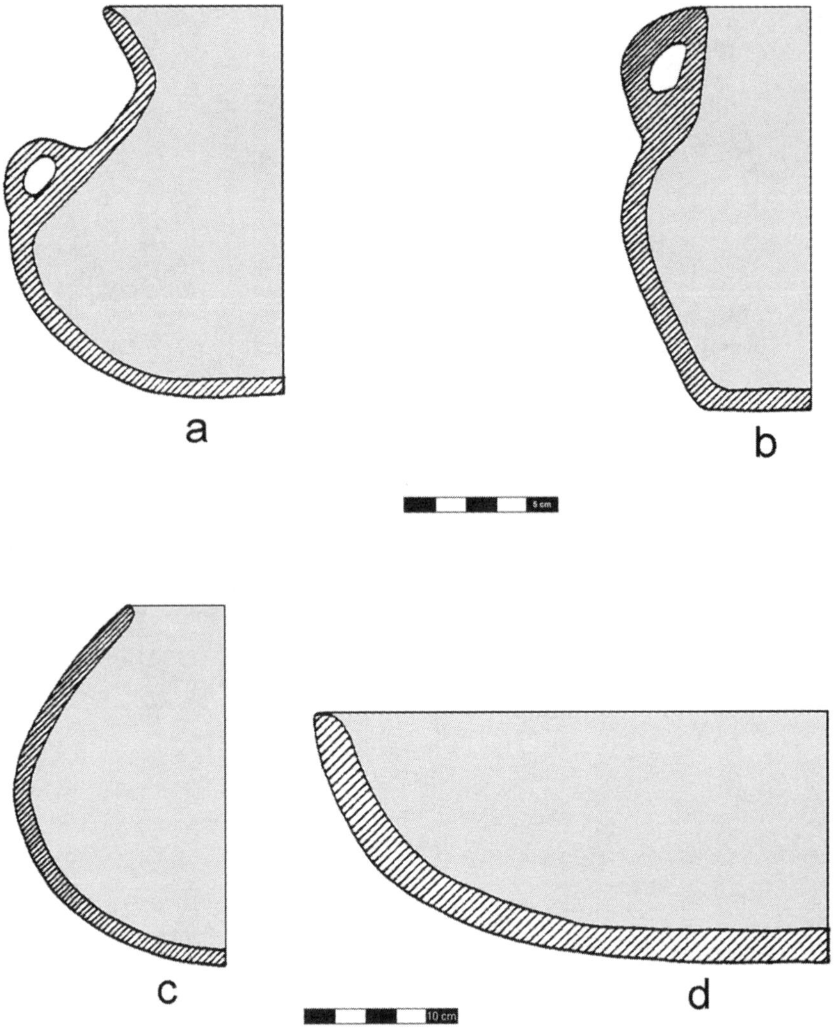

FIGURE 6.9. Additional examples of utility ware/plainware vessel profiles from Venadillo. (*a*) Globular jar with a wide flaring mouth and loop handles, with well-polished, dark buff slip, possibly of the type Polished Buff that is associated with Middle Chametla Polychrome (see Kelly 1938). (*b*) Pinkish buff jar with a straight neck and loop handles; black staining is probably a mineral deposit. Association: Burial 6. (*c*) Large orange-red-slipped olla; the upper half of the vessel slopes inward from a globular base to a constricted mouth. The surface, which is now worn and somewhat encrusted, is smoothed but not polished, except for some stick-polished marks. Association: Burial 10. (*d*) Large hemispherical bowl with an unpolished orange slip on the exterior and interior, which are poorly smoothed. It is asymmetrical in profile, and the rim is very uneven.

noted that Grosscup's (1964, 1976) analysis of the Amapa ceramics focused primarily on the recovered sherd assemblages, while that of Bell (1960) focused on whole or nearly whole vessels. Also, the reader will note that the descriptions of the Peñitas ceramics presented below are briefer than those for Chametla, Culiacán, and Amapa. Bordaz (1964:134–150) imposed a somewhat onerous classificatory system on the Peñitas sherd assemblage, assigning 158 entry numbers rather than named type designations. Nevertheless, he used typical technological and stylistic attributes—paste, temper, shape, surface finish, and decoration—to sort and define his entry categories. Bordaz's scheme is further exacerbated by Bell, who, with her detailed personal knowledge, classified the Peñitas ceramics found in the Marismas. She appears to have used generalized descriptors, such as "painted," "engraved," and "incised," without always identifying a specific Bordaz entry number.

FIGURE 6.10. Heavily white-slipped (washed), small composite jar from Burial 32 at Venadillo. Note the drips of slip on the body and rim.

Early Estero Period Decorated Ceramic Types

Early Chametla Polychrome (Kelly 1938:11–13, Figures 1–3, Plate 2)

Early Chametla Polychrome is an early type (Tierra del Padre phase). The paste is "uniformly granular," tan ("sand") colored, and often has a gray core. The interior and exterior surfaces are well smoothed and slipped. Red, white, and black are painted on a tan- to orange-slipped vessel. Designs are heavy and exclusively geometric, and most are angular elements that include triangles (often hatched), stepped diagonals, frets, and keys. Some curvilinear elements are also present (Figure 6.11). Kelly reports that three vessel forms are represented: convex bowls, platelike bowls, and recurved bowls. Grosscup (1964:290) indicates that Gavilán Polychrome (Gavilán phase) from Amapa is very similar to Early Chametla Polychrome.

Early Chametla Polychrome was identified in low frequencies at Juana Gomez, in Panales Test Mound B, and in the earthen mound portion of Mound C at Venadillo. Sweetman (1968, 1970:28) provides a summary description based on 435 sherds. The temper ranged from fine to medium, and the paste color varied from buff to light orange to red to dark gray with a gray core. A polished orange or creamy orange slip was present on both the exterior and interior surfaces. Sherd thickness ranged from 5 to 10 mm. No specific deviation from Kelly's (1938) description was mentioned.

Weil (1974:39–40) summarized four body sherds and two rim shreds of Early Chametla Polychrome. Paste colors fall within the yellow-brown class, and the temper ranges from fine to medium. Both surfaces of the specimens were slipped with a yellow-brown or dark brown slip; the surfaces were well smoothed. In terms of decoration, Weil's sherds were like those described by Sweetman (1968, 1970). Although no deviation from Kelly (1938) was specified, Weil did note that white paint alone was used for additional striping and that there were no examples of overpainting with dots on either red or black elements.

Early Black-banded Engraved (Kelly 1938:13–14, Figure 4a–c)

Chametla Black-banded Ware appears in three varieties, Early Engraved, Late Engraved, and Nonengraved. Kelly states that it is as nearly fundamental to Chametla ceramics as is Red-rimmed Utility Ware. She also notes that changes in vessel form accompany changes in decoration in the Black-banded Ware.

Early Black-banded Engraved (Figure 6.12) is an early type at Chametla, where it is sparse. In form and exterior decoration the type appears to be related to

FIGURE 6.11. Examples of Early Chametla Polychrome sherds from Juana Gomez.

FIGURE 6.12. Examples of Chametla Early Black-banded Engraved sherds from Juana Gomez.

FIGURE 6.13. Examples of Chametla Late Black-banded Engraved sherds from Juana Gomez.

recurved bowls of Early Chametla Polychrome. The exterior is covered with vertical or diagonal stripes, while the interior is covered in horizontal bands of black and red or just black. These bands are engraved, postfiring, in intricate geometric and sometimes curvilinear elements. The vessels are well polished. Grosscup (1976:254) notes the similarities between Amapa's Gavilán Polychrome and Early Black-banded Engraved but states that Gavilán Polychrome is rarely engraved and that if decorated, the black bands are decorated with white paint rather than engraving.

Grosscup (1964:286) further notes that Chametla Black-banded Engraved is a variety of Early Chametla Polychrome with designs engraved into a black band. At Chametla, this Black-banded Engraved type appears in the Tierra del Padre phase (Early Black-banded Engraved) and extends into the Baluarte phase (Late Black-banded Engraved).

In the Marismas, Early Black-banded Engraved appears in low frequencies at Juana Gomez, La Tarjea, Panales Test Mound B, and the earthen mound portion of Mound C at Venadillo. Several Black-banded Engraved sherds listed in the tables prepared by Bell are not specified as either Early or Late.

Sweetman (1968, 1970:22) described 20 sherds of Early Black-banded Engraved. The temper ranged from very fine to fine to medium, and paste color ranged from buff to reddish buff. Cores were gray or occasionally black. The sherds ranged from 5 to 8 mm thick. Sweetman notes that in the sample analyzed, engraving was limited to the black interior bands rather than to both black and red bands as described by Kelly (1938). None of the sites recorded by Weil (1974:39) during his survey produced any Early Black-banded Engraved.

Late Estero Period Types

Late Black-banded Engraved
(Kelly 1938:14–16, Figure 4d–f)

Kelly describes Late Black-banded Engraved (Figure 6.13) as a coarse and increasingly degenerate version of its predecessor. The red band and exterior polychrome are no longer present. Earlier, the exterior is decorated with either diagonal or vertical red and buff lines that are reminiscent of the decoration *mano colorado* on Red-rimmed Utility Ware. Late, the exterior bands are replaced by red and buff "splotches," a red band around the rim, or a red "wash." A black band is present just below the interior rim, and the remainder of the interior is filled with red-on-buff decorations that include heavy spirals, swastikas, or nebulous smears. The black band is wide initially but deceases in size over time. Similarly, the early engraving is well done like that of its predecessor, but it becomes less elaborate and well executed as the width of the band decreases.

This type was found at Juana Gomez, at La Tarjea, in Panales Test Mound B in the mid- to lower levels of the test units excavated, in the earthen mound portion of Mound C at Venadillo, and at Site 3 near Cuautla in low frequencies. It was generally more abundant at Venadillo than at the other sites.

Sweetman (1968, 1970:22) describes a sample of 79 Late Black-banded Engraved sherds from Juana Gomez. The temper of these ranged from fine to fine with coarse red inclusions to medium. Paste color was usually buff but sometimes also reddish brown, brown, or black; a black core was occasionally present. Sherd thickness varied from 4.6 to 9.0 mm. She notes no deviation from Kelly's (1938) description. However, she reports that in the earlier levels at Juana Gomez the red painting and

FIGURE 6.14. Examples of Chametla Black-banded Nonengraved sherds.

engraving are well executed, while in the later levels they are more poorly done. Sweetman also made a point of discussing a crudely made specimen with poorly executed incising. The sherd had black paste and was 6 mm thick. She suggested the possibility that this may have been an example of Sauer and Brand's (1932:64) Crude Incised type.

Weil (1974:38–39) recorded three rim sherds of Late Black-banded Engraved. The temper ranged from fine to medium, and the paste was within his yellow-brown class with a black core. The thickness of the sherds ranged from 4 to 8 mm. Weil notes no variation from Kelly's (1938) original description.

Black-banded Nonengraved (Kelly 1938:16, Figure 4h)

The Black-banded Nonengraved type developed out of Late Black-banded Engraved and partially overlaps with it temporally. In general, the loss of engraving is the main difference between the two types. The exterior is usually red, and over time the width of the black interior band decreases. In some instances, it disappears completely, with the result being a simple red-on-buff vessel.

Kelly (1938:16) makes special note that the Black-banded Ware (Figure 6.14) is the one ware that exhibits a full developmental sequence from early to late in the Chametla sequence. It is well made and well executed, having evolved out of Early Chametla Polychrome; she says that it is best technologically and most artistic of the Chametla ceramics. Kelly indicates that Sauer and Brand (1932:38–40) subsumed Black-banded Ware under their Chametla Polychrome.

Although limited in number, Black-banded Nonengraved appears to occur primarily in Late Estero period contexts when present, although it is mixed with other Early Estero and Middle Conchera period types in Mound B at Venadillo. However, as noted in Chapter 5, its association with Middle Conchera types is likely tied to the construction of the burial mound and the mixing of cultural deposits at the site. It also occurs at Juana Gomez, La Tarjea, and Panales Test Mound B.

Based on three sherds, Sweetman (1968, 1970:23) describes Black-banded Nonengraved as having very fine temper with a brown paste and a dark gray core. The sherds ranged in thickness from 6.0 to 6.5 mm. Weil (1974:41) describes three rim sherds and one body sherd of this type. The temper varied from fine to medium, and paste color varied within his yellow-brown class. Both surfaces were smoothed, and they were slipped; primary slip color also varied within the yellow-brown class. The exterior had a secondary slip that varied in color within the red-brown class. Neither Sweetman nor Weil reports any deviation from Kelly's (1938) description.

Middle Chametla Polychrome (Kelly 1938:16–18, Figure 5, Plate 5a–c)

The paste of the Middle Chametla Polychrome type is variable but generally granular, varying in color from tan to dark gray to a bright terra-cotta with a distinct gray core. This type is described as a black-and-red-on-cream ware that occasionally appears with white and orange overpaint on the red and black or red overpaint on the black. The decoration is almost exclusively on the exterior. Decorative elements are geometric, with a chevron and key dominating. Vessel rims are usually ticked with small red and black scallops. The decoration is typically in horizontal bands below the rim.

The surfaces of the vessels are well smoothed and polished and slipped with a cream to milky gray slip that occurs on both the interior and exterior. Kelly notes that both the slip and paint do not weather well; they become powdery and crack. She further indicates no apparent relationship with Early Chametla Polychrome and no overlap between the two types. However, Grosscup (1976:254) takes exception to this statement, observing that Middle Chametla Polychrome and its variant

FIGURE 6.15. Middle Chametla Polychrome Engraved vessels from Burials 3 and 4 at La Tarjea.

Middle Chametla Polychrome Engraved more closely resemble Early Chametla Polychrome than any other type in the Chametla sequence or any other type in the region. He further specifically suggests that is it likely that Middle Chametla Polychrome Engraved is derived from the Black-banded Engraved variants. Grosscup's explanation for this focuses on the apparent abandonment of the Chametla area and break in the ceramic sequence (as seen in the Marismas and at Amapa); he suggests that this break accounts for the lack of any transitional forms between Early Chametla Polychrome (including Black-banded Wares) and Middle Chametla Polychrome.

Middle Chametla Polychrome is reported from Juana Gomez, La Tarjea, Panales Test Mound B, Tecualilla (one sherd), and Mound B and the earthen portion of Mound C at Venadillo. Mound B is associated with the Middle Conchera period, and the presence of Middle Chametla Polychrome is likely intrusive because of mixing of cultural deposits.

Sweetman (1968, 1970:3) describes 28 sherds of Middle Chametla Polychrome from the Marismas. The temper ranged from very fine to medium; paste color ranged from buff to red to reddish brown to brown; and cores, when present, ranged in color from brick red to gray to black. Sherd thickness ranged from 5.5 to 8.0 mm. No specific variations from Kelly's (1938) description are reported.

Weil (1974:40) described a small assemblage of Middle Chametla Polychrome collected during his survey. The temper ranged from fine to medium, and paste colors fall within his yellow-brown class. Both surfaces were smoothed and are covered with yellow-brown or brown slip. No significant deviation from Kelly's description was noted, although the sherds Weil describes lacked the more intricate design elements reported by Kelly (1938:16). He suggests that this may be due to the size of the sherds collected from surface contexts.

Middle Chametla Polychrome Engraved (Kelly 1938:18, Plate 5d–i)

The Middle Chametla Polychrome Engraved type is Middle Chametla Polychrome with engraved geometric elements. The engraving is postfiring. This type varies somewhat in that it is always red-and-black-on-cream. There is no orange or white and no overpainting. Painted motifs are simpler because of the amount of engraving. It co-occurs with Middle Chametla Polychrome. Kelly notes that Sauer and Brand (1932) include this type in their Aztatlán Ware, although Kelly notes that there is some discrepancy between the Sauer and Brand description and the Chametla examples she describes.

Middle Chametla Polychrome Engraved (Figure 6.15) is not abundant, occurring only at four of the sites investigated during the Marismas project: Juana Gomez, La Tarjea, Panales Test Mound B, and Mound C at Venadillo, where, except for a single sherd, it is confined to the earthen portion of the mound.

Sweetman (1968, 1970:30) described 24 Middle Chametla Polychrome Engraved sherds from the Marismas. Temper ranged from fine to medium, and the paste color, from buff to tan to red to brick red to black. Gray and black cores were present. The sherds ranged from 3.0 to 8.5 mm thick. Both the interior and exterior surfaces were covered with a cream-colored slip, although some of the interiors were unslipped. No significant variation from Kelly's (1938) description of the type was noted; however, Sweetman did indicate that at least four of the 24 sherds were incised rather than engraved. Weil (1974:40–41) identified four sherds of Middle Chametla Polychrome Engraved. He reported no variation from Kelly's (1938) description.

FIGURE 6.16. Examples of Black-on-buff sherds.

Scalloped Rim (Kelly 1938:18, Plate 6a)

Scalloped Rim is a buff- to orange-colored type with circular pendants on the rim that usually extend from the rim to both the interior and exterior surfaces of the vessel; they usually alternate red and black. Paste color ranges from tan to terra-cotta; it is granular. The surface is slipped with color ranging from cream to ocher to orange-red, and the slip is sometimes thick enough to crack. Surfaces are smoothed and polished. This type is commonly found in association with Middle Chametla Polychrome. It occurs in low frequencies at Panales Test Mound B and Mound B at Venadillo. Some sherds appear in Middle Conchera period levels, but their presence is intrusive.

White-on-buff (Kelly 1938:18)

White-on-buff is a red-rimmed buffware with a polished red interior. The exterior is unslipped but decorated with "shadowy," usually concentric, arcs of lighter buff or white. The paste is granular and approaches that of plain Red-rimmed Utility Ware. A small number of such sherds are reported from Juana Gomez and Panales Test Mound B. Sweetman (1968, 1970:43) describes eight badly eroded sherds classified as White-on-buff. The temper was fine, and the paste color ranged from buff to brown to black; cores were light gray. The sherds were unslipped; one or the other surface was left plain, while the other was painted with concentric arcs. These sherds conformed to Kelly's (1938) description, although Sweetman states that none of the sherds appeared to exhibit the polished red interior described by Kelly. Sherd thickness ranged from 5.0 to 7.5 mm.

Black-on-buff (Kelly 1938:19, Plate 7g–i)

The Black-on-buff type appears to be a black-on-buff variant of Kelly's (1938:18–19) Red-rim Decorated, although with simpler design elements and less well polished. Paint on this unslipped buffware is in heavy black and done in bold, firm strokes. The paint is durable and often retains more luster than the vessel itself. Decoration occurs on either the exterior or interior within a band outlined with several black stripes. Design elements are geometric and usually angular, with a crosshatched triangle being the most common.

Grosscup (1976:256) notes a similarity between the Chametla Black-on-buff and the Amapa Botadero Black-on-buff, which is not abundant at Amapa and generally restricted to the Cerritos phase. Thus, both are contemporary. Grosscup further notes that these black-on-buffs resemble Tuxpan Red-on-orange and Lolandis Red Rim.

In the Marismas, Black-on-buff (Figure 6.16) is reported from Rincon de Panal and Chalpa, where it is listed as the Amapa variant. At Tecualilla, it is listed as a Chametla type and as an Amapa type, and at Venadillo it is listed as a Chametla type. Based on the distribution of Black-on-buff in the Marismas, it seems that the type may first appear in the Late Estero period and may persist into the Early Conchera period. Although not abundant, Black-on-buff appears to be more common in the Early Conchera period.

Red-rim Decorated (Kelly 1938:18–19, Plate 7a–f)

Red-rim Decorated, a good-quality buffware, is unslipped, although it sometimes appears that the vessels have a wash that is the same color as the paste. The color varies from buff to orange, and it is well polished. The most common form of decoration, which occurs on either the exterior or interior of vessels, is in a band below the rim, consisting of a red rim strip flanked by two narrow stripes. Below this is a series of continuous geometric elements including parallel stripes, zigzags, triangles, some scrolls, and circles, and below these are two to four narrow red stripes. Kelly notes that "conventionalized" zoomorphic elements are sometimes present in addition to the geomorphic elements. Furthermore, there are some examples of Red-rim Decorated with white paint, which usually occurs as a stripe below the rim. Kelly states that this "represents a perfect transition" to Sauer and Brand's Aztatlán Ware, lacking only engraving on the white band. She further notes that the Chametla Red-rim Decorated is the same as Decorated Red-rimmed Red-on-buff described by Sauer and Brand (1932:63–64), although their group also appears to have included Red-rimmed Utility Ware.

Red-rim Decorated is also known as Lolandis Red Rim (Kelley and Kelley 1971:125–127; Kelley and Winters 1960). In the revised Kelley and Winters (1960) phase sequence, based on work at the Guadiana Branch Chalchihuites Schroeder site (La Ferreria) (Kelley 1971) in highland Durango, this ceramic type is assigned to the Lolandis phase at Chametla. Lolandis Red Rim is interesting for several reasons. It is relatively abundant at the Schroeder site, and it appears rather suddenly during the Las Joyas phase (Foster 1995; Kelley 1971). However, it is not an indigenous Chalchihuites ceramic type (Kelley 1971:794–795; Kelley and Kelley 1971; Kelley and Winters 1960). It is further argued that its presence cannot be simply explained as a result of exchange. Kelley (1971; Kelley and Kelley 1971) argues that its abundance is likely the result of the migration of Nayarit or southern Sinaloa coastal peoples inland and into the Guadiana Valley (Grosscup 1976:256). The Kelleys cite the thin section study by Ward (1967:13–15), who observed that the paste-temper complex of Lolandis sherds from Durango was different from that of both local Chalchihuites wares and Lolandis sherds from Sinaloa; this was also seen as further evidence that the Durango Lolandis material did not represent trade wares from the west coast. Rather, the coastal potters had adapted their methods to local clay and temper sources. Although some might argue that the suggestion of an actual migration of coastal people into the Guadiana Valley is speculative at best, the timing of this fits well the archaeological data from the west coast and the geomorphic data from the Marismas in particular. As discussed in Chapters 2 and 4, at this time (AD 700–900) there appears to have been widespread abandonment of the west coast from at least as far south as the Amapa area north to the Chametla area. The Chalchihuites Las Joyas phase, the Lolandis phase at Chametla, the Tuxpan phase at Amapa, and the hiatus in the occupation of the Marismas between the Late Estero period and the Early Conchera period are generally contemporaneous.

Finally, Kelley and Kelley make a significant observation regarding the distribution and variation of this widespread ceramic type known as Red-rim Decorated/Lolandis Red Rim:

> Actually, the problem of the distribution and variation of the pottery type, which we have called "Lolandis Red-Rimmed," may have an even greater significance. What appears to various investigators to be roughly the same ware has now been noted and discussed or described by Sauer and Brand (1932, "Decorated Red-Rimmed Red-on-Buff"), Isabel Kelly (1938, "Red Rim Decorated"),... J. Charles Kelley and Howard D. Winters (1960, "Lolandis Red Rim"), George E. Fay (1959, "Tecuala Red-on-Buff"), Gordon Leonard Grosscup (1964, "Tuxpan Red-on-Orange"), and Stuart D. Scott (1968, "Tuxpan Red-on-Orange"), in addition to the present discussion. From the names applied alone, it will be seen that the ware is described in varying terms by the several researchers. This variation becomes even more apparent when the descriptions are examined. Thus, while Scott writes that the ware is unslipped, Kelly and Fay both note that it is usually unslipped but that some sherds are washed or self-slipped; and while Grosscup describes it as "lightly slipped," we have described it here as having "a fine textured slip." The picture that emerges is that of a distinctive ware, recognizable as such by several independent investigators; geographically widely distributed in Nayarit, Sinaloa, and Durango; and varying considerably in paste, slip, and possibly other criteria. However, the possibility exists that the apparent variation in characteristics of the ware itself represents instead erratic description based on the use of different descriptive and analytical criteria by the individual investigators [1971:126].

Red-rim Decorated (Figure 6.17) was recovered at Panales Test Mound B, Rincon de Panal, Chalpa, Tecualilla, and Mound B at Venadillo. Relatively speaking,

FIGURE 6.17. Examples of Red-rim Decorated.

FIGURE 6.18. Examples of Aztatlán Ware sherds (Chametla variety).

among the identified ceramic types, by period and in general, Red-rim Decorated was abundant.

Aztatlán Ware (Kelly 1938:19, Plate 7k–m; Sauer and Brand 1932:63, Plate 1–Panel A)

Sauer and Brand describe Aztatlán Ware as a red-rimmed, red-on-buff incised ware (Figure 6.18). It is unslipped, with horizontal red bands combined with mainly freehand scrolls and triangular and linear elements. These occur within zones outlined by incised parallel straight lines. In some instances, the space between the lines is filled with red or white paint. In other instances, painted and incised design elements are outlined in white paint. Decoration most commonly occurs on the exterior of the vessel. When present, interior decoration is limited to the occasional red line and even more rarely to some incising.

Kelly notes that Sauer and Brand's Aztatlán Ware is far more inclusive than her definition of the ware, with Sauer and Brand apparently having included all incised and engraved wares except for the Black-banded Ware. Kelly's Chametla Aztatlán Ware is essentially the same as her Red-rim Decorated. Aztatlán Ware at Chametla differs from Red-rim Decorated with a white band with sharply and well-executed incised geometric elements below the red rim. As described by Sauer and Brand these elements include freehand scrolls or triangular and linear elements within zones outlined by incised parallel straight lines. Other common elements include diagonal stripes, herringbone, frets, and zigzags; circles and scrolls are less frequent. Kelly states that several of these designs are combined and repeat to form a series of "stop" designs. Below the white band is a series of horizontal red stripes, and below these are sometimes additional decorations in red, most of which are indiscernible. However, a checkerboard pattern has been noted.

Regarding Aztatlán Ware and its analogous forms, Kelly states that Red-rim Decorated, Black-on-buff, and Aztatlán Ware form a related cluster that, in part, constitutes the Aztatlán complex. Of these three types, Red-rim Decorated is the most common and "probably the most fundamental" (Kelly 1938:19). Grosscup (1964:286) indicates that Aztatlán Ware is comparable to Botadero Incised at Amapa. Aztatlán Ware is also described by Kelly (1945) for the Culiacán area (see below). Aztatlán Ware was recovered in low frequencies at Juana Gomez, Panales Test Mound A, and Arriñitas and as a Culiacán variety at Rincon de Panal and Tecualilla.

Weil (1974:42) reports on a single sherd of Aztatlán Ware. It had fine temper, and the paste was within his yellow-brown color class. Both surfaces were smoothed and unslipped. Although Weil notes that the sherd conforms to Kelly's (1938) description of the (Chametla) type, he makes mention that the incising was well executed and suggests that this may be a Culiacán variety rather than a Chametla variety. He further notes that the incised design pattern was not similar to those illustrated in either the Chametla report (Kelly 1938) or Culiacán report (Kelly 1945).

Polished Buff (Kelly 1938:26)

Polished Buff ware has sand- to terra-cotta-colored coarse paste. The interiors and exteriors are slipped with colors that vary from rose-buff to ocher to cream, and the surfaces are well polished. This ware co-occurs with Middle Chametla Polychrome. Grosscup (1964:297) suggests some possible relationship to Amapa White.

Betty Bell, in her tabulations of decorated Marismas ceramics, only reports Polished Buff at Mound C, Venadillo, where it occurs in both the Late Estero period earthen mound and Middle Conchera period shell midden capping the mound. At Chametla, it is a Baluarte phase type, which is contemporaneous with the Late Estero period. Sweetman (1968, 1970:24), however, also reports Polished Buff at Juana Gomez. The lack of identification of this type at other sites may be due to the classificatory schemes employed by various analysts and the level of analysis completed for each of the sites. Polished Buff accounted for approximately one-third of the decorated ceramic assemblage from Mound C (Figure 6.19).

FIGURE 6.19. Probable Polished Buff jar from Burial 17A, Mound B, Venadillo (see Figure 6.8b).

Sweetman's (1968, 1970:24) description and discussion are based on an abundance of sherds classified as Polished Buff. The temper is reported as usually fine or medium, although very fine temper was also found. The color of the paste ranged from buff to cream to orange to rose to brown. Sweetman notes that occasionally black spots are present in the paste and that even within a single vessel, paste color can vary significantly. Reduced cores were often absent and when present, ranged from gray to black in color. Sherd thickness varied greatly; it was usually between 5 and 7 mm but sometimes as much as 9 mm; sherds 10, 12, and 15 mm thick were also recorded. Vessel surfaces were unslipped but polished.

Buff Exterior, Red Interior (Kelly 1938:26)

Buff Exterior, Red Interior is similar to Polished Buff but with a polished red interior of bowls and plates. As with Polished Buff, this type was identified at Juana Gomez and at Mound C, Venadillo. Sweetman (1968, 1970:43) described an assemblage of 25 Buff Exterior, Red

FIGURE 6.20. Examples of Red-rimmed Utility Ware sherds.

Interior sherds from Juana Gomez. The temper ranged from very fine to fine, the paste was buff, and the core was dark gray. A red polished slip was present on the interior as described by Kelly (1938). The exterior surface was polished as well. Sweetman notes that sherds with polished buff interiors and polished red exteriors were also present.

Red-rimmed Utility Ware (Kelly 1938:31)

Red-rimmed Utility Ware is the most abundant and most "fundamental" of the Chametla pottery types. Kelly states that it is a highly conservative style with only minor decorative changes (Figure 6.20). It dominated at the Tierra del Padre site, was noticeably less plentiful at the site of Cocoyolitos, and was sparse at the El Taste site. The sherds from Tierra del Padre had a reddish-colored paste that resulted in a pinkish hue to the sherd when the slip of the sherd was highly weathered.

The vessels are slipped on the interior and exterior, mainly with a buff slip; however, slips vary from gray-white to burnt orange. As indicated by the name, the ware exhibits a red rim. Other red decorations are sometimes present as well. Kelly describes what she calls "olla bowls" or large shallow bowls, which were often inverted and used as lids to burial urns.

Early versions of bowls are often decorated with solid red circles distributed across the interior below the rim (Kelly 1938:Figure 15a, Plate 13e). Because this decorative element is most often noted on sherds, the variability with which this design repeats itself is unknown. This circular design is replaced by a "marbleized surface" and *mano colorado* decorative techniques. The marbleized surface is actually striping, usually horizontal to slightly diagonal (sometimes vertical) (Kelly 1938:Plate 13c; see also Sauer and Brand 1932:Plate 1a, lower left), produced by the application of a thin buff or a light reddish wash on the buff surface of a vessel.

Grosscup (1976:220–221, 255) states that Amapa Red-on-buff is equivalent to the Chametla Red-rimmed Utility Ware; however, at Amapa there appears to have been a greater variety of designs represented—otherwise the two types are very similar in treatment and vessel form. One apparent significant difference is that none of the Amapa material exhibited the marbleized treatment. Grosscup further notes that the exteriors of some vessels are slipped red and that some olla and jar necks are painted red; he suggests that Kelly may have included such sherds in her Chametla redware.

Red-rimmed Utility Ware was recorded at multiple sites during survey (Weil 1974) and recovered at Panales Test Mound B, at Tecualilla, and in both Mound B and the earthen portion of Mound C at Venadillo. It appears in Late Estero contexts at Juana Gomez, Panales Test Mound B, and the earthen portion of Mound C at Venadillo. Its presence in Early Conchera contexts at Tecualilla and Mound B at Venadillo suggests either the persistence of the type or the mixing of temporal components.

Sweetman (1968, 1970:32) analyzed a significant sample of Red-rimmed Utility Ware sherds from the site of Juana Gomez (Figure 6.21). She notes considerable variation within this type. She reports that the temper

FIGURE 6.21. Red-rimmed Utility Ware bowl from Juana Gomez.

is sometimes fine but usually medium, with or without coarse red inclusions. Paste color is usually buff, but it also ranges from black to brown to brick red to salmon. The thickness of the specimens ranged from 4 to 14 mm; Sweetman reports that with the thicker vessels, the rim and shoulders are thicker than the body of the vessel. Slip colors range from cream to buff to orange. The simplest expression of this type is nothing more than a vessel decorated with a red rim. In addition to the painted red rim, decoration observed included vertical red lines on the interior with a plain exterior, horizontal lines in the interior with a plain exterior, red diagonal lines on the exterior (*mano colorado*) and plain interior, parallel lines on the interior with a plain exterior, marbleization, pendants hanging from the rim; pendants on the interior and diagonal lines on the exterior, red vertical lines on the exterior and horizontal lines on the interior, and splotches.

In her summary of her observations, Sweetman indicates that the quality of Red-rimmed Utility Ware seemed better in the lower levels (earlier) at Juana Gomez and that *mano colorado* and marbleization as decorative techniques were absent from sherds in the lower levels at that site. Other trends noted included an increase of orange-slipped sherds through time and a decrease of thick-rimmed cream-slipped sherds. Also, in the earlier levels, sherds of finely made vessels with red rims and orange slip dominated.

Weil (1974:37–38) was somewhat more structured in his classification of Red-rimmed Utility Ware recorded during his survey, identifying 13 varieties based on 55 sherds. Weil's categories, however, only included specimens that were decorated more extensively than those with just a red painted rim. Decoration occurred on both the interior and exterior surfaces; however, the specimens in his sample were too small to determine line orientation and the pattern of any line decoration or the presence of other design elements (e.g., splotches, dots). Slips and paste typically varied within the yellow-brown and dark brown classes, and decoration was most commonly within the red-brown class, although some within the yellow-brown class was also noted. Temper ranged from fine to medium. Interior surfaces ranged from medium smooth to smooth, and exterior surfaces were nearly always smooth. Sherd thickness varied from 4 to 12 mm.

FIGURE 6.22. Red-rimmed Utility Ware *mano colorado* from Burial 2 (B 2A and B 2B), Mound C, Venadillo.

Red-rimmed Utility Ware, *Mano Colorado* (Kelly 1938:31–32)

The mano colorado variety of Red-rimmed Utility Ware was produced when a series of streaks or stripes applied with a human hand covered in red paint were dragged across the surface of the vessel (Kelly 1938:Plate 13b, d, f; Figure 6.22). These variously appear as quick swipes, streaky blotches, or parallel rows. The width and size of these led Kelly to suggest that the potters were males. In some instances the interior of a bowl will be decorated using the marbleized technique, while the exterior is decorated with mano colorado. Grosscup (1976:255) states that this type of decoration is rare at Amapa and restricted to the Amapa phase.

The mano colorado variety of Red-rimmed Utility Ware was only recognized at Mound C at Venadillo, where it is found almost exclusively in the Late Estero earthen portion of the mound.

Baluarte Red-on-buff/Red-on-brown

Sweetman (1968:Plates XXII–XXIII, 1970:Plate 39) mentions, without describing, the presence of Baluarte Red-on-buff and illustrates two vessels. The specimen illustrated in Plate XXII was from a private collection and described as non-Chametla pottery.

Sweetman reports that the specimen illustrated in Plate XXIII was recovered from Tecualilla (Figure 6.23). However, no such type is identified at the site, although there is an unspecified red-on-buff listed. Unspecified red-on-buff was also identified at Panales Test Mound B and Rincon de Panal. Baluarte Red-on-buff is listed as a type for Mound A, Test Unit A, at Arriñitas. It may be a type name applied to a red-on-brown or red-on-buff, perhaps a miscellaneous red-on-buff from the Chametla area (Kelly 1938:23). Kelly (1938:Figure 16) illustrates a red-on-buff jar very similar to the Tecualilla specimen.

Polished Whiteware (Foy 1970:22–23; Sweetman 1968, 1970:44)

Sweetman describes a small assemblage of polished whiteware with a thick creamy white slip. She questioned whether the sherds were Amapa White. Foy also discusses a whiteware, questioning whether it was Cerritos White. Foy makes special mention of a white-slipped type, two of which were incised, that was found at Arriñitas. The slip was thick, and the sherds were polished. The paste of the sherds ranged from buff to orange to brownish red. She also noted that the Arriñitas sherds were similar to what Sweetman had described.

Grosscup (1976:222) describes Amapa White in detail, noting that both plain and incised sherds were present in his sample. A variety of incised geometric designs were present. Grosscup's (1964:287) description of Cerritos White is limited to the notation that the type is a heavily slipped whiteware. Neither Sweetman nor Foy chose to classify the white-slipped sherds they identified. Based on Grosscup's (1964, 1976) suggested phase associations of Amapa White and Cerritos White at Amapa, these white-slipped types may span the Late Estero period into the Middle Conchera period. It is possible that the small composite vessel illustrated above in Figure 6.10 is an example of this type, although it does not appear to be polished.

Amapa Red-on-cream (Grosscup 1976:222, Plate 2m, p–r)

The Amapa Red-on-cream type has a thick white to cream to gray slip like that of Amapa White. The paste varies in color from red to gray and often has a carbon streak; the temper is coarse. Incurved bowls are the typical vessel form, and the rims are sometimes thickened. Decoration consists of a variety of elements including red bands, red rims, dots, two horizontal bands that

FIGURE 6.23. Baluarte Red-on-buff jar from Tecualilla.

frame a sine curve, and diagonal lines. Amapa Red-on-cream was recovered at Arriñitas and Panales Test Mound B (Figure 6.24). One example (Figure 6.24, top row left) appears as a black-on-white sherd. It is likely that the vessel was fired in a neutral or reducing atmosphere where oxygen was not adequate or was lacking.

Sweetman (1968, 1970:20) described a sample of 14 Amapa Red-on-cream sherds from the Marismas. The temper ranged from very fine to fine and sometimes included coarse red inclusions. Paste color ranged from gray to buff to light brown to cream, and none of the sherds examined had a core. Both surfaces were white-slipped and well polished. Interior surfaces were undecorated except for an occasional red band on the rim. The exterior surfaces were decorated with simple geometric patterns.

Early Conchera Period Types

Cocoyolitos Polychrome
(Kelly 1938:23, Plate 9g–q)

Cocoyolitos Polychrome is a rather amorphous category of ceramics, a fact cited by both Kelly (1938) and Grosscup (1976). Kelly states that it is an "off-color polychrome style" (1938:23) that does not include any of her defined polychrome types. It appears that this type is defined on the basis of 15 sherds. The paste is fine and typically ranges in color from rose to brown. In addition to white overpaint on red and brown, some incising was also noted. Decoration is generally restricted to the surface, which is usually covered in its entirety. The interiors are unslipped but highly polished or smooth. Occasionally red stripes are painted in the interior. Grosscup (1964:288) indicates that Ixcuintla Polychrome, Iguanas

FIGURE 6.24. Examples of Amapa Red-on-cream.

Polychrome, and Cerritos Polychrome of the Amapa series fit within this typological category.

Temporally, Cocoyolitos Polychrome is associated with the Acaponeta phase at Chametla and the Ixcuintla and Cerritos phases at Amapa. In the Marismas, it is reported in small numbers at Panales Test Mound B (Amapa variety) and Tecualilla (Chametla variety).

Weil (1974:41) reports finding only a single sherd of Cocoyolitos Polychrome during his survey. The temper was fine, and the paste color was yellow-brown. The exterior surface had a yellow-brown slip, and the interior was painted; both surfaces were well polished. The sherd was 6.5 mm thick. Weil makes no specific mention of any deviation from Kelly's (1938) description.

Ribbed Ware (Kelly 1938:26)

Kelly briefly discusses a group of miscellaneous modeled wares at Chametla. Among these is a ribbed ware that includes both shallow and deep types. This ware (shallow) is reported in low numbers at Panales Test Mound B, Rincon de Panal, and Tecualilla.

Cerritos Polychrome (Bell 1960:60–63, Plate 15; Grosscup 1976:228–229, Plate 11h–s)

Cerritos Polychrome is an incised polychrome. The paste varies from red to gray, and a carbon streak is common. The surface is thinly slipped and polished, and it ranges from buff to brown-orange in color. Colors used in decoration include black, white, dark red, and rose. Grosscup states that Cerritos Polychrome and Gavilán Polychrome are the two most colorful polychromes in the Amapa ceramic assemblage. Decorative designs are very complex on this type. They are first outlined by incising and then completed by painting. Colors are typically dark red, rose, or white. More delicate designs are then applied using black, red, or white paint or sometimes incising. Grosscup notes that the more elaborate pieces take on the appearance of cloisonné. Design elements are both geometric and representational. The exterior base of the vessels is left undecorated. Grosscup notes that a "Coarse" variety of the type is also present, although it is rarely recovered. Its decoration is typical of Cerritos Polychrome, but the incisions tend to be wider and deeper, the designs are larger and less cluttered, the colors are darker, and the surface was polished after being incised, which results in the near obliteration of some of the incised lines. Grosscup suggests that this may be an idiosyncratic or perhaps regional variant.

Bell describes three annular-base whole vessels of Cerritos Polychrome, one of which has a base that is "terraced and pierced" (1960:60). Two of the specimens exhibit large central panels on the interior; each is surrounded by a running band. One panel is filled with an arrangement of geometric, curvilinear, and ray elements, and the other is filled with what may be a symbolic design.

Cerritos Polychrome was tentatively identified from the Well Unit at Arriñitas and in low numbers at Tecualilla (Figure 6.25). Sweetman (1968, 1970:27) described two sherds of Cerritos Polychrome. She noted no variation from Grosscup's (1964, 1976) subsequently published description of the type. Additionally, Sweetman (1968, 1970:45) describes three miscellaneous engraved sherds with very fine temper, orange-brown paste, and dark gray cores. The surfaces are unslipped. The interior surface was roughly smoothed while in a plastic state. The exterior was highly polished and engraved with curvilinear and straight lines. Sweetman notes that one sherd had a small patch of white paint with engraving through the paint. She suggests that these sherds may be Cerritos Polychrome.

FIGURE 6.25. Examples of Cerritos Polychrome from the Marismas Nacionales.

Botadero Incised (Grosscup 1976:226–228, Plate 10, Figure 1)

Botadero Incised (Figure 6.26) is an incised polychrome with a buff- to red-colored paste that frequently has a gray to black core. The vessels are lightly slipped and range from buff to orange-brown in color. Rims are painted red, and all designs are incised on the exterior surface. Red or white paint may fill portions of the incised designs. Bands or medallions covered with white paint often served as decorative fields; these are sometimes augmented by incised designs on the unpainted portions of the vessels. The designs were incised while the clay was still pliable. In some cases it appears that the incising occurred after the vessel was painted as there is paint pushed into the incisions; however, the vessel could have been painted after incising, with paint filling the incisions as it was applied. Designs, both geometric and representational, most frequently include frets, circles, and diagonals. The representational designs often depict feathered creatures or objects. An exterior white band outlined with red lines that border a red checkerboard pattern that covers the base of the vessel is typical of Botadero Incised bowls. Grosscup notes that such design characteristics are key attributes of the Aztatlán complex as defined by Kelly (1938).

Botadero Incised was recovered in low numbers from Panales Test Mound A, Panales Test Mound B, the Well Unit at Arriñitas, and Tecualilla. Sweetman (1968, 1970:25) describes 11 Botadero Incised sherds. The temper was very fine and sometimes included coarse red inclusions. The paste ranged from buff to orange or reddish brown to brown to black. A core was generally absent, but when present, it was light or dark gray. None of the sherds was slipped on the exterior or interior. The exteriors were decorated with red paint and a band of white paint that contained incised designs. Sweetman notes that incising on one sherd must have taken place in two stages, the first when the clay was still pliable and the second after the vessel had dried and hardened. The interior surface can be either rough or smooth. Sherd thickness ranged from 5 to 7 mm.

Sweetman classified the 11 sherds as Botadero Incised rather than Aztatlán Ware because the incised design elements were more curvilinear than those illustrated on the Aztatlán Ware at Chametla (Kelly 1938) and more like the Botadero Incised material from Amapa (Grosscup 1976). Sweetman describes one atypical sherd of Botadero Incised; it had incised white bands on both the exterior and the interior surfaces. Weil (1974) recovered two sherds of Botadero Incised; both had fine

FIGURE 6.26. Examples of Botadero Incised sherds.

temper, and the paste color fell within his yellow-brown class. Both were 6 mm thick. No variation from Grosscup's (1976) description is reported.

Botadero Black-on-buff (Bell 1960:50, Plate 9 Upper; Grosscup 1976:226, Plates 127a–d, 128)

Grosscup states that Botadero Black-on-buff sherds share many similarities with Tuxpan Red-on-orange and Botadero Red-on-white. It is typified by fine temper, and the paste color ranges from tan to cream. A gray to black core is common. Surface colors range from buff to yellow and cream; they are polished and likely lightly slipped. The thickness of vessel walls ranges from 6 to 9 mm, and bowl forms with simple rims are typical. Exterior decoration on bowls consists of a black rim and a decorative band below the rim bordered by black lines; black lines are common below the rim and above the design band, and several back lines often occur just below the design band. A wider black line below the other decorative elements serves to separate the upper decorated half of the bowl from the lower undecorated half of the bowl. Bowls with decorated interiors are plain on the exterior. Design elements include frets, lines, diagonal lines, triangular elements filled with small dots, interlocking designs, representational elements, and mythological creatures.

Bell describes five bowls of Botadero Black-on-buff. Three were shallow with flaring sides, and one was deep with straight sides; one was an annular-base *molcajete*. These vessels were unslipped and usually polished. The surfaces were yellowish buff in color. Two of the shallow bowls were decorated with geometric and curvilinear elements on the interior, and the other, with an animal design. The deep bowl was decorated with an exterior band of geometric elements, and the molcajete had an interior band of circles and an assortment of elements that may have been symbolic. The designs were painted in a very dark brown rather than a "dead" black.

In the Marismas, Botadero Black-on-buff was identified at Panales Test Mound A and Tecualilla. Sweetman (1968, 1970:24) describes a sample of seven Botadero Black-on-buff sherds. The temper was very fine and occasionally contained some coarse red inclusions. The paste color ranged from light buff to orange to brown, and if present, cores were either gray or black. She notes no variation from Grosscup's (1976) description, although she reports that if undecorated, the interior surface is roughly finished.

Mangos Engraved (Bell 1960:55–57, Plate 12; Grosscup 1976:233, Plate 13)

The Mangos Engraved type has a buff- to gray-colored paste with a carbon streak. The type is typically red-rimmed, with occasional red bands. Engraved designs occur on the undecorated (plain) exterior portion of the vessel below the rim. The designs are arranged around the vessel and commonly occur in panels or bands (Figure 6.27). Design elements include diamonds, triangles, knotted nets, and nucleated circles. Bell (1960), in her study of Amapa ceramics, identified a plain orange and a red-rimmed variety of Mangos Engraved. Grosscup accepted that the plain orange variety fell within the type. The plain orange variety studied by Bell consisted primarily of shallow bowls with flaring sides and some incurved bowls. There were also three bowls with incurved bodies and flaring rims; two of these were tripod vessels. One of these had incising that imitated gadrooning. The red-rimmed variety were all incurved bowls with an incised band on the exterior made of geometric and curvilinear elements or net-and-tassel elements; one specimen exhibited a band of repeated stylized bird elements. Three jars of the red-rimmed variety were tripods with rattle legs. Bell notes that two aberrant bowls assigned to the red-rimmed variety (similar in color and form) included traces of a geometric, negative paint design. The rims are painted dark red. Bell further indicates that all the vessels she examined appear to have a very thin, light orange slip.

Mangos Engraved is assigned to the Cerritos phase at Amapa. Mangos Engraved was identified in low numbers at Rincon de Panal, Tecualilla, and Cuautla Site 3.

FIGURE 6.27. Examples of plain and red-rimmed varieties of Mangos Engraved (adapted from Bell 1960).

Sentispac Red-on-buff (Bell 1960:48–50, Plate 8; Grosscup 1976:235, Plate 9g)

Sentispac Red-on-buff is a very rare type at Amapa, bearing resemblances to Sentispac Polychrome and to Cerritos White. A few sherds were scattered in Late period deposits at Amapa, and five vessels were recovered from burial contexts. The paste color ranges from reddish to tan, and a black core is present. The surface is slipped with color ranging from cream to light yellowish buff. Decorative elements consist of red bands on and below the rim and geometric, circular, cruciform motifs painted on both the exterior and interior of the vessel (Figure 6.28). The red decorative paint is often slightly smeared, indicating that the vessels were polished after being painted.

Bell describes one jar and four bowls of Sentispac Red-on-buff. The jar is a squat, flat-bottomed vessel with a flaring, relatively wide neck and low shoulder. Three of the bowls are deep with flaring sides, and the other is shallow with straight sides. Designs include bands and geometric, circular, and cruciform elements. All vessels were slipped, light yellow-buff in color with rust-red decoration. Fifteen sherds of Sentispac Red-on-buff were identified at Tecualilla; it was not represented elsewhere.

FIGURE 6.28. Examples of Sentispac Red-on-buff vessels from Amapa (adapted from Bell 1960).

Tuxpan Red-on-orange (Bell 1960:48, Plate 7 Lower; Grosscup 1964:300, 1976:237, Plates 9e, 16a–f)

Grosscup postulated that Tuxpan Red-on-orange was the diagnostic type for the proposed Tuxpan phase and that it persisted into the Cerritos phase at Amapa. The paste of Tuxpan Red-on-orange varies from red to dark gray in color, with a medium coarse paste. It comes in a variety of vessel forms including bowls, jars, ollas, and molcajetes. The vessels are painted in dark red that frequently has brown or purple tones. The paint is often smeared, indicating that the vessels were polished after painting. Decoration includes red rims and geometric elements within a band on either the interior or exterior walls of the vessel. Additionally, both "realistic" and geometric designs adorn the interior bottoms of the bowls. Bell reports that one of the two specimens she examined had a rabbit head painted on the interior bottom of the bowl that appeared to be a representation similar to the Aztec day-sign *tochtli*. Another she examined had a cross-banded design with small geometric and curvilinear areas filled with crosshatching (Figure 6.29).

In the Marismas the sites of Arriñitas, Panales Test Mound B, Rincon de Panal, Tecualilla, and Cuautla Site 2 yielded Tuxpan Red-on-orange. It generally appears to have co-occurred with Tuxpan Engraved, except at Venadillo, where no Tuxpan Red-on-orange was reported. Sweetman (1968, 1970:42) provides a description based on 27 sherds. Temper ranged from very fine to fine and sometimes included red inclusions. Paste color ranged from buff to orange to gray, and cores ranged from medium and dark gray to black. Sweetman indicates than none of the sherds were slipped while noting that Grosscup (1964) describes the type as lightly slipped. Other than this, Sweetman notes no variation from Grosscup's original description. She does, however, comment on Grosscup's observation that some vessels are decorated as Botadero Incised, Tuxpan Engraved, and Cerritos Polychrome on the exterior while the interior is decorated as Tuxpan Red-on-orange. In her analysis, Sweetman identified a similar example, although she classified the exterior decoration as Roblitos Polychrome.

Tuxpan Black-on-orange

The origin of the Tuxpan Black-on-orange type designation is unclear. Foy (1970:12) notes the presence of 12 sherds of this type at Tecualilla. However, in her subsequent sherd tallies for Tecualilla, Bell did not mention the type (see Chapter 5). Here it is assumed to be a variant of Tuxpan Red-on-orange. Neither Bell (1960) nor Grosscup (1964, 1976) mentions this type, although Grosscup (1976:237) notes that the red on Tuxpan Red-on-orange is typically a dark red.

Aguaruto Polychrome (Kelly 1945:29–30, Figures 7–8)

Aguaruto Polychrome is described as a black-and-red-on-buff type with boldly executed designs. It has a fine pinkish paste with a gray core. The surface is well smoothed but rarely exhibits polish. Vessels are unslipped, with the exterior ranging from yellow-tan to rosy tan in color. Pitting of the interior of bowls, due to weathering, was noted.

Rims lack a border or may have a simple red border that is sometimes bounded on the lower side by a black line. This most frequently occurs on sherds with exterior borders in black and red or sometimes on sherds with borderless rims. The principal designs occur in a band on the exterior surface. The band is most commonly divided into panels by vertical black lines. These panels

FIGURE 6.29. Examples of Tuxpan Red-on-orange sherds.

commonly contain solid red circles surrounded by small red dots, which alternate with panels that contain a small ellipse above a series of horizontal red lines or variations of this theme (Figure 6.30). The bottoms of vessels appear to be undecorated, but our information is too incomplete to be certain. Other design elements include tadpole-like figures and right-angle triangles similar to the Culiacán feathered arch.

Kelly states that its shape, bold open design, and nearly complete lack of black outlining, which gives the type an almost unfinished appearance, are the primary characteristics of Aguaruto Polychrome. She suggests a comparison between Aguaruto Polychrome and Sauer and Brand's (1932:64) Tacuichamona Polychrome. In the Marismas, only a few sherds of Aguaruto Polychrome were recovered, from Panales Test Mound B and Tecualilla.

Aztatlán Ware (Culiacán Variety) (Kelly 1945:27–29, Figure 6)

The Aztatlán Ware from Culiacán consists of two varieties, a fine ware that resembles Aztatlán Ware from the Chametla area (Fine—Figure 6.31) and a local version that is coarse and heavy (Heavy). It is not clear which is represented in the Marismas collection or why such a distinction was made. The paste is of "good quality," and the surface is slipped and well smoothed with some polishing. It has a red rim with either a single or three red stripes on the interior. Immediately below the exterior rim is the main decorative band, which is bounded above and below by parallel incised lines. Within this white or cream band are delicate and precisely incised geometric design elements that include diagonal wavy lines, paired diagonal lines, frets, stepped diagonals, scrolls, and circles. Kelly notes that the painted band tends to be smeared. Below the decorative band, a red checkerboard design from which pendant lines emanate covers the bottom exterior of the vessel.

The Heavy Aztatlán Ware is a coarser, heavier version of the Fine subtype (Figure 6.32). Kelly describes it as thick, cumbersome, indifferently smoothed, and hardly polished. The paste is somewhat coarser than that of the Fine version, which ironically Kelly found surprising, considering the overall crudeness of the ware. The paint is usually a rusty black or less often a maroon rather than red. White paint is present, but it is so badly preserved that nothing can be said of how it is used. The most common incised elements are parallel diagonal or vertical lines, scrolls, circles, and zigzags. The bottoms of the vessels are decorated in a checkerboard with pendant lines, as in the Fine version.

The Fine variety of Culiacán Aztatlán Ware was recovered from Rincon de Panal, and both the Fine and Heavy varieties were found at Tecualilla and Cuautla Site 3. These occurred in low numbers.

Culiacán Incised Ware (Kelly 1945:84–86, Figure 43)

Culiacán Incised Ware has a moderately fine paste that is sand-colored with a gray core. Surface colors vary from orange-tan to dun to rusty black. The exterior is undecorated, and the incising on the interior is not well executed. The bottom of the vessel is incised with

FIGURE 6.30. Examples of Aguaruto Polychrome sherds.

FIGURE 6.31. Examples of Culiacán Aztatlán Ware, Fine variety.

FIGURE 6.32. Examples of Culiacán Aztatlán Ware, Heavy variety.

FIGURE 6.33. Examples of Culiacán Incised Ware sherds.

crosshatching, which is usually more complex than a simple crosshatch, and above that, generally, the interior of the vessel is covered with two to four bands of incising. The band just above the crosshatching is narrow and often contains a series of eyelike elements. The uppermost band is bordered above and usually below by three parallel incised lines. This band is filled with hatched isosceles triangles that are pendants from the upper incised border. Decorative elements in the central sidewall bands are intricate and always geometric. They are most commonly curvilinear, and hatching and crosshatching are typically used to elaborate the designs (Figure 6.33). Kelly notes that the description on Culiacán Incised Ware is constrained by the lack of a large sample and the fact that most of the materials assigned to this ware are small sherds. The ware is similar to Aguaruto Incised, but Kelly depicts it as an inferior ware that is also simpler in design. One sherd of Culiacán Incised Ware was tentatively identified at Chalpa.

Culiacán Band Incised (Kelly 1945:100–102, Figure 53; Sauer and Brand 1932:63)

Culiacán Band Incised appears to be a variant of Culiacán Dun and Culiacán Fluted wares. Its paste is extremely fine and rosy tan to terra-cotta in color with a gray to black core that can approach the thickness of the sherd. The most common vessel form is the tripod bowl. The decoration is on the exterior in a band below the rim (Figure 6.34). Incising is well executed and better than on most Culiacán incised types. Hatched triangles running along the bounding bands are a common theme, as is hatching in general. Because of the size of the sherds, it is difficult to identify the variability of the design elements and patterns. Meanders and scrolls are used along with a number of "odd" motifs. There is some indication of bands being broken into a vertical band. Kelly notes that this is a relatively scarce ware and undoubtedly local. Culiacán Band Incised occurred in low frequencies at two Marismas sites, Rincon de Panal and Chalpa.

Cixto Incised (Foy 1970; Sweetman 1968)

The initial description of Cixto (Cisto) Incised was based on a single sherd from the site of Juana Gomez. Sweetman describes the type as having an orange paste, black core, and very coarse temper and ranging from 14 to 16 mm thick. It is unslipped, and the surface is uneven, having been roughly smoothed. The design, at the top of the sherd, is simple, consisting of deeply incised triangles, some of which are highlighted with red paint. Subsequently, Foy (1970) reports the presence of seven sherds of Cixto Incised at Tecualilla. She (1970:16) suggests a temporal range of AD 950–1250 for the type (Early Conchera into the Middle Conchera period). In subsequent publications, Sweetman (1970, 1974) does not discuss Cixto Incised; thus it is unclear whether she reclassified the sherd or whether she abandoned the tentative type.

Roblitos Polychrome

Sweetman (1968, 1970:36–38, 1974:74) also defined a second type, Roblitos Polychrome. Within this type she identified three named varieties and a group or category

FIGURE 6.34. Culiacán Band Incised sherd.

of miscellaneous sherds that also appear to fall within the type. These varieties were defined based on differences in design areas and motifs. The Variety I–Broad Red Interior Band description is based on 10 sherds. The variety is described as having an orange-brown to dark brown paste with no core or a black core and very fine temper with coarse red inclusions. Sherds ranged from 5 to 8 mm thick. Sweetman provides the following description of the decoration:

> The exterior is polished but not slipped, one sherd has a white band below the rim. The rim inclines towards the interior and is crossed with black lines. The rectangles between the lines are alternately paste color and red, the rim acting as a visual transition from the exterior to the interior. The interior has a broad undecorated light mauve-red to purple-red band immediately below the rim between 50–58 mm deep. Then follows a black line, and then a purple-red line with a line and dot motif. The interior is then decorated with a curvilinear design in black, purple-red, and orange paint on the base color [1968:71; see 1970:36].

Variety II–Stepped Fret is based on eight sherds and is described as having a brown paste with a black core and very fine temper. Sherds were 7 mm thick. It is unslipped. Sweetman gives the following description of the decoration (Figure 6.35):

> The interior is polished, with traces of red paint. The rim has patches of red, creamy orange, and white paint separated by black lines. The exterior has a band below the rim with a stepped fret as the main design. Colors include black, red, pinky-red, creamy orange and white. Occasionally red lines are painted over the other colors [1968:71; see 1970:36–38].

Variety III–Narrow Band is defined on the basis of six sherds. This variety has a brown to brownish black paste and a black core, and it is unslipped. Sweetman's description of decoration is as follows: "Interior either mauve-red or deep red, with two red lines below the rim. The exterior has a polychrome narrow band below the rim, design in pink, red, and white" (1968:72; see 1970:38).

Sweetman's (1968:72, 1970:38) Miscellaneous Sherds group within the Roblitos Polychrome type is based on 11 sherds. These have light brown, brown-orange, and brown paste. Cores are absent in some and dark gray or black in others. The sherds are unslipped. The decoration is described as follows: "The interior can be rough, painted red, or have a linear design. The exterior decorations include white, cream, pink, mauve-red, gray, and black" (Sweetman 1968:72).

Sweetman identified Roblitos Polychrome at Arriñitas. Subsequently, Davis (1969) identified Roblitos Polychrome at Panales Test Mound B, and Foy (1970) reports the type at Tecualilla. Foy (1970:6) notes that she

FIGURE 6.35. Roblitos Polychrome (Variety II).

did identify examples of Varieties II and III but made no distinction between varieties in her analysis. Bell seems to have considered Roblitos Polychrome an Amapa type in her tallies (see Chapter 5). However, Sweetman (1970: 102) notes that Betty Bell (via personal communication 1968) informed her that Roblitos Polychrome was also found at Peñitas and that a Roblitos Polychrome tecomate was on display at a local museum there.

Middle Conchera Period

Mazatlán Polychrome (Kelly 1938:19–21, Figure 8a–b, Plates 8f, 9b–d; Sauer and Brand 1932:61, Plate 1d)

Mazatlán Polychrome was first defined by Sauer and Brand. It is a red-on-buff with the red outlined in black. The examples from Chametla have a very fine tan paste with a gray core. Despite the fine paste, Kelly states that it has a "somewhat heavy, dead" feel to it (1938:20). It is completely covered with a cream-colored slip and has a good sheen (polish?) that resists wear. Design motifs are geometric and include triangles, stepped diagonals, and circles, again, all of which are outlined in black. The decoration is heavy and well executed (Figure 6.36). Mazatlán Polychrome was identified at Panales Test Mound B, Rincon de Panal, Chalpa, Tecualilla, and Cuautla Sites 1, 2, 3, and 4.

Weil (1974:43) reports two Mazatlán Polychrome sherds. The temper is fine, and the paste falls into his yellow-brown category. A yellow-brown slip is present on both surfaces, and both surfaces are well polished. The decoration includes angular geometric elements with arcs and circles and areas of red paint that are outlined with black paint. One of the sherds exhibited circular elements with leaflike striations. Both sherds measured 6 mm thick.

El Taste Polychrome (Kelly 1938:21–22, Figures 6–7, Plates 1a–b, 8b–e, 9e–f)

Kelly notes that the paste of El Taste Polychrome is identical to that of Mazatlán Polychrome, although El Taste Polychrome is a thinner type. Kelly recognized two varieties. The first is a red-on-gray/red-on-cream style (also identified as El Taste Red-on-gray and El Taste Red-on-cream—listed as such in Chapter 5) found on simple hemispherical or lipped (slightly flaring rim) bowls with decorated exteriors (Figures 6.37–6.38). Designs are red geometric lines on gray to cream backgrounds. The slightly flaring lips are often painted. Occasionally one or two thin black lines are added. Kelly states that this version could be confused with Red-rim Decorated; however, the painting on this variety of El Taste Polychrome is better executed.

The second variety is a true polychrome. It typically comes in the form of a globular bowl. The interior of the lip is decorated with red and black striping, with the black often being dotted with white paint. The interior of the bowl, other than the lip, is not decorated. The exterior of the lip is rose red. A narrow strip of orange to buff outlined in black and bearing "slight" overpainted designs in red is present below the lip. Below this is a wide rose-red band on which is painted an intricate conventionalized pattern in light magenta-rose, white, and orange-buff that is outlined in black. This band is bounded by black lines. Below this band is a black line from which teardrop-like elements extend toward the

FIGURE 6.36. Examples of Mazatlán Polychrome.

FIGURE 6.37. Examples of El Taste Polychrome Red-on-gray/cream sherds.

base of the bowl (Figure 6.39). Kelly notes that some sherds and several purchased specimens exhibit some variation in the paint colors used and the organization of the decorative elements.

Grosscup (1976:256) notes that El Taste Polychrome is similar to the Amapa type Iguanas Polychrome and that it is possible that some sherds classified at Iguanas Polychrome may actually have been El Taste Polychrome as a trade ware. In the Marismas, El Taste Polychrome was recovered at Panales Test Mound B, Rincon de Panal, Chalpa, Tecualilla, Mound B and the earthen portion of Mound C at Venadillo, and Cuautla Site 2. The only locality where it was relatively abundant was Mound C at Venadillo; otherwise it occurred in low numbers.

Weil (1974:43) describes 23 sherds of El Taste Polychrome. The temper varied from fine to medium, and the sherds were commonly slipped on both surfaces. Both the paste and the slip colors fall within his yellow-brown class. Interior surfaces are smooth, and the exterior surfaces vary from smooth to polished. Weil reports no significant variation in decoration from Kelly's (1938) original description. Sherd thickness varied from 4.5 to 9.0 mm.

El Taste Red-bordered
(Kelly 1938:22–23, Plates 8a, 9a)

The El Taste Red-bordered type typically occurs as a slightly flaring bowl. The paste of El Taste Red-bordered is the same as that of Mazatlán Polychrome and El Taste Polychrome, and it is slipped in a buff to gray color on both the exterior and interior. The edge of the rim is flattened, giving it a cross section that is squared in appearance. The edge of the rim is painted black and is sometimes ticked with white lines or with dots. Below the rim, on the interior of the bowl, is a broad red band that varies in color from mahogany to the El Taste rose red. The exterior below the rim is rarely painted. Kelly notes that Mazatlán Polychrome, El Taste Polychrome,

FIGURE 6.38. El Taste Polychrome Red-on-cream bowl.

FIGURE 6.39. Examples of El Taste Polychrome (top row specimens are from Venadillo).

FIGURE 6.40. El Taste Satin jar from Venadillo.

and El Taste Red-bordered are likely individual styles within a single ceramic ware.

In the Marismas, El Taste Red-bordered is reported at Panales Test Mound B, Rincon de Panal, Chalpa, Tecualilla, Mound B and the earthen portion of Mound C at Venadillo, and Cuautla Site 1. Frequencies between sites vary from low to relatively high.

El Taste Satin (Kelly 1938:26, Figure 10a, Plate 10m)

El Taste Satin has the same fine paste as El Taste Polychrome. Both the interior and exterior are slipped and highly polished. Colors range from gray to dun to orange (Figure 6.40). Small jars typically have a low-flaring rim like El Taste Polychrome, but the type also appears as simple hemispherical bowls.

Grosscup (1976:257) suggests that its presence or the presence of an analogous Amapa type may have gone unrecognized as any such sherds may have been simply categorized as a plainware. However, he notes that something very similar was recovered from mortuary contexts at Amapa.

In the Marismas, El Taste Satin was identified in very low frequencies at Panales Test Mound B, Rincon de Panal, Chalpa, Tecualilla, and Cuautla Site 1. At Mounds B and C at Venadillo, it accounted for fewer than 20 percent and just over 20 percent, respectively, of the decorated Chametla types identified. Weil (1974:43–44) describes a single sherd of El Taste Satin. The sherd had fine temper, both surfaces were slipped in a yellow-brown color, and both surfaces were polished. Weil mentions that the sherd exhibits an incised design that appears similar to the painted decoration found on El Taste Polychrome. Kelly (1938:26) makes no mention of decoration on El Taste Satin. It is possible that what Weil identified as El Taste Satin is actually a sherd of Sentispac Buff and the reported incising is an example of the imitation gadrooning found on that type.

Heavy (Deep) Ribbed (Kelly 1938:26)

As noted above, Kelly briefly discusses a group of miscellaneous modeled wares at Chametla that includes a shallow and deep ribbed type. A single sherd reported

FIGURE 6.41. Examples of Cerritos Engraved bowls from Amapa (adapted from Bell 1960).

as this Heavy (Deep) Ribbed type was recovered from Tecualilla.

Mazatlán Engraved

A single sherd of the Mazatlán Engraved type is reported from Chalpa, where it is listed as a Chametla type. The origin of this classification is unclear. It may be misidentified, or it may be an engraved or incised variant of Mazatlán Polychrome.

Botadero Red-on-white (Grosscup 1976:228, Plate 127k–p)

Botadero Red-on-white is similar to Botadero Black-on-buff and Tuxpan Red-on-orange. It has a gray-colored paste with a black core, and the temper is medium coarse (Type II—Grosscup 1976:216). Hemispherical and straight-sided bowls have red rims, simple geometric designs in bands below the exterior rim, and representational designs that often have small red dots painted on them. This type has been identified in low numbers at Tecualilla and Cuautla Site 3.

Cerritos Engraved (Bell 1960:52, Plate 10 Lower; Grosscup 1976:228, Plate 129a–e)

Grosscup's description of the Cerritos Engraved type is based on 12 whole vessels recovered from mortuary contexts; it was not commonly identified in the Amapa sherd assemblage. The paste ranges in color from buff to pink; it has a gray core; and the temper is coarse (Type III). Grosscup indicates that the surface was slipped white to gray, although Bell (1960) reports that only three of the 12 vessels she examined were slipped; the others were unslipped. She further noted that the slipped vessels had a clear white slip, while the surface color of the unslipped vessels was gray-white. Bell reports that the slipped bowls are decorated only with imitation gadrooning and small circles, while the unslipped bowls have an exterior band decorated with geometric and curvilinear elements, crosshatching, or a net-and-tassel design; many of these also exhibit imitation gadrooning (Figure 6.41). Grosscup notes that the designs are similar to those on Tuxpan Engraved and indicates that this appears to be a true engraved type. Cerritos Engraved was identified at Rincon de Panal and Tecualilla. It occurred in relatively low frequencies.

Ixcuintla Polychrome (Bell 1960:67, Plates 17–19; Grosscup 1976:231, Plates 135–153)

The Ixcuintla Polychrome type description is based on a large corpus of whole or nearly whole vessels that come in an array of forms ranging from bowls to jars to legged vessels to annular-base plates and jars to a ceramic effigy in the form of a human foot with a sandal. Molcajetes also occur. The paste of Ixcuintla Polychrome ranges from black to pink, and the temper, from medium coarse to coarse. The exterior surfaces are slipped in buff-orange or red and are well polished; the interiors are not slipped. The red slip may extend to near the base, or it may cover the entire exterior. Grosscup notes that a major variety consists of a red rim only. Bell notes that decoration is in red and white on the buff surface. Incising is sometimes present, but it is rare and not elaborate.

Bell notes that, in terms of decoration, this type is highly variable and suggests that multiple varieties may ultimately be derived from what is now treated as a single type. A variety of design elements are represented, including geometric, curvilinear, circular, zoomorphic, and possible symbolic forms; geometric elements are less frequently represented than the others. Birds and dogs are recognizable, as is a common motif thought to represent a coatimundi (Figure 6.42).

Marismas sites at which Ixcuintla Polychrome was identified include Rincon de Panal, Tecualilla, Mound C at Venadillo, and Cuautla Site 3. The type was not particularly abundant.

Ixcuintla White-on-orange (Bell 1960:52, Plate 10 Upper; Grosscup 1976:232)

Grosscup indicates that his description of the Ixcuintla White-on-orange type is based on Bell's analysis of six whole ceramic vessels (five shallow bowls and one plate) recovered from mortuary contexts and that sherds of this

FIGURE 6.42. Examples of Ixcuintla Polychrome vessels from Amapa (adapted from Bell 1960).

type are indistinguishable from body sherds of Ixcuintla Polychrome. The surface is light orange in color. Decoration is the same as that of Ixcuintla Polychrome. Bell indicates that bowls have geometric and curvilinear elements on the exterior and the plate has a stylized serpent head, which are executed in white paint (Figure 6.43). Ixcuintla White-on-orange was represented at Rincon de Panal and at Mound C at Venadillo.

Ixcuintla Engraved

The Ixcuintla Engraved type is not described by Bell (1960) or Grosscup (1964, 1976) for Amapa. It is listed as being present at the sites of Tecualilla and Cuautla Site 3. If identified by Bell, it is possibly an engraved or incised variant of Ixcuintla Polychrome or perhaps an engraved type in the vein of Cerritos Engraved. It is also possible that it is a local Marismas ceramic type.

Mangos Polychrome (Bell 1960:65, Plate 17 Upper; Grosscup 1976:233, Plates 157–159)

The Mangos Polychrome paste ranges from tan to buff in color, with a gray to black core; the temper is medium coarse. The surface is slipped and ranges from tan to buff in color. Bowl size varies but is moderately deep in most cases. Bell notes that occasionally there is a single red band of paint on the exterior edge of the bowl, but typically the rims are red. Decoration is restricted to the interior of the bowls, and commonly there are white bands

FIGURE 6.43. Example of an Ixcuintla White-on-orange bowl from Amapa (adapted from Bell 1960).

below the rim and near the base of the bowls that are outlined in red or black. Grosscup indicates that the design elements and layout are similar to those of Tuxpan Red-on-orange. Bell indicates that the interior design consists of a central panel bisected by two interlocked curvilinear elements or mostly filled with a stemmed spiral element that is pendant from the edge of the band or is quartered. The remaining areas are filled with geometric or curvilinear elements or, occasionally, a representation of an animal (Figure 6.44). A single sherd of Mangos Polychrome was identified at Tecualilla.

Otates Polychrome (Grosscup 1976:234, Plate 12h–l)

Grosscup states that Otates Polychrome is rare at Amapa. Its paste is tan to buff in color and lacks a carbon streak, and its temper is medium coarse (apparently similar to that of Santiago White-on-red). The interior surface is unslipped, smoothed, and gray in color, while the exterior surface is slipped with colors ranging from orange-red to purple-red. The slip tends to craze. The sherds suggest that a thin-walled olla with a thick-walled, outcurving rim is the predominant vessel type.

The design is made up of black lines outlined in white; the areas between black lines are filled with white, and white dots appear on black lines. Design patterns could not be determined, because of the small size of the sherds analyzed. Grosscup believes this to be an elaborate version of Santiago White-on-red, only one sherd of which is reported from the Marismas (Weil 1974:44), and he would have named it Santiago Polychrome if that name had not been applied to a type identified at Peñitas. A single sherd of Otates Polychrome is reported from Panales Test Mound B.

Peñitas Engraved (Bell 1960:58, Plate 13 Lower; Grosscup 1976:234, Plate 14f)

Peñitas Engraved paste color ranges from red to tan, with a black to gray core. The temper is coarse. This type appears as moderately shallow bowls with slightly incurved walls. Bell indicates that the interior and exterior bottoms, although well polished, are unslipped. The exterior sides exhibit a band of red slip through which designs are engraved. Grosscup states that this type is the same as Santiago Engraved except for the addition of one or more white painted bands, which occur on the exterior just below the rim, at the base of a decorated

FIGURE 6.44. Examples of Mangos Polychrome from Amapa (adapted from Bell 1960).

FIGURE 6.45. Peñitas Engraved bowl from Tecualilla.

area, between decorated areas, or just below the rim on the interior of the vessel. Bell indicates that the red band is typically divided into six panels. Decorative geometric and curvilinear elements occur in alternating panels. The alternate panels are filled with white paint. Peñitas Engraved was identified at Tecualilla (Figure 6.45) and at Mound C at Venadillo.

Santiago Engraved (Bell 1960:57–58, Plate 13 Upper; Grosscup 1976:234, Plate 14a–e)

The paste of Santiago Engraved is tan with a black to gray core, and the temper ranges from medium coarse to coarse. Some vessels are red slipped and highly polished, while others are orange and unslipped. Forms include bowls, tripod bowls and jars, and an annular-base jar. Designs include complex geometric and representational forms that are deeply engraved through the slip (Figure 6.46). Grosscup notes that in some cases a narrow band or medallion is left unslipped; the medallion may also be incised. Grosscup suggests that this type may have developed out of Mangos Engraved.

Bell's observations of vessels from Amapa cemeteries indicate that some of the bowls are slipped orange and the exteriors are covered with a wide band of red paint that covers most of the exterior body. Designs include geometric, curvilinear, and ray elements, as well as net designs that are arranged in bands. One of the bowls examined had four separate panels pendant from a narrow band. Several of the jars had rattle-tripod feet. Two jars exhibited band designs on the body and neck that used the same design elements and layouts as found on bowls. One jar had a wide band on the body that bore

FIGURE 6.46. Examples of Santiago Engraved sherds.

stylized human faces as well as a variety of curvilinear and ray elements. The designs on the neck were curvilinear or ray elements. One double jar with two legs on each body was among the vessel forms noted. Each of the bodies was decorated with three medallions engraved through red paint. The design elements used were probably symbolic.

In the Marismas, Santiago Engraved was recovered in relatively low numbers, from Rincon de Panal, Tecualilla, and Cuautla Site 1.

Sentispac Buff (Bell 1960:42, Plate 4 Lower)

Sentispac Buff is a highly polished, yellowish buff type that often appears in the form of a small jar with a flaring rim. Bell reports two globular high-necked jars with gadrooning on the body and a modeled band at the base of the neck above the body of the jar. Two such specimens were tripod jars with rattle legs.

A specimen recovered from Venadillo exhibited shallow linear grooves encircling the body just below the rim and the upper body of the jar (Figure 6.47). Below the shoulder the body of the jar is additionally decorated with very shallow, faint incising that imitates gadrooning. Two faintly incised lines encircle the vessel just below the neck, and near midway between the neck and the shoulder of the jar, a second band of three faintly incised lines is present. The decoration resembles the painted decoration found on some El Taste Polychrome.

Sentispac Buff was identified at Panales Test Mound B, Rincon de Panal, Chalpa, Tecualilla, and Mounds B and C at Venadillo.

Sentispac Polychrome (Grosscup 1976:235, Plate 9f)

The Sentispac Polychrome type was defined from a grab sample from the large site of Sentispac adjacent to the modern town of Sentispac and from sherds recovered during the excavations at Amapa. Grosscup notes that this is a distinctive but rare type at Amapa and suggests that it is not a local type.

This type has a red to tan paste with a gray core and medium coarse temper. Decoration consists of alternating narrow, horizontal lines of black or dark brown and red. Rectangular panels, possibly pendant from the lines, containing geometric patterns are also present. Bell notes, based on her analysis of a single whole vessel from a mortuary context, that the pendants extend from the lower band. Three sherds of this type were identified at Tecualilla.

Sentispac Black-on-buff

The origin of the Sentispac Black-on-buff identification is unclear. This designation is listed as a type at Tecualilla. It may be an aberrant Sentispac Red-on-buff or a variant of that type. As discussed above, a Chametla Black-on-buff has been identified at some Marismas sites.

Tuxpan Engraved (Bell 1960:52–55, Plate 11; Grosscup 1976:236–237, Plate 15)

Grosscup states that Tuxpan Engraved includes all incised monochrome sherds not classified as Cerritos Engraved (painted or slipped white) or Santiago Engraved (slipped red and engraved through the slip), and it does not include any of his Early period incised types. This is a somewhat catchall type, and Grosscup indicates that the category likely includes sherds of other engraved types including Mangos Engraved and Botadero Incised. The reader is directed to his description of the type for a full discussion of its variability. Bell states that Tuxpan Engraved comes in a variety of jar and bowl forms.

Because the type is a catchall, paste color is highly variable. Temper is most commonly medium coarse, although it includes coarse temper as well. Vessel form is highly variable between and within form categories. Bowls, jars, ollas, vases, and goblets are represented;

FIGURE 6.47. Sentispac Buff jar from Venadillo.

some have annular bases and tripod legs. Molcajetes are also found.

Designs are executed by incision, engraving, and excision (Figure 6.48). Red, white, and blue pigments are sometimes rubbed into the incisions postfiring. Bell notes that all bowls are decorated on the exterior, with bands that consist of a great variety of geometric, curvilinear, and net elements and net-and-tassel designs; many also have imitation gadrooning. Geometric elements include combinations of frets, scrolls, circles, crosses, and parallel lines. Grosscup notes that bowls commonly have an incised scalloped line at or just below their greatest diameters and that there are vertical lines extending from the point of each scallop; the scallops and pendant vertical line are frequently deep and broad enough that the incising could be called grooves or gadrooning. A knotted net that ends in a fringe is a common design element found on ollas.

Tuxpan Engraved was recovered at the Marismas sites of Arriñitas, Panales Test Mound B, Rincon de Panal, Tecualilla, Venadillo, and Cuautla Sites 3 and 4. In most cases, Tuxpan Engraved sherds occurred in relatively higher frequencies. Sweetman (1968, 1970:40) describes seven Tuxpan Engraved sherds. The temper was very fine and sometimes included coarse red inclusions. The temper ranged from buff to brown to reddish brown to black. Cores were buff, brown, and black. Sweetman noted no variation from Grosscup's original description.

Aguaruto Exterior Incised (Culiacán) (Kelly 1945:37, Figure 17)

Aguaruto Exterior Incised (Culiacán) is an unpainted type with a moderately coarse paste. The surface color varies from tan to gray to black. The incising is more variable; the incised lines are thinner and less even than those found on Aguaruto Interior Incised. Kelly indicates that the incised designs suggest affiliation to other Aztatlán wares. Designs include a chili-like motif, ellipses crossed by strands, guilloche, and arched panels below the main design band. According to Kelly, this is reminiscent of miscellaneous incised sherds associated with the Aztatlán complex at Chametla and Acaponeta. Only a few sherds of Aguaruto Exterior Incised were found at the site of Tecualilla.

Aguaruto Interior Incised (Kelly 1945:37, Figures 15–16)

Aguaruto Interior Incised is an unpainted ware with a surface color that ranges from tan to gray to black. It has a moderately coarse paste. The surface is smoothed and slightly polished. The incising is confined to the interior of the vessel, although there is occasionally the

FIGURE 6.48. Examples of Tuxpan Engraved sherds from the Marismas Nacionales.

presence of two or three incised lines on the exterior. The incisions are relatively wide; they appear to have been made with a blunt instrument. Aguaruto Interior Incised often appears as a tripod bowl with rattle supports. Three incised lines occur on the interior just below the rim. Generally, below these lines is an open field. The lower sides and bottom of the bowl are incised with a variety of decorative motifs and elements, some of which intrude into the open field above them. In some instances, the open field is replaced with a band of incised designs that include vertical lines, frets, and curvilinear elements. Crosshatching is present as a decorative element on the bottom of the bowl. A single sherd of Aguaruto Interior Incised was found at the site of Tecualilla.

Aztatlán Ware (Culiacán) (Kelly 1945:27–29, Figure 6)

The Culiacán Aztatlán Ware is divided into two groups: Fine Aztatlán Ware, which resembles its Chametla counterpart (see above), and Heavy Aztatlán Ware, which has no southern counterpart. Fine Aztatlán Ware is not abundant at the Culiacán localities. It appears mostly as medium bowls, and the sherds include one that appeared to be from a ceramic smoking pipe. The vessels are unslipped and well smoothed, and the paste is of "good" quality. The rim is red, and on the interior, below the rim, there are either one or three red stripes. On the exterior, immediately below the rim is the main design band, which is either white or cream. It is bounded by parallel incised lines and contains delicate and well-executed incised geometric lines that include paired diagonal lines, frets, step diagonals, or circles. Below this band, red checkerboarding is painted on the base of the vessel. Kelly suggests that it is likely that pendant red tapering lines finished off the decoration as was the case with the Culiacán Red-rim Decorated.

Heavy Aztatlán Ware is described as a coarser, heavier version of Fine Aztatlán Ware that is thick, cumbersome, not well smoothed, and only slightly polished. The paste is more granular but still fine given the overall heaviness and "untidy" finish of the ware. Rather than red paint, the paint is most often rusty black or a deep maroon; little in the way of white paint survived, but it appears to have been present. Incised designs occur in a band and consist of parallel diagonal or vertical lines, scrolls, circles, zigzags, and other elements. The painting and incising are not well executed. Below the incised band, the bowls are painted with checkerboard and pendant lines as with the Fine Aztatlán Ware.

In the Marismas, Culiacán-type Aztatlán Ware is reported at Rincon de Panal, Tecualilla, and Cuautla Site 3. It occurs in relatively high frequencies at Tecualilla.

Red-rim Decorated (Culiacán) (Kelly 1945:25–27, Figures 4–5)

Culiacán Red-rim Decorated is a rather thick ware with a fine paste that varies in color from buff to gray. A darkish core is sometimes present. The surface color ranges from cream to orange-tan; it is well smoothed and fairly well polished. The painted decoration is red bordering on maroon. On the exterior, below the red rim, are two narrow red stripes, and below these is the main design band. This band, commonly divided into panels by vertical or diagonal bars, contains various motifs including guilloche, one suggestive of the "Saw" of Culiacán Polychrome, and less frequently crosshatching, concentric diamonds, and other geometric elements. Below this band are three red stripes, and below these is checkerboarding that is followed by pendant tapering lines. The interior rim is also red, and below it are one or two additional red stripes. Only the site of Tecualilla produced a few sherds of Culiacán Red-rim Decorated.

Early Culiacán Polychrome (Kelly 1945:43–59, Figures 22–27, Plate 1a)

Early Culiacán Polychrome is one of three major groups of polychromes defined by Kelly at Culiacán, these being Early, Middle, and Late. She notes that all share identical paste, all are polished subsequent to painting, and all share certain "conventional" color combinations that include red-on-white, red-on-orange, black-on-gray, and white-on-black. In terms of Early Culiacán Polychrome, Kelly states that it is "remarkably" homogeneous in terms of paste, the uniformity of coloring, "the underpaint incision constant" (1945:58), and the brilliance of its finish, which she indicates makes Early Culiacán Polychrome readily distinguishable from the Middle and Late variants. The color of the paste, which is very fine, ranges from gray-tan to rose-tan, with a dark gray core. Underpaint incision (vessels are lightly incised before being painted) and rose pigment are, for the most part, confined to the Early variety, and although occurring in both Early and Middle polychromes, gray is "markedly stronger" in the Early variety as well.

Major decorative motifs and elements vary with other decorative elements, vessel form, interior treatment, and borders. In terms of design elements, the "Saw"—an interlocking step-fret coil—is the most common motif to appear on the Early polychrome, although it is more prevalent on Middle polychrome. Briefly, other motifs include a wavy panel (a rectangular panel vertically bisected by wavy or zigzag lines), a cross panel (a rectangular panel containing X elements with small circles between the arms), the funnel (a series of contiguous isosceles triangles—alternate triangles are usually colored and "heavily filled," while the intervening triangles may be a solid color and are not filled), the bar panel (a rectangular band bisected by pairs of vertical parallel lines), the guilloche (two dot-filled twisted lines), the fantail monster (a highly stylized and elaborate motif of a monster with a segmented body, angular appendages, and a fan-shaped tail), the feathered arc (an arch or full circle with the outer edge rimmed with small petal-like elements), the plume (an asymmetrical motif consisting of multiple plume-like or leaflike elements that rise vertically before bending to one side), the bird (depicted with a round body with a prominent beak and tail—also found as a bodiless, double-headed bird), the face (red band bordering a series of alternately inverted trapezoids containing conventionalized faces), and miscellaneous monsters. Again, these motifs and elements are executed in co-occurring color combinations. Kelly (1945:58–59) provides a succinct summary of the stylistic clusters that distinguish Early Culiacán Polychrome (Figure 6.49).

At the Marismas sites investigated, Early Culiacán Polychrome was identified at Panales Test Mound B, Chalpa, Tecualilla, and Mound B at Venadillo. Generally, relative frequencies of occurrence were low, except in one case at Tecualilla.

Navolato Polychrome (Kelly 1945:30–31, Figures 9–10)

Navolato Polychrome is a black-and-red-on-buff type. Decoratively, there is some overlap with Aguaruto Polychrome; however, designs on Navolato Polychrome are far more complicated and varied. Typically, each design motif is outlined in black. This makes the decoration look more extensive than that found on Aguaruto Polychrome, which appears to have greater open background space.

Decoration is almost exclusively confined to the exterior of vessels. Designs, which are executed in bold, free strokes, are horizontally distributed below the rim but not necessarily within panels as in some of the other Aztatlán complex types. A rim border is lacking on approximately half of the specimens. When present, some have a narrow red or black border stripe, while occasionally others have a series of stripes. The interior rims have one to three red stripes or a red band bordered by black lines. Motifs vary greatly; major ones include

FIGURE 6.49. Examples of Early Culiacán Polychrome sherds.

dome-shaped figures, concentric ellipses, a chili-like element on a vertical stem, step frets, curvilinear or angular step frets, and plumed or tailed elements. Additionally, Navolato Polychrome shares some elements with Aguaruto Polychrome. Most commonly these include vertical panels filled with parallel diagonal or horizontal lines or large solid circles. It also shares some designs with the Culiacán polychrome series, most notably the "Saw," some suggestion of the fantail monster, dot-dash filler, and circle-ellipse filler. Navolato Polychrome was identified in small numbers at Tecualilla, Mound B at Venadillo, and Cuautla Site 3.

Peñitas—Red Incised/Engraved (Entries 1–4, 7–8) (Bordaz 1964:151–156)

The paste of the Peñitas Red Incised/Engraved group or series of types is classified as Mitlán fine (Bordaz 1964: 145–148). This is a group of incised wares with the incised decoration variously being referred to as incised or engraved by Bordaz. Entry #1 of this grouping exhibits burnishing on the interior face and exterior base, or they are painted orange. The exterior wall and inside lip are painted red. Decoration is incised through the red paint, and one or two incised parallel lines are present near the edge of the interior lip. The lower edge of the red painted area is framed by one or two straight lines over a scalloped line. Several design layouts are typical. The first consists of a band engraved on the exterior lip and the aforementioned line or lines on the interior lip with the exterior body engraved, and the second is like the first with the addition of an engraved band between the lip and the engraved decorative band on the exterior body. At the lip, design motifs include individual or combinations of geometric or curvilinear elements. Step frets with triangular, spiral, or meander filler are most common. Designs on the body are arranged in panels and include large geometric and curvilinear elements, feathered serpents, jaguars, birds, and other naturalistic, stylized, or symbolic figures. Some lines exhibit traces of white paint, suggesting that the engraved line may have been filled with white paint to emphasize the decoration. Although Bordaz frequently refers to the incising as engraving, he does indicate that given the hardness of the pottery, incising likely took place before firing and in general, based on microscopic observation, after drying.

Bordaz notes that Entry #1 is very similar to Santiago Engraved at Amapa and shares design layouts and design elements with Aztatlán Polychrome, Cerro Isabel Engraved (subtypes A and C), and Guasave Polychrome at Guasave (Ekholm 1942).

Entry #2 is a version of Entry #1, which is a bowl with an incurved rim wall. Entry #3 is also the same, but with the addition of white paint decoration to the areas other than those painted red. Bands of white paint are present below the exterior rim; one is present on the interior lip; and they also occur in the decorative panels on the exterior or outlining engraved motifs. Bordaz indicates that white painted areas were engraved before or after the application of the paint. Entry #3 is the same as Amapa's Peñitas Engraved. Entry #4 is also the same as Entry #1, except in terms of vessel form. It also appears to be the same as Peñitas Engraved. Entry #7 only varies from Entry #1 in that red painting is restricted to a smaller portion of the surface of the vessel, thus leaving greater portions of the orange base exposed. Either the red decorated area or the unpainted orange base is engraved, but almost never both. Entry #7 shares the same affinities as Entry #1 with Santiago Engraved at Amapa and shares design layouts and design elements with Aztatlán Polychrome.

In this group, Entry #8 varies the greatest from Entry #1. The vessels are slipped buff or brown and are highly polished. In some instances, the surface is smudged black and highly polished. Designs are like those of Entry #1 with two exceptions: there is some decoration of the entire exterior of the vessel, and there is no engraved line on the surface of the interior lip. Tripods, molcajete plates, and annular-base vessels are represented. This type appears identical to Tuxpan Engraved, Fine Incised, Northern (Sauer and Brand 1932:64), and Aguaruto Exterior Incised. Design motifs are like some on Aguaruto Incised and Red Incised at Guasave (Ekholm 1942), some of the miscellaneous incised wares at Chametla (Kelly 1938), Red-rim Decorated, and both Fine and Heavy Aztatlán Ware at Culiacán (Kelly 1945).

Sherds thought to be of this group were identified at Rincon de Panal, Tecualilla, and Novillero Site 2. These occur in low frequencies.

Peñitas—Polychrome (Entries 43–49) (Bordaz 1964:174–180)

The paste of this group of Peñitas polychromes is Mitlán fine. The exterior of Entry #43 is painted red, with parallel and perpendicular black bands bordered in white and with white circles painted on the black bands. The rim band, both exterior and interior, is framed with a black band. The base of Entry #44 is well burnished, and the color of paint ranges from buff to orange. The interior of the lip is decorated with a series of small rectangles that are variously filled with red, dark brown, orange, and white paint. Each rectangle is separated by a thin vertical line of brown paint. These are then bordered at the bottom by a thin brown line that is overpainted with white dots. Frequently, an additional band, either red or brown, is present below the lower framing line. The rectangles do not extend much to the exterior face of the lip, and they are not bounded by a line. Entry #45 has a reddish slipped base, and sometimes it is only slipped on the interior. The lip band is the same as found on Entry #44, except that it covers equal portions of both the exterior and interior. In both cases, several red and brown parallel bands and lines that are overpainted with white dots are present below the band.

The paneled rim band of Entry #46 is the same as that of Entry #44. However, it differs in that the band is framed on both the exterior and interior by bands of brown paint that are frequently overpainted with white dots. The exterior below the band is painted with various geometric and curvilinear designs in red, brown, and white paint. The interior is decorated with fine linear designs in dark red-brown paint. Bordaz notes that the interior decoration found on this entry is similar to some found on Aztatlán Polychrome from Guasave.

The lip band on Entry #47 is the same as that of Entry #46. Only the exterior rim face is painted. The most common designs are executed in red, brown, orange, and white paint. The most common design elements include a line of dots and dashes usually painted immediately under the rim band on the painted surface of the vessel; bands or lines of red and brown paint (the brown paint is sometimes overpainted with white dots); or large, white painted interlocking curvilinear or rectangular scrolls framed with bands of red and brown paint that are overpainted with white S-shaped and reverse curve designs. The lip bands of Entry #48 and Entry #49 are the same as that of Entry #47. Only the exterior face of Entry #48 is decorated, and the most common design elements are the same as those for Entry #47. Entry #49 comes in the form of jars with the neck decorated in red, brown, and white paint over an orange base. The most common motif is a complex figure that includes simple curvilinear white scrolls or hooks.

Bordaz states that this group shares decorative layouts and designs with Cocoyolitos Polychrome from Chametla, Burrion Polychrome from Guasave, Iguanas Polychrome from Amapa, and Early, Middle, and Late polychromes from Culiacán. In the Marismas, sherds from this group were found at Panales Test Mound B, Rincon de Panal, Mound B at Venadillo, and Cuautla Sites 3 and 4.

Peñitas—Polychrome (Entries 50–54) (Bordaz 1964:180–182)

Bordaz states that another variety of Peñitas polychromes, also including Entry #104, is distinguished by the shape of the rim, the location of painted decoration, and decorative elements. All have Mitlán fine paste. They are all well burnished or slipped with colors ranging from yellowish to reddish buff; they have a red or brown band at the lip; and the edge of the lip is frequently overpainted with white at intervals (dashed?) or with a series of short, parallel dashed lines perpendicular to the rim. The body is painted with parallel bands of red and brown framed in white and overpainted with geometric and curvilinear elements. Common decorative elements include interlocking scrolls with circle filler, bands of ellipses and circles, and a series of dots and dot-dashes. Interlocking step-fret elements are also used.

Entry #50 is distinguished by rim decoration that is confined to the exterior and sometimes one or two thin, overlapping red or brown lines. Entry #51 exhibits exterior decoration only, except for occasional thin red, brown, and white bands or lines on the interior of the lip. Entry #52 is decorated on the interior only, as is Entry #53, which is distinguished by the absence of white paint. Bordaz, however, suggests that the white paint has simply eroded. Entry #54 is decorated on the exterior, with much of the decorated area consisting of two pairs of interlocking zigzag lines executed in red and white paint.

Bordaz found no similar types at Amapa, although there are some similarities in the treatment of the rim (white dashes and form) to El Taste Red-bordered at Chametla and the use of overpainted white dots and rows of dots and dashes at Culiacán and Guasave.

Other Peñitas Ceramic Categories

In addition to the Peñitas types discussed above, Bell lists (see Chapter 5) the following unspecified categories: Painted, Engraved, Incised, Red Incised, White-on-red, and Red-on-white, which appear to be generic types not identified specifically in the Bordaz typological scheme. Although these are unspecified in reference to Bordaz's (1964) entry numbers, there is little doubt that they are most likely Peñitas types given Bell's knowledge of Peñitas ceramics. Painted is most likely a reference to one of the Peñitas bichromes or polychromes, and Engraved and Incised (including Red Incised) likely fall within one of the 17 entries (##1, 2, 3, 4, 7, 8, 9, 10, 11, 22, 24, 66, 70, 90, 99, 105, 131) listed as exhibiting some form of engraving, incising, or gouging. It is worth noting that Bordaz (1964:267) observed striking similarities between the incised wares of Peñitas and those from Culhuacán (Brenner 1931) in the Valley of Mexico.

White-on-red is associated with three entries by Bordaz, Entries #79, #110, and #111. Bordaz (1964:207–208) indicates that Entries #110 and #111 are the same as Santiago White-on-red at Amapa (Bell 1960:53; Grosscup 1976:235). Bordaz does not seem to explicitly discuss a red-on-white type. Sweetman (1970:20) described a single white-on-red sherd she classified as Amapa White-on-red, which Grosscup (1976:223) describes as a tentative type based on four sherds. However, in her tabulation of decorated ceramic types for the Marismas, Bell identifies white-on-red sherds as Peñitas types. Sweetman (1970:44) also describes, based on four sherds, an unspecified white-on-red rim, which she suggests might be Amapa Polychrome. The temper of this group of sherds ranged from very fine to fine, and the paste color was either cream or buff. No cores were noted, and the sherds were unslipped. Both the interior and exterior surfaces were polished, and the sherds were decorated with a red band of paint on the exterior rim that was elaborated with white crosses. Sherd thickness ranged from 6 to 8 mm.

Late Conchera Period

As noted in Chapter 4, none of the sites investigated during the Marismas project appear to have been occupied during the Late Conchera period, which equates to the Santiago phase at Amapa and the La Quinta (Late Culiacán) phase at Culiacán—thus the lack of any decorated ceramics that are specifically assignable to this period.

Summary of Decorated Wares

The decorated ceramic assemblages, bichromes, and polychromes recovered during the Marismas project are dominated by previously established ceramic types from the sites of Chametla, Amapa, Culiacán, and Peñitas. Early, during the Estero periods and the Early to the mid–Middle Conchera period, Chametla types dominate. Amapa types are prevalent, with types from Peñitas and Culiacán being represented as well during the Middle and Late Conchera periods. Many of the untyped decorated wares appear to be variants of previously established types. Nevertheless, a number of the sherds undoubtedly represent either local, perhaps idiosyncratic types; local/regional types not previously recognized; or trade wares of unknown origin. Typological challenges are most likely related to the fact that most of the types are based on sherds rather than whole vessels (see Grosscup 1976:213–215), and since most of the Marismas ceramic material was recovered as sherds, this undoubtedly accounts for some of the issues regarding type or ware identification. Focusing the analysis on identifiable regional decorated ceramic types further complicates ceramic analysis in the area encompassing the Marismas. However, this is not a criticism of the work by previous analysts, for the goal was to identify regional types for ceramic cross-dating and establish a ceramic sequence that would allow chronological placement of the Marismas sites. Unfortunately, more comprehensive analyses were not completed, and thus our current understanding of the typological variability in the recovered decorated ceramic assemblages is limited. The work undertaken by the Marismas ceramicists did result in the identification of two provisional types, Roblitos Polychrome and Cixto Incised (Sweetman 1968, 1970, 1974).

Based on what was recovered it is clear that the prehistoric inhabitants of the Marismas produced and had access to an array of ceramic vessels. Vessel form, treatment, and adornment are highly variable (Figures 6.50–6.59). Vessel forms typically included bowls with flaring to recurved rims and jars with short and long necks (flaring and straight), some in the form of tripod vessels. It appears that forms typically associated with utility wares, such as tecomates and ollas, are, not surprisingly so, absent. Surfaces ranged from unpolished to highly

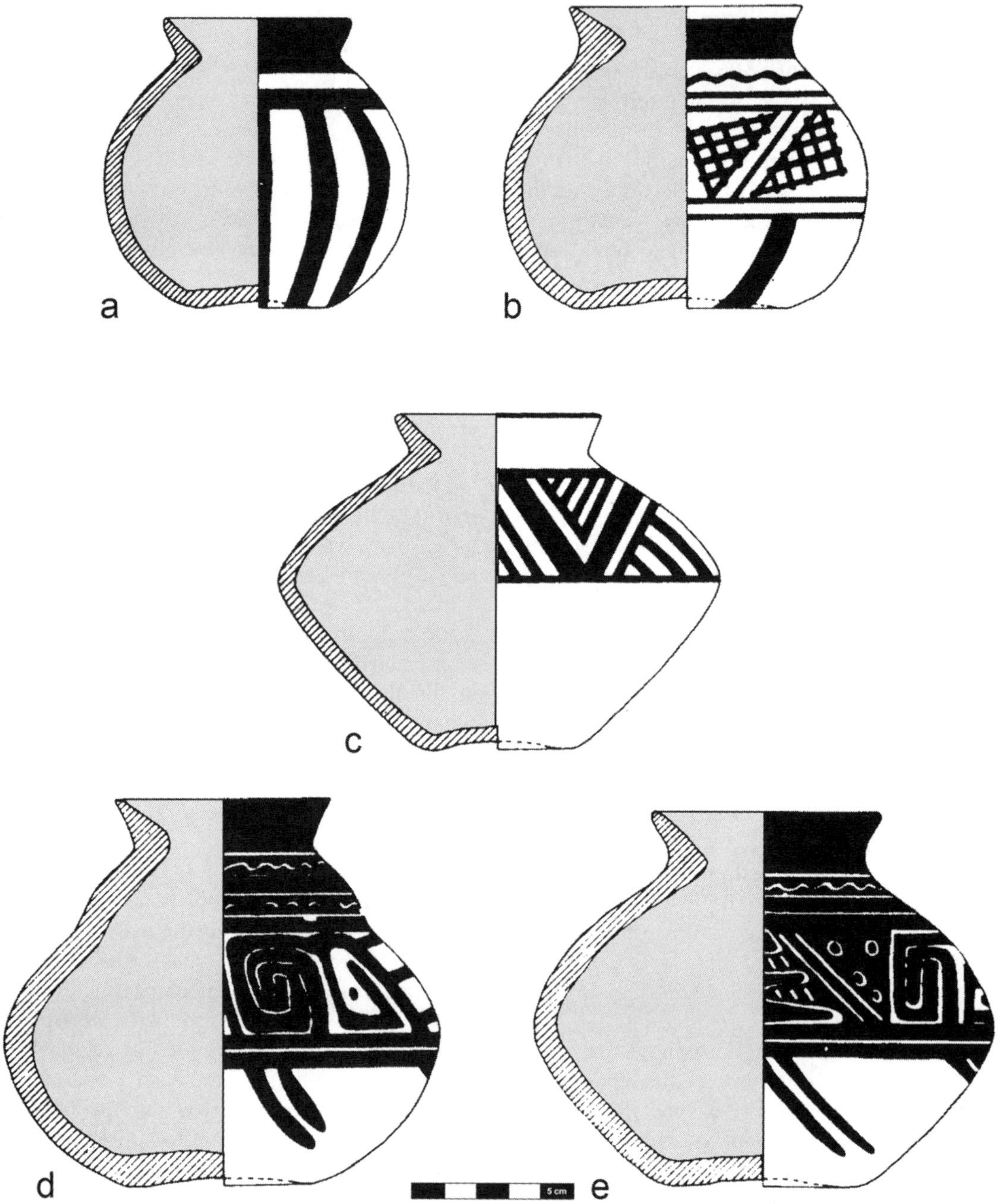

FIGURE 6.50. Decorated ceramics from La Tarjea. (*a*) Small ovoid jar with a very low, everted rim and fairly narrow mouth. The rim, exterior neck, and narrow band below it are painted red. Vertical red stripes, pendant from a lower red band, meet at the bottom of the vessel. Type: related to Chametla Red-rim Decorated? Association: Burial 6. (*b*) Small rounded jar with a low flaring rim and buff slip. The outer neck is painted red, with two narrow red stripes below. A band of opposed, crosshatched red triangles in pairs is painted around the center of the vessel. Extending down from this are three diagonal red lines that meet at what appears to have been a red painted base. Type: probably Chametla Red-rim Decorated. (*c*) Globular jar with a short flaring neck and wide mouth and orange slip. It has a painted red rim. Around the greatest diameter of the body a band consisting of hachured triangles attached to upper and lower framing lines is separated by a zigzag line. This decoration is dark red. Association: Burial 2. (*d*) Composite silhouette jar with a low flaring rim. This vessel is similar to (*e*) but with a more complex design. Alternating narrow bands of red and black are present below the neck, with the black incised. The vessel is buff-slipped and polished. There is a band of interlocking geometric elements in red and black with four pairs of diagonal red lines pendant from this band. Type: Middle Chametla Polychrome Engraved. Association: Burial 3. (*e*) Shouldered jar with a small mouth and flaring rim, buff-slipped, and polished. The neck and both outer and inner rim are painted red. The shoulder is covered with a band of geometric design elements in red and black paint; three of the black elements (elaborated triangles) are also incised. Four pairs of diagonal red lines are pendant from this band. Type: Middle Chametla Polychrome Engraved. Association: Burial 4.

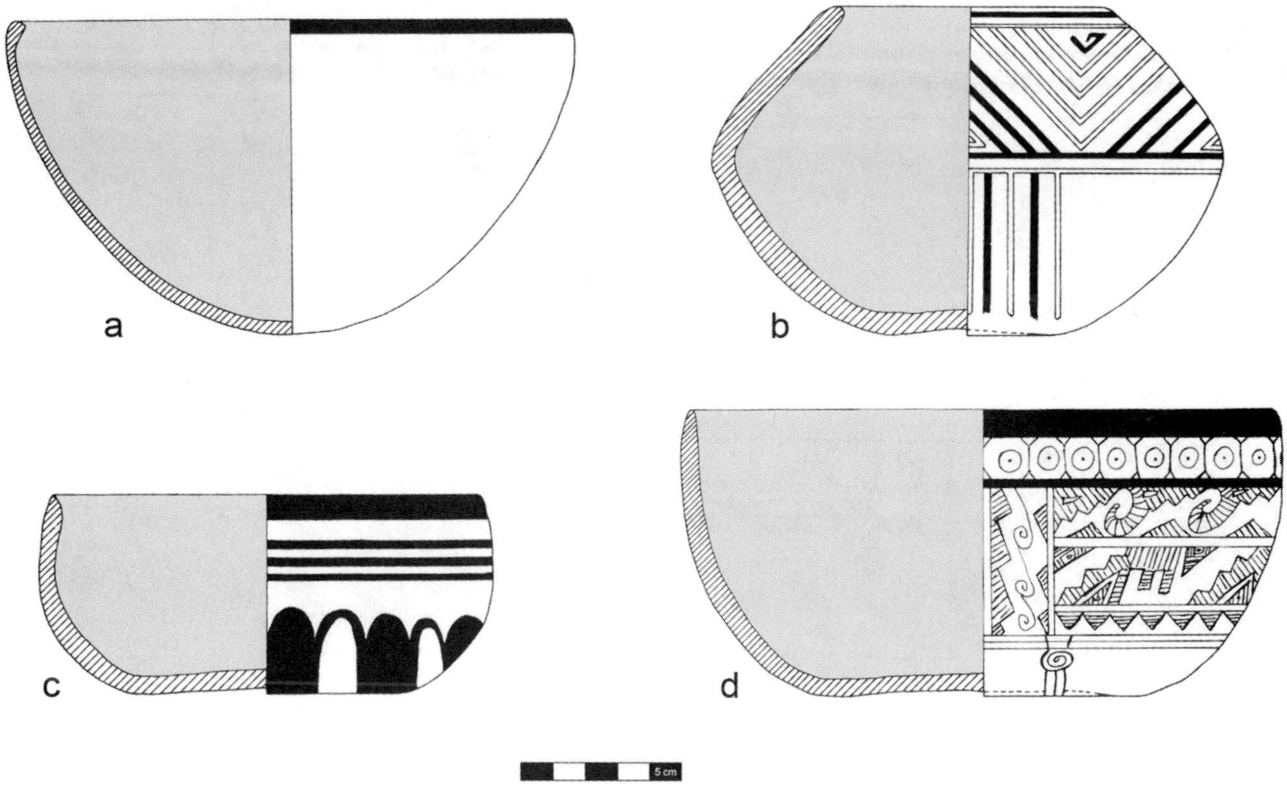

FIGURE 6.51. Decorated ceramics from Chalpa. (*a*) Deep utilitarian bowl. The interior surfaces are unsmoothed and unslipped. The rim and exterior of the rim are painted red. Type: Red-rimmed Utility Ware. Association: Burial 10. (*b*) Small shouldered globular bowl with an incurved mouth and concave base. The upper half of the bowl is decorated with black and red nested triangles on the slipped buff body. These occur within bands. The lower half of the bowl, below the shoulder, exhibits four panels of vertical alternating red and black stripes. The decorative design is carelessly executed. Association: Burial 18. (*c*) Simple open bowl, with a cream interior and red rim. There are red border stripes, with a design band of alternating red open and filled panels below. Type: El Taste Polychrome, Red-on-cream variant. Association: Burial 16. (*d*) Finely produced bowl with a cream interior. Polished bands of red paint (slip-like) and engraving are present on the yellow-brown (orange?) exterior of the vessel. The specimen is a complete hybrid, exhibiting some typical Middle Culiacán band incising but with a pendant that is a conventionalized motif of El Taste Polychrome. Its appearance is also similar to Santiago Engraved. Association: Burial 10.

polished. Decoration included painting, overpainting, incising, engraving, and the application of appliqués.

Finally, in her analysis of the Juana Gomez ceramics, Sweetman (1968:52) made note of several seemingly unusual sherds. She identified two sherds of early historic-period majolica, one from Level 1 and one from the surface of Mound 1. She tentatively identified these as San Luis Polychrome, a type dating to the 1600s and 1700s. Two sherds decorated in gray paint were recovered from Level 7, Mound 1. The sherds had fine temper; the paste was buff to dark gray; the exterior surfaces were polished and painted with black, red, and the aforementioned gray paint; the interior was plain; and they measured 5–10 mm thick. Two sherds of a very rough blackware were also noted; these each measured 9 mm thick. Two sherds with pink paint were also recovered. Sweetman states that similar sherds were noted in the shell midden capping Panales Test Mound A. She further notes that these two artifacts are separated by a considerable span of time; their origin remains a mystery.

Miscellaneous Types or Wares

Sweetman (1968, 1970) describes a number of miscellaneous types or wares; the descriptions are usually based on one or a small number of sherds that did not appear to conform to previously established ceramic types for the region. The origin of these types, whether local, regional, or trade wares, is not known.

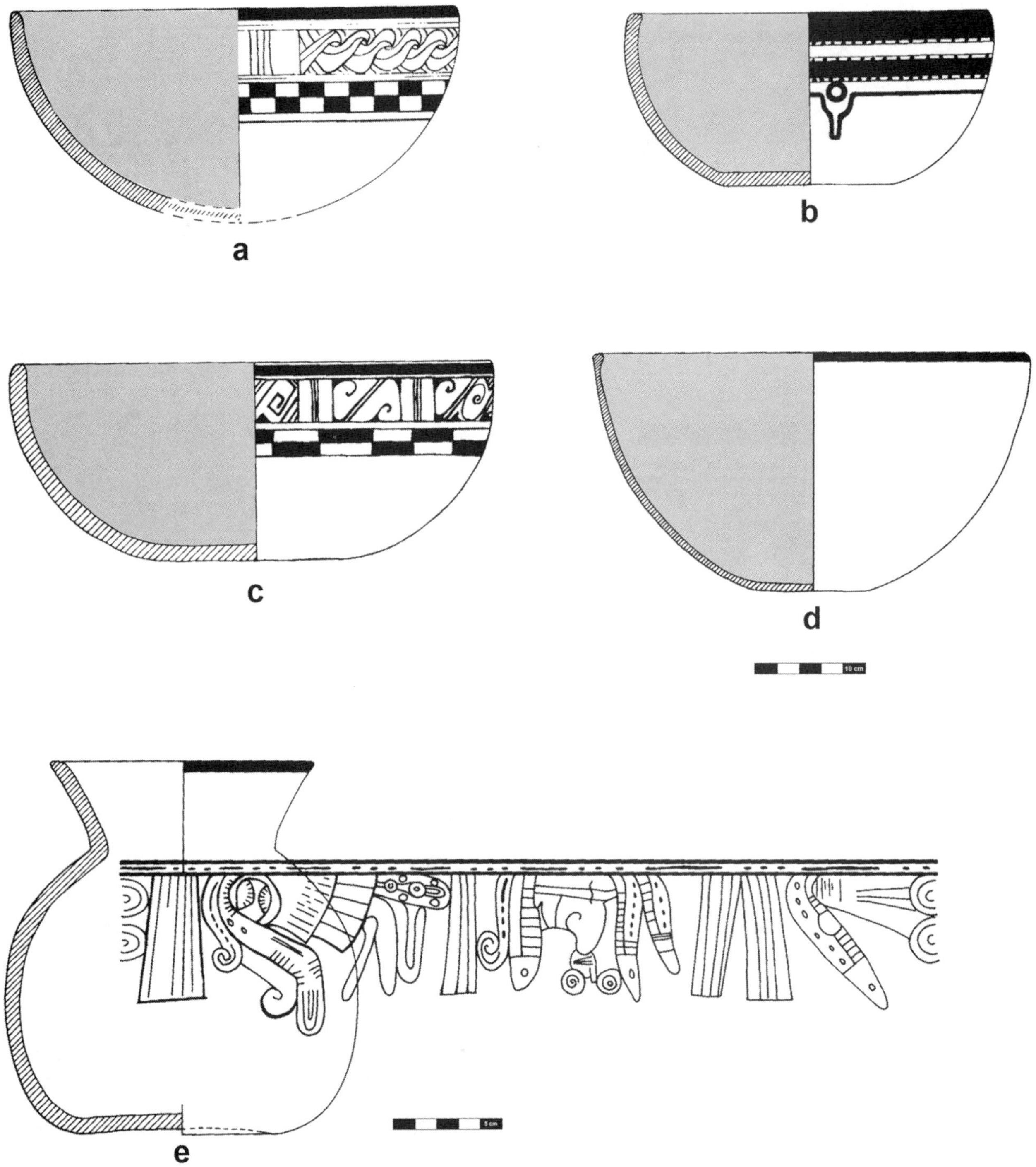

FIGURE 6.52. Examples of decorated ceramics from Tecualilla. (*a*) Bowl, a variant of Aztatlán Ware, unslipped orange, with a red rim, incised through a white band and, below that, checkerboard. (*b*) Bowl, orange-slipped, with alternate red and buff outlined in black, in a typical tassel motif, with white dots on the black lines. Type: El Taste Polychrome. (*c*) Bowl, with a lightly slipped brown-orange surface and geometric band decoration with a checkerboard edge and red rim. Type: Tuxpan Red-on-orange. (*d*) Red-rimmed bowl. (*e*) Jar, unslipped buff, painted in brown, black, and white, with a band with pendant feather ornaments and a pendant face.

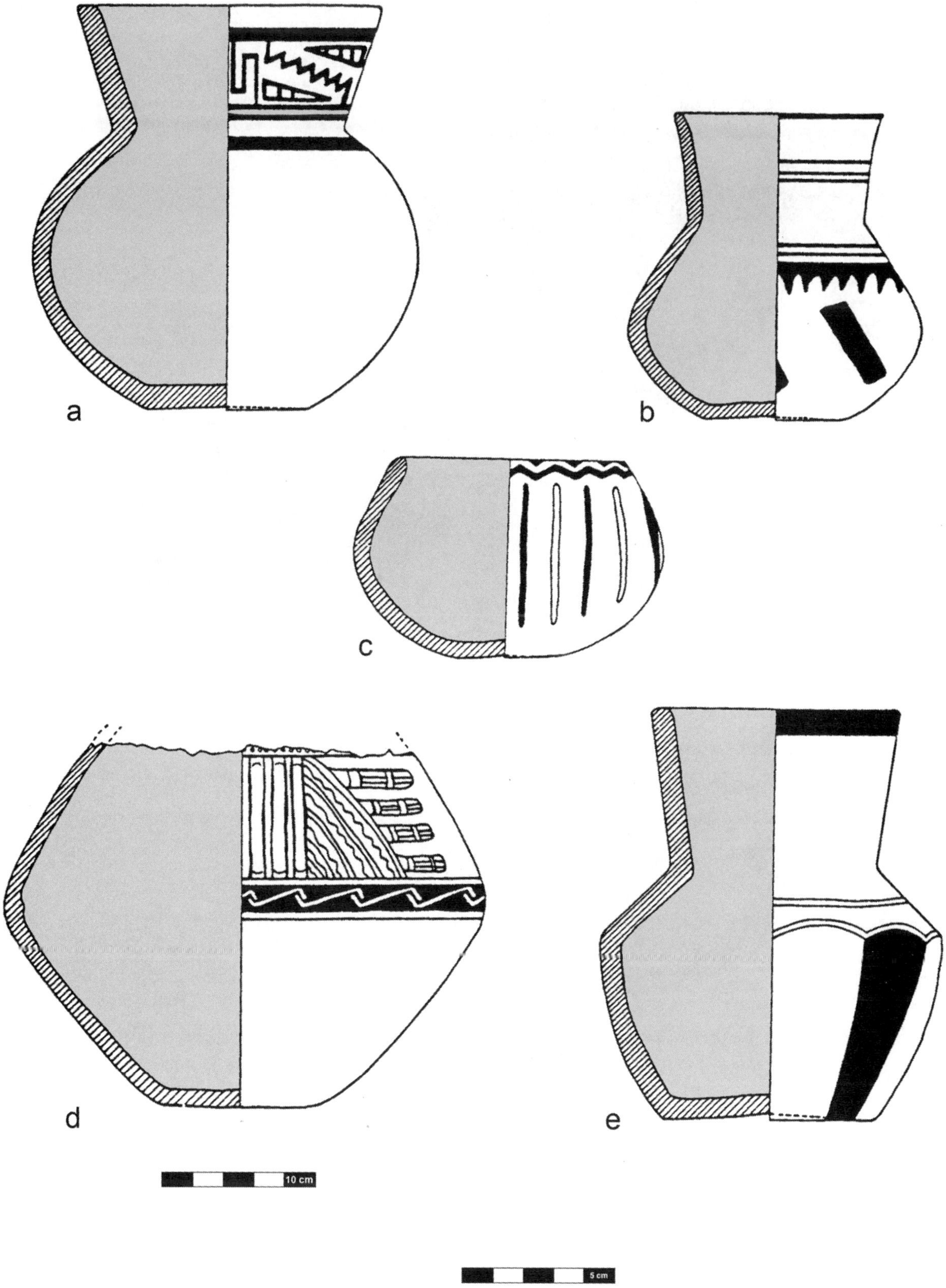

FIGURE 6.53. Additional examples of decorated ceramics from Tecualilla. (*a*) Jar, slipped and stick-polished buff, with a red geometric design on a high neck (see Kelly 1938:41 for an almost identical jar). (*b*) Small jar, slipped buff, with dark red diagonal lines and a band of small pendant triangles, stick-polished (see Bell 1971:708, Figure 10, for an almost identical bichrome jar from Amapa). (*c*) Small incurved bowl, with orange slip, red triangles and black scallops, and alternate red and black vertical lines. (*d*) Large spherical bowl. (*e*) Gadrooned jar, slipped buff, stick-polished, with alternate gadroons filled with red paint. The jar bears some resemblance to Kelly's El Taste Polychrome, Red-on-cream variant, but also shares characteristics of Sentispac Buff. It more strongly resembles general Sentispac Buff ware.

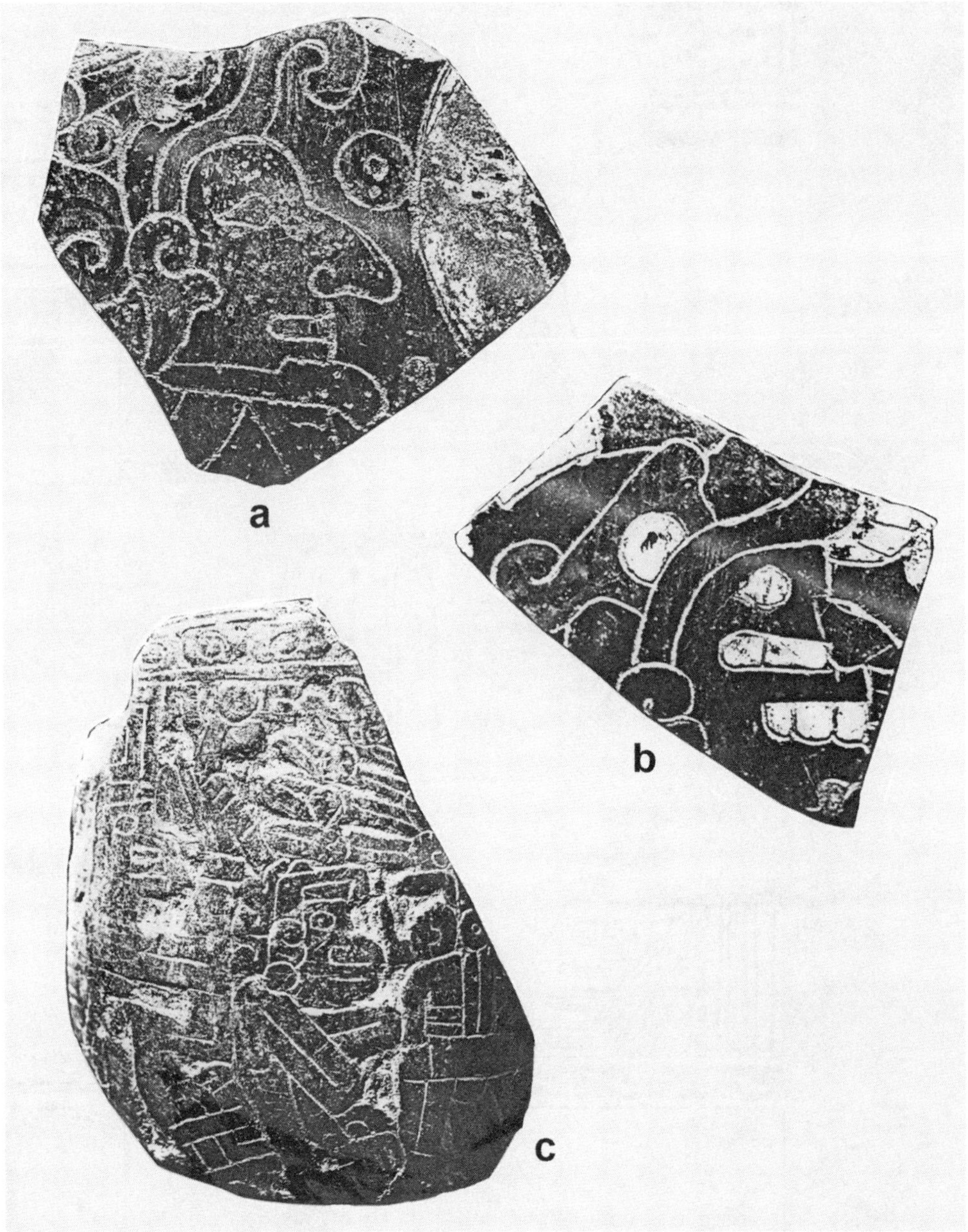

Figure 6.54. Incised sherds from Tecualilla. (*a*) Sherd showing a profile and portions of an elaborate headdress (to the left and above profile). Face painting is indicated by the row of dots and red stripe just forward of the ear. Colors are red, pink, orange (base color), and white. Unlike (b), on which the paint is durable and intact, the white on this sherd has largely flaked off. Type: Peñitas Polychrome (incised). (*b*) Sherd showing a profile with a nose plug and several large teeth. The figure appears to be wearing round earplugs, and portions of a headdress can be seen over the forehead and at the back of the head. The face is painted, half red, half orange, with a narrow red stripe on the orange. Colors are red, white, pink, and orange (base color); it is highly polished. Type: Peñitas Polychrome (incised). (*c*) Partial pipe bowl, polished black. The engraving is so deep that it almost appears to be champlevé. Within the deeply engraved designs, details are executed with fine incising. Shown is one fairly complete Mixteca-Puebla figure in profile with an elaborate headdress. A figure behind this is somewhat eroded, and the details are less clear. In front of the central figure is still another, of which the legs, part of the body, and part of an intricate headdress are visible. The band of concentric circles at the top is typical of Amapa and Peñitas. Type: Cocoyolitos style?

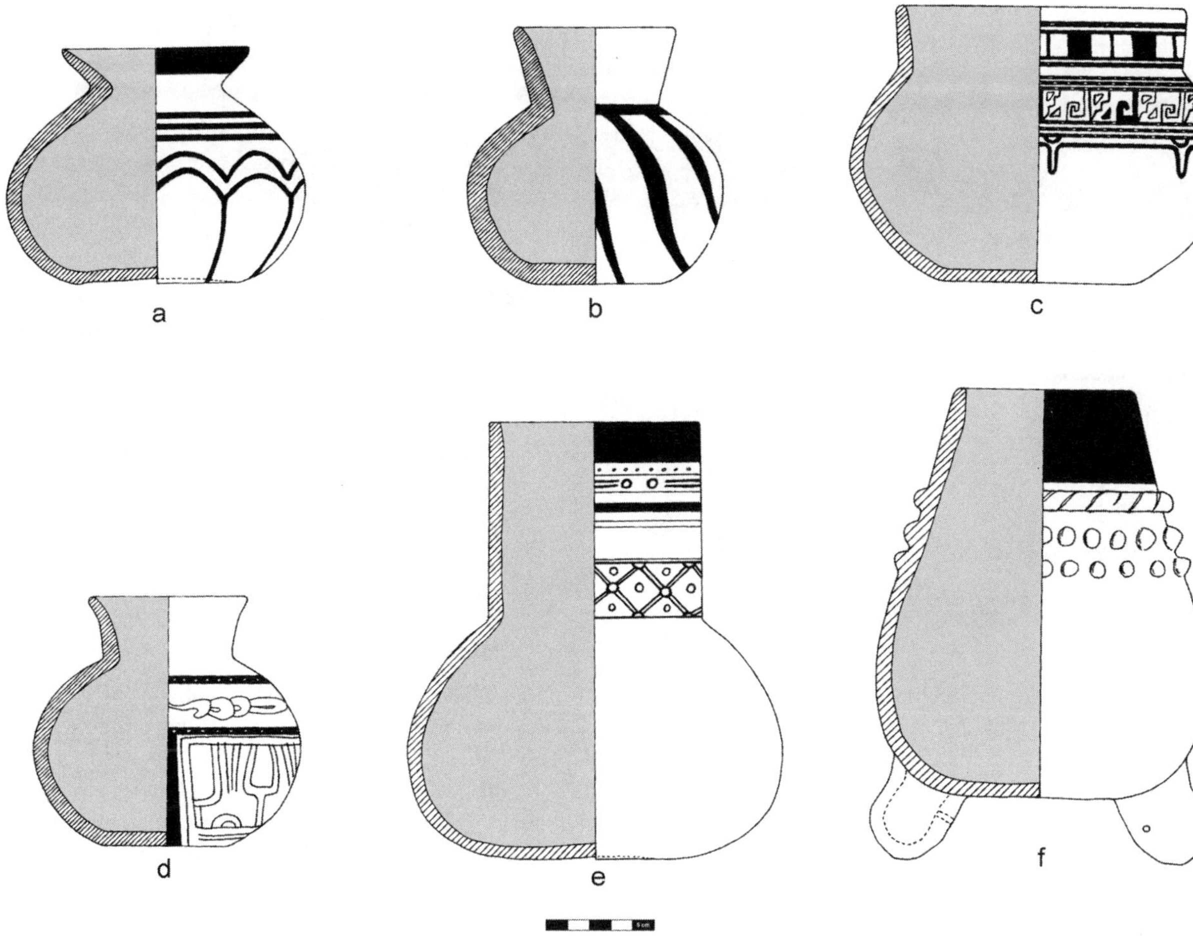

Figure 6.55. Examples of decorated ceramic vessels from Venadillo. (*a*) Red-on-buff/cream, slipped, globular jar with a slight shoulder and low, widely flaring rim. The outer rim is painted red; on the upper body there are three narrow red stripes, and on the lower body, there is imitation gadrooning in red paint (double line at top of panels, single line between). Type: El Taste Polychrome, Red-on-cream (or gray) variant. (*b*) Small rounded jar with a fairly high, flaring neck. It is slipped buff, apparently once polished but now much worn. There is deep diagonal grooving, with wide red stripes at the base of the neck and narrow red stripes on the neck itself. Type: may be related to El Taste Polychrome, Red-on-cream (or gray) variant. (*c*) Slightly shouldered bowl with a wide mouth and a short straight rim. It has alternating squares of red, orange, and white, with white overpainting on a black band on the upper portion of body. Panels consist of rectangular scrolls faced by stepped triangles. Colors include red, black, and white on an orange base, with white-dotted black lines above and below. Type: El Taste Polychrome: the vessel form is typical of El Taste Polychrome, but decoratively it is somewhat of a hybrid. (*d*) Orange-slipped jar with a slightly flared neck. There is a pad band on the upper body with black and white stylized monsters, framed by white-dotted black lines. The lower body has two solid red panels and two with red/orange designs—one checkerboard, one roughly arrow-shaped (Navolato)—separated by black bands with fragmentary white overpaint. Type: probably Early Culiacán Polychrome, though some panel elements resemble Navolato Polychrome. (*e*) Globular jar with a high neck. The body is slipped and polished buff. The upper portion of the neck has concentric lines of red, yellow, and black, with white overpainting on one wide black band and red painting on one buff band. At the base of the neck, there is a net pattern of red-on-buff base, with circles in each square of net. (*f*) Tripod jar, slipped and buff. The neck is painted red; it has an appliquéd fillet, which has been slashed diagonally. Both fillet and hobnails are painted orange.

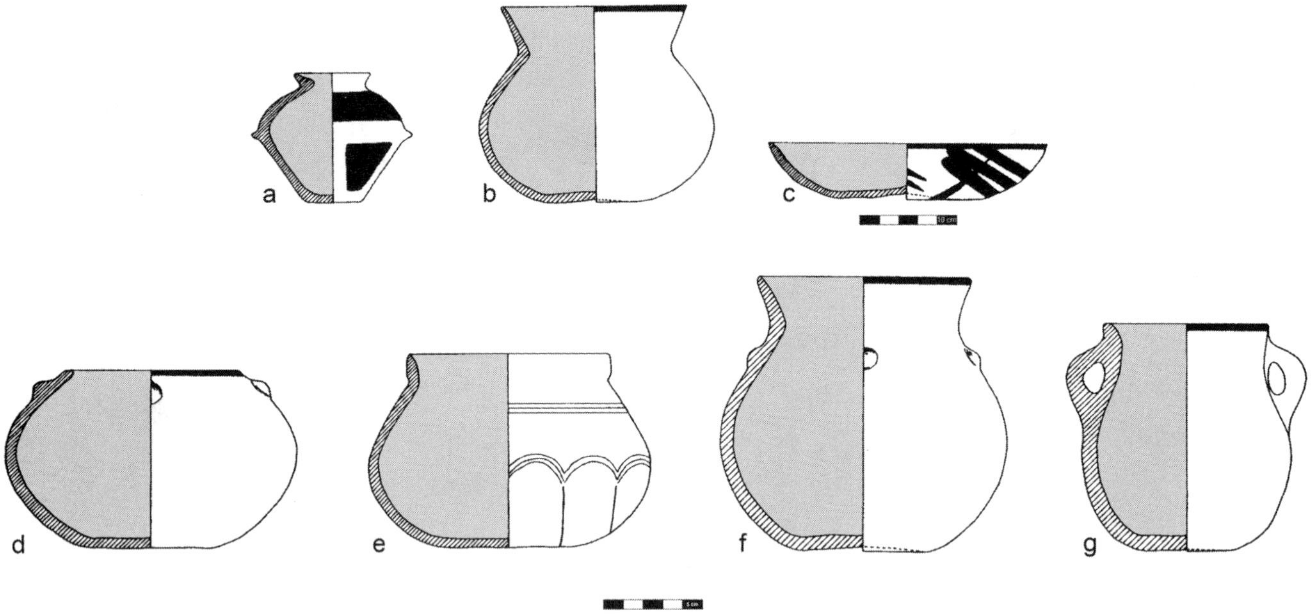

FIGURE 6.56. Additional examples of decorated ceramic vessels from Venadillo. (*a*) Small jar with a near-conical body below the neck and a low flaring rim. The upper body has a broad red painted band. The lower body is decorated with alternating red and white panels (two of each). (*b*) Globular jar with a slight shoulder. There is a thick, but eroded, white slip over the exterior and interior. It has a wide mouth and flaring rim, with traces of red paint on the rim. (*c*) Large, shallow bowl/lid for a burial urn. It is buff-slipped, with a red rim, *mano colorado* decoration on the exterior, and "marbleized" decoration in the form of stripes in a geometric arrangement on the interior. It conforms perfectly with Kelly's (1938:31) description of olla bowls in Red-rimmed Utility Ware, used as covers for burial ollas. (*d*) Incurved globular bowl with thick white slip inside and outside and a red painted rim. Four appliquéd nubbins are equally spaced just below the rim. It is much encrusted with mineralized soil deposits. Association: Burial 12. (*e*) Flattened globular jar with a low, slightly outcurved (not flared) rim and stick-polished imitation gadrooning. Type: Sentispac Buff. (*f*) Globular jar with a slight shoulder, thick white slip inside and outside, a flaring neck, and a red painted rim. Four small appliquéd nubbins are equally spaced just below the base of the neck. It appears to have been used in cooking or heating on a fire. (*g*) Jar with a slightly flaring neck and loop handles, very thick white slip over the exterior and interior, and a rim painted red.

Unnamed Red-on-orange-on-cream (Sweetman 1968, 1970:45)

Sweetman's description of an unnamed red-on-orange-on-cream ware is based on a single sherd with fine temper, brick-red paste, and no core. A cream-colored slip is present on both the interior and exterior surfaces. The exterior is decorated with a bold red and orange painted design on the cream slip. The interior is undecorated.

Unnamed Red-on-buff (Sweetman 1968, 1970:47)

Sweetman describes five unnamed red-on-buff sherds (Figure 6.59), which she indicates may represent two different types. The temper is very fine, the paste is buff-colored, and there is a gray core. Decoration can occur on both surfaces or just the exterior. Design elements include both rectilinear and curvilinear forms. Undecorated interior surfaces are roughly smoothed. Sherd thickness ranges from 5 to 8 mm.

Red-on-white-on-buff Ware (Sweetman 1968, 1970:48)

The Red-on-white-on-buff Ware description is based on three sherds with very fine temper and a paste that ranges from buff to light brown in color, with light gray cores. Slip colors vary from buff to orange. Decoration varies. The interiors of two sherds exhibit a painted brown rim; below this are two painted lines. One sherd has a plain exterior surface with a red rim. Some of the surfaces are elaborately decorated with broad white painted lines that form bands or a grid within which are painted fine curvilinear or rectilinear lines. Sweetman suggests that these may actually be sherds of Lolandis

FIGURE 6.57. Small effigy jar with a human head and face from Venadillo. It is unslipped buff and has a low red rim, and its bottom is painted red. Small loop handles serve as ears. It appears to have been made to be viewed upside down, that is, with the opening serving as the neck. It has a lightly modeled nose and forelock of hair; the eyes are merely rectangular gouges.

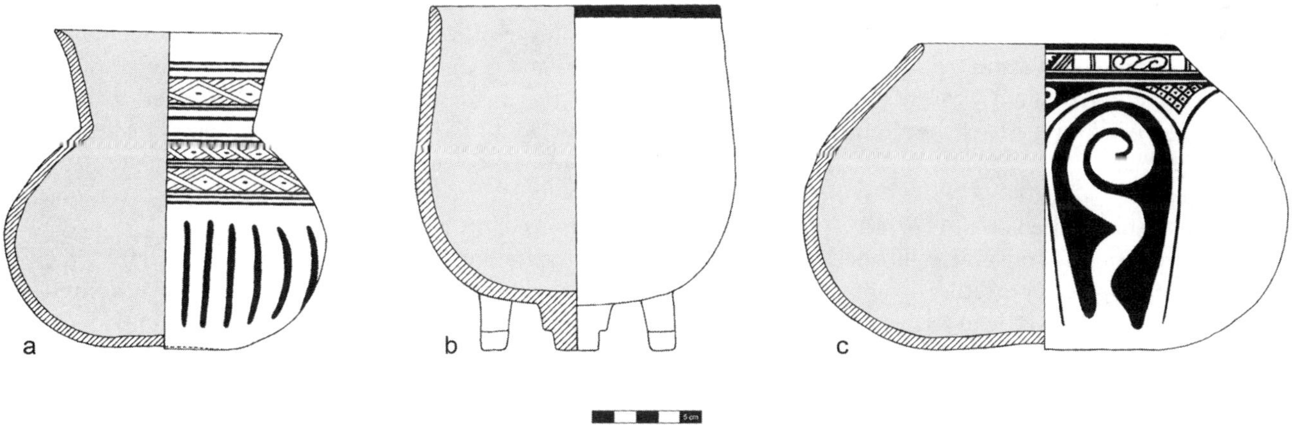

FIGURE 6.58. Decorated ceramics from Rincon de Panal, Panales Test Mound B, and Cuautla. (*a*) Red-on-buff jar with a wide, flaring, high neck from Rincon de Panal. One band is present on the neck, and the two on the upper body contain interlocking zigzags and hachured filler. Vertical red stripes adorn the lower body of the vessel, and it is highly polished. Type: unknown. Association: Burial 1. (*b*) Roughly cylindrical tripod jar with a rounded bottom, from Panales Test Mound B. It is slipped orange and unpolished, with shiny dark red paint around the rim. The slab tripod legs are terraced or stepped in form. Association: Burial 4? (*c*) Red-on-buff bowl with a constricted but slightly flaring orifice and no rim, from Cuautla. The vessel has a creamy buff slip, and it is decorated in a rose-colored red paint. In addition to the painted outline, each gadroon panel is slightly modeled at the top. Narrow bands of red are present around and slightly below the mouth of the vessel. Alternate gadroon-like panels contain a stylized element that is open at the bottom; the other panels contain no decoration. The areas between the lowermost edge band and the top of the gadrooning are filled with small series of diamond-like elements that contain dots within each individual element. Association: Burial 3A.

FIGURE 6.59. Examples of Sweetman's (1970) unnamed red-on-buff.

Red Rim Decorated. Sherd thickness ranges from 5.0 to 7.5 mm.

Red-on-orange Incised (Sweetman 1968, 1970:48)

The Red-on-orange Incised type description is based on a single sherd. The temper is very coarse, the paste color is orange, it has a black core, and it is unslipped. The interior is rough, and the exterior, which is smoothed but uneven, is incised with a series of deeply incised triangles. Sweetman states that "parts of the design are picked out in red paint" (1970:48). It is unclear what she intended to convey; perhaps she meant outlined or filled with red paint.

Miscellaneous Incised (Weil 1974:44)

Weil describes a sample of eight incised sherds that could not be typed or together do not constitute a type. This group varies greatly in terms of color, paste color, slipping, and surface finish. The incising is highly variable, and the design elements include hatched lines, "quasi-parallel" lines, and lines executed with varying degrees of precision.

CONCLUSION

In most respects, the study of prehistoric ceramic assemblages from the northwest Mexican coastal area is still in its infancy. The focus has primarily been on the elaborate and sometimes not-so-elaborate painted wares, while the plainwares have received relatively little attention. The early studies and typological schemes established by Isabel Kelly (1938, 1945, 1947) for Chametla and Culiacán, Gordon Ekholm (1942) for Guasave, Gordon Grosscup (1964, 1976) and Betty Bell (1960) for Amapa, and Jacques Bordaz (1964) are still widely used with little modification. This is in part due to the thoroughness of these studies. Nevertheless, these classificatory schemes are in need of reevaluation and refinement. As with most initial archaeological studies such as these, the limits of time and resources frequently result in inherent ambiguities and the generation of, as they say, more questions than answers. Unfortunately, the limited analyses conducted for the Marismas project have done little to expand or resolve current typological, temporal, and technological issues regarding the prehistoric ceramic traditions of the northwest Mexican coast.

Despite the very traditional approach utilized, the limited nature of the studies, and their ambiguities and those that abound in this discussion, the studies of ceramic assemblages from the Marismas have expanded our understanding of the geographic distribution of the types identified, expanded our understanding of the variability in decoration and form of the ceramic types produced in the area, broadened our understanding of the ebb and flow of external relationships between the occupants of the Marismas, and facilitated our understanding of the occupational history of the area. The changes in ceramic types over time seem to demonstrate the ebb and flow of economic and political power within the region as the influence of Chametla to the north waned and that of Amapa to the south grew.

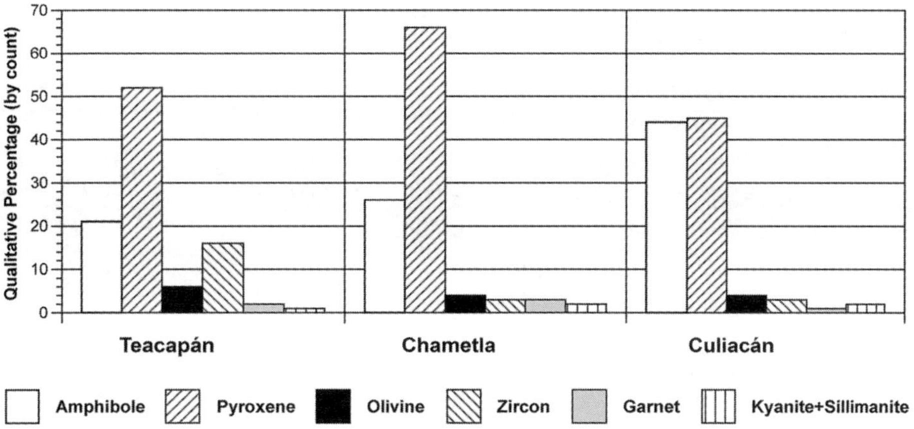

FIGURE 6.60. Percentages of dominant heavy minerals found in sherds from the Marismas Nacionales (Teacapán), Chametla, and Culiacán (adapted from Scott and Cazeau 1976:214).

In the intervening years since the Marismas project there have been significant advances in the study of prehistoric ceramics in terms of the social and economic context of ceramic types and production and exchange. Specifically, studies of production and exchange based on constituent analyses and the identification of clay and temper sources have advanced sociological and economic interpretations regarding the organization and structure of ancient societies. To date, such studies of west Mexican coastal ceramics have generally been few (e.g., Mountjoy et al. 1982). Cazeau (1972) and Scott and Cazeau (1976) began to address this issue with the study of heavy minerals in a sample of sherds from the Marismas, Chametla, and Culiacán. The dominant heavy minerals included amphibole, pyroxene, olivine, zircon, garnet, and kyanite + sillimanite. It was noted that the abundance of amphiboles increased from south to north (Marismas to Chametla to Culiacán), and of the five sherd samples processed from the Marismas and 118 from Chametla, it was noted that the total amphibole-pyroxene percentages were virtually identical (Figure 6.60; Table 6.2). This led Cazeau to ponder the question whether the Marismas and Chametla specimens had been produced at the same locality.

However, Scott and Cazeau (1976) compared sherds from the Marismas with data from several hundred sand samples from the Marismas area (Cottrell 1973). In the Marismas, the amphiboles and pyroxenes from the sand samples were primarily hornblende and augite, respectively. Scott and Cazeau indicate that the amphiboles and pyroxenes represented in the sherds occurred in the same frequency as in the Marismas sands, leading them to conclude that the analytical sample of ceramics was manufactured locally. They also analyzed sherds from Chametla and Culiacán; however, there is some ambiguity in the results. As with the sand samples for these areas, the sherds analyzed exhibited a dominance of amphiboles and pyroxenes. Nevertheless, Scott and Cazeau seem to suggest that the sherds analyzed represent vessels made in the Marismas.

Undoubtedly manufacture of ceramics occurred at sites within the Marismas. The question is, however, whether those vessels were primarily for local household use or whether they were ceramic vessels being produced for exchange. Considering the limited and preliminary nature of the Scott and Cazeau study, this issue cannot be resolved with either the archaeological or geologic data at hand. No evidence of ceramic production was identified during the Marismas excavations, but there is clear evidence that the inhabitants of the Marismas participated in extralocal exchange.

The late J. Charles Kelley (Kelley and Kelley 1971) discusses the results of a thin section study (Ward 1967) conducted on Lolandis Red Rim (Red-rim Decorated). As discussed above, Kelley (1971; Kelley and Winters 1960) found Lolandis Red Rim in quantity during his excavations at the Schroeder site (La Ferrería). The appearance of this type in the archaeological record at the Schroeder site was sudden. This fact together with its abundance led Kelley (1971, 1985) to suggest that west coast peoples may have actually migrated into the uplands of Durango sometime during the Las Joyas phase (AD 700–900) (Foster 1995, 2000), a time, we have argued (see Chapters 2 and 4), when much of the west coast from as far south as Amapa to Chametla (see also Grosscup 1976) was abandoned due to coastal inundation. Kelley found Ward's thin section study to be of particular interest and believed that it supported his argument for an immigration of west coast peoples. Kelley states, regarding Ward's study:

> She found that the Lolandis sherds from Durango shared a common paste-temper complex which was different from both the local Chalchihuites

Table 6.2. Percentages of Heavy Minerals Found in Sherds (Opaque Minerals Excluded).

	Teacapán Area					Chametla			Culiacán	
Minerals	#2	#3	#4	#5	#6	#118	#119a	#119b	#142	#144
Total amphibole	25	33	11	17	17	17	29	34	28	58
Hornblende	11	18	0	3	4	10	22	17	21	36
Actino-tremolite[a]	14	15	11	14	13	7	7	17	7	22
Total pyroxene	50	29	47	74	60	72	64	59	52	31
Clino-pyroxenes[b]	20	24	14	45	21	46	44	50	34	18
Ortho-pyroxenes[c]	30	5	33	29	39	26	20	9	18	13
Epidote	2	0	2	0	0	0	0	0	1	6
Olivine	14	5	13	3	2	7	5	2	7	2
Zircon	4	30	24	4	21	5	1	2	5	2
Garnet	2	0	0	0	0	0	0	0	1	0
Sillimanite	1	0	0	0	0	0	0	0	3	0
Kyanite	0	0	0	1	0	1	4	0	1	0

Note: Sherds are listed by sample number.
Source: Cazeau 1972.
[a] The amphiboles actinolite and tremolite combined.
[b] The clino-pyroxenes combine augite and diopside.
[c] The ortho-pyroxenes combine hypersthene and enstatite.

> wares and the Lolandis sherds from Sinaloa which she examined. There were similarities between Lolandis sherds from Durango and Sinaloa but the differences were too great to allow interpretation of the Durango sherds as trade wares. Apparently, only the migration of Lolandis potters to Durango, where they must have continued to make their own pottery with Sinaloa paste-temper formulae, but were forced to use local clays, can satisfactorily explain the situation [Kelley and Kelley 1971:126].

In other words, Lolandis Red Rim from the Schroeder site seemed to be a mix of west coast and Chalchihuites ceramic technologies.

A more recent study by Vidal (2011) is of relevance as well. She undertook the petrographic analysis of Mexican west coast ceramics recovered at Chalchihuites sites in the highlands of Durango. Sand samples were collected from localities along the Ríos Acaponeta, Baluarte, Presidio, and San Pedro on the west coast. Although the areas share many similarities in terms of the composition of the sands sampled, they are sufficiently distinct from one another. Six sherds (Middle Chametla Polychrome, Tuxpan Engraved, Aztatlán Ware, Red-rim Decorated, Iguanas Polychrome, and El Taste Polychrome) were processed for analysis. Vidal's results are interesting and generally confirm the west coast origin of the presumed west coast types found at the Durango sites. The Middle Chametla Polychrome sherd contains a sand composition that is present in the area between the Ríos Acaponeta and Baluarte; the Tuxpan Engraved sherd seems to have its greatest affinity to the Río Acaponeta sand sample, but the sand sample lacked the abundant amphiboles present in the sherd; the Red-rim Decorated sherd is most like the Río Acaponeta sand sample but also similar to the Río Baluarte sample; the Iguanas Polychrome sherd is most like the Río Baluarte sand sample; and the El Taste Polychrome sherd is most similar to the Río Acaponeta sand sample.

Interestingly, the Aztatlán Ware sherd is problematic. It shares an affinity with the Río Acaponeta sand sample; however, the Río Acaponeta sand sample lacks many constituents (quartzite, chalcedony, and goethite) present in the sherd. The three aforementioned constituents, present in sands from the Ríos San Pablo and Tunal in Durango, led Vidal (2011:104–110) to suggest that the sherd identified as Aztatlán Ware was a product of the Guadiana Chalchihuites area. A situation similar to that proposed by Kelley, a blending of west coast and Chalchihuites ceramic technologies, is a possibility. However, Vidal (2011:153) questions Kelley's (Kelley and Kelley 1971:125–127; Kelley and Winters 1960) suggestion that west coast peoples may have migrated into the uplands of Durango during the local Las Joyas phase, bringing with them a ceramic technology that incorporated local source materials to produced a ceramic ware in the tradition of the west coast. It is also possible that the Aztatlán sherd analyzed by Vidal represents a local version of the ware. Nevertheless, it is not clear why Vidal leans toward rejection of the possibility that there was a partial blending of west coast and Chalchihuites ceramic traditions that resulted in the production of local variants of ceramic types initially thought to have originated on the west coast, especially on the basis of a single sherd. It appears that Vidal is apparently unaware of Kelley's reference to Ward's thin section work.

As seen in the discussion above, it is clear that the use of petrographic techniques and more advanced geochemical methods such as electron microprobe analysis and inductively coupled plasma emission spectroscopy combined with the identification of potential clay and temper sources has the potential to greatly expand our current knowledge of the anthropology of Mexican west coast ceramics. However, it is going to require far more intensive and extensive study of ceramics and the identification of potential source localities.

As Grosscup stated 50 years ago, "There are major gaps in our knowledge of the ceramic archaeology of West Mexico" (1964:261; see also 1976:270). For the most part, those gaps remain, as the focus of ceramic studies remains on the identification of ceramic types to aid in establishing site chronology and the geographic distribution of those types. Numerous questions await further study regarding ceramic technological attributes (e.g., production methods, constituents), temporal variability in decoration and form within types and wares, spatial variability, and the distribution of types and wares. Some of these questions are tied to other issues such as the refinement of local and regional chronologies. In addition to the technological issues, numerous social and economic issues can be addressed through the study of ceramics. Can west Mexican ceramics be used to measure interaction between social units within and between communities, the political and social structure and complexity of households and communities, and the structure of regional, panregional, and long-distance exchange? Clearly, the archaeological ceramics of the west Mexican coast offer an array of challenges and opportunities.

CHAPTER 7

Artifacts

Michael S. Foster

Prehistoric west Mexico is known for its diverse and figurative material culture, which garners much attention from archaeologists, often beguiles art historians, and tends to overshadow other research initiatives such as the human ecological focus of the Marismas Nacionales investigations. Nevertheless, while perhaps more modest than other previously studied areas, in terms of material culture the Marismas Nacionales is no exception. In addition to the ceramic whole vessels and sherds, the Marismas excavations recovered numerous other ceramic artifacts, flaked- and ground-stone artifacts, and a small number of marine shell, bone, and copper artifacts. These are typical of previous archaeological excavations in the general region. From the Chametla area, for example, previous archaeological investigations yielded anthropomorphic and zoomorphic figurines, cups, ladles or spoons, stools or stands, spindle whorls, stamps, earspools, pipes, hollow ball rattles, and solid clay balls (Kelly 1938:45–61). A similar array of nonvessel ceramic artifacts was recovered from Culiacán. In addition to those artifact types listed for Chametla, excavations at Culiacán produced a pottery mask, seals/stamps, and whistles (Kelly 1945:121–141). Nonvessel ceramic artifacts from Amapa (Meighan, ed. 1976) included human figurines, zoomorphic figurine fragments, earspools, modeled and perforated discs (likely spindle whorls), whistles, stamps, palette-like ceramic slabs, and beads. Ground stone included manos, metates, pestles, axes and celts, labrets, beads and pendants, spindle whorls, abrading stones, polishing stones, and stone vessels. Flaked-stone artifacts included obsidian and chert blades, obsidian polyhedral cores, hammerstones, projectile points, other bifaces, scrapers (both formal and informal types), and marginally retouched implements made from both chert and obsidian. Marine shell artifacts included bracelets and pendants, and bone artifacts included awls and pins with carved designs (perhaps hairpins or pins to secure clothing resembling serapes or mantillas).

Copper artifacts were recovered from Culiacán and Amapa; no copper artifacts were recovered at the Chametla sites. Amapa copper artifacts included needles, bracelets and finger rings, a fishhook, bells, tweezers, and plaques.

As the summary presented above demonstrates, the array of artifact categories and types recovered during previous archaeological investigations in the region is considerable. Although sherds from ceramic vessels and whole or nearly whole ceramic vessels dominated the artifact assemblages recovered from the Marismas sites, a variety of other artifacts warrant further discussion. However, most of these artifacts were not subjected to intensive descriptive or metric analyses, and no functional or constituent (optical or geochemical) analyses were conducted. Furthermore, the lack of detailed studies, contextual information, and temporal control and the relatively few specimens in most of the artifact classes discussed below hinder an in-depth examination of temporal trends or patterns. Nevertheless, the following discussion provides descriptive and comparative details and illustrations of some of the artifacts recovered, which clearly situates the Marismas within the regional context and material archaeological traditions recognized in west Mexico.

Figurines

A small number of anthropomorphic ceramic figurines and figurine fragments, few of which have been described in detail or illustrated, were recovered during the investigations in the Marismas. These include solid handmade or modeled types as well as mold-made Mazapan-style figurines. Sites from which figurines or figurine fragments are reported include El Calón, Juana Gomez, Panales Test Mound A, Tecualilla, Cristo Rey, Venadillo, and El Mirador.

As discussed earlier in this volume (see Figure 5.7), the large shell mound site of El Calón yielded a number of anthropomorphic figurine fragments. Seventeen frag-

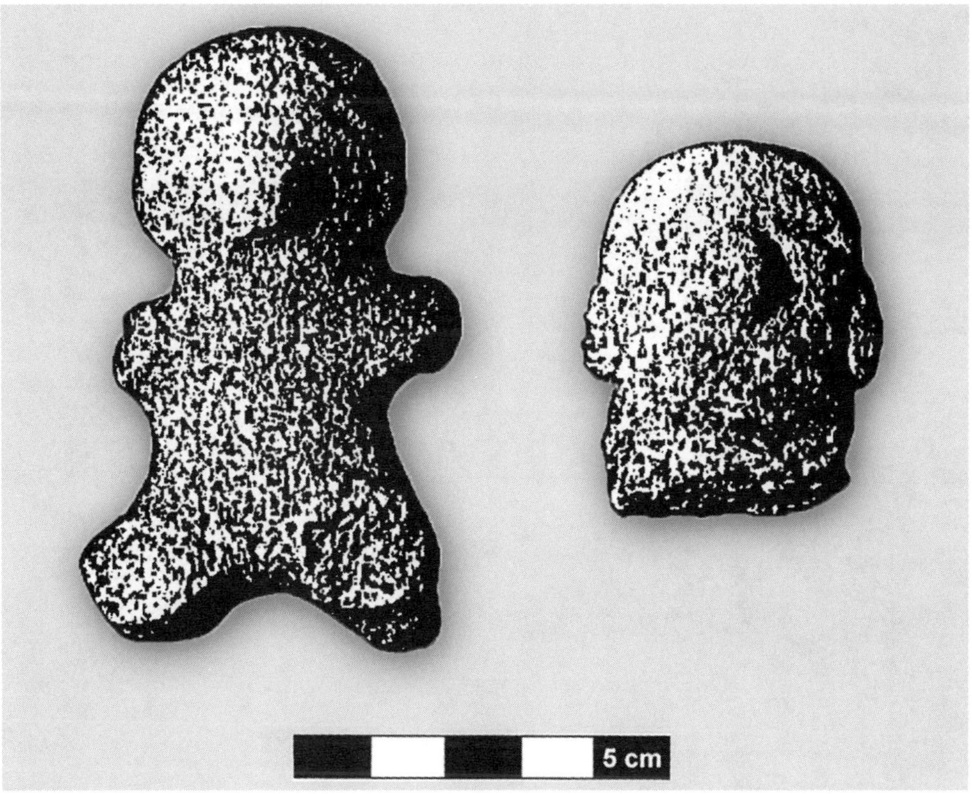

FIGURE 7.1. Examples of the El Calón figurines.

ments, mostly torsos and limbs, were recovered during the Marismas investigations. The pieces were solid and appeared to be hand-modeled. The paste was coarse, and very little detail beyond gross shape and size remained. The assemblage included one nearly complete specimen and the head of another (Figure 7.1). The nearly complete specimen appears in a seated position with its arms extended. There is a slight projection for a nose, but overall, the figurine lacks any significant anatomical detail. The right arm and feet/legs are missing.

The site of Tecualilla yielded three fragments of the distinctive Mazapan-style figurines, two heads and a body (Figure 7.2). Mazapan-style figurines are simple, low-relief figures that were made in a one-piece mold and are associated with many Early Postclassic Mesoamerican cultures. This style of figurine was first identified at Teotihuacán by Linne (2003). Subsequently, Acosta (1940) applied the name to a different style of molded figurine he found at Tula, Hidalgo. After Acosta's application of the name to the Tula specimens, the name was applied to similar mold-made Early Postclassic figurines in other parts of Mesoamerica.

The Tecualilla head fragments are typical Mazapan faces; they exhibit round earspools and the characteristic Mazapan-style headdresses. One specimen, Figure 7.2a, was light orange buff, and the other, Figure 7.2b, was dark gray in color. The slab body (Figure 7.2c) exhibited rudimentary feet. The arms and hands were molded in low relief, and the hands rest on the chest. The parallel horizontal bands across the body probably represent a textile robe or skirt. Mazapan figurines date to approximately AD 900–1200 in the Valley of Mexico. They were found at Amapa (Grosscup 1961; Meighan, ed. 1976), and Mountjoy (2000) reports them from sites in the Banderas Valley. Mountjoy (2000:100) further notes that, although Mazapan-style figurines are commonly associated with the Toltec culture of central Mexico, they are abundant in parts of the west Mexican coastal zone. He cites Stocker's (1983) observations that their abundance in west Mexican sites, both coastal and upland areas, has led some to suggest that the Mazapan-style figure tradition originated in west Mexico.

Level 1 of Test Unit (TU) 1 in Mound B at Venadillo yielded a curious cylindrical female figurine made from a light brown paste. The lower portion of the figurine is missing. Breasts are present, and the arms are represented by small bumps. A necklace or some form of neckband appears as an appliquéd piece of rolled clay, the mouth is pinched and protruding, and the eyes are appliquéd coffee bean–style eyes (Figure 7.3). Kelly (1945:Figure 64i) illustrates a cylindrical figurine from Culiacán, but the Venadillo specimen is not stylistically similar.

Eight figurine fragments were recovered from Mound C at Venadillo. Seven fragments are either arms or legs, and the other seems to be part of a head. All are from

FIGURE 7.2. (*a–c*) Mazapan-style figurine fragments from Tecualilla.

small solid figurines, but none were large enough to be diagnostic of any particular type. The majority of a slab figurine, with the arms and legs missing, was recovered from Level 6 in the earthen burial portion of the mound (Figure 7.4). It has a dark gray surface and is made of orange paste with a black core. It has coffee bean eyes, an appliqué nose, male genitals, and what seem to be earspools. Incising on the head may represent either hair or a headband, and there is a punctate design on the chest. Although the figurine was not directly associated with any of the burials recovered, its presence in the burial mound suggests that it may have been a funerary offering.

Other than the Mazapan-style figurines and perhaps several from El Calón, the figurines from the Marismas are not particularly similar to those described at Chametla, Amapa, Culiacán, or Guasave (Ekholm 1942; Grosscup 1961; Kelly 1938, 1945; Meighan 1976a; Santos Ramírez 2012b). As to whether these represent local varieties or are trade items, the former is suspected.

Beads

Three small anthropomorphic beads were recovered from an urn burial (Burial 4A and 4B) in Level 6 of Mound C at Venadillo (Figure 7.5). These are small sitting figures, buff in color, with appliquéd eyes and arms. They are 4 cm high, and each has a single notch in the back of the neck, suggesting that they may have been strung on a necklace. The individual in the urn was young and might have worn such a necklace.

A small, well-polished stone bead (Figure 7.6) was recovered from Burial 14, Mound A, at Chalpa. It is a light green material, and it is nearly symmetrical. The diameter of the bead varies from 1.5 to 1.6 cm. The diameter of the perforation varies from 0.59 to 0.66 cm in diameter.

A possible bead or pendant made from a dog phalange was recovered from Juana Gomez. It was burned, and the proximal end of the specimen had been cut off. It was polished, and two holes had been drilled through the anterior surface of the distal end (Wing 1968). A bone pendant made from the drilled astragalus (talus bone) of a dog was recovered from Tecualilla (Figure 7.7). Associated with Burial 3A at Cuautla were four drilled dog tooth beads. They were found in the area of the neck, indicating that they had been strung on a necklace.

Field notes from the excavations at Rincon de Panal mention a small "amulet." They indicate that the specimen is approximately 5.0 cm high/long, with a maximum diameter of 2.5 cm. It is cylindrical in form, with an open mouth and a sealed bottom, and based on the field sketch, it appears to have a slight taper to the bottom. There are two holes in the side of the specimen, suggesting that it may have been suspended, perhaps as

FIGURE 7.3. Cylindrical figurine from Mound B, Venadillo.

FIGURE 7.4. Slab figurine from Level 6, Mound C, Venadillo.

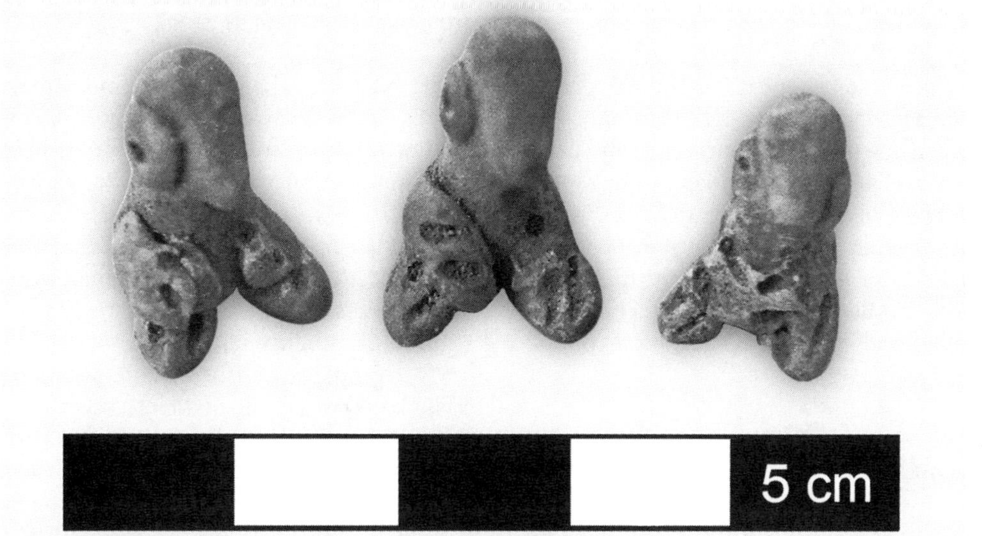

FIGURE 7.5. Anthropomorphic beads from Burial 4, Mound C, Venadillo.

FIGURE 7.6. Polished stone bead from Chalpa.

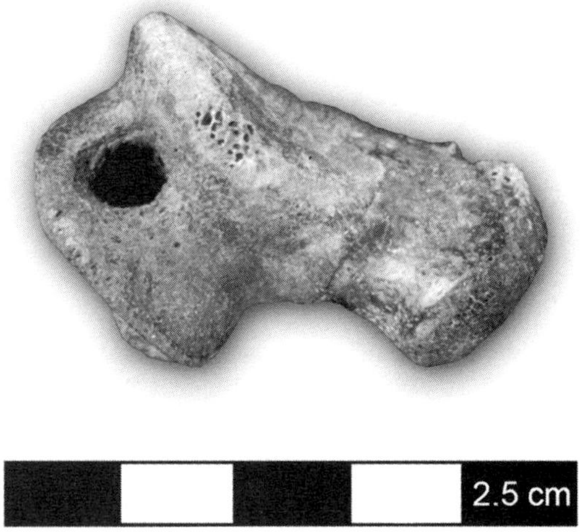

FIGURE 7.7. Pendant made from the astragalus of a dog from Tecualilla.

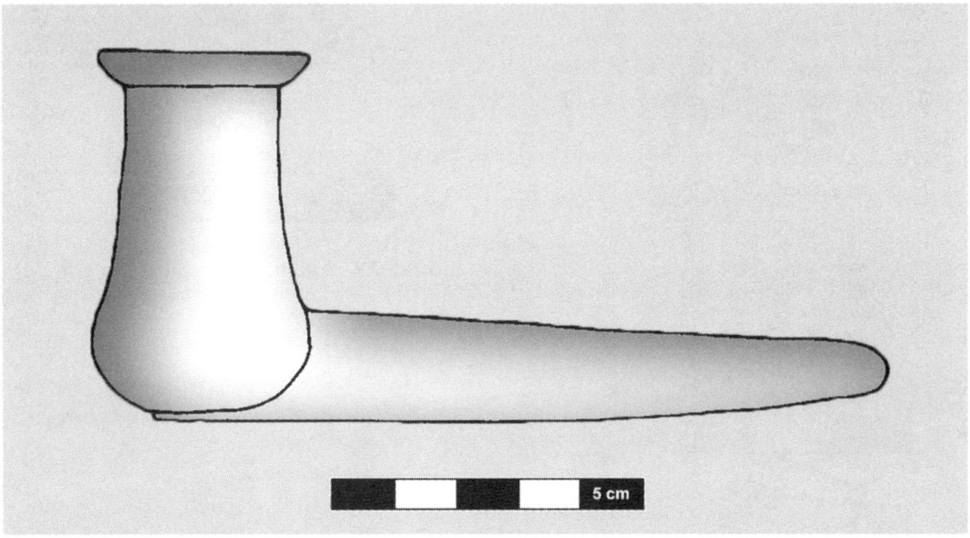

FIGURE 7.8. Polished buff, Cocoyolitos-style pipe from Burial 3, Rincon de Panal.

part of a necklace or some type of ceremonial regalia. It was described as a ceramic piece. The bottom surface (not specified as to either interior or exterior surface) exhibited a series of grooves, which sounds suggestive of a little grinding (i.e., *molcajete*-like) implement. The suspension holes were reported as being "punched" rather than drilled. Additionally, Gill (1971) notes the presence of bone beads with Burial 22 in Mound A and four red stone triangular beads found with Burial 1A in Mound B at Tecualilla.

PIPES

A small assemblage of fragmented and whole ceramic smoking pipes was recovered during the excavations. The fragmentary specimens were found in trash or other contexts, indicating that they had been discarded when broken, and the complete specimens were found as mortuary offerings. The whole and fragmentary specimens appear to be Cocoyolitos-style pipes (Kelly 1938:52–53) or Porter's (1956) type 3A, "sencillo sin soporte." Kelly found pipes concentrated at El Taste and Cocoyolitos A at Chametla and in upper levels at other sites. She concluded that pipes have an "indubitably late distribution" (1938:52). The local temporal association for a Cocoyolitos-style pipe in the Marismas sequence is the Middle Conchera phase.

A complete, polished buff Cocoyolitos-style pipe (Figure 7.8) was found in direct association with a flexed burial (Burial 3) at Rincon de Panal. Based on the drawing, the pipe measures 12.7 cm long; the pipe stem is 9.1 cm long, the bowl width is 3.5 cm; the bowl height is 5.8 cm; the bowl neck below the rim is 2.5 cm wide; and the rim is 3.4 cm wide.

Five pipes were recovered as funerary offerings from

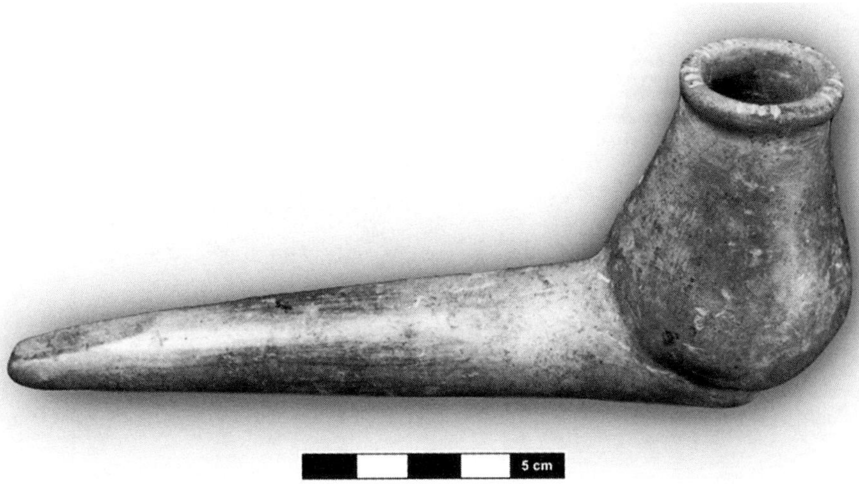

FIGURE 7.9. Buffware pipe from Chalpa.

TABLE 7.1. Ceramic Pipes from Cristo Rey.

Catalog Number	Specimen	Association	Length	Stem Length	Bowl Width	Bowl Height[a]	Bowl Neck Height	Bowl Neck Width[b]	Rim Width	Mouth Width
389	Figure 7.10a	Burial 22	16.5	13.3	3.2	4.5	1.3	2.7	3.5	2.5
385	Figure 7.10b	Burial 19	22.5	17.7	4.9	6.2	2.8	3.0	3.8	2.2
384	Figure 7.10c	Burial 19	28.9	21.3	7.7	5.4	1.7	3.4	4.5	3.3

Note: All measurements are exterior and in centimeters.
[a] Includes bowl and neck to the rim of the mouth.
[b] Measurement taken at midpoint on the neck below the mouth of the pipe.

the site of Chalpa. Although there is photodocumentation of one of the specimens (Figure 7.9), its association is unknown, and the other four are not described. The illustrated specimen is a polished buffware pipe that is decorated with five groups of three to four tick marks on the rim. It has a bulbous bowl that tapers toward its rim. The pipe measures 16.1 cm long; the stem is 11.75 cm long; the bowl width is 4.5 cm; the bowl height is 5.9 cm; the neck width just below the rim is 3.0 cm; the rim width is 3.1 cm; and the mouth is 2.1 cm wide. It is a Cocoyolitos-style pipe.

Three Cocoyolitos-style pipes were recovered from mortuary contexts in Mound E at Cristo Rey (Figure 7.10; Table 7.1). One (Figure 7.10a), from Burial 22, was well decorated with red paint and incising that occurred as zone-decorated red panels alternating with incised line panels. A second was an incised plainware pipe (Figure 7.10b) with a grooved bowl and finely incised lines on the stem and bowl neck. It was found with Burial 19. The third pipe (Figure 7.10c) was an undecorated plainware that was moderately polished. It too was found with Burial 19. These pipes support a Middle Conchera period date for Cristo Rey.

Twenty-two pipe fragments were recovered from Venadillo: five bowls, 16 stems, and one stem with a partial bowl. Two of the bowl fragments are so small that they can hardly be identified as pipe fragments. Both are the bases of the bowls at the point where the stem is attached; one is gray and fairly well polished, and the other is of a coarse paste and orange in color. Two of the remaining three bowls are complete. One is rather elongated, with a slightly flaring rim, straight sides, and a bulbous base; it is buff and slightly polished and 4.6 cm high, with an opening 3.5 cm wide. The second complete bowl is polished black, with a shallow incised design on the bulbous portion of the bowl; it has a slightly constricted neck with a flared rim. This piece is 3.7 cm high and 2.6 cm at the neck, and the maximum diameter is 3.4 cm. The other bowl fragment is gray and modeled as well as incised. The lip is not flared, but the bowl has been formed into a circular shape. One complete stem with a bowl fragment was recovered from one of the profile trenches. The piece was approximately 9.2 cm long and slightly polished buff; it has a flat bottom, possibly designed to prevent the pipe from rolling.

The 16 stem fragments were recovered from various levels and two trenches, and none could be identified as any particular style of pipe from Sinaloa. Mouthpieces, midsections, and portions that were attached to the bowls were all represented. Colors varied from orange to gray; some were polished, and others were not. There are no designs, either painted or incised, on any of the

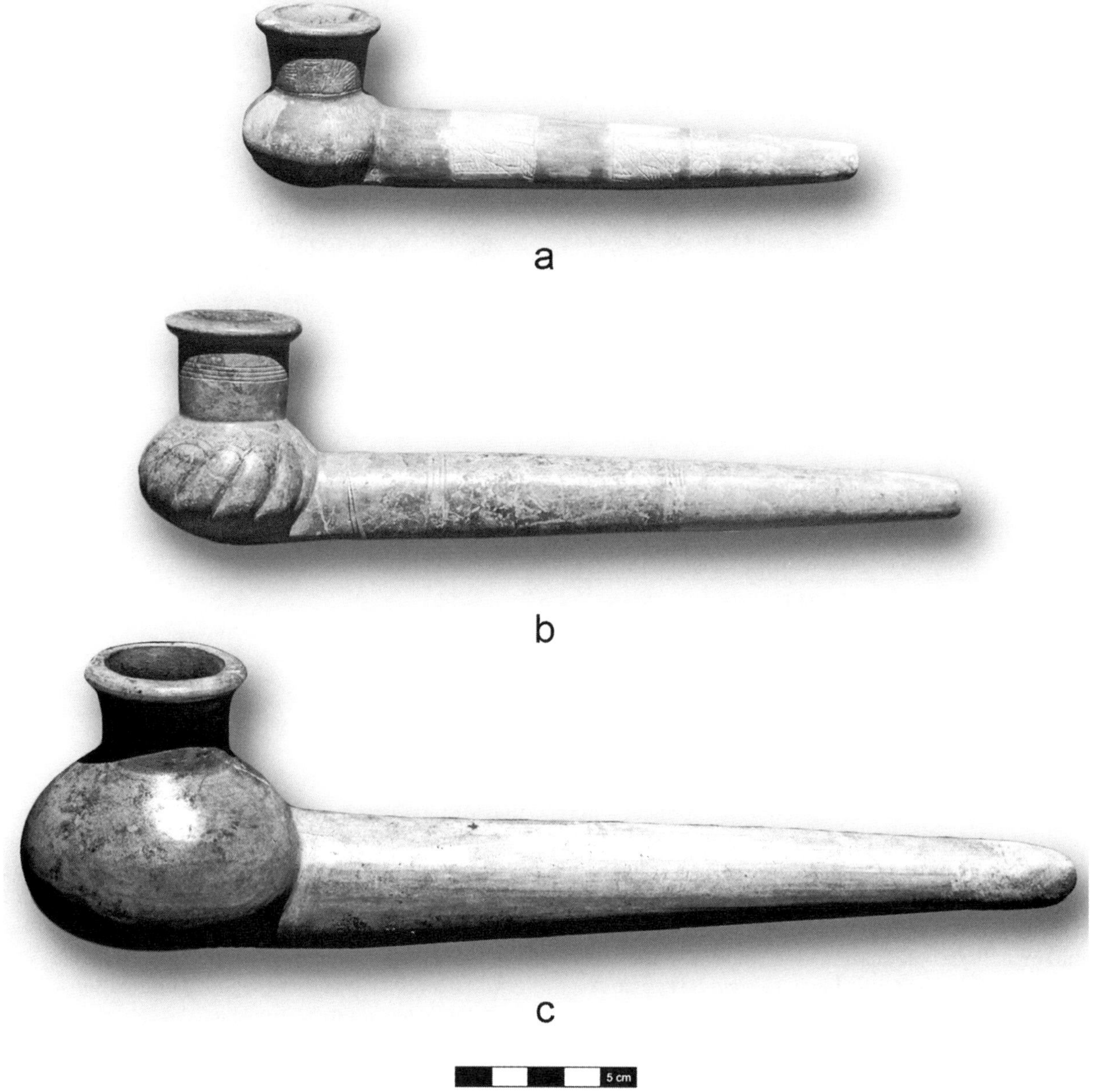

FIGURE 7.10. (*a–c*) Ceramic pipes from Cristo Rey.

stems. One unusual stem or tube section had a center hole 3.4 cm in diameter and is reminiscent of the cloud blower used in the American Southwest. It is of a very coarse gray paste, unslipped and unpolished.

None of the pipe fragments was exactly like any from Chametla, although some of the general shapes are similar to some of Kelly's (1938) descriptions. The bowl fragments are all different styles and possibly reflect local forms.

Recently, Mathiowetz stated the following regarding lobed pipes like those from the Marismas and the west Mexican coast:

> The occurrence of the distinctive lobe-like design around vessels also is manifested on another material culture object, the bowls of ceramic smoking pipes. Kelley (2000: 139) noted that the presence of ceramic pipes in the Aztatlán region grew in intensity from the Early to Late Postclassic. This conclusion is notable in that it suggests an increasing adherence to attendant ideologies tied to symbolic rain-making and cloud production [2011:562].

Mathiowetz further states,

FIGURE 7.11. (*a–c*) Whistles from Mound C, Venadillo.

> The lobed design noted around Aztatlán pipe bowls suggests that the pipes themselves served as symbolic open floral blossoms, such that the emanating smoke would be akin to the production and emanation of floral aroma. These clouds of floral smoke also symbolically represented the rain-filled clouds of ancestors [2011:563].

The Marismas borders the arid desert zone of northwest Mexico and the subtropics to the south. With a distinct dry season and agricultural success depending on the start of the monsoon, conjuring rain-filled clouds may have been an important ritual act.

WHISTLES

Three ceramic whistles were recovered from Mound C at Venadillo (Figure 7.11). Two are anthropomorphic in form, and the third has three projecting arms. The latter (Figure 7.11a) is similar to whistles from Amapa (Meighan 1976a:Plate 58) and a rattle described by Kelly (1945:129, Figure 62b) for Culiacán. It is unpolished and has an orange-colored paste.

The two anthropomorphic whistles were found in Level 6. One consisted only of a head with a hole in its top that may have been the mouthpiece. It is a slightly polished dark gray, and the eyes, nose ring, and turban on the head are appliqué (Figure 7.11b). It is almost identical to a whistle from Chametla (Kelly 1938:Plate 20m). The other whistle is a nearly complete seated figure, whose arms rest on his knees. The eyes are punched, and the nose is pinched out of the face, and there are appliquéd earspools. This figure is also wearing a turban that seems to be twisted and to have a tail that crosses the top of the head and hangs down toward the left ear. There is a large hole in the back of the figure and another in the top of the head. It is unpolished orange and is 4.5 cm high (Figure 7.11c).

Whistles, although not abundant, are not a rare artifact type found at sites in the region. As noted above, they are found at Chametla, Amapa, and Culiacán.

SPINDLE WHORLS

Spindle whorls are a common artifact found at sites on the west coast of Mexico and were recovered, although not in great abundance, at all but two of the Marismas sites investigated. They commonly occur as burial offerings and include plain and decorated ceramic and stone specimens. They appear in both Late Estero (mid–late Classic) and Early–Middle Conchera (Postclassic) period contexts. The most detailed information on spindle whorls from the Marismas comes from Mound C at Venadillo; data on nine complete or nearly complete specimens are available (Figure 7.12). All specimens appear to have been modeled around a spindle in order to create the spindle hole. No striations suggesting that the whorl was drilled after it was dried or fired were evident. Two are plain, one is decorated with red paint, and the rest are decorated by carving, incising/engraving, and punctation. In cross section, these range from biconical to spheroid to round.

Four of the small incised spindle whorls were decorated with small circular lines running completely around the whorl, diagonal lines, concentric circles, and triangles (Figure 7.12a–d). Color varies from buff to black. The medium-sized spindle whorls were incomplete and less elaborate than the others (Figure 7.12e–f). One spindle whorl (Figure 7.12e) was undecorated, orange in color, and slightly polished. A second one (Figure 7.12f) was neither incised nor punched but was decorated with diagonal light red bands on a buff background and moderately well polished, and a third whorl (Figure 7.12g) was crudely made of a coarse orange paste and was not slipped, polished, or otherwise decorated. The two large spindle whorls had a combination of

FIGURE 7.12. (*a–i*) Spindle whorls from Mound C, Venadillo.

punched and incised designs, which included concentric circles, bands entirely around the piece, figure 8s, dots, and lines (Figure 7.12h–i). Both were black (smudged?) and polished. One (Figure 7.12i) had white paint filling some of the small punctations. All, except the large plainware specimen (Figure 7.12g), come from the lower levels (Late Estero period) and are similar to Baluarte phase forms from Chametla (Kelly 1938:Figure 24) and Types 2b and 2c from Amapa (Meighan 1976a). Meighan notes that approximately one-half of the spindle whorls recovered at Amapa were made of stone, with the other half being ceramic. In addition to the modeled ceramic spindle whorls at Amapa (Meighan 1976a:76, Plate 52), modeled, perforated ceramic discs thought likely to be spindle whorls were also recovered. These are reminiscent of perforated sherd discs used as spindle whorls in the American Southwest (e.g., Stone and Foster 1994).

The Venadillo Mound C spindle whorls can be divided roughly into three groups based on size (Table 7.2): small, with a mean diameter of 2.50 cm and a mean height of 1.75 cm; medium, with a mean diameter of 3.36 cm and a mean height that could not be determined because of the fragmentary condition of the pieces; and large, with a mean diameter of 4.35 cm and a mean height of 3.50 cm. No data on the weight of each specimen were collected.

As noted above, spindle whorls were recovered from nearly every site investigated during the Marismas proj-

Table 7.2. Spindle Whorls from Mound C, Venadillo.

Catalog Number	Specimen	Provenience	Diameter (cm)	Height (cm)	Spindle Diameter (cm)
7210	Figure 7.12a	TU 4, Level 6	2.6	1.9	—
7215	Figure 7.12b	TU 2, Level 6	2.5	1.7	—
7206	Figure 7.12c	TU 3, Level 6	2.5	1.8	—
7189	Figure 7.12d	TU 4, Level 6	2.4	1.6	—
7214	Figure 7.12e	TU 2, Level 6	3.2	Incomplete	0.5
7184	Figure 7.12f	TU 2, Level 7	3.5	2.6	—
7607	Figure 7.12g	TU 2, Level 2	3.4	2.5	~1.0
7182	Figure 7.12h	TU 2, Level 5	4.6	3.5	—
7216	Figure 7.12i	TU 1, Level 5	4.1	3.5	0.9

Note: TU = Test Unit.

Figure 7.13. Spindle whorl decorated with incised dog forms from Tecualilla.

ect. One from the site of Tecualilla is of particular note. Images of five dogs (Figure 7.13) were incised on a spindle whorl that was also decorated with punctations filled with white paint. The spindle whorl itself was slipped brown with some red paint and well polished. No metric data are available for the specimen.

Another somewhat distinct spindle whorl was recovered from Burial 3A at Cuautla Site 3. It was a light-cream-colored stone, possible onyx, alabaster, or quartz. The whorl was biconical with flattened ends and appears similar to the Type 2 varieties at Amapa (Meighan 1976a: 78–80). Both ends have incised/carved double circles. There is a double incised band around the center of the whorl, and above and below the central band is a row of seven incised double circles with center dots (Figure 7.14). It measures approximately 2.4 cm in diameter.

As noted above, spindle whorls are a common artifact type recovered at sites on the Mexican west coast. They generally are made of stone or are modeled ceramic pieces, and they come in a variety of forms that are decorated by incising, engraving, or carving. Whorls serve as weights for spindles that are inserted in the central

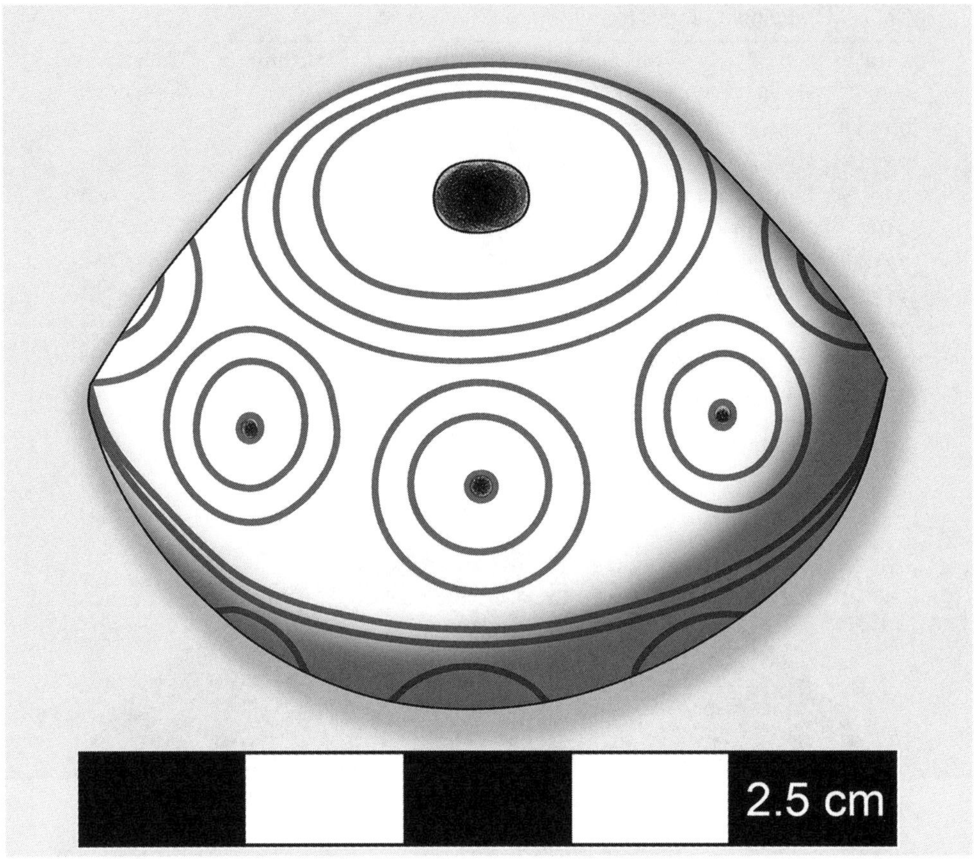

FIGURE 7.14. Illustration of the spindle whorl recovered from Burial 3A at Cuautla Site 3.

perforation of the whorl; they are used in the spinning of thread or yarn from fibers. The question often is, What types of fiber were spun? As discussed in Chapter 12, ethnohistorically, cotton was regularly reported as a crop grown on the west coast, and dress made from cotton cloth was commonly worn. Additionally, cotton cloth was often extracted as a tribute item.

In addition to cotton, agave fiber was extensively used in both prehistoric and ethnohistoric Mesoamerica and ethnographically across Mexico (e.g., Parsons 1972; Parsons and Parsons 1990). Parsons (1972) suggested, based on her study of Aztec whorls, that fiber type and yarn size were correlated to whorl size: small whorls were used to make fine cotton yarn that had relatively short, thin fibers, and larger whorls were used to make heavier, coarser thread from agave fiber, which is relatively long and thick. As seen from the sample of spindle whorls from Mound C at Venadillo, there is variability in the size, and presumably the weight, of spindle whorls from the Marismas.

Three studies of spindle whorls and thread/yarn size are of relevance here: Parsons and Parsons (1990), Stone (Stone and Foster 1994), and Teague (1998). Parsons and Parsons, as part of a general study of maguey utilization in central Mexico, experimented by having modern native weavers spin maguey fiber with archaeological spindle whorls; Stone studied an assemblage of 424 sherd disks, modeled whorls, and stone whorls from the large Hohokam site of Pueblo Grande in central Arizona; and Teague's research included whorls from across the Southwest and northwest Mexico. These studies support the hypothesis that whorl size (weight) and the length of the fiber of the material being spun are related to the thickness of the yarn/thread produced. However, the relationship between these two variables is not simple. These studies demonstrated that there is significant overlap in the thickness of the yarn that can be produced between whorls of various sizes. In other words, whorl size matters in the production of yarns, and in general, small spindle whorls cannot be used to produce heavy yarn. Large whorls, on the other hand, can produce a variety of thin to heavy yarns. Based on Stone's (Stone and Foster 1994:Figure 6.5) scatterplot, diameter in centimeters against weight in grams, of modeled whorls, it is estimated that the Venadillo Mound C whorls discussed above weighed between 3 and 30 g. This suggests that fine to heavy yarn or thread was being spun at the site. Parsons and Parsons's (1990) study of maguey fiber production indicates that larger whorls, in excess of 10 to as much as 75 g, were used to spin maguey fiber. Thus, it

FIGURE 7.15. Hollow ceramic ball from Mound C, Venadillo.

appears that the Marismas whorls were more likely to have been used in the spinning of cotton fiber, although maguey fibers could have been spun with some of the larger whorls.

Additionally, based on Stone's (Stone and Foster 1994) and Teague's (1998) studies, the apparent lack of clustering of a given size of whorl suggests a lack of specialization in terms of the size of yarn being produced. In other words, the Venadillo spinners were producing yarns or threads of various sizes rather than of a particular size. Whether this was the case across the Marismas is not known. Experienced spinners, no doubt, could to some degree control the width of spun fiber through subtle adjustments in the mechanical and physical process of spinning, and that spinning equipment was potentially customized or reflected the preferences of individual spinners. The lack of standardization in spinning equipment, therefore, also implies that spinning was not regulated necessarily beyond the level of individual spinners or their households and that the shared conventions of appropriate yarn width were approximated by each spinner on her own using personalized tool kits and equipment.

This is supported by the fact that in the 20 cases in which spindle whorls are reported as funerary offerings, 18 of those were females; two, a child and an adolescent, were of undetermined gender. None of the male burials reported had spindle whorls buried with them. Thus, it appears that spinning and probably weaving were not just female-related tasks; it is also reasonable to assume that tool kits for spinning and weaving may have been highly personalized or so closely identified with the deceased that they could not be passed on to living descendants or associates, who might be less inclined to use them for both practical purposes and spiritual reasons.

Ceramic Ball Rattle

A hollow ceramic ball (Figure 7.15) was found in association with Burial 8 in Level 8, Mound C, at Venadillo. It is brown-slipped and polished, measuring 5.7 cm in diameter. It contained some small beads or pebbles that

FIGURE 7.16. Examples of prismatic obsidian blades from Mound C, Venadillo.

allowed it to function as a rattle that could have been held in the hand and shaken during dancing or ceremonial activity. Kelly (1938:51) reports small hollow ceramic ball rattles and globes from Chametla. These measured 3.5–5.0 cm in diameter; the Venadillo specimen measured 6.26 cm in diameter.

Net Sinker

Level 4 in Mound C at Venadillo produced a modified sherd that may have been a net sinker. It was roughly figure 8 in shape with two notches near one end and made of a ceramic vessel with an orange paste with a dark core. It appears that the vessel from which the sherd was derived was fired at a low temperature. Coe (1961) recovered similar notched sherds, which he believed to be net sinkers, at La Victoria in Guatemala, and Stark (2001:214, Figure 10m–p) describes a small collection of such sherds at Mixtequilla, Veracruz.

Flaked and Ground Stone

Although flaked and ground stone were reported at all sites investigated, the most intensive analysis of any flaked and ground stone recovered from the Marismas is that of the assemblage from Mound C at Venadillo (Foster 1984). Thus, the following discussion focuses on the Mound C artifacts while discussing specimens from other sites where detailed descriptions are available.

Blades and Flakes

Most flaked-stone material was derived from obsidian. Some quartz and granite flakes and debris were also present. The most abundant type of flaked-stone artifact was the blade (Figure 7.16). The 41 blade fragments recovered from Mound C at Venadillo included platform ends, midsections, and distal sections that were either triangular or trapezoidal in cross section. Nearly all the blades show some signs of use or some form of edge damage. Small flakes are missing from most edges, but only a few show signs of retouching. The widths of the blade fragments varied from 0.6 to 1.2 cm, and their lengths, from 1.0 to 6.6 cm. Polyhedral cores were present, but those I know of were small and not abundant. The size may have been the result of the cores being reduced as flakes were removed; however, it is probable that many of the blades may have been obtained through exchange. In addition to the blades, 48 obsidian flakes, 11 quartz flakes, and two granite flakes were recovered. None exhibited any evidence of use.

Projectile Points and Bifaces

One nearly complete projectile point (spearpoint?), 8.6 cm long (Figure 7.17), and 10 point fragments were recovered from Mound C at Venadillo. All are made of obsidian and exhibit bifacial flaking. About half the fragments seem to be from stemmed points.

FIGURE 7.17. Obsidian bifaces from Venadillo.

A large stemmed obsidian projectile point measuring approximately 7 cm in length was recovered from Level 6 in TU A of Mound 1 at Juana Gomez. A large side-notched obsidian projectile point was recovered from TU 2, Level 4 (75–100 cm), at Cuautla Site 3 (Figure 7.18). It measured 7.4 cm long, with a blade length of 6.5 cm, shoulder width of 3.3 cm, stem width of 2.5 cm, and basal width of 3.4 cm. The blade was slightly asymmetrical, the shoulders were slightly rounded, and the base was slightly concave to straight. It was well made. This style of large side-notched points is represented at Amapa (there classified as knives—Meighan 1976d:Plates 94–95) and elsewhere on the west coast and the uplands of west Mexico (e.g., Lister 1949; Meighan and Foote 1968) along with stemmed and corner-notched points; these appear to date to the Postclassic period.

The role of obsidian in the precolumbian cultures of Mesoamerica was widespread, and it has been extensively studied. With the use of various sourcing techniques, the study of local, regional, and panregional obsidian procurement and exchange has become commonplace in recent years, including in west and northwest Mexico (e.g., Darling 1998; Glascock 2007; Glascock et al. 2010; Ohnersorgen et al. 2012; Pierce 2012). Glascock and others (2010) have completed a comprehensive study of obsidian sources in Jalisco, Nayarit, and Zacatecas; 26 sources, two of which—La Lobera, Jalisco, and Huitzila, Zacatecas (Darling 1993; Darling and Glascock 1998; Darling and Hayashida 1995)—are neighboring sources, are described. The quality of these sources ranges from excellent to poor, with nearly half being excellent and all but three of the rest being rated as good/medium. The La Joya, Jalisco, source, located between Etzatlán and Magdalena, is the largest in Mesoamerica (Weigand and Spence 1982). In addition, the largest obsidian workshop, Las Cuevas, located to date in western Mexico occurs near La Joya. Many of the source areas identified have nearby specialized workshops that produced prismatic blades. La Joya blades have been identified at Amapa (Ericson and Kimberlin 1977) and visually at Guasave (Spence et al. 1980).

Glascock and others (2010) state that the west Mexican obsidians come in an array of colors, from green to gray to blue to black as well as some red, yellow, white, orange, and mixtures of colors. Some of the obsidians have gold and silver tinges, and some have a multicolored, iridescent sheen that produces a rainbow-like appearance, thus rainbow obsidian. The aforementioned La Joya source produces rainbow obsidian as well as dark green obsidian. Despite the color variability found at La Joya, the compositional profile of the La Joya obsidians is uniform. Glascock and others (2010) analyzed more than 1,000 obsidian artifacts with X-ray fluorescence and neutron activation analysis; 27 unique compositional groups were identified as a result. None of the obsidian artifacts from the Marismas were submitted for constituent analyses and sourcing.

FIGURE 7.18. Illustration of the obsidian projectile point from Cuautla Site 3.

However, a sample of 39 obsidian artifacts from four Marismas sites was visually assessed and sourced based on color by Dr. Michael Spence (Table 7.3). Twenty-six of the specimens were identified as being from the La Joya source. A second source, an unknown green source possibly from Nayarit, was also identified. The source of the two remaining specimens was unidentified. Although some have questioned visual sourcing, Pierce (2012) notes that Ohnersorgen and his colleagues (2012) observe that many of the volcanic sources in western Mexico produce obsidians of distinct hues and that these obsidians can be sorted through macroscopic analysis with relatively high accuracy. Although hardly as definitive as geochemical techniques, visual sourcing is a potentially useful preliminary tool.

Manos and Metates

Mound C produced nine mano fragments, plano-convex in cross section; one was a surface find, and eight came from the upper levels of Mound C. Three mano types are represented. Two fragments are from the overhanging type as described by Kelly (1938:62, Figure 28b, 1945:154, Figure 75a) at Chametla and Culiacán. One fragment was from the overhanging end itself, and the other is a body fragment with a portion of the overhanging end. Both were made from granite. These would have been used on straight-sided slab-style metates. Three fragments were from small circular one-handed manos, all from granite as well. One has a rough surface that had been slightly smoothed from use; the surface may have been pecked in order to sharpen the mano and increase its grinding effectiveness. One of the other fragments had some beveling, probably from shaping, on its upper edge, but it shows no indication of extensive use.

Five of the fragments, three midsections and two ends, all granite, were from two-handed manos. Three showed signs of extensive use; they showed light beveling toward the edges and exhibited use of both faces.

Five metate fragments were also recovered from the shell cap of Mound C. All are small and seem to be from center portions of metates. One fragment was made from vesicular basalt, and the others were granite.

Pestles

Two small vesicular basalt pestles were recovered from Level 6 in Mound C at Venadillo. They are semirectangular in shape with rounded edges, and both are 6.5 cm long.

Smoothing Stone

A small piece of pumice, plano-convex in cross section, was recovered. It is thought to be a smoothing or polishing stone showing extensive use.

Bone

Only three bone artifacts were recovered during the Marismas excavations. As noted above, a possible pendent made of a dog phalange was recovered from Juana Gomez (Wing 1968) and another made from the astragalus of a dog was recovered from Tecualilla; both were drilled for suspension. The third worked bone specimen was found in Mound C at Venadillo near the surface of the west trench. It is a musical rasp (Figure 7.19) carved from the jawbone of a caiman (*Crocodilus acutus*). Meighan (ed. 1976:Plate 112t) reports a rasp made from a notched dog mandible at Amapa; it appears, based on its illustration, similar to the Marismas specimen. Kelly (1945:141–142, Figure 71a) recovered a notched deer antler rasp at Culiacán. Human long-bone rasps that are common at Postclassic period sites in central Mexico are generally unreported from coastal west Mexico. The site of Tizapan el Alto on the south shore of Lake Chapala in Jalisco yielded two examples of rasps made from human bone (Meighan and Foote 1968:Plate 23j–k). Mathiowetz

Table 7.3. Probable Sources for Obsidian Artifacts from the Marismas Nacionales.

Site	Provenience	Artifact Type	Source
Rincon de Panal	TU C, Level 4 (160–200 cm)	13 prismatic blades	La Joya, Jalisco
		1 prismatic blade	Unknown green source (Nayarit?)
		1 side-notched projectile point made from a prismatic blade	La Joya, Jalisco
	A to C Level (0–40 cm)	1 prismatic blade	La Joya, Jalisco
Tecualilla, Feature A	TU E, Level 5	1 prismatic blade (2 fragments)	Unknown green source (Nayarit?)
	TU F, Level 6	1 prismatic blade	La Joya, Jalisco
	TU G, Level 4, Lot 4	1 projectile point	Unknown
		1 flake	Unknown
		4 prismatic blades	La Joya, Jalisco
	TU G, Level 6, Lot 7	2 prismatic blades	La Joya, Jalisco
		3 prismatic blades	Unknown green source (Nayarit?)
	TU H, Level 7, Lot 11	1 prismatic blade	La Joya, Jalisco
Panales Test Mound B	Cut 1, earthen mound, Level 1	5 flakes/debitage	Unknown green source (Nayarit?)
Chalpa, Feature B	Level 4	3 prismatic blades	La Joya, Jalisco
		1 flake/debitage	Unknown green source (Nayarit?)

Note: Visual identifications made by Dr. Michael Spence. TU = Test Unit.

Figure 7.19. Top and bottom views of a rasp made from the jawbone of a caiman (*Crocodilus acutus*) from Mound C, Venadillo. Holes are tooth sockets.

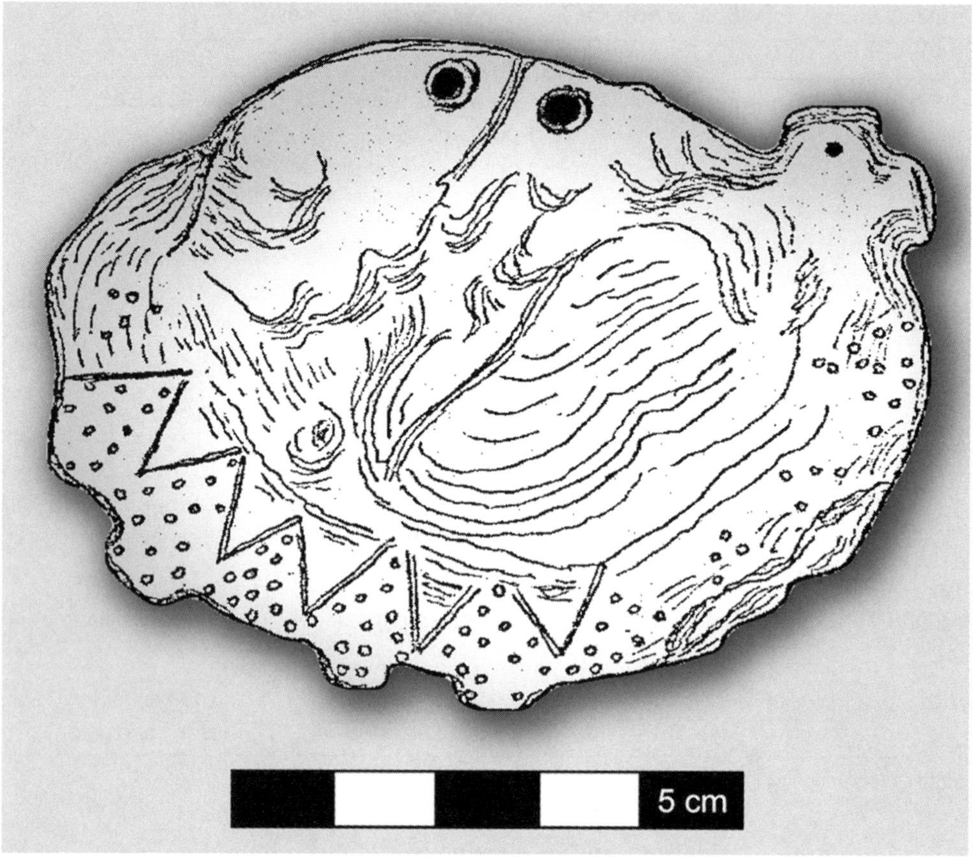

FIGURE 7.20. Shell pendant from Chalpa (adapted from Shenkel 1972).

(2011:266), citing the work of Ramírez de Swartz and others (2005:316) at La Peña, Jalisco, in the Sayula Basin not far from Tizapan el Alto, notes the recovery of another notched bone rasp. However, he does not specify whether the rasp was of human or animal bone. Mathiowetz (2011:254–287), in his study of human bone rasps found in elite burials at Paquimé in northern Chihuahua, provides a detailed discussion of the symbolism of musical rasps and their use in ritual in Mesoamerica and the prehistoric American Southwest. However, the lack of context for the Marismas specimen (recovered from the west trench of Mound C at Venadillo) limits interpreting its ritual or symbolic significant. Nevertheless, it may well have been used in community-wide ceremonies by people of local status, or it may have been nothing more than a household musical instrument used in household activities or household rituals.

There is a surprising lack of worked bone and shell in the Marismas. It is possible that much of it was traded out of the area, but the abundance of both materials makes it seem odd that so little was found. The preservation of bone and shell is generally good throughout the area, so the infrequent occurrence of bone and shell artifacts cannot be entirely the result of poor preservation.

SHELL

One of the very few marine shell artifacts reported from the Marismas sites is attributed to Chalpa (Shenkel 1972). The specimen was in the possession of a local resident who said that he had recovered it from the site. It is a pendant or gorget made from the shell of *Pinctada mazatlanica* (*Avincula barbata*; see Keen 1971:79). The species is an offshore shallow-water species that ranges from Baja California to Peru. Shenkel suggests that it is a zoomorphic representation—a "fat multi-legged turkey-like creature" (1972:78; Figure 7.20)—measuring 8.4 cm long and 6.5 cm wide. The following is Shenkel's description:

> Spaced equidistant along the bottom edge are 10 open bottomed engraved triangles with their apexes oriented toward the center of the pendent. These triangles vary between 10 and 13 mm wide and 8 to 11 mm tall. The zigzag space that they create along the bottom edge is filled with randomly spaced small holes slightly over a millimeter in diameter and less than 0.2 mm deep. Projecting from about the 1:00 o'clock position on the oval is the head, a rounded,

FIGURE 7.21. Burial 2 at Cuautla Site 3, with shell bracelets on the arm and cowry shell offering (lower right edge).

> bilobed stylization possibly representing a head and beak. One hole 2 mm in diameter and over 1 mm deep representing an eye on the head portion. Two suspensory holes 3 mm in diameter are located on either side of the 12:00 o'clock position 7.2 mm apart and 2 mm from the upper edge. Wear patterns on the edges of these holes indicate that the suspensory thong crossed the front of the pendent. The angle of the wear marks from the holes on the reverse side of the pendent indicate that it was worn close to the throat on a thong of about 50 cm in length. The remaining features of the pendant are three intact and one battered projections 7 mm long by 3 mm high in the 6:00, 6:30, 8:00, and 8:30 o'clock positions. These may represent feet, feathers, or both [1972:78–80].

Two shell bracelets/armlets along with a whole, unmodified cowry shell (cf. *Cypraea cervinetta*) and a whole *Oliva* sp. shell were recovered from Burial 3A at Cuautla Site 3 (Figure 7.21). The genus of mollusk from which the bracelets were made is not reported in the field notes; however, based on the photograph, they appear to be produced from *Glycymeris* or *Laevicardium*, a large bivalve species. The lower bracelet has a maximum diameter of 5.4 cm and an aperture diameter measuring 4.1 cm. The upper bracelet measures 5.9 cm in diameter, with an aperture of 4.4 cm. The cowry shell is 8.0 cm long.

In some instances, shell may have been present in burials or as funerary offerings; however, this association is ambiguous. Based on a review of field notes and burial forms, it is not clear whether shells were deliberately placed as offerings or whether their presence represents a mixing of deposits or is otherwise fortuitous.

At Amapa, 17 shell bracelet fragments were recovered from four separate graves; all are reported as being made from *Glycymeris* cf. *gigantea* (Meighan 1976e). Kelly (1945:144–146, Figure 72) reports a small assemblage of marine shell ornaments that is dominated by shell bracelets but includes beads and pendants. What appears to be a *Laevicardium* bracelet or armlet is illustrated as her Figure 72a. Ekholm (1942:109–111, Figure 21) discusses and illustrates shell artifacts recovered at Guasave. These include carved beads and pendants, whole

FIGURE 7.22. Small copper ax from Chalpa.

shell beads and pendants, shell plaques (tesserae?), and bracelets, some of which have had the umbo areas carved into faces. Some of the illustrated bracelets appear to be plain, thin *Glycymeris* bracelets.

Copper

Western Mexico, including the West Mexican Coastal Plain, has a long history of production and use of copper (e.g., Hosler 1994; Mountjoy 2000; Mountjoy and Torres 1985; Pendergast 1962a, 1962b). Metallurgy has been regarded as a Postclassic time marker in Mesoamerica, and it is a hallmark of the late Aztatlán tradition. There is general agreement that metallurgical technology came to west Mexico from Central or South America (Hosler 1994:87; Meighan 1969; Mountjoy 1969). But while there is little doubt that it originated elsewhere, the question of when it appeared in western Mesoamerica continues to be debated. A date around the start of the Postclassic period has generally been accepted for many years, but there is now increasing evidence that metallurgy may have appeared earlier in west Mexico than elsewhere in Mesoamerica. Mountjoy (Mountjoy and Torres 1985), using Meighan's (1976b) chronology for Amapa (which here is rejected in favor of Grosscup's [1976]), has suggested that copper artifacts appear as early as AD 600 in west Mexico. Mountjoy further argues: "The origin and early development of metal working on the west Mexican coast appears intimately related to the Aztatlán tradition" (2000:97). Hosler (1994) has divided west Mexican copper metallurgy into two periods, Periods 1 and 2. Period 1, AD 600–1200/1300, copper artifact types include bells (the most commonly made artifact type), open rings, perforated eye needles, beam design tweezers, axes, awls, and fishhooks. These include items made from low-arsenic copper-arsenic alloys. Period 2, AD 1200/1300–Spanish contact, saw a significant advancement in the technology and skill of west Mexican metallurgists. It is during this time that a variety of copper alloys, including bronze and some copper and silver and copper and gold, appear. These allowed artisans to refine and elaborate the designs of the items being produced. As a result, the shape and size of copper bells varied greatly. Other metal items included ornamental tweezers, open rings, and sheet metal body ornaments. The alloys also allowed the production of hardened cutting, carving, and piercing tools such as axes, awls with one or both ends pointed, narrow blades, and needles. Although data are limited, Hosler (1994:132) indicates that very few of the copper items from the Sinaloa area appear to represent Period 2 copper technology and that Period 1 technology appears to have persisted rather than being replaced in Period 2.

The Marismas sites of Arriñitas, Chalpa, Tecualilla, Cristo Rey, and Venadillo produced copper artifacts. No evidence (e.g., kilns, tools) of copper production was identified during the Marismas excavations, and it is assumed that the items were obtained through exchange. The Marismas copper artifacts were recovered from a burial mound, but none in direct association with a burial; thus, their use as funerary offerings is unclear. Some of these items could have been placed as offerings, disturbed, and redeposited during subsequent interments. Other items may have been lost or dropped, eventually mixing into the fill of burial mounds. No gold or silver artifacts were recovered.

Gill (1974:88) recovered what he believed to be a small (miniature) copper ax (Figure 7.22) in Feature C at Chalpa, measuring approximately 2.63 cm wide by

FIGURE 7.23. Copper from Mound C, Venadillo.

5.95 cm long. Based on the photograph, one edge (right) appears to be damaged, and the other (left) appears partially beveled. In form, it is unlike axes or celts with flaring bits that are reported elsewhere from the west Mexican coastal zone (e.g., Mountjoy and Torres 1985:Figures 5.3, 5.4e). Perhaps the piece was axe-money (used as tribute or in trade), although it does not resemble in form or technology the axe-monies described elsewhere in west Mexico (Hosler 1994:166–168, Figure 5.27).

Two pieces of copper were recovered from Tecualilla, one from TU F and the other from TU G. The specimen from TU F (Level 9) was a small lump of copper, possibly a small ingot, measuring 1.87 × 2.81 × 0.98 cm. It weighed 23.49 g and was smoothed or ground. The function of this specimen is unknown. The second copper artifact recovered appears to have been a pin or needle. It was heavily corroded, measured 8.16 cm long and 0.22 cm thick, and weighed 1.72 g.

A possible copper hook (fishhook?) was recovered at Cristo Rey, and the site of Arriñitas produced two small copper bells; however, these were not described. An amorphous piece of heavily corroded copper was recovered from Level 5 at Mound C at Venadillo (Figure 7.23). The maximum dimensions of this Y-shaped artifact were 5.4 × 6.0 mm; neither its original shape nor its function can be determined.

Summary and Discussion

Although analyses of non-ceramic vessel artifacts recovered during the Marismas Nacionales excavations were limited, the preceding discussion provides insight into the array of material culture associated with the prehistoric inhabitants of the area; it also provides insight into resources, both raw and finished products, and patterns of acquisition and exchange. Several aspects of the Marismas non-ceramic vessel artifact assemblage as a whole are curious, if not compelling. In general, the non-ceramic vessel artifact assemblage is small, and there is low variability within many of the recovered artifact classes. In part, this may be an outcome of focusing excavations on burial mounds and shell middens and the lack of excavations between mounds and middens that might have resulted in the identification of domestic structures and other features. It may also be due to recovery techniques (e.g., lack of screening or fine screening). However, these patterns cannot be totally attributed to methodological issues. Low levels of artifact diversity among various artifact classes may be an indication of the socioeconomic status of the households and individuals occupying the area and the social significance of prehistoric Marismas settlements. These communities and their inhabitants may have simply lacked the opportunity, social status (or responsibilities), or resources to acquire and own an array of material culture. The absence of material culture other than ceramic vessels is further evident in burial assemblages. As discussed in Chapter 8, few items were found in association with human burials, and few of these appear to have been made or acquired solely for use as funerary offerings.

One notable characteristic of the Marismas artifact assemblages is the lack of utilitarian or ornamental marine shell artifacts in spite of their proximity to open beaches where shells could be collected. A shell pendant or gorget reportedly from Chalpa and the two shell bracelets and whole shell funerary offerings from Cuautla are the only marine shell artifacts described as a result of the Marismas excavations. Shells were sometimes reported as funerary offerings by Gill and his students, but the descriptions are vague, and it is not clear whether these are individual whole shells or pieces of shell. Their status as funerary offerings is also highly ambiguous as they could be pieces or individual shells that occurred as mound fill. Marine shell ornaments and artifacts are found at sites along the west coast of Mexico and in adjacent uplands (e.g., Ekholm 1942; Feldman 1976; Kelly 1938; Meighan, ed. 1976), and marine shell was a commodity that was widely traded (e.g., Bradley 1993; Tower 1945; Trubitt 2003). However, no evidence—tools or shell debris—of shell ornament and jewelry manufacture was identified at any of the Marismas sites, although it is possible that limited, household production of shell ornaments could have occurred.

The Marismas sites produced a small assemblage of bells, whistles, rattles, and a rasp. Similar types of artifacts have been recovered from other west Mexican coastal sites. The presence of these items indicates that sound or music was incorporated in ritual. Whether this was done at the household or community level is unclear; such items could also have been taken to ceremonies at regional centers or shrine sites such as El Calón.

Another aspect of ritual activity is expressed by figurines. Gender-specific and gender-neutral specimens are represented in the recovered assemblage. The figurines recovered from El Calón during the Marismas project are simple, modeled, gender-neutral specimens that exhibit no evidence of clothing or personal adornment. The Mazapan figurine fragments, as is typical, display evidence of earspools, headdresses, and clothing. The Venadillo Mound B specimen appears to be a female with a necklace or perhaps a collar to a tunic or dress. The Venadillo Mound C specimen may be exhibiting a headdress and a medallion or breastplate on his chest. The lower half of his body is nude. The presence of figurines at El Calón and in burial mounds at sites suggests use both at public shrines and in local ceremonies, possibly in domestic rituals, before being used as funerary offerings, discarded, or lost.

Items of personal adornment were not abundant. Ceramic and shell beads, shell bracelets, and a shell pendant were recovered. It is possible that some of the amorphous pieces of copper recovered, in their original state, were also items of personal adornment. The copper pin recovered at Tecualilla could have been used to fasten clothing or perhaps as a hairpin. Most were found in mortuary contexts and likely represent personal items or items placed on or with the body. With the possible exception of the one small green stone bead recovered at Chalpa, none of the items identified as personal adornment appear to be of exotic material. In addition to jewelry, the presence of spindle whorls suggests that cotton and perhaps maguey were being woven locally. Cotton could have been woven into belts, headdresses, ponchos, blouses, kilts, or loincloths for everyday or ceremonial use.

As discussed, the spindle whorls recovered came in a variety of sizes, indicating that thread or yarn of varying widths was produced. One of the interesting aspects of the spindle whorls is their stylistic similarity and seemingly high degree of uniformity across sites along the west Mexican coast. Although ceramic spindle whorls would not necessarily take a highly skilled craftsperson to manufacture and could have been produced at the community or household level, this uniformity is suggestive of mass production and exchange. Funerary evidence suggests that spinning and (by extension) weaving were activities dominated by females and that variation in whorl size and the presence of spindle whorls may indicate that women maintained their own tool kits for producing yarn, which were subsequently interred with them. The presence of spindle whorls in the latter Classic period and in the Postclassic period suggests that cotton growing and yarn production were features of the Marismas occupation, consistent with later ethnohistoric evidence of cotton growing and the production of cotton cloth on the west coast at contact (see Chapter 13).

The smoking pipes recovered, both whole and fragmentary, generally appear to be Cocoyolitos-style pipes and date to the Postclassic period. The whole pipes were found in association with human burials and likely represent personal items. Although occasionally found with males, the majority of ceramic smoking pipes recovered were typically found in association with female burials, and in several instances, multiple pipes were present within a single burial.

There are 74 species of native tobacco in the Americas (see Winter 2000a, 2000b, for a detailed discussion of the use, natural history, and distribution of tobacco in the Americas). Winter (2000a:108) indicates that the Tarahumara and Huichol historically raised *Nicotiana rustica* and still use it for ritual purposes, although modern forms of tobacco are also grown and used. In addition to the Tarahumara and Huichol, other groups in adjacent areas, the Cora and Tepehuan, are reported to use tobacco for ritual purposes as well as for pleasure (e.g., Díaz-Romo and Alvarez 2000; Lumholtz 1902; Pennington 1963, 1969). Whether the pipes found in burial contexts were personal items or were placed as part of the mortuary process, for example, ritual smoking at the time of the interment and placement with the body for smoking in the afterlife, is unclear. It is also possible that pipes were used in curing ceremonies. Also unknown is whether the pipes were produced locally or obtained through exchange. Pipe making, as opposed to the making of spindle whorls, likely required more technological skill and craftsmanship.

A variety of flaked and ground stone was recovered. The presence of obsidian, quartzite, and granitic debitage and the presence of cores indicate that flaked-stone tools, including prismatic blades, were produced and maintained at the Marismas sites. Although the evidence is limited and tentative, it is likely that most of the obsidian was obtained from sources in highland Jalisco, specifically the La Joya source and the nearby Las Cuevas workshop. However, other sources may be represented, and it is possible that even obsidian from the central Mexican Pachuca (Hidalgo) source made its way to some of the Marismas sites. The presence of obsidian cores indicates that some blade manufacturing occurred locally, but it is also likely that blades themselves, as well as bifaces and projectile points, were obtained through exchange. The alluvial valleys of the nearby Sierra Madre Occidental were likely sources for a variety of lithic materials for both flaked- and ground-stone implements. Granitic material, rhyolite, and vesicular basalt are among the types of material that could have been procured from the river valleys cutting through the piedmont bordering the Marismas. Cobbles, based on size and shape, requiring minimal modification for use could have been collected for grinding implements such as manos and pestles or for the making of mauls or axes. Larger pieces of stone could have been collected for shaping metates, mortars, or other grinding implements. As with some of the flaked stone and other items discussed in this chapter, finished ground-stone products such as formal, well-shaped metates, manos, mortars, and pestles were also obtained through exchange. No lithic workshops were identified, and evidence of stone-working implements was limited to a few probable hammerstones; however, this may have been because of the limited scale of the excavations at any given site.

Clearly, the prehistoric occupants made and acquired an array of tools and items to sustain their domestic and ritual realms. As discussed in Chapter 8, they made great use of the raw biotic resources of the estuary, coastal, and marine ecosystems to nourish themselves. These environments also provided biotic raw materials used to build houses, make tools, and decorate themselves. However, because of the perishable nature of many of these types of materials, we can only speculate about the array of materials used and items made. Nevertheless, a variety of man-made lithic, mineral, and ceramic items were preserved in the archaeological record of the area. These indicate the use of local, near-local, and extralocal raw resources as well as the procurement, either through exchange or by direct acquisition, of finished products. The mechanisms through which nonlocally produced items and nonlocally available raw materials were acquired are unclear based on the data available. Several, but not necessarily mutually exclusive, models of exchange have been offered to explain how products and raw resources were managed and moved across the west Mexican landscape, especially during the panregional Postclassic Aztatlán period (ca. AD 900–1350/1400).

The Aztatlán tradition was a widespread material culture and ideological complex of socioeconomic and information networks that linked not only the cultures of west Mexico to one another but those cultures to central Mexico and elsewhere in northern Mexico (Foster 1999; Kelley 1986, 2000; Mountjoy 2000; Publ 1985, 1986; Sauer and Brand 1932). The Aztatlán phenomenon has variously been referred to as an archaeological culture, tradition, and horizon. Initially, the research focus was on widespread similarities in material culture, especially ceramic decoration, from one area to the other on the west Mexican coast and the apparent and suggested ties to the Mixteca-Puebla culture of central Mexico (e.g., Ekholm 1942; Kelley 1986, 2000; Sauer and Brand 1932). As discussed in Chapter 12, the Spanish encountered a number of native provinces—Culiacán, Chametla, Aztatlán, and Sentispac—along the west Mexican coast from northern Nayarit to northern Sinaloa. These political entities were culturally complex, exhibiting large ceremonial and market towns with high population densities, large armies, agricultural and fishing industries, and extensive trade networks. Tribute-exacting caciques administered these provinces.

As suggested by the archaeology of the Marismas and previous investigations, the west coast was divided by such political entities before the arrival of the Spanish (see also Publ 1986, 1990). Mountjoy (2000:95) has asserted that there was at least one Aztatlán tradition ceremonial/civic center established near the center of every large coastal river valley between Tomatlán, Jalisco, and the northern border of Sinaloa. In terms of the Marismas villagers, perhaps San Felipe Aztatán (Garduño and Gámez Eternod 2005) was such a center.

It has been demonstrated archaeologically that a great variety of finished goods and raw material moved along the trade corridors that extended up and down the coast and well up into the uplands of west and northwest Mexico and the central highlands. Obsidian, copper, ceramics, figurines, ornaments, cotton cloth, salt, dried fish and shellfish, feathers, and shell were undoubtedly only a portion of the items that moved back and forth in this system. Kelley (1986, 1995, 2000) has argued that the mercantile system, with entrepreneurial long-distance traders transporting their goods to gateway or emerging gateway communities, was responsible for the development of the Aztatlán tradition. Although economics was seen as the prime mover, these traders were also the conduit for the spread of ideological and social change. By the time of the Spanish invasion, the Aztatlán tradition had collapsed, and ties to central Mexico had been cut by the rise of the Tarascan Empire (Kelley 1993). However, it is clear that the ethnohistoric cultures of the area were functioning remnants of the Aztatlán system. The Spanish observed the widespread trade and the movement of goods along the coast (see Chapter 12) and saw that the region was a route for the movement of goods to and from the American Southwest (e.g., Kelley 1986, 1993; Riley 1987, 2005).

Others have opted to argue that prehistoric goods and raw materials moved through west and northwest Mexico through systems of prestige exchange—that is, long-distance exchange was maintained by a social elite or groups of elites residing in various political subdivisions (Darling 1998:391). Such exchange builds alliances between groups, and because the commodities generally exchanged are often luxury or exotic items or resources, they are symbols of authority and status to those who obtain and manage them. However, it is likely that neither the mercantile model nor the prestige exchange model fully accounts for all exchange processes undertaken by the people of the Marismas and the west Mexican coast in general.

The structure and dynamics of intercommunity exchange are complex and multitiered (McAnany 1987). Prestige and exotic items undoubtedly were acquired and exchanged in different spheres than utilitarian and low-value items. To what degree, if any, the elite and mercantilists controlled access and the redistribution of utilitarian and low-value items is uncertain in terms of the materials recovered at the Marismas sites. Certainly there was exchange within extended households and between households at the local level, and clearly there was acquisition of finished products and nonlocal resources and products by the Marismas populations. Nevertheless, the limits of extant data on resource zones and the acquisition of resources, the production and distribution of products at primary and secondary civic/ceremonial centers, and inter- and intracommunity exchange hinder a more comprehensive understanding of prehistoric economic and political organization along the west Mexican coast. Regardless, as can be seen by the material culture described in this chapter, be it through long-distance traders trying to enrich themselves, power-hungry local elites and aspiring elites trying to establish and maintain power, or farmers and fisherfolk trying to carve out a life for themselves and their families, an array of products were acquired, traded, made, and consumed by the prehistoric inhabitants of the Marismas and the adjacent West Mexican Coastal Plain.

CHAPTER 8

Diet and Subsistence

Michael S. Foster

Diet and subsistence are often considered together in archaeology, although they represent two distinct aspects of a culture's economy and adaptation to the environment (Spielmann 1990). Dietary studies focus on the identification of the foods eaten, the seasonal availability of foods, and the nutritional benefits of the foods consumed. The focus of subsistence studies is on the acquisition of food resources, their allocation, and their preparation.

The Marismas Nacionales is a resource-laden environment; arable land and a variety of terrestrial and aquatic plant and animal species are present within this extensive mangrove-estuarine system. The subsistence potential of the area is further enhanced by the fact that it is an ecotone (Odum 1971); as such, the Marismas estuary-lagoon-mangrove system is a transitional zone or environment between two diverse environments. To the east lie the piedmont and the western flank of the Sierra Madre Occidental, and to the west, the Pacific Ocean. Ecotones by definition contain a greater diversity of species than either of the adjoining environmental zones (Odum 1971:157–159). The terrestrial species common to the lowland and foothill zones are at home in the Marismas, while the saline estuary offers the mollusks, shrimp, fish, birds, and mammals that inhabit oceanic shoreline environments on a seasonal and annual basis.

Prehistorically, access to the estuary and ocean from most of the sites on the Teacapán Peninsula and adjacent coastal plain was facilitated by the use of established and maintained trails cut through the thickets and mangroves. The resources of the nearby piedmont, however, would have required significantly more effort to access because of the distance and, especially during the rainy season, the dense vegetation. Nevertheless, the potential edible biomass in the foothills was a perfect complement to seasonal estuarine, marine, and coastal food resources. Life along this coastal ecotone was bountiful and encouraged centuries of human occupation and use.

A variety of faunal specimens representing prehistoric dietary remains were recovered or documented during the Marismas project. The most frequently observed and most abundant were mollusks. Shell middens dot the landscape, and scatters of shell representing food remains were a common occurrence at the habitation sites investigated. In addition to mollusks, the remains of a variety of terrestrial and aquatic fauna were recovered. In all, the project identified 5,039 faunal specimens, exclusive of molluscan remains (counted as number of individual specimens [NISP]). Faunal assemblages were recovered from the nine sites and ranged from 187 NISP at Juana Gomez to 864 NISP at Rincon de Panal, although the total number of specimens recovered from the site of Venadillo exceeded 1,000. The marine shell and vertebrate bones recovered or otherwise documented attest to the collecting and hunting of nondomesticated animal species from a variety of habitats within and adjacent to the Marismas, including the estuary, lagoonal system, and ocean.

The recovery of botanical remains was remarkably slim by comparison with faunal remains. A few pollen grains of cucurbits and maize were identified in sediment samples collected for geomorphic and paleoenvironmental reconstructions (see Chapter 2). Pollen samples from archaeological contexts yielded no information regarding prehistoric plant collection or agricultural practices, and no macro archaeobotanical remains were recovered. Nevertheless, the dietary remains recovered do provide insights into general subsistence and land-use patterns, plus the variety of species procured and consumed by the prehistoric inhabitants of the Marismas.

Additionally, dietary and subsistence information may be gleaned from the recovered artifact assemblages, particularly the ground- and flaked-stone tools that appear to be associated with food procurement and processing. The presence of grinding implements suggests the processing of plant foods, both cultigens and wild species, and the presence of projectile points may be indicative of hunting. One notched sherd recovered may be a net sinker used for fishing.

TABLE 8.1. Archaeomollusks from the Marismas Nacionales.

Taxon	Common Name	Niche
Bivalvia		
Arcidae		
Anadara grandis	Ark shell	Sa-B
Anadara tuberculosa		Sa-B
Noetia reversa	Ark shell	Sb
Mytilidae		
Choromytilus palliopunctatus	Mussel	R
Pteriidae		
Pteria sterna	Wing oyster	R
Ostreidae		
Ostrea corteziensis	Oyster	B
Ostrea palmula	Oyster	B
Crassatellidae		
Pododesmus cepio	Jingle shell	Sa
Corbiculidae		
Polymesoda (*Neocyrena*) sp.	Marsh clam	B
Veneridae		
Anomalocardia subrugosa	Venus clam	Sa
Chione gnidia	Venus clam	Sa-X
Chione subrugosa		Sa
Protothaca grata	Venus clam	Sa-X
Tivela byronensis	Venus clam	Sa-X
Tivela delessertii	Venus clam	Sa
Tivela planulata	Venus clam	Sa
Tivela stultorum	Venus clam	Sa-Sb
Mactridae		
Mulinia pallida		Sa-Sb
Tellinidae		
Rangia medica		B
Donacidae		
Donax gouldii	Bean clam	Sa-Sb
Donax punctatostriatus	Bean clam	Sa-Sb
Gastropoda		
Neritidae		
Neritina sp.	Nerite	Sa
Littorinidae		
Littorina fasciata	Periwinkle	Sa
Turritellidae		
Turritella anactor	Turret shell	X
Turritella gonostoma	Turret shell	X
Turritella leucostoma	Turret shell	X
Potamididae		
Cerithidea montagnei	Horn shell	B
Cerithidea valida	Horn shell	B
Calyptraeidae		
Crepidula striolata		R
Naticidae		
Natica chemnitzii	Moon shell	Sa
Polinices uber	Moon shell	Sb-X
Muricidae		
Hexaplex brassica		X
Hexaplex (Muricanthus) nigritus	Rock shell	Sb
Thaididae		
Thais biserialis	Dye shell	R
Thais kiosquiformis		B-R
Buccinidae		
Northia northiae	Whelk	Sb
Melongenidae		
Melongena patula		Sa
Nassariidae		
Nassarius luteostoma	Nassa mud snail	Sa
Fasciolariidae		
Opeatostoma pseudodon	Tulip shell	Sb
Olividae		
Agaronia propatula		Sa
Agaronia testacea	Olive shell	Sa
Oliva sp.	Olive shell	Sa
Olivella volutella		Sa-Sb
Cancellariidae		
Cancellaria solida	Nutmeg shell	X
Terrestrial Gastropoda		
Polygyridae		
Polygyra texasiana		L
Spiraxidae		
Streptostyla turgidula		L

Note: B = low-salinity lagoon and mangrove environment (brackish); L = land; R = intertidal and shallow rocky shore assemblage; Sa = intertidal sand beaches and sand flats to 10 m; Sb = sand and sand/mud, 11 to 26 m, nearshore; X = intermediate shelf assemblage, 27 to 65 m.

Source: Taxonomic order (family) based on Keen 1971.

Diet

Mollusks

Linear and mounded shell middens representing food refuse are prevalent archaeological features on the Marismas landscape (Shenkel 1969a, 1971a, 1974). Marine shell refuse was also found in buried archaeological contexts in proximity to hearths and scattered in feature fill. It is clear that the prehistoric occupants of the area were well acquainted with mollusks as a source of food. The Marismas is not unique in this regard. Shell middens, dating from the Middle Archaic period to the early contact period, are common along the coast of west Mexico from the northern desert of Sonora to the tropical lowlands of southern Chiapas (e.g., Bowen 1976; Foster et al. 2008; Mountjoy 2000; Voorhies et al. 2002). The most impressive shell middens in the Marismas are the linear oyster middens found along portions of the estuary; however, these are thought to mostly date to the early contact period and the Spanish occupation of west Mexico (see Chapter 12). Nevertheless, oysters were important prehistorically. Oysters (*Ostrea* spp.) and clams, *Tivela byronensis* in particular, are the most common taxa represented in the prehistoric middens of the area. Numerous other species are also represented in the middens, but in relatively small numbers, and although many are edible, their dietary significance is not always clear. Members of 10 endemic families (14 genera, 21 species) of marine Bivalvia and 14 families (17 genera, 22 species) of Gastropoda were indentified, as were two species of terrestrial gastropods (Table 8.1) and barnacles (a crustacean). None of the taxa identified appear to be exotic or nonlocal.

Shell middens are an invaluable source of information on prehistoric diet, land use, and environmental information, and they often contain quantities and varieties of nonmolluscan faunal remains (e.g., Meighan 1970). However, despite the abundance of shell refuse and the seeming importance of mollusks as a food, the amount of edible biomass contributed to a diet by mollusks is often overestimated. As Voorhies (2000; see also Osborn 1977a, 1977b) observes, the abundance and concentration of shell is a reflection of mollusk processing and disposal. As middens are specialized use loci, it is often unclear how long they took to accumulate; they can represent numerous processing and consumption episodes during which only a relatively small quantity of mollusks was consumed.

Nutritional Value of Mollusks

Many of the mollusks listed in Table 8.1 may represent incidentally collected individuals or individuals collected for purposes other than consumption. Oysters and Venus clams, the most commonly represented mollusks found in the Marismas shell middens, and a number of other taxa present are known to have been consumed prehistorically, ethnohistorically, and ethnographically. Waselkov (1987:117–123) and Claassen (1998:183–187) present useful summaries of the nutritional value of mollusks, although, as Waselkov notes, published data on molluscan nutritional values are limited and vary greatly from one source to another. General observations indicate that mollusks are good sources of protein, they have low fat content, they are high in omega-3 fatty acids, they have caloric values ranging from 54 to 234 kcal per 100 g of wet flesh, and their carbohydrates range from 0 g to 8.9 g per individual. They are also good sources of vitamin B_{12} and iron and contain large amounts of thiamine, iodine, and trace minerals but are lacking in vitamin C (Waselkov 1987:122). The amount of protein, carbohydrates, and fat in mollusks varies seasonally, by sex, and by environmental circumstances (Claassen 1998:183, Table 1.5). The amount of protein available in mollusks is slightly less than or equal to the amount found in most fish; however, they generally contain less protein than terrestrial mammals and birds while equaling or exceeding the protein value of most plant foods. It is of particular note that oysters and several other species contain high-quality protein as well as sufficient amounts of essential amino acids (Waselkov 1987, citing Borgstrom 1962). Claassen (1998:183) reports that the amount of protein available in mollusks ranges from 5.5 g to 22.0 g (8–77 percent of the tissue) depending on the species. She further notes, citing Erlandson's (1988:105) minimum daily requirement of 40 g of protein for humans, that 100 g of mollusks could provide 14 to 55 percent of that requirement. Nevertheless, although moderated by physical activity, the consumption of large quantities of protein over several days holds the potential for protein poisoning, which can result in illness or in extreme cases even death (e.g., Claassen 1998:183; Noli and Avery 1988). Thus, the idea that mollusks could serve as the main or sole source of food for extended periods is contrary to human nutrition and survival.

Although extended use of mollusks as the primary source of dietary protein to the near exclusion of other foods holds the potential for illness, it is clear that mollusks are an important food. The question is, What role did mollusks play in the diet of the prehistoric inhabitants of the Marismas? Were they a supplementary food—were mollusks collected to compensate for the lack of other sources of protein—or were they a complementary food—were they a regular and significant part of the diet? Mollusks are commonly thought of as a dietary supplement or even a starvation food. However, in nutrient-rich environments such as estuaries like the Marismas estuary-lagoon system, mollusks are often concentrated in abundance, and they are available most or all of the year in predictable locations. Collectors are therefore almost assured of successfully fulfilling their subsistence needs during lean times or as needed (Waselkov 1987:110, 123).

Seasonality

There is no direct information (oxygen isotopes, incremental growth studies) regarding when the prehistoric inhabitants of the Marismas collected mollusks. This is further complicated by the fact that there is little information on mollusk exploitation and consumption by horticultural or agricultural people, although Claassen (1986, 1998:178–179) suggests that mollusks may have served as a seasonal staple for prehistoric food producers. Most data are derived from studies of hunter-gatherers (e.g., Bowdler 1976; Meehan 1982). However, as discussed above, in a setting such as the Marismas, mollusks were likely available year-round and therefore could have been collected as needed while, perhaps, being more heavily exploited during the late winter and spring dry season, when agricultural stores were low and wild plant resources were diminished or unavailable.

A number of other factors undoubtedly influenced when and how intensively mollusks were collected. Late spring and summer, before the onset of the summer monsoons, are oppressively hot in the Marismas, and if mollusks were collected for consumption, it would have been necessary to consume them quickly before they spoiled or preserve them through drying, salting, or smoking. Another issue that affects mollusks in the summer is the occasional red tide, which makes mollusks poisonous to humans and kills mollusks off. During the 1974 field season (June–July), the oyster population in the estuary was sufficiently reestablished that harvesting them was allowed, and the Marismas project cooks would regularly have oysters available in the late afternoon. Also, during the summer, children were observed opportunistically collecting and eating *Donax*—a small clam easily collected from the beaches along the mouth of the estuary and along the ocean—raw in the same fashion as one eats sunflower seeds. Water temperature may also affect the availability of certain mollusks. As water temperatures cool, some mollusks migrate to deeper, warmer waters or burrow deeper into substrates (Waselkov 1987:111). However, Waselkov notes that the impact of temperature change more likely influences species selection rather than the overall use of mollusks. Additionally, the salinity of brackish water environments such as the Teacapán Estuary affects the availability of species such as the oyster. Historically and during the modern era, freshwater from flooding during the rainy season has resulted in the extirpation of oyster and shrimp populations in the estuary. This undoubtedly occurred prehistorically as well. Meehan (1977, 1982) notes that such an event killed off most of the mollusks in an estuary exploited by the Anbarra, Gidjingali speakers of Australia; they responded by catching more fish than they had during the previous year.

Crustaceans

Crustaceans (Arthropoda/Crustacea) were represented by the remains of crabs. Members of Brachyura (crabs) are reported only at the site of Venadillo. It is likely that their use as a source of food is underrepresented in the reported faunal data as they would have been ever present in the area.

Vertebrate Fauna

Vertebrate faunal remains were recovered from middens, from the fill of burial mounds, or from other excavated contexts at nine sites. These included remains representing foods, the complete or partial (disturbed) skeletons of animals deliberately buried in the burial mounds, and the remains of animals that happened to die or were killed in or on the feature. Deliberately buried animals were usually easy to distinguish, as a majority of the skeleton of the animal was found articulated or spatially clustered. A number of isolated elements of dog were recovered; these are generally thought to represent disturbed and redeposited remains of deliberately buried individuals. Although dogs were widely used as a food throughout Mesoamerica (e.g., Emery 2001:256; Flannery 1976; Wing 1978), there is only limited evidence that they were butchered and consumed by the prehistoric occupants of the Marismas.

Faunal remains were generally collected as encountered or recovered during the screening of excavated fill through ¼-inch screen. The only systematic effort to collect faunal remains occurred during the excavation of Mound C at Venadillo (Cumbaa 1973; Foster 1984), where faunal remains were collected from four 2-×-2-m excavation units. All excavated fill was screened through ¼-inch screen, and all faunal elements present were collected by level. Bulk grab samples of the screened sediment were then collected from two of the excavation units for water screening and flotation; a fine-mesh screen—18 squares to an inch—was used for recovery. This resulted in the recovery and identification of 26 taxa not recovered using a ¼-inch screen (Cumbaa 1973:67). It is not clear that all these specimens, mainly small reptiles and amphibians, represent food remains; Cumbaa (1973) suggested that the small fish might represent bait rather than food remains.

The recovered faunal assemblages were, for the most part, quite similar, although some notable differences exist. Bone was classified as subsistence fauna if it was an isolated or fragmentary element or if it was burned.

Nevertheless, interpretation of some of the remains, especially those of armadillo, snake, rodents, lizards, and amphibians, is problematic since they lacked definitive characteristics that indicate they represent food remains. As noted above they may represent natural deaths or individuals that were killed inadvertently, or perhaps they represent agricultural or household pests that were killed but not consumed as food. The lack of reported or identified contextual information, such as evidence of burrows or dens, somewhat hinders the interpretation of such remains. Regardless, archaeological and ethnographic data indicate that many of these animals were systematically or opportunistically hunted and consumed.

A number of vertebrate species are represented at all of the sites. These include dogs (*Canis familiaris*), opossum (*Didelphis virginiana*), birds and turtles, and catfish (Ariidae). Present in at least two-thirds of the sites are armadillo (*Dasypus novemcinctus*), rabbit (*Sylvilagus* sp.), small rodents, deer (*Odocoileus virginianus*), iguana (Iguanidae), snook (*Centropomus* sp.), and snapper (*Lutjanus* sp.). Turkey (*Meleagris gallopavo*) and musk turtle (Kinosternidae) are also commonly found. Bone fragments of aquatic animals constitute approximately 60 to 80 percent of the overall sample considered to represent food remains.

Quantification of the assemblages was based on the number of identified specimens per taxon, reported as the NISP, and the minimum number of individuals (MNI) was also calculated. A total of 5,038 NISP representing 93 taxa (65 nonoverlapping) were observed; this resulted in the identification of at least four families and 14 genera of mammals, three families and six genera of birds, two families and two genera of amphibians, four families and 18 genera of reptiles, 10 families and 14 genera of fish, and a number of unidentified invertebrates (crabs) (Table 8.2). In most instances, no more than one or two individuals of a given genera or species are represented at a given site. Dogs, apparently because of their presence as burials in burial mounds, were the most frequently recovered fauna.

Nutritional Value of Vertebrate Fauna

The terrestrial and nonmolluscan aquatic fauna exploited by the prehistoric Marismas inhabitants provided a variety of nutrients—protein, vitamins, and minerals—to supplement a presumed agriculturally based diet. Deer is a good source of protein, although it is high in cholesterol and saturated fat; it is also a good source of iron, phosphorus, selenium, zinc, copper, niacin, and vitamins B_2 and B_{12}. Rabbit meat is high in protein and low in fat, sodium, and saturated fatty acids. Although rabbit meat is also high in cholesterol, it is a good source of vitamin B_{12}, niacin, iron, phosphorus, and selenium. Fish in general are good sources of protein and omega-3 fatty acids, are low in fat, and provide vitamins A and B as well as calcium.

Faunal Remains at Marismas Project Sites

Juana Gomez

The site of Juana Gomez is the northernmost site investigated during the Marismas project. It is located inland from the Teacapán Peninsula approximately 12 to 13 km north-northwest of the Laguna Agua Grande and about 16 km east of the coast, although today tidal marshes and flats lie about 6 km to the west. It lies outside the estuary-lagoon system that forms the heart of the Marismas; the archaeological components investigated date between AD 300 and 700, to the Early and Late Estero periods. A total of 187 vertebrate specimens were examined from Juana Gomez; 182 of those are identified as taxa that are likely food sources. Five dog elements recovered represent a single individual thought to be a burial (Table 8.3). At least 18 individuals, including the one dog, are represented in the remains; no MNI could be determined for the 134 fish bone fragments recovered. Aquatic animal remains (n = 153) account for 84.1 percent of the NISP, while terrestrial animal remains (n = 28) account for the remaining 15.4 percent, although the amount of available meat represented by the terrestrial taxa likely constitutes a much greater percentage of the total available meat represented. The relative abundance of aquatic species indicates that fish provided a significant amount of the animal protein consumed, despite the fact that the site was some distance from estuarine or marine habitats. Although duck was likely a food source, birds do not appear to have contributed significantly to the diet.

As previously noted, Juana Gomez produced a bone artifact made from a dog phalange. The specimen was burned, its proximal end had been cut off, it was polished, and two holes had been drilled through the anterior surface of the distal end (Wing 1968).

Arriñitas

Arriñitas is located on the east bank of an arm of Teacapán Estuary; the site dates to AD 900–1050, the Early Conchera period. Subsistence fauna account for 54.2 percent of the NISP recovered (Table 8.4). Mammals, including opossum, armadillo, rabbits, rodents, and probably deer, account for 20.1 percent of the total identifiable subsistence NISP; birds, including turkey, for 18.2 percent; reptiles, including iguana, turtles, snakes, and frogs, for 33.7 percent; and fish, including shark, sea catfish, snook, grouper, snapper, and corvina, for just over 27 percent. Based on the MNI, terrestrial taxa made up a greater portion of the diet at Arriñitas. However, in terms of available meat, fish likely provided near-equivalent amounts. Two dog burials were also recovered.

TABLE 8.2. Identified Archaeofauna from Sites in the Marismas Nacionales.

Taxon	Common Name	Habitat
Mammals		
Didelphis virginiana	Opossum	Terrestrial
Soricidae	Shrews	Terrestrial
cf. *Notiosorex crawfordi*	Crawford's desert shrew	Terrestrial
Leporidae	Hares and rabbits	Terrestrial
Sylvilagus sp.	Rabbit	Terrestrial
Rodentia	Unidentified rodents	Terrestrial
Thomomys umbrinus	Pocket gopher	Terrestrial
Oryzomys sp.	Rice rat	Terrestrial
Sigmodon hispidus	Cotton rat	Terrestrial
Peromyscus cf. *boylii*	Brush rat	Terrestrial
Cetacean	Sea mammals	Aquatic
Canis familiaris	Domestic dog	Terrestrial
Procyon lotor	Raccoon	Terrestrial
Felis concolor	Mountain lion	Terrestrial
Felis wiedii	Margay	Terrestrial
Odocoileus virginianus	White-tailed deer	Terrestrial
Artiodactyla	Cloven-hoofed mammals	Terrestrial
Mammalia	Unidentified mammals	
Birds		
Casmerodius albus	Great common egret	Aquatic/shore
Eudocimus albus	White ibis	Aquatic/shore
Anas sp.	Duck	Aquatic/shore
Anatidae	Ducks and geese	Aquatic/shore
Gallinula chloropus	Common gallinule	Aquatic/shore
Meleagris gallopavo	Turkey	Terrestrial
Zenaida sp.	Dove	Terrestrial
Columbidae	Doves and pigeons	Terrestrial
Ardeidae	Herons and bitterns	Aquatic/shore
Passeriformes	Perching birds	Terrestrial
Aves	Unidentified birds	
Reptiles		
Crocodilus acutus	American crocodile	Aquatic
Crocodylidae	Crocodilians	Aquatic
Kinosternon sp.	Musk turtle	Aquatic
Kinosternidae	Musk turtles	Aquatic
Chrysemys sp.	Pond turtle	Aquatic
Chelonia mydas	Green turtle	Aquatic
Cheloniidae	Sea turtles	Aquatic
Testudines	Turtles	Aquatic
Squamata	Scaled reptiles	Terrestrial
Phyllodactylus cf. *lanei*	Leaf-toed gecko	Terrestrial
Iguana iguana	Green iguana	Terrestrial
Iguanidae	Iguanid lizards	Terrestrial
Anolis cf. *nebulosus*	Anole	Terrestrial
Colubridae	Nonvenomous snakes	
Natrix valida	Water snake	Terrestrial
Rhadinaea hesperia	Slender snake	Terrestrial
Masticophis flagellum	Coachwhip	Terrestrial
Oxybelis acuminatus	Vine snake	Terrestrial
Drymarchon corais	Indigo snake	Terrestrial
Coluber constrictor	Boa	Terrestrial

TABLE 8.2. (cont'd.) Identified Archaeofauna from Sites in the Marismas Nacionales.

Taxon	Common Name	Habitat
Lampropeltis triangulum	King snake	Terrestrial
Leptodeira sp.	Cat-eyed snake	Terrestrial
Rhinocheilus cf. *antonii*	Mexican long-nosed snake	Terrestrial
Tantilla bogerti	Black-headed snake	Terrestrial
Trimorphodon paucimaculatus	Sinaloa lyre snake	Terrestrial
Serpentes	Snakes	Terrestrial
Amphibians		
Anura	Frogs and toads	Aquatic
Scaphiopus cf. *couchii*	Spadefoot toad	Terrestrial
Bufo sp.	Toad	Aquatic
Hylidae	Tree frogs	Terrestrial
Microhylidae	Narrow-mouthed frogs	Terrestrial
Amphibia	Unidentified amphibians	Aquatic
Fish		
Chondrichthyes	Sharks	Aquatic
Carcharhinus sp.	Requiem shark	Aquatic
Galeocerdo cuvier	Tiger shark	Aquatic
Aetobatis narinari	Spotted eagle ray	Aquatic
Osteichthyes	Unidentified fish	Aquatic
Elops affinis	Machete	Aquatic
Clupeidae	Herring	Aquatic
Ariidae	Sea catfish	Aquatic
Arius steindachneri[a]	Sea catfish	Aquatic
Bagre pinnimaculatus[a]	Gaff-topsail catfish	Aquatic
Batrachoididae	Toadfish	Aquatic
Centropomus sp.	Snook	Aquatic
Serranidae	Sea bass	Aquatic
Epinephelus sp.	Grouper	Aquatic
Carangidae	Jacks	Aquatic
Caranx sp.	Jack	Aquatic
Trachinotus sp.	Pompano	Aquatic
Lutjanidae	Snappers	Aquatic
Lutjanus sp.	Snapper	Aquatic
Gerridae	Mojarras	Aquatic
Gerres sp.	Mojarra	Aquatic
Pomadasyidae	Grunts	Aquatic
Pomadasys macracanthus[a]	Grunt	Aquatic
Haemulon sp.	Grunt	Aquatic
Sparidae	Porgies	Aquatic
Calamus sp.	Porgy	Aquatic
Sciaenidae	Drums	Aquatic
Cynoscion sp.	Corvina	Aquatic
Cynoscio stolzmanni[a]	Corvina	Aquatic
Mugil sp.	Mullet	Aquatic
Polydactylus sp.	Threadfin	Aquatic
Eleotridae	Sleepers	Aquatic
Eleotris sp.	Sleeper	Aquatic
Sphoeroides sp.	Puffer	Aquatic
Invertebrates		
Brachyura	Crabs	Semiterrestrial

Note: Total: 92 taxa (65 nonoverlapping).

[a] Otolith identified at La Tarjea, recovered from Levels 6 and 7, Test Unit A, Mound A (Feldman 1972).

TABLE 8.3. Juana Gomez Archaeofaunal Assemblage.

Taxon	Common Name	Habitat	Number of Individual Specimens (NISP)	Percent NISP	Minimum Number of Individuals
Mammals					
Didelphis virginiana	Opossum	Terrestrial	1	0.5	1
Dasypus novemcinctus	Armadillo	Terrestrial	5	2.7	1
Leporidae	Hares and rabbits	Terrestrial	1	0.5	1
Rodentia	Rodents	Terrestrial	8	4.3	2
Canis familiaris	Dog	Terrestrial	5	2.7	1
Odocoileus virginianus	White-tailed deer	Terrestrial	7	3.7	1
Birds					
Anas sp.	Duck	Aquatic/shore	1	0.5	1
Aves	Birds	—	1	0.5	1
Amphibians					
Bufo sp.	Toad	Aquatic	3	1.6	1
Reptiles					
Testudines	Turtles	Aquatic	1	0.5	1
Serpentes	Snakes	Terrestrial	1	0.5	1
Fish					
Chondrichthyes	Sharks	Aquatic	4	2.1	1
Osteichthyes	Fish	Aquatic	134	71.7	—
Ariidae	Sea catfish	Aquatic	6	3.2	1
Centropomus sp.	Snook	Aquatic	2	1.1	1
Lutjanus sp.	Snapper	Aquatic	2	1.1	1
Mugil sp.	Mullet	Aquatic	4	2.1	1
Eleotridae	Sleepers	Aquatic	1	0.5	1
Subsistence Fauna Totals					
Habitat undetermined			1	0.6	1
Aquatic subsistence fauna			158	86.8	9
Terrestrial subsistence fauna			23	12.6	7
Total subsistence fauna			182	98.33[a]	17
Assemblage Totals					
Aquatic animals			158	84.49[a]	9
Terrestrial animals			28	15.50[a]	8
Habitat undetermined			1	0.1[a]	1
Burials/offerings (dogs)			5	2.7[a]	1
Total recovered			193		18

[a] Percentage of total NISP.

Panales

Panales is an island in the Teacapán Estuary. Excavations were carried out at three loci on Panales: Panales Test Mound A (Pearson 1968), Panales Test Mound B (Shenkel 1969a), and Rincon de Panal (Taylor and Rautenkranz 1971). Although it is not specified, it is believed that the faunal assemblage generally ascribed to Panales and discussed here was recovered during the Test Mound B excavations. Shenkel (1969a:33–41) reports the nearly complete skeletons of two dogs and other faunal remains, and Pearson (1968) only makes passing mention of "nonhuman" bone and shell. Test Mound B consisted of an earthen mound capped with a shell midden; the earthen mound, based on associated ceramics, dates to AD 500–700 (Late Estero period), and the shell cap dates to AD 900–1350 (Early and Middle Conchera periods).

Subsistence fauna account for 41 percent of the NISP (Table 8.5). More than half of the NISP (59.1 percent) of

TABLE 8.4. Arriñitas Archaeofaunal Assemblage.

Taxon	Common Name	Habitat	Number of Individual Specimens (NISP)	Percent NISP	Minimum Number of Individuals
Mammals					
Didelphis virginiana	Opossum	Terrestrial	9	1.6	2
Dasypus novemcinctus	Armadillo	Terrestrial	118	20.8	2
Leporidae	Hares and rabbits	Terrestrial	2	0.4	—
Sylvilagus sp.	Rabbit	Terrestrial	7	1.2	2
Rodentia	Rodents	Terrestrial	12	2.1	4
Thomomys umbrinus	Pocket gopher	Terrestrial	1	0.2	1
Oryzomys sp.	Rice rat	Terrestrial	2	0.4	1
Sigmodon hispidus	Cotton rat	Terrestrial	8	1.4	4
Canis familiaris	Dog	Terrestrial	35	6.2	2
Procyon lotor	Raccoon	Terrestrial	7	1.2	1
Odocoileus virginianus	White-tailed deer	Terrestrial	17	3.0	1
Birds					
Casmerodius albus	Great common egret	Aquatic/shore	2	0.4	1
Eudocimus albus	White ibis	Aquatic/shore	5	0.9	1
Meleagris gallopavo	Turkey	Terrestrial	2	0.4	1
Ardeidae	Herons and bitterns	Aquatic	7	1.2	2
Aves	Unidentified birds	—	47	8.5	6
Reptiles					
Kinosternon sp.	Musk turtle	Aquatic	2	0.4	1
Chrysemys sp.	Pond turtle	Aquatic	9	1.6	1
Testudines	Turtles	Aquatic	34	6.0	2
Iguana iguana	Green iguana	Terrestrial	4	0.7	1
Iguanidae	Iguanas	Terrestrial	1	0.3	
Drymarchon corais	Indigo snake	Terrestrial	106	18.7[a]	1
Serpentes	Snakes	Terrestrial	35	6.2	1
Amphibians					
Anura	Toads and frogs	Aquatic	1	0.2	1
Fish					
Chondrichthyes	Shark	Aquatic	3	0.2	1
Osteichthyes	Fish	Aquatic	43	8.6	—
Ariidae	Sea catfish	Aquatic	32	5.7	5
Centropomus sp.	Snook	Aquatic	3	0.5	2
Epinephelus sp.	Grouper	Aquatic	6	1.1	1
Lutjanus sp.	Snapper	Aquatic	5	0.9	2
Cynoscion sp.	Corvina	Aquatic	1	0.2	1
Subsistence Fauna Totals					
Habitat undetermined			47	15.3	
Aquatic subsistence fauna			153	49.8	21
Terrestrial subsistence fauna			107	34.9	22
Total subsistence fauna			307	52.2[a]	43
Assemblage Totals					
Aquatic animals			153	28.0[a]	21
Terrestrial animals			366	64.7[a]	24
Habitat undetermined			47	8.3[a]	6
Burial/offerings (dogs)			35	6.2[a]	2
Intrusive			224	39.6[a]	
Total recovered			825		45

[a] Percentage of total NISP.

TABLE 8.5. Panales Archaeofaunal Assemblage.

Taxon	Common Name	Habitat	Number of Individual Specimens (NISP)	Percent NISP	Minimum Number of Individuals
Mammals					
Didelphis novemcinctus	Opossum	Terrestrial	1	0.4	1
Sylvilagus sp.	Rabbit	Terrestrial	1	0.4	1
Canis familiaris	Dog	Terrestrial	150	59.0	4
Odocoileus virginianus	White-tailed deer	Terrestrial	10	3.9	1
Birds					
Meleagris gallopavo	Turkey	Terrestrial	1	0.4	1
Reptiles					
Chelonia mydas	Green turtle	Aquatic	1	0.4	1
Cheloniidae	Sea turtles	Aquatic	4	1.6	—
Testudines	Turtles	Aquatic	9	3.5	—
Iguanidae	Iguanas	Terrestrial	7	2.8	1
Fish					
Osteichthyes	Fish	Aquatic	41	16.2	
Ariidae	Sea catfish	Aquatic	8	3.2	2
Caranx sp.	Jack	Aquatic	1	0.4	1
Trachinotus sp.	Pompano	Aquatic	1	0.4	1
Lutjanus sp.	Snapper	Aquatic	6	2.4	1
Sciaenidae	Drums	Aquatic	10	3.9	1
Sphoeroides sp.	Puffer	Aquatic	3	1.2	2
Subsistence Fauna Totals					
Aquatic fauna			84	80.8	9
Terrestrial fauna			20	19.2	5
Total subsistence fauna			104	40.9[a]	15
Assemblage Totals					
Aquatic fauna			84	33.1[a]	9
Terrestrial fauna			170	66.9[a]	9
Burials/offerings (dogs)			150	59.1[a]	4
Total recovered			254		18

[a] Percentage of total NISP.

the Panales archaeofaunal assemblage comes from four dog burials. In terms of subsistence taxa, most notable are turkey and both green and sea turtles. Opossum, rabbit, and deer account for 11.5 percent of the subsistence fauna NISP; turkey, for 0.9 percent; iguana, turtle, and sea turtles, for 20.2 percent; and fish, including sea catfish, jack, pompano, snappers, drum, and puffer, for 68.3 percent.

Rincon de Panal

Rincon de Panal, located on the island of Panales, dates between AD 900 and 1350, the Middle Conchera period. Dog remains from presumed burials account for nearly 48.8 percent of the total NISP. This is followed by armadillo. However, 263 pieces, 30.4 percent of the NISP, of armadillo bone from a single individual were recovered. Armadillos are known to burrow and live in dens; it is possible that this represents an animal that died in its den rather than the carcass of a hunted animal or a burial offering. Although armadillos were hunted and are represented in the archaeofaunal assemblages at other Marismas sites, this individual is considered to be intrusive.

Subsistence fauna account for 20.7 percent of the NISP (Table 8.6). Opossum, rabbit, rodents, raccoon, and deer were among the mammal remains that account for 15.6 percent of the subsistence fauna NISP; birds including turkey and dove account for 8.4 percent; reptiles

TABLE 8.6. Rincon de Panal Archaeofaunal Assemblage.

Taxon	Common Name	Habitat	Number of Individual Specimens (NISP)	Percent NISP	Minimum Number of Individuals
Mammals					
Didelphis virginiana	Opossum	Terrestrial	4	0.5	1
Dasypus novemcinctus	Armadillo	Terrestrial	263	30.4	1
Leporidae	Hares and rabbits	Terrestrial	1	0.1	1
Rodentia	Rodents	Terrestrial	7	0.8	3
Sigmodon hispidus	Cotton rat	Terrestrial	1	0.1	1
Canis familiaris	Dog	Terrestrial	422	48.5	9
Procyon lotor	Raccoon	Terrestrial	7	0.8	1
Felis wiedii	Margay	Terrestrial	3	0.3	1
Odocoileus virginianus	White-tailed deer	Terrestrial	3	0.3	1
Mammalia	Mammals	Terrestrial	2	0.2	—
Birds					
Meleagris gallopavo	Turkey	Terrestrial	6	0.7	2
Columbidae	Doves and pigeons	Aquatic	2	0.2	1
Aves	Birds	Aquatic	7	0.8	3
Reptiles					
Crocodylidae	Crocodilians	Aquatic	6	0.7	2
Kinosternidae	Musk turtles	Aquatic	1	0.1	1
Chrysemys sp.	Pond turtle	Aquatic	3	0.3	1
Testudines	Turtles	Aquatic	3	0.3	
Iguana iguana	Green iguana	Terrestrial	2	0.2	1
Serpentes	Snakes	Terrestrial	2	0.2	1
Fish					
Osteichthyes	Fish	Aquatic	79	9.1	—
Ariidae	Sea catfish	Aquatic	21	2.4	4
Centropomus sp.	Snook	Aquatic	5	0.6	2
Lutjanus sp.	Snapper	Aquatic	11	1.3	2
Haemulon sp.	Grunt	Aquatic	2	0.2	1
Mugil sp	Mullet	Aquatic	1	0.1	1
Subsistence Fauna Totals					
Aquatic subsistence fauna			141	78.8	14
Terrestrial subsistence fauna			38	21.2	17
Total subsistence fauna			179	20.7[a]	31
Assemblage Totals					
Aquatic fauna			141	16.3[a]	18
Terrestrial fauna			723	83.7[a]	23
Burials/offerings (dogs)			422	48.8[a]	9
Intrusive (armadillo)			263	30.4[a]	1
Total recovered			1549		41

[a] Percentage of total NISP.

including caiman, iguana, and snake account for 9.5 percent; and fish including sea catfish, snook, snapper, grunt, and mullet account for 66.5 percent. Based on the NISP for subsistence fauna, fish represent the majority of the remains; however, the amount of meat represented may not significantly outweigh that represented by terrestrial taxa identified.

Of particular note among the mammalian remains were those of a *Felis wiedii* or margay, a small spotted cat similar to an ocelot. Whether this animal was captured for its pelt or meat or whether it was kept as a pet is unknown. Margay pelts, like ocelot, jaguar, or puma pelts, would certainly have been valued.

Chalpa

Chalpa is located adjacent to one of the arms of the Teacapán Estuary. The site dates to the Early and Middle Conchera periods, AD 900–1350. The faunal assemblage from Chalpa is heavily skewed by faunal remains recovered as burials or as funerary offerings. These account for 91 percent (639 elements) of the NISP and include six deer skulls and the remains of 16 dogs and four raccoons. The six deer skulls with antlers were buried, but no lower jaws were present. These skulls were used as burial offerings presumably after the edible and usable parts (hide, tendons, and long bones) of the deer had been consumed, although it is possible that ritualistic requirements precluded the use of the animals and that only the heads were taken. It is unclear why the mandibles were removed and whether they were discarded or put to some other use. In all, remains of eight deer are represented at the site, representing a potentially significant amount of edible biomass. The dogs account for nearly 74.1 percent of the faunal specimens identified.

Subsistence remains account for just over 10 percent of the NISP at Chalpa (Table 8.7). For those specimens identified as subsistence- or likely subsistence-related remains, terrestrial species, including opossum, rodents, and probably deer, account for 37 percent of the NISP identified as subsistence fauna; birds including turkey account for 9.6 percent; reptiles including caiman, iguana, and turtle account for nearly 8.2 percent; and fish—sea catfish, snapper, porgy, drums, and mullet—account for 52.1 percent. The MNI could not be determined for two of the unidentified bird bone and 18 unidentified fish bone fragments recovered.

The taxa represented at Chalpa vary from those at Juana Gomez in that iguana and caiman are represented; the presence of caiman is undoubtedly a reflection of the proximity of Chalpa to the Teacapán Estuary. Also represented, perhaps as a domesticated species, is turkey. The Spanish report that indigenous groups on the west coast of Mexico at contact were keeping turkeys (Emery 2001; Sauer 1932; see Chapter 12). Today, wild turkeys are present in the uplands of the Sierra Madre Occidental to the east of the Marismas, and their range may have been broader prehistorically. Two of the birds represented, herons/bitterns and white ibis, may have been taken for their plumage. It is also likely that the hides of deer and other mammals and perhaps crocodiles and caimans were retained and used.

Tecualilla

Tecualilla is located adjacent to an arm of the Teacapán Estuary; it dates from AD 900 to 1350, the latter Early Conchera and Middle Conchera periods. A relatively greater variety of taxa is represented in the faunal remains from Tecualilla (Table 8.8). Subsistence fauna account for 76.5 percent of the NISP. Opossum, armadillo, mountain lion, rabbit, rodents, deer, and an unidentified sea mammal account for 20.9 percent of the total subsistence NISP represented. Birds including herons or bitterns, duck, turkey, common gallinule, and dove account for 6.2 percent. Amphibians (frogs) appear in minor amounts (0.3 percent). However, collectively, reptiles including caiman, iguana, turtle, and boa constrictor account for 15.8 percent. Fish including shark, sea catfish, snook, grouper, jack, snapper, mojarras, grunt, corvina, drum, mullet, sleeper, toadfish, and a ray account for 56.5 percent. The remains of *Felis concolor*—mountain lion—may represent an animal hunted for both its pelt and its meat. The interesting aspect of the Tecualilla faunal assemblage is the presence of sea turtle, a ray, and an unidentified sea mammal. Although their primary habitat is the ocean, these animals are known to make their way into the estuary, and they undoubtedly did so in the past. Once in the estuary, they may have been more easily captured. Six dog burials were also recovered. One horse bone appeared in this assemblage and is certainly intrusive.

Cristo Rey

Cristo Rey is located on the eastern side of the Teacapán Peninsula and appears to date to the Middle Conchera period (AD 1050–1350). The recovered Cristo Rey faunal assemblage is relatively small compared with those from the other sites investigated. Dogs from burials make up the majority of the materials; seven individuals represent 86.8 percent of the total NISP. Subsistence fauna account for 13.2 percent of the total NISP (Table 8.9). The remains of opossum and rabbit account for 8.2 percent of the subsistence NISP; birds, 2 percent; turtles, 30.6 percent; and fish including sea catfish and snapper, just 59.2 percent. Fish appear to account for the majority of the meat represented by the subsistence NISP.

Venadillo

Venadillo is located near the east-center edge of the Teacapán Peninsula within a kilometer of the estuary. The Venadillo faunal assemblage was the largest and most

Table 8.7. Chalpa Archaeofaunal Assemblage.

Taxon	Common Name	Habitat	Number of Individual Specimens (NISP)	Percent NISP	Minimum Number of Individuals
Mammals					
Didelphis virginiana	Opossum	Terrestrial	5	0.7	1
Thomomys umbrinus	Pocket gopher	Terrestrial	6	0.9	1
Canis familiaris	Dog	Terrestrial	501	71.4	16
Procyon lotor	Raccoon	Terrestrial	94	13.4	4
Odocoileus virginianus	White-tailed deer	Terrestrial	44	6.3	8
Mammalia	Mammals	Terrestrial	1	0.1	1
Birds					
Ardeidae	Herons and bitterns	Shore/aquatic	1	0.1	1
Eudocimus albus	White ibis	Shore/aquatic	1	0.1	1
Meleagris gallopavo	Turkey	Terrestrial	3	0.4	1
Aves	Birds	—	2	0.3	—
Reptiles					
Crocodylidae	Crocodiles, caimans, alligators	Aquatic	1	0.1	1
Testudines	Turtles	Aquatic	3	0.4	1
Iguanidae	Iguanas	Terrestrial	2	0.3	1
Fish					
Ariidae	Sea catfish	Aquatic	1	0.1	1
Lutjanus sp.	Snapper	Aquatic	11	1.6	1
Calamus sp.	Porgy	Aquatic	6	0.9	1
Mugil sp.	Mullet	Aquatic	1	0.1	1
Sciaenidae	Drums	Aquatic	1	0.1	1
Osteichthyes	Fish	Aquatic	18	2.6	—
Subsistence Fauna Totals					
Habitat undetermined			2	2.7	
Aquatic subsistence fauna			44	60.3	9
Terrestrial subsistence fauna			27	38.0	5
Total subsistence fauna			73	10.4[a]	14
Assemblage Totals					
Aquatic animals			44	6.3[a]	9
Terrestrial animals			656	93.4[a]	33
Habitat undetermined			2	0.3[a]	
Burials/offerings (dogs, raccoons, deer)			601	85.6[a]	26
Total recovered			702		42

[a] Percentage of total NISP.

varied recovered during investigations in the Marismas (Table 8.10) and reflects the intensity of recovery methods used during the excavation of Mound C (Cumbaa 1973; Foster 1984). Unlike for the previous sites, the NISP and MNI were not determined for all the taxa identified at Venadillo, and in some cases, the occurrence of a taxon is noted as simply present. The earthen portion of Mound C dates to the Late Estero period (AD 450–700), and the shell midden capping the mound dates to the Middle Conchera period (AD 1100–1400), as does Mound B.

Dog was the most commonly recovered taxon; the majority of these remains are believed to represent burials. In total, 861 elements representing 27 individuals were recovered from Venadillo. Of these, 477 elements representing 15 individuals were from Mound C. Six

TABLE 8.8. Tecualilla Archaeofaunal Assemblage.

Taxon	Common Name	Habitat	Number of Individual Specimens (NISP)	Percent NISP	Minimum Number of Individuals
Mammals					
Didelphis virginiana	Opossum	Terrestrial	5	0.6	2
Dasypus novemcinctus	Armadillo	Terrestrial	34	4.1	2
Sylvilagus sp.	Rabbit	Terrestrial	13	1.6	5
Rodentia	Rodents	Terrestrial	10	1.2	4
Sigmodon hispidus	Cotton rat	Terrestrial	6	0.7	3
Cetacean	Sea mammals	Aquatic	5	0.6	1
Canis familiaris	Dog	Terrestrial	193	23.5	6
Procyon lotor	Raccoon	Terrestrial	2	0.2	2
Felis concolor	Mountain lion	Terrestrial	1	0.1	1
Odocoileus virginianus	White-tailed deer	Terrestrial	21	2.6	4
Artiodactyla	Cloven-hoofed mammals	Terrestrial	1	0.1	—
Mammalia	Mammals	Terrestrial	35	4.3	—
Birds					
Anas sp.	Duck	Aquatic/shore	1	0.1	1
Anatidae	Ducks and geese	Aquatic/shore	1	0.1	—
Gallinula chloropus	Common gallinule	Aquatic/shore	1	0.1	1
Meleagris gallopavo	Turkey	Terrestrial	2	0.2	1
Zenaida sp.	Dove	Terrestrial	12	1.5	1
Columbidae	Doves and pigeons	Terrestrial	5	0.6	1
Ardeidae	Herons and bitterns	Aquatic/shore	3	0.4	3
Aves	Unidentified birds	—	14	1.7	—
Reptiles					
Crocodylidae	Crocodilians	Aquatic	3	0.4	2
Kinosternon sp.	Musk turtle	Aquatic	3	0.4	—
Kinosternidae	Turtles	Aquatic	13	1.6	4
Chrysemys sp.	Pond turtle	Aquatic	8	1.0	4
Cheloniidae	Sea turtles	Aquatic	2	0.2	1
Testudines	Turtles	Aquatic	63	8.7	—
Iguanidae	Iguanas	Terrestrial	6	0.7	3
Coluber constrictor	Boa	Terrestrial	1	0.1	1
Amphibians					
Anura	Frogs and toads	Aquatic	1	0.1	1
Bufo sp.	Toad	Aquatic	1	0.1	1
Fish					
Chondrichthyes	Sharks	Aquatic	1	0.1	1
Aetobatis narinari	Spotted eagle ray	Aquatic	1	0.1	1
Osteichthyes	Fish	Aquatic	256	31.2	—
Ariidae	Sea catfish	Aquatic	53	6.5	10
Batracoididae	Toadfish	Aquatic	1	0.1	1
Centropomus sp.	Snook	Aquatic	7	0.9	3
Epinephelus sp.	Grouper	Aquatic	2	0.2	1
Caranx sp.	Jack	Aquatic	5	0.6	2
Lutjanus sp.	Snapper	Aquatic	16	2.0	4
Gerridae	Mojarras	Aquatic	1	0.1	1
Haemulon sp.	Grunt	Aquatic	1	0.1	1
Sciaenidae	Drums	Aquatic	1	0.1	—

Table 8.8. (cont'd.) Tecualilla Archaeofaunal Assemblage.

Taxon	Common Name	Habitat	Number of Individual Specimens (NISP)	Percent NISP	Minimum Number of Individuals
Cynoscion sp.	Corvina	Aquatic	1	0.1	1
Mugil sp.	Mullet	Aquatic	2	0.2	1
Eleotridae	Sleepers	Aquatic	4	0.5	2
Eleotris sp.	Sleeper	Aquatic	2	0.2	1
Subsistence Fauna Totals					
Habitat undetermined			14	2.2	—
Aquatic subsistence fauna			459	73.2	49
Terrestrial subsistence fauna			154	24.6	30
Total subsistence fauna			627	76.5[a]	79
Assemblage Totals					
Aquatic fauna			459	56.0[a]	49
Terrestrial fauna			347	42.3[a]	33
Habitat undetermined			14	1.7[a]	
Burials/offerings			193	23.5[a]	6
Total recovered			820		85

[a] Percentage of total NISP.

Table 8.9. Cristo Rey Archaeofaunal Assemblage.

Taxon	Common Name	Habitat	Number of Individual Specimens (NISP)	Percent NISP	Minimum Number of Individuals
Mammals					
Didelphis virginiana	Opossum	Terrestrial	3	0.8	1
Sylvilagus sp.	Rabbit	Terrestrial	1	0.3	1
Canis familiaris	Dog	Terrestrial	321	86.8	7
Birds					
Aves	Unidentified birds	—	1	0.3	—
Reptiles					
Chrysemys sp.	Pond turtle	Aquatic	10	2.7	1
Testudines	Turtles	Aquatic	5	1.4	—
Fish					
Ariidae	Sea catfish	Aquatic	5	1.4	2
Lutjanus sp.	Snapper	Aquatic	1	0.3	1
Osteichthyes	Fish	Aquatic	23	6.2	—
Subsistence Fauna Totals					
Habitat undetermined			1	2.0	—
Aquatic subsistence fauna			44	89.8	4
Terrestrial subsistence fauna			4	8.2	2
Total subsistence fauna			49	13.2[a]	6
Assemblage Totals					
Aquatic fauna			44	11.9[a]	4
Terrestrial fauna			325	88.8[a]	9
Undetermined habitat			1	0.2[a]	
Burials/offerings (dogs)			321	98.8	7
Total recovered			370		13

[a] Percent of total NISP.

TABLE 8.10. Venadillo Archaeofaunal Assemblage.

Taxon	Common Name	Habitat	Number of Individual Specimens (NISP)[a]	Percent NISP[a]	Minimum Number of Individuals[a]
Mammals					
Didelphis sp.[b]	Opossum	Terrestrial	4		1
cf. *Notiosorex crawfordi*[c]	Crawford's desert shrew	Terrestrial	Present		—
Dasypus novemcinctus[b]	Armadillo	Terrestrial	7		2
Sylvilagus sp.[b]	Rabbit	Terrestrial	9		2
Rodentia[b]	Unidentified rodents	Terrestrial	10		2
Oryzomys sp.[b]	Rice rat	Terrestrial	Present		—
Sigmodon hispidus[b]	Cotton rat	Terrestrial	Present		—
cf. *Peromyscus boylii*[c]	Brush rat	Terrestrial	Present		—
Canis familiaris[b]	Domestic dog	Terrestrial	384		12
Procyon lotor[b]	Raccoon	Terrestrial	6		1
Odocoileus virginianus[b]	White-tailed deer	Terrestrial	16		1
Mammalia	Unidentified mammals	Terrestrial	10		—
Birds					
Anatidae[b]	Ducks and geese	Aquatic/shore	Present		—
Ardeidae[b]	Herons and bitterns	Aquatic/shore	Present		—
Passeriformes[c]	Perching birds	Terrestrial	Present		—
Aves	Unidentified birds	—	12		2
Reptiles					
Crocodilus acutus[c]	American crocodile	Aquatic	Present		—
Crocodylidae	Crocodilians	Aquatic	7		1
Kinosternidae[b]	Musk turtles	Aquatic	4		2
Chrysemys sp.	Pond turtle	Aquatic	1		1
Testudines	Turtles	Aquatic	6		1
cf. *Phyllodactylus lanei*[c]	Leaf-toed gecko	Terrestrial	Present		—
Iguana iguana	Green iguana	Terrestrial	17		1
cf. *Anolis nebulosus*[c]	Anole	Terrestrial	Present		—
Colubridae[c]	Nonvenomous snakes	Terrestrial	Present		—
Natrix valida[c]	Water snake	Terrestrial	Present		—
Rhadinaea hesperia[c]	Slender snake	Terrestrial	Present		—
Masticophis flagellum[c]	Coachwhip	Terrestrial	Present		—
Oxybelis acuminatus[c]	Vine snake	Terrestrial	Present		—
Coluber constrictor[b]	Boa	Terrestrial	Present		—
Lampropeltis triangulum[c]	King snake	Terrestrial	Present		—
Leptodeira sp.[c]	Cat-eyed snake	Terrestrial	Present		—
Rhinocheilus cf. *antonii*[c]	Mexican long-nosed snake	Terrestrial	Present		—
Tantilla bogerti[c]	Black-headed snake	Terrestrial	Present		—
Trimorphodon paucimaculatus[c]	Sinaloa lyre snake	Terrestrial	Present		—
Iguanidae	Iguanid lizards	Terrestrial	2		—
Serpentes	Snakes	Terrestrial	5		1
Amphibians					
cf. *Scaphiopus couchii*[c]	Spadefoot toad	Aquatic	Present		—
Bufo sp.[b]	Toad	Aquatic	2		1
Hylidae[c]	Tree frogs	Terrestrial	Present		—
Microhylidae[c]	Narrow-mouthed frogs	Aquatic	Present		—

TABLE 8.10. (cont'd.) Venadillo Archaeofaunal Assemblage.

Taxon	Common Name	Habitat	Number of Individual Specimens (NISP)[a]	Percent NISP[a]	Minimum Number of Individuals[a]
Fish					
Chondrichthyes	Sharks	Aquatic	7		1
Carcharhinus sp.[c]	Requiem shark	Aquatic	Present		—
Galeocerdo cuvieri[c]	Tiger shark	Aquatic	Present		—
Osteichthyes	Unidentified fish	Aquatic	155		—
Elops affinis[c]	Machete	Aquatic	Present		—
Clupeidae[c]	Herring	Aquatic	Present		—
Ariidae[b]	Sea catfish	Aquatic	13		2
Centropomus sp.[b]	Snook	Aquatic	4		1
Serranidae[c]	Sea bass	Aquatic	Present		—
Epinephelus sp.	Grouper	Aquatic	2		1
Carangidae[c]	Jacks	Aquatic	Present		—
Trachinotus sp.[b]	Pompano	Aquatic	Present		—
Lutjanidae[c]	Snappers	Aquatic	Present		—
Gerres sp.[c]	Mojarra	Aquatic	Present		—
Pomadasyidae[c]	Grunts	Aquatic	Present		—
Haemulon sp.[b]	Grunt	Aquatic	Present		—
Sparidae[c]	Porgies	Aquatic	Present		—
Sciaenidae[b]	Drums	Aquatic	Present		—
Mugil sp.[b]	Mullet	Aquatic	18		1
Polydactylus sp.[c]	Threadfin	Aquatic	Present		—
Eleotridae	Sleepers	Aquatic	1		1
Sphaeroides sp.	Puffer	Aquatic	1		1
Invertebrates					
Brachyura[c]	Crabs	Semiterrestrial	72		21
Subsistence Fauna Totals[a]					
Habitat undetermined			12	3.1	2
Aquatic subsistence fauna			221	56.5	14
Terrestrial subsistence fauna			158	40.4	11
Total subsistence fauna			391	50.5[d]	27
Assemblage Totals[a]					
Aquatic fauna			221	28.5[d]	14
Terrestrial fauna			542	69.9[d]	44
Habitat undetermined			12	1.6[d]	2
Burials/offerings (dogs)			384	49.5[d]	12
Total recovered			775		60

[a] Only for collected NISP and minimum number of individuals data.
[b] Large-mesh screen (1/4-inch) recovery, Mound C.
[c] Small-mesh screen recovery, Mound C (Cumbaa 1973:73, Table 8).
[d] Percent of total NISP for collected NISP data.

of the 477 pieces of dog bone were recovered from the Mound C shell midden (Levels 1–4). The remainder, found in the earthen burial mound (Levels 5–8), appears to represent intentional burials, and some were associated with the human burials. The presence of the dog bone in the shell midden is thought to represent food remains (Cumbaa 1973:76). This is the only evidence suggesting that dogs were consumed. The use of dog in Mesoamerica as a source of food was common (e.g., Wing 1978), but dog remains from contexts that suggest they were regularly eaten by the inhabitants of the Marismas are rare.

TABLE 8.11. Archaeofauna from Large-Mesh Recovery at Mound C, Venadillo.

Level and Fauna	Number of Individual Specimens (NISP)	Percentage of NISP by Level	Minimum Number of Individuals
Shell Midden			
Level 1			
Crab	—		—
Fish	—		—
Reptile/amphibian	—		—
Bird	—		—
Small mammal	1	50.0	1
Dog	—		—
Deer	1	50.0	1
NISP subtotal	2	100	
Level 2			
Crab	—		—
Fish	3	12.5	3
Reptile/amphibian	7	29.2	3
Bird	—		—
Small mammal	10	41.6	5
Dog	1	4.2	1
Deer	3	12.5	1
NISP subtotal	24	100	
Level 3			
Crab	—		—
Fish	11	26.8	7
Reptile/amphibian	11	26.8	4
Bird	2	4.9	1
Small mammal	—		—
Dog	4	9.8	2
Deer	13	31.7	2
NISP subtotal	41	100	
Level 4			
Crab	33	36.3	9
Fish	28	30.8	9
Reptile/amphibian	4	4.4	3
Bird	2	2.2	2
Small mammal	8	8.8	3
Dog	—		—
Deer	16	18.6	3
NISP subtotal	91	100	
Totals Recovered			
Nondog	153	96.8	57
Dog	5	3.2	3
Total	158	100	60

Level and Fauna	Number of Individual Specimens (NISP)	Percentage of NISP by Level	Minimum Number of Individuals
Earthen Burial Mound			
Level 5			
Crab	28	29.2	7
Fish	27	28.1	7
Reptile/amphibian	14	14.6	5
Bird	14	14.6	4
Small mammal	12	12.5	4
Dog	100		2
Deer	1	1.0	1
NISP subtotal[a]	196	100	
Level 6			
Crab	9	16.7	3
Fish	15	28.8	5
Reptile/amphibian	10	18.5	4
Bird	3	5.6	3
Small mammal	16	29.6	8
Dog	350		8
Deer	1	1.9	1
NISP subtotal[a]	54	100	
Level 7			
Crab	—		—
Fish	15	68.2	11
Reptile/amphibian	1	4.5	1
Bird	—		—
Small mammal	3	13.6	1
Dog	14		1
Deer	3	13.6	1
NISP subtotal[a]	22	100	
Level 8			
Crab	1	16.7	1
Fish	2	33.3	2
Reptile/amphibian	—		—
Bird	1	16.7	1
Small mammal	2	33.3	2
Dog	8		1
Deer	—		—
NISP subtotal[a]	6	100	
Totals Recovered			
Nondog	178	26.5	72
Dog	472	72.6	12
Total	650	100	84

Source: Cumbaa 1973:Figure 6.

[a] NISP total excludes dogs, which are considered burial remains.

With the exception of crab, most other taxa are represented by one or two individuals. As many as 21 individual crabs are represented. Several species of crab were commonly observed on the site during excavations, and they are plentiful on the shores at the mouth of the estuary and ocean.

In Mound C deer are represented by 48 NISP or 10 MNI. These remains potentially signify the largest amount of meat represented in the Mound C assemblage. The next most commonly represented taxa are fish, with a total NISP of 267 and an MNI of 45. In his discussion of the Mound C faunal assemblage, Cumbaa (1973) suggests that the presence of some of the smaller reptiles and amphibians may be incidental. However, citing Thomas (1971), Cumbaa ultimately concludes that the smaller reptile and amphibian material from Mound C, for the most part, was probably food remains because of the types and limited number of elements represented. Cumbaa also notes the presence of numerous extremely small fish bone and interprets this as evidence of a mass collecting technique.

Although a greater number of taxa were identified as a result of the recovery of the Mound C material, the results are not vastly different from those from other sites. Fish, deer, reptiles, and small mammals constitute the bulk of the food remains. Additionally, crab was identified, and Venadillo is the only site where dog remains are thought to represent food remains. Bird remains were not abundant, and turkey was not identified.

Mound C is the only example of a feature excavated during the Marismas project where a stratigraphic sample of faunal remains provides some comparative temporal information regarding faunal exploitation. As noted above, the earlier earthen burial mound dates between AD 450 and 700, and the shell midden covering it dates from AD 1100 to 1400. However, inferences regarding the distribution, number, and variety of nonmolluscan fauna within the two temporally distinct archaeofaunal assemblages are tempered by the limits of the representativeness of the recovered sample. Additionally, Cumbaa (1973) combined the Mound C data into seven broad groups or generalized taxa (crab, fish, reptiles/amphibians, birds, small mammals, dog, and deer), and this information is limited to the remains recovered from the screening of the excavated fill through ¼-inch mesh. It does not include the results from fine screening and flotation (Table 8.11).

The reason for the lack of nonmolluscan faunal remains in Level 1 is unclear. Less volume of midden material was excavated for Level 1, and the lack of faunal bone may be a sampling error. It is also possible that the last dumping/depositional episodes at the center of the mound simply consisted of mostly shell with little faunal bone.

The highest numbers of individual nonmolluscan animals (MNI) represented occur in Levels 4 ($n = 29$) and 5 ($n = 28$) (excluding dog remains in Level 5). These levels represent the interface between the shell midden and the earthen burial mound. The high number of animal remains may be a reflection of the initial use of the area as a refuse midden. Also, despite the assertion that many of the small mammals, reptiles, and amphibians present likely represent food remains, middens are often home to a variety of burrowing and scurrying creatures. Although it may be expected that such remains be articulated, they could have become disarticulated and scattered during trash deposition events or subsequent bioturbation.

The faunal remains in the burial mound probably represent redeposited refuse from the general area of the mound that was mixed in as human remains were buried. Like some of the remains in the shell midden cap, some of the small mammals, reptiles, and amphibians represented may be the remains of creatures inhabiting dens within the mound. Recent animal burrowing was noted, although evidence for previous burrowing was not readily identifiable. The number of individual animals represented in the burial mound generally equals or exceeds that of those present in the midden. This suggests that at some point the area was used for the disposal of household trash. It is also unclear whether trash disposal at the locus continued even after Mound C became a burial mound.

Crab was primarily found in Levels 4, 5, and 6, and they were most abundant in Levels 4 and 5. This may be the result of the initial dumping of trash on the burial mound. However, it is possible that some of the crab remains are natural occurrences representing the death of the animal or the results of predation by birds and other animals. In terms of human predation, these remains may be concentrated as a result of the processing and consumption of a number of individual crabs during a single event.

Other fauna are more evenly distributed across multiple stratigraphic layers. Fish are represented in all levels; reptiles and amphibians are present in all but Levels 1 and 8; birds are present in Levels 3, 4, 5, 6, and 8; small mammals are found in Levels 1–2 and 4–8; and deer are found in Levels 1–8. Dog remains believed to be burials account for 72.6 percent of the NISP in the earthen burial mound. Otherwise, as Cumbaa (1973:76) notes, with the exception of the marine shell, the faunal assemblages from the Mound C shell midden and the burial mound were generally similar in both the taxa and the number of NISP represented (see Table 8.11). Deer, fish, small mammals, and reptiles were represented and appear to have been staples of the diet.

TABLE 8.12. El Mirador Archaeofaunal Assemblage.

Taxon	Common Name	Habitat	Number of Individual Specimens (NISP)	Percent NISP	Minimum Number of Individuals
Mammals					
Didelphis virginiana	Opossum	Terrestrial	1	0.2	1
Dasypus novemcinctus	Armadillo	Terrestrial	2	0.4	1
Canis familiaris	Dog	Terrestrial	426	85.4	10
Sylvilagus sp.	Rabbit	Terrestrial	1	0.2	1
Rodentia	Rodents	Terrestrial	2	0.4	2
Odocoileus virginianus	White-tailed deer	Terrestrial	7	1.4	3
Mammalia	Mammals	—	3	0.6	—
Birds					
Aves	Birds	—	3	0.6	2
Reptiles					
Kinosternidae	Musk turtles	Aquatic	1	0.2	1
Chrysemys sp.	Pond turtle	Aquatic	6	1.2	1
Amphibians					
Anura	Frogs and toads	Aquatic	1	0.2	—
Bufo sp.	Toad	Aquatic	4	0.8	1
Fish					
Chondrichthyes	Sharks	Aquatic	1	0.2	1
Osteichthyes	Fish	Aquatic	33	8.8	—
Ariidae	Sea catfish	Aquatic	5	1.0	2
Centropomus sp.	Snook	Aquatic	2	0.4	1
Caranx sp.	Jack	Aquatic	2	0.4	1
Subsistence Fauna Totals					
Habitat undetermined			6	8.1	2+[a]
Aquatic subsistence fauna			55	74.3	8
Terrestrial subsistence fauna			13	18.6	10
Total subsistence fauna			74	13.6[b]	18
Assemblage Totals					
Aquatic fauna			55	11.0[b]	8
Terrestrial fauna			439	88.8[b]	20
Habitat undetermined			6	1.2[b]	
Burials/offerings			426	85.2[b]	10
Total recovered			926		28

[a] Not included in total subsistence fauna minimum number of individuals calculation.
[b] Percentage of total NISP.

El Mirador

El Mirador was located on the Teacapán Peninsula within 1 km of the estuary. Based on the ceramic assemblage recovered, the site dates between AD 1050 and 1350, the Middle Conchera period. The El Mirador faunal assemblage is dominated by dog remains; 10 individuals were recovered as burials and account for 85.2 percent of the total NISP (Table 8.12). Subsistence fauna account for the remaining 14.8 percent. The remains of opossum, armadillo, rabbit, deer, and rodents account for 21.6 percent of the subsistence NISP fauna; birds, 4.1 percent; turtles, 9.5 percent; amphibians (frogs), 6.6 percent; and fish including shark, sea catfish, snook, and jack, 58.1 percent. Although fish make up the majority of the NISP, they may account for less than half of the consumable meat represented in the overall assemblage.

Plant Foods

No macrobotanical specimens, wild or domesticated, were recovered, and the results of pollen analyses in reference to economically important plant foods are ambiguous. Little economic plant pollen was identified, and the contexts in which they occur add no significant insight into the variety of wild plants and cultigens used as foods, the intensity of their use, their contributions to diets, or their seasonality. Undoubtedly, the ancient Marismas inhabitants were as well acquainted with the wild plant foods available to them as they were with the fauna they exploited. The archaeological and ethnographic literatures are replete with examples of wild plant use by traditional agriculturalists. Peoples exploited wild flora not only for their food products but for a variety of other purposes as well, including medicines, fibers, clothing, implements such as bowls and ladles, weapons, construction materials, fuels, insect repellents, poisons and stupefaction agents, and dyes (e.g., Beals 1945; Coe and Diehl 1980; Felger and Moser 1985; Nietschmann 1973). Wild plant foods likely made up a regular portion of daily diets and may have contributed greater proportions during times of agricultural stress, especially in the early summer with the onset of the monsoon, when plants would have sprouted and the new shoots, young leaves, and flowers appeared. By late summer into the fall, some plants would have yielded ripe fruits that could have been harvested and consumed or prepared for storage. Plants with dietary potential that occur in the area include agave, cacti, composites, legumes, and spurge, to name a few.

Agriculture

It is assumed that the main dietary staples of the prehistoric Marismas inhabitants were corn, beans, and squash. By the time the coastal plain was occupied by ceramic-producing populations, agriculture was long established throughout Mesoamerica as well as the neighboring American Southwest as the primary mode of subsistence for sedentary populations. Ethnohistorically, the Spanish report that agriculture was practiced along the west coast of Mexico (Sauer 1932; see Chapter 12). However, the only direct evidence for cultigens resulting from the Marismas project comes from a few grains of pollen from cucurbits and maize recovered from cores taken near El Calón (see Chapter 2). Other evidence for a heavy reliance on agriculture comes from the artifact assemblages recovered from the sites investigated. Ground-stone implements likely used in the processing of agricultural produce include manos, metates, and pestles. Although mostly fragmentary, these implements were commonly recovered artifacts. The manos—usually two-hand types including overhanging manos—and metates were often well shaped and presumably were used to process domesticated crops from cultivated fields (e.g., Hard 1986).

Additionally, the recovery of spindle whorls from sites in the Marismas suggests that cotton was being grown and spun. Ethnohistorically, the Mexican west coast was known as a source of cotton textiles, and cotton was a commonly reported crop (Mota y Escobar 1940; Riley 1987; Sauer 1932; see Chapter 12). Furthermore, archaeological and ethnographic evidence from Mexico and the American Southwest indicates that cottonseeds were parched and consumed as food. Cottonseed has a relatively high oil content (Karen R. Adams, personal communication 2010). Cottonseeds were recovered in bowls with burials (food?) and in storage jars in rooms at Casas Grandes (Di Peso et al. 1974:8:314–315). Ethnographically, people in Tepotzlán, Morelos, once used cottonseed for food (Lewis 1951:85), and in the neighboring American Southwest, cottonseeds are reported as food both archaeologically and ethnographically (e.g., Castetter and Bell 1942; Russell 1908). It is also possible that agave fiber was being spun with the recovered spindle whorls (e.g., Parsons 1972; Parsons and Parsons 1990) and that agave was a source of *aguamiel* (honey water), pulque, or food for the prehistoric inhabitants of the Marismas (e.g., Anderies et al. 2008; Parsons and Parsons 1990).

Subsistence

As discussed in Chapter 12, at the time of Spanish contact, the native inhabitants of the Mexican west coast grew a variety of crops, may have domesticated turkeys and other birds, and hunted and captured a variety of terrestrial and marine fauna. Based on the data from the Marismas project, it is clear that such subsistence practices were long established in the region. Because of the lack of direct evidence such as macrobotanical and pollen remains, agriculture and the use of wild plant foods are assumed. However, their use is strongly suggested by the presence of grinding tools and other artifacts that would have been used in either their processing or their capture. It is further assumed that the agricultural diet consisted mainly of corn, beans, and cucurbits.

The ethnohistoric accounts list a variety of wild fauna being used as food, and those lists are nearly identical to the archaeological evidence of animals captured and hunted by the earlier inhabitants of the Marismas. One of the more notable exceptions is the lack of evidence for the use of shrimp in the archaeological record of the Marismas. During early colonial times, shrimp are reported as being caught for household consumption and for export to the interior of Mexico (see Chapter 12).

The wild animals and plants were taken from a variety of microenvironments or biotopes within or adjacent to the estuarine-lagoon-mangrove system that composes the Marismas. Their collection undoubtedly required the use of a variety of techniques, although the only direct evidence of their procurement and processing is limited to flaked- and ground-stone artifacts recovered that may have been used in hunting and food processing. The following discussion relies on modern ethnographic accounts in an attempt to gain insights into possible subsistence techniques, both agriculture and hunting and collecting, that may have been used to acquire the foodstuffs necessary for daily survival by the prehistoric inhabitants of the Marismas.

Microenvironments

Although the natural setting and biodiversity of the Marismas are summarized in Chapter 2, they are briefly reviewed here in order to present the context for the discussion of subsistence, related resource procurement, and land use. Following Coe and Flannery (1967; see also Rossman 1975; Voorhies 1976; Zarky 1976), the region that encompasses the Marismas can be subdivided into a series of microenvironments—areas that environmentally support a particular ecological community of plants and animals—which were exploited for the biotic, geologic, and mineral resources they contained. These include the ocean; the beach and low scrub; the estuary and lagoon system; the mangrove forest; the monte; salt playas; and the piedmont, foothills, and uplands of the western flank of the Sierra Madre Occidental.

Ocean

The Pacific Ocean adjacent to the northwest Mexican coast contains a diversity and abundance of mammals, mollusks, crustaceans, reptiles, and fish suitable for human consumption and use (e.g., Hubbs and Roden 1964). Some of these resources could be taken near shore using nets, spears, fishhooks, and traps or weirs. In addition to the variety of fauna potentially available, it is also possible that the prehistoric peoples of the Marismas used various seaweeds, which can be collected as drift from the beach or gathered at low tide, for both food and utilitarian purposes (e.g., Abbott and Williamson 1974; Felger and Moser 1985; Uphof 1968).

Beach and Low Scrub

There is a narrow strip of beach and border area that extends along the edge of the Teacapán Estuary near its mouth and along the oceanfront above and below the mouth of the estuary. A variety of viny weeds and small bushes grow at the upper margins of the beach area, and in some places coconut groves are present. Lizards, snakes, and iguanas can be seen making their way along the beach and into the adjacent vegetation. Clams, gastropods, and some crabs are the primary food resources present in this zone.

Estuary and Lagoon System

The estuary, which is a tidal estuary and lagoon system, is the source of a great many fish and mollusks and in terms of food, is potentially the richest of the microenvironments represented. Today, as they undoubtedly did during prehistoric times, many of the taxa present in the ocean make their way into the Teacapán Estuary and are often captured there. Oysters are the predominant species of mollusk present. The brackish water of the estuary system is ideal for the survival of the oyster and shrimp, although their presence and quantities are affected by the salinity of the estuary. Red snapper, robalo, mullet, snook, and other fish are nearly always to be found in the estuary, and shrimp seem to be most abundant during August and September (Wing 1969:98).

Mangrove Forest

The shores of the estuary are covered with a thick forest of red, white, and black mangroves that are less than 20 m in height with silt roots (Wagner 1964:260). These roots form a maze through which movement by people is nearly impossible. The floor of the forest is often inundated by tidal saltwater and consists of a black, sticky mud inhabited by small crabs. The trees provide nesting areas for waterfowl, and the wood was, historically, used for building materials.

Monte

The monte area is sometimes characterized as savanna or low deciduous forest (Hardy and McDiarmid 1969:57; Wagner 1964:252). The wildlife that inhabits or intrudes into the monte includes iguanas, armadillos, snakes, foxes, and deer. Most of the archaeological sites within the Marismas are found within this zone, and it is, as it is today, where farming would have occurred.

Salt Playas

Interspersed within and occurring at the edge of the monte are salt playas. These low-lying areas are seasonally wet and dry and for the most part infertile, except for some grasses and the crabs that survive in the salt ponds. These areas have been used in recent times as salt sources for preserving fish (Wing 1969:102), and it is reasonable to assume that this was the case prehistorically.

Piedmont, Foothills, and Uplands of the Western Flank of the Sierra Madre Occidental

The piedmont is the apron of sloping alluvium and lower foothills along the western slope of the Sierra Madre Occidental. West (1964) identifies this area as part of the "extratropical highlands." As elevation rises, the thorn and tropical deciduous forests of the lowlands give way to a chaparral and coastal scrub association, which in turn gives way to pine and oak forests above 1,500 m.

Table 8.13. Distribution of Marismas Archaeological Mollusk Taxa by Niche.

Mollusk Type	B	B-R	Sa	Sa-B	Sa-Sb	Sa-X	Sb	Sb-X	X	R
Gastropods	2 (9%)	1 (4%)	8 (35%)		2 (9%)		2 (9%)	1 (4%)	5 (22%)	2 (9%)
Bivalves	4 (19%)		5 (24%)	2 (<10%)	3 (14%)	3 (14%)	2 (<10%)			2 (<10%)
Total	6 (14%)	1 (2%)	13 (30%)	2 (5%)	5 (11%)	3 (7%)	4 (9%)	1 (2%)	5 (11%)	4 (9%)

Note: B = low-salinity lagoon and mangrove environment (brackish); R = intertidal and shallow rocky shore; Sa = intertidal sand beaches and sand flats to 10 m; Sb = sand and sand/mud, 11 to 26 m, nearshore; X = intermediate shelf, 27 to 65 m.

The prehistoric inhabitants of the Marismas may not have ventured far up the western flanks of the Sierra, preferring to hunt and collect in the lower elevations of the piedmont where biodiversity is high, with the nonmarine fauna of the lower foothills consisting of the same taxa found on the adjacent coastal plain.

Collecting, Hunting, and Processing

Mollusk Collection

The abundance of shell middens in the Marismas is a clear indication that mollusks played a role in the local diet. Nevertheless, it is difficult to assess the overall contribution mollusks made to the local diet. Claassen (1998:178) notes that there are few data that pertain to the role of mollusks in the diets of horticultural and agricultural societies, and Waselkov (1987:136–138) observes that as societies become more dependent on agriculture, their consumption of mollusks declines. One of the most comprehensive studies of ethnographic shellfishing is that of Meehan (1982). She reports that among the Anbarra, Gidjingali speakers of Arnhem Land of northern Australia, mollusks only accounted for 6 to 17 percent of the total caloric intake. The Anbarra subsist on a combination of hunting and European foods. Mollusks are also seen as a starvation food and as a supplement to the diet during times of dietary stress. The abundance of this resource in the Marismas undoubtedly varied seasonally and from year to year, as did their consumption by the prehistoric inhabitants of the area. Nevertheless, despite the fact that the prehistoric Marismas populations were undoubtedly agriculturalists, they consumed significant quantities of mollusks. Whether this was on a seasonal basis or on a regular basis as a supplement to the diet is unclear.

Forty-four marine gastropod and bivalve species have been identified in the archaeological middens in the Marismas (see Table 8.1, Table 8.13). Twenty-one percent of the mollusk species identified are associated with brackish water habitats (estuaries or lagoons). In terms of prehistoric subsistence, the important brackish water species include oysters (*Ostrea* spp.), ark (*Anadara* spp.), and rangia (*Rangia medica*), all of which are bivalves.

The majority of the species represented in the middens, 47 percent, inhabit intertidal sand beaches and flats to a depth of 10 m. These include the Venus clams (*Tivela* spp., *Chione* spp., and *Protothaca* spp.) and the Pacific crown conch (*Melongena patula*). These would have been available along sandy shores at the mouth of the estuary and the shores of the ocean. Some of the nearshore species such as the black murex (*Hexaplex nigritus*), a gastropod, were widely consumed prehistorically. These would be captured at low tide or when they were washed up onshore. It is not known to what degree the other taxa contributed to the diet; it is possible that some of the species may have been collected inadvertently or for their colorful shells.

Mollusk behavior certainly shaped human collecting or harvesting practices. During low tide exposed beaches and any mud- or sand flats could be inspected for trails or siphon holes, and shellfish could be dug up using a stick or by sifting the sand or mud with one's feet or fingers (Waselkov 1987). Oysters attached themselves to the roots of the mangroves along the edge of the estuary or, perhaps, to fossil shell beds that were present. Those on the mangrove roots may have been exposed or partially exposed and easily seen during low tide. Regardless, they could have easily been pulled or knocked off their roosts using a stick or a rock. *Anadara*, *Raginaa*, and other bivalve mollusks would have been available on the beaches at the mouth of the estuary and along the shores of the estuary and lagoons. It is also possible that some of the species were obtained by diving or with a dragnet. Waselkov (1987) reports that some deeper-water species, as well as sea urchins, were collected from canoes by the use of pronged poles or spears, while in other cases scoops or rakes were used to mass harvest mollusks. Nevertheless, it is worth noting that Coe and Flannery (1967:80) observe that many intertidal and deepwater species that often occur in low frequencies in middens, or at sites in general, may represent beach shell gathered for utilitarian, ornamental, or ceremonial purposes.

Even though mollusks were collected and consumed in quantity, it is most likely that they were collected by hand or with a digging stick. Once extracted, they were placed in baskets or nets for transport to a processing locus or back to the village for processing and consumption. Many of the middens are close to the shoreline or estuary, indicating that the meat was extracted there and then taken back to the village for further processing and consumption. Although there are accumulations of shell at many of the habitation sites, the volumes of these middens are often considerably less than the volume of those at presumed processing loci.

Felger and Moser (1985) describe most of the above procurement techniques as methods used by the Seri of the central Sonoran coast for collecting mollusks. They further indicate that women are generally the ones participating in mollusk collecting, although men and children help as well (see also Claassen 1998). Meehan (1982:119) indicates that among the Anbarra, shell collecting was a woman's occupation and that women were very efficient foragers. However, both females and males and the young and old participated in the activity at various times during the year, with differing levels of effort. Meehan observes that group members were well aware of who were good food producers and who were not. The abilities of good female food producers were recognized to be equal to the skill of good male producers. Lazy people were recognized as such. Female task groups included women and girls up to the age of 15 who were unmarried or did not have any children vs. women who were married, had children (married or unmarried), or were widowed.

It is often assumed, because mollusks are small packages of protein in a hard shell living in restricted habitats, that mollusk collection requires more effort than it is worth (Waselkov 1987). Meehan's (1982:90–102) study of the Anbarra also presents one of the few discussions of time spent collecting mollusks relative to dietary return. One individual collected 9 to 19 kg of mollusks during three trips that lasted from 50 to 85 minutes each. Another individual collected 8 to 16 kg of mollusks during three trips that lasted 50 to 110 minutes each. In one period of two hours and 45 minutes, one person collected 26.2 kg of mollusks, and at one location, a group of women and children collected a total of 97 kg of mollusks in 80 minutes. This time included collecting and often the washing of shells before they were cooked.

Meehan (1982:125) observed that in the year of her study, females gathered 85 percent (3,235 kg) of the *Tape hiantina* (a favored clam), with each female averaging about 77 kg, compared with 19 kg collected by each male. Males, on the other hand, collected mollusks less often, but their per-day average was 11 kg, compared with 8.5 kg per day per female. On an annual basis, females' average trek was 22 km, as compared with 7 km for males; however, the daily average for males was 7 km, compared with 2 km for females, with the average amount of time spent collecting by males being seven hours, compared with two hours for females. Meehan further notes that while waiting for the tide to reach the desired level, men often fished or carried out other foraging activities. The amount of mollusks collected by boys was comparatively small, with older boys generally collecting mollusks incidentally while engaged in other activities. Although small boys accompanied their mothers to the shell beds, they usually played or otherwise did not participate in collecting mollusks (Meehan 1982). Bird and Bird (2000) observed that among the Meriam foragers (Melanesian islanders in the Torres Strait), children were effective collectors but that they encountered high-ranked taxa at a lower rate than adults. As a consequence they ended up exploiting a wider variety of taxa, mostly of lower value, which they then delivered whole to the residence. Bird and Bird note that age-linked differences in mollusk predation may be difficult to identify archaeologically and that a disproportional presence of low-value taxa may be taken as an indication of subsistence intensification.

Based on the Anbarra data (Meehan 1982), if we assume that at a minimum the meat in the shells collected accounted for 15 to 20 percent of specimen weight, returns in fresh meat range from 1.2 to 5.25 kg of meat per individual foray. The group of Anbarra women and children that collected 97 kg of mollusks in 80 minutes likely produced a yield of at least 14.5 to 19.5 kg of meat. Thus, despite reservations that mollusk collection may not be a viable expenditure of energy for dietary return, on the contrary, it appears that mollusk foraging activities are well worth the effort.

Several methods of processing mollusks can be inferred from the Marismas archaeological record. Most of the clamshells observed occur in a disarticulated state. As bivalve mollusks cook, they open up, and after the meat is extracted the shells are discarded and can quickly become disarticulated. The *Tivela* shell middens that dot the landscape appear to be processing loci where the meat was extracted for further processing such as drying or smoking. It appears that in these cases, the shell was pried open, the meat was extracted, and the shell was discarded. The majority of the shells do not appear to have been exposed to heat, although they could have been boiled or steamed. At several sites, excavations exposed food-processing areas with hearths and scattered shell in proximity (see Figure 5.40). It is assumed that mollusks were cooked at these loci and the shell was discarded in the area of the hearth. Additionally, the fired-glazed clay balls (see Chapter 6) reported at several Marismas sites are often found in these contexts, suggesting that they may have been used in the mollusk-cooking process.

Meehan (1982) observed that after being collected, mollusks were piled on the beach and covered with grass and sticks. A fire was set, and the mollusks were quickly cooked. The mollusks were eaten there, or the cooked meat was taken elsewhere and consumed (Claassen 1998; Waselkov 1987). Felger and Moser (1985) indicate that the Seri of the central Sonoran coast cook mollusks by steaming them, placing them around hot coals, boiling them, or cooking them on a grill; they also eat them raw. Larger species, such as black murex, are cooked by burning sticks on top of them, in much the same fashion as the Anbarra cooked clams. During the Marismas fieldwork, it was observed that the small bivalve *Donax gouldii* was collected from the ocean beach by children at play and eaten like one eats sunflower seeds—pried open with the teeth, tongue, and/or fingers; the meat sucked out; and the shell spit out. They are also said to form the basis of an excellent soup.

Fishing

With the possible exception of some of the probable brackish water sleepers (eleotrids), none of the fish identified at sites in the Marismas are freshwater or low-salinity species. Snook (*Centropomus* spp.), snapper (Lutjanidae), catfish (Ariidae), and mullets (*Mugil* spp.) are saltwater/brackish water species and survive well in estuaries, while others such as cartilaginous fish (Osteichthyes), mojarras (Gerridae), grunts (Pomadasyidae), croakers/drums (Sciaenidae), toadfish (Batrachoididae), jacks (Carangidae), and groupers (*Epinephelus* spp.) are more restricted to the high-saline waters at the mouths of estuaries and of the ocean (Voorhies et al. 2002). Although not abundant (probably because of the lack of systematic recovery techniques) in the faunal assemblages recovered, fish bones were common and occurred at every site investigated. As noted in Chapter 12, many of the villages encountered by the first Spanish in the area were referred to as fishing villages, and fish were caught in quantities not only for household or local use (immediate consumption or preserved for later consumption) but for exchange inland or as tribute offerings. Ethnohistorically, fish, from both the estuaries and the ocean, contributed significantly to local diets, and it is highly likely that the same can be said for the earlier inhabitants of the Marismas area. We can infer that the large production of salt mentioned in the Spanish sources was intended as a preservation agent for the local supply, as well as for the exportation of fish and oysters (see Chapter 12).

Wing (1969) collected 22 species of fish during her 1968 studies in the Marismas and lists some 86 species that inhabit the estuary, lagoon, and ocean adjacent to the region. Fish were an important resource for those living in the Marismas in 1968, as they are today. A variety of techniques are used to capture them (Wing 1969). Both seines and cast nest are used, as are harpoons and hooks, and stationary traps were used to catch both fish and shrimp. The seines are about 2 m wide and up to 50 m in length and are used primarily for catching snapper, corvina, and robalo. Smaller mesh seines are used for shrimping; some of these are bag-like in appearance and are essentially purse seines. Larger mesh nets are used as gill nets to catch mullet.

Two types of cast nest are used. One has a bag along its lower edge, is 2 to 4 m in length, and when used for fishing, has a two-finger mesh. A variety of this type of net is used to catch shrimp; it is shorter, no longer than 2 m in length, and has a one-finger mesh. The second type of cast net, called a *carete*, is used solely for shrimp. It is a net with a hole at its center. Strings or straps are tied to the edge of the net, and when the net is retrieved, these are pulled through the center hole causing the net to close and form a bag that catches everything within (Wing 1969).

Weirs and stationary traps were also used to catch fish and shrimp. The *tepeste* is a stick barricade that runs from the shore out toward the middle of the channel where there is an opening through which fish could be guided. Near this opening is a platform from which one could harpoon fish. The stationary traps are made by driving mangrove stakes into the bottom of the channel to form a C-shaped fence, the opening of which is oriented downstream. This fence is about 2 m in length, and it is attached to two arms or extensions that run to both banks of the channel. This facility is bounded upstream and downstream by a fence of palm logs that act to protect the trap from the force of the current (Wing 1969). Nets and traps may have been used in conjunction with or through the use of stupefying agents (e.g., Coe and Diehl 1980:117).

The aforementioned harpoons came in two types. One was a double-pronged type with barbed points that were hafted on a pole that was about 2 m long. This type of harpoon was usually used to take fish trapped in weirs. The other type of harpoon was a single-pointed type that was commonly used in hunting caiman (Wing 1969).

Although no fishhooks were recovered during the Marismas excavations, a copper fishhook was recovered at Amapa (Meighan 1976f:114, Plate 101), and more recently, shell and copper fishhooks have been reported from Punta Mita at northernmost tip of Bahía de Banderas northwest of Puerto Vallarta (González Barajas and Beltrán Medina 2007). Pendergast (1962a) reports the presence of copper fishhooks from sites in the uplands of west and northwest Mexico. Thus, the use of fishhooks prehistorically in the region appears to have been widespread.

These fishing techniques and the devices employed are most certainly modern versions of those used prehistorically. In addition to fish, sharks, rays, and an unidentified marine mammal were also represented in the archaeological record. Today, sharks and dolphins occasionally make their way into the estuary; it is likely that if present prehistorically, these were taken using harpoons, although they could have been netted or trapped depending on their size. The ray could also have been taken in the estuary or in the shallow waters along the ocean beach.

Sea turtle (Cheloniidae) remains were also recovered archaeologically; it is likely that these were taken in shallow water or captured on the beach, although they probably occasionally made it into the estuary, where they could have been captured by hand, netted, or harpooned. The capture of a green turtle (*Chelonia mydas*) at the mouth of the Teacapán Estuary by some men was observed during the summer of 1974. They removed only the fins and associated muscles before tossing the animal back into the ocean to drown. Nietschmann (1973:211–212) reports that the modern Miskito Indians of coastal Nicaragua practice live storage of green turtles for up to two weeks by keeping them in the shade. He further notes that older informants stated that turtle crawls used to be built in villages and that these were used to keep turtles for months by feeding them mangrove leaves until the turtles were needed for food. Nietschmann (1973:164) reports that about 46 kg of edible meat is available from a 90-kg (medium-size) green turtle (green turtles taken ranged in weight from 14 to 180 kg). In addition to sea turtles, Coe and Flannery (1967:75) note that modern villagers living on the coast of Guatemala occasionally catch river turtles as well. These turtles were also kept alive until eaten. No mention is made of whether their eggs were collected.

Fish were a valuable resource to the prehistoric inhabitants of the Marismas; many of the species represented—snook, snapper, and jack—can reach or exceed a meter in length. Although the modern inhabitants of the Marismas use a variety of techniques to capture estuary and marine species, it is not known what methods were employed by the prehistoric inhabitants of the area. The only insight comes from Venadillo, where a notched sherd, possibly a net sinker, was recovered. Similar artifacts are reported from La Victoria, Guatemala (Coe 1961), and Nicaragua (Healy 1980; McCafferty 2008). Other evidence from the site of Venadillo may also suggest the use of nets for fishing. As previously discussed, the high number of very small fish bones recovered from Mound C led Cumbaa (1973) to conclude that this was evidence of a mass collecting technique such as netting. Regardless, the common occurrence of fish remains as well as those of other marine and estuary species indicates that the ancient peoples of the area were adept at capturing the creatures that inhabited the estuary and nearby ocean.

Ethnographic studies indicate that many groups of people supplement their diet with fish on a year-round basis (e.g., Kelly 1977; Pennington 1963, 1969), although Nietschmann (1973:160) points out that among the Miskito, fishing is more commonly practiced during the wet season when other meats are more difficult to capture. Nietschmann cites Bennett's (1959) reference to the Chocó of Panama, who avoided the forest and preferred fishing and agricultural production during the wet season. Coe and Diehl (1980:117) note that robalo (snook) is favored and that the peak of the fishing season from September to the end of December occurs when the river is sediment-laden and the fish cannot see nets being cast from boats. When the waters are clear, fishing more frequently occurs at night, when fish feed at the surface and can be caught more easily.

Once caught, fish are gutted and roasted or buried in hot coals; cooking takes about 20 minutes, depending on size. When cooked, the scales, skin, and fins are easily peeled off. Boiling is sometimes used, and the meat was likely cut into strips to be dried, salted and dried, or smoked for storage. Dried fish is usually boiled for consumption (Kelly 1977; Pennington 1963, 1969, 1980). In some areas fish are soaked in brine, which retards bacterial growth during the drying or smoking process. Drying fish in the Marismas area may have been somewhat problematic because of the ambient humidity, especially during the wet season. Smoking may have been the preferred alternative, although salt may have been available for brining. Nevertheless, ethnohistoric data for the region make no reference to either smoking or drying fish, although salting fish for preservation is mentioned (see Chapter 12). Specifically, Mota y Escobar (1940) reports that at Chametla fish not consumed immediately were preserved by salting.

Hunting

A variety of large and small terrestrial fauna are represented in the animal remains recovered, the largest being deer and the smallest being rodents and amphibians. Deer (*Odocoileus* sp.) remains were present at eight of the sites in small quantities of individuals. A total of eight individual deer were represented at Chalpa, of which six were crania with antlers attached that were placed in a burial mound as apparent mortuary offerings.

In their study of modern human ecology on the southern coast of Guatemala, Coe and Flannery (1967:73–74) note that deer avoid mangrove forest because the maze of dense silt roots and marshy soil make movement difficult and that they are also greatly bothered by bloodsucking insects such as mosquitoes and gnats. They are much more at home on the savanna (monte), especially during the wet season, where they could hide in thickets and browse on grasses; they would also have access to agricultural fields where they could browse on the shoots and leaves of growing plants (Coe and Diehl 1980). The piedmont would provide an even more suitable habitat for deer, but traversing the coastal plain, especially during the wet season, to the Sierra piedmont represents an expenditure of considerable energy. Today, in the Marismas, deer are occasionally hunted locally (Wing 1969). Coe and Diehl (1980:102) report that ethnographically deer are often hunted with dogs and either by an individual or by a group of hunters.

With the exception of the site of Chalpa, the lack of faunal elements indicates that deer contributed only a small portion of the prehistoric diet at the sites investigated. It is, however, possible that deer could have been killed and butchered at the kill site and their meat, perhaps boneless, was brought back to the village. In all likelihood, they were hunted with a spear or bow or trapped individually on the monte. Surprisingly, no awls, needles, or other utilitarian implements made from deer long bone were recovered.

Rabbits (Lagomorpha) identified to genus were cottontails (*Sylvilagus* sp.) and were represented at eight of the nine sites from which animal remains were recovered. However, with the exception of Tecualilla, where five individuals were represented, only one individual was represented per site. The low number of individuals recovered suggests that they were not hunted intensively or in community drives. Rabbits probably lived on the monte to take advantage of hiding places and foods, and they may have been agricultural pests. They are usually captured by individual hunters or could have been clubbed or stoned, with some effort by people tending fields or garden plots. Dogs may also have been used to hunt them.

Armadillos (*Dasypus novemcinctus*) were represented at six sites. Coe and Diehl (1980:103–105) report that some modern inhabitants of coastal Veracruz consider nine-banded armadillo meat to be a delicacy, especially when it is roasted in its own shell. Armadillos are burrowing animals and live in areas that are not usually prone to flooding. Today, dogs are used to discover burrows; the animal is then dug out and dispatched (Coe and Diehl 1980).

The remains of iguana (*Iguana iguana*) were recovered from five sites. Iguanas occur in low numbers and do not appear to have contributed significantly to the prehistoric diet of the area. Wing (1969) notes that both *Ctenosaura pectinata*—the Mexican spiny-tailed iguana—and *Iguana iguana*—the green iguana—are abundant in the Marismas today; however, the modern inhabitants of the area do not make much use of them for food, although a soup made from *Ctenosaura* is said to have great medicinal use in the treatment of dysentery. In contrast, Coe and Flannery (1967:75) report that both black (*Ctenosaura similis*—black spiny-tailed iguana) and green iguanas are eaten in coastal Guatemala. Hunting/collecting occurs in midafternoon when iguanas seem to gather. Green iguanas, which are arboreal, are frequently captured alive, bound, and taken back to the house and kept alive until they are butchered or sold. In February and March female green iguanas descend from the trees to lay their eggs. When on the ground they can be hunted with dogs. The eggs are a prized foodstuff and are considered tastier than the meat. Pregnant females are sometimes opened up so that the eggs can be extracted. The iguanas are then sewn up and released in the hope that they may actually survive and be capable of reproducing the following year. Coe and Flannery (1967) observed green iguana nests with up to 33 eggs. Green iguanas inhabit river- and estuary banks. They are diurnal, arboreal, and nonaggressive.

In contrast, black iguanas are mostly terrestrial and considerably more aggressive, thus they must be hunted with care. Once subdued, they are brought back to the village alive, but their mouths are sewn shut to avoid their bite. The iguanas are hunted with dogs that run them down and shake them into a semiconscious state before they are taken, or they are clubbed by young boys. Black iguanas are all white meat and are said to taste better than green iguanas. The eggs of the black iguana are not extensively sought by modern peoples. Coe and Flannery (1967) note that along with green iguana, fish, and so-called mouthless crabs (Gecareinidae), black iguana made up most of the meat consumed by the family living at the Salina La Blanca site when excavations occurred. Nietschmann (1973:Table 20) notes that a 5-kg iguana (*Iguana* sp.) yields nearly 3 kg of meat. It is interesting, considering the density of modern iguana populations (both black and green) in the Marismas, that if common prehistorically, they were not more intensively used.

Rodent (Rodentia) remains are a common occurrence in archaeological contexts. Although it is probable that some of these remains are natural (animals living and dying in the mounds), based on the number of specimens represented and the types of elements recovered, Cumbaa (1973) argues that rodents were consumed prehistorically in the Marismas. Ethnographic data indicate that rodents are taken using a variety of methods, most of which are not observable in the archaeological record. They are clubbed or stabbed with sticks, stoned, trapped or snared, taken from their burrows, or hunted with a bow and arrow. These animals could have been hunted or trapped in agricultural fields or as part of other activities such as firewood or wild plant collecting (e.g., Szuter 1991). Their preparation for consumption requires little effort. They are most often roasted on a spit without the removal of their skins or evisceration; their feet, ears, and tails burn off, and once they are cooked, their skins are pulled off easily before they are eaten, guts, bones, and all. Any bones not easily crushed by chewing are spit out or otherwise discarded (Szuter 1991).

Crab remains were only recovered from the site of Venadillo. Among the species of crab found in the area today are fiddler crabs (*Uca* spp.) and red crabs (*Goniopsis* spp.). Fiddler crabs inhabit salt marshes, where they play an important role in the salt marsh ecology by consuming decaying plant and animal matter before they in turn become food for a variety of animals that inhabit marshes including humans. Among the Miskito they are known as the *cangrejo de mangle*, the mangrove crab (Nietschmann 1973). Coe and Flannery (1967:77) recovered crab remains at Salina La Blanca, mostly the mouthless or blue land crab (*Cardisoma crassum*), which were commonly eaten by the modern inhabitants of the site. They use small wooden box traps, and children are frequently the main collectors. The crabs are roasted or boiled.

Domesticated Animals

Dogs (*Canis familiaris*) appear to be the only species unequivocally domesticated by the prehistoric inhabitants of the Marismas (Wing 1968, 2013). Dog remains were most often recovered as intentional burials. They may have been kept as pets and for hunting; and they may have been sacrificed to accompany human burials and serve as guides or companions in the afterlife, or if they died of old age or otherwise, they were simply buried out of respect. In some instance, such as at Mound C at Venadillo, where a dog appears to have been placed on a human burial urn, it appears that the animal was killed to accompany the burial.

Although most of the dog remains were recovered from burial mounds and associated with human burials, isolated elements and fragments in midden contexts may represent food remains (Cumbaa 1973). As previously discussed, dogs could be easily maintained by feeding them scraps of food or by letting them forage for themselves. Since dogs live on-site and are domestic, it would require little expenditure of energy to capture them for butchering (e.g., Coe and Diehl 1980; Stark 1977). However, the lack of dog remains in Marismas sites clearly identifiable as food remains suggests that there was not a heavy dependence on them as a source of meat. Recently, however, Manrique (2004) reports that dog bone clearly exposed to heat was recovered from a midden deposit at San Felipe Aztatán, an apparent major Aztatlán center on the Río Acaponeta about 37 km southeast of Teacapán. The implication is that the bone represents food remains. The early Spanish invaders report that dogs were found in most villages and found in numbers (see Chapter 12). P. Scott (Chapter 12) reports that Indian allies of Gonzalo López (Guzmán 1963:84) took as many of the dogs as they could as a mobile source of food.

At least 67 individual dogs were identified, and most of these are thought to represent intentional burials (Wing 2013). The majority of the remains were isolated or partially articulated skeletal elements, although 13 articulated skeletons were recovered. As discussed above, only at the site of Venadillo, in the Mound C shell midden, is there some evidence to suggest that dogs may have been eaten. The sites of Panales, Tecualilla, and Chalpa produced the most complete specimens. Other sites that produced dog remains include Arriñitas, Cristo Rey, El Mirador, Rincon de Panal, and Venadillo.

Wing (2013) has completed both quantitative and qualitative analyses of dog crania, dentition, and postcranial skeletons for interpretive, descriptive, and comparative purposes (Table 8.14). The following discussion is based on Dr. Wing's recent study and her generous sharing of data. In additional to the metric observations, the analyses included the assessment of snout length (Degerbøl 1961:39), the presence or absence of premolars (Allen 1920; Drögemüller et al. 2008; Wright 1960), and the assessment of damage to the canines and incisors (Wing 1968, 2013).

In terms of size, the dogs represented at the site of Panales were the largest in all respects. There was some overlap between Panales and Chalpa, but the mean sizes of the two populations were distinct (Wing 2013). In terms of dentition, no premolars were absent from the Panales specimens. Wing (2013) notes that missing premolars in dogs are often the result of a congenital condition, although they can also be absent as a result of disease or accident. Congenital absence of the first lower premolar is typical in many breeds of dogs, but the condition varies at different rates in different populations. Among the Mexican hairless (*xolotizcuintli*), the absence or abnormality of the entire premolar row is characteristic. None of the Panales dentaries examined exhibited missing premolars, and 36 percent of the canines and incisors examined exhibited damage. The overlapping index for the Panales sample was calculated to be 76 to 90, which indicates no overlapping of the upper premolars and longer snout lengths (Wing 2013).

The dog remains from Chalpa are interesting in that two different varieties may be represented. The Chalpa remains were recovered from two areas of Mound 3. Excavations near the center of Mound A, Test Unit A, resulted in the recovery of dog burials as well as burials of raccoons. Excavations were also completed at the periphery of Mound 3, in excavation units designated Test Unit B and the north extension of Square A. These excavations also resulted in the recovery of dog remains, as well as those of raccoons and the previously discussed deer skulls with antlers. Although considerable overlap in size between the dog remains recovered near the center of the mound and those recovered from the periphery of the mound was noted, the dogs recovered from the periphery were generally smaller than those represented from the center. In addition to being smaller, the Chalpa specimens exhibit missing premolars and more tooth damage than the Panales specimens (Wing 2013). Eight complete dentaries were examined from Chalpa; the first lower premolar was missing from four specimens, and the first two were missing in a single specimen. None of the specimens was lacking the entire premolar row, and no abnormally shaped teeth were observed. Some 78 percent of the canines and incisors from the sample of remains from the center of Mound 3 exhibited damage, while 75 percent of those from the periphery of the mound exhibited damage. An overlapping index could only be determined for a single specimen from the center of Mound 3; it was 79–80, indicating no crowding. The index for those from the periphery of the mound was 93–100, indicating slightly greater crowding (Wing 2013). Additionally, the postcranial skeletons of the dogs from Chalpa are relatively shorter than those of the Panales dogs. The significance of the size difference in the two dog populations represented in Mound 3 is unclear; it is possible that temporal differences are represented, or it is possible that two distinct dog populations are represented.

The sample sizes from Tecualilla and the other sites are too small for similar comparative analyses. Nevertheless, based on available data, the Tecualilla dogs appear to be of similar size to those from the periphery of Mound 3 at Chalpa. In five of the six Tecualilla cases available for study, there was evidence of damaged canines and incisors. One of the canid specimens from Tecualilla was a dog's canine tooth that had been made into a pendant by drilling the root of the tooth, presumably so that it could be strung.

Wing (2013) calculated live weight estimates based on the width of the occipital condyles of the Marismas dogs and comparative materials and data (Reitz and Wing 2008). Estimated live weights for the Marismas dogs are calculated to range from about 7 to 14 kg (roughly 15 to 32 lbs).

What the Marismas dogs looked like remains a point of speculation. A spindle whorl recovered from Tecualilla is incised with a representation of a dog with erect ears, a tail that curves up over its back, and a spotted coat (see Figure 7.13). Wing (2013) notes the similarity of this representation with the ceramic dogs, the so-called Colima dogs, that are found in the Preclassic period shaft tombs of west Mexico (e.g., Schöndube 1998). Wing further notes that the Colima dogs are often depicted with humanlike dentition, in that the tooth lines are even, lacking protruding canines. She observes that the smashing of the canines and incisors found in the Marismas dogs would result in a similar appearance. It should be noted that a similar pattern of damage was also noted in the raccoon burials recovered from the Marismas (Wing 1968, 2013).

At the time of the Spanish conquest, a variety (colors, shapes, and sizes) of dogs were described (Allen 1920). Based on her study of the Marismas remains, Wing (2013) believes that the common Indian dog is most likely represented by the Panales remains while two varieties of small Indian dogs are represented at Chalpa. She further concludes, because of the lack of remains congenitally missing more than two premolars, that the Mexican hairless is not represented in any of the Marismas remains. One of her concluding comments is that the dogs (as well as the raccoons), as reported by early Spanish chroniclers, may have served to guide the deceased to the land of the dead.

Wing (1968, 2013) also observes that the presence of raccoons in the burial mounds across the Marismas seems to mirror that of dogs. She suggests that raccoons were considered the equal of dogs and kept as pets. Specifically, she refers to Allen's (1920) argument that what some early Spanish chroniclers believed to be dogs or referred to as dogs (*ixcuintlepozotli*) were actually raccoons. Her argument that raccoons were of equal importance to dogs is further supported by the fact that many of the raccoon dentaries examined exhibited the same type of damage to the canines and incisors as was prevalent in the dogs.

Turkeys were represented at four sites, with no more than one or two individuals per site. In addition to dogs, it is possible that turkeys were also domesticated. The ethnohistoric literature indicates that turkeys were kept by people of the northwest Mexican coast at the time of European contact (Sauer 1932), although Beals (1932) reports that turkeys were not widely kept at the time of the Spanish arrival. The Marismas is, today, near the range of wild turkeys in northwestern Mexico. Based on the limited number of individuals at some sites and their absence at other sites, it does not appear that turkeys were raised or hunted in numbers. No turkey eggshell was observed at any of the sites investigated. Breitburg (1993) indicates that records of precontact use of turkey in Mesoamerica are spotty and the only intensive reported use comes from the Valley of Mexico, where thousands of birds were commonly exacted as tribute by Montezuma. Otherwise, turkey does not appear to have been commonly used as food. Coe and Diehl (1980) indicate that some people did not keep turkeys, because turkeys consume too much maize. If turkeys did not inhabit the coastal plain but, rather, the piedmont of the Sierra, hunting them would have required a significant expenditure of effort. However, they may have been taken while hunting for other animals such as deer. Recently, Manrique (2004) reports turkey remains at San Felipe Aztatán.

TABLE 8.14. Summary of Measurements of Dog Remains (in Millimeters).

Measurement and Site	*N*	M	Observed Range	SD	Coefficient of Variability
Cranial Width					
Panales	1	55.1			
Tecualilla	1	49.2			
Chalpa	4	50.1	47.0–52.8	2.4014	4.79
Venadillo	1	52.2			
Cristo Rey	1	50.5			
Occipital Condyle Width					
Panales	1	34.0			
Rincon de Panal	3	32.7	30.7–35.5	2.5146	8.69
Chalpa	3	32.6	29.4–36.4	3.5473	10.88
Tecualilla	1	32.9			
Cristo Rey	2	30.3	29.5–31.1	1.1314	3.73
Venadillo	1	35.9			
El Mirador	1	34.0			
P^4 Alveolar Length					
Arriñitas	1	18.2			
Panales	3	19.1	18.8–19.5	0.3606	1.89
Rincon de Panal	9	16.2	14.0–18.4	1.0064	6.21
Chalpa	18	18.2	14.3–20.0	1.225	8.12
Tecualilla	3	18.2	18.6–18.7	0.5568	3.05
Cristo Rey	8	16.7	15.2–18.4	1.2317	8.38
Venadillo	12	18.1	16.2–18.8	0.5413	3.17
El Mirador	15	16.6	15.5–18.5	0.6668	4.02
P^4–M^2 Alveolar Length					
Panales	1	34.5			
Rincon de Panal	5	29.9	28.1–32.0	1.8727	6.26
Chalpa	9	30.1	26.9–33.0	1.7649	5.86
Cristo Rey	2	28.3	26.5–28.1	1.1314	4.14
Venadillo	1	33.3			
El Mirador	1	31.6			
M_1 Alveolar Length					
Arriñitas	2	20.95	20.9–21.0		
Panales	2	21.4	21.1–21.7	0.4243	1.98
Rincon de Panal	10	19.8	18.5–21.4	0.9947	5.02
Chalpa	20	19.9	18.0–21.5	0.944	4.74
Tecualilla	9	20.5	18.0–23.0	1.7624	8.60
Cristo Rey	6	19.5	18.2–21.3	1.0172	5.22
Venadillo	11	19.7	18.4–21.0	0.796	4.04
El Mirador	13	19.7	18.5–21.2	1.0216	5.19
Height of Dentary at M_1					
Panales	3	22.7	22.0–23.0	0.5774	2.54
Rincon de Panal	6	19.98	18.1–22.2	1.8841	9.43
Chalpa	11	19.5	16.7–21.6	1.3344	6.84
Cristo Rey	3	20.8	20.6–21.0	0.2082	1.00

TABLE 8.14. (cont'd.) Summary of Measurements of Dog Remains (in Millimeters).

Measurement and Site	*N*	M	Observed Range	SD	Coefficient of Variability
Humerus Length					
Panales	3	140.7	138.0–144.0	3.0551	2.17
Rincon de Panal	2	118.9	113.7–124.0	8.2832	6.13
Chalpa	2	110.8	99.6–122.0	15.8392	14.29
Venadillo	2	118.5	107.0–130.0	16.263	13.72
El Mirador	3	114.3	105.0–125.0	10.066	8.81
Humerus Distal Width					
Panales	3	29.3	28.3–31.6	2.1595	8.37
Rincon de Panal	3	24.6	23.1–26.7	1.8583	8.55
Chalpa	5	25.7	22.0–30.1	2.9912	11.64
Tecualilla	1	22.3			
Cristo Rey	2	25.0	23.5–26.5	2.1213	8.49
Venadillo	7	26.8	23.1–30.5	2.8266	10.55
El Mirador	7	25.9	23.5–28.1	1.7462	6.74
Radius Length					
Panales	1	140.0			
Chalpa	4	115.7	101.6–121.8	9.4366	8.16
Venadillo	3	123.3	120.0–125.0	2.8868	2.34
El Mirador	2	114.5	104.0–125.0	14.849	12.97
Cristo Rey	2	134.5	130.0–139.0	6.364	4.73
Rincon de Panal	3	115.1	108.0–124.0	8.1504	8.08
Greatest Width of Femur Head					
Panales	3	16.4	14.5–18.8	1.6921	10.32
Rincon de Panal	4	14.0	13.4–14.4	0.45	3.21
Chalpa	6	15.5	14.0–16.8	0.9832	6.34
Cristo Rey	2	16.8	16.4–18.2	0.5657	3.37
Tibia Length					
Panales	3	158.7	155.0–160.0	2.5166	1.60
Rincon de Panal	4	129.6	120.0–143.0	9.6683	8.46
Cristo Rey	1	140.0			
Chalpa	2	126.1	114.1–138.0	16.899	13.40
Venadillo	1	142.0			
El Mirador	2	136.5	133.0–140.0	4.9497	3.63
Calcaneum Length					
Panales	1	40.0			
Rincon de Panal	2	33.9	31.8–35.9	2.8991	8.55
Chalpa	1	41.9			
Tecualilla	4	36.5	35.0–38.3	1.6781	4.60
Cristo Rey	3	36.4	34.7–39.0	2.2869	6.28
Venadillo	3	34.4	31.7–38.2	2.7502	8.99
El Mirador	8	35.5	33.5–38.1	1.4755	4.16
Astragalus					
Rincon de Panal	5	20.3			
Cristo Rey	3	21.4	19.0–21.2	0.8591	4.23
Venadillo	1	22.5	20.8–22.6	0.6532	3.01
El Mirador	6	21.7	19.0–24.2	2.4007	11.22

Farming

Traditional Mexican agriculture is based on three primary plants: corn (*Zea mays*), beans (*Phaseolus* spp.), and squash (*Cucurbit* spp.). Undoubtedly, as it is today, a form of swidden agricultural was practiced in the Marismas area prehistorically. The agricultural cycle in the area is dictated by the local climate, which is marked by a distinct dry season that is followed by a rainy season. As reported in Chapter 2, the Marismas lies between the southern boundary of the arid subtropical climate of central and southern Sonora and the wet and dry tropical climate jungles of the west Mexican coast. The annual rainfall at Teacapán ranges between 800 and 900 mm, and it can vary dramatically from year to year. Typically the rains begin in late June or early July and last through late September or early October. Field preparation involves the cutting of vegetation in April and May. The cut vegetation is allowed to dry, and it is then burned prior to planting. Coe and Diehl (1980:77) note that on the coast of the Gulf of Mexico planting does not occur until there has been at least one good rain. This is done so that there is sufficient soil moisture for seeds to sprout. This is the case in the Marismas as well, where today corn is planted immediately after the start of the rainy season in July. Two to three grains of a hybrid maize are dropped in holes that are about a meter apart (Wing 1969).

Following Coe and Diehl (1980:74), field size varies from 1 to 4 ha, and 1 ha is usually sufficient to provide a family with enough maize for a year. Additionally, a single-hectare field is small enough to be worked by one individual without outside help. Larger fields may have been managed by larger task groups consisting of extended family members or cooperative labor groups.

On the Gulf Coast, as maize approaches maturity, all the stalks, except for those that have ears that are going to be eaten as corn on the cob, are bent or doubled over (Coe and Diehl 1980:78). This is done to keep the rain out and to prevent the husks of mature ears from opening and exposing the maize to pests and, once the stalk is bent, the plant is left alone to dry for eight to 15 days. Once thoroughly dried, the ears are picked, and the best are kept for seed corn or household consumption. The remainder is used to feed animals, to trade, or to sell.

At harvest, the dry corn is shucked with a *piscador*, which is made from a split deer metapodial or preferably, today, an iron spike. The grains are removed from the cob in the field, where they are sacked. Wing (1969) observed at Chalpa a 4-ha field that produced from 4,500 to 9,000 kg of corn. This approximates and exceeds the observations by Coe and Diehl (1980) for the Gulf Coast, who also observed that a family of 12 used 24 to 36 sacks (2,160–3,240 kg) of corn a year and another family of five adults and four children consumed about 16 sacks (1,460 kg) a year. Fifty to 75 percent again as much corn is used for livestock feed (Wing 1969).

In the Marismas area after the harvest the fields are left fallow until the next planting. These fields are used for agriculture for seven years before being left for pasturage for three years, before they are once again prepared and used for cultivation (Wing 1969).

Coe and Diehl's (1980:82–83) study indicates that beans are a major component in the diet in their study area. There, for the most part, people prefer to plant beans other than black beans in fields separate from maize and from black beans and either early in the year or in the late summer to early winter, so they are not growing during the rainy season. Coe and Diehl (1980: 83) indicate that a variety of squash and gourds are planted and at about the same time as maize. Sometimes squash seeds are placed in the same hole with maize, but because the maize stalks shade the growing squash, most people prefer to plant them in separate fields. Beans are planted in the cornfields in October, and both maize and beans are harvested in December (Wing 1969:103).

In the Marismas, maize and beans were the major crops grown, along with a variety of other crops including squashes and gourds, chilies, cabbage, peanuts, tomatoes, sugarcane, watermelon, guava, avocado, mango, papaya, citrus, and fruit trees. Coconut palms are found near Teacapán and along the coast, but none grow very far up the estuary (Wing 1969).

Summary

The Marismas Nacionales is a resource-rich environment replete with a variety of faunal and floral resources capable of sustaining human populations. Many of these resources also served utilitarian purposes by providing raw material for tools, ornamentation, and shelter. The most abundant animal remains, recovered and observed, are those of mollusks. Shell middens are numerous and attest to the large quantities of mollusks extracted from the area (Shenkel 1969a, 1971a, 1974). Species of *Tivela* and oyster appear to have been favored, although numerous other taxa are also represented. *Tivela* are intertidal beach taxa, while oysters are at home in the estuary; thus, their presence is indicative of the exploitation of both the beach/ocean and estuary biotopes. It is not clear whether the dominance of these two taxa represents a dietary preference or indicates that these taxa were the most common, abundant, and productive and the most easily gathered. They were probably collected by hand, possibly by task groups consisting of females and children. Additionally, mollusks were probably a relatively dependable resource because of their year-round availability and their presence in known locations. Those taxa, like oysters, living in the estuary were subject to changes in

TABLE 8.15. Summary of Archaeofaunal Data, Number of Individual Specimens.

Site	Terrestrial Food Animals		Aquatic Food Animals		Habitat Undetermined	Total Subsistence Fauna	Burials and Offerings	Intrusive Animals	Total Recovered
	Number	Percent	Number	Percent					
Juana Gomez	23	12.6	158	86.8	1	182	5	—	187
Arriñitas	107	34.9	153	49.8	47	307	35	224	566
Panales	20	19.2	84	80.8	—	104	150	—	254
Rincon de Panal	38	21.2	141	78.8	—	179	422	263	864
Chalpa	27	36.9	44	60.3	2	73	629	—	702
Tecualilla	154	24.6	459	73.2	14	627	193	—	820
Cristo Rey	4	8.2	44	89.8	1	49	321	—	370
Venadillo[a]	158	40.4	221	56.5	12	391	384	—	775
El Mirador	13	18.6	55	74.3	6	74	426	—	500
Total	544		1,359		83	1,986	2,565	487	5,038

[a] Numbers and percentages are only for number of individual specimens for which Venadillo data were calculated (excludes those species simply noted as present).

the salinity of the estuary and may have been less abundant during the rainy season, when large amounts of freshwater flowed into the estuary, reducing its salinity and its suitability as a brackish water habitat. As previously noted, flooding of the estuary in modern times caused the local extinction of oysters and, although not represented in the recovered faunal assemblages, shrimp. Although mollusk remains are abundant, it is not clear whether they were a dietary staple or they were a supplemental food.

In addition to mollusks, the archaeofaunal assemblages recovered include a variety of crab, fish, amphibians, reptiles, birds, rabbits, deer, and perhaps dogs that were eaten. Fishing was probably done primarily in the estuary and lagoons rather than in the ocean. Fish and marine mammals could have been speared, trapped, or netted. The largest land animal represented is deer, but it does not appear to have been common in the area. At the site of Chalpa, excavated deer skulls with their antlers were associated with human burials, indicating that deer had ceremonial significance as well. Dr. J. Andrew Darling (personal communication 2012) suggested that perhaps these antlered skulls were used as part of a costume (ceremonial/dance) in the fashion of a Yaqui deer dancer. Human occupation of the coastal plain may have resulted in overhunting early on or caused deer to move out of the area, although, as discussed previously, deer dislike the mosquitoes and gnats that abound in the Marismas. None of the taxa represented are exotic, although some, such as the mountain lion, margay, and turkey, may have been more at home in the piedmont bordering the east edge of the wetlands and estuary. Although varied, most of the taxa identified generally occur in low numbers at all the sites from which animal remains were recovered. However, there was little systematic collection of faunal remains, and our understanding of the number of individuals captured, and the variety of taxa used, is undoubtedly limited because of sampling error.

In general, there appears to be relatively little variation in the faunal assemblages from the sites investigated, and they tend to be more alike than dissimilar. This similarity also seems to have persisted over time. Nevertheless, two factors likely influenced the availability of faunal resources and procurement patterns, seasonal variation and annual variation in the climate. With the onset of the monsoon in early summer, the monte becomes lush with new growth, and the available plant biomass rapidly increases. During this time, game animals would have concentrated in the monte to consume this vegetation and take advantage of available water. During the dry season, these animals would have moved to locations where water and vegetation were more readily available (e.g., Coe and Diehl 1980:101). Despite the lack of any direct data (pollen, archaeobotanical remains), it is assumed that agriculture supplied the bulk of the caloric intake for the Classic and Postclassic period inhabitants of the Marismas and that the diet was primarily corn, beans, and squash and other cultigens, supplemented with hunted marine and terrestrial fauna, mollusks, and wild plants.

As shown in Tables 8.15 and 8.16, NISP and MNI counts indicate, in most cases, that the majority of the fauna represented at each site were derived from the estuary-lagoon system. Only at Juana Gomez is the MNI equally distributed between terrestrial and aquatic fauna. This is not unexpected considering the site is the most inland of those investigated. At Arriñitas the MNI

TABLE 8.16. Summary of Archaeofaunal Data, Minimum Number of Individuals (MNI).

Site	Terrestrial Food Animals	Aquatic Food Animals	Subtotal	Total Subsistence Fauna*	Burials and Offerings	Intrusive Animals	Total Recovered
MNI							
Juana Gomez	7	9	16	17	1	—	18
Arriñitas	22	21	43	43	2	3	48
Panales	5	9	14	14	4	—	18
Rincon de Panal	17	14	31	31	9	—	40
Chalpa	5	9	14	14	26	—	40
Tecualilla	30	49	79	79	6	1	86
Cristo Rey	2	4	6	6	7	—	13
Venadillo	11	14	25	27	12	—	39
El Mirador	10	8	18	18	10	—	28
Percent MNI							
Juana Gomez	33.9	50.0		94.4	5.6		18
Arriñitas	45.8	43.8		89.6	4.2	6.3	48
Panales	27.8	50.0		77.8	22.2	—	18
Rincon de Panal	42.5	35.0		77.5	22.5	—	40
Chalpa	12.5	22.5		35.0	65.0	—	40
Tecualilla	34.9	60.0		91.8	7.0	1.2	86
Cristo Rey	15.4	30.8		46.2	53.8	—	13
Venadillo	22.8	35.9		69.2	30.8	—	39
El Mirador	35.7	28.6		64.3	35.7	—	28

* Total may include species from undetermined habitats.

count for terrestrial fauna exceeds that of aquatic fauna, although the NISP for aquatic fauna is considerably higher than for terrestrial fauna. This emphasis on foods available in the estuary and lagoons is even more evident when the role of mollusks is also considered. Although there were undoubtedly seasonal and annual fluctuations in the availability of certain taxa of mollusks, the estuary was a dependable source of food throughout the year, especially during the dry season, when terrestrial fauna may have moved to other locations seeking freshwater and vegetation.

Additionally, most of the taxa represented inhabited microenvironments in proximity to the sites investigated, and it would have required relatively little effort to access these resources. If not present on the site or in immediate proximity, most of the fauna were available within 1 to 3 km—and at most 5 km—of the site. Zarky (1976) notes, in an analysis of early Formative period sites in the Ocós area of Guatemala, that catchment radii of 5 km resulted in a great deal of overlapping catchment areas and suggests that a catchment zone with a 0.5-km radius was likely more typical of the area exploited. It is also likely that all agricultural lands were within 1 km to 3 km from the villages. Although agricultural fields were likely used exclusively by a given village, it is also possible that villages had their own collecting and hunting territories, although considering the circumscribed nature of the Marismas and the proximity of sites to one another, these were shared resource areas.

Although not evident through recovered pollen or macrobotanical remains, agriculture is indicated by the presence of manos and metates at the sites investigated and the fact that agriculture had long been established as the primary source of food for most populations living in Mesoamerica at the times the Marismas was occupied. Undoubtedly, a version of swidden agriculture was practiced. Fields would have had to be on elevated lands where they were less likely to be flooded or affected by the water table. As noted in Chapters 2 and 4, the Marismas appears to have been abandoned or occupied on an extremely limited basis between roughly AD 700 and 1000 because of the subsidence or inundation of the coastal plain. The swampy conditions during this period would have made agriculture difficult if not impossible. If some land was available, ground moisture may have been such that newly planted seeds would have rotted or crops would not have reached maturity.

In summary, subsistence data, as well as other archaeological evidence such as the presence of burial mounds at village sites, indicate that the ancient peoples of the west Mexican coast occupied the Marismas on a year-round basis during the Early Classic period and for most of the Postclassic period. They survived on a diverse diet that included mollusks, fish, a variety of terrestrial fauna, and probably agriculture. In general, the nonagricultural subsistence pattern appears to have been oriented to the harvesting of estuarine fauna. Through time there appears to have been little variation in the overall hunting and collecting activities of the prehistoric inhabitants of the Marismas. Nevertheless, despite it being a diverse and rich environment, there was undoubtedly seasonal and annual variation in the resources available that affected what and how available resources were harvested and their contribution to the diet. The dietary and subsistence data from the Marismas reflect a pattern seen ethnohistorically elsewhere and in estuarine environments of ancient Mesoamerica (e.g., Coe and Diehl 1980; Coe and Flannery 1967; Flannery 1976; Nietschmann 1973; Stark 1977; Voorhies et al. 2002; Wing 1978; Zarky 1976).

Acknowledgments

I would like to thank Dr. Elizabeth Wing for providing the data presented in Tables 8.14–8.16. I would also like to thank her for a critical review of this chapter. Despite the intervening years, Dr. Wing's study of the archaeofauna and dog remains from the Marismas Nacionales project and that of her student Dr. Stephen L. Cumbaa remain the most extensive studies of archaeofaunal remains from the west Mexican coastal zone. Any shortcomings, misinterpretations, or errors are solely mine.

CHAPTER 9

Mortuary Practices

Michael S. Foster

Mortuary studies in American archaeology have a long and varied history. Historically, the study of human burial practices was primarily descriptive, with emphases on preservation of the body or skeleton, its treatment, the grave, and, if present, associated funerary offerings. Early studies also focused on the philosophical and religious determinants that influenced the treatment and disposal of the body. By the late 1960s, the focus shifted to the role of social organization as the main determinant of variation of mortuary practices within societies and how mortuary practices were influenced by the circumstances of death (e.g., Binford 1971; Braun 1979; Brown 1971; Carr 1995:109–119, 188; O'Shea 1981; Tainter 1978).

Recently, Carr (1995) undertook a review of the history of mortuary studies in American archaeology and a study of mortuary practices. Based on an extensive survey of the Human Relations Area Files, he reviewed 31 nonstate societies in an effort to determine the range of factors affecting mortuary practices. He designed his study to parallel Binford's (1971) cross-cultural study so that the results could be compared and integrated. In the process, 46 mortuary practices and 29 possible causes were coded and evaluated, with a resulting 1,887 observations (Carr 1995:123–147).

Carr (1995:106–107) states that the primary goal of his effort was to substantiate the role of philosophical-religious beliefs in the study of mortuary remains and practices. Second, he wanted to test the validity of the propositions employed by archaeologists to reconstruct social organization from mortuary remains. Carr (1995: 188–195) concludes that although determinants of mortuary practices are more complex and multivariate than generally assumed in many American mortuary studies, it is possible to reconstruct past behavior because mortuary practices and remains are symbolic and can be related in a semideterminant way to their referent meanings. When considered together, social and philosophical-religious factors often outweigh physical and circumstantial factors as the primary determinants of mortuary practice cross-culturally.

For example, Carr (1995:188–189) observed that philosophical-religious and social factors (ideation and worldview) were as influential as social organization, if not more so, noting that they were five to 10 times more influential than circumstantial and physical factors (which, in turn, were more influential than environmental factors) (Mitchell 1994a). Philosophical-religious factors declined in importance with an increase in social complexity from band-level hunter-gatherers through complex hunter-gatherers to horticultural tribes. However, social factors were found to dominate in societies with petty hierarchies, while philosophical-religious factors reasserted themselves in paramount chiefdoms. When compared with philosophical-religious and social determinants, circumstantial and physical factors were less influential generally in mobile societies than in most sedentary societies.

In regard to social persona and its representation, with increasing social complexity the affects of personal identity declined systematically, while horizontal social distinctions increased in funerary contexts (Carr 1995: 191). Carr (1995:190) notes that the primary dimensions of social persona and social organization—age, sex, horizontal and vertical social position, personal identity, and social classification at death—were frequently identifiable. Of these factors, in mortuary expressions age and status (horizontal and vertical position) were more influential than sex, personal identity, and the circumstances of death regardless of sociopolitical complexity (Carr 1995:191).

The philosophical-religious factors, according to Carr, that most frequently determined mortuary practices were beliefs about universal orders and symbols. Another philosophical-religious determinant was the relationship between the soul and the treatment of the corpse (Hertz 1907). Regional cemetery location was strongly influenced by cosmology and similar beliefs, while grave location was strongly influenced by philosophical-religious and social organizational determinants. Other apparent philosophical-religious factors included body orientation, body position, and the arrangement of burial goods in the grave.

Finally, Carr's study de-emphasizes social organization as a primary determinant of mortuary behavior, even though it is still seen as influential. Location within a cemetery was the most frequently recognized indicator of horizontal social relationships reflecting kinship or affinity (Carr 1995; Goldstein 1976; Mitchell 1994a). No relationship between horizontal social positions and grave orientation was noted, a point that also contradicted earlier studies (e.g., Binford 1971). Vertical social position was found to be manifested in the energy expended on the burial, the elaborateness of the burial facility, and the quantity of funerary offerings. However, the quality of funerary offerings was not a factor (Carr 1995; Mitchell 1994a; Tainter 1978).

The Marismas Nacionales Mortuary Assemblage

The remains of 248 individuals were recovered and analyzed as a result of the Marismas Nacionales project (Table 9.1). These include complete and fragmentary inhumations, secondary burials, possible ceremonial or ritualistic skull burials, and possibly the burial of body parts other than skulls. The majority of these burials were recovered from burial mounds.

The primary burials come in a variety of forms including extended and flexed inhumations resting on their backs, on their sides, or seated. The secondary burials were recovered as either bundle or urn burials. Also identified was a secondary mass burial. As previously discussed, the recovery of human remains was a primary research goal associated with the project's bioanthropological studies (Gill 1968, 1971; see Chapters 10 and 11). Although the resulting data on burial types and customs were secondary to the bioanthropology, they nevertheless provide insight into human social organization, the economy, and the worldview of the prehistoric inhabitants of the Marismas.

Carr's (1995) study provides the general, theoretical framework for the following discussion. Mortuary studies in west Mexican archaeology are limited by comparison with other more well-documented areas in Mesoamerica, thus emphasizing the importance of the Marismas funerary assemblage. Most studies are simply descriptions of burials, burial types, and burial goods in site reports where there is little subsequent discussion of their potential philosophical-religious or social significance. Cabrero G. (1995) recently published a summary

Table 9.1. Number of Individuals Recovered by Site.

Site	Feature Total	Site Total
Juana Gomez		2
La Tarjea		14
Feature A	10	
Feature B	4	
Panales Test Mound A		1[a]
Panales Test Mound B		5
Rincon de Panal		7
Chalpa		69
Feature A	25	
Feature B	1	
Feature C	11	
Feature D	32	
Tecualilla		52
Feature A	39	
Feature B	13	
Cristo Rey		30
Feature A	7	
Feature B	5	
Feature D	2	
Feature E	16	
Venadillo		53
Feature B	40	
Feature C	13	
El Mirador		9
Cuautla		6
Total		248

[a] Two burials were recovered from Panales Test Mound A; however, only one was analyzed, and because of the lack of a description of Burial 2, it is not included in the final burial count or further discussed.

of mortuary practices from west and northwest Mexico, whereas Carpenter's (1996) analysis of the burials from Guasave, Sinaloa (Ekholm 1942), and Ravesloot's (1988) study of the Paquimé burials from Chihuahua are somewhat more monograph treatments of mortuary behavior observed at these two archaeological sites. This study builds on the earlier work of Gill (1971) and his students (Schutte 1973). Descriptions of each burial recovered during the Marismas project are presented online on the publisher's homepage for this book.

Terminology

The following terms and categories are used to describe the Marismas burials. Condition refers to the completeness of the skeletal remains: 90–100 percent skeletal presence is complete; 70–90 percent is nearly complete (and in good condition); 50–75 percent is incomplete. A skeleton is classified as fragmentary if it is less than 50 percent complete. The state of completeness appears to be affected by four primary factors: soil moisture and chemistry; postdepositional disturbances such as intrusions of subsequent burials and animal burrowing; cultural practices such as the burying of individual skeletal elements such as skulls and arms or the incomplete collection of skeletal remains allowed to deflesh before secondary burial (e.g., bundle or urn burial); and vandalism.

Burial type describes the method and manner of interment; both primary and secondary burials are represented (Gill 1971). Primary burial types include extended and flexed inhumations. Extended burials are those that are fully extended with the long axis of the skeleton in a straight line and with the legs and arms in straight or near fully extended positions. These usually occur in a supine position—extended on their backs. Flexed types include semiflexed burials, with the legs bent at the knees and with the femurs at an angle of 90 degrees or more to the vertebral column; flexed burials, with the femurs at an angle of less than 90 degrees to the vertebral column; and tightly flexed burials, with the knees pressed to or as close as possible to the torso. Gill (1971) sees semiflexed burials as a variant of extended burials. Although these terms are generally applied with consistency in the field notes and site reports, there may be some variance in the reporting of flexed burials, with the term *flexed* being used to describe flexed as well as tightly flexed burials at some of the sites.

Secondary burials include bundle and urn burials. Both are represented by disarticulated skeletal elements—in the case of bundle burials, collected and bound together by either tying or wrapping the bones into a compact package. It is also possible that the disarticulated bones were collected and simply placed in an appropriately sized pit in the burial mound. Urn burials contain disarticulated skeletal remains of an individual or multiple individuals. It is assumed that, lacking evidence to the contrary, a corpse was left exposed to decompose to the point where the connective tissues and tendons no longer held the skeleton together or it could be easily broken so that the remains could be placed in a large urn. There is some evidence that corpses were cut into pieces for placement in a bundle. In addition to primary and secondary burial types, skeletal elements, mainly skulls, may have been buried individually.

Age estimates follow Gill's (1971) previous studies. The age categories are newborn (death at or near the time of the birth), infant (0 to 2 years), child (2+ to 12 years), adolescent (12+ to 18 years for females and 20 years for males), young adult (18+ to 35 years for females and 20+ to 35 years for males), mid-adult (35+ to 50 years), and old adult (50+ years). In some cases, the assigned age of an individual was adult with no further qualification (young, mid, or older). In the following discussion, such individuals are included in the mid-adult category. See Gill (1971) for the methods of sex and age determination.

The following provides a summary of the number and character of mortuary remains and the conditions of their recovery and documentation by site, followed by a general discussion of the Marismas Nacionales assemblage with observations regarding prehistoric funerary practices and mortuary behavior on the coast of west Mexico. Funerary offerings are items believed to have been intentionally buried with the individual as an accompaniment in death. For the descriptions of various culturally induced cranial and dental modifications represented in the Marismas burials, see Chapter 10.

Juana Gomez

Only two human burials were recovered from excavations at Juana Gomez, an Early to Late Estero period site; both were from the Gully test units. Burial 1 may have been a poorly preserved bundle burial of a young adult male. All skeletal elements were disarticulated, and there was no evidence that the remains had been dismembered using cutting tools or by physical means such as pulling or tearing. Instead, many of the bones were broken and found in a jumble. Based on field notes and subsequent descriptions of the burial, it is not clear whether it is a bundle burial or a heavily disturbed primary burial. No funerary offerings were associated with B 1, and there is no information regarding orientation or other treatment.

Burial 2 was an extended inhumation of a young female in a supine position with her arms extended at her sides; the orientation of the body, the direction of the head and the long axis of the spinal column, was not specified on the burial record form. No funerary offerings occurred in association with the burial; however, some small rocks partially encircled the crown of the skull. Although no associated funerary offerings are listed with the burial, an obsidian flake described as an obsidian point was found at the middle of the right forearm. Another stone was located on the inside of the upper right thigh. Although certainly suggestive, especially the stones at the top of the cranium, it is not certain whether these items are part of the mortuary treatment or whether they are fortuitous occurrences.

La Tarjea

Fourteen burials were recovered at La Tarjea, a Late Estero period site. Details regarding most of the human burials from La Tarjea are limited because of the poor and fragmentary nature of the remains; no field forms were available for review, and no osteological analyses of the remains were undertaken. Additionally, excavations at La Tarjea were limited by time and bad weather; thus not all the burials encountered were fully exposed.

Rucker (1969) provides a summary of the burials on which this discussion is based.

Ten burials were recovered from Mound A, and four, from Mound B. Due to poor preservation of the human skeletal material, observations regarding mortuary treatment based on age and sex are not possible. Burial type could be determined or suggested for only nine of the 14 burials recovered. In total three from Mound A were extended, three appeared in semiflexed positions, and one appeared in a tightly flexed position. Burials in Mound B include one extended burial, one tightly flexed burial, and one (B 3) for which the burial type could not be determined with certainty, though it may have been buried in a seated position. Thus, extended burials account for 29 percent of the excavated population; semiflexed burials, for 21 percent; flexed burials, for 14 percent; and undetermined, for 36 percent. The variability in burial types at La Tarjea is likely a reflection of philosophical-religious beliefs (e.g., Carr 1995). It is also possible that the variety of burial types seen at La Tarjea has temporal significance, although it is not apparent based on the archaeological data at hand. An unidentified Early Conchera period component could be present at La Tarjea, therefore explaining the presence of the extended burials or suggesting that the La Tarjea burials represent a transition when the old burial form (extended) was being replaced by a new form (flexed).

Burial 10 from Mound A is one of the few instances recorded during the Marismas excavations where an actual grave pit was identified and there was evidence of possible postinterment ritual associated with the burial of an individual. The profile of a straight-sided, presumably rectangular grave was exposed. In virtually all other cases, because burials were in burial mounds that appear to have grown by accretion with the addition of burials over time, evidence of grave pits was virtually nonexistent. The presence of an ash lens above and partially collapsed into the grave pit indicates that a postinterment burial ritual involving a fire over the grave may have taken place. Several fires appear to have been built and allowed to burn within a relatively short period of time. Whether these fires were associated with the mourning process, or the transformation from the living world to the realm of the dead, or recognition of the individual's status within the community is unknown.

When it could be determined, five of the 10 individuals were buried with their head oriented to the north or in a northerly direction. Two were to the east; one, to the south; and one, to the west.

Arriñitas

Excavations at Arriñitas produced only a few isolated human teeth and skeletal remains. No burials were identified.

Panales Test Mound A

Two human burials were recovered from Panales Mound A, a feature with both Late Estero and Early to Middle Conchera components. Both the burials were secondary bundle burials, but only B 1 was analyzed. Its sex was undetermined, but it was probably a female, mid-adult in age. The skull was facedown, and both the skull and mandible were placed atop the pile of bones, which included only short sections of long bones, a thoracic vertebra and other vertebral fragments, and part of the left scapula. A section of left femur, measuring 20 cm long, was broken at both ends. All other long-bone sections were cut or sawed smooth on at least one end (Gill 1971: 80). This suggests that the postcranial skeleton had been cut up for packaging in the burial bundle. No funerary offerings were found in association with Burial 1.

Panales Test Mound B

Five burials were recovered from Panales Test Mound B, also a feature with both Late Estero and Early to Middle Conchera components. The burials recovered from Panales Test Mound B were generally in poor condition. No grave pits were identified, and no evidence of postinterment rituals was found. The sex for only B 1 could be determined with certainty: it was male. Burial 2 was classified as probably female; the sex of the three remaining burials could not be determined. The male was a young adult, as was the probable female. A third young adult was represented along with one adolescent and one child.

Burial type could only be determined for the male and probable female; both were extended. The male's head was oriented to the north, and the probable female's was oriented to the southwest. Burial 2, the probable female, may have had a broken metate associated with it as a funerary offering. The broken metate, approximately two-thirds complete, was located about 30 cm southwest of the skull (Figure 9.1). Gill (1974) did not believe that the metate was associated as a funerary offering. However, if the metate fragment was placed with the burial and if the individual was a female, the metate may represent a gender-specific offering or household item routinely used by the deceased. No other funerary offerings were found in association with B 2.

A variety of funerary offerings were found with B 3 and B 4. Burial 3, an adolescent, had two vessels, beads representing one or two necklaces, and five spindle whorls. The presence of the spindle whorls suggests that the individual was a female, as they are exclusively found with females. The question is whether the individual was a weaver or whether a close relative, perhaps the individual's mother, was a weaver and the spindle whorls represent a personal gift given in grief or in anticipation of the individual's future occupation or perhaps occupation in

FIGURE 9.1. Burial 2, Panales Test Mound B, showing the location of the metate in relationship to the burial.

the afterlife. The beads may have been personal adornment. There are no detailed descriptions of the ceramic vessels found in association with the burial. Because of the lack of descriptions, the significance of the vessels cannot be assessed. Nevertheless, B 3 at Panales Test Mound B has more funerary offerings than typically recovered with burials in the Marismas.

Burial 4 was that of a very young child buried with a carved red bead and a polished bone necklace piece; these may represent one of two necklaces or bracelets buried with the child. It is likely that these were items of personal adornment or a gift from a grieving parent. The tripod vessel buried with the individual represents something of value, but it is not known whether it was a household item buried with the child or whether it was made or obtained specifically as a funerary offering.

Rincon de Panal

Seven burials were recovered from the T-shaped mound tested at Rincon de Panal, an Early through Middle Conchera period site. There are discrepancies between the preliminary report (Taylor and Rautenkranz 1971) on the fieldwork and the subsequent osteological analyses (Schutte 1973) regarding the treatment of the burials and the number and types of funerary offerings. Thus, the reports on the fieldwork and field notes are relied on for describing treatment and funerary offerings, and the osteological analyses, for age and sex data.

Three of the individuals recovered from Rincon de Panal were identified as male; two, as female; and one, as possibly male. The sex of the seventh could not be determined. Five of the individuals were young adults, one was a mid-adult, and the age of the seventh could not be determined. Four of the burials were in flexed or semiflexed positions. Of these, two were male, and one was a female; and the individual whose sex could not be determined was buried in a semiflexed position. One of the males (B 3) was in a seated position with the legs drawn tightly to the chest. The burial type for the probable male could not be determined because of poor preservation.

Funerary offerings were found with five of the individuals—two males, one probable male, a female, and the individual of undetermined sex. Among the funerary offerings associated with B 1 was an elaborately decorated, unclassified red-on-buff-type ceramic vessel (see Figure 6.58a), along with another ceramic vessel and 10 spiral marine shells. A ceramic, plain buff, Cocoyolitos-style smoking pipe (see Figure 7.8) was recovered with B 3; it had been placed on his chest. Three other burials had single occurrences of plainware and several crudely made ceramic vessels.

Burials 1 through 6, three (possibly four) males and two females, exhibited various forms of tabular erect cranial modification. With the exception of B 2, all were intermediate forms of deformation, with B 5 and B 6 exhibiting a bilobed variety of intermediate deformation. Burial 2 exhibited a bilobed variety of plano-lambdoid deformation. It is notable that this appears to be the oldest individual identified among the seven recovered, suggesting perhaps that this more extreme form of cranial modification was associated with someone of higher socioeconomic status and that because of this status he had better access to resources that resulted in greater longevity. If this is true, his status did not afford him any consideration in terms of funerary offerings, of which there were none, though this could be an artifact of the extent of the recovery effort. Finally, in six cases where it could be determined, all the burials were oriented to the east or northeast. Based on the limited sample size, no difference in the orientation of males vs. females was apparent.

Chalpa

Investigations at Chalpa resulted in the recovery of remains of 69 individuals from Mounds A, C, and D (Gill 1971, 1974). The occupation at Chalpa spanned the Early and Middle Conchera periods. Many of the remains were fragmentary and likely represent disturbed burials.

TABLE 9.2. Age Distribution by Sex for Burials from Mound A, Chalpa.

Age	Male	Female	Probable Male[a]	Total	Percent Male	Percent Female
Adolescent	1	1		2	50	50
Young adult	3	3	1	6	50	50
Mid-adult/adult	2	3		5	40	60
Old adult		3		3		100
Total	6	10		16	37.5	62.5

[a] Not included in totals for percentages.

TABLE 9.3. Body Orientation by Sex for Burials from Mound A, Chalpa.

	Body Orientation				
Sex	South	Southwest	West	East	Northwest
Male	1	1	1		
Female	2	3	3	1	
Undetermined			2		
Probable male					1
Total	3	4	6	1	1

Mound A

The remains of 25 individuals were recovered from Mound A. The condition of these ranged from complete to fragmentary, with both males and females represented, and ages ranged from newborn to old adults. Of the 16 burials where the sex could be determined with certainty, six were males, and 10 were females. One probable male was identified, and the sex of eight of the individuals could not be determined. The Mound A burials include one newborn, six infants, one child, two adolescents, seven young adults, five mid-adults (including two specified as adult with no further assessment), and three old adults.

Nearly one-third of the individuals recovered were newborns, infants, or children. Of these, the sexes of the newborn, the six infants, and the one child could not be determined, and for those whose sex could be determined with certainty, females account for a higher percentage than males (Table 9.2).

Extended, flexed, semiflexed, and seated burial types were represented. Burial type for nine of the burials, mostly the newborn, infants, and children, could not be determined. Two infants and the probable male were buried in extended positions (19 percent); three females were buried in semiflexed positions (19 percent); one male, three females, and one child were buried in flexed positions (31 percent); and two males and three females were buried in seated flexed positions (31 percent). In all, nine females were buried in some form of flexed position: three semiflexed, three flexed, and three in a seated flexed position. Male burial types, including the probable male, exhibited slightly more variability, with one male buried in a semiflexed position, two in seated flexed positions, and the probable male in an extended position.

Orientation of the body was determined for 15 of the burials (Table 9.3). Body orientations to the west dominate, followed by the southwest.

Funerary offerings were found with 14 (56 percent) of the 25 burials; three males and nine females had funerary offerings with them. The probable male also had a funerary offering. Three of the burials identified as male and one identified as female lacked funerary offerings, and except for one El Taste Polychrome bowl found with a child, no funerary offerings were found with remains identified as the newborn, infants, or the other children. Funerary offerings were not abundant; the number of items ranged from one to as many as five items. The number of funerary offerings found with nine females ranged from one to five items and averaged 2.3 items. Ceramic vessels were the most common found with the female burials, and ceramic smoking pipes were found with three individuals, as were spindle whorls. In one case a pipe and a spindle whorl co-occurred (B 10), and it was noted that the pipe had been used. Spindle whorls were exclusively associated with females in the Feature A burial assemblage, and pipes were found only with females with a single exception, B 12A, a male. The female in B 2 was interred with three animal burials. Other items of note include an obsidian projectile point found with the female in B 8.

The number of funerary offerings associated with the three males with offerings ranged from one to three, with most being ceramic vessels. Several of the male burials

FIGURE 9.2. Mass secondary inhumation, Burial 1, Mound C, Chalpa.

also appear to have had items of personal adornment buried with them. Burial 6 had two shell beads and a possible earring associated with it, and the probable male burial (B 14) had a green stone bead. As noted above, B 12A, a male, was buried with a smoking pipe.

Of the six individuals identified as males, only one (B 12A) exhibited cranial deformation; he was accompanied by a funerary offering. Six of the 10 females identified had cranial modification, and of these five were also accompanied by funerary offerings. The presence of cranial modification could not be determined for seven of the eight individuals of undetermined sex; one exhibited no evidence of cranial modification.

Two of the males and the probable male exhibited dental modifications; all three were found with associated funerary offerings. Two of the females exhibited dental modification, and both had funerary offerings; two females lacked evidence of dental modification, and both had funerary offerings. The presence or absence of dental modification could not be determined for six of the females. Only one male and one female identified in the Mound A remains exhibited both cranial and dental modification as well as funerary offerings.

Mound C

Two burials containing 11 individuals, one a mass burial of 10 individuals and a burial of a single individual, were recovered from Mound C at Chalpa. Burial 1 consisted of the fragmentary remains of 10 individuals: seven adults, two children, and one newborn (Figure 9.2). The sexes of the children and newborn could not be determined. The adults were represented by four males, two females, and one probable female.

Burial 1 appears as an unusual secondary mass inhumation. Four of the burials were represented by nearly complete skulls that had been placed in a cluster with the faces toward the center and pointed slightly downward (Figure 9.3). Piled on top of the skulls were their mandibles and postcranial elements, on top of which were placed the remains of the six other individuals. Although no grave pit was identified, the burial was confined to an area measuring 92 × 101 cm. The postcranial elements resting on the mandibles were completely mixed, and it appeared that the material had been tightly packed in the grave. No funerary offerings were associated with Burial 1.

The second burial in Feature C, B 2, was a complete young adult female. She was flexed on her back, accompanied by a large olla and a small dish, and exhibited both cranial and dental modifications.

Mound D

The remains of at least 32 individuals were recovered from Mound D. Unfortunately, many of these were badly disturbed and highly fragmentary. Many of the Mound D individuals were identified postfieldwork during laboratory studies, thus expanding the sample size believed to have been recovered during excavations. Some of the damage may have been done with the interment of other burials in the mound or as the result of looting. Children, subadults, and adults were represented, with ages ranging from newborn to old adult. A pair of the remains,

FIGURE 9.3. Four human skulls that were part of the mass secondary inhumation, Burial 1, Mound C, Chalpa.

B 11A and B 11B, appear to represent premature infants, possibly a multiple birth (twins). These two individuals were buried in proximity, and if not stillborn, they probably died shortly after birth. Burial 14B was that of an infant also less than one year in age. The remains of two children and one subadult were also present. Three of the females were mid- to old adults at the time of death, and one individual tentatively identified as male was an old adult.

Sex could be determined with certainty for six males and eight females; three possible males and two possible females were also identified. Sex could not be determined for any of the newborns or children represented.

Because of the fragmentary nature of the remains, the burial type for only eight of the 32 individuals from Mound D could be identified. One male and three female burials were flexed on their backs; one probable male and one female were buried in flexed positions; one infant was buried in a semiflexed position; and one child was buried in a seated position. Semiflexed burials dominated the assemblage, accounting for 56 percent of the identifiable burial types, followed by flexed burials (33 percent) and the single seated burial (11 percent).

The orientation of the body could be determined for only seven of the burials. Three females were oriented to the northwest; one, a seated child of undetermined sex, to the west; one male, to the south; one female, to the southwest; and one probable male, to the north.

Only 11 of the Mound D burials could be assessed for the presence or absence of funerary offerings. Because of mixing, the association of funerary offerings with B 8 and B 9C is uncertain; the offering with B 8 could be associated with one of the burials from the B 9 group, as could the offering with B 9C. Noting these caveats, nine of the individuals were accompanied by offerings, and two were not. One of the burials lacking funerary offerings was an adult male, and the other was an adult tentatively identified as a male. Of those with funerary offerings, two were male, four were female, and the sex of the remaining three could not be determined. The number of items per burial ranged from one to four when present; five of the individuals had a single item, three had two, and one had four. The individual (B 10A) with the most funerary offerings was an elderly adult female. Accompanying her were a spindle whorl, a smoking pipe, a broken olla, and an obsidian projectile point. However, this burial may have intruded into an earlier burial, and thus the association of the projectile point with B 10A is tenuous.

The presence or absence of cranial modification could be determined for only eight of the individuals recovered from Mound D. Two were male, and three were female; the remaining three individuals were either stillborn fetuses or newborns. In only four instances could the presence or absence of dental modification be determined among the adults; two males and one female lacked dental modification, and only a single female had modified dentition. Based on the evidence available, there was no co-occurrence of cranial and dental modifications in the Mound D burials; however, as noted above, the remains were generally fragmentary and incomplete. In three instances, the presence of cranial modification was found with individuals who also had associated funerary offerings: one male (B 12A) and two females (B 6 and B 10A). One male (B 4A) with cranial deformation exhibited no evidence of dental modification and no funerary offerings.

Tecualilla

Two burial mounds, Mounds A and B, were investigated at the site of Tecualilla, an Early through Middle Conchera period site (Gill 1971, 1974). The excavations in Mound A were conducted over a three-year period and produced the remains of 39 individuals. Excavations in

TABLE 9.4. Body Orientation of Burials from Mound A, Tequalilla.

	Body Orientation							
Sex/Age	North	Northeast	East	South	Southwest	West	Northwest	Undetermined
Male	1	2	2	1		5	2	4
Probable male			1					1
Female	1			2	1	2		1
Probable female	1							1
Infant[a]						1		
Child[a]						1		4
Undetermined								2
Total	3	2	3	3	1	9	2	13

[a] The sex of these individuals could not be determined.

Mound B resulted in the recovery of 13 individuals. The condition of these ranged from complete to fragmentary, both males and females were represented, and ages ranged from newborn to old adult. Extended, flexed, and semiflexed burial types were represented, and funerary offerings were found with 24 of the burials.

Mound A

The Mound A population was mostly adults, although adolescents and children were represented. Seventeen (47 percent) of the individuals were males; two were probable males; seven (19 percent) were females; two were probable females; and the sexes of eight of the individuals could not be determined. With one exception, a fragmentary old adult, the burials of undetermined sex were either adolescents or children.

The burial types found in Mound A exhibited some variability; these included extended, semiflexed, seated, and secondary burials. Burial type could not be determined for nine (25 percent) of the 36 individuals represented in Mound A. Twelve (33 percent) of the burials were extended (including one incomplete but likely extended), 11 of which were extended on their backs and one that was extended on its right side. Seated burials were the next most common burial type, represented by seven (19 percent) individuals; four (11 percent) were semiflexed burials, three of which were on their backs and one that was on its left side; and four (11 percent) were secondary burials. One of the secondary burials (B 3B) is described as a secondary inhumation, but Gill (1971:68) does not further describe the burial or explain this particular burial type. Other secondary burials consisted of two bundle burials and one isolated skull. Also, Volkmann and Gill (1970:54) indicate that in the case of B 23, the front of the rib cage had been opened and a decorated vessel had been placed in the left side of the chest cavity.

Of the 12 extended burials, seven were male, one was a probable male, two were female, and one was a probable female; the sex of one individual could not be determined. Ten were adults (eight males, two females); one, an incomplete infant; and one, an adolescent male. Burials 21A and 21B, respectively, an old male and a young adult female, were buried extended on their right sides in the same grave. Five of the seated burials were adult males, and one was an adult female; one seated burial was that of an adolescent. The four semiflexed burials were those of three adult females and a child of undetermined sex. These were classified as semiflexed because their knees were slightly bent. All three of the secondary burials were males, as was the isolated skull.

Body orientation could be determined for 23 of the burials (Table 9.4). The most common direction was west, with none of the bodies oriented to the southeast. Thirty-nine percent of the males were aligned to the west. Thirty-three percent of the females were also aligned to the west, and 33 percent were aligned to the south. One infant and one child of undetermined sex were also oriented to the west.

The presence or absence of funerary offerings could not be determined for seven of the individuals from Mound A. Of the remaining 29 individuals, 17 were buried with funerary offerings, 11 were not, and one case is uncertain. An animal burial was found in proximity to B 16, the incomplete remains of an infant; however, the direct association of the animal burial with B 16 could not be established. Fifteen cases in which sex could be established exhibited funerary offerings, including 12 males, two females, and one probable female. One child and one adolescent were also buried with funerary offerings. The number of items per individual ranged from one to four and included plain and decorated ceramic vessels, a carved shell pendant, shell beads, obsidian blades, spindle whorls, projectile points, and an animal burial. In two cases spindle whorls were found with females, and three spindle whorls were found with the burial (B 12C) of an adolescent of undetermined sex. As previously stated, spindle whorls were most commonly found with females among sites in the Marismas and, therefore, were likely associated with female activities or actions, such as spinning, or perhaps a mother's gift to her deceased child. The obsidian blades and a projectile point occurred as funerary offerings only with males in Mound A.

In Mound A, funerary offerings were more commonly found with males, including nine cases in which ceramic vessels were present vs. only three for females. Males often had multiple vessels buried with them, as well, consisting of decorated and plainware vessels. The three females only had one plainware vessel each. In the four cases where some form of ornamentation or jewelry (bone beads, pectorals, shell beads) occurred, all were male.

When the presence or absence could be determined, 10 of 15 males had some form of cranial modification, as did five of six females. One adolescent of undetermined sex also exhibited cranial modification. The presence or absence of dental modification could be determined for 13 individuals; four out of five males exhibited evidence of dental modification, as did one of three females. There were five additional cases where no evidence of dental modification was identified; all were subadults (infant, child, or adolescent). There were three cases, all male (B 10, B 12A, and B 14A), where the co-occurrence of cranial and dental modification was present. Two females and an adolescent of undetermined sex exhibited cranial modification but no evidence of dental modification; one male exhibited dental modification but no cranial modification, and another male lacked any modification to the skull or dentition.

Evaluation of the association of funerary offerings, cranial modification, and dental modification is hampered by the significant number of Mound A cases lacking information on one or more of the three variables; only eight of the 36 burials could be assessed on all three of the characteristics. Three male burials (B 10, B 12A, and B 14A) exhibited a co-occurrence of funerary offerings, cranial modification, and dental modification. Two burials, B 12C and B 17, respectively, an adolescent of undetermined sex and a young adult female, exhibited funerary offerings and cranial modification but no evidence of dental modification. A male bundle burial (B 18) had no associated funerary offerings and only exhibited evidence of dental modification; B 21B, a female buried together with a male (B 21A), lacked burial offerings and exhibited cranial modification only; and B 23, a male, had associated funerary offerings but no evidence of either cranial or dental modification. For 28 of the Mound A burials the presence or absence of funerary offerings, cranial modification, or dental modification could not be determined.

Mound B

Thirteen individuals were recovered from Mound B at Tecualilla. All were either incomplete or in poor condition. The sex of the individuals could only be determined with certainty in seven cases: five males and two females, although one additional burial was probably male and another was a probable female. Adults account for 69 percent of the Mound B population. Two adolescents, one child, and a newborn were also represented. Because the remains were so incomplete burial type could be determined for only seven individuals. These include an extended burial of a mid-adult male positioned facedown; two seated burials, one adolescent male and one adult female; a secondary bundle burial; a possible second secondary burial; and two urn burials. One purported urn burial (B 1D) contained a skull and some partial skeletal material. A small vessel, presumably an offering, was found inside the urn. The second similarly purported urn burial (B 1H) was also fragmentary. A third highly questionable urn burial (B 1B) was badly disturbed, and it was reported as containing only partial skeletal remains, the legs, which are described as articulated. Because of their poor preservation and fragmentary nature, little can be said of how these two, possibly three, urn burials compare with other urn burials found in the Marismas or in neighboring areas.

The orientation of the body could be determined for only three of the 13 Mound B individuals. Two males were oriented south, and one female was oriented southwest.

The poor preservation of the human remains from Mound B also limits assessment of the distribution of funerary offerings relative to cranial and dental modifications. One adolescent male (B 1A), exhibiting dental modification and no cranial deformation, may have had a necklace or bracelet consisting of four red stone beads; however, the association is ambiguous. There were three other burials with funerary offerings present; however, the presence or absence of either cranial or dental modification could not be determined. A child burial (B 2B), for which cranial deformation was unidentifiable, lacked funerary offerings and dental modification. The presence or absence of funerary offerings, cranial modification, or dental modification for the remaining eight burials could not be determined.

Cristo Rey

Excavations in Mounds A, B, D, and E at the site of Cristo Rey, a Middle Conchera period site, resulted in the recovery of 30 individuals: 11 males, two possible males, five females, one possible female, and 11 of undetermined sex (Schutte 1973). Adults, juveniles, and children were represented, with ages at death ranging from child to old adult. Primary burial types included tightly flexed,

flexed, semiflexed, and secondary burial types, and the secondary burial types include both bundle and urn burials. Burial types for three of the 30 burials could not be determined, because of condition.

Mound A

Seven individuals were recovered from Mound A; six were classified as fragmentary, and one, as incomplete. The sex of only two individuals (B 6 and B 7) could be determined. Both were males; one was a young adult, and one was a mid-adult. The remaining individuals included three children, one adolescent, and one young adult; their sexes were undetermined.

Four of the individuals recovered from Mound A were buried in a tightly flexed position, one was in a flexed position, and one was a bundle burial. The orientation of the body could be determined for only four of the individuals: B 2, a male, was oriented south; B 6, a male, was oriented west; B 7, a male, was oriented northeast; and B 9, a child of undetermined sex, was oriented east. Burial 9, a tightly flexed child, was accompanied by a whole plain vessel that was located in the chest area. Schutte (1973:23) speculated that perhaps the chest had been opened and the vessel had been placed within the chest cavity. Volkmann and Gill (1970:54) had previously suggested that such an event was represented in B 23 at Tecualilla. However, B 9 at Cristo Rey was too poorly preserved to determine whether this was the case. It is more likely that the vessel had been placed on or very near the individual and subsequently collapsed into the chest cavity area as the body decayed.

Five of the individuals were accompanied by funerary offerings. The number of offerings ranged from one to four. Burials 6 and 7 each had a ceramic vessel, the latter identified as plainware. Three spindle whorls and a polychrome vessel were found in association with B 2, an adolescent of undetermined sex. Two children (B 9 and B 10) were each buried with a plainware vessel, and B 10 also had two beads and possibly two sherds. The third child had no funerary offerings.

Funerary offerings and cranial or dental modifications co-occur in two of the Mound A burials. Burial 6 was buried with a ceramic vessel and exhibited both cranial and dental modification, and B 7 was buried with a ceramic vessel and exhibited cranial modification only.

Mound B

Six individuals were recovered from Mound B; all the remains were in fragmentary condition. The age of B 1 could not be determined. Burial 13 was a child, B 3 was an adolescent, and B 4, B 5/11, and B 12 were all young adults. The sex of only one of these individuals, B 12, a male, could be determined with certainty. Burials 3 and 5/11 were identified tentatively as males.

Burial 1 was an urn burial, although the urn and remains were fragmentary. Burial 3 was a bundle burial, B 4 was flexed, B 12 was tightly flexed, and B 13 was flexed. Burial 5/11 was so fragmentary that the burial type could not be determined. The orientation of the bodies of two of the five individuals could be determined; B 4, a young adult of undetermined sex, was oriented southwest, and B 12, a male, was oriented northwest.

Two of the Mound B burials, B 4 and B 5/11, had associated funerary offerings or possible offerings; however, the presence or absence of cranial or dental modifications could not be determined for either. Burial 12 had a whole ceramic vessel in association; the presence or absence of cranial modification could not be determined, and no dental modification was present.

Mound D

The remains of two individuals were recovered from Mound D. Burial 14 was seated in a semiflexed position. The fragmentary remains were those of a young adult of undetermined sex. The body was oriented to the northwest; it lacked funerary offerings and exhibited dental modification.

Burial 15 may be the remnant of a heavily disturbed bundle burial. The remains were those of a female and were fragmentary and incomplete. The skull was face-down, and an obsidian blade fragment found nearby was reported as a funerary offering.

Mound E

All 16 burials in Mound E at Cristo Rey were fragmentary or incomplete. Age categories included three children, eight young adults, one mid-adult, two classified simply as adult, and two old adults. Eight of the burials were those of males, five were female, and the sex of three could not be determined.

Flexed burials were the predominate burial type: four semiflexed, two flexed, eight tightly flexed, and one bundle burial. Two burials, one female and one child of indeterminate sex, were disturbed and scattered with no specific burial type recognized. Five of the males were tightly flexed, one was flexed, and two were semiflexed. Two of the females were tightly flexed, one was flexed, and one was semiflexed. One child of undetermined sex was also buried in a semiflexed position. The bundle burial was that of a child of undetermined sex. Flexed burials account of 13 (93 percent) of the 14 cases where burial type could be determined.

Orientation of the body could be determined for 14 of the burials (Table 9.5); the orientation for one child and for one individual of undetermined sex could not be determined. East- and west-oriented burials were the most frequent, accounting for 43 percent and 36 percent, respectively; south- and southwest-oriented burials account for 21 percent.

Seven (44 percent) of the 16 Mound E burials had

TABLE 9.5. Body Orientation of Burials from Mound E, Cristo Rey.

	Body Orientation			
Sex/Age	Southwest	South	East	West
Male	1	1	4	3
Female		1		2
Child			2	
Total	1	2	6	5

TABLE 9.6. Body Orientation of Burials from Mound B, Venadillo.

	Body Orientation				
Sex/Age	East	West	Northwest	South	Southwest
Male	3	2	3	2	1
Female	1	6	2	2	1
Infant [a]	1				
Child [a]				1	
Total	5	8	5	5	2

[a] The sex of these individuals could not be determined.

associated funerary offerings. A child burial (B 25B) was found in spatial association with B 25A, an old adult male. However, it seems likely that B 25B represents an intrusion into B 25A. The number of funerary offerings per burial ranged from one to four. Burial 19, a male, was accompanied by two ceramic smoking pipes, one vessel, and one obsidian projectile point; B 22, a female, was accompanied by one vessel (see Figure 6.7d) and two spindle whorls; B 23, a male, had a studded vessel (see Figures 5.37 and 6.7a); B 25A, a male, was accompanied by a crocodile tooth; B 27, a male, was found with a whole vessel (see Figure 6.7c); B 28, a female, was probably associated with a smoking pipe (the burial was scattered); and B 30, a female, was accompanied by one vessel (see Figure 6.7e), one spindle whorl, one smoking pipe, and possibly one obsidian flake.

Eight (50 percent) of the 16 burials from Mound E at Cristo Rey exhibit cranial modification. There were four instances of individuals with a co-occurrence of funerary offerings and cranial modifications; three were males (B 19, B 23, and B 25A), and one (B 30) was a female. No evidence of dental modification is reported for B 19; the presence or absence of dental modification could not be determined for B 23, B 25A, and B 30. Four of the burials lacking funerary offerings (B 18, B 21, B 24, and B 26) are also reported to exhibit cranial deformation; the presence or absence of dental modification could not be determined for these burials. Burial 27 was found with a whole ceramic vessel and lacked any dental modifications; the presence or absence of cranial modification could not be determined. Burial 16 lacked funerary offerings and exhibited no evidence of cranial deformation; the presence or absence of dental modification could not be determined.

Venadillo

The remains of 53 individuals were recovered from the excavations at Venadillo, 40 from Mound B, a Middle Conchera period feature, and 13 from the earthen burial mound, an Early to Late Estero period feature, below the shell midden that formed Mound C. Schutte and Speas (1973) note that many of the Mound C burials, although superimposed, intruded into one another, resulting in a mixing of remains.

Mound B

Infants through old adults were represented in the burials from Mound B. These include four infants, three children, one adolescent, 17 young adults, five mid-adults, nine old adults, and one simply classified as an adult. The largest group consisted of young adults (43 percent), followed by old adults (23 percent). Of the 40 individuals represented, 15 were male, and 15 were female. Nine individuals were of undetermined sex, seven were either infants or children, one was an adolescent, and one was an old adult. One individual (B 22) is an adult reported as a probable female.

Burial type could not be determined for nine of the individuals recovered from Mound B. Flexed burials account for the majority of the 32 burials where burial type could be identified; 23 (72 percent) flexed burials were represented, along with one semiflexed burial. Three seated (flexed) burials were also found, along with two bundle burials, one extended burial, and an isolated skull and mandible (B 23). The skull, probably from a disturbed burial, was found directly below and touching B 19, although it is possible that it represents a funerary or other type of votive offering. Burials 5A and 5B, those of a young adult female and an infant or very young child, are bundle burials, which were interred in close proximity to each other. It is unclear whether they were buried together. However, similarities in interment suggest not only that they were buried together but that they may be mother and child.

Orientation of the body could only be determined for 25 of the burials in Mound B at Venadillo (Table 9.6). Eight of the individuals were oriented to the west; five each, to the east, south, and northwest; and two, to the southwest. The sample size is such that inferences are nearly meaningless. Males were most commonly oriented to the east or northwest, while females were most commonly oriented to the west.

Funerary offerings were found with 23 of the individuals from Mound B. Three of these individuals are described as having olla fragments associated with them, but these are not described in detail, and it is not clear whether these are remnants of ceramic vessels or sherds

FIGURE 9.4. Urn burial before exposing its contents, Burial 2, Mound C, Venadillo.

placed as offerings or simply refuse in the fill of the burial mound. Nevertheless, these are counted as funerary offerings in the following discussion. Eleven (73 percent) of the 15 males and 11 (69 percent) of the 16 females had funerary offerings associated with them. Of the nine individuals of undetermined sex, eight lacked funerary objects, and the ninth had olla fragments. The number of items ranged from one to four and included plainware and polychrome vessels, spindle whorls, a figurine head, beads, an obsidian projectile point, an obsidian blade, and an animal burial. The percentage of burials with funerary offerings was relatively high compared with the other Marismas sites investigated, as well as the number of items per burial. Ten vessels found with the Venadillo Mound B burials (B 2, B 3, B 6, B 9, B 12, B 17A, B 25, B 26A, B 26B, and B 32A) are illustrated in Chapter 6, Figures 6.8 and 6.56d.

Cranial modification was observed on 16 of the individuals from Mound B; two individuals lacked cranial modification, and there were 22 individuals for which the presence or absence of cranial modification could not be determined or was not reported. Only one Mound B burial (B 23) exhibited dental modification; there were 11 burials that lacked dental modification, and its presence or absence could not be determined or was not reported for 28 of the burials. The single individual, B 23, with dental modification also exhibited cranial modification; as noted above, this individual was represented by only a skull and mandible and may have been associated with B 19 and B 20. Seventeen cases provided information on the co-occurrence of funerary offerings and cranial modification. There were 13 cases of individuals with funerary offerings and cranial modification; seven were male, and six were female. There were two individuals, B 8A and B 24 (both females), who exhibited cranial modification but lacked funerary offerings, and there were two individuals, B 7 and B 15 (a female and a male, respectively), with funerary offerings and no cranial modification.

Mound C

The remains of 13 individuals recovered from Mound C at Venadillo were in poor condition, and some were highly fragmentary and incomplete. Infants, adolescents, young adults, mid-adults, and an old adult were represented. The sexes of eight of the individuals could be determined; two were males, and six were females.

Burial type was evident in nine cases. Two were urn burials (B 2A and B 2B, and B 4A and B 4B), each of which contained the remains of two individuals. Five of the individuals, one male, two females, and two of undetermined sex, were buried in flexed positions. The burial type for three infants and one female was not identified, because of poor preservation and disturbance.

The incomplete remains of two young adults, one male (B 2B) and one female (B 2A), were recovered from a single burial urn covered by an inverted dish (*mano colorado*—see Figure 6.22). The burial urn, its lid, and its contents were crushed when the vessel collapsed in situ (Figure 9.4). The remains consisted of the skulls, long bones, ribs, scapulae, vertebrae, and phalanges. The skeletons appeared to have been allowed to decompose before the bones were collected for burial, and no evidence of cut marks or scraping was identified to suggest that the remains were dismembered and cleaned prior to burial. The male skeleton lacked its long bones, while

FIGURE 9.5. Exposed contents of Burial 2 (B 2A and B 2B), Mound C, Venadillo.

FIGURE 9.6. Urn burial before exposing its contents, Burial 4, Mound C, Venadillo.

those of the female were present. The ribs, scapulae, vertebrae, and phalanges collected were placed in the bottom of the urn, and the long bones and skulls were placed on top (Figure 9.5). In addition to the human bone in the urn, a bone from a caiman was also present and may have been included inadvertently when the human bones were gathered for burial, again suggesting that the bodies had been allowed to deflesh prior to burial.

Burial 4 was another burial urn with the remains of an adult female (B 4A) and an infant (B 4B). The urn and its contents were badly crushed (Figures 9.6 and 9.7). The burial urn, like Burial 2, had a dish inverted over its mouth. The bone was too highly fragmented and disturbed to determine whether the remains had been defleshed or otherwise disarticulated before being placed in the urn for burial. Recovered from within the urn were three small, anthropomorphic beads in a sitting

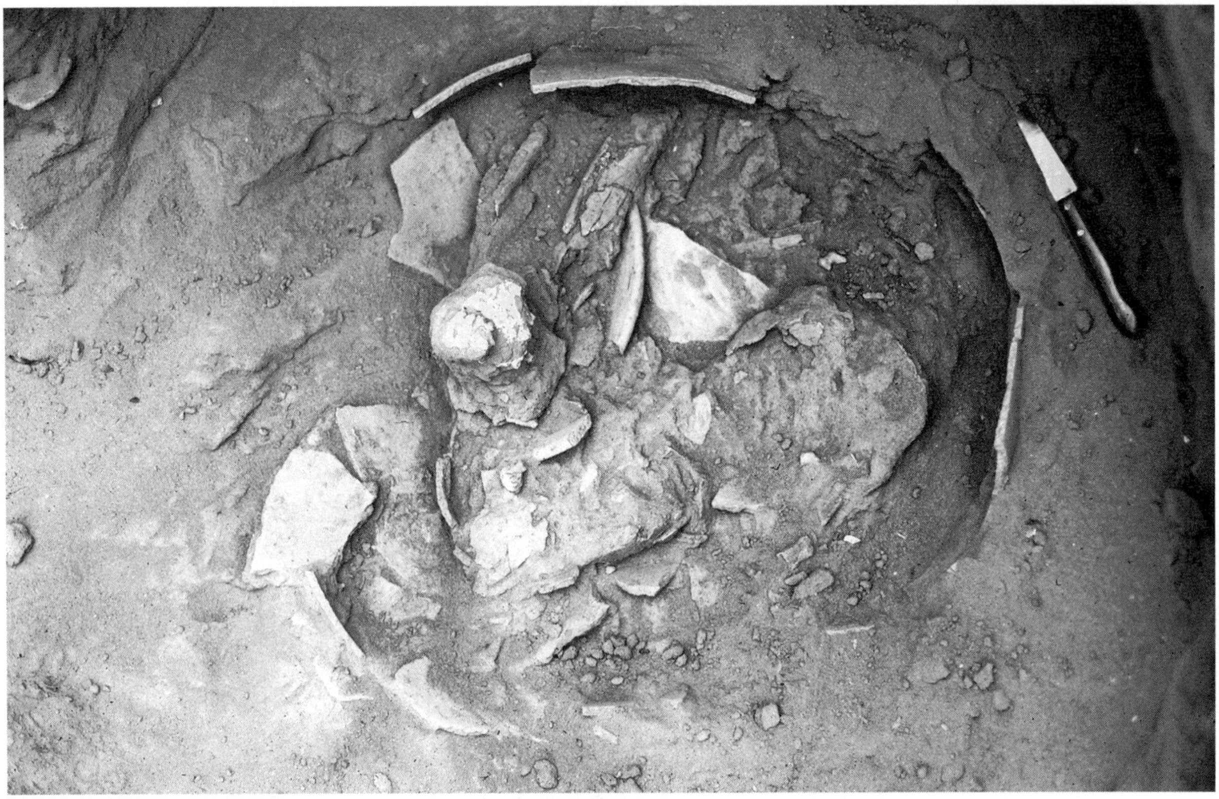

FIGURE 9.7. Exposed contents of Burial 4 (B 4A and B 4B), Mound C, Venadillo.

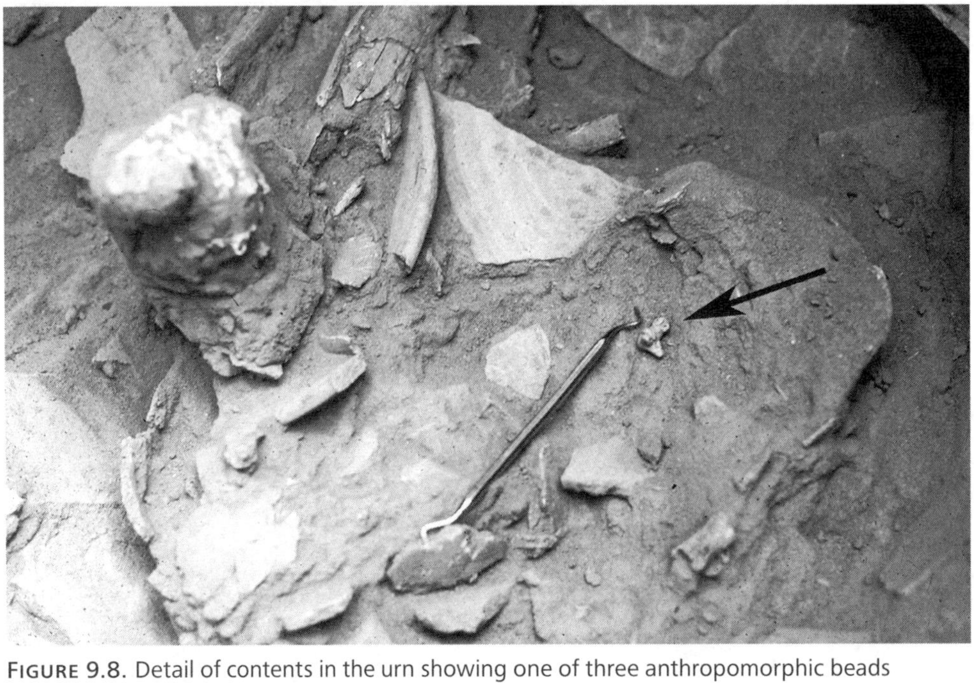

FIGURE 9.8. Detail of contents in the urn showing one of three anthropomorphic beads recovered from within, Burial 4 (B 4A and B 4B), Mound C, Venadillo.

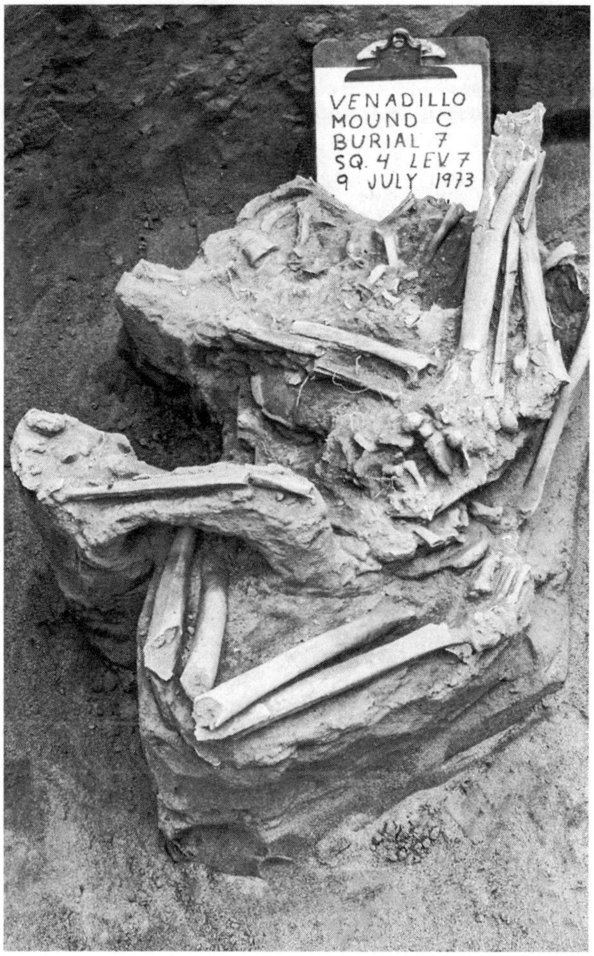

FIGURE 9.9. Possible multiple burial, Burial 7 (B 7A and B 7B), Mound C, Venadillo.

position (Figure 9.8; see Figure 7.5). Each bead had a small notch on the back of its neck, indicating that they may have been strung from a necklace worn by or placed with the deceased. Additionally, the fragmented skeletal remains of a dog were found on or just above the urn in a position indicating that it may have been killed and placed on the urn as a companion for the journey to the afterlife.

In addition to the urn burials, B 7A and B 7B may represent a multiple burial of a young adult male and an old adult female (Figure 9.9). The male remains were incomplete, and the female's were complete, although the bone was poorly preserved. Both were buried in a flexed position, and one appears to have been placed on top of the other.

The orientation of the body could be determined for five of the individuals. One female was oriented to the south, and one, to the north, as well as one male who was oriented to the east. One individual of undetermined sex was oriented to the south, and one was oriented to the southwest.

Most of the 13 Mound C burials lacked funerary offerings. As described above, three anthropometric beads were found with Burial 4, the urn burial containing the remains of an adult female and an infant. It is not clear with which individual the beads, presumably from a necklace, are associated. Additionally a dog appears to have been placed on or near the urn as a guardian or companion. One adult, sex undetermined, had a round ceramic ball rattle found in association as a funerary offering (see Figure 7.15).

The preservation of the skeletal material from Mound C was such that the presence or absence of cranial modification could not be determined for any of the individuals. As for the presence or absence of dental modification, B 2B, the male in the urn burial, exhibited no evidence of dental modification; nor did B 3, an adolescent female. The presence of dental modifications for the remaining 11 burials from Mound C was undetermined. The two burials exhibiting no evidence of dental modification also lacked funerary offerings.

El Mirador

The remains of nine individuals were found at El Mirador, a Middle Conchera period site. Eight burials were recovered from Mound A, and one was recovered from Mound B.

Mound A

Two of the individuals recovered from Mound A were children, five were young adults, and one was a mid-adult. The sexes of the eight individuals from Mound A could not be determined with certainty. Four individuals, including a child, were classified as undetermined; three, as possible females; and one, as a possible male.

Burials 1 and 2 were found within a single burial urn. Both skeletons were badly crushed, and the bone was fragmentary and poorly preserved. Burial 1 was a child, and B 2 appeared to be the remains of a young adult of undetermined sex (probably female). The urn consisted of a redware bowl covered with an inverted hemispherical bowl. Two flexed burials, B 6 (tightly flexed) and B 7, were encountered, along with one extended burial (B 8). Burial type could not be determined for the three remaining individuals.

Body orientation was determined for three of the eight Mound A individuals; however, their sex could not be identified. Burial 6 was oriented to the northwest; B 7, to the west; and B 8, to the north.

No funerary offerings accompanied any of the burials, with two possible exceptions. A metate fragment was found near B 6; two pieces of obsidian were found under and above the right knee of B 8, and a cluster of adobe balls was found about 30 cm above its feet. It is uncertain whether these items are funerary offerings.

FIGURE 9.10. Burial 3A, Cuautla Site 3.

The presence or absence of cranial modification could not be determined for any of the eight Mound A burials. One individual (B 5A) exhibited dental modification; evidence for dental modification for the remaining seven individuals could not be identified.

Mound B

A single burial (B 4) was recovered from Mound B. The remains were highly fragmentary and poorly preserved, consisting only of a few fragments of cranium and long bones. The individual was a young adult, possibly a male. Burial type and orientation of the body could not be determined, and no funerary offerings were present.

Cuautla Site 3

Several sites in the vicinity of Cuautla, Nayarit, approximately 25 km south of the tip of the Teacapán Peninsula, were arbitrarily selected for limited testing to determine their age and cultural associations. The recovered artifacts and human remains date to the Middle Conchera period. Excavations at Site 3, although limited in scale, resulted in the recovery of the remains of six individuals. The remains were either fragmentary or incomplete. Of the six, the sexes of only two, one male and one female, could be determined. Four were young adults, one was a mid-adult, and one was a child.

Burial types for four of the individuals were unidentified because of the fragmentary and scattered nature of the remains. The male (B 4) was buried in an extended supine position, and the female (B 3A) was in a semiflexed position on her right side. Burial 2 was highly fragmentary; the field notes suggest that it may have been a bundle burial. Orientation of the body could not be determined for four of the individuals. The male (B 4) was oriented east, and the female (B 3A) was oriented to the northeast.

Funerary offerings were found with only one individual, the female (B 3A), which was one of the more elaborate burials recovered during the Marismas excavations. Burial 3A was that of a mid-adult buried in a semiflexed position. Although bone preservation was poor, it was possible to document the burial. The individual was placed on her right side with the knees bent at a near 90-degree angle. The right arm was bent with the hand under the skull, and the left arm was straight with the left hand at the knees. The long axis of the body was oriented northeast-southwest, with the head to the north. The burial was resting on coarse beach sand and was covered by a finer beach sand (Figure 9.10).

Immediately next to the skull, near the forehead, was a small red-on-buff jar (see Figure 6.58c). Next to it was an incised spindle whorl made from quartz (see Figure 7.14). Also present were *Olivella* cf. *dama* shells and a medium-size cowry (*Cypraea* sp.) shell. These were also found near the head. Four canid teeth were recovered as well. These had been drilled through the roots and were likely strung on a necklace buried with the individual. On the lower left arm were two shell bracelets (species not specified but likely either *Glycymeris* sp. or *Laevicardium elatum*).

Three of the Cuautla Site 3 burials lacked funerary offerings, including the male. Because of the disturbed nature of the burials, the presence or absence of funerary offerings could not be determined for two individuals of undetermined sex.

Cranial modification also could not be established for any of the six Cuautla Site 3 burials, including B 3A, which featured the female with abundant funerary offerings. Burial 4, the young adult male, exhibited an extreme form (C-4) of dental modification on all four maxillary incisors; no funerary offerings were found in association with this burial.

Discussion

Combined, the 248 individuals recovered during the Marismas project were and are one of the largest samples of human skeletal remains from northwestern Mesoamerica. As can be seen from the summaries above and

TABLE 9.7. Burial Type (Number of Individuals) by Site.

Site	Burial Type									
	Extended	Semiflexed	Flexed[a]	Seated[b]	Urn	Bundle	Mass Secondary Burial	Secondary	Undetermined	Total
Juana Gomez	1					1				2
La Tarjea	4	3	2						5	14
Panales Test Mound A						1				1
Panales Test Mound B	2								3	5
Rincon de Panal		2	2	1		1			1	7
Chalpa	4	8	9	6			10		32	69
Tecualilla	12	4		9	3	3		2	19	52
Cristo Rey		4	16	1	1	3			5	30
Venadillo	1	1	28	3	4	2		1	13	53
El Mirador	1		2		2				4	9
Cuautla	1	1							4	6
Total	26	23	59	20	10	11	10	3	86	248
Percent of total	10.5	9.3	23.8	8.1	4.0	4.4	4.0	1.2	34.7	
Percent of identified burial types (*n* = 162)	16.0	14.2	36.4	12.3	6.2	6.8	6.2	1.9		

[a] This category includes those burials classified as flexed and tightly flexed. Tightly flexed burials occurred at both Tecualilla and Cristo Rey.
[b] This category includes flexed burials in seated positions as well as burials in seated positions with their legs extended.

the following two chapters, the recovery of these remains has provided an array of cultural and bioanthropological data.

However, it is important to remember that the condition of the skeletal material ranged from complete or nearly complete to highly fragmentary and scattered, with the bone itself being in very good condition to highly friable and splintered. Although some mounds produced relatively large numbers of individuals, many of those individuals were in incomplete or fragmentary condition and were disturbed postdepositionally. Considering the frequency with which disturbed and fragmentary remains were encountered, there appears to have been little concern regarding disturbing previously interred individuals. As a result, the sample sizes from which inferences regarding mortuary behavior and treatment can be drawn tend to be small. Cabrero (1995) makes a similar observation regarding the burial assemblages she reviewed in her survey of mortuary practices across west and northwest Mesoamerica. In reviewing this chapter, J. Andrew Darling (personal communication 2014) made the comment that a lot of the bone, both in the Marismas and in west and northwest Mexico in general, is reported in poor condition. He questions whether this was because of manipulation through cleaning, some kind of charneling, bundling, or disturbance of the interred that resulted in skeletal material being pushed aside or taken out and cleaned before reburial.

Burial Types and Treatment

A variety of burial types were identified during the Marismas excavations, including the primary types extended, semiflexed, and flexed and the secondary types bundle, urn, and mass inhumation (Table 9.7). Although a burial type may dominate in a specific mound, typically multiple burial types are represented. The question that arises is whether the presence of multiple burial types represents repeated use of the mound over an extended period—indicating temporal changes in mortuary practices—or differences in kin, social, or religious preferences. Gill (1971, 1974) notes that virtually all the burial types identified were present early, during the Classic period (Early and Late Estero periods). At the time, he noted the lack of urn burials, although several probable urn burials were thought to be present at Tecualilla. Subsequent work at Venadillo (Foster 1974, 1984) identified urn burials in Mound C, which is dated to the latter portion of the Early and Late Estero periods, and at El Mirador, a Middle Conchera period site. Thus, all varieties of burial types identified in the Marismas appear to span the entire prehistoric occupational sequence discussed in this volume.

Extended Inhumations

Extended burials were typically in a supine position with the face up. In some cases the arms were placed with the hands crossed at the wrists over the lower torso, and the feet were crossed. In one case, B 4A from Mound B at Tecualilla, a mid-adult male, appears to have been buried facedown; however, the burial was incomplete, with the torso area missing. Gill (1971:75) observes that the body was in a somewhat "sprawling" posture and that it appeared as if it had been positioned "casually" in a large burial pit.

Extended burials account for 16 percent of those where burial type could be determined. The burial type extends through the Marismas sequence and is represented at all the sites investigated with the exception of Cristo Rey. Gill (1971:67, 95) has long argued that extended burials occur most commonly early and are generally replaced late in the sequence by flexed burials, based on data from Tecualilla, where they are most commonly found in the earliest levels (Early Conchera) (no flexed burials were recovered from the site).

This burial type is widespread across western and northwestern Mesoamerica (Cabrero 1995:152–153). They commonly occur with other burial types, and their relative percentage compared with other burial types varies through time. During the Preclassic period they are the dominant, if not the only, burial type present in mortuary assemblages. Cabrero (1995) further observes that during the Postclassic period, between AD 1100 and 1400, extended burials reemerged as the predominant burial form across west and northwest Mesoamerica, although there are sites where flexed burials are common. This does not appear to pertain to the Marismas, where flexed burials dominated during the Postclassic. Cabrero appears to be well aware of issues regarding sample size and the representativeness of the data she reviewed for her survey of mortuary customs across the region.

Seated Burials

Seated burials account for 12.3 percent of the burials where burial types could be determined. Seated burials include flexed burials placed in a seated position and seated burials where the legs are extended with the arms along the side of the body and the hands next to the femurs or placed in the lap. It is not clear whether these are variations within a single burial type or whether two distinct burial types are represented. Gill (1971:67–74) suggests that the seated burials with their legs extended are variants of the extended burial type. The majority of the seated burials with their legs extended come from Tecualilla. Gill (1974:94) reports that this burial form was also identified at La Isla del Ídolo, Veracruz, but at the time had not been reported elsewhere in Mesoamerica.

Flexed seated burials are discussed as a separate burial type here because they are flexed and easily could be considered a variant of the flexed form of burial. Seated burials were identified at five of the 11 sites from which burials were recovered. A sixth possible seated burial, B 6, was identified at La Tarjea (Late Estero period). Its classification, however, is ambiguous since it is described as possibly upright but was classified as an undetermined burial type. Flexed seated burials are found as late as the Middle Conchera period.

Flexed Inhumations

The degree to which bodies were buried in a flexed position varied from slightly flexed to very tightly flexed. Semiflexed burials account for 14.2 percent of the burial types that could be identified, and flexed burials account for 36.4 percent; together these categories account for nearly a third of the burials. As discussed above, some if not all of the semiflexed burials may be extended inhumations with the knees only slightly bent to fit a grave that was not sufficiently long to accommodate the full length of the body (Gill 1971:75). By the latter Middle Conchera period, his late Postclassic period, Gill (1971: 99, 1974, 1985) asserts that the dominant form of burial is flexed, which coincides with the proposed migration of new peoples from the south, based on evidence for the introduction of new genetic traits into the Marismas populations ca. AD 1200 (see Chapter 11). New arrivals intermarried and interbred with the local population but did not displace it. As well, they introduced burial forms including flexed burial.

The flexing of bodies for burial, as opposed to extending them, not only represents differences in ideology or perhaps personal preference; it may also say something about the practicality of burial and the duration of the mortuary process. Rao (2013) indicates that, based on data from the Bahamas—an environment not all that dissimilar to the Marismas—rigor mortis begins one to two hours after death and takes an additional one to two hours to develop. It lasts 12 to 18 hours in the Jamaican summer and 18 to 24 hours in the winter. Also, if the onset of rigor occurs quickly, it passes quickly; and if delayed, it lasts longer. A number of factors affect the timing and duration of rigor; these include age, nature of the death, muscular condition, and atmospheric conditions. If the body was not immediately placed in a flexed position shortly after death, it is likely that the flexing of the body for burial would have to wait until it came out of rigor. If this was the case, the burial process might have taken several days to complete. However, if the body was flexed before the onset of rigor and the goal was to bury the body while rigid to facilitate handling, the burial process may have been of short duration, with grieving and other mortuary activities following the burial.

Flexed burials appear through the entire Marismas sequence but are most common in the middle to late Postclassic period (Middle Conchera period). At Cristo Rey and at Venadillo Mound B (both date to the Middle Conchera period), flexed burials dominate and are present to the exclusion or near exclusion of extended burials.

Secondary Mass Inhumation

Burial 1 in Mound C at Chalpa is unique among the burials recovered during the Marismas project, consisting of the disarticulated remains of 10 individuals. Gill describes the burial as follows:

> Evidently the bodies within the mass grave had been allowed to reach an advanced state of decomposition before burial, since no two bones were found articulated (except for cranial bones). Of course, the possibility of a reburial cannot be totally discounted; even though, if this were the case, many bones were apparently lost in the process and never reburied. That is, some of the specimens within the burial were found to be exceedingly fragmentary. No grave goods were included. Mandibles were not articulated with the skulls, and some of the bones were found in fragmentary condition. The burial consisted of two levels. The uppermost level contained mandibles, a few skull fragments and numerous long bones and other post-cranial material, completely mixed and packed tightly within an oval 3.0 × 3.5 foot grave. Directly below the assortment of mandibles and post-cranial material, four skulls were found placed in a tight ring, their faces inward and tilted down. No such pattern has been seen anywhere else within the estuary [1971:58–61].

Gill offered no interpretation or explanation of this burial, although he generally discounted the possibility that it represented a reinterment of individuals disturbed during the burial of another individual. He notes that no other burials occurred in proximity to B 1 and that it was unlikely that the burial of a single individual would have disturbed the remains of 10 previously buried individuals. Furthermore, Gill indicates that the mass burial occurred in an area of Mound C that contained few other burials.

One possible explanation comes from Gill's (1971:209) observation that most of the individuals in the mass burial, he emphasizes *most*, exhibit cribra orbitalia (lesions localized on the orbital roof). However, it is not clear whether Gill is referring only to the four skulls or to other cranial fragments found in the mass burial. Cribra orbitalia is seen as typically arising from chronic iron-deficiency anemia. However, recently, researchers (Walker et al. 2009) have associated the condition with a vitamin B12–deficient diet, a condition exacerbated by a lack of animal protein, childhood disease, and poor sanitation. They point out that other factors such as environmental and cultural stresses would further contribute to the lack of vitamin B12, thus aggravating the condition. Gill (1974:93) makes further note of the condition and suggests that these individuals may have been victims of disease and that their remains may have been left exposed for an extended period, resulting in incomplete skeletal presence and poor preservation condition. It is possible that this group of individuals was sufficiently diseased that they were quarantined or otherwise socially circumscribed. As they died, their bodies were placed together and allowed to decompose prior to the burial of what skeletal remains were still available.

Perhaps the four individuals represented by the skulls died at or near the same time. Two were males, one older adult and one young adult; one was a young adult female; and one was a child of undetermined sex. The cause of death is unknown. It is not unreasonable to suggest that the individuals were a family unit; perhaps the young adults were father and mother to the child, and the older adult was a male relative on either the father's or mother's side. It is also possible that the older adult was the father to the three other individuals. If their deaths occurred within a short period of time, perhaps their bodies were laid out to decompose in something akin to a charnel house or family plot, either of which may have been previously used and contained the partial remains of previously deceased individuals. When the time came to collect the remains for burial, the remains of the previously deceased were collected, either inadvertently or deliberately, for interment. Upon interment, the skulls of the four kin, because of their completeness and perhaps out of respect for the head/skull, were placed in the bottom of the grave pit face-to-face, again out of respect and concern for their togetherness in the afterlife. Alternatively, it is possible that all 10 individuals died at the same time and that they were all laid out in the open to decompose prior to burial. Over time it is possible that the bodies were scavenged by animals and that sufficient time passed that some of the skeletalized remains became more decomposed than others, thus when collected not all bodies were complete, as some of the bones were more degraded than others.

Although Gill (1971) does not believe that the remains of the 10 individuals represent reburial of remains disturbed during the burial of other individuals in Mound C, nevertheless it is possible that the remains were recovered from another burial mound and were reinterred in Mound C. It is also possible that the burial represents a ritualistic burial beyond the level of a familial or household unit: a burial that was part of a ritualistic killing of the individuals. However, apparently no evidence of trauma or dismemberment of the bodies was identified. Thus, this burial type remains unexplained, and any explanation is a point of speculation. Nevertheless, bone beds containing a number of human remains have been identified in other parts of northwest Mesoamerica (e.g., Cabrero 1995; Nelson et al. 1992), but none appear to approximate B 1 at Chalpa.

Bundle Burials

Bundle burials are a form of secondary burial and account for 6.8 percent of the identified burial types; in all, 11 bundle burials were recovered from six of the sites investigated. By definition, bundle burials are the packaged remains of an individual. The body required dismemberment, or it would have been allowed to decompose to a point where body parts could be pulled or easily cut from the corpse or the skeleton became nearly or completely disarticulated. The skeletal material would have been collected and bound with cordage or woven material such as cloth or a *petate*, which also decomposed in the ground.

Bundle burials are akin to urn burials in that defleshed and disarticulated remains are packaged for burial. Both represent an investment in time beyond digging a grave. Bodies had to be laid out to decompose, and presumably there was at least some effort to protect the remains to prevent them from being scavenged and lost. This might have included placement of the body in a charnel house or perhaps elevated on a scaffold of some type. Many of the bundle burials were poorly preserved. However, the Panales Test Mound A bundle burial exhibited evidence that it had been cut into pieces as well as probably pulled apart. No such indications were noted for the other bundle burials. Other evidence for the treatment of bundle burials comes from Rincon de Panal, where Taylor and Rautenkranz (1971) report that the soil surrounding the burial was slightly discolored. They suggest that this may have been from the material in which the burial was wrapped when interred.

Bundle burials, although relatively few, appear as a burial type from early to late in the Marismas sequence. They do not appear to be a widespread burial type in west and northwest Mesoamerica (Cabrero 1995).

It is also of note that Schutte (1973:23) reports that "numerous" dog bundle burials were recovered, mostly from the upper levels, from Mounds A and B at Cristo Rey. The bones were fragmented and tightly packed together as if packaged; there was little or no articulation of elements. There is no further discussion or description of the burials beyond Schutte's brief mention, and this is the only report of such dog burials at sites investigated during the Marismas project.

Urn Burials

Urn burials or possible urn burials were found at four sites, Tecualilla, Cristo Rey, Venadillo, and El Mirador. The three urn burials from Tecualilla were in poor condition, and all were recovered from Feature B. The identification of B 1B as an urn burial is very tentative; it was badly disturbed and scattered. The second urn burial, B 1D, contained only a skull and some fragmentary bone. Also found within the urn was a small ceramic bowl. The third urn burial, B 1H, is not described in any detail by Gill (1971) or on the field forms. Gill's (1971:76, 1974:94) general description simply states that each contained adult remains and that the ollas containing the remains did not have dishes inverted over their mouths.

The two urn burials from Mound C at Venadillo, as described above, each contained the remains of two individuals, while the two at El Mirador each contained a single individual. These four urn burials were much like those described at Chametla (Kelly 1938), except the second urn burial at El Mirador lacked a dish inverted over the mouth of the olla containing the remains.

In general, the Marismas burial urns contained disarticulated remains. The remains found in the urns at Tecualilla contained incomplete remains, and it did not appear that the remains had been placed in the urns in an orderly manner. However, the Tecualilla urns were highly fragmentary and disturbed. Burial 2 (B 2A and B 2B) from Mound C at Venadillo was the best preserved of the Marismas urn burials. It contained the remains of a young adult male and young adult female, perhaps a couple or brother and sister. The skeletal material from the two individuals appears to have been placed in the burial container in an orderly and considerate manner. Burial 4 (B 4A and B 4B) from Mound C was badly crushed, and it was not possible to determine how the skeletal material had been placed in the burial urn. The two urn burials from El Mirador were badly crushed as well.

Urn burials are not an unusual burial feature on the West Mexican Coastal Plain. In addition to those found at Chametla, Culiacán, and Guasave before the Marismas project, subsequent discoveries of urn burials have occurred in the region (see Chapter 2). In the Marismas, the number of individuals buried in urn burials made up 4 percent of the population when burial types could be identified (10 individuals in eight burial urns). The burial urns, the actual ceramic vessels holding the remains, appear considerably larger than the largest of the ceramic vessels, utilitarian and decorated, recovered as either whole vessels or sherds from nonburial contexts.

FIGURE 9.11. Burial 21, Mound B, Venadillo.

This suggests that the burial urns themselves were made specifically as repositories for human remains. Whether they were made within the household or by a specialist within the community or obtained through exchange is unknown. Regardless, the urns likely represent an economic or labor investment that is atypical of general mortuary practices in the Marismas.

In contrast the use of burial urns was widespread in the precolumbian Americas, and the urns commonly held cremated remains. Cremation was infrequent or nonexistent among the precolumbian cultures of the west Mexican coast. Rather, urns contained the disarticulated remains of individuals whose bodies appear to have been left, as discussed above, to decompose and skeletonize. It is not clear why urn burial was a chosen alternative to inhumation or even bundle burials. Certainly, the placement of the skeletal remains in an urn subsequent to the defleshing of the body represents a belief regarding the treatment of the remains and the souls of individuals that differs from beliefs regarding the direct interment of the body in the earth. In the cases from Venadillo Mound C, it appears that related individuals were placed in the urns: a male and a female in one case and a female and an infant in another. In these two instances, the goal may have been simply to keep family members together for their passage to the underworld.

Isolated Skulls and Skeletal Remains

The burial of isolated skulls and body parts was not a widespread occurrence among the Marismas burials. An isolated skull, B 18, from Mound A at Tecualilla is one of the few examples. The remains consisted of a skull and mandible. Isolated skulls within burial mounds and burial contexts are known from other sites along the northwest coast of Mexico (e.g., Cabrero 1995; Ekholm 1942; Lister 1955). Ekholm (1942:120) suggested that these may represent trophy skulls. The B 18 skull occurred near B 17, but its association with B 17 is ambiguous, as is its context in general. However, if B 18 is associated with B 17, a female, that may preclude it from being a trophy skull, assuming that a trophy skull is the head of an enemy and that males were more likely to be warriors. It is possible that the skull represents special treatment of a venerated family member or other kin. Although no headless skeletons were recovered from Mound A, there was sufficient disturbance to suggest that B 18 may not necessarily represent the interment of an isolated skull but, rather, the skull may have been redeposited separately when the postcranial skeleton was disturbed.

Burial 23 from Mound B at Venadillo was also recorded as an isolated skull and mandible of a young adult male located directly below Burial 19. The association of the two burials is unclear. Based on the field form and subsequent description of the burial, it is likely that B 23 represents a disturbed burial rather than a trophy skull buried with B 19.

Burial 21 from Mound B at Venadillo was curious. It consisted solely of the lower left arm and hand, the right leg and foot, and the left foot. As can be seen in Figure 9.11, the left arm appears to have been severed at the elbow, leaving the radius and ulna intact. The right leg appears to have been severed just above the knee, and the left foot, at the ankle. No other skeletal elements, fragmentary or whole, that could have been part of B 21 were found. It is possible that, if disturbed by another burial, the left leg could have been removed; however, the lower left arm was undisturbed, and it would seem that some elements of the torso (ribs, spine) should have been present. It is noteworthy that the remains are in relatively correct anatomical position for an individual buried in a semiflexed position on his or her right side. Nevertheless, considering the relative lack of disturbance of the remains present, the burial does not unambiguously appear to have been disturbed by another burial. Additionally, no burial was found in proximity that may have caused any disturbance to B 21.

If this is not a disturbed primary burial, what, then, does it represent? One plausible explanation is that these remains represent the primary burial of severed body parts. If in fact these were severed body parts, the question remains, For what purpose were they taken from an individual? Whether this was done for ritual purposes cannot be determined from the remains themselves. It is also possible that these represent all that was left of an individual postmortem. The three body parts apparently were not decomposed sufficiently to become disarticulated during the burial process. Whatever the explanation, as noted above, one thing that cannot be explained easily is that the remains appear to be all that was left of an individual buried in a semiflexed position on the right side. It is possible that these severed or otherwise dismembered remains were placed in positions that approximated anatomically a semiflexed burial because of prevailing mortuary customs and respect for the individual.

Age

Carr (1995:191) observed that age and social position were often found to be important factors affecting mortuary practices in societies at all levels of sociopolitical complexity. The question is, What was considered old in the precolumbian societies that occupied the Marismas? In his study of mortality rates at various Marismas sites, Gill (see Chapter 10) notes that overall mortality rates for infants (including newborns—0–2 years) and young adults (20–29 years) are the highest, followed by the 50+, 30–39, and 40–49 age groups, respectively. Unfortunately Gill used age categories for his mortality rate study that differ from the categories he (1971) employed to assign an individual to an age group (e.g., infant, child, adolescent, etc.), which are also used in this discussion (see above). Nevertheless, it is clear from his study that the majority of adults died between the ages of 20 and 49 years and that the relative number of individuals who lived beyond 50 years was small.

Weigand and Weigand's (1991) study of death and mourning among modern Huichol Indians of neighboring Jalisco provides a starting point for this discussion. According to Weigand and Weigand (1991:66–68), among the Huichol outward expressions of mourning are brief since the spirits of ancestral humans are close by and influence the course of daily life in ways that benefit both the ancestors and the living. The most immediate gestures of grief are expressed in references to suicide, particularly among women after the death of a mother or child; however, suicides are rarely carried out.

It is also very important for the immediate survivors to understand and agree upon the cause of death so that this can be conveyed in a convincing way to other members of the community. In other words the cause of death affects the pattern of the grieving process (Carr 1995; Weigand and Weigand 1991:66). If the death is accidental or due to natural causes (and unequivocally not associated with sorcery), crying, wailing, sadness, and expressions of suicide are the typical forms of grieving, which are later followed by chants and animal sacrifices during lineage ceremonies. If sorcery is thought to be the cause of death, the typical forms of sorrow can be accompanied by deep anger and threats of vengeance.

Mourning is expressed at two levels, within the lineage and to a lesser degree within the community (the larger sociopolitical unit within which the lineage functioned), and the degree of mourning is expressed in terms of the value of the individual. The Weigands (1991:67) indicate that old people, especially old males with singing and recitation abilities, are deeply mourned unless they had become mentally incapacitated. The significance of such a loss resides in the ceremonial importance of these individuals because of their direct connection with ancestral humans and ancient ancestors, a connection that bonded the individual to the ancestors during a lifetime of religious dedication. The ceremonial value, and to a lesser degree the emotional value, of the very old to the lineage is highly significant. The loss of the very old in terms of their economic value to the lineage is negligible. The loss of old individuals is felt to a lesser degree than for the very old at the ceremonial level as well as the economic and emotional levels. Also of high value within the lineage were mature men and women of working and reproductive age. The Weigands (1991:67) note that in addition to emotional attachments, the economic impact of the loss of such individuals is openly acknowledged. It is during this stage of life when the ceremonial value of individuals, presumably males (although not explicitly stated by the Weigands), starts to develop and become important. Those with no ceremonial or economic value to the lineage and, therefore, mourned the least across the lineage are young children (presumably infants and very young children), even though there is great emotional impact at the loss of the young.

The ceremonial, economic, and to a lesser degree emotional value of the loss of the very old, old, mature, and young to the community parallels that of the lineage. Again, the loss of the very old has a significant impact on the ceremonial component at the community level, while the loss of old individuals significantly impacts the community's political structure. The impact of the loss of the very old on the political structure of the community is notable but less than that of the old, who, inferentially, are more engaged in the community's political structure as established leaders and decision makers. The loss of the old is mourned for their value economically and for emotional reasons, and the loss of the mature is mourned for their political and economic value within the community. The loss of the mature for ceremonial considerations is diminished, as they would have just been developing their value in the ceremonial realm. At the community level the loss of a mature individual is felt, but the impact is limited. The loss of the young at the community level is greatest at the economic level, presumably because they participate in food production. Their loss is also noted in the political realm. The loss of a youth or a child has negligible value for the community in the ceremonial realm, as does the loss of a child in the political, economic, or emotional realms.

Unfortunately, the Weigands (1991) provided little regarding how an individual's value at the lineage and community levels is manifested in the treatment and burial of the dead. As Carr (1995) observed, many of the determinants of the treatment of an individual upon death are manifested archaeologically.

It was possible to estimate the age of 163 of the 248 individuals recovered (Table 9.8). The question posed here is, Were any of the mounds used exclusively or nearly so to bury the dead of a particular age group? The fact that none of the burial mounds were completely excavated and sample sizes are generally small per mound makes it difficult to answer this question. Only four of the sites or mounds produced sufficiently large samples that can be used to assess this question. These are Mounds A and D at Chalpa, Mound A at Tecualilla, and Mound B at Venadillo. Mound A at Chalpa has a comparatively higher number of infants represented, but in general all age groups are represented. All age groups are also represented in Mound D at Chalpa, Mound A at Tecualilla, and Mound B at Venadillo. As cited above, based on Gill's discussion of mortality rates, the distributions of age groups in the four mounds with relatively larger sample sizes are what would be expected (Figure 9.12). This evidence does not suggest that mound burial was limited to old or very old individuals.

If the elderly and very old were venerated socially or politically or both by the family and community within the societies that occupied the Marismas during precolumbian times, is this manifested in how the body was treated in death? As previously discussed, little information is available regarding grave types and any rituals or other activities directly associated with the grave and the body; the only significant exception is B 10 from La Tarjea, where there is evidence of presumed ceremonial fires at the grave. Unfortunately, the skeletal remains of the individual were not assessed for age or sex.

As can be seen in Table 9.9, the three burial types (extended, semiflexed, and flexed) that account for the majority of burials when burial type could be identified are represented in all age groups, with the exception of the fetal or newborn categories. Young, mid-, and old adults were most commonly buried in flexed positions. Here, this is interpreted as being a reflection of the prevailing mode of burial at a specific period in regional prehistory, although family or local beliefs and traditions may have determined the position of the body.

Gill (1971:39) raised the possibility that, based on the excavations at Tecualilla, it was possible that the lower parts of the mounds had been reserved for adolescents and adults, with infants and children being buried nearer the top of the mounds. However, he also suggested that it was possible that there was a simple change in the function of the burial mounds, with the infants and children being added later. Nevertheless, these suggestions remained points of speculation, as they are still.

Regarding infant burials, the majority of infants were buried in an extended position, although infant remains were also recovered as semiflexed, flexed, and bundle burials, as well as one individual being recovered from an urn burial. Gill (1971:57) notes that extended inhumations of infants were generally typical of such burials from the Marismas, and he cites Vaillant for an explanation:

> In the burials of small children and babies, the position of the limbs is of little consequence. The space occupied by the head and body is so that an extended or contracted disposal of the limbs requires little alteration of the pit [Vaillant 1935:181].

Gill (1971:87) further proposed that, although speaking of the presence or absence of funerary offerings, children who had reached two years of age were treated like adults. However, a review of Table 9.9 suggests that, in terms of burial type, infants as well as children were generally treated the same as adults.

Children, adolescents, and adults of all age groups are represented as seated burials, while within this group only adolescents and adults of all age groups are

Table 9.8. Age Group (Number of Individuals) by Site.

Site	Age Group								
	Fetal/ Newborn	Infant	Child	Adolescent	Young Adult	Mid-Adult/ Adult	Old Adult	Undetermined Age	Total
Juana Gomez					2				2
La Tarjea									
Mound A								10	10
Mound B			1					3	4
Panales Test Mound A						1			1
Panales Test Mound B			1	1	3				5
Rincon de Panal					5	2			7
Chalpa									
Mound A	1	6	1	2	7	6	2		25
Mound B							1		1
Mound C	1		2		3	4	1		11
Mound D	3	2	1	4	4	13	5		32
Tecualilla									
Mound A		1	6	2	14	8	7	1	39
Mound B	1		1	2		9			13
Cristo Rey									
Mound A				1	2				3
Mound B			3	1	3	1		1	9
Mound D					2				2
Mound E			3		8	3	2		16
Venadillo									
Mound B	2	2	3	2	15	8	8		40
Mound C		4		1	3	4	1		13
El Mirador, Mound A			2		6	1			9
Cuautla			1		4	1			6
Total	8	15	25	16	81	61	27	15	248
Percent of total	3.2	6.0	10.1	6.5	32.7	24.6	10.9	6.0	100
Percent of identified age groups ($n = 233$)	3.4	6.45	10.7	6.9	34.8	26.2	11.6		100

Note: Newborn, death at or near the time of the birth; infant, 0–2 years; child, 2+ to 12 years; adolescent, 12+ to 18 years for females and 20 years for males; young adult, 18+ to 35 years for females and 20+ to 35 years for males; mid-adult, 35+ to 50 years; and old adult, 50+ years.

sometimes found seated with their legs extended. The age groups infant through old adult are also all represented by bundle burials.

The urn burials clearly represent a departure from "normal" or prevailing burial types. The burial urns, which are the largest class of ceramic vessels produced, themselves represent an investment of material wealth. Additionally, the fact that the remains of the individuals within them were dismembered and/or allowed to decompose before they were collected and placed in the urn indicates additional investment in an extended mortuary process. An infant, a child, three young adults, and four mid-adults/adults were represented in the urn burials. The fact that in one case a young male and female, and in another an adult female and an infant, were found in one urn suggests that age was not a factor in the burial of these individuals. It is also suggested here that, in general, age was not a determining factor affecting whether an individual was buried in any of the urns recovered during the Marismas project.

The secondary mass inhumation identified at Chalpa appears to be an aberrant burial type. It consisted of the remains of a newborn, a child, young adults, mid-adults, and an old adult. What this burial represents is unclear. Based on the fact that five of the seven age groups defined are represented, it does not appear that age was a determining factor in the composition of the burial; however, since it is a secondary burial, it could represent the reinterment of disturbed burials from another location. In the absence of additional information, it is

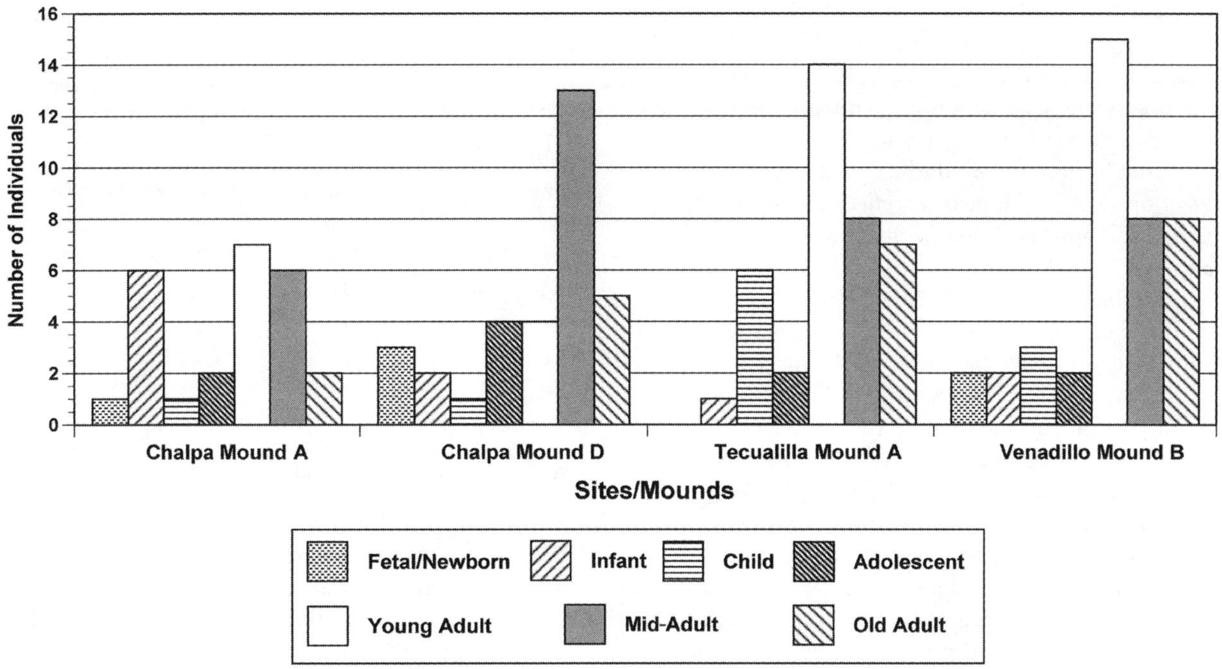

FIGURE 9.12. Age profiles of selected mounds at Chalpa, Tecualilla, and Venadillo.

TABLE 9.9. Age Group (Number of Individuals) by Burial Type.

	Age Group									
Burial Type	Fetal/ Newborn	Infant	Child	Adolescent	Young Adult	Mid-Adult/ Adult	Old Adult	Undetermined Age	Total	Percent of Total
Extended		5	1	1	12	2	2	3	26	10.5
Semiflexed		1	2	2	6	6	2	3	22	8.9
Flexed[a]		1	4	3	28	12	10	2	60	24.2
Seated			1	3	6	5	5		20	8.1
Urn		1	1		3	4		1	10	4.0
Bundle burial		1	1	2	4	1	2		11	4.4
Secondary mass inhumation	1		2		2	4	1		10	4.0
Isolated body part					1				1	0.4
Secondary					2	1			3	1.2
Undetermined	5	9	12	5	17	26	4	7	85	34.3
Total	6	18	24	16	81	61	26	16	248	
Percent of total	2.4	7.3	9.7	6.5	32.7	24.5	10.5	6.4		100

Note: Fetal/newborn, death at or near the time of the birth; infant, 0–2 years; child, 2+ to 12 years; adolescent, 12+ to 18 years for females and 20 years for males; young adult, 18+ to 35 years for females and 20+ to 35 years for males; mid-adult/adult, 35+ to 50 years; and old adult, 50+ years.

[a] Includes both flexed and tightly flexed burials.

possible that age may have been a determinant in the treatment of the individuals at the time of their initial interment.

In summary, prevailing burial practices were likely the primary determinant of how the body was positioned within the grave regardless of age. However, as observed ethnographically for the Huichol, it is reasonable to assume that age and circumstances at death particularly of a loved one influenced the manner and intensity of expressions of grief that may not be immediately apparent in the archaeological record, if at all. While there was variability in body positioning throughout the Marismas sequence, prevailing trends seem to have dictated the practice of extended burials during the Early and Late Estero into the Early Conchera periods, followed by flexed burials in the Middle Conchera period. Body position does not appear to reflect the age of the deceased and most likely follows more widespread cultural patterns.

On the other hand, if the aged were venerated, it might be anticipated that funerary offerings were more commonly buried with older individuals as an indication of their special place within the family or the community. Table 9.10 summarizes the presence or absence of funerary offerings by age group. It does not include the secondary mass inhumation from Chalpa, instances of uncertain association of funerary offerings, or cases where funerary offerings were present but the age of the individual could not be determined. As for the two urn burials from Mound C at Venadillo, each of the two individuals within each urn was counted as having a funerary offering. Also, 12 of the 14 burials from La Tarjea were accompanied by funerary offerings; however, no age determinations were made for any of these burials.

TABLE 9.10. Presence or Absence of Funerary Offerings by Age Group.

Age Group	Cases Present	Cases Absent	Total Cases
Fetal/newborn	1	—	1
Infant	4	11	15
Child	6	9	15
Adolescent	11	2	13
Young adult	41	24	65
Mid-adult/adult	20	1	21
Old adult	15	14	29
Total	98	61	159
Percent of total	62	38	100

Note: Excludes individuals of undetermined age.

Of the 159 cases for which the presence or absence of funerary offerings was determined with an absolute or high degree of certainty, 62 percent of the individuals had funerary offerings, and 38 percent lacked them. Based on Table 9.10, infants and children were less likely to be accompanied by funerary offerings; the fetal/newborn category sample size is too small to be of interpretive value. In his initial study of burials from the Marismas, primarily the sites of Chalpa and Tecualilla, Gill (1971, 1974) noted that none of the infants recovered had associated funerary offerings, although subsequent work resulted in the finding of one infant—B 4B, in the urn burial of an adult female and an infant—with three beads. However, it is important to remember that nearly all remains identified as infants were badly preserved and often disturbed. Gill further notes that children, in terms of funerary offerings, appear to be treated like adults. Nevertheless, a review of Table 9.10 suggests that children are less likely (40 percent) to have associated funerary offerings. This appears to represent a marked increase over infants, where only 27 percent were found with funerary offerings, suggesting that at the family and perhaps the community levels, individuals between the ages of two and 12 years were viewed as having greater social and perhaps economic value.

Eighty-five percent of the adolescents were found with offerings, as were 63 percent of young adults, 95 percent of mid-adults/adults, and 52 percent of old adults. Although this parallels the mortality rates reported by Gill in Chapter 10, that is, there was a greater number of mid-adults/adults, it is not surprising that a higher number of associated funerary offerings is found with this group. Nevertheless, this high frequency of funerary offerings also likely reflects the social and economic value of the individuals both within the family and within the community. As discussed by the Weigands (1991), Huichol adolescents and young adults through time become more socially, economically, and politically important as they become more effective producers and as they take on the social and political responsibilities of their families, kin groups, and community. By the time they ascend to adulthood (mid-adults/adults) these roles reach their apex within the family and community, in terms of their social, economic, and political value. If it is assumed that funerary offerings are a reflection of expected individual achievement, it is not surprising that age-appropriate investment in offerings is consistent with the age of the deceased.

As might also be expected, there is a decline in the frequency with which funerary offerings were placed with the oldest adults. Only 52 percent are found with offerings, while 48 percent were buried without offerings. If, as the Weigands (1991) indicate, the elderly were primarily venerated for their ceremonial value and ancestral or institutional knowledge, the lack of associated funerary offerings may indicate that their value within the family and community decreased as age affected their abilities to participate in these realms. Furthermore, their value economically became negligible as they became more consumer than producer. Thus, the fact that a relatively high number of older adults were buried without funerary offerings may be an expression of their diminished ceremonial, political, and economic capacity and perhaps even emotional value to their families and communities.

TABLE 9.11. Sex (Number of Individuals) by Site.

Site	Sex			
	Male	Female	Undetermined	Total
Juana Gomez	1	1		2
La Tarjea				
Mound A			10	10
Mound B			4	4
Panales Test Mound A			1	1
Panales Test Mound B	1		4	5
Rincon de Panal	3	2	2	7
Chalpa				
Mound A	6	10	9	25
Mound B			1	1
Mound C	3	2	6	11
Mound D	5	8	19	32
Tecualilla				
Mound A	18	7	14	39
Mound B	5	2	6	13
Cristo Rey				
Mound A	2		5	7
Mound B	1		4	5
Mound D			2	2
Mound E	8	5	3	16
Venadillo				
Mound B	15	15	10	40
Mound C	2	5	6	13
El Mirador, Mound A			9	9
Cuautla	1	1	4	6
Total	71	58	119	248
Percent of total	29	23	48	100
Percent of identified sex (n = 129)	55	45		100

Sex

The sexes of 129 of the 248 individuals recovered were determined. This discussion does not include possible or probable assignments or individuals of undetermined sex. Table 9.11 was generated to evaluate the proposition that sex influenced where an individual was buried; in other words, Were there any mounds that were reserved exclusively for males or females? The high number of individuals of undetermined sex within each mound and in some cases small sample size render a definitive conclusion impossible. In most instances, one sex slightly outnumbers the other. The widest disparities between males and females occur in Mound A at Chalpa, where females outnumber males 10 to 6, and in Mound A at Tecualilla, where males outnumber females 18 to 7. The Tecualilla Mound A data are suggestive; the ratio of males to females in the sample for which sex was determined is 2.5:1. However, with a high number of individuals of undetermined sex, the significance of the prominence of males remains uncertain.

The number of individuals for which sex and the presence or absence of funerary offerings could be determined is 108: 58 males and 50 females (Table 9.12). As indicated, both males and females were more commonly buried with funerary offerings than not, and it appears that females were buried with offerings slightly more frequently than males. More specifically, if the data in Table 9.12 are assumed to be at least somewhat representative, it appears that females buried in Mound A at Chalpa, males and females in Mound B at Venadillo, and to a slightly lesser degree males in Mound A at Tecualilla were more typically buried with funerary offerings than not.

There are likely multiple explanations for this—perhaps the most parsimonious being that the inclusion of grave goods was simply a cultural tradition practiced at the family level. This may be reflected in the fact that an individual was more likely to be buried with offerings than not. But if Mound A at Chalpa and Mound A at Tecualilla are examples, the presence of funerary offerings may be linked to sex. Gill (1971, 1974) noted that at Chalpa females appeared to be more commonly buried with offerings than males, while at Tecualilla the opposite was true. As will be discussed further, Gill's observations on the quality or value of the offerings associated with these two groups led him to speculate not only that burials with offerings were higher-"status" burials but that they may reflect local differences in social organization. Specifically, Chalpa was a matrilineal society, while Tecualilla, where male interments included more grave goods of presumably higher material value, was a patrilineal society.

Orientation

Carr (1995) observed that in nearly all of the cases in his cross-cultural study, the primary determinant of body orientation was based in the philosophical-religious realm and often linked to the perceived direction of the afterworld. The orientation of the body for burial was determined for 122 individuals (70 percent) recovered from the Marismas (Table 9.13), as opposed to 126 individuals whose orientation could not be established due to poor preservation or burial type (urns, bundle, seated upright). Burial orientation, like many other aspects of burial practices in the Marismas, exhibits considerable variability. Overall, west was the direction toward which bodies were most commonly oriented (26 percent), followed by east (20 percent), south (16 percent), northwest (14 percent), southwest (12 percent), north (8 percent), northeast (2 percent), and southeast (2 percent).

TABLE 9.12. Sex and the Presence or Absence of Funerary Offerings by Site.

	Sex and Presence/Absence of Funerary Offerings			
Site	Male Present	Male Absent	Female Present	Female Absent
Juana Gomez		1		
La Tarjea [a]				
Panales Test Mound A				
Panales Test Mound B		1		
Rincon de Panal	2	1	1	1
Chalpa				
Mound A	3	3	9	1
Mound B				
Mound C [b]			1	
Mound D	2	1	4	
Tecualilla				
Mound A	11	6	2	4
Mound B			1	
Cristo Rey				
Mound A	3			
Mound B	1			
Mound D				
Mound E	4	4	3	2
Venadillo				
Mound B	10	2	11	4
Mound C		2	2	3
El Mirador, Mound A				
Cuautla		1	1	
Total	36	22	35	15
Percent of sex total (males: $n = 58$; females: $n = 50$)	62	38	70	30

Note: Excludes individuals for which sex or the presence or absence of funerary offerings was not determined with a high degree of certainty.

[a] Eleven of the 14 individuals recovered from La Tarjea were accompanied by funerary offerings; however, the sex of these 14 individuals was not determined.

[b] Excludes the 10 individuals in the mass secondary inhumation from Mound C at Chalpa.

It cannot be known whether orientation at the time of burial was meant to approximate a cardinal direction or whether those burials oriented between cardinal points were oriented purposely. While west was the most frequent orientation, 26 percent of burials were deviations off of west (northwest and southwest), suggesting that a westward orientation could be as high as 52 percent of burials.

Males were buried to either the east or the west 54 percent of the time, 27 percent to the east and 27 percent to the west. Females were most frequently oriented to the west (33 percent); 18 percent were oriented to the south, and 15 percent were oriented to the southwest and northwest each. In all, 66 percent of the females were oriented to the west or in a westerly direction. This perhaps suggests that the burial of females with their heads oriented in a westerly direction was rooted in some pervasive fundamental belief regarding the significance of the west and its relation to females.

The cardinal directions, associated with deities and colors, played significant roles in Mesoamerican ideology (e.g., Mathiowetz 2011; Riley 1963). The east was often seen as the most important because of the rising sun (the sun being reborn) and its association with fertility, birth, rebirth of the soul, creation, light, rain, corn, and spiritual power, with the west being the underworld. However, the old dilemma regarding burial orientation hangs over any interpretation of directionality regarding the orientation of the body: Was the head oriented to the east (facing away from the east) because of the importance of that direction or to the west so that the face was oriented to the east so the individual could see the rising sun? Furthermore, the world of the Mesoamerican, as well as that of neighboring Southwestern peoples, was divided into fourths or into eights. As discussed above, the Marismas burials were oriented in an array of directions, with the cardinal directions apparently being the most important, especially west, followed by east and then south and north.

Patterning in the orientation of the body based on the sex of the individual is reported at other sites in western Mexico. One of the larger samples with good data comes from Tizapan el Alto, located on the south shore of Lake Chapala in the uplands of Jalisco, where 52 burials were recorded; of these 43 (77 percent) were oriented either east or west (Meighan and Foote 1968:44). Interestingly, Meighan and Foote (1968:44) note that most of the burials were within 10 degrees of the respective cardinal direction. Three of the other burials for which orientation could be determined were oriented to the southwest, two were to the southeast, and one was to the northeast. Also, the Tizapan el Alto burials exhibit strong patterning with 19 (86.4 percent) of the 22 individuals identified as males oriented to the east and six (85.7 percent) of the seven individuals identified as females oriented to the west (Meighan and Foote 1968:44). The site is dated between AD 900 and 1250, making it contemporaneous with the Early and Middle Conchera periods of the Marismas. The burials likely date to the Tizapan complex, which Meighan and Foote (1968:36) date to AD 1100–1250.

To the north of the Marismas, a burial mound near Guasave yielded 166 complete and 21 fragmentary skeletons (Ekholm 1942). In a reanalysis of the Guasave mortuary practices, Carpenter (1996) identified, based on Ekholm's unpublished field notes, 196 individuals; 14 of these were only partially excavated or were isolated elements from presumably disturbed burials. In all, Carpenter used data from 182 individuals in his restudy of the Guasave burials; these included 72 burials from the Huatabampo component and 104 burials from the Guasave component of the mound as identified by Carpenter (1996). However, it was not possible to identify the sex of many of these individuals, because of poor skeletal preservation. Of the 46 individuals whose sex was identified for the Huatabampo period, 22 males and 20 females were oriented to the south; one male and one female, to the southeast; and two males, to the west. Overall, when sex was not included, 59 Huatabampo period inhumations were oriented to the south, four were oriented to the southeast, and three were oriented to the west. Sex and orientation were determined for 35 Guasave period individuals; of these, 18 males and 15 females were oriented north, one male was oriented to the northeast, and one female was oriented to the northwest (Carpenter 1996).

By comparison, the Marismas burials exhibit the same general variability in the orientation of burials as seen at Tizapan el Alto but a far greater degree of variability in orientation than found at Guasave. The lack of east/west-oriented burials at Guasave (see Ekholm 1942: Figure 22) is both interesting and perplexing.

Funerary Offerings

As discussed above, when the age of the individual could be determined, funerary offerings were found associated with all age groups, from newborn to old adult, and for the 159 individuals within this sample, funerary offerings were present 62 percent of the time. Offerings were less frequently represented in the categories of newborn, infant, and child and more prevalent in the adolescent through mid-adult categories; they were present with old adults just over half of the time.

Overall, when sex could be identified, funerary offerings were associated with 36 (62 percent) of the 58 individuals identified as males and 35 (70 percent) of the 50 females. As discussed above, there were instances in individual mounds when offerings present either with males or with females appeared to outnumber significantly those with the opposite sex.

Assessing the significance of the funerary offerings associated with burials from the Marismas project is hampered by the lack of detailed descriptions and the lack of consistency in the descriptions of burial assemblages over the life of the project. This is further complicated by the fact that although cumulatively a large number of burials were recovered from the Marismas, many of the sites produced small numbers of individuals. In recent times, investigators have employed a methodology—grave lot values (GLV)—that assigns value to individual artifacts within a funerary assemblage in an effort to discover patterns that indicate differences in ranking and prestige among a site's inhabitants. McGuire's (1987, 1992) study of burials at the large Hohokam site La Ciudad in the Phoenix Basin of central Arizona brought this technique to the forefront of mortuary studies in the American Southwest. Each funerary artifact was evaluated for its context, origin, and labor investment. Subsequently, the technique was employed, supported with a variety of univariate and multivariate statistical techniques, at the Hohokam primate towns of Los Muertos and Pueblo Grande (Brunson 1989; Mitchell 1994b; Mitchell and Brunson-Hadley 2001). Recently, Carpenter (1996) employed the method in his reanalysis of the Guasave burials. He used GLV to the near exclusion of all other statistical techniques with the exception of a limited application of chi-square analysis.

Following Carpenter's study, and perhaps to his dismay, McGuire (2001:40) disavowed the use of GLV as an analytic tool, seeing it as overly simplistic and

Table 9.13. Body Orientation of Burials from the Marismas Nacionales.

Site and Sex	Body Orientation								Total	Site (Mound) Total
	North	Northeast	East	Southeast	South	Southwest	West	Northwest		
La Tarjea										8
Sex unknown	3		2		2			1	8	
Panales Test Mound B										2
Male	1								1	
Sex unknown						1			1	
Rincon de Panal										6
Male			3						3	
Female			1					1	2	
Sex unknown								1	1	
Chalpa										24
Mound A										(13)
Male					1	1	1		3	
Female					1	3	3		7	
Sex unknown						1	2		3	
Mound C										(1)
Female	1								1	
Mound D										(10)
Male					1	2			3	
Female						1		3	4	
Sex unknown	1		1				1		3	
Tecualilla										27
Mound A										(24)
Male	1		2		1		6	3	13	
Female	1				2	1	2		6	
Sex unknown	1		2				2		5	
Mound B										(3)
Male					1	1			2	
Female					1				1	
Cristo Rey										21
Mound A										(4)
Male		1					1		2	
Sex unknown			1		1				2	
Mound B										(2)
Male								1	1	
Sex unknown						1			1	
Mound D										(1)
Sex unknown								1	1	
Mound E										(14)
Male			4		1		3		8	
Female					1	1	2		4	
Sex unknown			2						2	
Venadillo										30
Mound B										(25)
Male			3	1	2		2	3	11	
Female			1	1	2		6	2	12	
Sex unknown			1		1				2	

TABLE 9.13. (cont'd.) Body Orientation of Burials from the Marismas Nacionales.

	Body Orientation									Site (Mound)
Site and Sex	North	Northeast	East	Southeast	South	Southwest	West	Northwest	Total	Total
Mound C										(5)
Male			1						1	
Female	1				1				2	
Sex unknown					1	1			2	
El Mirador, Mound A										2
Sex unknown							1	1	2	
Cuautla										2
Male		1							1	
Female			1						1	
Total	10	2	25	2	20	14	32	17	122	122 (122)
Percent of total (*n* = 122)	8.0	2.0	20.0	2.0	16.0	12.0	26.0	14.0	100	
Total males	2	2	13	1	7	4	13	7	49	
Percent of total males (*n* = 49)	4.0	4.0	27.0	2.0	14.0	8.0	27.0	14.0	100	
Total females	3	—	3	1	7	6	13	6	39	
Percent of total females (*n* = 39)	8.0	—	8.0	3.0	18.0	15.0	33.0	15.0	100	
Total sex unknown	5	—	9	—	6	4	6	4	34	
Percent of total sex unknown (*n* = 34)	14.7		26.5		17.6	11.8	17.6	11.8	100	

Note: Excludes burials for which the orientation of the body could not be determined (including urn burials, secondary mass inhumations, and secondary burials). If, for example, an individual was listed as north-northwest, that individual was classified as northwest in this table.

ignoring the considerable complexity that potentially dictates mortuary treatment. McGuire, ironically, observed that studies subsequent to his own appeared to employ the technique blindly or to the exclusion of other analytic or interpretive approaches. What led McGuire to this epiphany was the recognition that order identified in the data was a construct of the technique rather than the discovery of actual patterning in the data. While a GLV analysis was contemplated for the Marismas data set, understanding the aforementioned limitations of the descriptive data regarding funerary offering assemblages, no statistically based analyses were attempted, and instead, the following empirically based discussion is offered.

A variety of funerary offerings were recovered with the Marismas burials. Ceramic vessels, both decorated (polychromes, incised, bichromes) and plainware, were the most common. Spindle whorls, ceramic smoking pipes, effigy vessels, obsidian blades and projectile points, and ceramic, shell, and animal tooth beads were among the other reported artifact types. Interestingly, items of personal adornment such as necklaces, bracelets, ear- and nose rings, finger rings, and earspools are either rare or not present. Summaries of funerary offerings by burial are also presented in Chapter 5 and online (see above).

Offering Complexity

This discussion focuses on the size (number of items), variability, and quality of funerary offerings recovered from the Marismas burials. The size of funerary offering assemblages ranged from a single item to as many as nine (Table 9.14). Beads likely representing a single necklace were counted as a single item. Of the 111 cases where funerary offerings occur, individuals were buried with a single item 50.5 percent of the time, and burials with two items occurred 18.9 percent of the time (Table 9.15). With an average number of 2.02 items per individual across all Marismas burials with funerary offerings, 69.4 percent (one or two items per burial) of the Marismas burials fall below the average for all burials. Burials with three items account for 16.2 percent, and those with four items account for 11.4 percent. Burials with more than four items account for only 2.7 percent of the burials;

TABLE 9.14. Number of Funerary Offerings by Site.

Site	Number of Individuals with Funerary Offerings	Range of Funerary Offerings per Burial	Total Number of Funerary Offerings	Average Number of Funerary Offerings per Individual
Juana Gomez[a]				
La Tarjea	9	1–3	11	1.2
Panales Test Mound A				
Panales Test Mound B	3	1–9	13	4.3
Rincon de Panal	5	1–3	7	1.4
Chalpa				
Mound A	14	1–4	28	2.0
Mound B				
Mound C	1		2	2.0
Mound D	10	1–4	18	1.8
Subtotal	25	1–4	48	1.9
Tecualilla				
Mound A	18	1–5	41	2.3
Mound B	5	1–3	8	1.6
Subtotal	23	1–5	49	2.1
Cristo Rey				
Mound A	5	1–4	10	2.0
Mound B	5	1	5	1.0
Mound D	1	1	1	1.0
Mound E	7	1–4	15	2.1
Subtotal	18	1–4	31	1.7
Venadillo				
Mound B	22	1–4	53	2.4
Mound C	3	1–2	4	1.3
Subtotal	25	1–4	57	2.3
El Mirador, Mound A[b]	2	1	2	1.0
Cuautla	1	6	6	6.0
Total	111	1–8	224	2.02

[a] Burial 2 at Juana Gomez had small stones partially encircling the skull. An obsidian flake and another small stone were present, although of uncertain association as funerary offerings.
[b] The El Mirador burials were in poor condition, and the association of funerary offerings listed here is uncertain.

there was also one case with five, one with six, and one with nine (0.9 percent each). Another perspective is that burials with one or two items account for 69.4 percent of the cases; burials with three or four items, 27.6 percent; and burials with more than five items, 2.7 percent.

Is the variability seen in the number of funerary offerings significant? Of the 248 sets of human remains recovered, 111 (44.8 percent) individuals (males, females, and individuals of undetermined sex) were found to have associated funerary offerings. Seventy-three of the individuals are reported as not having associated funerary offerings; the remainder (64 individuals—including the 10 individuals in the secondary mass inhumation at Chalpa) include those for which the presence or absence of funerary offerings could not be determined or those with associated artifacts that are not clearly funerary offerings. The data, where the presence or absence of funerary offerings could be determined, suggest that offerings were more likely to be placed in the grave than not, and as discussed above, adults of varying ages are significantly more likely to have associated offerings than are newborns, infants, or children. Again, the presence of funerary offerings is suggestive of social, economic, and ritual or political position within a family or household or within the community. The data, despite their limitation and ambiguity, suggest that at least in death there were "haves" and "have-nots," and among the "haves" there were two groups, those with one to four

Table 9.15. Occurrence of Funerary Offerings per Burial.

Site	Number of Funerary Offerings							Total
	1	2	3	4	5	6	9	
La Tarjea	8		1					9
Panales Test Mound B	1		1				1	3
Rincon de Panal	4		1					5
Chalpa								
Mound A	5	5	3	1				14
Mound B		1						1
Mound D	5	3	1	1				10
Tecualilla								
Mound A	6	6	2	3	1			18
Mound B	3	1	1					5
Cristo Rey								
Mound A	3		1	1				5
Mound B	5							5
Mound D	1							1
Mound E	4		1	2				7
Venadillo								
Mound B	7	4	6	5				22
Mound C	2	1						3
El Mirador	2							2
Cuautla						1		1
Total	56	21	18	13	1	1	1	111
Percent of total	50.5	18.9	16.2	11.7	0.9	0.9	0.9	100

items in their funerary assemblages and those with five to as many as nine items in their funerary assemblages.

Two of the three individuals with five or more offerings, B 3 ($n = 9$) from Panales Test Mound B and B 3A ($n = 6$) from Cuautla, come from very small burial populations. Burial 12C ($n = 5$) from Mound A at Tecualilla is from a mound from which 36 burials were recovered. The Panales Test Mound B and Tecualilla burials are interesting. Panales Burial 3 is that of an adolescent of undetermined sex who was buried with two vessels, six small white beads (counted as a single item), 16 tooth-shaped beads (counted as a single item), and five spindle whorls. Burial 12C was that of a child of undetermined sex buried with two plainware vessels and three spindle whorls. It should be noted that Gill (1971) indicates that the association of the funerary offerings with B 12C is probable but somewhat uncertain. Assuming that the association of the items with B 12C is sound, it is interesting that spindle whorls account for the majority of funerary offerings in each case and that they are associated with a child (B 12C) and an adolescent (B 3). Spindle whorls, when the sex of the individuals can be determined, are exclusively found with females.

Although the number of items with each of these burials is twice as much or more than the average per burial with offerings, the variability in artifact class (e.g., ceramic vessels, other ceramic artifacts, pipes, dog burials, ground stone, obsidian points, obsidian flakes, etc.) is low. In the case of B 3 from Panales Test Mound B the artifact classes represented include ceramic vessels, personal adornment (two, possibly one, necklaces or bracelets), and spindle whorls. In the case of B 12C from Tecualilla, only ceramic vessels and spindle whorls are represented. Burial 3A from Cuautla exhibits the greatest variability, with four artifact classes represented, including a vessel, a spindle whorl, three items of personal adornment (a dog tooth necklace and two shell bracelets), and a whole cowry shell. Other burials with four artifact classes represented in the funerary offerings include B 10A from Mound D at Chalpa, B 30 from Mound E at Cristo Rey, and B 16 and B 26A from Mound B at Venadillo. As can be seen in Table 9.16, two-thirds of the burials from the Marismas only had one artifact class represented in the funerary offerings, and 27 percent had two; only 12 percent had three or four artifact classes represented.

Table 9.17 summarizes artifact classes (or types) by sex at sites in the Marismas. Gill (1971, 1974) argued that at Chalpa the females had more lavish funerary offerings, while at Tecualilla the opposite was true for males. At Chalpa nine different artifact classes are associated with females, while only four were found with males. At Tecualilla, six artifact classes are associated with males, and only two are associated with females. Two other sites, Cristo Rey and Venadillo, have relatively larger samples. At Cristo Rey, five artifact classes are present with both males and females; however, at Venadillo 11 artifact classes are associated with the females, and seven are associated with the males. Overall, ceramic vessels are the most common offering, accounting for 47.3 percent of the individual artifacts recovered as funerary offerings. Spindle whorls were the next most common, followed by smoking pipes and personal adornment

Table 9.16. Number of Artifact Classes Represented in Burials from the Marismas Nacionales.

	Number of Artifact Classes				
Site	1	2	3	4	Total
La Tarjea	10				10
Panales Test Mound B	1		2		3
Rincon de Panal	4	1			5
Chalpa	14	7	3	1	25
Tecualilla	13	10	1		24
Cristo Rey	12	2	2	1	17
Venadillo	16	6		2	24
El Mirador	1	1			2
Cuautla				1	1
Total	71	27	8	5	111
Percent of total	64	24	7	5	100

Table 9.17. Occurrences of Artifact Classes/Types by Sex in Burials from the Marismas Nacionales.

Site and Artifact Class/Type	Male	Female	Sex Unknown	Total
La Tarjea				
Polychrome vessels			4	4
Decorated vessels (bichromes and appliquéd)			2	2
Plainware vessels (plain and monochrome)			1	1
Effigy jars			2	2
Unspecified types of ceramic vessels			3	3
Panales Test Mound B				
Plainware vessels (plain and monochrome)			1	1
Spindle whorls			2	2
Smoking pipes			5	5
Beads/pendants (necklaces and bracelets)			4	4
Ground stone			1	1
Rincon de Panal				
Decorated vessels (bichromes and appliquéd)			1	1
Plainware vessels (plain and monochrome)			1	1
Unspecified types of ceramic vessels			2	2
Smoking pipes			1	1
Beads/pendants (necklaces and bracelets)			1	1
Chalpa				
Decorated vessels (bichromes and appliquéd)	1	5	2	8
Plainware vessels (plain and monochrome)		5	2	7
Unspecified types of ceramic vessels	3	5		8
Spindle whorls		5		5
Smoking pipes	1	4		5
Ceramic stamps		1		1
Beads/pendants (necklaces and bracelets)	1		4	5
Obsidian projectile points		2		2
Flaked stone			1	1
Unmodified stone			1	1
Worked shell		1		1
Animal burials		3	1	4

TABLE 9.17. (cont'd.) Occurrences of Artifact Classes/Types by Sex in Burials from the Marismas Nacionales.

Site and Artifact Class/Type	Male	Female	Sex Unknown	Total
Tecualilla				
Polychrome vessels	1			1
Decorated vessels (bichromes and appliquéd)	4		1	5
Plainware vessels (plain and monochrome)	11	3	4	18
Unspecified types of ceramic vessels	1			1
Spindle whorls		5	3	8
Beads/pendants (necklaces and bracelets)	5		1	6
Obsidian blades/flakes	5		3	8
Animal burials			2	2
Cristo Rey				
Polychrome vessels			1	1
Decorated vessels (bichromes and appliquéd)		1		1
Plainware vessels (plain and monochrome)	2		3	5
Effigy jars				
Unspecified types of ceramic vessels	3	2		5
Spindle whorls		3	4	7
Smoking pipes	2	2		4
Sherds			1	1
Beads/pendants (necklaces and bracelets)			1	1
Obsidian projectile points	1			1
Obsidian blades/flakes		1	1	2
Green-glazed rocks			2	2
Crocodile teeth	1			1
Venadillo				
Polychrome vessels	1	1		2
Decorated vessels (bichromes and appliquéd)	1	1		2
Plainware vessels (plain and monochrome)	7	3		10
Unspecified types of ceramic vessels	10	12		22
Spindle whorls		4		4
Figurines		1		1
Sherds	2		1	3
Beads/pendants (necklaces and bracelets)	1	2		3
Obsidian projectile points		1		1
Obsidian blades/flakes		1		1
Ground stone	1			1
Dog teeth		1		1
Green-glazed rocks		1		1
Ceramic rattles			1	1
Animal burials		1		1
El Mirador				
Flaked stone			1	1
Ground stone			1	1
Adobe balls			1	1
Cuautla				
Decorated vessels (bichromes and appliquéd)		1		1
Spindle whorls (stone)		1		1
Sherds				1
Beads/pendants (necklaces and bracelets)		1		
Shell bracelets		2		2
Whole shells		1		1
Total	65	83	74	222

such as necklaces and bracelets (beads, pendants, pectorals, and shell bracelets).

Polychrome Vessels

Polychrome vessels were only specifically identified in six cases, representing a total of eight specimens (five from individuals of unknown sex, two from males, and one from females). Because of the sample size, inferences regarding occurrence by sex cannot be made. However, these were considered high-value items because of their production costs, and they make up only 7 percent of the ceramic vessels recovered as funerary offerings from the Marismas burials. The polychrome vessels recovered appear to be types or variants of types that are typically found in the area of southern Sinaloa and northern Nayarit; they could have been made locally or acquired through exchange at local markets or subregional or regional centers.

Decorated Ceramics

The category of decorated ceramics is not well defined in burial descriptions or in the field notes. Here it is assumed that it includes painted wares (bichromes), vessels with appliqués, well-polished and incised wares, and perhaps handled vessels or modeled vessels of various forms. These too were likely higher-value items because of production costs. Ceramics classified as decorated account for 17.5 percent of the ceramic vessels recovered. These too could have been made locally or acquired through exchange at local markets or subregional or regional centers.

Plainwares

The plainware category includes primarily utilitarian vessels that were either unslipped or slipped, smoothed, and either not polished or not highly polished. Plainwares account for 36.8 percent of the ceramic vessels recovered from burials. Ceramics of unspecified type or ware, most likely plainwares, account for 36 percent of ceramics recovered from burials. Combined, those ceramics identified as plainwares and those identified as unspecified types or wares account for 73 percent of the ceramics recovered as funerary offerings. Like both polychromes and other decorated vessels, the plainwares could have been made locally or acquired through exchange at local markets or subregional or regional centers.

Effigy Vessels

Only two effigy jars, both from La Tarjea, were recovered as funerary offerings. Not described in detail, one is reported to be a representation of a turtle, and the other, a frog or toad. Again, because of production costs these may have been high-value items. It is of note that both represent aquatic animals; perhaps they hold ritual significance and were tied to either water or rain.

Spindle Whorls

Spindle whorls clearly appear to be a sex-related artifact type when it comes to funerary offerings. In 18 of the 20 cases where spindle whorls were recovered as offerings, all were with females. The two exceptions accompanied the remains of a child and an adolescent, both of undetermined sex. This is likely indicative of women as weavers within the family and community.

Ceramic Smoking Pipes

As discussed, pipes are most commonly found with females. Of the nine cases where they were found as funerary offerings, three were with males, and six were with females. The pipes placed as funerary offerings may suggest a ritual or curing activity or role on the part of the deceased. As discussed in Chapter 7, smoking and breath are commonly associated with rituals carried out to ensure the arrival of clouds and rain-bringing winds (Mathiowetz 2011). The abundance of pipe fragments in nonmortuary contexts suggests that smoking was also a recreational activity as well.

Personal Adornment

There were 23 instances in which items of personal adornment were recovered. Artifact classes or types recovered included shell, ceramic, stone, and dog tooth beads; shell pectorals; and shell bracelets. As noted above, multiple beads and pendants found together likely formed a single necklace or bracelet and were counted as a single item. Because of the lack of detailed descriptions, it is difficult to assess the value of these items and interpret their significance as funerary offerings. These may have been personal ornaments worn by the individuals during their life or personal items presented by family members or other survivors to adorn the body for burial and the afterlife. Among the most elaborate items recovered were two shell bracelets and four dog tooth beads with B 3A from Cuautla.

Obsidian

Obsidian artifacts were recorded as funerary offerings. Five blades or flakes and one projectile were found in association with males, and two flakes or blades and three projectile points were found with females. The apparent association of obsidian projectile points with females in addition to males is interesting. The material would have been nonlocal and likely procured through exchange at a market or with a trader. It is assumed that the obsidian projectile points had material and ideological value because of the lack of locally available volcanic glass and the skill needed to produce them.

Ground Stone

Although present, ground-stone artifacts occur rarely as funerary offerings. Burial 2 at Panales Test Mound B produced a broken metate that appears to be associated as a funerary offering. The individual was likely a female, although sex could not be determined with certainty. This suggests a specific gendered association linking women with the processing and preparation of foodstuffs. A possible metate fragment is also listed as a funerary offering for a burial of undetermined sex from El Mirador. Likewise, Burial 26B from Mound B at Venadillo produced a ¾-grooved ax in association with a male, again suggesting a potentially gendered relationship. Perhaps the investment in time and energy to produce ground-stone implements and their utilitarian nature made them undesirable as funerary offerings. These may have been seen as expensive offerings because of production or acquisition costs. There appears to have been little elaboration within the ground stone industry in the Marismas; there are no ground-stone bowls, effigies, or other such items that may have been viewed as more appropriate funerary offerings than a metate, mano, or ax.

Animal Burials

Dogs and raccoons were deliberately buried in the Marismas burial mounds, and a number of these were found in direct association with human burials, indicating that they had been placed in association as part of the mortuary program. Seven animal burials are listed in association with a human burial, one of which included three animals in a single human burial. Wing (2013) in her extensive study of dog remains from the Marismas cites Sahagun (Anderson and Dibble 1952:41), noting that dogs were important in that they carried the souls of their masters across the place of the nine rivers in the land of the dead. It is interesting that there may have been some perceived equivalence between dogs and raccoons and their potential role in the afterlife.

Miscellaneous Items

Various types of funerary offerings are represented by a single or very limited number of occurrences. These include a cylindrical ceramic stamp, sherds, an unmodified stone, a crocodile tooth, a round ceramic rattle, a whole shell, and the so-called green-glazed rock. These occur as the only item in a funerary assemblage, or they are found in assemblages consisting of several items. Generally, they appear to be low-value items that were added to the grave as at least some token in recognition of the individual's passing.

Summary

In summary, funerary offering assemblages are not particularly large in the Marismas, with four or fewer items making up the vast majority. Of the 248 individuals represented, nearly 45 percent were accompanied by funerary offerings. However, this picture is not necessarily a reflection of reality in terms of the number of individuals actually buried with or without offerings. Many of the burials were disturbed, and many were incomplete. Nevertheless, it appears that two groups of individuals are represented: those with funerary offerings and those without. Ceramic vessels, polychromes, decorated types, and plainwares, were the most common funerary offerings, with nondecorated wares (plainwares and those of unspecified type or decoration) appearing most frequently. As noted above, the polychromes account for about 7 percent of the ceramic vessels recovered as funerary offerings, and other decorated wares, almost 18 percent. Together, these two categories represent 25 percent of funerary ceramics recovered. As discussed in Chapter 6, where tallies comparing decorated wares and plainwares are available, decorated wares represented about 15 percent of the ceramic assemblage recovered from a site or a specific mound. Thus, at 25 percent of the ceramic vessel assemblage from burials, decorated ceramics appear disproportionately frequent compared with their representation in overall site ceramic assemblages. If the decorated wares were considered to be of higher value, their presence may indicate that individuals buried with decorated wares were socially or economically advantaged over others in the community. Ceramic vessels, decorated or plain, dominate the funerary assemblages in terms of investment, and their presence crosscuts all age groups with the exception of newborns, infants, and children; and not surprisingly, they are most commonly found with mid-adults, who represent the largest segment of the population recovered in the burial mounds. The presence of ceramic vessel funerary offerings also crosscuts sex.

Spindle whorls occur only with females, and ceramic smoking pipes are more frequently found with females. Items of personal adornment occur with both males and females, as well as a variety of other artifact types. No one funerary assemblage stands out as exceedingly rich or lavish, nor is there an inordinate number of items in any one assemblage that sets it apart significantly from the others. The items present seem to represent things taken from the household of the deceased or personal items buried on or with the individual. Some appear to represent tokens offered simply as gestures of respect. Most, if not all, of the offerings are of local origin, that is, in use or owned by household members at the time of the deceased's passing, and there is no evidence to suggest that items used as offerings were produced specifically for that purpose.

Funerary Offerings, Cranial Modification, and Dental Modification

If funerary offerings represent some measure of social differentiation in the past life of the deceased, are funerary offerings correlated with the presence or absence of cranial and dental modifications, and if so, might these culturally induced modifications to the human body also be indicative of social standing? Table 9.18 presents a summary comparison of the co-occurrence of funerary offerings and cranial and dental modifications by site and sex. Cases where the sex of an individual could not be determined are excluded, as are cases where the presence or absence of both cranial and dental modifications could not be determined. Thus, this discussion is limited to data from five sites, Chalpa, Tecualilla, Cristo Rey, Venadillo, and a single case from Cuautla. In all, 77 cases—those with funerary offerings and either cranial or dental modification or both—are assessed.

Seventeen cases from Chalpa—11 from Mound A, one from Mound C, and five from Mound D—were suitable for study. Three males from Mound A were found to have both funerary offerings and dental modification, but only one of these also had cranial deformation. The presence or absence cranial modification for two could not be determined. Of the eight females from Mound A, six had funerary offerings and exhibited cranial deformation; two lacked information regarding its presence or absence. Only two females exhibited dental modification, and only one exhibited both cranial and dental modifications. The presence or absence of dental modification could not be established for four of the Mound A females. A single case, a female with funerary offerings and both cranial and dental modifications, was recorded for Mound C at Chalpa.

Five burials from Mound D at Chalpa had evidence of funerary offerings and cranial or dental modifications. One male lacked funerary offerings and evidence of dental modification, but the presence of cranial modification could not be determined. Another male had associated funerary offerings and exhibited cranial deformation but lacked dental modification. All three females in the sample had funerary offerings; two also exhibited cranial modification, but the presence or absence of dental modification could not be determined. The presence or absence of cranial modification was unidentifiable for the third female, and her teeth were not modified.

Overall, at Chalpa 16 (94 percent) of 17 individuals had funerary offerings. These include four (23.5 percent) males and 12 (70.6 percent) females with offerings and one (5.9 percent) male lacking offerings. Of these, 11 (68.7 percent) also exhibited cranial modification, and six (37.5 percent) exhibited dental modification. Five (3.13 percent) individuals exhibited no dental modifications, whereas in three (18.8 percent) cases, both cranial and dental modifications were present. There were only two examples (12.5 percent) where cranial modification was present and dental modification was lacking. No examples were identified featuring dental modification and no cranial deformation. The Chalpa sample suggests a link between the presence of funerary offerings and cranial modification. However, an association between offerings and dental modification does not appear as strong, with six instances in which they co-occur and five in which offerings were present and dental modifications were lacking.

Tecualilla Mounds A and B yielded 22 cases: 15 males and six females from Mound A and one male from Mound B. Of the 10 males with funerary offerings, eight exhibited cranial modification, and of these three exhibited both cranial and dental modification. One of the males with offerings exhibited no cranial or dental modification; two exhibited no cranial modification, but the presence or absence of dental modification could not be determined for them. The presence or absence of dental modification could not be determined for the remaining four with both offerings and cranial deformation. The single male from Mound B with funerary offerings lacked evidence for either the presence or absence of cranial modification but did exhibit dental modification.

Thirteen (59.1 percent) of the 22 individuals in the Tecualilla sample had funerary offerings; nine (40.9 percent) lacked them. These include 11 (50 percent) males and two (9.1 percent) females with offerings and five (22.7 percent) males and four (18.2 percent) females lacking offerings. Of the 13 individuals with offerings, eight (61.5 percent) exhibited cranial modification, and four (30.8 percent), dental modification; there were three (23.1 percent) who had offerings and exhibited both cranial and dental modification. For six cases (66.7 percent) out of the nine that lacked offerings cranial modification was present; and of these one lacked dental modification. Two (22.2 percent) of the individuals lacking offerings also lacked cranial modification; one of these exhibited dental modification, but the presence or absence of dental modification could not be determined for the other. One female lacking funerary offerings also displayed dental modification, but the presence or absence of cranial modification could not be determined; likewise, the single male from Mound B also exhibited dental modification, but the presence or absence of cranial modification could not be determined.

The Tecualilla sample indicates that, at a ratio of 2.25:1, burials with funerary offerings and cranial modification (n = 9, 69.2 percent) occurred more frequently than individuals (n = 4, 30.8 percent) with offerings lacking cranial modification. As at Chalpa, this seems to suggest that individuals with cranial deformation were more likely to receive funerary offerings. The presence or absence of dental modification could not be determined for nearly 60 percent of the Tecualilla sample, and only three (23.1 percent) of the individuals with funerary offerings also exhibited dental modifications. In this instance, however, the same three individuals also exhibited cranial deformation. The limitations of sample size for dental modifications render inferences regarding a link between funerary offerings and dental modification mute, but it is suggestive that those with cranial deformation not only have a higher likelihood of having funerary goods but may also have dental modifications as well.

Cristo Rey yielded a total of 16 cases from Mounds A, B, and E. The two males from Mound A with funerary offerings both exhibited cranial modification; one also exhibited dental modification, and the other's dentition was not modified. The single male from Mound B was accompanied by funerary offerings but lacked any evidence of cranial or dental modification. Thirteen individuals from Mound E were suitable for study: eight males and five females. Two of the males lacked offerings as well as any evidence for cranial or dental modification. Two other males also lacked offerings, but both exhibited cranial modification; one of these also exhibited dental modification, and the other lacked evidence for dental modification. Two males with offerings exhibited cranial modification but lacked dental modification, and the two remaining males with funerary offerings lacked any evidence of either cranial or dental modification.

Ten (62.5 percent) of the 16 individuals in the Cristo Rey sample were accompanied by funerary offerings; six (37.5 percent) lacked offerings. These include seven (43.8 percent) males and three (18.7 percent) females with offerings and four (25 percent) males and two (12.5 percent) females lacking offerings. Four (40 percent) of the individuals with offerings displayed cranial modification, and six (60 percent) lacked cranial modification. Two (33.3 percent) of the six individuals lacking offerings exhibited cranial modifications, while four (66.7 percent) did not. Only a single male with offerings exhibited both cranial and dental modifications. Fourteen (87.5 percent) of the 16 individuals from Cristo Rey lacked dental modification, and its presence for one individual could not be determined. At Cristo Rey the association between funerary offerings and cranial modification appears less than that at either Chalpa or Tecualilla. Based on the Cristo Rey sample as employed here, dental modification appears not to have been as widely practiced as at either Chalpa or Tecualilla, and there is virtually no link among the presence of funerary offerings, cranial modification, and dental modification.

Mound B at Venadillo produced 21 cases, nine males and 12 females, that can be evaluated. One male with offerings exhibited both cranial and dental modifications; three with offerings and cranial modification exhibited no modification of their dentition; and the remaining five with offerings exhibited cranial modification, but the presence or absence of dental modification could not be determined.

Four of the females in the sample lacked offerings; of these, three exhibited cranial modification, but the presence or absence of cranial modification for the fourth could not be determined. One of the four with offerings exhibited cranial modification but lacked dental modification, one for which the presence or absence of cranial modification could not be determined exhibited no dental modification, and the presence or absence of dental modification for the other two could not be determined. Five females with offerings exhibited cranial modifications but no evidence of modifications to their dentition. Two females with offerings exhibited cranial modification, but the presence or absence of dental modifications could not be determined; and one with offerings lacked cranial modification, but the presence or absence of dental modification could not be determined.

Seventeen (81 percent) of the 21 individuals in the Venadillo sample were buried with funerary offerings, and four (19 percent) were not. These include nine (42.9 percent) males and eight (38.1 percent) females with offerings and four (19 percent) females lacking offerings. Of the 17 individuals with offerings, 16 (94.1 percent) displayed cranial modification; and only one lacked any modification. Only one of the individuals with offerings exhibited both cranial and dental modification. Seven (41.2 percent) individuals with offerings and cranial modifications lacked dental modifications, and the presence or absence of dental modifications could not be determined for seven other individuals with offerings and cranial modifications. Of the four individuals lacking offerings, three exhibited cranial deformation; its presence or absence could not be determined for the fourth individual. Dental modifications were present on only one individual with an offering, and that individual also displayed cranial modification. Eight (47.1 percent) of the 17 individuals with funerary offerings lacked dental modifications. The presence or absence of dental modifications could not be determined for 10 individuals; of these seven (70 percent) had offerings and displayed cranial modifications, one (10 percent) had offerings but lacked cranial modification, and two (20 percent) lacked offerings but displayed cranial modification.

The sample from Venadillo suggests a strong link at 94 percent between funerary offerings and the presence of cranial modifications. Few individuals with offerings displayed both cranial and dental modifications, and in

TABLE 9.18. Summary of the Presence or Absence of Funerary Offerings, Cranial Modification, and Dental Modification by Sex in Burials from the Marismas Nacionales.

Site and Burial Number	Sex	Funerary Offerings	Cranial Modification	Dental Modification
Chalpa				
Mound A				
B 6	Male	Present	Unknown	Present
B 12A	Male	Present	Present	Present
B 19	Male	Present	Unknown	Present
B 2	Female	Present	Present	Present
B 3	Female	Present	Present	None
B 5A	Female	Present	Present	Unknown
B 8	Female	Present	Unknown	Present
B 10	Female	Present	Unknown	None
B 15	Female	Present	Present	Unknown
B 18	Female	Present	Present	Unknown
B 20	Female	Present	Present	Unknown
Mound C				
B 2	Female	Present	Present	Present
Mound D				
B 4A	Male	None	Unknown	None
B 12A	Male	Present	Present	None
B 6	Female	Present	Present	Unknown
B 10A	Female	Present	Present	Unknown
B 13A	Female	Present	Unknown	None
Tecualilla				
Mound A				
B 1	Male	Present	Present	Unknown
B 2	Male	Present	None	Unknown
B 3A	Male	None	Present	Unknown
B 7	Male	Present	None	Unknown
B 8	Male	None	Present	Unknown
B 9	Male	Present	Present	Unknown
B 10	Male	Present	Present	Present
B 12A	Male	Present	Present	Present
B 14A	Male	Present	Present	Present
B 18	Male	None	None	Present
B 20	Male	None	None	Unknown
B 21A	Male	None	Present	Unknown
B 22	Male	Present	Present	Unknown
B 23	Male	Present	None	None
B 25	Male	Present	Present	Unknown
B 11	Female	None	Present	Unknown
B 13	Female	None	Present	Unknown
B 17	Female	Present	Present	None
B 19	Female	None	Unknown	Present
B 21B	Female	None	Present	None
B 24	Female	Present	None	Unknown
Mound B				
B 1A	Male	Present	Unknown	Present

Table 9.18. (cont'd.) Summary of the Presence or Absence of Funerary Offerings, Cranial Modification, and Dental Modification by Sex in Burials from the Marismas Nacionales.

Site and Burial Number	Sex	Funerary Offerings	Cranial Modification	Dental Modification
Cristo Rey				
Mound A				
B 6	Male	Present	Present	Present
B 7	Male	Present	Present	None
Mound B				
B 12	Male	Present	None	None
Mound E				
B 17	Male	None	None	None
B 18	Male	None	Present	Unknown
B 19	Male	Present	None	None
B 23	Male	Present	Present	None
B 24	Male	None	Present	None
B 25A	Male	Present	Present	None
B 26	Male	None	None	None
B 27	Male	Present	None	None
B 16	Female	None	None	None
B 22	Female	Present	None	None
B 28	Female	Present	None	None
B 29	Female	None	None	None
B 30	Female	Present	None	None
Venadillo				
Mound B				
B 3	Male	Present	Present	Unknown
B 11	Male	Present	Present	Unknown
B 17A	Male	Present	Present	None
B 18	Male	Present	Present	None
B 19B	Male	Present	Present	None
B 23	Male	Present	Present	Present
B 20	Male	Present	Present	Unknown
B 26B	Male	Present	Present	Unknown
B 27	Male	Present	Present	Unknown
B 2	Female	Present	Present	None
B 4B	Female	None	Unknown	None
B 5A	Female	Present	Present	Unknown
B 6	Female	Present	Present	Unknown
B 7	Female	Present	None	Unknown
B 8A	Female	None	Present	None
B 16	Female	Present	Present	None
B 24	Female	None	Present	Unknown
B 25	Female	Present	Present	None
B 26A	Female	Present	Present	None
B 29	Female	Present	Present	None
B 32B	Female	None	Present	Unknown
Cuautla				
B4	Male	None	Unknown	Present

fact there is only a single case. However, the presence or absence of dental modification could not be determined for nearly half of the sample, and where they could be identified dental modifications were absent from nearly half. This suggests that there is little if any linkage between the presence of funerary offerings and dental modification at Venadillo.

The single individual from Cuautla for which information is available lacked offerings but exhibited dental modifications. Unfortunately, cranial deformation could not be determined on this individual.

Summary

Vertical and Horizontal Ranking

The goal of this chapter has been to assess mortuary behavior as manifested in the Marismas burials for linkages to social organization and ideation (worldview). As discussed above, great effort is often exerted in an effort to identify social persona and reconstruct social organization through the analysis of mortuary evidence and funerary offerings. As Carr (1995:190) notes, the primary dimensions of social persona and social organization that are frequently identifiable in mortuary practices and skeletal remains are age, sex, social position (horizontal and vertical), personal identity, and the circumstantial social classification of the deceased at death.

In his studies of the Marismas burials, Gill (1971, 1974) drew a direct link between social status and the presence or absence of funerary offerings and the presumed richness or elaborateness of those offerings. This was a common approach within the methodological and theoretical paradigm at that time. Funerary offerings were seen as indicators of wealth and status. Although Gill was cognizant that the line between status and the presence or absence of funerary offerings was not necessarily a direct one, he nevertheless thought of those with funerary offerings, especially those with larger and apparently richer mortuary assemblages, as "elites."

Clearly, there were individuals who were buried without accompanying funerary offerings and those who were buried with offerings. Based on the available data it appears that age influenced whether funerary offerings were placed with an individual. Newborns, infants, and children were less likely to receive accompanying grave goods, a trend noted by Gill (1971, 1974). The occurrence of funerary offerings increases with age from adolescents through mid-adults and adults and declines with older adults. However, this is not necessarily significant, in that this trend simply parallels the mortality profile of the individuals buried in the mounds. For example, the more mid-adults present in the assemblage, generally the higher the frequency of funerary offerings.

There is a continuous gradation in the number of funerary offerings from none to as many as nine. Nevertheless, burials with one or two items account for 69.4 percent of the cases; burials with three or four items, 27.6 percent; and burials with more than five items, only 2.7 percent. Most of the funerary offerings were ceramic vessels, and most of these were plain utilitarian wares. Decorated ceramics were present but did not make up a high proportion of the ceramics found in funerary assemblages. In addition, empirically speaking, most of the nonceramic artifacts found as offerings appear to have been domestic or personal items that were offered simply as expressions of bereavement. Although present, items of personal adornment were relatively few.

Two artifact types distinguish themselves from other types of items used as funerary offerings: ceramic smoking pipes and spindle whorls. Smoking pipes occur most frequently with females, and spindle whorls occur exclusively with females. Smoking pipes occurred a third of the time with males and two-thirds of the time with females. The significance of their presence with females over males is unclear, as is their function. As discussed, Mathiowetz (2011) links the lobed pipes of the west Mexican coast to the calling of rain-bearing clouds. It is likely that the presence of pipes indicates socioreligious involvement in ritual activities during the lifetime of the individual. Thus, the distribution of pipes between males and females may be indicative of females' participation or preeminence in rituals related to rain and fertility. Whether these activities extended beyond the limits of the community in which they resided is unknown.

Spindle whorls appear unequivocally as sex-specific items and were certainly important, nonperishable components of weaving tool kits. As many as five spindle whorls were found with a single individual. Their exclusive association with females indicates that weaving was a female task. Mathiowetz (2011), citing McCarty and Matson's (1975:207) translation of the work of Fray Antonio Arias de Saavedra, reports that seventeenth-century observations of the Cora indicate that women typically were buried with items they used in life, including spindles. More recently, an entire weaving kit including a backstrap loom, battens, thread, and spun wool was buried with a Huichol woman (Fikes 2010:175). The presence of this artifact type may signify community-wide as well as family recognition of the economic importance of those individuals buried with spindle whorls.

The Marismas burial data do not indicate the presence of a highly ranked society, although several possible statuses were indicated through the analysis of funerary offerings and mortuary treatment based on age and sex. No clear qualitative differences in funerary offerings or mortuary treatments suggest sharp sociopolitical differences within the Marismas communities investigated; there is no evidence of centralized leadership, nor is there evidence of inherited rank. Funerary offerings generally consisted of ceramic vessels, personal adornment, tools, perhaps some ceremonial-related items (pipes), and miscellaneous items that reflect personal use, gifts or tokens of bereavement, and perhaps membership, prestige, and authority. However, not all people were buried with such offerings. Although there was variability in the distribution of funerary offerings, correspondence of funerary offerings and cranial and dental modifications, burial types, and age- and sex-related treatment, ranking does not seem to have been significant. Instead, these patterns appear to imply that social standing was based on age, and an individual's accomplishments within the family and perhaps within the Marismas community had a greater influence on material expressions of grief at the time of burial. This is not to say that the inhabitants of the Marismas communities were not part of a larger and structured sociopolitical system with a centralized authority. But if this were the case, it was not expressed in the form of localized mortuary practice.

Worldview

There is little doubt that the Marismas populations actively participated in a ceremonial calendar centered on the agricultural cycle and perhaps fishing (see Chapter 12). Burials were or tended to be oriented to the cardinal directions, with the east and the west being equally the most important. This was likely tied to fertility and the rains and the resting place of the soul (e.g., Mathiowetz 2011). The presence of individuals buried with smoking pipes that may have been tied to the making of clouds (smoke) and the bringing of rain (e.g., Mathiowetz 2011) suggests that these individuals had ceremonial responsibilities within the community that, in turn, served to define them personally.

It is noted that several effigy vessels (including a frog or toad and a turtle) were recovered from burials and that these represented species linked to aquatic habitats and water. Little else recovered from the burials overtly conveys any suggestion of ideology, symbolically or technologically.

Ancestor worship or veneration of ancestors may be manifested in the presence of isolated skulls. However, whether these represent actual isolated interments, funerary offerings, trophies, or disturbed burials is unclear. The Marismas examples were not treated with red ocher, nor was only the front half of the skull found, as reported at Guasave (Ekholm 1942:120). Despite the suggestion that the Guasave examples might be trophy skulls, an understanding of their significance remains elusive.

A shift in burial types from predominately extended to predominately flexed burials sometime around AD 1200 is a clear indication of a change in worldview that was accompanied by the apparent arrival of new peoples into the area (Gill 1971, 1974; see Chapter 11). Local ideological shifts notwithstanding (see Mathiowetz 2011), the new arrivals to the Marismas did not extinguish the old ways completely, and earlier forms of burial continued to be practiced. In conclusion, the treatment of the 248 individuals recovered from the Marismas is a reflection of prevailing—and changing—social, political, and religious ideologies within the families and within the larger socioeconomic and political structure in which those families lived their daily lives.

CHAPTER 10

Paleopathology and Cultural Modification of Postclassic Period Skeletal Remains

George W. Gill

The West Mexican Coastal Plain has long been recognized as an important zone of prehistoric occupation in greater Mesoamerica (Gorenstein and Foster 2000; Sauer and Brand 1932). This region also served as a conduit for the movement of both population elements and cultural goods and thus became a link between the American Southwest and tropical, lowland Mexico (Ekholm 1942; Kelley 2000; Riley 2005). The role of this area as a corridor for the movement and interaction of peoples and its constricted nature, being bounded on the east by the Sierra Madre Occidental and on the west by the Pacific Ocean, helped produce a particular population dynamic and genetic history (see Chapter 11). In addition to the information obtained on genetically affected traits and population interactions, study of the human skeletons recovered from the Marismas Nacionales has also yielded insights into patterns in mortality, disease, and injury, as well as cultural modifications of the skeleton (e.g., cranial and dental modifications). It is clear that the recovery and analysis of human remains from the Marismas have greatly enhanced our overall understanding of the region's prehistory, to include not only human adaptation to a coastal-estuarine environment but also prehistoric social organization and worldview.

Most of the data discussed below are derived from the Postclassic (Conchera) period sites of Tecualilla, Chalpa, Venadillo Mound B, Rincon de Panal, and Cristo Rey. The dating, temporal placement, and geographic locations of these sites are presented in Chapters 4 and 5 of this volume. The remains of 248 individuals were excavated during the Marismas project from 1968 to 1974, and the majority of these remains come from those same five sites. In general, the skeletal remains recovered from these sites are not only more numerous but also better preserved than those from the Classic (Estero) period samples. Nevertheless, the information from Classic period sites is discussed when possible. Overall sample sizes when subdivided by sex and age impose some limits on interpretation and discussion, as does a lack of precise temporal association in some instances. So, methodologically, in order to assess trends in the data, samples from multiple sites have been pooled by period, generally the early Postclassic period vs. the late Postclassic period. Sample sizes for each site group, or compilation of site samples, will be listed within each individual table, figure, or discussion section.

Age at Death Distributions

The early Postclassic period sites of Panales (Mounds A and B) and Tecualilla (Early Conchera—Middle Conchera) show similar age at death distributions and are therefore pooled in Figure 10.1 to increase sample size ($n = 44$). The most remarkable finding from these sites is the value for young adult males (ages 20–29). Even though a preponderance of male individuals from Tecualilla Mound A (the site that provides the largest portion of this sample) could bias these results, the very high value for young adults in general from all other Marismas sites suggests that this trend is real. It is significant that the deepest burial levels in Feature A at Tecualilla seem to have been commemorated with a "marker mound" constructed on top of the original mound later in the Postclassic period. This rapidly constructed upper portion of Mound A contained very few burials. Its primary purpose may have been to support a temple or other votive structure. The burials recovered from the lower levels of the Tecualilla Mound A site clearly do not represent a balanced population sample. These burials appear to be the remains of higher-status individuals as indicated by the quality and quantity of grave goods. More than artificially inflating the young adult male death rate, these burial practices seem to have deflated the infant age at death values. The 6.8 percent infant age at death figure shown in Figure 10.1 is about half of that indicated by the other sites. It seems

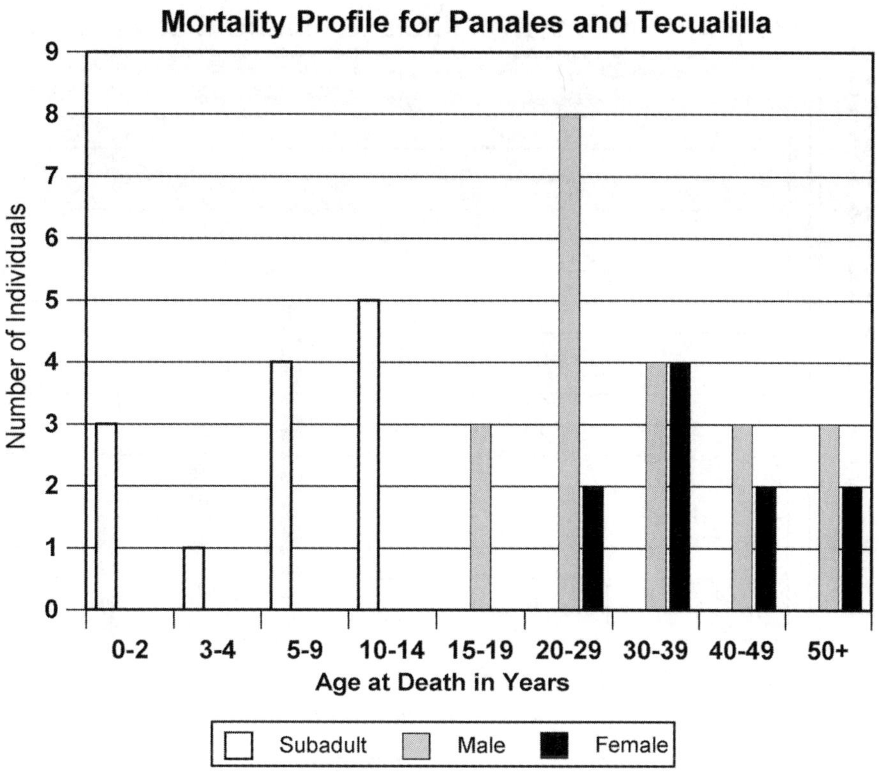

FIGURE 10.1. Mortality profile for Panales and Tecualilla.

that infants were simply not buried within the lower levels at Tecualilla Mound A, where the highest-status burials occur. The few burials of infants and small children at that site were interred somewhat later in the upper, marker mound levels. Milner et al. (2008) note that when such occurrences are encountered archaeologically they constitute one of the many challenges in paleodemography in discerning actual mortality from age at death values.

Another threat to the efficacy of interpreting mortality from age at death values has been known since the early 1980s. Sattenspiel and Harpending (1983) have demonstrated that fertility affects age at death distributions as much as actual mortality does. The higher the population birthrate, the lower the mean age at death. In other words, as stated by McCaa, "High fertility leads to 'young' populations" (2002:95). Particularly average life expectancy calculations are virtually meaningless for archaeological populations if temporal stability (regarding birth, death, migration, etc.) is not known to exist. For Marismas populations many important factors are not known, so life expectancy calculations have not been made. Also, in some cases sample sizes are small. It is believed, however, that variations in individual age at death categories between site populations probably do have meaning.

Much more normal population profiles seem to be represented by the Chalpa mounds and Venadillo Mound B (Figure 10.2). Here, if it is assumed that the presence or absence of funerary offerings reflects differential social standing within the local population, these clearly show a mix of higher- and lower-status burials, balanced sex ratios, and individuals of all ages. The samples from these sites are larger as well (total n = 86). Infant numbers are high (over 20 percent), and the young adult values also remain high. Burial practices had changed by the time these later Postclassic sites were occupied, and it is possible that cultural changes had actually led to a higher infant death rate. It is also possible that cultural changes had merely led to more infants being buried along with the high-status adults.

The most recent Marismas sites of Rincon de Panal (Early Conchera–Middle Conchera) and Cristo Rey (Middle Conchera) show similar mortality curves and have likewise been pooled (Figure 10.3). Even when these two site groups are combined the total sample size is small (n = 29). This may be the reason for a lack of infant specimens, even though two other possibilities that cannot be ruled out are (1) a truly low infant death rate and (2) differential burial treatment regarding infants. The high age at death values for young adults continued at these two sites.

Figure 10.4 shows age at death profiles for all Postclassic period Marismas sites combined (n = 159). The

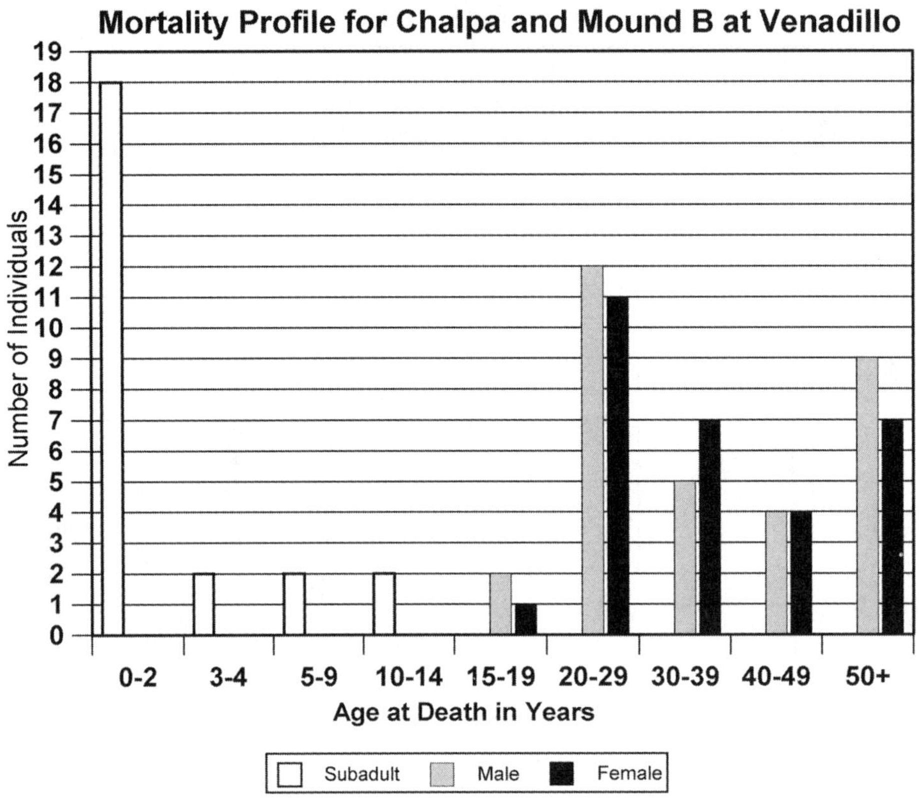

FIGURE 10.2. Mortality profile for Chalpa and Mound B, Venadillo.

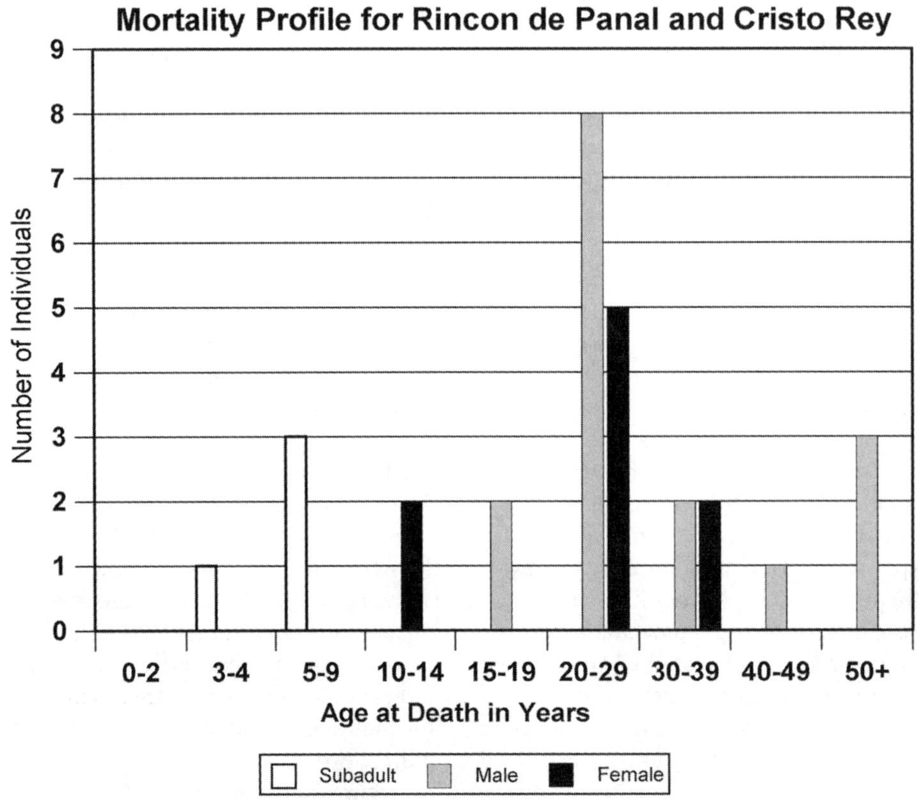

FIGURE 10.3. Mortality profile for Rincon de Panal and Cristo Rey.

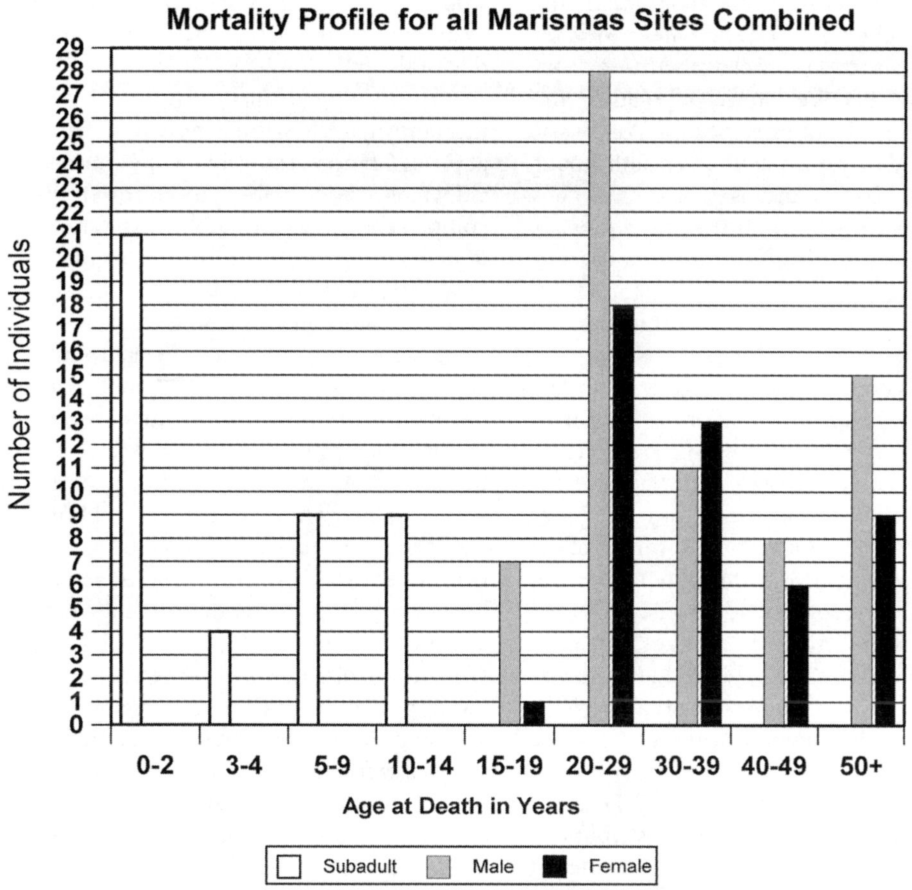

FIGURE 10.4. Mortality profile for Panales, Tecualilla, Chalpa, Venadillo Mound B, Rincon de Panal, and Cristo Rey combined.

infant numbers remain high at 13.2 percent, as does the young adult (20–29 years) death rate of 28.9 percent.

Studies of mortality rates in Mesoamerica, especially in proximity to the Marismas, are limited. Thus even though separated temporally and spatially from our Postclassic Marismas sites, the Preclassic village site of El Arbolillo (900–500 BC) in the Valley of Mexico is of comparative interest, in that it shows a similar mortality profile (Vaillant 1935). The El Arbolillo sample ($n = 58$), for instance, shows an infant age of death figure of 13.8 percent, nearly identical to the Marismas value. If all age categories of children, as shown in Figure 10.4, are combined, this produces a rate among children within the Marismas of 18.9 percent. This figure matches Vaillant's upper value for all of his samples from central Mexico (between 9.8 percent and 18.9 percent).

The place where the prehistoric Marismas Nacionales differs most in age at death from the El Arbolillo sample is in the young adult death rate. Only 15–20 percent of those in central Mexican populations died between the ages of 18 and 35 years, but by comparison fully 28.9 percent of the Marismas young adults died in an even shorter span between 19 and 30 years. The El Arbolillo sample shows a correspondingly higher mortality in the middle age group—roughly 34–52 percent, as opposed to the Marismas value of 23.9 percent for the age groups of 30–39 years and 40–49 years combined.

The percentage of older-aged individuals (50+ years) is 15.1 percent (see Figure 10.4). Once again, this is comparable to the value attained in the El Arbolillo sample (10 percent to nearly 25 percent). It is clearly the difference in the balance between the young adult and middle-aged adult death rates that stands out between the Preclassic period central Mexicans and the Postclassic Marismas populations.

Kreger (2010) in a recent paleodemographic and paleopathological study of 309 Postclassic human skeletons from the Cholula site, a central Mexican urban center in the state of Puebla, found an age at death distribution more divergent from the Marismas values than that at El Arbolillo. At Cholula the infant numbers were quite high, like those in the Marismas and at El Arbolillo, but Kreger found a very low young adult death rate for both males and females. She also found low values for the midlife categories (31–35 years, 36–40 years, and 41–45 years). Her highest age at death values were for older-aged individuals (50+ years of age). This is a very unusual pattern for any prehistoric population and suggests not only a peaceful existence for Postclassic Cholula but also good nutrition and sound health among these precolumbian Mesoamericans.

A large Postclassic site from northwestern Mexico, Casas Grandes (Paquimé) in northern Chihuahua, constitutes a better example for comparison with the Marismas sites than the central Mexican ones just mentioned due to its closer proximity to the Marismas Nacionales. Thus it might be expected to show an age at death distribution more closely aligned to that of the Marismas. Such is the case according to the results of studies by both Benfer (1968) and Di Peso and others (1974). Benfer (1968) in his study of 654 skeletons from the Casas Grandes site found high infant values, as with all of the other studies of precolumbian Mexican populations, but more importantly he discovered additional mortality peaks for both males and females in their 20s, just as we have seen within the Marismas populations. Benfer was surprised that the mortality figures for urban agriculturalists at Casas Grandes compared closely with those from hunter-gatherer populations he had studied from central Texas, 600 mi east of Casas Grandes. He attributes this close similarity in age at death distributions to similar conditions between the two groups in spite of subsistence differences and geographic separation. He points to the possibilities of complications surrounding childbirth in the case of females and warfare in the case of the high mortality for young males.

Di Peso and others (1974:8:326) also discuss mortality at Casas Grandes; they examined mortality by period and also by age and sex categories. They list the infant values at all periods at 13.9 percent, with much lower figures for juveniles and adolescents and then a high value again for young adults (19–35 years) of 20.4 percent. Their charts show a somewhat higher death rate for the mid-adults (36–50 years) than we have found within the Marismas and a correspondingly lower number at the older adult stage (50+ years). Their figure of 6.3 percent of adults over 50 years is noticeably lower than the Marismas value of 15.1 percent.

Di Peso and others also provide a valuable comparison of male vs. female numbers for the different periods, for each of the three adult age categories of young adult, mid-adult, and older adult. In the two prehistoric periods for which they had sufficient specimens for comparison (Viejo period: AD 800–1200; Medio period: AD 1200–1400) their findings are very similar to ours for the Marismas sites. That is, the high young adult death rate at Casas Grandes is even higher for males than for females, as within our Marismas samples, and their somewhat low representation of older adults is fairly even between the two sexes, a situation also similar to that of the precolumbian Marismas.

A possible explanation for the high young adult mortality in the Marismas, as with some interpretations at Casas Grandes, is warfare. For example, Scheiber and Gill (1996) and Scheiber (2006, 2008) note, from their compilation of age at death statistics on Northwestern Plains Indians from various periods, that during times of warfare (as evidenced by projectile points buried in bone) the mortality for young adults (especially males) increases while that for middle-aged adults declines.

Comparative data from west and northwest Mexico beyond Cases Grandes are limited. Nevertheless, much of value has come from the Casas Grandes site over the years, and an important recent study has focused on fragmentary human remains and burials from that site (Casserino 2009). Previous studies and interpretations of the site's skeletal remains, as just mentioned, have invoked speculation regarding warfare and even a massacre (Benfer 1968; Di Peso et al. 1974:2:612–613; LeBlanc 1999:252). Casserino's analysis, on the other hand, has led him to the conclusion that the trauma seen and element representations present are the result of ritual sacrifice and cannibalism. Based on Casserino's discussion, the Casas Grandes data in general, and the condition and treatment of the Marismas skeletal remains, the high rate of mortality seen among young adults does not in general appear to be associated with ritual sacrifice and cannibalism, although there is clear evidence of ritualistic burial of skulls and other body parts (see Chapter 9). At this point it is not entirely clear to what degree warfare, intrasocietal or domestic conflict, birthrate, disease, or even accident produced the age at death distributions observed in the Marismas material.

Skeletal Injury

Skeletal injuries of various kinds are evidenced on the bones from the Marismas Nacionales. Few of these injuries appear to have resulted in death, however, which is perhaps somewhat surprising considering the high mortality among young adults (suggesting at least some warfare). However, a full understanding of the extent of skeletal trauma and its effect on mortality is greatly hampered by the lack of preservation and missing bone elements of many of the individuals recovered.

The Marismas evidence suggests that the earlier Postclassic populations sustained more injury than those from later times. Tecualilla males, for instance, have the highest injury rate, with 53 percent showing some evidence of skeletal injury. Forty percent of the females from that site show somewhat similar evidence. Skeletal injuries were much less common at Chalpa, judging from a frequency of 22 percent of male specimens showing injury and a rate of 27 percent among females. Adult sample sizes are not great at either Tecualilla ($n = 24$) or Chalpa ($n = 32$), but they are sufficient to allow cautious inference. Venadillo Mound B reveals even less trauma, with only five adults (three males, two females) showing skeletal injury (of a total of 27 adults).

The smaller sample sizes from the late sites of Rincon de Panal and Cristo Rey (which are partially contemporaneous) preclude a meaningful comparison with earlier sites that have the largest samples. Five instances of skeletal injury have been noted, however, from those sites (mostly from Rincon de Panal) and consist of a healed cut mark to a fibula (female), a fracture of a left zygomatic (male), rib fractures (male), and two lower leg fractures.

The most dramatic cases, where death appears to have occurred as a direct result of the injuries, come from Tecualilla and Chalpa. An adult male from Tecualilla, Feature B, Burial 1E, shows a depressed fracture of the frontal bone with no signs of healing. The fragmentary nature of this skeleton unfortunately precludes examination for additional evidence of injury.

A young adult male from Chalpa, Feature D (B 4A), shows a fracture of the frontal bone. It had healed only very slightly before death. The crushing blow to the frontal area resulted in an override fracture with an 11-mm-wide angular fragment of bone overlapping the rest. The fracture occurred at a point 25 mm above nasion, with death following before much remodeling of the bone had taken place.

Depressed fractures of the cranium are evidenced in other cases as well, but in these instances the fractures do show signs of healing. From Chalpa, Feature D, a middle-aged male (B 5) shows a slight frontal depression that had healed. An old female from Feature A at Chalpa (B 18) shows a similar frontal depression, also healed and partially remodeled. Radiographic analysis reveals that she had also sustained a fracture to the left tibia, which had healed but with subsequent periostitis. The subsequent remodeling had resulted in a platelike deposit on the surface of the tibia covering nearly one-fifth of the length of the diaphysis. This woman had also sustained a rib fracture, which was in the process of healing at the time of death.

A young adult female from Chalpa, Feature C (B 1E), shows a healed depressed fracture of the skull near point bregma. One male from Venadillo Mound B shows a similar defect to the left frontal area.

An adult male from Tecualilla (B 8) shows two depressed fractures of the posterior left parietal. He also shows a rib partly crushed and healed and periostitis affecting all leg bones. A Tecualilla female (B 24) also shows a healed depressed fracture of the occipital.

Other injuries that would seem to be the result of interpersonal violence are not so common among these Marismas populations, but one such possibility does come from Tecualilla. Burial 18, a young adult male, exhibits a small fracture of the right nasal bone. The injury had healed but with a slight protuberance at the fracture site. An additional possible example of violence is a young adult female from Chalpa (B 20), who shows evidence of multiple injuries. Her sternum had been broken but well mended by the time of death. Periostitis of one of her tibiae could also have been the result of that previous injury.

Two Colles-like "fall" fractures to distal radii have been described among the Marismas samples. An old female from Chalpa, Feature D (B 10A), shows a healed fracture of the left radius with the characteristically reflected distal end. An adult male from Tecualilla, Feature B (B 4A), shows an almost identical fracture (also of the left radius). These are not true Colles' fractures, in that the ulna is not involved, only the distal radius.

A young Tecualilla male (B 10) reveals an unusual injury to the left radius. He suffered the effects of an avulsed ligamentous attachment (a forced detachment) to the proximal end of that bone (Ellis R. Kerley, personal communication 1970). A slight periostitis is also evidenced at the site of injury. He also shows periostitis along the surface of the left tibia.

Another young adult male from the same mound, Mound A, at Tecualilla (B 4) reveals effects of multiple traumas. A crushing injury to a distal hand phalanx, probably during childhood, resulted in the dorsal relocation of the articular surface of the joint. Also, an earlier injury to the distal end of the right fibula resulted in subsequent ankylosis with the tibia, as well as severe infection to the area (roughened bone surface). This individual also shows inflammatory response in the hip and shoulder joints and along one-half of the right ulna.

A female from Tecualilla Mound A (B 13) shows a broken thumb. The proximal phalanx had healed but in misalignment. An old male from the same mound (B 21A) had apparently suffered injury to the sacrum. The caudal tip of the sacrum was broken and healed out of alignment. Another old male from the same site (B 22) shows a previously broken and healed fifth metatarsal. An adolescent male (B 25) from the same mound site shows a healed break of the distal big toe phalanx.

From the adjacent Feature B mound at Tecualilla an adult female skeleton (B 1B) exhibits inflammatory periostitis of the left ankle, evidenced by a buildup of reactive bone on the distal left tibia and fibula. It is likely that this resulted from trauma since lipping is seen on some tarsal bones and the big toe phalanges. These phalanges show a definite osteoarthritis with characteristic eburnation of the joint surfaces. These arthritic changes appear to

TABLE 10.1. Distribution of Carious Permanent Teeth of Children by Tooth Type and Age.

Tooth Type	5–9 Years Age Group			10–14 Years Age Group		
	Number of Teeth	Number of Carious Teeth	Percent Carious	Number of Teeth	Number of Carious Teeth	Percent Carious
Maxilla						
Incisors	3	0	0.0	15	0	0.0
Canines	0	0	0.0	8	0	0.0
Premolars	0	0	0.0	14	0	0.0
Molars	5	0	0.0	16	0	0.0
Subtotal	8	0	0.0	53	0	0.0
Mandible						
Incisors	2	0	0.0	12	0	0.0
Canines	1	0	0.0	7	0	0.0
Premolars	2	0	0.0	15	0	0.0
Molars	5	0	0.0	17	1	5.9
Subtotal	10	0	0.0	51	1	2.0
Total	18	0	0.0	104	1	1.0

TABLE 10.2. Distribution of Carious Permanent Teeth of Adolescents and Young Adults by Tooth Type, Age, and Sex.

Age Group and Tooth Type	Males			Females			Total		
	Number of Teeth	Number of Carious Teeth	Percent Carious	Number of Teeth	Number of Carious Teeth	Percent Carious	Number of Teeth	Number of Carious Teeth	Percent Carious
15–19 Years Age Group									
Maxilla									
Incisors	6	0	0.0	4	0	0.0	10	0	0.0
Canines	5	0	0.0	2	0	0.0	7	0	0.0
Premolars	7	0	0.0	4	0	0.0	11	0	0.0
Molars	8	0	0.0	4	0	0.0	12	0	0.0
Subtotal	26	0	0.01	14	0	0.0	40	0	0.0
Mandible									
Incisors	6	0	0.0	4	0	0.0	10	0	0.0
Canines	4	0	0.0	2	0	0.0	6	0	0.0
Premolars	8	0	0.0	4	0	0.0	12	0	0.0
Molars	8	1	12.5	3	0	0.0	11	1	9.0
Subtotal	26	1	3.8	13	0	0.0	39	1	2.6
Total	52	1	1.9	27	0	0.0	79	1	1.3
20–29 Years Age Group									
Maxilla									
Incisors	69	0	0.0	49	3	6.1	118	3	2.5
Canines	42	4	9.5	33	2	6.1	75	6	8.0
Premolars	89	6	6.7	69	3	4.3	158	9	5.7
Molars	126	16	12.7	77	14	18.2	203	30	14.8
Subtotal	326	26	8.0	228	22	9.6	554	48	8.7
Mandible									
Incisors	75	4	5.3	40	4	10.0	115	8	7.0
Canines	43	1	2.3	33	2	6.1	76	3	3.9
Premolars	89	6	6.7	67	4	6.0	156	10	6.4
Molars	135	25	18.5	80	17	21.3	215	42	19.5
Subtotal	342	36	10.5	220	27	12.3	562	63	11.2
Total	668	62	9.3	448	9	10.9	1,116	111	10.0

have been the result of trauma (Kerley, personal communication 1970).

Very few comparative data on skeletal injury exist within the osteological reports from other west Mexican archaeological sites. The skeletal analyses focus largely upon osteometrics, cranial modification (skull deformation), dental modification (tooth decoration), and certain disease conditions but offer very little information on skeletal injury. Casserino's (2009) study on the evidence for violence at Casas Grandes is a notable exception. Many of his findings, however, do not closely parallel the situation within the Marismas. For instance, a large number of the sets of remains at Casas Grandes were noninterred skeletons, and 52 percent of those show postmortem modification and/or burning of the bone, assumed by Casserino to be related to cannibalism. He describes the damage to the noninterred specimens, as well as to 10 percent of the interred skeletons, as extensive fracturing and splitting of long bones, chipping and abrasions, cut marks to the cranium and postcranial elements (with a focus on articular areas and muscle attachment sites), and burning/charring mainly affecting cranial elements. None of these features has been found on any of the skeletons from within the Marismas Nacionales.

Dental Pathology

Examination of the permanent dentitions of all specimens from Postclassic Marismas sites yields a sample of 2,083 teeth. Table 10.1 shows the results of the study of the subadult children. Even though only 18 permanent teeth are represented in the 5–9 years age group, it is probably significant that none of these teeth reveal carious lesions. Furthermore, only one carious tooth (a molar) is to be found among the 104 preserved teeth of the 10–14 year olds. So, it appears that young children within the Marismas during Postclassic times had healthy teeth.

Things remained about the same during the adolescent years (Table 10.2). Here again only one individual (a male) shows a carious tooth. As in the case of the subadult children it is a mandibular molar.

Conditions clearly changed during the young adult years (20–29 years). The frequency of caries increases from 1.3 percent to 10.0 percent for this age group; however, it does not seem to increase over the next 10 years of life, as suggested by the identical 10.0 percent frequency for the 30–39 years age group (Table 10.3). As expected the older individuals (over 40 years of age) show a significant increase in the occurrence of caries (to 17.5 percent). The fact that the increase is largely within the female sample may have to do with the greater age of some of the female specimens.

The Table 10.4 comparisons tend to suggest that a more important consideration (than either the sex of the person or the age of the adult individual) is the temporal placement of the site group. It is seen that the earlier Postclassic sites of Tecualilla, Chalpa, and Venadillo produce only about half the frequency of caries (8.9 percent with the three site samples combined) as do the later sites of Rincon de Panal and Cristo Rey (18.1 percent with those two site samples combined). Since caries frequencies increase with carbohydrate intake, there may have been a greater reliance on agriculture at some sites, while at others, such as Tecualilla, Chalpa, and Venadillo, the agricultural diet was more heavily supplemented with animal protein. Maybe the larger amount of deer bone found at Tecualilla is meaningful with regard to actual subsistence activities there. Certainly these dental data would tend to support this position.

Table 10.5 shows that the overall caries frequency for the Marismas of 11.0 percent fits well within worldwide agricultural economies of prehistoric times. The only Marismas frequencies that might suggest a mixed economy of agriculture with significant hunting and gathering are those from the earliest Postclassic sites of Tecualilla and Venadillo and perhaps Chalpa as well.

Arthritis, Periostitis, Osteomyelitis, and Cranial Hyperostosis

Arthritis

Osteoarthritis, including both osteophytosis of the spinal column and lipping of the joint surfaces of long bones, is quite common within the skeletal assemblage from the Marismas Nacionales. Apparently rheumatoid arthritis also occurred in prehistoric times judging from the examples of ankylosing spondylitis (and ankylosis of some bones of the feet) as well as cases of the severe destruction of entire joints with reactive pannus occurring before death. Figure 10.5 provides examples of both forms of arthritis.

The percentages of adult skeletons showing arthritic changes differ noticeably among the various Marismas populations. Regarding osteoarthritis the frequency seems to show no particular trend through time but does seem to reflect regional variation. At Tecualilla, for instance, 37.5 percent of the adult skeletons reveal some degree of arthritic change in the spinal column, bony joints, or both. Yet Chalpa shows only 19 percent, and Venadillo, Mound B, 28 percent. Rincon de Panal provides a sample of adult postcranial bones too small ($n = 3$) to say anything (except that all three do show a slight degree of arthritic change). The Cristo Rey sample of nine studiable adults, which shows a 44.4 percent frequency, is likewise too small to allow much confidence in the results.

By contrast the rheumatoid arthritis frequencies appear to show a trend through time. The Tecualilla sample

TABLE 10.3. Distribution of Carious Teeth of Adults by Tooth Type, Age, and Sex.

Age Group and Tooth Type	Males			Females			Total		
	Number of Teeth	Number of Carious Teeth	Percent Carious	Number of Teeth	Number of Carious Teeth	Percent Carious	Number of Teeth	Number of Carious Teeth	Percent Carious
30–39 Years Age Group									
Maxilla									
Incisors	9	1	11.1	14	2	14.3	23	3	13.0
Canines	5	0	0.0	8	0	0.0	13	0	0.0
Premolars	15	3	20.0	15	1	6.7	30	4	13.3
Molars	21	6	28.6	19	0	0.0	40	6	15.0
Subtotal	50	10	20.0	56	3	5.4	106	13	12.3
Mandible									
Incisors	20	0	0.0	11	0	0.0	31	0	0.0
Canines	9	0	0.0	7	1	14.3	16	1	6.3
Premolars	19	0	0.0	14	1	7.1	33	1	3.0
Molars	25	6	24.0	18	2	11.1	43	8	18.6
Subtotal	73	6	8.2	50	4	8.0	123	10	8.1
Total	123	16	13.0	106	7	6.6	229	23	10.0
40+ Years Age Group									
Maxilla									
Incisors	22	0	0.0	24	6	25.0	46	6	13.0
Canines	19	2	10.5	15	6	40.0	34	8	23.5
Premolars	54	8	14.8	32	6	18.8	86	14	16.3
Molars	57	11	19.3	31	7	22.6	88	18	20.5
Subtotal	152	21	13.8	102	25	24.5	254	46	18.1
Mandible									
Incisors	38	1	2.6	25	4	16.0	63	5	7.9
Canines	26	4	15.4	19	1	5.3	45	5	11.1
Premolars	57	9	15.8	34	9	26.5	91	18	19.9
Molars	63	11	17.5	21	9	42.9	84	20	23.8
Subtotal	184	25	13.6	99	23	23.2	283	48	17.0
Total	336	46	13.7	201	48	23.9	537	94	17.5

shows a 29.2 percent rheumatoid arthritis frequency, while Chalpa shows half that (13.6 percent), and Venadillo, Mound B, only 4 percent. The small samples from Rincon de Panal and Cristo Rey provide no examples of rheumatoid arthritis. Since genetics plays a role in the development of rheumatoid arthritis, it is possible that the declining frequencies through time are another indicator of gene flow into the Marismas region (see Chapter 11).

Periostitis and Osteomyelitis

Due to injury and infection the skeletal populations within the Marismas exhibit both periostitis and osteomyelitis. At Tecualilla 23.3 percent of all individuals show at least some periostitis on the long bones. At Chalpa 26.5 percent show the same conditions, and at Venadillo, Mound B, 22.7 percent of the adults reveal either periostitis or osteomyelitis. Of the 22 Venadillo Mound B adults with most of the long bones represented, five show clear signs of infection. Yet only one individual (B 4A), a middle-age male (40–50 years), shows a definite (and severe) case of osteomyelitis. The entire left tibia of this individual was affected by the disease, and almost the entire interior of the diaphysis had formed a giant sequestrum. A thin outer layer of reactive bone is also recognizable on that tibia. This same specimen reveals a rather severe periostitis of the left ulna with slight involvement of the left radius. His left femur also shows an abnormally flattened head, 62 mm in diameter, as opposed to his normal right femoral head of 47 mm. The neck of that femur had shortened as well, to about half normal length (23 mm as opposed to a 42-mm length on the right femoral neck). The abnormal left femur looks to be healthy and normal apart from the head-neck defor-

TABLE 10.4. Summary of Caries Frequency Data by Site.

Site and Sex/Age	Number of Individuals	Number of Teeth	Number of Carious Teeth	Percent Carious
Tecualilla	29	609	50	8.2
Male	16	367	32	
Female	8	167	18	
Young	5	75	0	
Chalpa	33	540	57	10.6
Male	14	244	21	
Female	15	270	36	
Young	4	26	0	
Venadillo	29	442	34	7.7
Male	15	226	22	
Female	14	216	12	
Young	0	0	0	
Rincon de Panal	6	144	21	14.6
Male	4	91	1	
Female	2	53	20	
Young	0	0	0	
Cristo Rey	21	348	68	19.5
Male	10	207	46	
Female	7	92	20	
Young	4	49	2	
Total	118	2,083	230	11.0

mity, which may indicate altered biomechanical stresses as a result of the severe osteomyelitis in the adjacent long bone. A spondylolisthesis (fully detached L5 neural arch) may be another result of abnormal stresses to his hip and lower back.

One other male from Venadillo Mound B (B 18) shows a possible osteomyelitis of the right femur. The diaphysis is quite malformed (flat and wide) when compared with the normal left femur. Of the two Venadillo Mound B females who show periostitis, one of these (B 25) had a more severe case, which affected the right ulna at what appears to have been the site of a former fracture. This acute periostitis may have been progressing to an actual osteomyelitis at the time of death (at about age 29 years).

Of the four adult males from Rincon de Panal with well-preserved long bones, none reveal signs of infection. Of the two females recovered, one (B 6) does show periostitis of the left tibia and fibula following possible fracture.

As mentioned above, some of these cases of periostitis and osteomyelitis within the Marismas population appear to have developed subsequent to bone fractures or other injuries. Other cases show no such connection. The majority affected show rather slight localized periostitis along areas of muscle attachment or elsewhere along the shaft. The tibia, fibula, and ulna are common sites of affliction.

According to Mann and Hunt (2005), periostitis results from inflammation that can be caused by either infectious activity or direct trauma such as a blow to the bone that lifts the periosteum from the bone surface, stimulating bone deposition. Infectious activity can be introduced to the site directly by injury or indirectly by blood-borne infection (Mann and Hunt 2005; White et al. 2012). In the case of the Marismas skeletons it appears that both traumatic episodes and blood-borne disease causes were involved.

Osteomyelitis is a more serious condition of the bone. It involves deep bone infection (even the marrow cavity) and generally results from a deep penetrating injury or less often a serious blood-borne infection (Mann and Hunt 2005; Ortner 2003; White et al. 2012).

Cranial Hyperostosis

A quite common condition among adult specimens from all Marismas sites is cranial hyperostosis, a thickening of the cranium. Figure 10.6 provides an example of this condition. Various degrees of cranial thickening occur, and even though mid-parietal metrics were collected on

TABLE 10.5. Comparative Caries Frequencies.

Population	Location	Context	Number of Teeth	Percent Carious
Examples of Hunting and Gathering Societies				
Aborigine	Australia	Skeletal	10,561	1.6
Northwest Coast	Alaska and British Columbia	Skeletal	5,500	0.4
Blossom site (CA-SJO-68)	Central California	1000 BC	289	2.4
Indian	California	Skeletal	438	1.6
Northwest Plains Indian	Wyoming and Montana	Skeletal	704	2.6
Examples of Mixed-Economy Societies				
Jomon	Central Japan	1000 BC	1,377	8.6
Papuan	New Guinea	Living	6,142	3.8
Polynesian	Pacific Islands	Skeletal	1,166	1.7
Polynesian	Paupahan	Living	11,072	9.3
Central Plains Indian	Nebraska	Skeletal	2,445	5.3
Examples of Agriculturally Based Societies				
Mongoloid	China, Japan, Korea	Skeletal	826	4.1
Mongoloid	China	Living	129,634	7.7
Mongoloid	Central China	Living	20,778	2.1
Chinese	China	Living	95,830	2.5
Yayoi	Japan	200 BC	1,206	19.7
Japanese	Japan		76,893	18.0
South America	Bolivia etc.	Skeletal	3,295	6.3
Pachacamac	Peru	Skeletal	2,727	15.5
Peruvian	Peru	Skeletal		15.4
Maya (Xcambó, Mexico)	Northern Yucatán	AD 250–550		
Males			189	7.4
Females			208	21.2
Maya (Xcambó, Mexico)	Northern Yucatán	AD 550–750		
Males			1,234	14.0
Females			1,089	27.4
Ayalan	Ecuador	500 BC–AD 1730	2,395	10.1
Easter Island	Pacific	AD 1600–1800	1,019	27.1
Mexican Indian[a]	Coastal west Mexico	AD 1000–1350	2,083	11.0

Source: All data, apart from the present study, from Cucina et al. 2011; Owsley et al. 1985; Turner 1979; Zitt 2008.
[a] Data from the present study.

some adult specimens (which appeared to show thickening beyond normal), in most instances measurements were taken only at the right and left asteria. Most range from 8 mm to 12 mm in thickness, with normal crania generally measuring 6 mm or less. No infants or children show the condition.

Cranial hyperostosis has been observed at other prehistoric sites in Mexico, but the precise cause or causes are unknown. In all cases the cranial bones appear to be uniformly thickened throughout due to hyperostosis of the diploe, without detectable thickening of the inner or outer tables (Gill 1971). This same condition has been reported by Gavan (1949) among members of the Toliman complex from Jalisco and by Johnston (1968) on a specimen from Tikal. Also, in 1969, while at the American Museum of Natural History in New York City I examined 23 skulls collected by Ekholm (1942) from Guasave, Sinaloa. Three of these clearly show thickened crania.

In 1970 Ellis R. Kerley examined several of the cranial hyperostosis cases from the Tecualilla and Chalpa sites. It was his impression that these might possibly represent normal, nonpathological crania. The entirely normal appearance of the skulls and the continuous nature of the trait, ranging from normally thin skulls to very thick, with varying degrees between, led him to at least consider this possibility (see Gill 1971:202). Based upon examination of the Tikal specimen, however, both

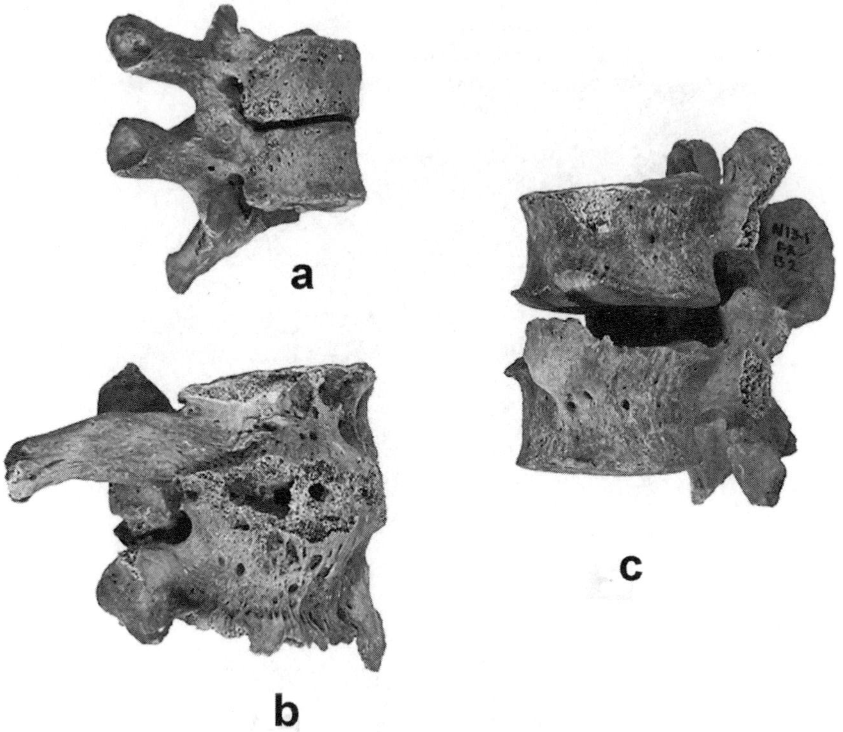

FIGURE 10.5. Examples of arthritic vertebrae from Feature A, Tecualilla. The (*a*) thoracic vertebrae and (*b*) lumbar vertebrae of the Burial 22 male exhibit what is most likely rheumatoid arthritis with ankylosing spondylitis. (*c*) The lumbar vertebrae of the Burial 2 male show a slight, probably subclinical osteoarthritis (with osteophyte).

Johnston (1968) and Lawrence Angel of the Smithsonian Institution preferred a pathological explanation for the condition. They did reject thalassemia as a cause, however, as do I with regard to these Marismas cases. Thalassemia is an inherited chronic anemia resulting from faulty hemoglobin production. It causes hyperostosis of the cranial diploe but produces an abnormal appearance to the basic structure of the diploe not seen in the Marismas cases. Talavera and Gelover (2008) in their study of Aztatlán skeletons from southern Sinaloa and northern Nayarit encountered cranial hyperostosis and attribute it to dietary deficiencies, most likely a lack of iron absorption. Lack of vitamin A, C, or D can contribute, as can diarrhea, which, according to these authors, impedes the assimilation of iron.

Within the skeletal populations of the Postclassic Marismas Nacionales, males tend to show a higher incidence of cranial hyperostosis than females. At Tecualilla 58.5 percent of males show thickening, while only 12.5 percent of females show this condition. At Chalpa 76.9 percent of males exhibit hyperostosis, as compared with 46.7 percent of females. Venadillo Mound B seems to show less sexual dimorphism, with 37.5 percent of males and 33.3 percent of females exhibiting thickened crania. The condition of the skeletal remains at Venadillo, however, allowed only 20 skulls to be reliably evaluated. Fewer yet could be accurately assessed at Rincon de Panal ($n = 3$) and Cristo Rey ($n = 5$), where frequencies of 33.3 percent and 60.0 percent were found, respectively, with sexes combined. Small sample sizes from the late sites simply prohibit assessment of any possible temporal trends over the Postclassic period.

Meighan's (ed. 1976) work at Amapa, Nayarit, a Postclassic site near the Marismas Nacionales, produced fragmentary human remains of 10 individuals. Four of these specimens have preserved crania that allowed assessment of cranial hyperostosis and other pathological conditions, as well as other cranial characteristics. Bernor (1976) conducted a basic osteological inventory and assessment of the Amapa sample, and he and I at different times examined four of the better-preserved skulls from the series. These four were shipped to my laboratory for study by Meighan (Gill 1976). Within this small sample Bernor and I both independently identified a single case of cranial hyperostosis (B 1, No. 3701). The individual, a male of approximately 35 years of age, reveals a slight but unmistakable thickening of the cranial diploe. The type of cranial thickening in this case is identical to the examples identified within the Marismas skeletal series.

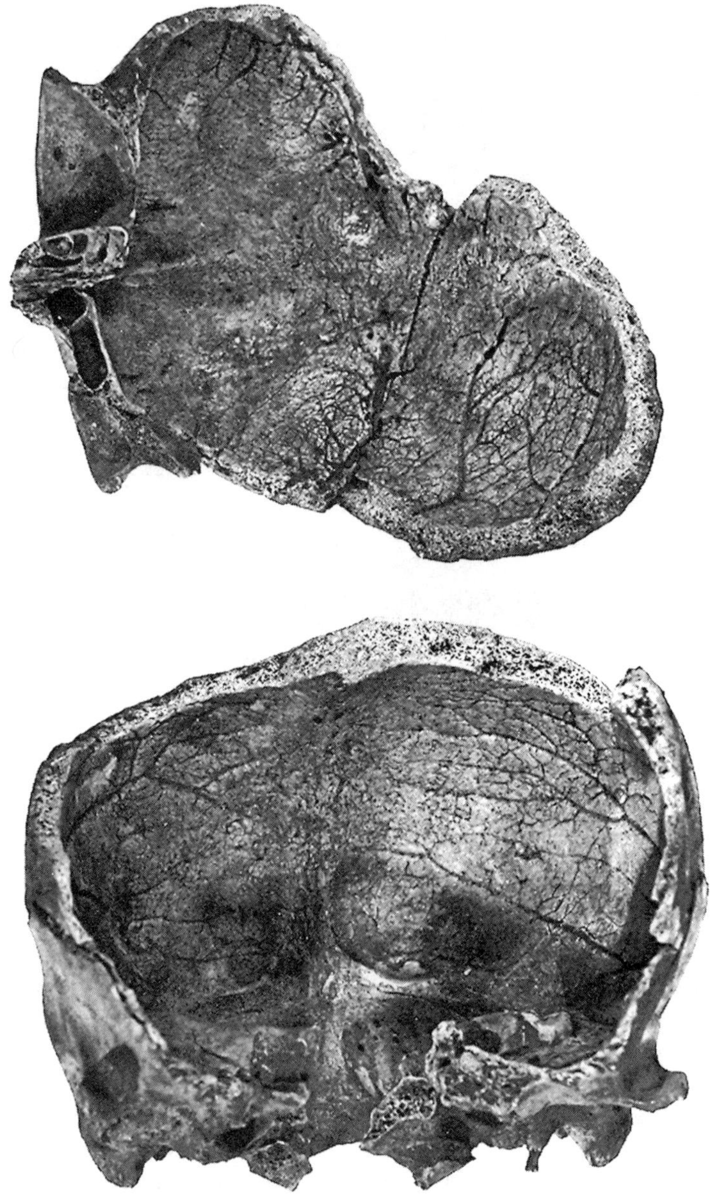

Figure 10.6. Example of extreme nonspecific cranial hyperostosis of an adult male from Feature A, Chalpa (salvage skull 3).

Evidence for Precolumbian Tuberculosis

Evidence for prehistoric tuberculosis in the Americas has been considered by physical anthropologists and physicians since the early part of the last century (Hrdlička 1909). Over the first part of this 100+-year period a healthy skepticism tended to predominate thinking, often for very good reasons (Buikstra 1981). Beginning in Peru and Mexico, however, some compelling evidence from both prehistoric skeletons and mummies began to accumulate. By 1970, Hurtado (1970:75) accepted without reservation the existence of tuberculosis in prehistoric Mexico. By this time I had encountered five cases from the Marismas sites of Chalpa and Tecualilla exhibiting massive destruction to the vertebral column from low-grade, chronic granulomatous infections strongly suggesting tuberculosis (Gill 1971:210–214). Over the next decade came additional strong skeletal evidence from across North America, north of Mexico. Skeletal evidence was presented from geographic areas as widely dispersed as west-central Illinois (Buikstra and Cook 1981), southern Ontario (Hartney 1981), and central North Dakota (Williams and Snortland-Coles 1986). Numerous skeptics remained, however, until after 1990, when sufficient samples of the tubercle bacillus, *Mycobacterium tuberculosis*, were isolated from Peruvian

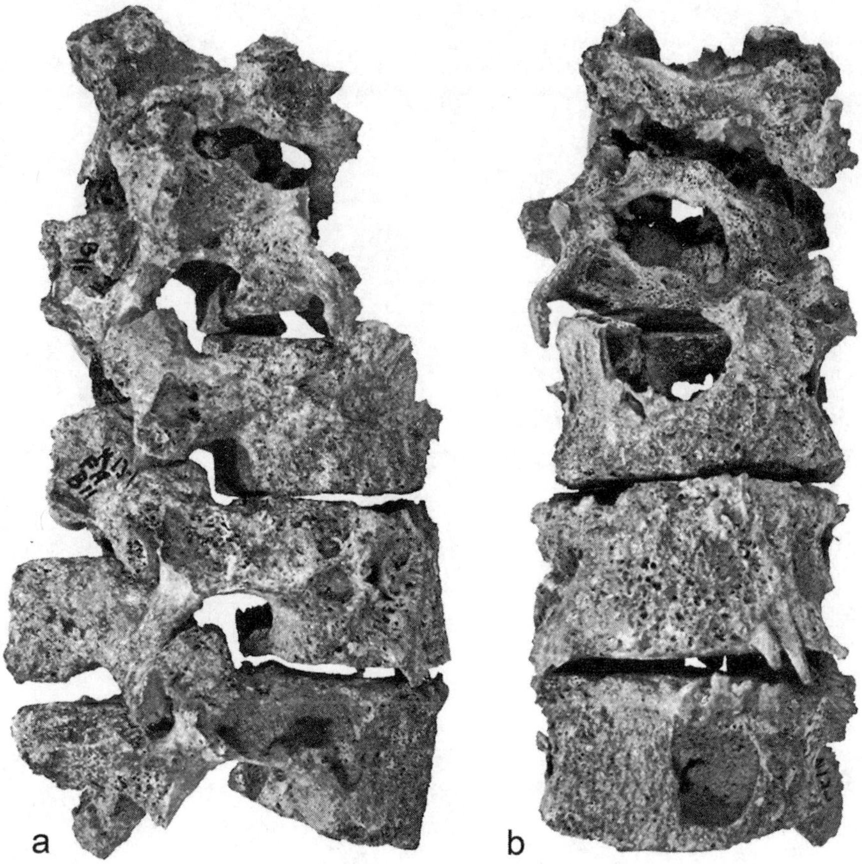

FIGURE 10.7. (*a*) Lateral view and (*b*) ventral view of chronic, low-grade granulomatous infection (probable Pott's disease) evidenced in the spinal column of an elderly female from Burial 11, Feature A, Tecualilla.

mummies of established prehistoric age to sway the majority of researchers (Prat and Souza 2003; Salo et al. 1994). Further support came with the work of Salo et al. (1994), with the first published molecular diagnosis using polymerase chain reaction techniques for micro bacteria's ancient DNA in the same Peruvian material. This was followed by a second molecular diagnosis published by Arriaza et al. (1995). Today, then, the existence of precolumbian tuberculosis in the New World is no longer in question, but there is still importance to the documenting of specific prehistoric cases in order to understand better the frequency, severity, and temporal depth of the disease within various geographic regions of the Americas.

Figure 10.7 shows the lumbar spine of an old female from Tecualilla (B 11) likely suffering from tuberculosis of the skeleton (Pott's disease) at the time of death. The photographs clearly show that the low-grade infection and destruction centered in the twelfth thoracic and lumbar vertebrae, as is the case with Pott's disease. Little or no fusion of the weakened vertebrae had occurred in this case. Figure 10.8 provides a more detailed view of one of the Burial 11 lumbar vertebrae.

Another case, almost identical to the Burial 11 situation at Tecualilla, comes from Chalpa, Feature A. It is the skeleton of a middle-aged female (B 5A) that shows the same kind of destruction caused by chronic, granulomatous infection. In this specimen, just as in the case of Tecualilla Burial 11, the lower thoracic and lumbar vertebrae show large eroded areas. Additionally, in this Chalpa case, collapsing centra with subsequent fusion caused a considerable shortening and kyphosis (humpback condition) of the spinal column. Figure 10.9 shows two of these collapsed and fused thoracic vertebrae. Also the distal end of a first rib and the ends of two long bones were mildly affected by the destructive processes.

A middle-aged male specimen from Feature D at Chalpa (B 5) shows evidence as well of having been afflicted with a very similar granulomatous infection. Tremendous areas are eroded from the vertebral centra in the thoracic and lumbar areas of this specimen. Posterior fusion of five of the thoracic vertebrae had also occurred, possibly as a response to collapsing centra.

A young adult female skeleton from Tecualilla (B 17) also shows evidence of chronic, low-grade infection. The manifestations do resemble Pott's disease, but the pattern is less compelling in this example (Figure 10.10) than the

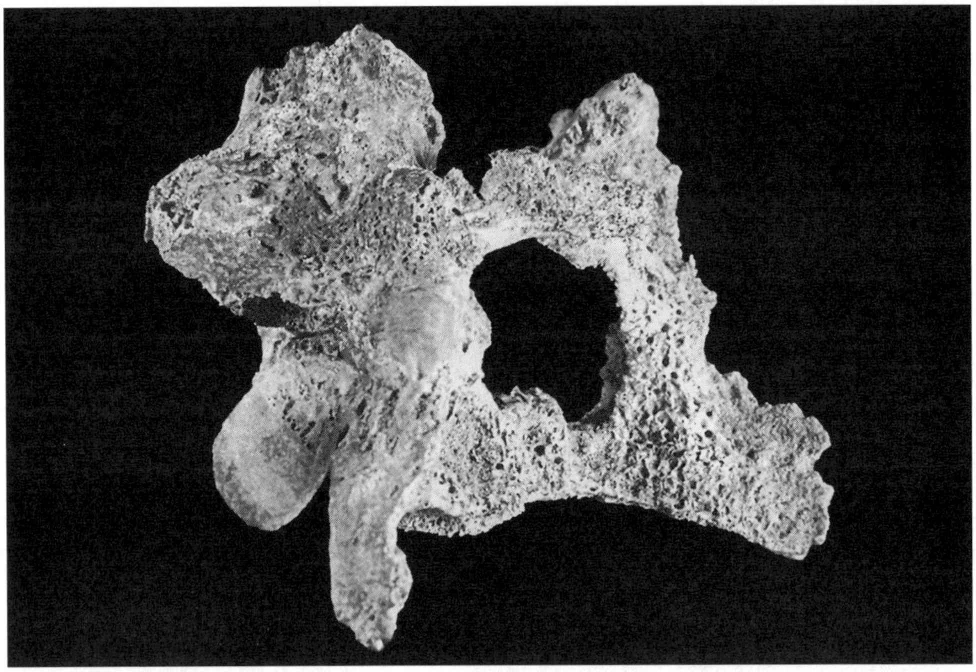

FIGURE 10.8. Lumbar vertebra from the Burial 11 spinal column, Feature A, Tecualilla.

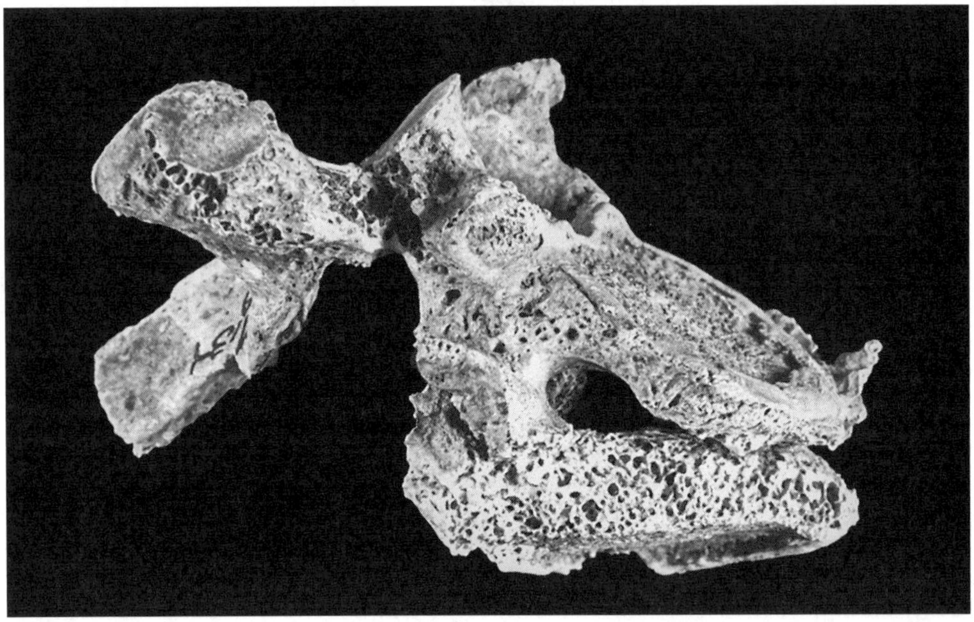

FIGURE 10.9. Collapsed and fused thoracic vertebrae of an adult female from Burial 5A, Feature A, Chalpa.

Burial 11 case from Tecualilla and the Burial 5 example from Chalpa, Feature D. This Tecualilla female reveals three lumbar vertebrae with areas of destruction. The proximal end of a fibula and the sacrum also show effects of disease.

No indications of massive destruction to the vertebral column have been observed on skeletons from Venadillo, Rincon de Panal, or Cristo Rey even remotely resembling the Tecualilla and Chalpa cases. Of course, it should be kept in mind that sample sizes from these last three sites are not as great.

Skeletal Anomalies

Four postcranial skeletal anomalies have been identified within the Postclassic period skeletal samples of the Marismas. These are sternal foramen, olecranon foramen, spina bifida occulta, and the fusion of the first coccygeal vertebra to the sacrum (which creates six seg-

FIGURE 10.10. Lumbar vertebra of a young female from Burial 17, Feature A, Tecualilla.

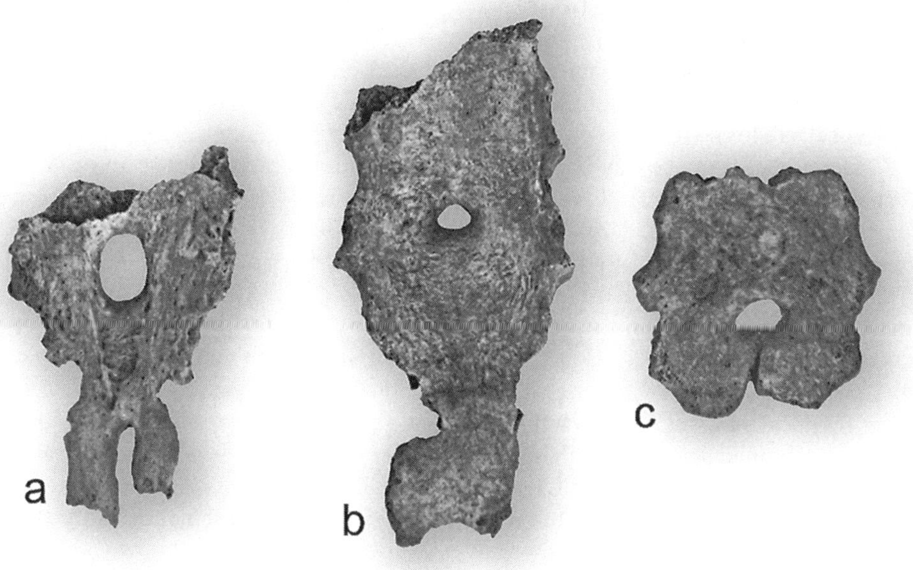

FIGURE 10.11. Broken distal ends of three sternal blades from Tecualilla: (*a*) Burial 1, Feature A; (*b*) Burial 10, Feature A; (*c*) Burial 1A, Feature B.

ments of the sacrum rather than the normal five). Figure 10.11 shows the rather rare anomaly of the sternal foramen. It occurs in 1 percent of U.S. whites and is assumed to be genetically determined by a rare recessive allele. This obvious perforation through the lower portion of the sternal blade is so rare in all major human populations (Verna et al. 2013) that some forensic anthropologists and pathologists have never seen an example of this trait, and as a result it is occasionally mistaken for gunshot trauma or some other projectile trauma (Byers 2011).

The trait occurs in elevated frequencies only in small, isolated populations that possess a higher-than-normal coefficient of inbreeding. It is surprising, and probably quite meaningful, that the 15 studiable sternal blades from Tecualilla show a 33.3 percent frequency

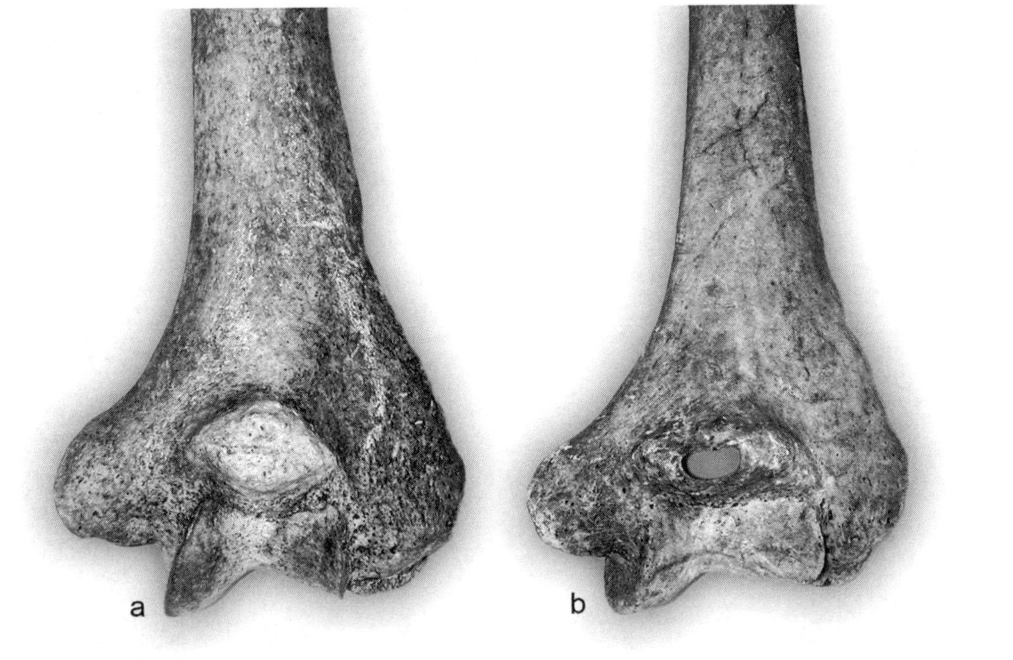

FIGURE 10.12. (*a*) Distal humerus without an olecranon foramen and (*b*) a distal humerus with a foramen.

of occurrence of this generally rare anomaly. The 12.5 percent frequency at Chalpa is also important. Unfortunately the fairly small sample size at Chalpa ($n = 8$) precludes a definitive statement as to whether the frequency had actually declined at the Chalpa site compared with the earlier high frequency at Tecualilla. Probably it had, since no examples at all are found from the more recent sites of Venadillo Mound B, Rincon de Panal, and Cristo Rey ($n = 9$ sternal blades, total from all three sites). It is tempting to believe that the gradual decline through time indicated by these data is real and to attribute it to the increased gene flow suggested by all other skeletal data (craniometrics, dental metrics, and discrete trait analysis—see Chapter 11).

Figure 10.12 shows an olecranon foramen of the olecranon fossa of the distal humerus. This is a trait common to females in many human populations but is found less often among the males in most, if not all, human groups. The perforation appears merely to be the result of reduced robusticity of some individuals (especially females) in the area of the elbow joint. Most males have a thin partition of bone in that area of the humerus between the radial fossa anteriorly and the olecranon fossa posteriorly. In many females, however, the robusticity at that location is so reduced that the thin partition of bone gives way to an actual perforation. The presence of this opening during forearm extension helps allow for hyperextension at the elbow joint. This is because the olecranon process of the ulna (tip of the elbow) not only enters the fossa but can penetrate part of the way through it.

Both the sternal foramen and olecranon foramen are listed in tables in Chapter 11. Only basic tabulations are provided, however, in that chapter.

Figure 10.13 illustrates the frequency of occurrence of the olecranon foramen at all five Postclassic sites. Besides the strong sexual dimorphism for the expression of this trait, so often reflected in other *Homo sapiens* populations, there does seem to be a decline through time with this trait as well, at least from the earlier Postclassic sites of Tecualilla and Chalpa (where the female frequency is 75 percent or more and the male frequency is a little over 20 percent) on through the later Postclassic sites of Venadillo Mound B, Rincon de Panal, and Cristo Rey (where the female frequency is 50 percent or less and the male frequency is probably less than 8 percent).

The cases of spina bifida noted in this study are not instances of fully developed, clinical spina bifida. They are all examples of spina bifida occulta, a less severe, generally subclinical manifestation of this condition. Its causes are not precisely known but are thought to consist of a combination of genetic and environmental factors, as in the cases of fully manifested, clinical spina bifida. For the purposes of this study neural arches that reveal approximately one-third incomplete closure (20 mm or more open arch) have been classified as spina bifida occulta cases, while those with less are considered within the normal range for neural arch closure.

The cases of fusion of the first coccygeal vertebra to the caudal extremity of the sacrum (surface of S5) produce an elongated sacrum with six segments. So does the fusion of the fifth lumbar vertebra (L5) to the cra-

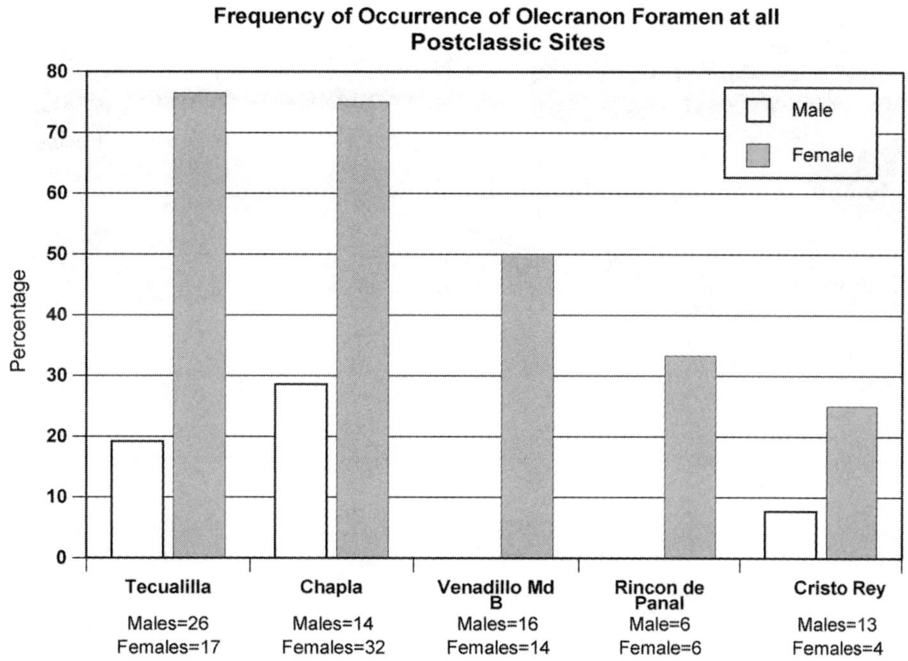

FIGURE 10.13. Frequency of occurrence of the olecranon foramen at Tecualilla, Chalpa, Venadillo Mound B, Rincon de Panal, and Cristo Rey.

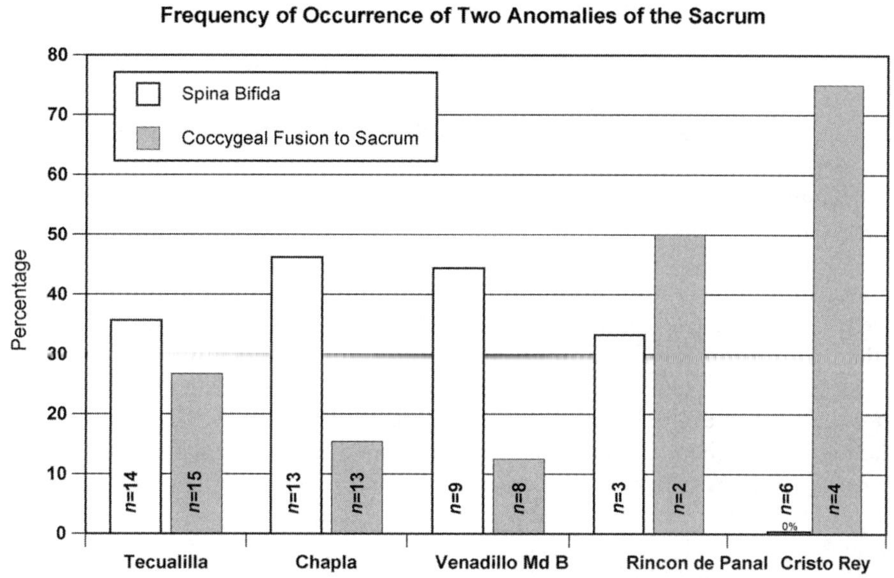

FIGURE 10.14. Frequency of occurrence of two anomalies of the sacrum at Tecualilla, Chalpa, Venadillo Mound B, Rincon de Panal, and Cristo Rey.

nial surface of the sacrum (S1), but no instances of this latter condition have been found among the Marismas skeletons. All six-segment sacra consist, in other words, of fusions of the first coccygeal segment to S5. This trait is not as well studied as the other anomalies mentioned here but seems to be at least in part controlled by genetic factors and to develop early in adult life. Earlier studies of the six-segment sacra (Gill 1971) pooled sexes since there appeared to be no sex difference in expression. In the recent analysis of data from Venadillo, Rincon de Panal, and Cristo Rey the same pooling was done, but it may be of interest that within the relatively small samples of preserved sacra from these three sites all of the ones with six segments ($n = 5$) were from male skeletons.

Figure 10.14 shows the frequencies of both of these sacral anomalies at all sites. Even discounting to some

FIGURE 10.15. Swelling of the right tibia of an adult male showing osteitis from Burial 20, Tecualilla.

extent the likely inflated frequencies of coccygeal fusions at Rincon de Panal and Cristo Rey, where sample sizes are quite small, no clear pattern seems to emerge. A decline in coccygeal fusions from Tecualilla to Chalpa on down to the Venadillo site could be real but is hard to confirm with such small samples. Regarding spina bifida occulta, however, it seems even more evident that no clear pattern of change exists. Rather, a consistent pattern (of about 33 percent frequency or more) remains through all periods.

Osteitis and Culture

Some cases of osteitis (bone swelling) of tibiae have been observed that do not appear to be the result of disease. Figure 10.15 illustrates one of these tibiae from an older adult male (55–65 years) from Tecualilla (B 20). The surface of the bone in these cases is quite smooth rather than uneven, as would be the case in instances of periostitis or osteomyelitis. Also, the grooved impressions characteristic of the increased vascularity in diseased bones are totally lacking in these cases.

The two best-documented examples within the Marismas skeletal sample are both males, and they embrace a temporal span from the early Postclassic period site of Tecualilla to the most recent site of Cristo Rey. The Cristo Rey example is described as a swelling of the diaphysis of the left tibia of a young male (B 19), 26–33 years of age (Schutte 1973). It resembles quite closely the example from the older male at Tecualilla (B 20).

From the beginning of the Marismas skeletal project, during my dissertation research on the Tecualilla and Chalpa samples (Gill 1971), Dr. Ellis R. Kerley served as my paleopathology consultant. He believed that these particular cases of osteitis represented a cultural phenomenon rather than disease, perhaps a bodily reaction to some kind of constriction around the leg. Such a constriction would reduce normal circulation to the lower leg and thus cause such an osseous response (Gill 1971). He asked whether some kind of ligature was worn by the prehistoric inhabitants of the region. I responded that none had ever been identified and they would likely not preserve in the archaeological record in that tropical area. Subsequently the project historian, Patricia Scott, during archival research in Mexico City, located drawings made in 1530 during the Spanish conquest of the Marismas region known as the Guzmán Entrada (Figure 10.16). As may be seen from this detailed illustration, there appear to be ligatures present on the fallen warrior at the bottom of the drawing. These ligatures, midway between the knee and ankle, are in precisely the same location as the tibial swellings on our male specimens from Tecualilla and Cristo Rey. Other early drawings of the battles of conquest in this region exist, and they also reveal ligatures on warriors similar to those shown in Figure 10.16. Some of the other drawings show objects hanging from the ligatures, which precludes the interpretation of these adornments as merely body paint or tattoos. Thus, it appears that Dr. Kerley's assessment was quite likely correct, that the swellings of these tibiae represent the result of ligatures and not disease-caused bone pathology. It may also be of interest that the Tecualilla male comes from the high-status burial mound at Tecualilla (Feature A) and the Cristo Rey male with that same condition is associated with the greatest amount of burial offerings of any of the Cristo Rey burials (two pipes, an obsidian point, and a vessel), thus probably representing a high-status burial as well.

Intentionally Produced Cultural Modifications

Two kinds of cultural modification, artificial cranial and dental modifications, were widely practiced in the precolumbian Marismas. Both of these were widely practiced elsewhere in Mesoamerica as well and within other regions of the precolumbian Americas.

FIGURE 10.16. Plate 69, Lienzo de Tlaxcala, depicting the Guzmán Entrada, showing a warrior with ligatures.

Artificial Cranial Modification

A total of 81 studiable crania come from the Postclassic sites of the Marismas, and of these nearly 89 percent show definite indications of artificial cranial modification. The Dembo and Imbelloni (1938) classificatory system, a system widely used in the study of Mesoamerican crania (Comas 1960; Romano 1965; Saul 1972), is adopted in this study. Some contemporary studies of artificial cranial modification classify skulls into one of only two broad categories of modification: fronto-occipital (a result of anteroposterior compression) or circumferential (a result of constriction to the entire circumference of the cranial vault) (Jimenez et al. 2012; Khonsari et al. 2013). For this study such a basic classificatory system is inadequate due to the wide variety of forms exhibited within the Marismas and the potential cultural meaning within the observed variety.

The Dembo and Imbelloni system distinguishes the same two major categories of artificial cranial deformity: the tabular (fronto-occipital), which shows flat surfaces caused by boards, and the annular (circumferential), caused by a soft wrap (which leaves no flattened areas on the skull). Both of these primary categories, however, are further subtyped by Dembo and Imbelloni as erect or oblique, depending upon the angle of the vertical axis of the cranial vault (see Comas 1960 for further description).

No true annular forms exist within the Marismas skeletal samples, and none show the oblique subtype. All are, in other words, of the tabular erect type of modification. This is the oldest and most widely distributed modification type in Mesoamerica. In fact it accounts for nearly every example of artificially deformed skull in all of Mexico. The annular (circumferential) form comes largely from Peru. Ganot Rodríguez and Peschard Fernández (1997), in a thorough survey of 227 known artificially deformed skulls from prehistoric sites in western and northern Mexico, find all but two to be tabular erect. The two exceptions are both of the tabular oblique subtype (one from El Zape site, Durango, and one from El Salto site, Durango).

The tabular erect skulls from the Marismas sites come in four common forms:

1. Extreme. These crania show marked deformity of both the frontal and occipital areas, with maximum compression of these two bones.

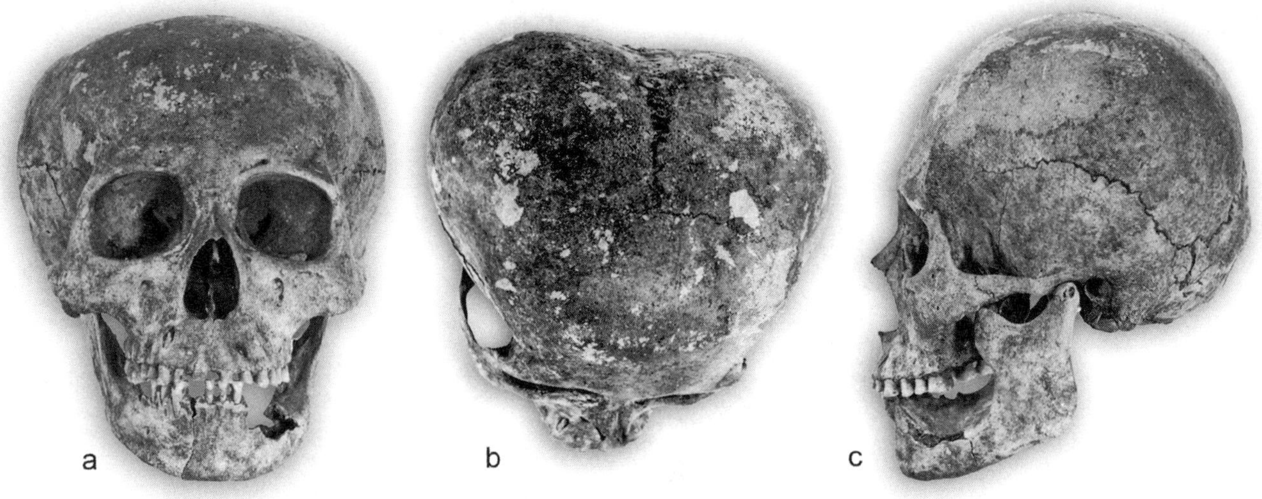

FIGURE 10.17. Tabular erect form of cranial modification with bilobed effect, Burial 3, Feature A, Chalpa: (*a*) frontal view; (*b*) top view; (*c*) left lateral view.

2. Intermediate. Fronto-occipital compression has occurred, but neither the frontal nor the occipital areas show a marked degree of flattening.
3. Plano-lambdoid. A well-flattened occipital region is evident in this form, and yet the frontal area shows no clear traces of compression.
4. Plano-frontal. The fourth category has not been observed in the Marismas skeletal samples but is described as showing a smooth, flat frontal area, with little modification to the occipital.

All four forms can be subdivided into four varieties: bilobed, trilobed, pseudocircular, and paralleled (parallelepiped). Of these, only the trilobed variety has not been identified among the Marismas samples. Examples of the other three have, and they are defined as follows:

1. Paralleled. This skull variety results in a cube-shaped cranium (with a flatter lambdoid area in the plano-lambdoid cases than in the intermediate).
2. Bilobed. This is a divided, sometimes heart-shaped cranium, caused at least in part by a sagittal head binding.
3. Pseudocircular. A high, circular-shaped cranium characterizes this variety. It is produced by a combination of soft wrapping and boards, and it results in a flat occipital region.

Figure 10.17 shows the most common form of cranial modification found in the Postclassic period Marismas Nacionales, a tabular erect intermediate skull with bilobed effect. Figure 10.18 shows the two less common pseudocircular and paralleled forms, and Figure 10.19 illustrates the exotic but even more rare extreme form.

Figures 10.20 and 10.21 show the distribution of the forms and varieties encountered within the Marismas Nacionales. These graphic depictions suggest that some patterns were consistent throughout the Postclassic period. For instance, the intermediate form of modification (both with and without the bilobed effect) is the most prevalent. The plano-lambdoid form with bilobed effect is also relatively common throughout the temporal span of the Postclassic period.

Perhaps not as evident from the figures (which combine sites) is the fact that artificial cranial modification seems to have increased slightly after the earliest Postclassic phase. For instance, fewer than 75 percent of the Tecualilla skulls show artificial modification, while approximately 95 percent of those from all sites after Tecualilla exhibit evidence of the practice. Perhaps a greater specialization in the forms produced occurred through time as well. At least a complete loss of the paralleled variety occurred after Tecualilla, and there may have been a loss of both the extreme form and the pseudocircular variety after Venadillo. The sample sizes from Rincon de Panal and Cristo Rey are too small, however, to confirm this.

The distribution of the various styles by sex is completely equal. A greater effort may have been made, however, in the application of the practice in general with baby girls than with infant boys. For instance, of the six undeformed skulls from Tecualilla, all are male. Also, the most extreme hyperbrachycranic indices (which result from more diligently applied forces) from both Tecualilla and Chalpa are largely among females.

No changes in the relative frequency of forms or the prevalence of the custom in general can be demonstrated from the Chalpa site on through the later site groups. Therefore, the noticeable cultural changes with regard

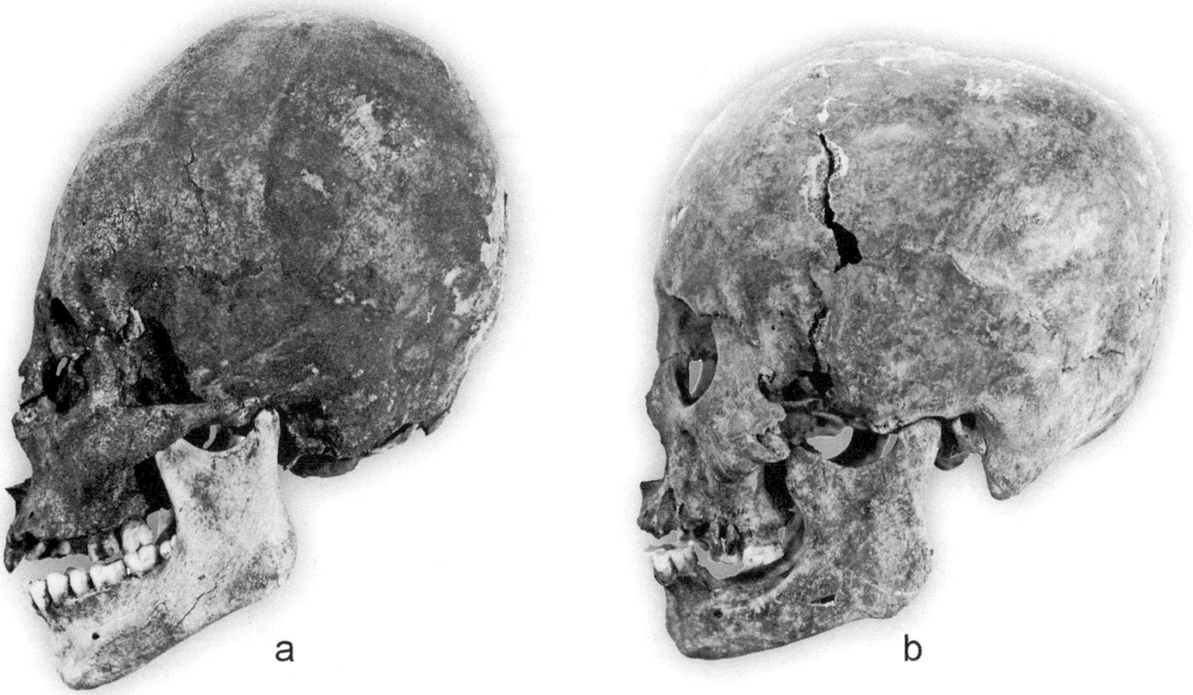

FIGURE 10.18. Lateral views of the less common cranial modification forms: (*a*) pseudocircular variety, Burial 2, Feature A, Chalpa; (*b*) paralleled variety, Burial 11, Feature A, Tecualilla.

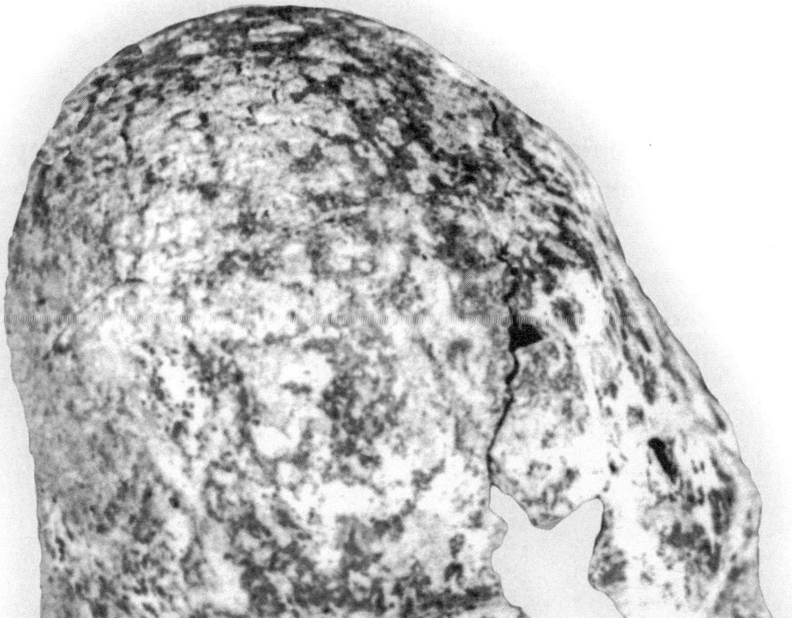

FIGURE 10.19. Right lateral view of the extreme form of cranial modification revealing concave depressions in both the frontal and occipital regions, Burial 14A, Feature A, Tecualilla.

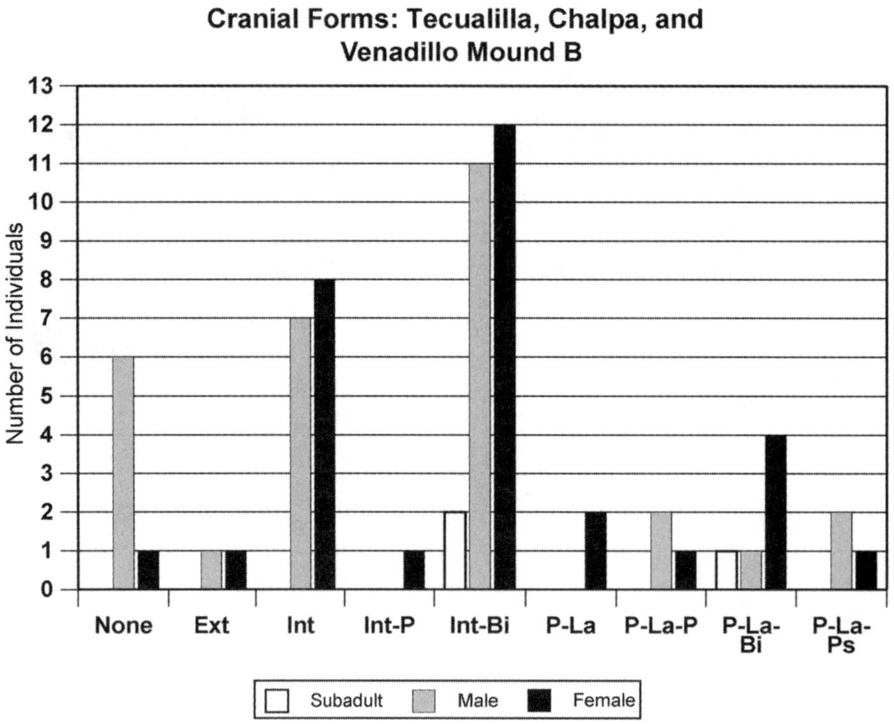

FIGURE 10.20. Frequency of modified cranial forms from the early sites Tecualilla, Chalpa, and Venadillo Mound B (pooled). Ext = extreme; Int = intermediate; Int-P = intermediate paralleled; Int-Bi = intermediate bilobed; P-La = plano-lambdoid; P-La-P = plano-lambdoid paralleled; P-La-Bi = plano-lambdoid bilobed; P-La-Ps = plano-lambdoid pseudocircular.

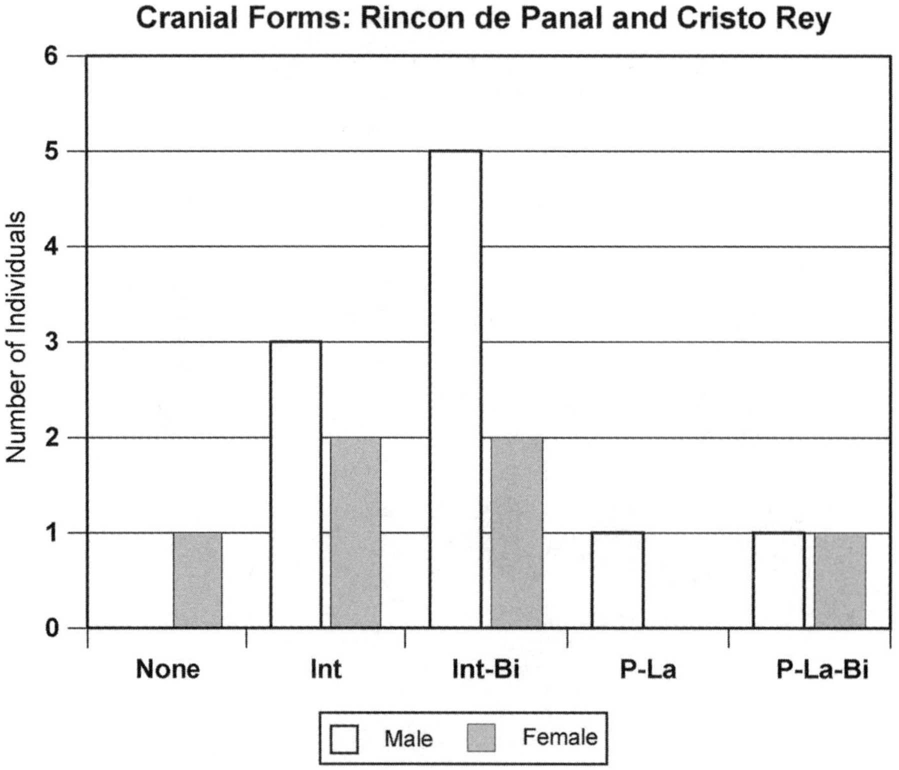

FIGURE 10.21. Frequency of modified cranial forms from the more recent sites Rincon de Panal and Cristo Rey (pooled). Int = intermediate; Int-Bi = intermediate bilobed; P-La = plano-lambdoid; P-La-Bi = plano-lambdoid bilobed.

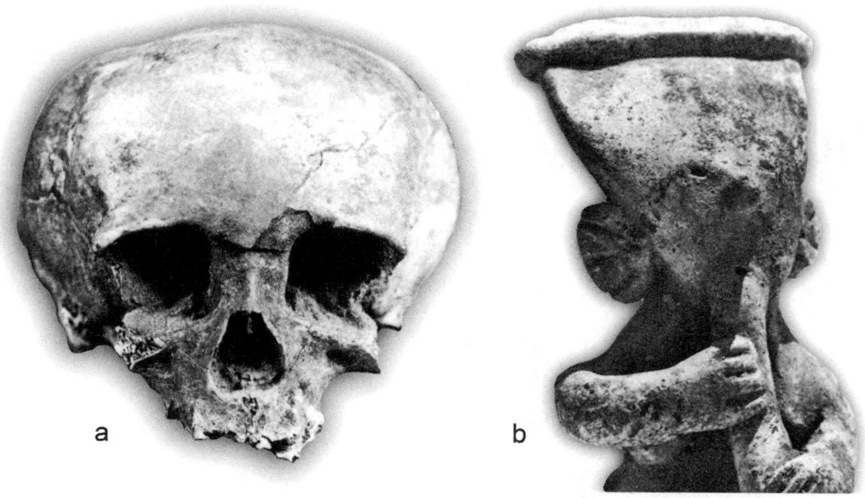

FIGURE 10.22. Comparison of (*a*) a common form of modified cranium with (*b*) a common figurine style from West Mexico.

to this practice appear to have been between Tecualilla, where 23.5 percent of the deformed crania are of the paralleled variety, and Chalpa, where no paralleled examples are found. It is probably meaningful that the demise of the paralleled variety of cranial modification (about AD 1250) coincides with the first evidence of genetic change in the region, as discussed in Chapter 11, and an abrupt change from extended primary burials to flexed interments (see Gill 1974, 1985; see Chapter 9).

Examination of the practice of cranial modification at other sites in northwestern Mexico provides some interesting comparisons with the findings here but also poses some challenges due to the poor preservation of crania at many of the sites. Within the Amapa sample, for instance, there are only two studiable crania (Nos. 3700 and 3703). Both do show clear signs of cranial modification (Gill 1976:199) and in fact compare well with the Marismas crania. One, a female, reveals an intermediate form of modification of the bilobed variety. The other, a male, shows the extreme form without the bilobed effect.

At Culiacán, Sinaloa, Hulse (1945) describes 252 burials, but due in part to poor preservation, only 33 adult crania could be assessed with regard to cranial length and width. Even though the emphasis of his study was upon genetically affected skeletal traits, he does mention the "universal practice" of cranial modification practiced by the group. No details are given, however, as to the forms of modification.

At Guasave, Glascock (1980) mentions poor preservation of the crania as a hindrance both to her assessment of the frequency of occurrence of artificial cranial modification and to the classification of individual crania. She states that 40 percent of the skeletal sample reveals definite cranial modification but that this figure could be much higher, "but for the poor preservation of more than half of the skeletal remains" (1980:85). Even though only two of the Guasave crania are described by Glascock as complete, others are sufficiently preserved to allow her to classify 11 within the Guasave series. Of these she describes three of the extreme form (as seen at Tecualilla and Amapa), six as intermediate, one as plano-lambdoid, and one as plano-frontal. Four of the six that show an intermediate form of modification are also bilobed, like some of those at the Chalpa site.

Di Peso et al. (1974) mention that the fragmentation and poor preservation of most of the crania from Casas Grandes preclude accurate assessment of the extent of the practice of artificial cranial modification. They state that at least 12.8 percent of the 681 studied individuals exhibited undeniable signs of modification and that a better-preserved subset of crania, studied by osteologists on the project, showed a 50 percent frequency of artificially deformed skulls. Researchers at that site attribute almost all of the modification to simple cradleboard practices. The type and degree of fronto-occipital modification that they describe does support this. They mention only three cases from the Viejo period that clearly represent a purposeful attempt to shape the cranium. This constitutes a significant departure from the situation at the Marismas sites, and at Guasave and Amapa as well, where clearly purposeful efforts were made to achieve specific cranial shapes on many if not most of the crania.

Early in the excavation and analysis of the Marismas skeletal sample, parallels were noticed between three common cranial shapes (e.g., extreme, bilobed, pseudo-circular) and art styles of clay figurines from the same region and period (Gill 1969a). Figure 10.22 shows one of these comparisons. A hypothesis of meaningful congruence between the shapes of the human skulls and the

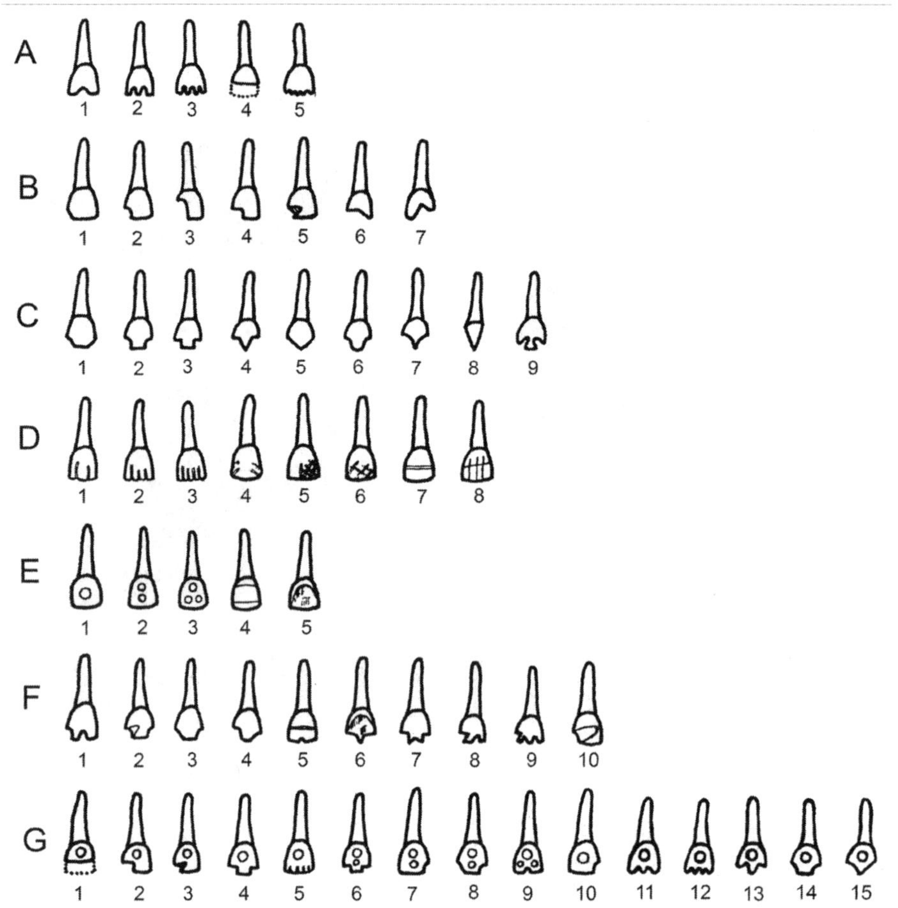

FIGURE 10.23. Romero's chart of precolumbian dental modifications (adapted from Romero 1970).

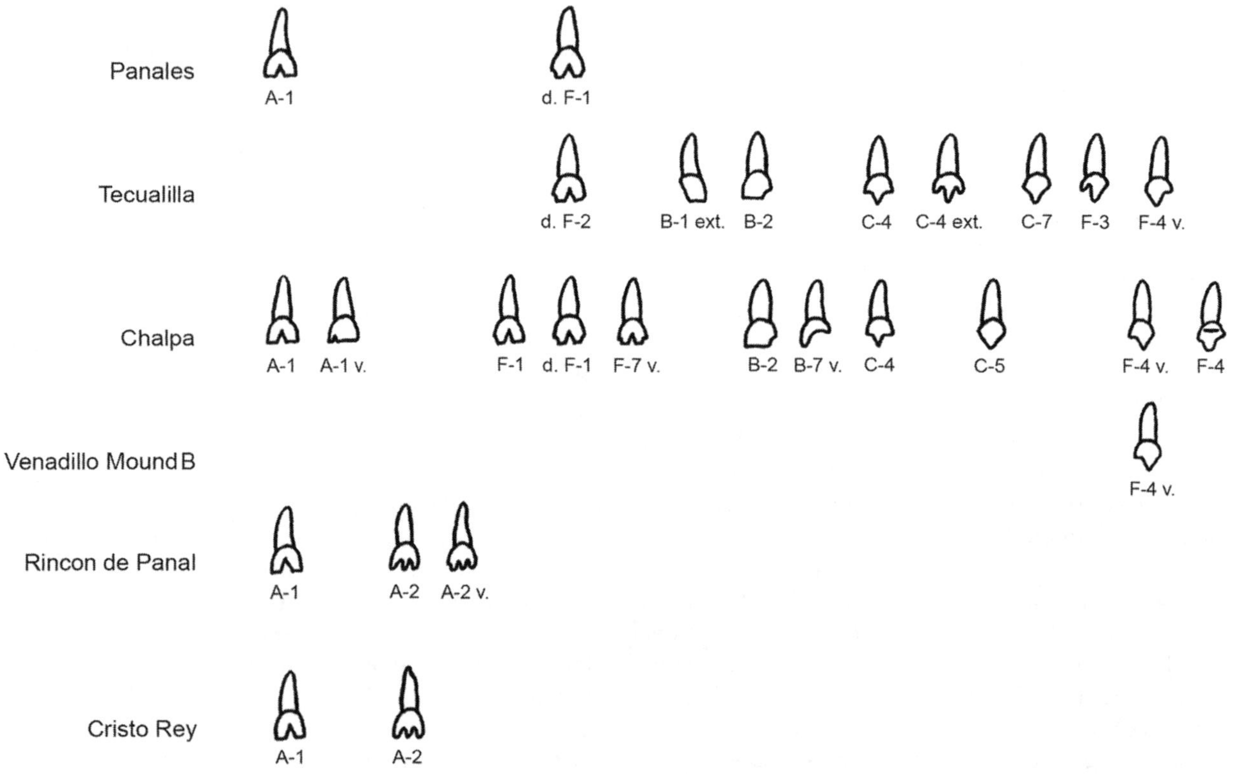

FIGURE 10.24. All forms of dental modification from the Marismas Nacionales (Gill 1985).

clay figurine head shapes was formulated and tested in the summer of 1970 (see Gill and Prince 1970). A total of 247 clay figurines from regional museums in Sinaloa and Nayarit were examined, and subsequent analysis demonstrated that a significant correlation did exist between the most common skull forms in the Marismas and the most prevalent head shapes represented on the clay figurines (Gill 1977).

All human skulls and nearly all of the clay figurines were found to be of the same four basic artistic styles (Gill 1977, 1985). Even the percentages in some cases were found to be similar between the number of crania of a certain type and the figurines of a corresponding shape. This would seem to suggest a greater cultural significance for certain stylistic motifs and also that this perception was shared by those shaping the infants' heads and those creating the clay figurines. Since many of the clay figurines are more extreme than the skulls, and at times even seem to represent hairstyles and headdresses, it is clear that the figurines were not always intended to be mere copies of actual human head shapes. Instead, both the skulls and the figurines appear to represent manifestations of the same conceptual artistic ideal (Gill 1977, 1985). Limits in the plasticity of the human cranium probably mean that the figurines come closer to the original aesthetic ideal in the minds of the aboriginal artisans than do the actual skulls. However, both were being fashioned in Postclassic times according to extremely similar artistic styles, and these probably symbolized very similar concepts.

Dental Modification

Quite commonly the upper incisors and canine teeth of skeletons from the Marismas Nacionales show carving. Figure 10.23 illustrates the various forms of decorated teeth found throughout the Americas as presented by Romero (1970). This study uses the Romero system in classifying all dental modifications. Many of the forms depicted in Romero's classification (and a few variants from them) have been found at the Marismas sites. However, no examples of inlay or incising of labial surfaces have been found at any of our sites. Furthermore, only modification of the maxillary teeth (no mandibular teeth) was practiced at least by Postclassic times. All forms found at the six Postclassic sites of the Marismas Nacionales are shown in Figure 10.24.

All forms of decorated teeth from the six sites are either notched or pointed. Of the 17 forms identified, seven are actual variants of more common forms within Romero's system. Each of these cases–A-1 variant, A-2 variant, double F-1, F-7 variant, B-7 variant, C-4 extreme, and F-4 variant—was reviewed personally with Romero in order to ensure a correct classification (Gill 1985:196). In the cases of the A-1 variant and F-7 variant, it seems clear that the artisan was attempting to produce aesthetic effects quite different from those of the more common variations of the same form (as illustrated by Romero). This becomes even more apparent when these variant forms are viewed in the context of the full dental pattern produced by the artist (Figure 10.25). So, as may be seen in Figure 10.25, the effect of the A-1 variant form in pattern "b," for instance, is clearly different from the A-1 effect of pattern "a." The A-1 variant and F-7 variant forms have been fully described and illustrated previously (Gill 1985:196–203).

Some cases from Tecualilla and Chalpa (see Figure 10.24) show radical degrees of application of both notches and points, with large portions of the original tooth having been removed. This underscores the importance of this practice within this prehistoric culture. So does the fact that nearly 45 percent of studiable individuals 15 years of age or older from all six Postclassic sites show some form of decoration.

Also revealing is a tabulation of the frequency of patterns by site. The earliest Postclassic sites of Panales, Tecualilla, and Chalpa reveal frequencies of the occurrence of dental modification of 66.7 percent, 62.5 percent, and 61.5 percent, respectively. The later sites of Venadillo Mound B, Rincon de Panal, and Cristo Rey, on the other hand, show much lower frequencies of 11.1 percent, 33.3 percent, and 16.7 percent occurrence of this practice, respectively. So, overall, the number of people with decorated teeth averaged about 63 percent at the earliest Postclassic sites and then declined to about 20 percent a century or so later at the three late Postclassic sites.

Decline in the practice of dental modification during the Postclassic period is evidenced in two additional ways beyond the decline in frequency. The more extreme forms also appear to be found less often at the late sites. At Chalpa and Tecualilla, for instance, between 25 percent (Chalpa) and 80 percent (Tecualilla) of the decorated teeth show the more extreme, pointed forms. At the two most recent sites of Rincon de Panal and Cristo Rey, on the other hand, none of the modified teeth reveal extreme forms. Likewise, the number of teeth involved in the modification practice within each maxillary tooth row declines during this same period. Only the upper central incisors are affected in half of the cases at the three most recent sites, while at the early Postclassic sites of Panales, Tecualilla, and Chalpa 93 percent of individuals with carved teeth show involvement beyond the central incisors. These findings are consistent with what Romero (1958:Chart 12) found in Michoacán and the Valley of Mexico, where he documented more extreme forms and a greater variety of forms at earlier temporal horizons.

One of the last forms of dental modification to survive in precolumbian Mexico is the rather subdued A-2

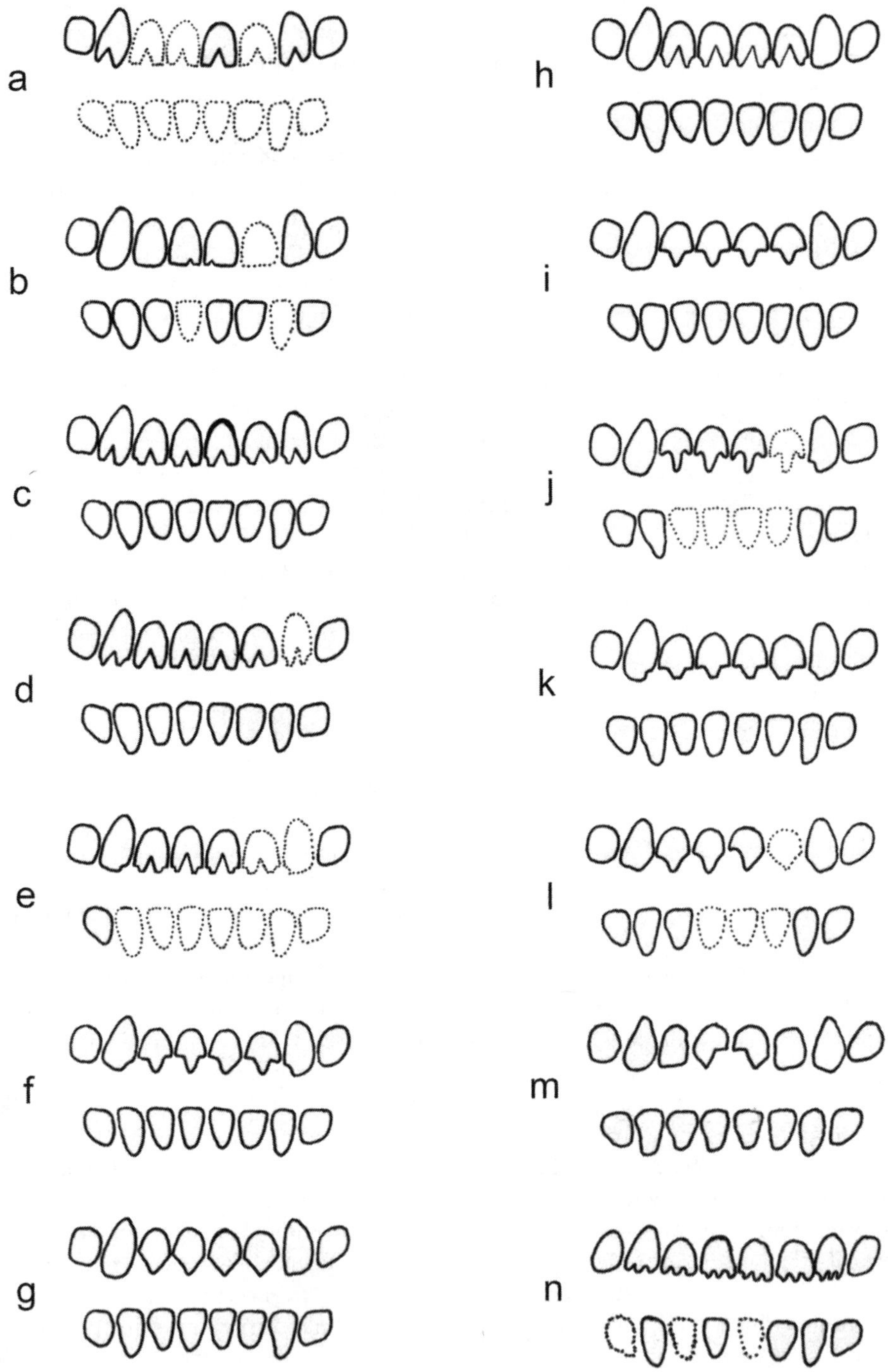

FIGURE 10.25. *(a–n)* Dental patterns from the Marismas Nacionales (Gill 1985).

style, and this is true at the Marismas sites as well. In fact, it first appears quite late, at only the Postclassic period sites of Rincon de Panal and Cristo Rey.

In examining other trends and correlations it is of interest that neither frequency of occurrence nor style of decoration correlates with sex. As the percentages of males and females showing dental decoration were examined, the numbers were seen to be precisely the same (Gill 1985:199). In a preliminary assessment, the correlation of the presence or absence of funerary offerings and dental modification was tested (Foster and Gill 1977); no significant correlation was found. This is consistent with earlier findings in Mexico and elsewhere in Central America (Smith 1972; Stewart 1978).

Dental modification in the Marismas appears to have served as a rite of passage or initiation into adulthood (Gill 1971, 1985). Age of application was just about 15 years of age for both boys and girls judging from our skeletal evidence. Those below that age show no sign of decorative practices, and those at that age show signs of very recent application of the techniques (no wear etc.). This is consistent with the findings elsewhere in greater Mesoamerica (Smith 1972; Stewart 1941; Sweet et al. 1963).

Geographic distributions of the various forms of carved teeth provide considerable information on routes of diffusion, rates of movement of these ideas, and also the geographic limits of particular forms. For instance, the C-4 form found at Tecualilla and Chalpa is an early southern style. It appeared in the Yucatán by AD 400–500 and is found in many other parts of Mexico after that time (Romero 1958). However, its presence at the sites of Tecualilla and Chalpa may constitute the most northerly limit of this form (Gill 1985). Glascock (1980) describes nine examples of dental modification from Guasave and classifies some of the incisors as C-6. Careful examination of her illustrations, however, shows that a few of these could be heavily worn examples of the C-4 type. In a survey of 53 adult crania from southern Sinaloa and northern Nayarit, Talavera and Gelover (2008) found only three to have dental modifications. They show the A-1 form, and at least one reveals the F-1 type as well.

As Romero (1958) points out, the notched A-1 and A-2 forms are the most widespread in the Americas. Also the A-1 form is the most widely distributed temporally among our Marismas sites (Gill 1985:207). It is represented at all sites except Venadillo Mound B. The A-1 form has been tracked by Romero (1958) almost continuously from Jersey County, Illinois, in North America to Chile and Bolivia in South America, and it bridges a temporal span of some 1,100 years, from AD 500 at Teotihuacán to AD 1600 in Illinois (Gill 1985:207). The A-2 form, according to Romero (1958), was the last to disappear in the Valley of Mexico, around AD 1450. It was the last surviving form in Oaxaca and Veracruz (lasting until AD 1600) and along with the A-1 form was the last to die out in several other places as well, including Illinois (sometime after AD 1600). So, perhaps not so surprisingly, the A-1 and A-2 forms are also the last two forms found at the Marismas sites. They are in fact the only forms found at both Rincon de Panal and Cristo Rey, suggesting an age of disappearance similar to that in the Valley of Mexico. Ganot Rodríguez and Peschard Fernández (1997), in discussing dental modifications at the Postclassic site of El Cañón del Molino, Durango, also report only these two forms (A-1 and A-2) from a single cranium at that site, from which 34 crania were recovered.

Amapa (Meighan, ed. 1976) produced seven decorated incisor teeth from a small skeletal sample of 10 individuals (Bernor 1976). Five of those teeth show the F-1 form, and the other two, the C-4 form. At least one of those two shows the extreme C-4 design (Gill 1976). Amapa not only is close geographically to the Marismas Nacionales but is contemporaneous with the earlier Postclassic Marismas sites. The more extreme form of C-4 shown by the single Amapa incisor does not have as wide a distribution as the normal C-4 does. It has been found at Tecualilla but not Chalpa or the other Marismas sites. Romero (1970) unfortunately does not distinguish between the normal C-4 and the extreme C-4 variant. Fastlicht and Romero (1951:Plate 8) do, however, illustrate the extreme C-4 variant. This variant has also been photographed and described by Long and Wire (1966:38, Figure 20c) from Barra de Navidad, Jalisco. Unfortunately they misclassified these upper incisors as C-7 forms, but their photograph of the specimen shows a clear C-4 extreme. Other teeth showing the extreme C-4 design, identical to those from Tecualilla and Amapa, have been recovered from a site near Sentispac, Nayarit (Gill 1976). These teeth were within the collections of the late Hector Galvéz (Instituto Nacional de Antropología e Historia, Mexico). So, the extreme C-4 tooth form appears to have been a well-established style in the region of northern coastal Nayarit in the early Postclassic period.

In sum, perhaps the most interesting information to emerge from this study of dental modifications in the Marismas is the strong suggestion of decline of the practice in three different ways from the earliest Postclassic sites to the later ones. Downward trends regarding the prevalence of the overall practice as well as the severity of the forms applied seem quite compelling from our evidence. Perhaps less convincing is the argument that fewer teeth within the dental arcade were involved through the passing of time. Yet additional evidence comes to support this last idea from Guasave. The burial zone at Guasave, north of the Marismas sites, is dated older than our Postclassic sites (Ekholm 1942; Glascock 1980; Romero 1958), and interestingly the decorated sets of dentitions there show not only involvement of maxillary incisors and canines, like some from Tecualilla and Chalpa, but also involvement of maxillary premolars. Furthermore mandibular teeth at Guasave are often carved as well, a condition not found at even our earliest Marismas sites. Additionally, the Guasave specimens show many of the extreme, pointed forms of carved teeth and very few notches other than the widespread A-1 form (Glascock 1980). This is to be expected if the three trends in the decline of this practice suggested by our Marismas data are genuine changes in cultural behavior rather than fortuitous occurrences within the database resulting from small sample sizes (particularly from the late sites of Rincon de Panal and Cristo Rey). Thus the information from Guasave strengthens the argument for genuine changes in cultural behavior during the Postclassic period within the Marismas Nacionales.

Summary and Conclusions

An extended program of research focusing on the study of disease pathology, trauma, skeletal anomalies, and age at death distributions has produced a sizable quantity of significant information on the lifeways of the prehistoric inhabitants of the Marismas Nacionales. Equally important insights have been gained from examining the effects of cultural activities on the skeleton, including both the purposeful modifications (artificial cranial modification and dental modification) and the accidental effects of cultural practices (such as inflammation of the tibiae seemingly caused by ligatures).

The discovery of a high death rate among young adult males at most of the Postclassic sites certainly suggests high-risk activity among these males. The answer would seem to lie with warfare or at least some form of interpersonal conflict. Severe perimortem injuries to two males (one from Tecualilla and one from Chalpa), and to none of the females, support this, as does the higher injury rate in general within the male samples. Chalpa is the one exception, where the male injury rate is only 22 percent while the female rate is 27 percent. At Tecualilla, on the other hand, the male rate is 53 percent, and the female rate is 40 percent. The high rates for both sexes at Tecualilla probably signal a period of conflict, while the overall lower rates at Chalpa, particularly among males, may indicate a more peaceful time.

Dental pathology suggests a number of interesting things. First, the teeth of children and adolescent individuals in the precolumbian Marismas were quite healthy. A moderately high caries frequency of 11 percent is found among the adults, which is quite consistent with an agricultural economy. More hunting seems to have occurred during the earlier Postclassic, judging from the much lower caries rate among the adults at Tecualilla, Chalpa, and Venadillo (8.9 percent), while rates twice that high are found at the later sites of Rincon de Panal and Cristo Rey (18.1 percent). The earlier sites may, in fact, represent a period of mixed economy (agriculture supplemented with significant hunting and gathering) rather than a predominantly agricultural economy. An expanded sample of late Postclassic dentitions would be quite helpful in settling this question of a potential shift in economic subsistence activity.

Nearly 89 percent of the crania from the Marismas reveal the practice of artificial cranial modification. All examples are of the tabular, erect type of deformation. The practice seems to have gained even more momentum after Tecualilla. The combined frequency for the sites occurring after Tecualilla is actually 95 percent. The three forms and three varieties found within the Marismas appear to be equally distributed between males and females and also appear to show little change through time. Yet the "paralleled" variety of deformation seems to have disappeared after Tecualilla. Also, no extreme forms of deformation are encountered after Venadillo; nor were examples of the pseudocircular variety. This may indicate greater specialization within this practice through time, or it might only mean that the reduced sample sizes from the later sites have fortuitously omitted certain examples of the practice.

It is perhaps meaningful that the apparent demise of the paralleled variety of deformation (about AD 1250) coincides with the first evidence of genetic change in the region and also an abrupt change in burial practices from the extended, primary burial form to the flexed. A few of the extreme dental modifications appear to have disappeared at this time as well. Certainly, important changes at various levels of society appear to have been under way by AD 1250.

Nearly all examples of the most extreme hyperbrachycranic cranial indices are within the female samples of artificially deformed crania. Furthermore, most skulls showing no cultural modification are found within the male samples. All of this suggests that even though no styles of artificial deformation were gender-specific, the prehistoric society did place more emphasis on the practice (both extent and intensity of application) for infant girls than for infant boys. Since an aspect of this cultural practice appears to have been to enhance physical appearances, perhaps society placed slightly greater value on this achieved beauty for females than for males.

Clear parallels have been demonstrated between the shapes of actual skulls from these Postclassic sites and the shapes of heads of clay figurines from the same region and period. It is clear that the clay figurines were not always intended to be mere copies of actual human heads, since some of the figurines are more extreme than the skulls. In fact, at times they seem to represent hairstyles and headdresses and not a human head shape at all. Instead, the two separate artists' media (skulls, on the one hand, and figurines, on the other) appear to represent discrete manifestations of the same conceptual ideal. Both symbolize the same or similar concepts, but in general neither is intended as a copy of the other.

Dental modification within the precolumbian Marismas appears to have served as a rite of passage into adulthood, with an age of application of just 15 for both boys and girls. The same forms and varieties were applied to both sexes and to those of various levels within the social hierarchy. Strong evidence also exists for a decline in dental modification practices through time in three different ways: (1) less prevalence of the overall practice, (2) reduced severity of the forms applied, and (3) the involvement of fewer teeth within each individual's dental arcade. Very clearly the later Postclassic sites show fewer teeth involved; much milder, less extreme forms applied; and a much larger number of individuals showing no signs of dental decoration at all. In this region of coastal west Mexico at least, the practice of dental modification was clearly in decline well before Spanish contact. The opposite seems to be true for cranial deformation. It seems, rather, to have been increasing in overall popularity up until the time of conquest.

Two examples of osteitis of the tibia have been noted on adult males who may have been warriors. Historical evidence in the form of artists' drawings from the time of Spanish conquest suggests that these mysterious swellings (of otherwise normal, healthy bone) might be the result of decorative ligatures worn tightly around the lower leg. Drawings depicting the 1530 Guzmán Entrada actually show such ligatures on the lower legs of warriors in the region of the Marismas Nacionales.

Migration and subsequent gene flow into the Marismas region appears to have lessened regional inbreeding. This is indicated by dramatic declines from the earliest to the latest sites in certain genetically influenced anomalies and pathological conditions. Rheumatoid arthritis, for instance, drops from an adult frequency at Tecualilla of 29.2 percent to 13.6 percent at Chalpa, 4 percent at Venadillo, and 0 percent at the last two sites. Furthermore, the generally quite rare anomaly of the sternal foramen likewise drops in frequency from an incredibly high 33.3 percent at Tecualilla to 12.5 percent at Chalpa and 0 percent at Venadillo, Rincon de Panal, and Cristo Rey. Also, the olecranon foramen among female adults within the Marismas Nacionales drops from the high frequency at Tecualilla of 75 percent to a frequency of 50 percent, or less, at the later sites.

Tuberculosis of the skeleton (Pott's disease) has not been extensively documented in precolumbian Mexico, yet a few rather clear cases have been demonstrated from our Postclassic period sites within the Marismas. At least three adult skeletons from the Tecualilla and Chalpa sites reveal the classic pattern of destruction to the spinal column characteristic of Pott's disease. This helps shed light on prehistoric demographic patterns within the region, to include not only coastal migration and contacts but population densities and interactions.

Information from various osteological sources has been presented here, all the way from mortality, injury, and disease to cultural practices, aesthetics, and cultural change. Many insights into the adaptations and lifeways of these precolumbian coastal people have been gained. Many more, however, await the accumulation of additional skeletal information from particularly the late Postclassic temporal span and also the integration of the skeletal information presented here with other archaeological, cultural, and natural science information. In these ways, eventually an even fuller picture will emerge of life in the prehistoric past of the Marismas Nacionales of coastal west Mexico.

Acknowledgments

Many people and various institutions provided valuable assistance to me during the 46-year period of the osteological research summarized in Chapters 10 and 11 of this volume. The two people most responsible for my involvement in the Marismas Nacionales project from 1968 to 1974 were Dr. Stuart D. Scott, the Marismas project principal investigator, and Dr. William M. Bass, my doctoral dissertation adviser at the University of Kansas (KU). The initial inspiration and planning of the overall project to include the concept of burial recovery and skeletal analysis came from Dr. Scott. His support of me and my assistants and students in all phases of my studies went far beyond that of the normal principal investigator. It led not only to the success of my work but to a close colleague relationship and eventual lifelong friendship.

When Stuart Scott requested a human osteologist from KU to join his State University of New York at Buffalo archaeology team in 1968, Bill Bass chose me, a first-year graduate student in physical anthropology, to fill the position. Furthermore, following my first field season of burial excavations in the Marismas, and my decision to pursue this work as a dissertation project, Dr. Bass assisted me in obtaining a National Science Foundation Dissertation Grant (GS-2497) for continued fieldwork. He not only inspired and encouraged my interest in biological anthropology, even as an undergraduate student, but continued to support me as I moved through the KU doctoral program. I want to sincerely thank Dr. Bass for his inspiration, help, and encouragement. A better mentor, colleague, and lifelong friend could not be found. Dr. Ellis R. Kerley was also on my doctoral committee at KU. He provided valuable guidance during the analysis of paleopathologies found in the Marismas sample.

During my graduate student years at KU a number of fellow students joined the osteology part of this project and provided valuable assistance. The two who did the most were Dr. William S. Lyon, then a fellow graduate student at KU, and Susan Prince Warren, a KU undergraduate. Bill Lyon was supported by my National Science Foundation grant for the 1969 field season and worked with me at Tecualilla and Chalpa. Susan traveled with me in 1970 to museums in western Mexico to study the ceramic figurines that are compared here with the artificially deformed human crania from the Marismas sites.

During those important early years of field research and collection curation at KU many additional individuals provided important guidance and assistance. Among those are the late Drs. T. Dale Stewart, Smithsonian Institution, and Georg Neumann, Indiana University, as well as then–fellow graduate student Dr. Richard Jantz. Family members who provided project assistance well beyond the usual were my father, Dr. George L. Gill (laboratory photography); my father-in-law, P. Craig Livesay; and my wife during those years, Carol Gill Richmond (Spanish translation and typing of all report manuscripts and my dissertation). Anthropologist colleagues in Mexico, particularly Dr. Javier Romero, Instituto Nacional de Antropología e Historia, and Dr. Arturo Romano, former director of the Museo Nacional de Antropología in Mexico City, provided photographs and published material that assisted greatly in the interpretation of research results. Many others in Mexico rendered invaluable assistance over our years of field research. Landowners Juan and Tomás Fonseca and their family members made my work at Chalpa a most pleasant and memorable experience. My gratitude is also extended to their cousin Martín Fonseca and his family, who owned the land at Tecualilla.

With my acceptance of a faculty position at the University of Wyoming (UW) in 1971 and the completion of my doctoral work, skeletal collections from Chalpa and Tecualilla were moved to Wyoming, and a new phase of Marismas osteological research began. Two of my earliest master's students, Dr. James Schutte and Henry W. Case, completed their thesis projects on Marismas skeletal material. Schutte and Case, along with a number of other undergraduate and graduate students, traveled to west Mexico for field and laboratory work. Of particular help were Dr. Douglas W. Owsley, Susan Weigner Brown, Marla Speas Wold, and John Leedy. Important data compilations and statistical analyses were accomplished as student class projects by Laura Jurgenson and Phillip Lanum and by Dr. Marcel Kornfeld and Dr. Kim Smiley. I wish to thank all of these former UW students for their important contributions.

More recently, another group of professionals and other individuals facilitated the completion of my two chapters. The most important of these is the editor of this volume, Dr. Michael S. Foster. Mike directed the field work at Venadillo, and he has been a major facilitator for many other aspects of the Marismas project, doing more than any of us in recent years to pull together the various parts of this multidisciplinary study and see its publication. We all owe him a sincere debt of thanks.

I also want to thank Dr. Rick Weathermon, senior research scientist, Department of Anthropology, UW, for his assistance in compiling the drafts of graphics for my Chapters (10 and 11) and for his review of and comments on Chapter 10. Thanks also go to Mike Foster for finalizing the graphics for publication. Keith Kanbe, office associate, Department of Anthropology, UW, prepared the electronic versions of my manuscripts, and his word processing skills and assistance are greatly appreciated. UW anthropologists Dr. Melissa Murphy and Dr. James Ahern reviewed my chapters; their valuable suggestions and time are greatly appreciated.

Last, I want to thank my wife, Denise Ann Gill, for her considerable patience and support during the final research and writing phases of this work. This wrap-up effort has taken a considerable amount of time from my early retirement and space from within our home. I also appreciate the patience and understanding of my four children and seven grandchildren. Even though they live away from home, a few more visits and enjoyable times could have taken place had Granddad's mind not been off in the swamps of coastal west Mexico!

CHAPTER 11

Skeletal Biology and Population Affinities

George W. Gill and Henry W. Case

The area encompassing the Marismas Nacionales is rich with evidence of prehistoric human activity. During the Mesoamerican Classic period and more especially the Postclassic period this extensive lagoonal and estuarine system was particularly active, in terms of both population dynamics and cultural changes (Gill 1971, 1974, 1977, 1985; Gill and Case 1983; Schutte 1973; Scott and Foster 2000). As previously discussed, the large coastal plain of which the Marismas Nacionales is a part forms a long, relatively narrow corridor between the American Southwest to the north and the tropical lowlands and west Mexican uplands to the south, a corridor that has long been seen as a route of contact linking the two regions (e.g., Ekholm 1942; Foster 1999; Kelley 2000; Mathiowetz 2011; Sauer and Brand 1932).

Although exact avenues of population movement and cultural diffusion within the west Mexican coastal zone and between the American Southwest and west and central Mexico are not well defined, there is mounting and unequivocal evidence that such contacts occurred. Nevertheless, some (e.g., Carpenter 1996; Smith and Heath-Smith 1980) have argued that such movements in the past have been greatly overemphasized. They deemphasize migrations, diffusion, and the "waves of influence" model, especially as it applies to the Mixteca-Puebla art style. They believe that certain Postclassic changes in west Mexico and other regions peripheral to the central highland region were more a result of interdependent exchange and communication networks than of diffusion from a nuclear core area (e.g., Smith and Heath-Smith 1980:15–16). Vaillant's earlier view is that "there is evidence that much of the distribution [of Mixteca-Puebla culture] was accomplished by *movements of peoples*; the movement seems datable as between 1100–1300" (1940:300; emphasis added). It is the contention here that hypothetical, and to some extent disputed, Postclassic "movements of peoples" might better be documented or refuted on osteological grounds than from the study of ceramic art styles, which could, of course, spread either with or without the actual movement of people. It is partly to that end that the present human osteology study was begun.

Figure 5.1 shows the locations of sites in the Marismas that produced the human skeletal sample for this study. The Classic period samples are from Panales and the earthen portion of Mound C at Venadillo; these have been previously reported (Gill 1984). The Postclassic sites including Tecualilla, Chalpa, Venadillo Mound B, Rincon de Panal, and Cristo Rey produced far more skeletons, and they constitute the nucleus of the skeletal sample discussed here. The few or very poorly preserved human remains from the two Classic period sites of Juana Gomez and La Tarjea are not included in this study.

As mentioned in Chapter 10, of the 248 human skeletons analyzed as a result of the Marismas project, the majority come from the five Postclassic sites cited above. Adolescents, children, and infants have, of course, been omitted from the metric tabulations and comparisons in this study, and adult sample sizes for each site group will be listed by site sample within each subsequent individual table.

Dating of the burial sites is treated fully in Chapter 4 (see Figure 4.1). Table 11.1, however, provides an abbreviated Postclassic burial site chronology for the Marismas (Early Conchera–Middle Conchera).

The chronological sequence of the above list of sites appears to be secure, based upon not only radiocarbon dates and ceramic evidence but also changes in burial practices and other cultural patterns. The exact dates for the beginning and termination of the placement of burials at these sites, however, are not precisely known.

For more than 25 years following the completion of excavations in the Marismas Nacionales, University of Wyoming students, former students, and the two authors of this chapter have been involved in the analysis of craniometric, odontometric, anthroposcopic, and other skeletal data. Although the results of these studies have

TABLE 11.1. Postclassic Burial Site Chronology for the Marismas Nacionales.

Site	Date
Cristo Rey	AD 1100–1400
Rincon de Panal	AD 900–1350
Mound B, Venadillo	AD 1150–1300
Chalpa	AD 900–1300
Tecualilla (burial zone)	AD 1000–1200

formed parts of master's theses, conference presentations, and other broader osteological and dental studies, few of these studies have been published. It will be the purpose of this chapter to integrate the more important findings of these various studies into a cohesive whole, which will describe many physical characteristics of the prehistoric inhabitants of the Marismas Nacionales and shed light on origins and relationships. The directions and timing of population migrations and the presence and extent of gene flow between certain groups are also discussed.

METHODS

The osteological and odontological analyses within this study have been undertaken from four approaches: osteometrics, odontometrics, evaluation of discrete nonmetric traits (Inca bones, extra sutures, etc.), and anthroposcopic approaches to continuous nonmetric cranial characteristics (sizes of supraorbital ridges, nasal profiles, etc.).

Measurements and Observations

Fifty-seven cranial and postcranial measurements were taken on each of the adult and late adolescent specimens. Thirty-one nonmetric observations were made on the presence or absence of certain discrete traits of mostly the cranium. Twenty-nine of these are ones utilized by Jantz (1970) for the Plains Arikara. All 31 are illustrated and defined in Gill (1971) and reviewed more recently by Rogers (2008). Also, Appendix B provides illustrations of most of the ones utilized in this study. In addition to these discrete observations, 17 nondiscrete, continuous traits were examined, and these are also fully described and in some cases graphically illustrated in Gill (1971). Two dental measurements, mesiodistal crown diameter and buccolingual (labiolingual) crown diameter, were taken on all adult dentitions from the six Postclassic site samples. Also, 100 laboratory photographs were taken of crania, teeth, and postcranial bones.

Scope of the Study

The scope of examination in a descriptive skeletal and dental analysis such as this one should indeed be broad since even though heritability calculations have been made for certain craniometric traits (Carson 2006), knowledge of the exact genetic bases for these characteristics remains limited. This is especially true when there is equal interest in sociocultural and other environmental effects upon the skeleton. Certainly the latter has been the case since the earliest years of the present study. Regarding the former point, that is, the limited knowledge of the exact genetics behind skeletal traits, the problem exists not only with regard to the quantitative polygenetic traits but with the qualitative discrete ones as well. All appear to be closely related to genotype, and some certainly more intimately than others (Carson 2006:175; Dobzhansky 1962:84; Johannsdottir et al. 2005: 204–205), yet known alleles have been identified for almost none of them. The broad-spectrum approach, therefore, utilizing many different variables, appears to be the soundest approach.

Of the 59 measurements taken in this study, 49 of the most meaningful ones have been selected for presentation or for the calculation of indices presented in the tables. All 31 of the discrete observations are listed in subsequent tables, and all 17 observations on the nonmetric, continuous traits are discussed.

Techniques

The measurements selected for this investigation, and listed in the metric tables, have been defined by a number of human osteologists (Bass 2005; W. Howells, personal communication 1968; Howells 1973; Olivier 1969). Most of these measurements are standard among all osteologists, and some are even self-explanatory with regard to measurement technique, since the cranial landmarks are contained within the name itself. However, some of the measurements are not self-explanatory, and the techniques of the various investigators differ. For instance, cranial length and breadth were taken according to Bass (2005) and Howells (personal communication 1968; 1973), who define opisthocranion as the most distant (posterior) point on the skull excluding the external occipital protuberance. Olivier (1969) makes no provision for the exclusion of the occipital protuberance, which in the case of a number of deliberately modified crania forms the most posterior point. Thus, the more common approach of Bass and Howells has been used here, and the external occipital protuberance has been excluded.

The techniques of Olivier (1969), on the other hand, have been used for palatal length and breadth. The palatal measurements are of the external palate, termed length and breadth of the alveolar arch by Olivier. Interorbital breadth has been taken from the maxillofrontal points and not dacryon, and biorbital breadth was taken as a maximum diameter along the lateral margins of the orbit and not from fixed ectoconchian points as defined by Howells (1973). Prosthion has been considered here the most anterior-inferior point in the midline of the upper alveolar process (Giles and Elliot 1962). Therefore, the nasion-prosthion measurement of upper facial height is taken to the most inferior point in the midline of the alveolus, often given the separate name alveolare or infradentale (Bass 2005; Howells 1973). All of these measurements were taken with standard osteometric spreading or sliding calipers. All cranial measurements were either taken directly by or verified by Gill, except for those taken on the Venadillo Mound B material, which were collected by Susan Weigner Brown.

All dental metrics were taken by Case with a Boley gauge (to 0.1 mm). The beaks were sharpened to facilitate interproximal placement of the instrument. As mentioned above, mesiodistal and buccolingual (labiolingual) crown diameters were taken on each permanent tooth. These were taken using the methods of Moorrees (1957). For a more detailed discussion of these methods and the reasons for using them, one may consult Case (1976).

Since right and left teeth of the same type are highly correlated, measurements were taken from only the left side of the dental arch, unless a left tooth could not be measured. Then the right counterpart was substituted. Many teeth in the Marismas sample could not be measured due to caries, chipped enamel, and dental mutilation. Most adult dentitions from the Marismas samples show some wear. If it was felt that more than a slight reduction resulted from attrition, the tooth was not measured.

Corpal length of the mandible, gonial angle, and height of the ascending ramus were all taken with a Swiss-made mandibulometer (G. P. M. Gneupel). This is particularly important to keep in mind with regard to corpal length, which is a considerably greater diameter when taken directly along the full length of the horizontal ramus of the mandible using sliding calipers.

Auricular height and all measurements from porion were taken with the Western Reserve Head Spanner. Palatal depth was taken with a standard osteometric palometer, with the lateral wings of the instrument placed between the first and second molars, according to standard procedure.

Cranial capacity was calculated on the original Tecualilla and Chalpa samples in 1971 using direct measurement with a volumetric cylinder (Length Bonnett Engineers) and mustard seed (Gill 1971:104). There are advantages for conducting this tedious exercise on the crania since some comparative skeletal samples from western Mexico and the American Southwest, particularly in earlier studies, were measured by these same techniques. In the present study, however, recalculations were made using the formulas of Lee and Pearson presented in Olivier (1969). This not only has made the assessments easier but has increased sample sizes (since more fragmentary crania can be utilized). Cranial module was also calculated, and it consists of an average of cranial length, breadth, and basion-bregma height (Bass 2005).

Measurements of postcranial bones were taken with a standard osteometric board and sliding calipers. Also, a flexible metric tape was used for mid-shaft perimeter measurements on femurs and humeri. Perimeter measurements of these two bones, in conjunction with measurements of maximum length (for both bones) and physiological or oblique length (femur), allow for the calculation of indices of robusticity (Olivier 1969). These have been calculated for each adult individual, and in cases where the paired bones have yielded different indices of robusticity, an average of the two results has been used. The methods of Olivier (1969) have also been used for the calculation of the sacral index (index of sacral breadth) and pelvic index (pelvic height index).

Craniofacial indices have been calculated for both males and females from each major site according to the methods outlined in Bass (2005). Statistical analysis of craniofacial metrics has been undertaken by various investigators utilizing both Penrose (1954) distance (CH^2), shape (CZ^2), and size (CQ^2) and principal component scores. The Penrose tabulations from cranial metrics are the product of a class project undertaken by former University of Wyoming students under the supervision of Case; for this study 15 variables were chosen by Gill for their reliability in showing genetic relationships. These are minimum frontal breadth, bizygomatic breadth, nasal height, nasal breadth, left orbital height, left orbital breadth, maxillofrontal breadth, biorbital breadth (maximum), biauricular breadth, cheek height, porion-nasion, porion-subnasale, porion-prosthion, mandibular symphysis height, and breadth of the ascending ramus of the mandible. Since these measurements were obtained on more males than females, some female site comparisons cannot be made. A total of 56 males from Tecualilla, Chalpa, Venadillo Mound B, Rincon de Panal, and Cristo Rey and 37 females from the first three of these sites were included.

The principal component scores were computed by Case using the SAS routines PROC CORR and PROC MATRIX. The principal components analysis has compared 42 Marismas skeletons among themselves and with other Mexican Indian groups (Genovés 1970; Romano 1970) and the Pecos Pueblo (Hooton 1930). Since the Marismas skeletal series is small, the variance-covariance matrix was computed from a wide range of Native American skeletons, some of which have been measured by the authors and others that were taken from the literature. In addition to the Marismas series, skeletons used in the variance-covariance matrix include the individual skeletons reported in Genovés (1970), Romano (1970), Gill (1978), Miller and Gill (1980), Bass and

TABLE 11.2. Indices of Tecualilla Males.

Index	*N*	Mean	SD	Range
Cranial (Unmodified) [a]				
Cranial index	5	80.77	3.99	75.27–84.27
Cranial module	5	152.20	2.27	150.33–155.67
Height-length	5	75.22	2.73	71.27–78.08
Height-breadth	5	93.31	5.59	88.44–102.92
Upper facial	4	54.29	2.02	52.86–57.78
Total facial	3	89.68	4.71	85.71–96.30
Nasal index	5	50.38	5.52	43.40–60.38
Orbital index	4	78.90	3.82	72.34–81.82
Cranial (Modified)				
Cranial index	10	96.70	8.52	86.04–112.00
Cranial module	9	152.59	3.33	148.33–157.33
Upper facial	7	49.39	0.95	47.52–50.67
Total facial	7	82.37	2.42	80.24–86.67
Nasal index	6	49.82	2.68	47.05–54.90
Orbital index	6	79.42	5.08	72.74–85.37
Postcranial				
Robusticity (humerus)	13	20.76	1.47	18.60–23.48
Robusticity (femur)	14	20.14	0.45	19.28–20.61
Sacral index	10	104.29	5.61	95.50–115.00
Pelvic index	12	77.19	3.21	72.26–82.33

[a] These unmodified male skulls constitute the only significant sample of unmodified crania from anywhere in the Marismas Nacionales and are therefore listed separately.

TABLE 11.3. Indices of Chalpa Males.

Index	*N*	Mean	SD	Range
Craniofacial [a]				
Cranial index	6	97.54	2.96	83.83–101.84
Cranial module	3	149.56	3.50	144.67–152.67
Upper facial	2	53.68	3.67	50.00–57.35
Total facial	2	87.60	5.78	81.82–93.38
Nasal index	7	47.36	1.63	46.15–51.02
Orbital index	5	87.18	6.94	85.71–97.56
Postcranial				
Robusticity (humerus)	3	21.48	0.21	21.21–21.54
Robusticity (femur)	4	19.96	0.40	19.27–20.17
Sacral index	3	106.64	10.04	99.09–120.83
Pelvic index	1	79.05		

[a] All mean craniofacial indices listed here are based upon indices from skulls showing at least some evidence of artificial deformation.

TABLE 11.4. Indices of Tecualilla and Chalpa Females.

	Tecualilla				Chalpa			
Index	*N*	Mean	SD	Range	*N*	Mean	SD	Range
Craniofacial [a]								
Cranial index	6	99.10	10.82	83.83–116.55	13	98.85	10.31	84.52–125.17
Cranial module	5	148.47	4.21	141.67–153.00	12	148.97	3.46	142.62–154.67
Upper facial	4	50.96	5.35	44.06–56.35	8	49.81	2.78	46.04–54.55
Total facial	4	81.56	8.58	71.33–90.40	8	83.78	3.52	77.70–89.15
Nasal index	5	51.70	2.45	50.00–56.52	8	48.30	4.90	41.18–55.10
Orbital index	6	80.15	5.51	71.74–87.50	10	84.30	6.41	76.74–98.70
Postcranial								
Robusticity (humerus)	6	19.47	1.78	16.18–22.22	9	20.13	1.14	18.46–22.10
Robusticity (femur)	6	18.54	1.28	16.75–20.48	7	19.17	0.95	18.10–20.58
Sacral index	2	116.28	7.63	108.65–123.91	4	108.64	5.38	100.90–112.61
Pelvic index	4	76.46	4.18	69.29–79.27	4	74.04	3.98	71.22–82.87

[a] The mean craniofacial indices listed here are based upon skulls evidencing at least some artificial deformation.

Lacy (1963), Bass and Barlow (1964), Birkby and Bass (1963), and Webb and Snow (1981), as well as seven unpublished Northwestern Plains Indian skeletons (from Wyoming) that were measured by Case.

The high percentage of modified crania in the Marismas sample precludes the use of many cranial vault measurements for principal components analysis. So, in order to compare the Marismas skeletal sample with other Mexican Indian groups it has been necessary to limit the present analysis to six of the original 15 variables. These are minimum frontal breadth, bizygomatic breadth, nasal height, nasal width, left orbital height, and left orbital breadth.

The statistical methods used in the odontometric studies include (1) descriptive statistics calculated for each site group, (2) distributions for each tooth measurement compared between site groups in pairwise fashion by the Mann-Whitney *U* test, and (3) Penrose's distance (CH^2), shape (CZ^2), and size (CQ^2) calculated between pairs of Marismas populations and between each Marismas population and other New World groups. The Penrose comparisons are based upon 16 odontometric variables. These 16 characters used in the calculations are the mesiodistal and buccolingual diameters of both maxillary and mandibular P1, P2, M1, and M2.

Skeletal Population Analysis

As mentioned earlier, the Classic period skeletal remains from the Marismas are all exceedingly fragmentary; not a single measurable human bone was recovered from the sites of Juana Gomez, La Tarjea, or Panales. Some dentitions from Panales were adequate for study, but these are temporally more recent and may have been early Postclassic in age. The one exception from the Classic period is Mound C at Venadillo, which produced four fragmentary but measurable adult skeletons (Appendix C; see also Gill 1984). Comments on these Classic period remains will be integrated into the discussion of the much more numerous Postclassic period skeletons.

Cranial Metrics of the Postclassic Skeletons

Appendix D presents craniometric data for all Postclassic period skeletal populations of the Marismas Nacionales. These are drawn from the metrics presented by Gill (1971) and Schutte (1973), as well as data files on unpublished Marismas skeletal samples such as Venadillo Mound B.

For descriptions of the exact burial mound features from which each Tecualilla and Chalpa skeleton has come, one may consult Gill (1971:106–111), and for Rincon de Panal and Cristo Rey the reader is referred to Schutte (1973:16–33). All Postclassic specimens from Venadillo come from Mound B.

Table 11.5. Indices of Venadillo Mound B Males.

Index	*N*	Mean	SD	Range
Craniofacial [a]				
Cranial index	9	96.83	2.56	91.95–101.21
Cranial module	8	151.29	5.36	139.33–156.67
Upper facial	1	54.42		
Total facial	1	83.67		
Nasal index	5	45.83	2.46	43.65–50.00
Orbital index	5	84.24	2.60	80.85–86.96
Postcranial				
Robusticity (humerus)	9	21.08	1.36	18.34–22.85
Robusticity (femur)	8	19.86	0.80	18.95–20.97
Sacral index	3	108.63	7.53	100.95–116.00
Pelvic index	1	77.74		

[a] A single individual in this sample shows a normal, unmodified cranium, and all others show signs of artificial cranial deformation. The cranial index figures for the single unmodified specimen are omitted from this table and are discussed elsewhere.

Sample number unfortunately differs from one variable to the next within the tables of measurements and indices since some specimens are well preserved and others are so fragmentary that they produced only one or two reliable measurements. All measurements listed in the tables in Appendixes C and D are given in millimeters except gonial angle, which expresses degrees from the perpendicular.

Cranial Form

A review of most of the cranial length and breadth means from the Appendix D tables, as well as cranial indices from Tables 11.2–11.8, shows the effects of artificial cranial deformation. Only the five unmodified male crania from Tecualilla (Table 11.2), a single male from Venadillo, and two females (Venadillo and Cristo Rey) give an indication of the normal genetic predisposition of the population with regard to cranial dimensions. As shown in Table 11.2 the five Tecualilla males with unmodified skulls average 80.77 in cranial index. The single unmodified male from Venadillo reveals a mesocranic index of 78.92, well inside the range of variation for Tecualilla males. When all six of these Marismas males are pooled, half are high mesocranic in cranial form, and half are brachycranic; and their mean cranial index is 80.45, almost on the 80.00 line of separation between mesocrania and brachycrania. The two unmodified female skulls from the Marismas suggest the same. The one from Venadillo calculates to 80.95, and the female from Cristo Rey is 81.11. An unmodified male cranium in a private collection in Sinaloa that is supposed to have

TABLE 11.6. Indices of Venadillo Mound B Females.

Index	N	Mean	SD	Range
Craniofacial[a]				
Cranial index	8	99.06	6.05	85.98–104.38
Cranial module	4	146.67	4.23	142.33–152.33
Upper facial index	3	52.07	2.10	49.65–53.38
Total facial index	3	86.15	1.93	83.92–87.30
Nasal index	4	56.71	2.92	53.33–60.47
Orbital index	5	81.74	3.56	78.57–85.71
Postcranial				
Robusticity (humerus)	5	20.95	1.78	19.17–23.77
Robusticity (femur)	5	19.86	1.90	18.34–22.97
Sacral index	2	117.34	4.74	113.98–120.69
Pelvic index	2	74.59	8.01	68.92–80.25

[a] One specimen in this sample possesses an unmodified skull, and all others show signs of artificial cranial deformation. The cranial index figures for the single unmodified specimen are omitted from this table and are discussed separately.

originally come from the Chalpa site (Feature A) is also slightly brachycranic (83.83), and one of the Tecualilla females that shows little, if any, deformity has an index of 83.82. The five unmodified male crania from Tecualilla show medium or high cranial length-height indices, and the one from Venadillo is also medium (70.27). Yet, when breadth-height calculations are examined, the male population is seen to be more variable, from low to high, but with a very medium average. Both females lack cranial height measurements and therefore cannot be compared.

Our findings, then, indicate a reasonable degree of population variation regarding the genetic predisposition for cranial form between high medium (mesocranic) and broad (brachycranic). Of course, the possibility of slight, nondetectable artificial cranial deformation having affected these skulls, particularly the crania showing brachycranic indices, cannot be ignored. As may be observed from the tables, the mean cranial indices for all modified groups of both sexes are hyperbrachycranic, ranging from 88.45 to 99.38. The means for the female samples slightly exceed those for the males from the same sites, indicating that artificial deformation was somewhat more extreme among the females of the population. This conforms to the impression gained through visual observation of the same crania.

Figures 11.1 and 11.2 show representative adult male and female crania from the Marismas Nacionales. The male in Figure 11.1a reveals an unmodified cranium. The other male, in Figure 11.1b, and both females in Figure 11.2 show the more typical condition of artificially modified crania. The two males in Figure 11.3 show cranial and facial characteristics that deviate from the average within the Marismas, giving some indication of the range of variation within these populations. The mesocranic male in Figure 11.3b (Burial 2) is one of the five unmodified Tecualilla males that constitute the sample reflected in Table 11.2. This specimen also shows sharp facial features quite unusual for the Postclassic period Marismas and some other traits that will be discussed more in subsequent sections. The heavier featured brachycranic male in Figure 11.3a does show mild cranial deformation.

Cranial Size

Cranial module is remarkably similar for all groups tested. For the females the value averages about 148 mm, and for males the value is closer to 152 mm. The smaller value of 149.56 mm for the Chalpa males is probably

TABLE 11.7. Indices of Rincon de Panal and Cristo Rey Males.

	Rincon de Panal				Cristo Rey			
Index	N	Mean	SD	Range	N	Mean	SD	Range
Craniofacial[a]								
Cranial index	3	94.55	1.01	93.93–95.71	5	97.70	3.86	92.97–102.55
Cranial module	2	151.84	1.65	150.67–153.00	5	153.33	4.70	148.67–160.67
Upper facial	1	51.88			1	51.95		
Total facial	2	85.63	1.74	84.40–86.86	1	87.30		
Nasal index	4	50.74	2.13	49.51–53.92	3	49.94	5.26	44.00–54.00
Orbital index	4	84.14	2.34	82.22–87.50	3	81.68	6.28	85.55–88.10
Postcranial								
Robusticity (humerus)	3	19.93	1.51	18.86–21.66	4	21.54	0.93	18.73–24.32
Robusticity (femur)	3	19.10	0.83	18.56–20.06	3	20.07	1.32	18.74–21.38
Sacral index	1	115.15						
Pelvic index	1	78.68						

[a] The mean craniofacial indices listed here are based upon indices from skulls evidencing at least some artificial deformation.

TABLE 11.8. Indices of Rincon de Panal and Cristo Rey Females.

	Rincon de Panal				Cristo Rey			
Index	*N*	Mean	SD	Range	*N*	Mean	SD	Range
Craniofacial [a]								
Cranial index	1	99.38	0.88		1	88.48		
Cranial module	1	149.33			1	147.33		
Upper facial	1	47.21						
Nasal index	1	53.33						
Orbital index	1	82.93						
Postcranial [b]								
Robusticity (humerus)	1	19.86			2	20.09	0.13	20.00–20.18
Robusticity (femur)					3	20.83	2.57	18.42–23.54
Sacral index	1	115.69						

[a] All crania except one from Cristo Rey show at least some evidence of deformation. The cranial index figures for the one unmodified specimen are omitted from this table and discussed in the text of this report. Total facial index is omitted because none of the specimens were complete enough to yield that index.
[b] Pelvic indices are not listed here since none were obtainable from females at these sites.

the result of the small sample size. Cranial capacity, as expected, is also quite similar between the various populations of the Marismas Nacionales. Using the Lee and Pearson method of calculating cranial capacity (Olivier 1969), the eight measureable males from Tecualilla show an average of 1,432 cc. This is considerably larger than the 1,374 cc average (1,355 cc unmodified and 1,387 cc modified) reported among Tecualilla males in 1971 (Gill 1971: 119), when a direct volumetric measure was used. The difference is due in part to the greater sample size today but more so to the different techniques. Each Tecualilla and Chalpa specimen with cranial capacity directly measured in the earlier study (Gill 1971) has been reanalyzed using the Lee and Pearson method. Males generally show from 50 cc to 100 cc greater capacity when the Lee and Pearson calculations are used, and females average about 25 cc more. The greater distortion of the Lee and Pearson method for males may be due to the thicker crania and heavy bony excrescences of the Marismas males, which artificially inflate results based upon exterior measurement. The Lee and Pearson figures are used in this study, however, since these are the only figures available on three of the Marismas populations (Venadillo, Rincon de Panal, and Cristo Rey).

Using the Lee and Pearson formula results for cranial capacity estimate produces the following male averages: Tecualilla, 1,432 cc ($n = 8$); Chalpa, 1,389 cc ($n = 4$); Venadillo Mound B, 1,437 cc ($n = 7$); Rincon de Panal, 1,447 cc ($n = 2$); and Cristo Rey, 1,467 cc ($n = 5$). Female capacities are considerably less and average as follows: Tecualilla, 1,327 cc ($n = 3$); Chalpa, 1,313 cc ($n = 10$); Venadillo Mound B, 1,300 ($n = 4$); and Rincon de Panal, 1,324 cc ($n = 1$). No measurable female cranium was recovered from the Cristo Rey site. Again, the smaller-than-expected average for Chalpa males is most likely the result of small sample size.

When cranial capacity figures from all Postclassic sites are pooled by sex group, which certainly appears justifiable due to the very slight variations by site group, a 123 cc difference between the sexes (males: 1,436 cc, $n = 26$; females: 1,313 cc, $n = 18$) was found. Even considering the slightly more inflated figures for the males caused by the shortcomings of the Lee and Pearson formulas, this difference is significant and supports other osteological indications of strong sexual dimorphism among the Marismas populations.

Auricular height is not as variable among the tested samples as basion-bregma height, but both show fairly wide individual variation within and between populations. As may be observed from the Appendix D tables, auricular height means for both males and females center around 120–121 mm, except for the very small sample of Chalpa males, which averages less than 118 mm, and the surprisingly low mean from Venadillo Mound B of approximately 113.8 mm for both sexes. Most Cristo Rey and Rincon de Panal samples are too small to judge auricular height, but the few values obtainable suggest that they may have averaged as low as the Venadillo population.

Facial Form

Our findings seem to support those of Giles and Bleibtreu (1961) in that deformation of the skull does not appear to affect significantly those measurements and indices restricted solely to the region of the face. Comparison of the mean facial indices of recovered unmodified specimens with those of the modified ones from the same sites (largely testable only among Tecualilla males)

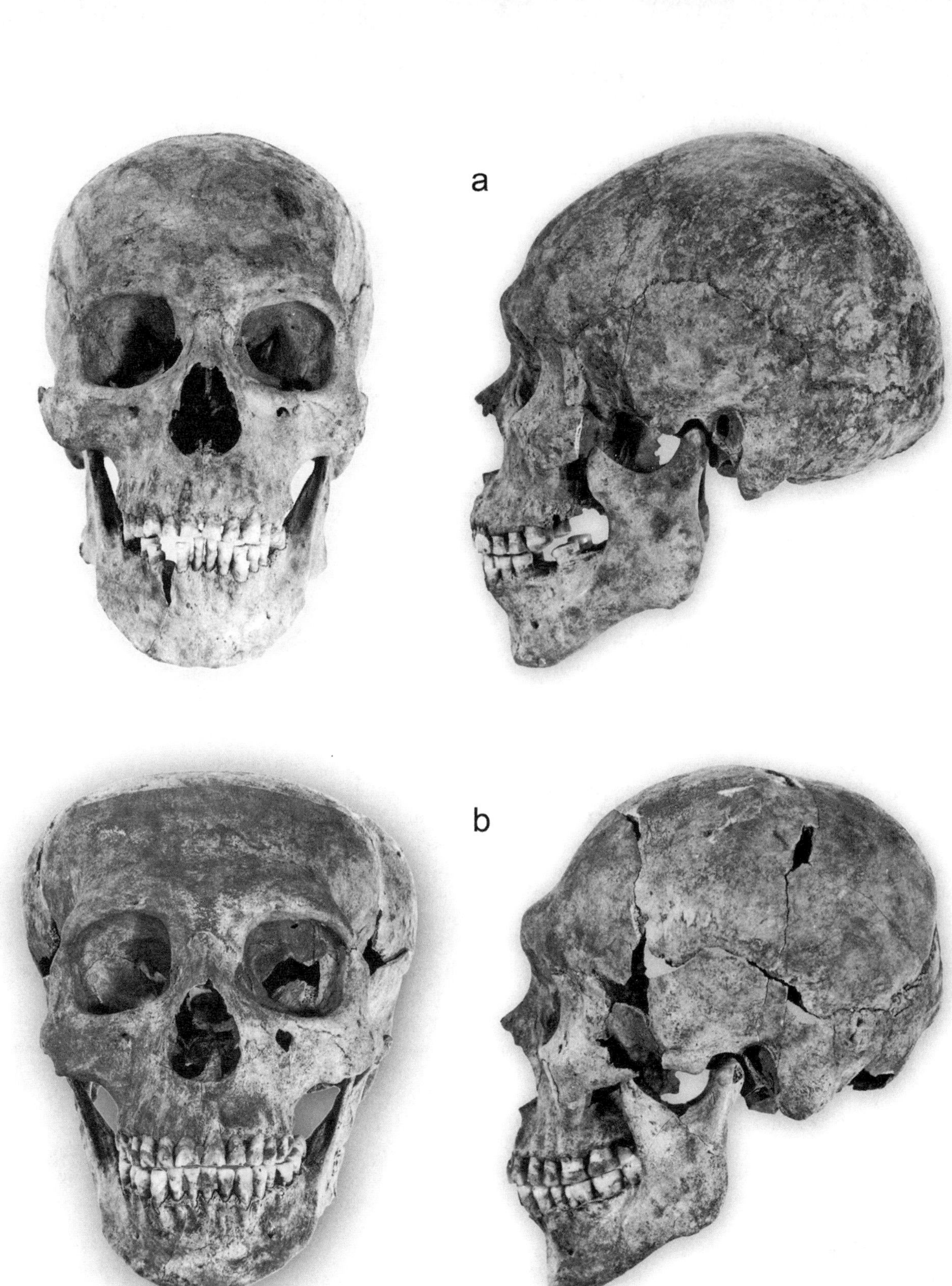

FIGURE 11.1. Male crania from (*a*) Tecualilla and (*b*) Chalpa showing a common pattern of morphology within the Marismas Nacionales. Both reveal rugged faces with moderate alveolar prognathism. Cranium a is from Burial 7, Mound A, Tecualilla, and Cranium b is from Burial 12A, Mound A, Chalpa. The Chalpa skull (*b*) exhibits artificial cranial deformation, while the one from Tecualilla (*a*) does not.

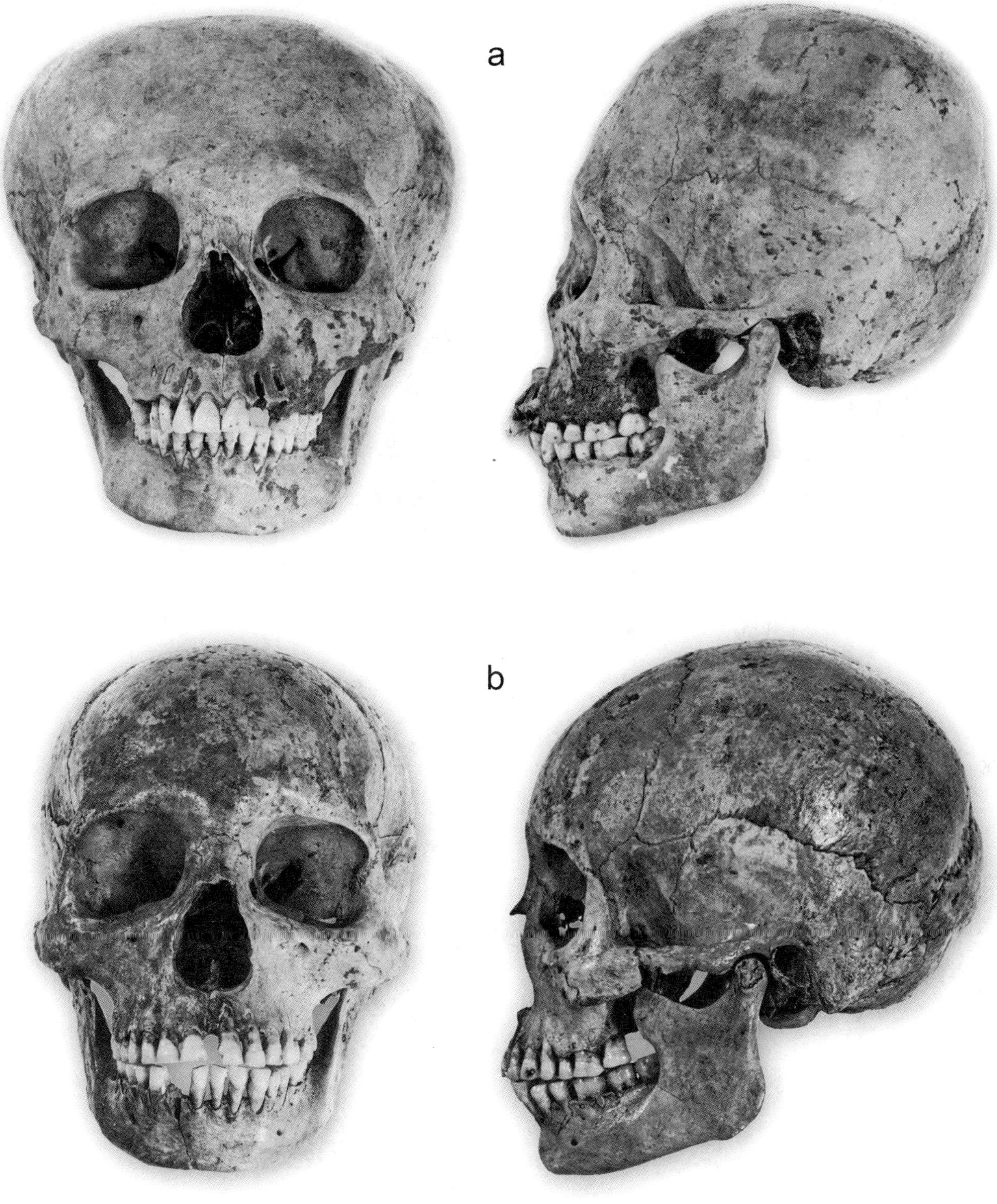

FIGURE 11.2. Female crania from (a) Tecualilla and (b) Chalpa showing common patterns of morphology within the Marismas Nacionales. Short faces and wide interorbital breadths are common among the females. Cranium a is from Burial 17, Mound A, Tecualilla, and exhibits obvious artificial cranial deformation, and Cranium b is from Burial 20, Mound A, Chalpa, and shows much less deformation.

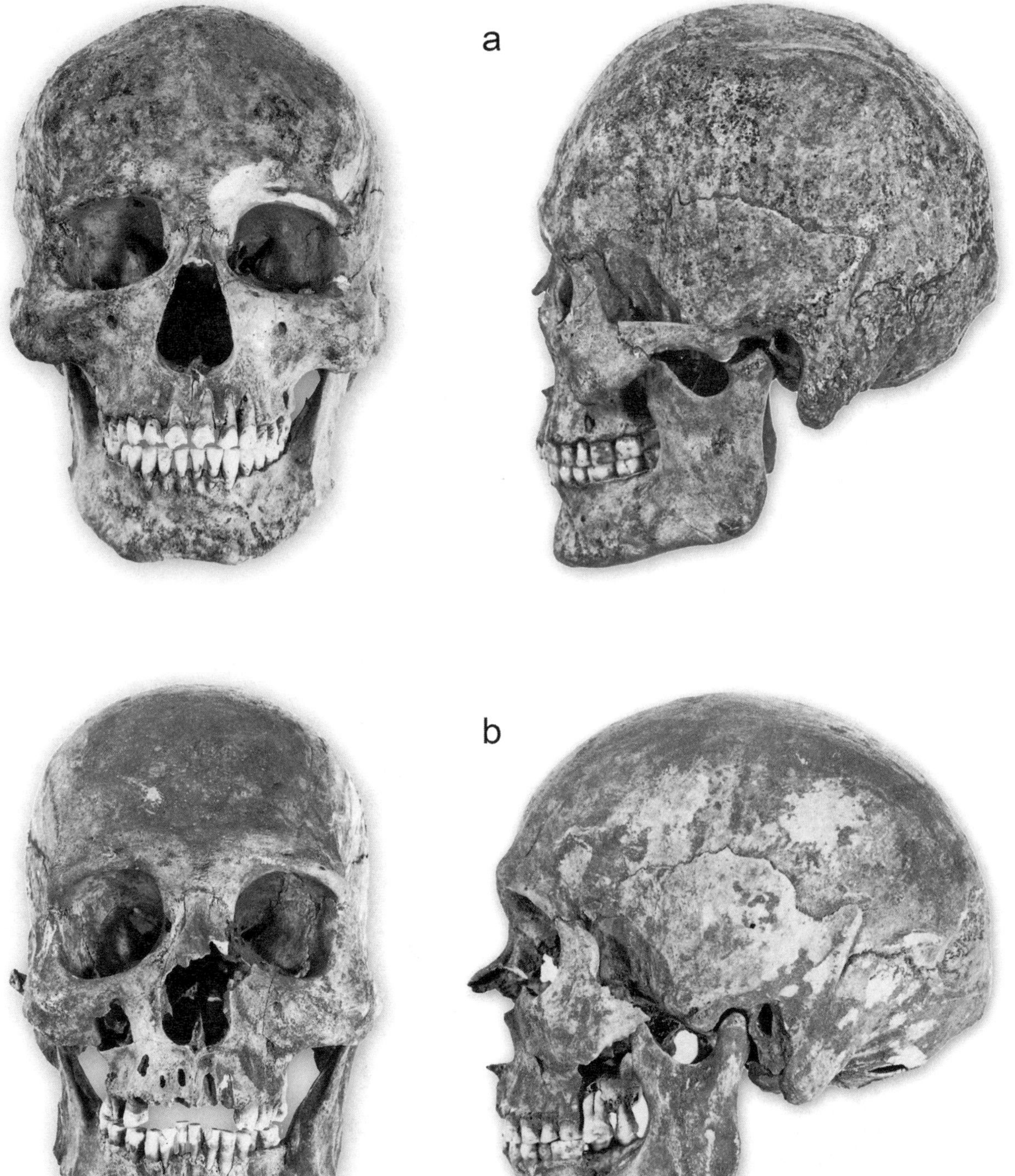

FIGURE 11.3. (*a–b*) Two male crania from Mound A at Tecualilla showing morphological variation within the Marismas Nacionales. These two males illustrate deviations from the more common morphological patterns, in the direction of (*a*) wider cheekbones and heavier jaws and (*b*) a more prominent nasal skeleton and chin with reduced alveolar prognathism. Cranium a is from Burial 10 and exhibits a slight artificial deformation, and Cranium b is from Burial 2 and shows no sign of deformation.

reveals no noticeable differences. Unfortunately even the best samples from Tecualilla are a little too small for statistical testing; especially problematic is the small sample of unmodified individuals.

The facial form of these prehistoric peoples ranged from medium to broad. The mean upper facial index of most of the Marismas samples is mesene. Where it deviates from this, as in the case of Chalpa females and the modified sample of Tecualilla males, it is toward a high euryene value (slightly broad). A few slender faces can be found among the Tecualilla and Chalpa skeletons, but certainly not enough of them to sway the average significantly in that direction.

When the mandible is included for the calculation of total facial index, the faces appear relatively shorter and broader, particularly among females, who show noticeably shorter symphyseal heights. Two of the three female samples (Tecualilla and Chalpa) show euryprosopic or broad total facial indices, while the Venadillo females' average is noticeably higher; most male samples also show higher mesoprosopic values. Here again, the only narrow faces to be found come from the earlier Postclassic sites of Tecualilla and Chalpa.

Nasal Dimensions and Orbital Shapes

Nose form among these populations also tends toward medium to broad, even though the individual variation within each group is great and the averages for two of the male samples actually reach narrow proportions. Both Chalpa and Venadillo Mound B males average leptorrhine, but the other four male samples yield mesorrhine means. The greater number of narrow, leptorrhine noses among both males and females at Chalpa and among the Venadillo males is noticeable. Interestingly, the narrow-nosed individuals from those populations (which are temporally more recent than Tecualilla) possessed other physical characteristics that set them apart from the earlier Tecualilla people and the majority of their contemporaries as well. These trait correlations and their possible meaning will be discussed later in more detail.

Nasal index is one of the two indices available for comparison from the small Classic period sample. Unfortunately, as shown in Appendix C, Table C.2, the two females that could be measured reveal very different indices (one very narrow and the other a rather wide medium).

Perhaps more striking than the temporal differences in nose form are the sex differences. Females from the Marismas Nacionales clearly had shorter, broader noses, throughout all periods (with the possible exception of the Classic period females). The greatest sexual dimorphism is seen at Venadillo Mound B, where the male skulls reveal a leptorrhine average and the females, a platyrrhine one. The small sample size of Venadillo females ($n = 3$) could account in part for the dramatic contrast, but within the larger Tecualilla and Chalpa samples the same general trend is clearly evident.

Another trait that differs somewhat between the Tecualilla skeletons and the later ones is orbital form. All Tecualilla groups average 79 or 80 in orbital index, well into the chamaeconch range according to Bass (2005). Most of the Chalpa specimens, as well as those from Venadillo Mound B and the other later sites, show orbits identical to those from Tecualilla; however, a few individual crania in these younger samples show a different orbital form and index. These few are clearly higher in form and contribute to higher index averages, which in most instances push these samples into the mesoconch range. None of the sample means from the temporally later groups are as low as those from Tecualilla (or from the Classic period Venadillo Mound C site). As may be seen from the tables, they run from nearly 82 (Venadillo females) to over 87 (Chalpa males), depending upon how many individuals in each sample show the "new" trait of higher, more rounded orbits.

As may be seen from the Appendix D tables, the mean biorbital breadth is rather consistently around 98–99 mm for the Marismas females. Males show a somewhat larger dimension and means that are less consistent between samples. Interorbital breadth is particularly interesting. At all Marismas sites females show interorbital dimensions equal to or greater than those of the males, despite the overall smaller size of the female face. This wide spacing of the female eyes was obvious from visual examination of the Marismas crania as they were lifted from the burials and is now adequately borne out by the osteometric results.

The Palate and Mandible

Form of the palate among the various Postclassic period Marismas inhabitants is very consistent. Indices of the superior alveolar arch show broad or brachyuranic palates among all groups. Length of the palate appears to decrease slightly through time, judging at least from the male samples (Tecualilla, longest; Cristo Rey, shortest). Depth of palate is also greatest among the Tecualilla males, who average nearly 17 mm. Other males average nearer the 14–16 mm range.

The male skeletal populations also show a decrease in symphyseal height of the mandible from the large-faced Tecualilla individuals, at approximately 37 mm, to Cristo Rey at only 34.3 mm in height. Dimensions of the female mandibles, in contrast, appear to hold rather consistently from site to site.

Other dimensions of the mandible, such as ramus height and breadth, bigonial diameter, and bicondylar diameter, show little if any variation by site group and not much sexual dimorphism either. Some of the corpal length values for Rincon de Panal and Cristo Rey (Appendix D, Tables D.6 and D.7) collected by Schutte (1973) appear unusually large when compared with those of the other groups. Error is the suspected explanation of the large standard deviation in these cases and the lack of consistency with other mandibular and maxillary dimensions. In checking Schutte's original notes a number of estimates and measurements from a single side of the mandible were found, as if sliding calipers were used on some broken specimens. If this was done without cognizance of the large differences in results between the caliper method and the mandibulometer method of deriving corpal length, it could have the effect of inflating some values and creating the large standard deviations found in his data, and it is suggested that the reader disregard the corpal length values in those two tables.

Males from the Marismas Nacionales do reveal gonial angles of the mandible slightly more acute than those of females. No differences in angle can be discerned, however, between the various site groups.

The Midface

Cheek height is somewhat greater among males, and it is suggested from the data that the Tecualilla males were a little more robust in this area of the face than any of the later ones. No systematic analysis of facial prognathism has been undertaken. However, visual observation and facial measurements taken from porion seem to suggest a degree of upper facial (alveolar) prognathism in excess of the average for Native Americans generally.

Postcranial Osteometry

Living Stature

Stature calculations according to the Genovés (1967) formulas reveal an average for the Marismas males of about 5 ft, 4½ inches. The 14 Tecualilla males averaged 163.56 cm, with the tallest at 169.47 cm (about 5 ft, 6¾ inches) and the shortest at 160.44 cm (5 ft, 3¼ inches). Chalpa males were nearly identical in size, judging from the five skeletons with measurable tibias. The Chalpa male mean is 163.62 cm, and the range is from 160.64 cm (5 ft, 3¼ inches) to 168.00 cm (almost 5 ft, 6¼ inches).

Measurable tibias for females at Tecualilla and Chalpa are even fewer (six from Tecualilla and five from Chalpa). Tecualilla females averaged 151.86 cm (a little less than 5 ft) in height, and those from Chalpa, 155.34 cm (about 5 ft, 1 inch). The tallest females from both sites were 158.39 cm at Tecualilla and 158.92 cm at Chalpa (about 5 ft, 2½ inches). The shortest Tecualilla female was 146.42 cm (less than 4 ft, 10 inches), and the shortest Chalpa woman was 149.41 cm (nearer 4 ft, 11 inches).

Individuals at Venadillo Mound B averaged about an inch shorter than those from the Tecualilla and Chalpa sites, judging from the 14 measureable adult skeletons studied. The nine Postclassic Venadillo males averaged 161.36 cm (about 5 ft, 3½ inches), and the five females, 150.83 cm (4 ft, 11½ inches). The tallest statures at Venadillo Mound B are the same as those from Tecualilla and Chalpa, that is, 168.74 cm (about 5 ft, 6½ inches) for the tallest male and 158.09 cm (5 ft, 2¼ inches) for the tallest female. The shortest individuals from Venadillo, however, are about 2 inches shorter than the shortest members of the Tecualilla and Chalpa groups. The shortest Venadillo male measures 155.93 cm (5 ft, 1½ inches), and the shortest female was 143.97 cm (4 ft, 8½ inches).

Very few stature calculations are available from the sites of Rincon de Panal and Cristo Rey. All four measurable males from these sites (two from each) fall within the ranges established by the other Marismas populations, yielding a mean of 166.06 cm (5 ft, 5½ inches). The single female stature obtained was calculated from the preserved fibula of a Cristo Rey female and is 151.5 cm (just less than 5 ft), quite typical for females of the prehistoric Marismas Nacionales.

Shoulder Breadth

Shoulder breadth among these prehistoric people appears to have been rather great relative to overall stature, judging from the maximum length of the clavicle. This would seem to be particularly true of the earlier Postclassic populations from Tecualilla and Chalpa. Also, sexual dimorphism with regard to this trait was apparently as great as in Europeans. Olivier mentions that the clavicle length of French females is 90 percent that of French males. The mean clavicle length from the Marismas females (132.4 mm) is approximately 90 percent that of the males (148.0 mm). Examination of ranges, as well as these means, indicates that Olivier's scale for the determination of sex (based upon European data) should probably be adjusted slightly for these west Mexican Indians. This is not what was stated earlier (Gill 1971) based upon a smaller sample of only Tecualilla and Chalpa skeletons, but now that samples are larger, the picture becomes clearer. That is not surprising since the overall body size of the Marismas people is showing itself (through several of our skeletal studies) to be noticeably smaller than that of Europeans. The male range in clavicle length is 135–158 mm, and that for the females is 121–149 mm. Distribution charts are helpful (Figure 11.4) and suggest that for this west Mexican Indian population Olivier's chart for sex determination from the clavicle should be adjusted downward 1 mm for males and 2 mm for females. Olivier's scale lists all clavicles above 150 mm in length as male and all those below 138 mm as female. The Marismas data suggest that for these populations a measurement above 149 mm better indicates a definite male and a reading below 136 mm more likely means definite female.

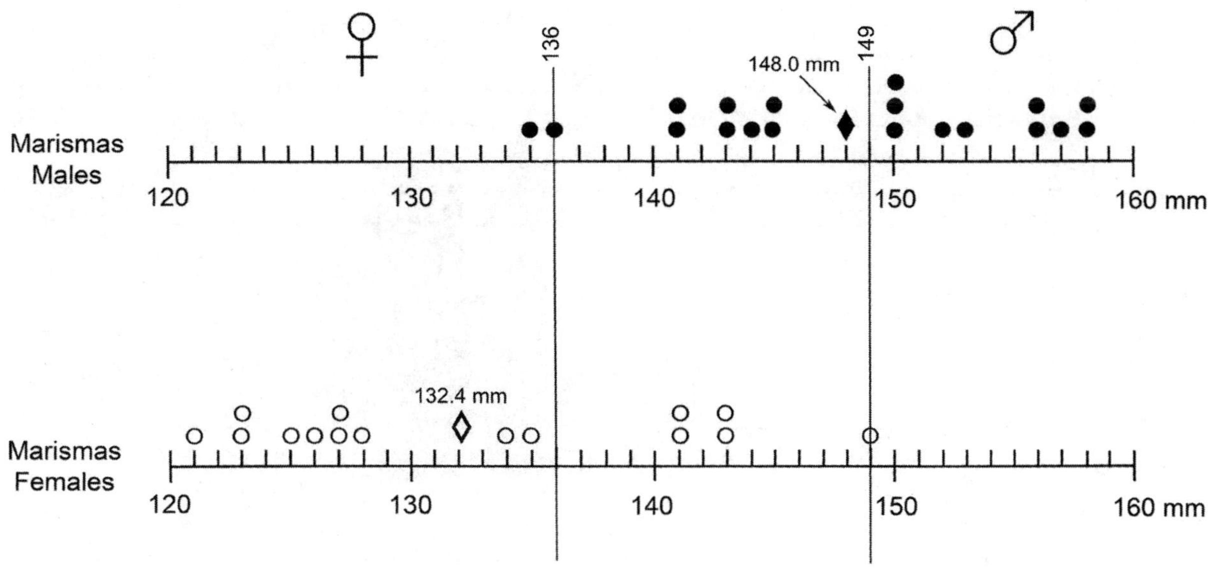

FIGURE 11.4. Clavicle lengths of adults from the Postclassic Marismas Nacionales sites.

As shown in Table 11.9, applying this adjusted scale to the individual skeletons within the Marismas sample produces a better classification result (at least for males) than does the utilization of the unmodified Olivier scale (which is often used by forensic anthropologists today for all human populations). These data would indicate that even though Olivier's European scale seems to work well for sexing the remains of modern American whites and blacks, perhaps for Mesoamerican Indian populations, and presumably other small-statured populations, our adjusted scale might be a better one to use. A further testing of this adjusted scale could, and probably should, be made in the future on a larger sample of Mesoamerican Indian skeletal remains in order to better test its validity and perhaps further adjust the cutoff values.

TABLE 11.9. Sex Identification Accuracy from Clavicle Length of Adults from Postclassic Marismas Nacionales Sites.

Sex Identification and Scale	Misclassified	Not Classifiable	Correctly Classified
Males			
Gill scale	1	8	10
Olivier scale	2	10	7
Females			
Gill scale	0	5	10
Olivier scale	0	5	10

Note: Review of this table reveals the increased accuracy in sexing Marismas adults when using the modified scale shown in Figure 11.4.

Femoral Head Diameter

Femoral head diameters also indicate that standard scales of sex determination from the femur (based upon U.S. and British white samples) must be modified somewhat to be applicable to these Marismas populations. The range in femoral head diameter for the males at all Marismas sites is 41–50 mm, and the mean is 45.5 mm. The range for the females is 36–44 mm, and the mean is precisely 41 mm. These means and ranges show that femoral head diameter among these populations averages about 3 mm less than among American whites and blacks (Byers 2011; Krogman and Iscan 1986; Steele and Bramblett 1988; Stewart 1979). The widely used Stewart (1979) scale for modern U.S. whites is shown in Table 11.10.

A slight modification of a similar scale (originally proposed by Krogman [1962]) was suggested by Gill (1971:126) for sex determination within the Marismas populations. This was based upon femurs from only 38 specimens from Tecualilla and Chalpa. Today that sample stands at 67 individuals from all Marismas sites, and the Gill (1971) scale still provides the best possible

TABLE 11.10. Scales for Sex Identification from Femoral Head Diameter (in Millimeters).

Scale	Population	Female	Female?	Sex?	Male?	Male
Stewart (1979)	Modern U.S. whites	<42.5	42.5–43.5	43.5–46.5	46.5–47.5	>47.5
Gill (1971)	Aboriginal west Mexicans	<40.5	40.5–42.5	42.5–43.5	43.5–44.5	>44.5

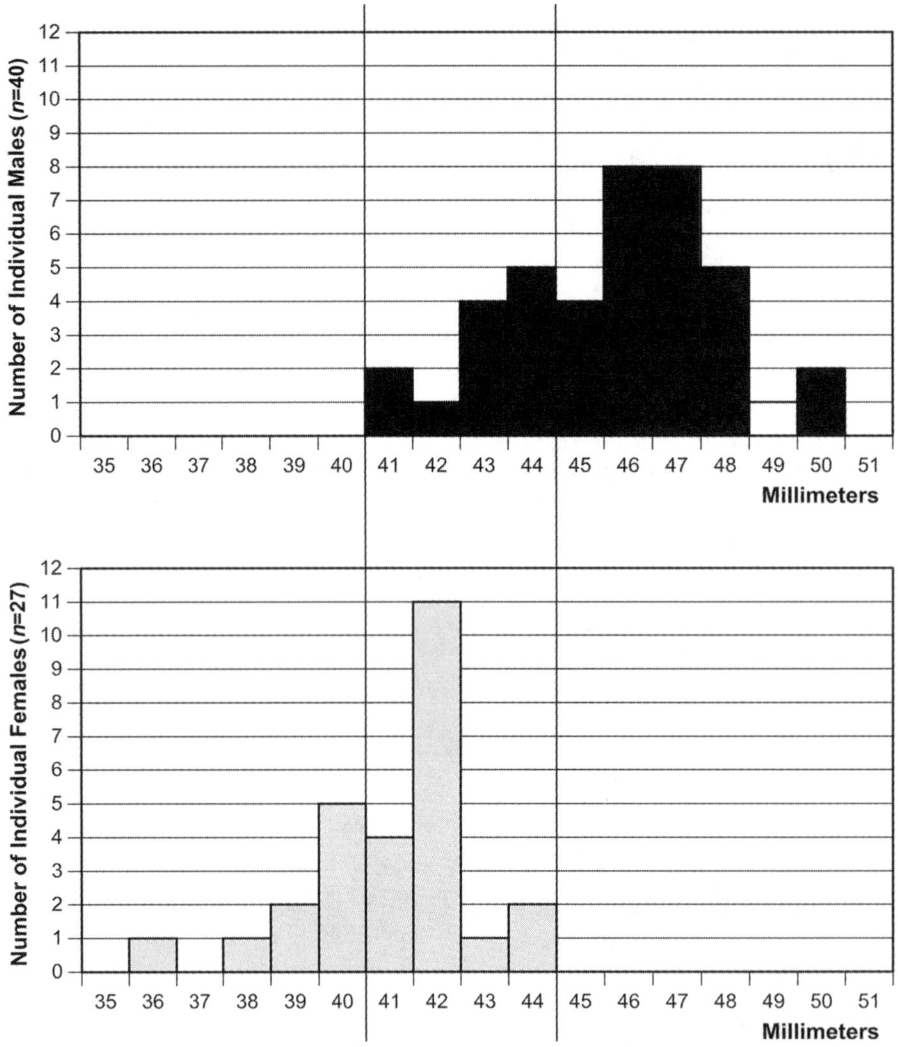

FIGURE 11.5. Femoral head heights of adults from the Postclassic Marismas Nacionales sites.

cutoff values for the total sample. That scale devised for aboriginal west Mexicans is also shown in Table 11.10.

Figure 11.5 shows plot points for each of the 67 Marismas skeletons for which measurable femurs exist. The adjusted cutoff values are also shown. As can be seen from Figure 11.5, 37 individuals from the Marismas are correctly sexed using this scale, and none are misidentified with regard to sex. Had the Stewart scale (used today indiscriminately by most forensic anthropologists for all North American populations) been utilized here, only 32 individuals would have been correctly sexed, and three males would have actually been misidentified as female.

Vertical Head Height of the Humerus

Examination of the vertical height of the head of the humerus shows, as with the femoral head diameters, an average dimension of approximately 2 mm less than that for American whites and blacks. Figure 11.6 is taken from Stewart (1979) and shows that a reading above 45 mm indicates male and one below 43 mm is very likely female in those two populations. A plot of the values for the males ($n = 30$) and females ($n = 20$) from the Marismas Nacionales is shown in Figure 11.7. Among these west Mexican Indians, a diameter above 43 mm indicates male, and a diameter below 41 mm is almost certainly female.

The mean height of the head for the male sample from the Marismas is 44.9 mm, and for females it is 38.8 mm. This sex difference of 6.1 mm is almost identical to that found by Dwight (1905) in a sample of 400 adult whites (difference of 6.09 mm) in a classic study in the early days of forensic physical anthropology (Krogman and Iscan 1986; Stewart 1979).

Robusticity

Robusticity calculations have been made according to the procedures of Olivier (1969), as mentioned earlier. Results have been compiled on all well-preserved adult femurs and humeri from all Marismas sites. As reflected in the tables of indices, femoral robusticity seems to run

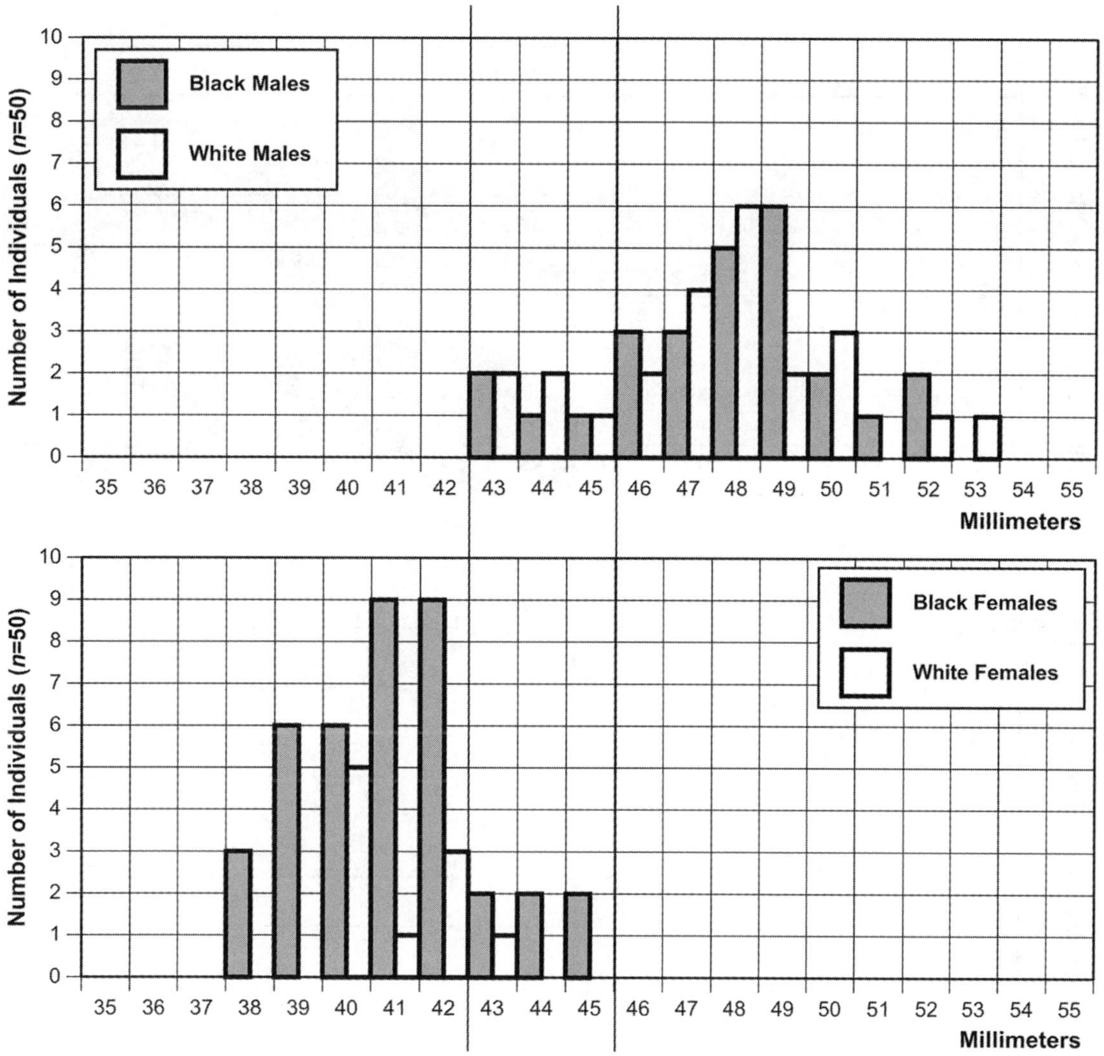

FIGURE 11.6. Vertical humeral head heights of American blacks and whites from the Terry Collection (after Stewart 1979).

slightly higher for males; the males average approximately 20, and the females, 19. This value is the same for Europeans (Olivier 1969). Robusticity of the humerus runs slightly higher (males average about 21; females, about 20). This is slightly higher than expected for some groups of whites but about that expected for many Mongoloid populations (Olivier 1969).

Pelvic Dimensions

Examination of pelvic dimensions, as shown in Tables 11.2–11.7, shows values fairly typical of other Native American groups. Sacral indices for the males at Tecualilla, Chalpa, and Venadillo Mound B are 104.3, 106.6, and 108.6, respectively (Rincon provides a single male value, and Cristo Rey, none). These sacral index values are a bit higher than the 100 average expected for Mongoloid males generally but are below the 110 value expected for whites. The Marismas females, as expected, show higher values than the males. Our values for pelvic index indicate results similar to those for other Mongoloid populations and whites. White females show values for this index that cluster around 74, and males, nearer 79 (Olivier 1969:253). The Marismas populations are tending very much toward those same values, as may be seen in the tables.

Anthroposcopic Characteristics

Observations have been made on 31 discrete traits of the skeleton (29 cranial and two postcranial) and 10 nondiscrete, continuous traits. These 41 nonmetric traits are all defined and illustrated in Gill (1971), and some useful illustrations for this chapter are provided in Appendix B. Rogers (2008) also defines and illustrates most of the same 31 discrete traits utilized in this study (all except mandibular and postcranial traits).

Discrete Trait Observations

The 29 discrete cranial traits utilized in this study, and illustrated in Appendix B, were originally drawn from an inventory prepared by Jantz (1970) for his skeletal

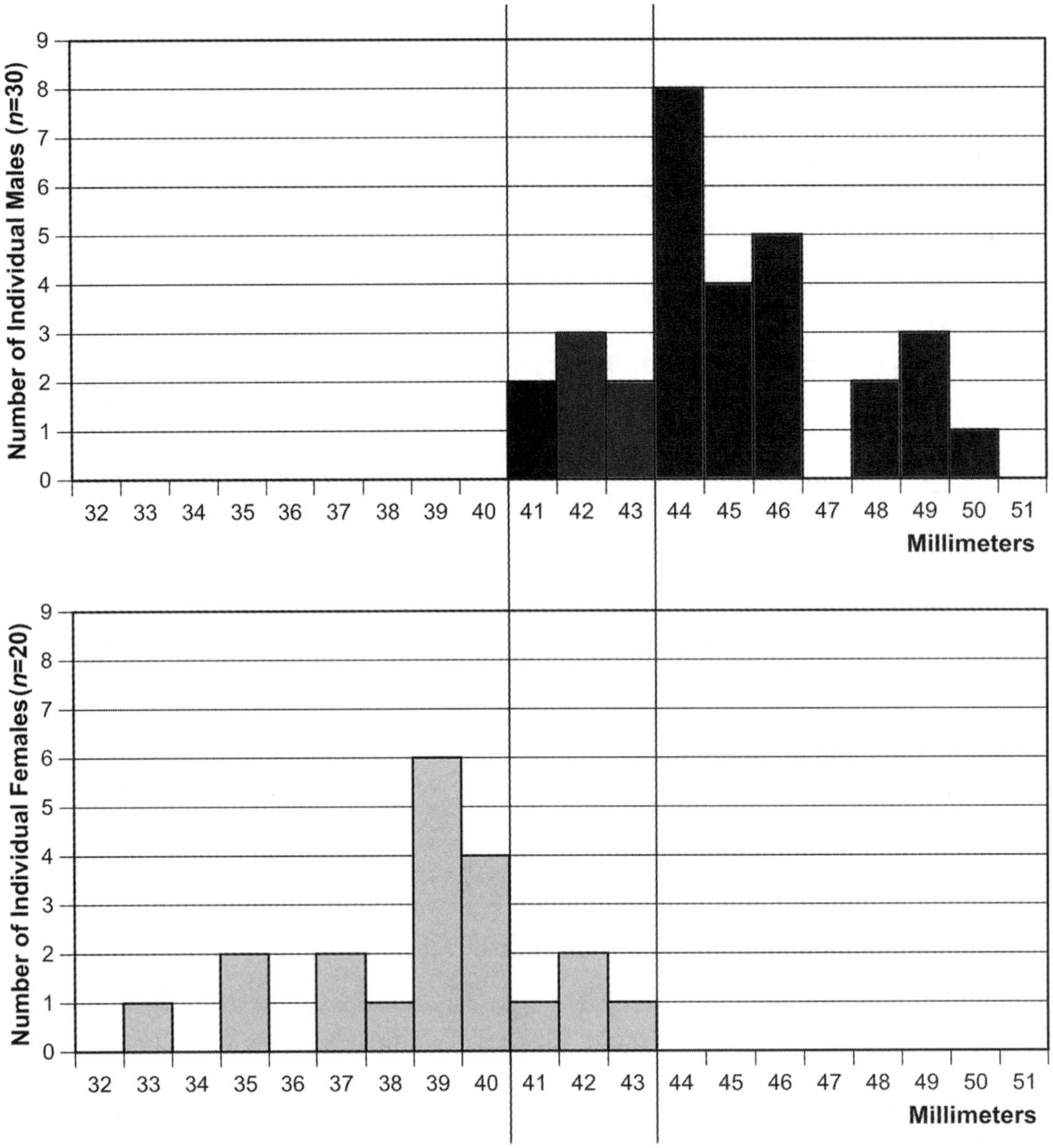

FIGURE 11.7. Humeral head diameters of adults from the Postclassic Marismas Nacionales sites.

analysis of the Northern Plains Arikara. Many of these same traits are also defined and illustrated by El-Najjar and McWilliams (1978) and others. Postcranial traits are less often studied, but two useful ones included in this study are the perforation of the olecranon fossa of the humerus (olecranon foramen) and the perforation of the blade of the sternum (sternal foramen).

The data collected on all 31 traits are summarized in Table E.1 in Appendix E and in Tables F.1–F.4 in Appendix F. The single table in Appendix E provides the only information available on discrete trait frequencies from the Classic period. Unfortunately the paucity of data from these five fragmentary individuals tells us very little. It is perhaps of interest, however, that two out of the three studiable female olecranon fossae do show perforations (for a frequency of 66.7 percent). This tendency, as will be discussed below, continues within the Marismas female population throughout the Postclassic period.

Much more can be said about the Postclassic period skeletal samples shown in Appendix F. Sample sizes are large enough that meaningful patterns can be discerned and statistical tests could be applied to look for both sex differences (Table 11.11) and site group differences (Table 11.12). Examination of the Appendix F tables suggests a certain degree of genetic relationship, and therefore assumed genetic continuity through time, for the five site groups. This is revealed best by noticeable frequencies of some discrete traits that are often quite rare or totally lacking in some human populations. For instance,

TABLE 11.11. Sex Differences in the Occurrence of Nonmetric Traits.

Trait	Males (n = 60)		Females (n = 50)		Difference
	Number	Percent	Number	Percent	Percent
Cranial					
Lambdoid ossicle—medial	68/75	90.7	50/56	89.3	1.4
Lambdoid ossicle—lateral	61/69	88.4	41/56	73.2	15.2
Parietal foramen	35/72	48.6	38/59	64.4	15.8
Mastoid foramen	58/78	74.4	31/54	57.4	17.0
Mastoid foramen exsutural	45/74	60.8	20/54	37.0	23.8*
Coronal ossicle	35/52	48.6	20/74	27.0	21.6**
Epiteric bone	4/40	10.0	2/48	4.2	5.8
Mastoid sutural ossicle	20/54	37.0	14/49	28.6	8.4
Parietal notch bone	6/66	9.1	4/59	6.8	2.3
Asterion ossicle	14/65	21.5	13/55	23.6	2.1
Tympanic dehiscence	19/93	20.4	16/82	19.5	0.9
Double anterior condylar canal	14/73	19.2	8/50	16.0	3.2
Accessory lesser palatine foramen	32/52	61.5	19/37	51.4	10.1
Supraorbital foramen	59/94	62.8	29/70	41.4	21.4
Frontal foramen	17/90	18.9	10/75	13.3	5.6
Structure into the infraorbital foramen	24/72	33.3	20/50	40.0	6.7
Accessory infraorbital foramen	9/61	14.8	3/43	7.0	7.8
Mylohyoid bridge	13/81	16.1	6/78	7.7	8.4
Accessory mental foramina	6/92	6.5	3/79	3.8	2.7
Epactal bone	17/38	44.7	10/31	32.3	12.7
Inca bone	6/44	13.6	0/32	0.0	13.6
Sagittal ossicle	16/29	55.2	15/28	53.6	1.6
Palatine torus	22/39	56.4	13/27	48.2	8.2
Pharangeal fossa	10/37	27.0	5/28	17.9	9.1
Superior sagittal sinus—left	17/42	40.5	10/38	26.3	14.2
Frontotemporal articulation	1/53	1.9	0/49	0.0	1.9
Metopic suture	1/52	1.9	1/38	2.6	0.7
Bregmatic bone	0/39	0.0	0/34	0.0	0.0
Mandibular torus	3/96	3.1	0/87	0.0	3.1
Postcranial					
Perforated olecranon fossa	10/72	13.9	47/71	66.2	52.3**
Perforated sternal blade	5/19	26.3	1/13	7.7	18.6

* $p \leq .05$; ** $p \leq .01$, according to chi-square using Yates's correction for continuity.

epiteric bones, parietal notch bones, and pharyngeal fossae are found throughout the Marismas region and throughout the entire Postclassic temporal span. It is also noteworthy that a number of Inca bones are observed from at least the earliest Postclassic sites of Tecualilla, Chalpa, and Venadillo Mound B. As may be discerned from Table 11.12, sample sizes from Rincon de Panal and Cristo Rey are insufficient to reliably test for the possible continued presence of that trait in later Postclassic times. The occurrence of olecranon foramina in the male populations of the Marismas, throughout the Postclassic period, may be noteworthy as well.

Table 11.11 indicates those discrete traits that show a statistically significant sex difference in frequency. Interestingly, the Marismas male population shows a greater number of sutural ossicles than do the females for all of the major cranial sutures (lambdoid, coronal, and sagittal), but this difference reaches the level of statistical significance only in the case of the coronal ossicle frequencies. Most craniofacial foramina are also more common among the males (except for parietal foramina), but only in the case of exsutural mastoid foramina does this difference reach the level of statistical significance. Not surprisingly the perforation of the olecranon fossa was several times more frequent among the females of the Marismas than among the males (66.2 percent for

TABLE 11.12. Pairwise Comparison of Identifiable Nonmetric Traits in Five Marismas Nacionales Sites by Chi-Square or Fisher's Exact Test.

	Site Pair									
Trait	1–2	1–3	1–4	1–5	2–3	2–4	2–5	3–4	3–5	4–5
Cranial										
Parietal foramen					$p \leq .05$					
Parietal notch bone				$p \leq .01$		$p \leq .05$				
Asterion Ossicle	$p \leq .01$						$p \leq .01$			
Tympanic dehiscence	$p \leq .01$			$p \leq .05$	$p \leq .01$			$p \leq .01$		
Frontal foramen	$p \leq .05$	$p \leq .05$				$p \leq .01$	$p \leq .01$	$p \leq .01$	$p \leq .01$	
Accessory infraorbital foramen						$p \leq .01$				
Mylohyoid bridge				$p \leq .05$			$p \leq .05$			$p \leq .05$
Epactal bone						$p \leq .01$				
Palatine torus								$p \leq .05$		
Pharangeal fossa									$p \leq .05$	
Postcranial										
Perforated sternal blade					$p \leq .01$	$p \leq .05$	$p \leq .05$			

Note: 1 = Tecualilla; 2 = Chalpa; 3 = Venadillo; 4 = Rincon de Panal; 5 = Cristo Rey. Fisher's exact test was used for combined samples of 30 or less. Chi-square, using Yates's correction for continuity, was employed on larger samples.

females, as compared with 13.9 percent for males). This is typical within *Homo sapiens* in general and may relate to the overall greater robusticity of the male humerus (White et al. 2012).

Table 11.12 reflects the results of a pairwise statistical comparison of these same discrete traits by site group. Many of these traits show no difference from the earliest to the latest Postclassic sites, but a few do show differences between the various Marismas populations. Occurrences of tympanic dehiscences (foramen of Huschke) and frontal foramina seem to differ the most from one Marismas site group to the next. As the degree of similarity and difference among site groups is surveyed, it does in general seem to support the idea of somewhat closer genetic relationships of the earlier Postclassic period sites (Tecualilla, Chalpa, Venadillo) with each other than with the two later sites (Rincon de Panal and Cristo Rey). Conversely, even less difference is seen between the two late sites of Rincon de Panal and Cristo Rey as they are compared with each other. Consistent with this is the fact that the biggest differences can be seen between Chalpa (an earlier site) and Rincon de Panal (a later site) and between Venadillo (earlier site) and Cristo Rey (later site). The one exception to this is the fact that no meaningful differences are discerned between Tecualilla and Rincon de Panal, even though those sites could be separated by as much as a century or more.

Nondiscrete, Continuous Trait Observations

The nonmetric continuous traits considered within this study are orbital form and inclination, form of the palate, nasal profile, the degree of nasal spine and sill development, form of the external auditory meatus, development of the supraorbital ridge, chin form and projection, and two other traits of the mandible (genial tubercle development and the degree of gonial eversion). These traits, like the discrete traits discussed above, are defined and fully described by Gill (1971). Illustrations of the differing degrees of nasal sill development are also provided by Gill and Deeds (2008). Assessment of these traits has provided excellent information not only for yielding a more complete basic description of the prehistoric people of the Marismas Nacionales but also for providing evidence of gene flow and temporal changes.

Examination of orbital shape and inclination shows a great deal of homogeneity at Tecualilla. Of the 24 orbits observable on both males and females (17 male, seven female), 23 exhibit a rhomboid shape with either slight ($n = 21$) or medium ($n = 2$) inclination. One female demonstrates square orbits (which, by definition, are ones without any orbital inclination).

At Chalpa more variation is seen in orbital form, even though most specimens show identical shapes to those found at Tecualilla. Of the 25 studiable orbits from Chalpa (11 male, 14 female), two reveal a round form (one male, one female), and two females show square orbits. All others, however, exhibit the usual rhomboid shape with slight inclination (one male with medium inclination).

Even though the skulls from Venadillo Mound B show great similarity to those at Tecualilla and Chalpa (with many showing rhomboid orbits with slight inclination), three differences were observed. First, twice as many females from this site (four of the total of eight) show square orbits, instead of the usual rhomboid form, and even one male from this site shows square orbits.

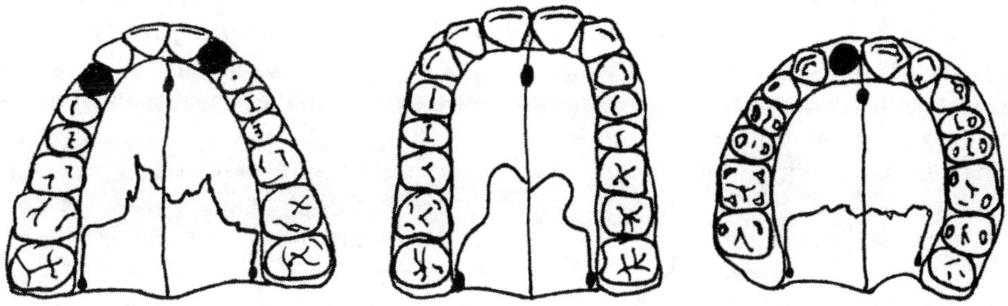

FIGURE 11.8. Human palate shapes: (*left*) parabolic; (*center*) hyperbolic; (*right*) elliptic (adapted from Gill 1998; Olivier 1969; Rawlings 2002).

Also one female reveals an elliptic orbital form, the only instance within the entire Marismas region. Half of the males also show medium, rather than slight, orbital inclination, which is an obvious increase from Tecualilla and Chalpa.

Unfortunately the Rincon de Panal and Cristo Rey site skeletal samples are so small that further trends are hard to track. The 10 individuals studiable from those sites (unfortunately only one female from Rincon and none from Cristo Rey) show the rhomboid orbital form typical of particularly the male samples from the earlier Marismas sites. Noteworthy, however, is that two males (one from Rincon and one from Cristo Rey) show pronounced orbital inclination. None of the earlier Postclassic site samples have individuals with pronounced orbital inclination. Three others (one Rincon male and two from Cristo Rey) show medium inclination, suggesting similarity to Venadillo Mound B in this regard.

So, temporal shifts in orbital shape do appear to have occurred at the Postclassic sites investigated, with an increase through time of the angle of inclination of the orbit among males, from very few medium cases at Tecualilla and Chalpa to about half of all males at Venadillo, and the later sites of Rincon de Panal and Cristo Rey even have examples of pronounced inclination. It would also seem that an increase in the frequency of square orbits among females occurred between Venadillo and the two population samples from the earlier sites of Tecualilla and Chalpa.

The only characteristic of orbital form that differs with sex is the degree of inclination. All but one of the Marismas individuals showing zero inclination (square orbits) are females, and all 12 of the specimens exhibiting medium or pronounced inclination are males. It also appears that this sexual dimorphism increases through time within these Marismas skeletal samples, with female inclination lessening after Tecualilla and male inclination increasing after Venadillo.

Form of the palate reveals some clear differences by site group. Figure 11.8 shows the three shapes of the human palate described by physical anthropologists (Brues 1990; Byers 2011; Gill 1971; Olivier 1969) and used frequently by forensic anthropologists (Byers 2011; Gill 1998; Rawlings 2002; Rhine 1990) to assist in the assessment of ancestry. Basically, parabolic palates are common to people of European descent (as well as Polynesians); hyperbolic palates, to Negroid and Australoid peoples; and elliptic palates, to Native Americans (and also other Mongoloid peoples). Research by Rawlings (2002) shows North American Indians to possess a frequency of elliptic palates of 42.7 percent. Chapman and Gill (1993) have shown that Native Americans from North and South America (combined samples) show approximately 48 percent elliptic palates and 50 percent parabolic. The present study is one of the few focusing on Mesoamerican skeletal samples.

As in nearly every other study of the human palate, this study finds no sex differences in the frequency of occurrence of these forms. Also, as is often the case with other studies of late prehistoric Native American skeletal samples, no occurrences of the hyperbolic palate form were found. The only possible case (a female from Tecualilla) showed significant dental loss, alveolar resorption, and alteration in the alveolar margins (and was therefore excluded from this study).

Tecualilla and Chalpa samples (combined, $n = 41$) show nearly identical frequencies of the elliptic vs. parabolic forms (83 percent elliptic, 17 percent parabolic). The small Cristo Rey sample ($n = 7$) exhibits a very comparable pattern (86 percent elliptic, 14 percent parabolic). In contrast to these, the Venadillo Mound B population ($n = 16$) shows almost the reverse (32 percent elliptic, 68 percent parabolic). This trend is reflected also by the small Rincon de Panal sample ($n = 5$), which shows entirely parabolic palates. A pooling of all site groups ($n = 76$) yields a frequency of 68 percent elliptic palates and 32 percent parabolic ones.

Shape of the external auditory meatus is very consistent within all Marismas populations. A total of 92 individuals possess studiable external auditory meatuses, and of these, 89 are elliptic in form (two are round, and one reveals a very narrow slit shape). The only slit-like meatus comes from Cristo Rey, the most recent site. The two round shapes are from Tecualilla and Venadillo. Two specimens from Chalpa had almost round openings as well, and two from the small Rincon sample were almost slits. So, if there is any change at all in the Marismas temporally regarding this trait, there may be a tendency for a few wider meatuses to occur from the earlier sites and a few narrower ones from the later populations.

The nasal skeleton of the Marismas populations shows some variation, but it seems to be more within population groups than between them. Concavo-convex nasal profiles are certainly the rule, which should not be surprising for a Mesoamerican Indian region such as the Marismas Nacionales. Of the 38 well-preserved nasal skeletons (a few from all site groups), 34 are concavo-convex, and only four are straight in profile. None are concave, and none are fully convex. Three of the straight noses are from the Chalpa site (one male, two females), and one is from Venadillo (a female). Assessment of the nasal spines ($n = 45$) reveals a preponderance of medium ($n = 27$), with only five small spines (mostly Tecualilla and Cristo Rey) and 13 large ones (none from Venadillo).

The picture of nasal sill development is somewhat more complex and seems to indicate some sex difference as well as variation by site group. A total absence of the nasal sill is a condition occasionally found at all Marismas sites except Chalpa (and Rincon de Panal, which only produced five nasal sills well-enough preserved to be evaluated). Of the total ($n = 72$), 17 percent ($n = 12$) reveal a total absence of sill. Males and females are equally represented in this sample numerically ($n =$ 6 each), but percentages for the females are somewhat higher. A total of 32 (18 males, 14 females) show dull sills, and 20 (15 males, five females) show medium ones. Of the eight sharp sills (none from Venadillo), five are found on males. Looked at another way, 55 percent of the nasal sills within the male sample are rated either absent or dull, and yet 71 percent of female sills are so rated. The fact that three times as many males as females show a medium amount of sill development is probably significant, yet sample sizes are too small for a compelling statement about this. A study by Gill and Deeds (2008) among Northwestern Plains Indians shows considerable variation in nasal sill development temporally (more medium and sharp sills with more recent populations) but little sex difference.

No differences are detectable in chin form among the various Marismas site groups, but not so surprisingly the form does differ markedly between the sexes. No truly bilateral chins can be shown to exist among females at any of the sites ($n = 41$), but among males ($n =$ 45) bilateral chin form is actually the most commonly encountered type (51 percent show this shape). Only 12 median chins (27 percent) are to be found within the pooled male sample, while 10 (22 percent) are median-bilateral. Yet 29 females (71 percent) show median chins, 12 (29 percent) show median-bilateral ones, and none reflect the male-like bilateral condition.

Regarding the degree of projection of the chin, by contrast, there are no noticeable sex differences. Some slight differences by site are suggested, however. At Tecualilla, for instance, no negative chins are found ($n = 15$ males, $n = 8$ females). All show neutral chins, a slight projection, or a medium amount of projection. At Chalpa ($n = 21$ total), however, four individuals (two males, two females) reveal negative chins. All other Chalpa mandibles show either neutral chins or a slight degree of projection. Of the 22 specimens from Venadillo, two show negative chins, eight have neutral ones, 10 have a slight projection, and two show a medium amount of chin projection. A pooling of the two small skeletal populations of Rincon de Panal ($n = 5$) and Cristo Rey ($n = 14$) yields three negative chins (two at Rincon, one at Cristo Rey), three neutral, eight with a slight projection, and five medium. So it appears as though negative chin development entered the Marismas Nacionales around AD 1300 as evidenced at Chalpa and remained as a skeletal variation within the region from that time onward (evidenced at all subsequent sites).

Gonial angle eversion is somewhat more pronounced, on average, at Tecualilla and Cristo Rey than at Chalpa, Venadillo, and Rincon de Panal. Most of the males from Tecualilla (10 of 15) show at least a small degree of gonial angle eversion (but none of the eight females show any at all). At Cristo Rey nine of 13 show at least a small degree as well. At Chalpa, by contrast, only five of 23 individuals show any gonial angle eversion, and 14 individuals show clear examples of inversion at gonion. Of the 25 cases at Venadillo, nine individuals show inversion as well, but a slightly higher number than at Chalpa ($n = 12$) do show at least some eversion. The small sample from Rincon de Panal ($n = 4$) shows a pattern similar to that at Venadillo and Chalpa, with all individuals showing a lack of gonial angle eversion (one shows slight eversion on one side). At every site some individual specimens show gonial angle inversion, and at no site do very many reveal pronounced degrees of eversion (one Tecualilla male, one Chalpa male, and no individuals from any other sites). Even a medium amount of eversion is rare (three males from Tecualilla, one male and three females from Chalpa, one male and one female from Venadillo, and no individuals from either Rincon de Panal or Cristo Rey).

Genial tubercles seem not to vary at all by sex and perhaps not significantly by population either. It is noteworthy, however, that of the 87 adult mandibles examined from all Marismas sites, only one showed the "pit" condition (a total lack of tubercles). This was a male from the Chalpa site. Most others show either small or medium tubercles, and some do show large ones (7 of 23 at Tecualilla, 5 of 22 at Chalpa, 1 of 24 at Venadillo, and 2 of the 18 from Rincon and Cristo Rey). If anything, there may be a slight drop in frequency through time of the large tubercles. Tecualilla and Chalpa show 27 percent large tubercles, and yet the other sites combined reveal only 7 percent large genial tubercles.

An examination of supraorbital ridge size shows a pattern quite typical of Native Americans in general. Males show well-developed brows with a rating of "medium" for nearly all. An occasional instance of truly large browridges (typical of males of Australoid populations or archaic sapiens groups) can be seen. Within the Marismas populations two of these cases occurred, both from the earliest two Postclassic period sites (one Tecualilla male, one Chalpa male). Nearly all females, on the other hand, show a "slight" degree of ridge development. Only two females (both from late sites) show absolutely no browridge at all (one Venadillo Mound B female, one Cristo Rey female). Pooling the adult males from all sites ($n = 53$) reveals 43 specimens (81 percent) with medium browridge development, two (mentioned above) with truly large brows (4 percent), and eight (15 percent) with slight browridge development. The pooled female sample ($n = 37$) shows 34 specimens (92 percent) with slight browridge development, only one female (from Venadillo Mound B) with a medium brow (3 percent), and two (mentioned above) with no browridge at all (5 percent). If there is any temporal trend at all within the Marismas populations, it is toward a slight reduction in browridge size through time, but to demonstrate this conclusively larger sample sizes would be necessary from the two most recent sites, Rincon de Panal ($n = 6$) and Cristo Rey ($n = 9$).

Dental Traits

The dental metrics mentioned above (mesiodistal and buccolingual crown diameters) have been presented previously for all site groups within the Marismas Nacionales along with complete descriptive statistics (Case 1976). Distributions of individual measurements by site are repeated here in Appendix G. Male and female tooth dimensions were combined in Case's original study in most instances since no statistically significant differences ($p < .05$) were demonstrated by the Mann-Whitney *U* test (for pooled sites or by individual sites). Canine measurements and the labiolingual crown diameters of the maxillary central incisor teeth are the exceptions. These metrics do show sexual dimorphism that Case (1976) found to be statistically significant at the .05 level. Tables G.1–G.4 in Appendix G provide mean crown diameters for all categories of teeth for each of the major site groups within the Marismas Nacionales.

The principal diagnostic traits of the Marismas dentitions, as regards Native American populations in general, are the buccolingual crown diameters of the maxillary second and third molars and the mandibular second premolar. These are quite small for Native American groups. Also important is the large mesiodistal crown diameter of the maxillary third molar. The mesiodistal diameters of the remaining maxillary teeth, as well as the mandibular anterior teeth, show no special characteristics and generally occupy midway positions in the range of Native American populations. Buccolingual crown diameters among the Marismas inhabitants tend to be small in relation to those of other groups, and only the Aztec and Aleut have an appreciable number of dimensions smaller than these populations.

Within the Marismas the Cristo Rey and Rincon de Panal populations are similar, and both have smaller means than those found at the other sites. The greatest similarities, however, are found among the populations from Tecualilla, Chalpa, and Venadillo, and the small Classic period sample from Panales seems to resemble these groups as well.

Population Distance

Population distance calculations have been undertaken from three different approaches: (1) Penrose's distance (CH^2), shape (CZ^2), and size (CQ^2) tabulations for each pair of Marismas populations based upon 15 craniometric variables (mentioned in the "Methods" section above); (2) Penrose's distance, shape, and size computed for each pair of Marismas populations using 16 dental characteristics (see "Methods"); and (3) computation of principal component scores for several Marismas males and females, based upon six of the original 15 cranial measurements. This examination includes comparisons with a wide range of Southwest and Mexican samples.

The Penrose tabulations in Table H.1 of Appendix H are the product of the class project mentioned in the "Methods" section above, in which 15 craniometric variables were selected. As may be seen from this table these metrics were obtained on more males than females, so some site comparisons for females could not be made. A total of 56 males from Tecualilla, Chalpa, Venadillo Mound B, Rincon de Panal, and Cristo Rey and 27 females from the first three of these sites were included in this study. As may be seen from the numerical values, a ranking of biological distance exists that fits very well the other observations from crania that Tecualilla and Chalpa, and to some extent Venadillo Mound B, form a closely related group of populations, with Rincon de Panal and Cristo Rey exhibiting a somewhat more distant relationship.

Further support for the relationships of these site groups comes from an analysis of the odontometric data

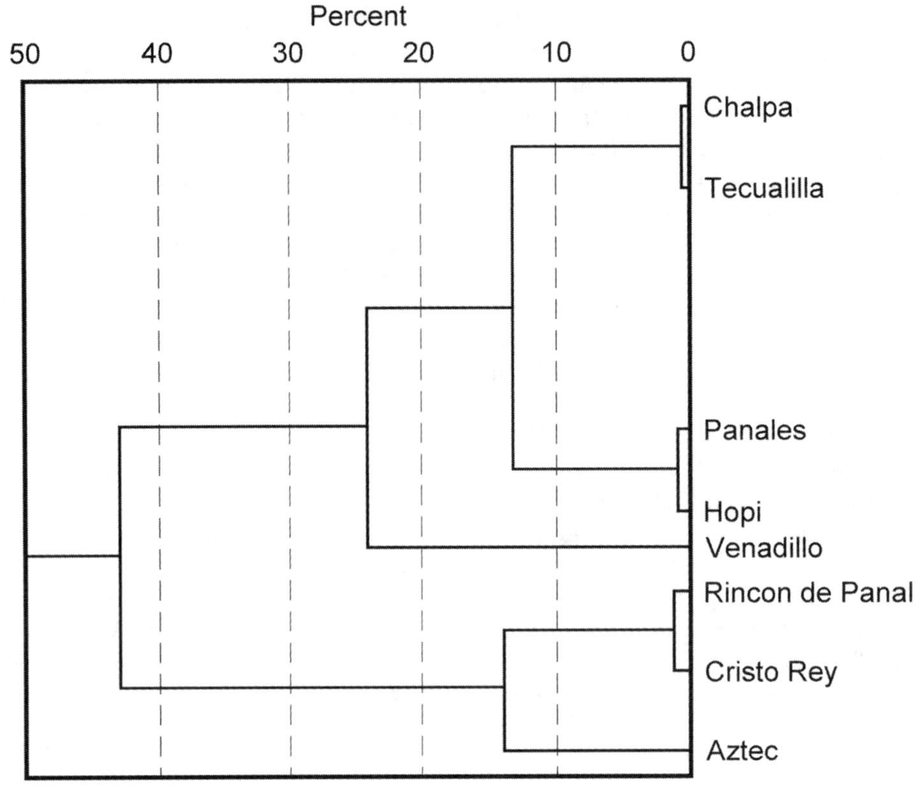

FIGURE 11.9. Dendrogram of the Marismas Nacionales site populations and the Hopi and Aztec. Based on 0/0 (0/0 = $[CQ^2 \times 100]/CH^2$), an analysis of 16 odontometric variables.

(Case 1976). Table H.2 of Appendix H shows the results of Penrose's distance, shape, and size computed for each pair of Marismas populations using the 16 dental characters listed above. This comparison also shows Chalpa and Tecualilla to be the most similar populations, with Venadillo resembling each of these site groups as well.

The pairwise comparisons that show reasonably small distance (CH^2) values involve the populations of Tecualilla, Chalpa, Venadillo, and to a lesser extent Panales. The larger distances involve the above four groups when compared with Rincon de Panal and Cristo Rey. Rincon de Panal and Cristo Rey show greater affinity to each other than to the other site groups, but this relationship is not close. The distance estimates confirm the impressions received from the univariate tests that two groups are present: one consists of Panales, Tecualilla, Chalpa, and Venadillo, and the second is made up of Rincon de Panal and Cristo Rey.

Pooled Marismas Nacionales samples have also been compared with several Native American populations using Penrose's distance, shape, and size (with the same 16 dental variables). Both univariate analysis and Penrose's distance (CH^2) demonstrate biological affinities of the Marismas groups with the Hopi and Aztec (Case 1976:220). Figure 11.9 is a dendrogram based upon the Penrose results that includes the Marismas site groups and the Hopi and Aztec. It shows the Hopi clustering more with the earliest Postclassic site populations, that is, Group 1, which consists of Tecualilla, Chalpa, and Venadillo (as well as the few remains from Panales), and the Aztec clustering with the later ones, Group 2, which is made up of Rincon de Panal and Cristo Rey.

In the odontometric study Case also observed that "Chalpa occupies a position between Tecualilla and the Aztec which may suggest a slight influence from the South" (1976:223). This fits well our observations based upon craniometrics, some discrete cranial traits, and anthroposcopy that some southern traits are discernible within the Chalpa sample and within the later site samples as well. Not only are Rincon de Panal and Cristo Rey similar to each other, but, as previously discussed, both resemble the Aztec more than do the other Marismas populations. The cluster of these two, however, is not as tight as the one formed by the other four Marismas populations (Group 1). Yet, despite the large distances separating these two Marismas groups, Rincon de Panal and Cristo Rey generally tend to show closer affinities to the other Marismas populations than they do to other Native American groups.

Unfortunately, the Guasave site and the Aztec are the only Mexican Indian populations for which good comparative dental data exist. Of all comparative populations, Guasave (Glascock 1980), in general, is the most distant from the Marismas populations. The Aztec, on the other hand, show little divergence from the Marismas groups. The Aztecs appear to occupy a position midway between Panales, Tecualilla, Chalpa, and Venadillo (Group 1), on the one hand, and Rincon and Cristo Rey (Group 2), on the other. When compared with individual Marismas populations, the Aztec are closest to the Venadillo population. A visual inspection of the Mexican populations compared with those of the American Southwest (Dahlberg 1963) suggests a close resemblance of Guasave to the Pima, whereas the Marismas populations are more similar to the Hopi.

Last, principal components analysis has been used to compare 42 Marismas skeletons among themselves, with other Mexican Indian groups, and with the Pecos Pueblo. The figures presented in Appendix I show the principal component scores of the individual skeletons from the Marismas Nacionales and for the comparative samples. Most of the plots of comparative scores in those figures represent individual skeletons, but a few of the ones taken from Genovés (1970) are average values of two or more specimens.

The first principal component (PC1), which accounts for 58.7 percent of the variability, is easily interpreted as a size component with high loadings on bizygomatic breadth and minimum frontal breadth. The second principal component (PC2) contributes 17.5 percent to the total variability and contrasts the minimum frontal breadth with nasal height and to a lesser extent bizygomatic breadth. The former shows a high negative loading, while the latter two have positive loadings.

For males the plot shows considerable overlapping among the Marismas sites and of the Marismas populations with the comparative sample. Although the full range of variability for the Marismas sites is unknown, there does seem to be some nonrandom sorting of the Marismas skeletons according to size. Tecualilla and Venadillo Mound B tend to cluster at the high end of the scale, whereas Chalpa, Rincon de Panal, and Cristo Rey have lower values. First principal component scores for the Marismas skeletons are generally higher than those for the comparative sample. Only Santa Maria (preceramic) and Seri (northern) score high enough to be comparable with Tecualilla and Venadillo Mound B. Chalpa, Rincon de Panal, and Cristo Rey generally occupy midway positions between Tecualilla and Venadillo Mound B, on the one hand, and the bulk of the comparative Mexican Indian sample, on the other. For females, the Marismas sites are separated by the second principal component, while the first is unimportant. In sum, the Marismas individuals differ from the remaining Mexican Indians by size in this study utilizing six facial measurements.

Summary and Conclusions

After many years now of excavating burials, measuring long bones, skulls, and teeth, visually assessing cranial morphology, and analyzing these skeletal remains of the Marismas Nacionales statistically, a much sharper picture of biological relationships, migration, and population change has come into focus. The Marismas data suggest that over the roughly 1,000-year span, a long period of cultural and populational stability existed. The prehistoric inhabitants of the region in the earlier part of the time span (represented by bones and teeth from the Classic period sites of Venadillo Mound C and Panales and from the earliest of the Postclassic sites, Tecualilla) were a homogeneous and possibly endogamous people representing a very early population element of central and western Mexico. In fact, the famous Paleoindian cranium from the Valley of Mexico, Tepexpan Man, is nearly identical metrically to the sample of unmodified male skulls from the Tecualilla site. Beginning around AD 1200, however, new traits began to appear, first at the Chalpa site and then with increasing frequency through Venadillo Mound B, Rincon de Panal, and Cristo Rey. Some of the data, however, suggest a good deal of genetic continuity as well.

Many traits remained stable through time, and analysis of the Postclassic skeletal sample (barely adequate at the latest sites) does reflect this. The picture that emerges, with a focus on the characteristics that remained constant throughout the centuries, is a rather complete one. We have a good view of stature, cranial and facial form, dental dimensions, and a number of more subtle features of the cranium and postcranial skeleton. The population was of relatively short stature (on a worldwide scale), with a male average of 5 ft, 4½ inches, and the females averaging about 5 ft tall.

According to Genovés (1970), this is typical of populations in Mexico at this latitude. Males reveal a cranial size average of 1,436 cc, and females, 1,313 cc. Judging from the few unmodified crania (mostly from Tecualilla), these prehistoric people were genetically inclined toward a fairly medium head form (high mesocranic to slightly brachycranic) that was moderate in height.

The face form of these people was medium to broad, with very few individuals with long narrow faces. The few that occur are from the earlier sites. In total facial height (including the mandible) the females had noticeably shorter faces. The women also had shorter noses and wider spacing between the eyes than the men, thus providing some sexual dimorphism in facial morphology as well as in size. Nose form in general for this population was medium, but with a lot of individual variation and with males more often inclining toward longer, narrow noses and females inclining more often toward shorter, broader ones. A fairly strong sexual dimorphism is also exhibited in shoulder breadth (assessed through clavicle lengths) and the size of joint surfaces (through femoral and humeral head diameters), which parallels the condition among the quite sexually dimorphic white and black populations of today, but on a slightly smaller overall scale.

Like most Native American populations these prehistoric people of the Marismas Nacionales had both elliptic and parabolic shapes of the palate. Instead of the nearly equal distribution of about 50 percent of each form, as some Native American populations exhibit, these people had more than twice as many elliptic palates (68 percent) as parabolic (32 percent). The Marismas male samples also show more bilateral chins (51 percent) than expected among Native American (and other Mongoloid) peoples.

Some other traits varied much more through time than the characteristics just mentioned. Eye orbits, for instance, are low and wide at Tecualilla and the earlier sites but average medium a little later in time due to the presence of a few high, rounded orbits. A few longer palates also exist at the earlier sites, as well as greater cheek height dimensions and larger symphyseal height measurements for some mandibles.

A slight reduction in browridge size seems to have occurred through time, as well as subtle changes in the angle of inclination of the orbits. Males show an increase in the angle of inclination of the superior margin of the orbit through time, while females show an increase in the frequency of square orbits (i.e., those with a total lack of inclination), thus leading to a picture of increasing sexual dimorphism through time with regard to that particular trait. Negative chin projection first appears at Chalpa (probably about AD 1200) and remains within the Marismas population from that time onward.

Population distance statistics from both craniometrics and odontometrics suggest a clustering of the Marismas site populations into two groups. Group 1 consists of Tecualilla, Chalpa, and Venadillo (and the earlier site of Panales), and Group 2 consists of Rincon de Panal and Cristo Rey. Group 1 shows affinity with the Hopi and other Puebloan populations of the American Southwest, while Group 2 shows a closer affinity to the Aztec. Some of the best evidence for at least some genetic continuity comes from the discrete trait analysis. Certain somewhat unusual discrete characteristics, both cranial and postcranial, persist throughout time at all sites within the Marismas Nacionales.

An important question regarding this interesting balance found in the skeletal data between traits suggesting continuity and those suggesting change is, What precise forces account for it? It could be migration and gene flow, genetic drift, natural selection, or migration and total replacement (if by a reasonably closely related population). The directional patterning of some of the changes such as the decreasing size through time of teeth, cheekbones, palates, and some other features would seem to eliminate genetic drift as a possibility. This might suggest, in fact, natural selection since the pattern for hundreds of thousands of years within the hominid line has been precisely this kind of change, that is, a reduction in face size, tooth size, and general robusticity. On a wider scale within Mexico, and the rest of the Americas for that matter, natural selection might very well be the ultimate causation (and therefore explanation) of the patterns of skeletal change observed within the Marismas Nacionales, but it is certainly not the proximate (more immediate) cause. The changes are simply too rapid and too irregular in their patterning (by site group) to be due to natural selection alone. Also, sudden changes in burial practices occurring at the same time that the first biological changes appear would not be the prediction from a within-locale natural selection model. Changes such as this suggest either migration and genetic exchange or migration and replacement.

Another possible approach to the variation described here is to view its source as more geographic and less temporal—that is, representing contrasting genetics of different tribal or village elements living more or less contemporaneously within the Marismas. Due to the very close geographic proximity of all of the Marismas sites, however, this interpretation appears less likely. Also, small but clear population differences can be demonstrated between most site groups, and the biological variation does appear to increase proportionately with the degree of temporal separation and not the degree (or extent) of geographic separation. In fact, the geographic barriers of river channels, and possible political boundaries within the Marismas, fail to produce any such comparable correlation with the biological variations.

So, some sort of migration and subsequent biological change appears to be the best interpretation for the variation in the Marismas skeletal data. The number and nature of the characteristics that show continuity within the Marismas would seem to virtually preclude the "migration and replacement" model and favor much more the "migration and gene flow" interpretation. The craniometric and odontometric tests probably shed less light on this issue than does a close inspection of the discrete trait occurrences. The continuation of some of these rather unusual physical traits throughout the Postclassic period, in conjunction with the introduction of new traits, appears explained best by migration followed by settlement and intermarriage, not replacement. Some of the data suggest that some gene continuity would be almost a certainty.

All observations, measurements, and statistical analyses made on both skulls and teeth reveal that the people of the Marismas Nacionales for a long time were a genetically isolated element having biological affinities with populations in the north of Mexico and the southwestern United States. Particularly the Hopi and other genetically conservative Puebloan populations show strong ties with Marismas populations of the Classic and early Postclassic periods. This early population in the Marismas represents a very old population element in northern and central Mexico, as represented by the Tepexpan remains and termed years ago by Georg Neumann the "Hanid" variety within the Americas. Then at about AD 1200, when the first changes in burial practices are evidenced in the Marismas region (Gill 1971, 1974, 1985; see Chapter 9), a few Aztec-like physical traits are evidenced, beginning at Chalpa. As time went on, the genes from the intrusive element became more frequent, and some populations (e.g., Cristo Rey) became, with regard to certain dental characteristics, at least, more like the intrusive population than the original population.

Since the Marismas data suggest noticeable genetic change in a very finite coastal region over a rather short span of a few centuries, these findings may have bearing on the broader question of Postclassic culture change in Mexico. They stand in support of the traditional interpretations of Vaillant (1940) and others as mentioned in the introduction of this chapter, who tended to view Postclassic cultural diffusion as being "accomplished by movements of peoples... between AD 1100–1300" (1940: 300). This stands counter to the thinking of Smith and Heath-Smith (1980), also mentioned above, which sees such movements of past peoples has having been greatly overemphasized. As stated earlier, Smith and Heath-Smith wish to de-emphasize migrations and the entire "waves of influence" model. They (1980:15–16) believe that certain Postclassic period changes in west Mexico and other regions peripheral to the central highland region were more the result of interdependent exchange and communication networks than diffusion from a nuclear core area. Perhaps their hypothesis will hold for certain kinds of culture change, but as far as the Marismas data are concerned they support much more readily a picture of movement of peoples. Summarizing, population migration followed by settlement and intermarriage with the local indigenous people is the interpretation that is best supported by the osteological and odontological data. This interpretation seems consistent with the archaeological findings of the Marismas project as well as what has been argued for the wider region (e.g., Kelley 1986, 2000).

CHAPTER 12

The Marismas Nacionales

A Historical View of the Conquest and Early Postconquest Periods

Patricia Kay Scott

As described in sixteenth- and seventeenth-century documentary sources, the Marismas Nacionales was a bleak place, plagued by a variety of noxious spiders, bats, scorpions, reptiles, and biting insects such as mosquitoes and ticks. The rivers, estuaries, and lagoons were dangerously full of caiman, and sharks were a frequent impediment to coastal fishermen. The conquistadores and early missionaries witnessed violent storms, saw devastating floods, and endured seasons of excessive heat. Yet the Marismas is described by the conquerors as an area of abundant natural resources, which helped to sustain the elaborate preconquest culture of the well-populated coastal area and the trail that passed through the eastern margin of the Marismas, serving as a link in trade between central Mexico and the American Southwest (Meighan 1971:754–755).

Nuño Beltrán de Guzmán claimed the area of the Marismas Nacionales as part of Nueva Galicia during his 1530–1531 conquest of west Mexico. Guzmán and his forces perceived that the majority of the coastal population between the Río Santiago and the Río Chametla (Río Baluarte) lived within three native social-political divisions, referred to in this chapter as Centispác, Aztatlán, and Chametla (Figure 12.1). These *provincias*, as the Spanish initially referred to them, shared overlapping tribal units, common languages, many cultural likenesses, and common enemies. The conquistadores and other early chroniclers list particular settlements for these entities and describe leaders for each.

The following information about the Marismas Nacionales during the contact period is extracted and compiled from eyewitness accounts and other Spanish colonial documents and published ethnohistoric syntheses spanning the sixteenth to the eighteenth century. An overview of sources is followed by a comprehensive review of each of the three provincias identified above—Centispác, Aztatlán, and Chametla. An in-depth description of settlements by place-name is given under each provincia in a format similar to the historical geographies of Gerhard for other portions of colonial Mexico, which were published in the 1970s and 1980s and later revised and reprinted in 1993 (see 1993a, 1993b, 1993c). The chapter concludes with summarizing comments and observations related to political organization, environmental factors such as changing hydrology, and trade and an interpretive discussion of a possible link of El Calón, and the Marismas in general, to a seventeenth-century account of the cosmology of the Chora (Cora) Naylita (Nayarita, Náyari).

Ethnohistoric Sources and Eyewitness Accounts

Much of the primary information on the contact history of west Mexico has been expertly summarized and/or interpreted in the writings of Sauer and Brand (1932), Sauer (1932, 1934, 1935), López Portillo y Weber (1935), Kelly (1938), Grosscup (1964), Mecham (1968), Grimes and Hinton (1969), Spicer (1969), Brand (1971), Meighan (1971, 1974), Cline (1972a, 1972b), Gerhard (1972), and Anguiano (1976), but little attention, still, has been given directly to the region encompassing the marine estuaries of the Marismas Nacionales. The research for this chapter was undertaken in the late 1960s and early 1970s. Resources in the archives and libraries of Mexico City and Guadalajara provided much information. During the time frame of my research, Marina Anguiano undertook an extensive review of the conquest of this area, and in 1976 she completed "Nayarit: Costa y Altiplanicie en el Momento del Contacto, Ecologia y Fuentes" as an M.A. thesis at the Universidad Nacional Autónoma de México. Readers interested in the conquest and early postconquest history of the Marismas and areas east and south of it in Nayarit should consult this massive and more comprehensive study. Her extensive information related

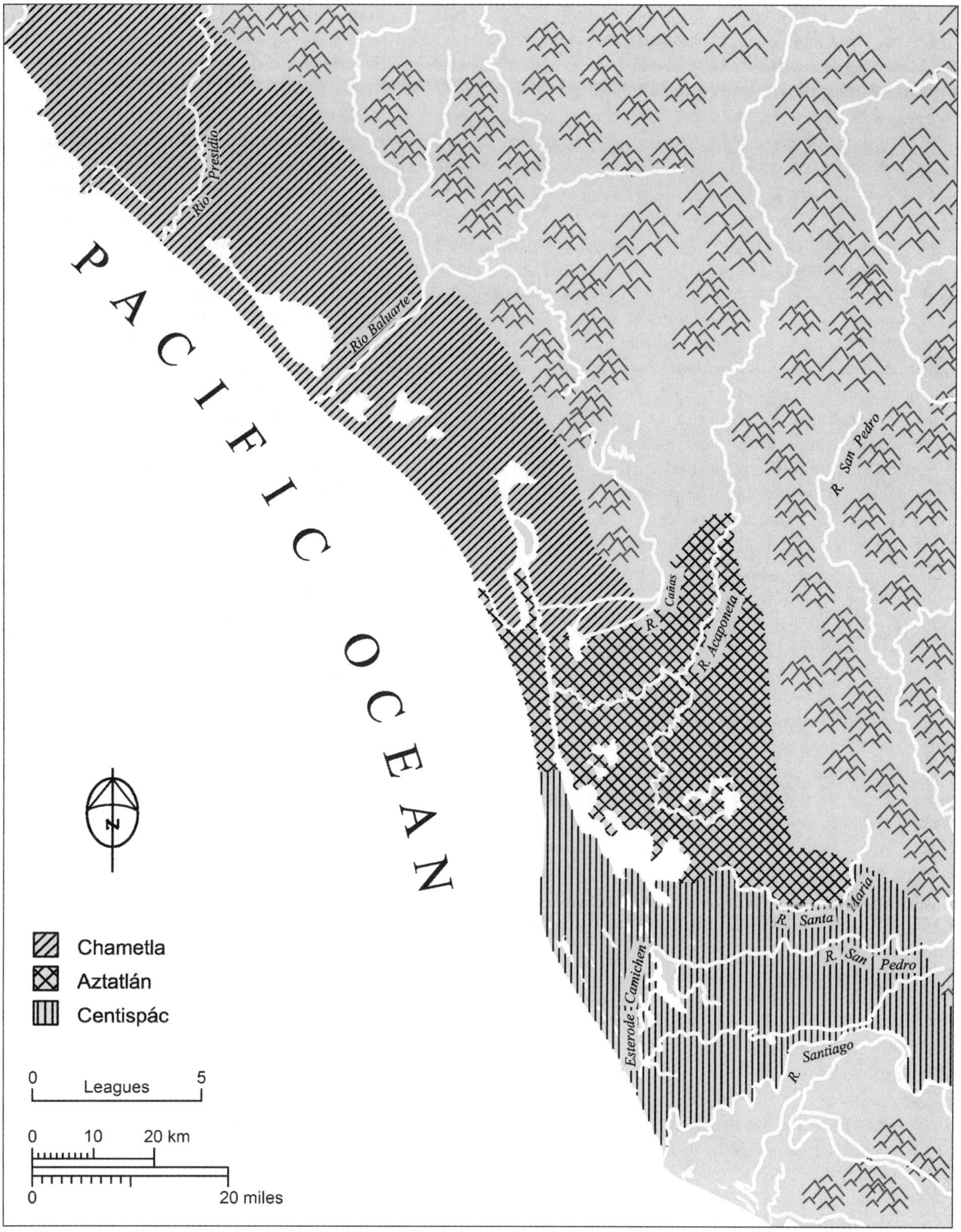

FIGURE 12.1. Approximate territorial areas of Centispác, Aztatlán, and Chametla at the time of conquest. The part of Chametla that extended beyond the Río Presidio to the Río Piaxtla is not shown.

to the conquistador's records and those of Tello (1945, 1974) and Mota Padilla (1973) are especially relevant to the Marismas research.

The earliest primary sources and descriptions of the Marismas Nacionales come from the colonial record of the conquest of west Mexico, including accounts by Nuño Beltrán de Guzmán (1955), Gonzalo López (1963), García del Pilar (1955, 1963), Juan de Sámano (1963), Pedro de Carranza (1963), and four anonymous conquistadores (*Relación Anónima Cuarta* 1955; *Relación Anónima Primera* 1963; *Relación Anónima Segunda* 1963; *Relación Anónima Tercera* 1955) who were with Guzmán as he claimed the area in 1530–1531 in the name of Spain. A somewhat later source, *Las Relaciones Geográficas*, collected from 1578 to 1586, provides very little information about the Marismas area. However, other Spanish colonial descriptions offer abundant ethnohistoric insights from the sixteenth, seventeenth, and eighteenth centuries, including the published writings of Alonso Ponce (1968), Alonso de la Mota y Escobar (1940), Domingo Lázaro de Arregui (1946), Antonio Tello (1945, 1974), Antonio Arias de Saavedra, Matias de Mota López y Padilla (Mota Padilla 1973), and Joseph Antonio Villa-Señor y Sánchez (1952).

Fray Alonso Ponce visited pueblos and Franciscan convents from 1584 to 1588 and was in Acaponeta in 1587. Details of his visits were recorded for New Spain's Comisario General Fray Antonio de Ciudad Rodrigo. Ponce's original account was entitled *Tratado curioso y docto de las grandezas de la Nueva Espana: Relación breve y verdadera de algunas cosas de las muchas que sucedieron al fray Alonso Ponce en las provincias de la Nueva España, siendo comisario general de aquellas partes*. This was first published in 1872. Fray Alonso de la Mota y Escobar was the Franciscan bishop of Guadalajara from 1598 to 1607. From 1602 into 1605 he undertook pastoral visits to other missions, and ca. 1604 he produced the *Descripción Geográfica de los Reinos de Nueva Galicia, Nueva Vizcaya y Nuevo León*, in which he described the natural history, politics, economy, and religious aspects of the communities he visited, including portions of the Marismas.

Domingo Lázaro de Arregui was in the Acaponeta region as early as 1617. His *Descripción de la Nueva Galica* was finished in 1621. It was produced for the *presidente de la audiencia* at a time when the creation of the Obispado of Durango (established in 1622) was being considered. Arregui's manuscript included a map depicting portions of the Marismas study area. José María Muriá (2000:259) analyzed Arregui's work and notes that his information seems mostly to be original, that is, not taken from earlier sources other than one other seventeenth-century writer.

Fray Antonio Tello headed several Franciscan Jalisco missions and visited the Marismas area sometime between the 1620s and 1650s. He described his visits in his *Crónica Miscelánea en que se trata de la Conquista Espiritual y Temporal de la Sancta Provincia de Xalisco en el Nuevo Reino de Galicia y Nueva Vizcaya y Descubrimiento del Nuevo México*, which he finished in 1653. Some of that work was lost, but the recovered portion was first published in 1891. His writings are at times problematic, and for ethnohistorians his *Crónica* can be a confusing mix of what he observed during his travels and information he took from earlier accounts (see Anguiano 1976:265–278; Brand 1971:651–652; Van Horne 1936 for discussions of the value of Tello's reports).

Fray Antonio Arias de Saavedra was the administrator at the Franciscan Mission de Nuestra Señora de la Asunción in Acaponeta from 1656 to 1672. Prior to becoming the administrator, he was there for three years as the associate priest. During the three decades he was in Acaponeta, he observed the social and religious lives of Indian groups living nearby and in the Sierra Madre Occidental bordering Acaponeta. In his own words, the "mountain range was inhabited by infidels, Indian renegades, and apostates" whom he hoped to convert. His mid-seventeenth-century ethnographic knowledge was incorporated into a 1673 report entitled *Información Acerca del Estado de Nayarit*. It was first published in 1899 in the *Colección de Documentos Ineditos, Históricos y Etnográficos, Acerca de la Sierra de ese Nombre* compiled by Alberto Santoscoy. Arias de Saavedra's report has been translated into English by Dan S. Matson (McCarty and Matson 1975). The account is especially important in that it sheds light on the relationships between the indigenous peoples in the mountains and in the coastal areas, including the Marismas research area. Arias de Saavedra's account also details Indian religious ceremonies, one of which he observed at the mouth of the Río Cañas where it flows into the Marismas (see discussions of Acaponeta and Olita below).

Matías Angel de Mota López y Padilla, often cited as Mota y Padilla or Mota Padilla, was born in Guadalajara and later became an eighteenth-century governmental official and historian. About 1741 he completed the work known as *Conquista del Reino de la Nueva Galicia* (see Mota Padilla 1742). This extensive survey includes sixteenth- and seventeenth-century economic, social, and political information compiled from earlier accounts, as well as eighteenth-century information (see Anguiano 1976:279–315 for an extensive review of Mota Padilla).

Joseph Antonio Villa-Señor y Sánchez was an eighteenth-century Mexican geographer and historian. He was officially commissioned in 1742 to determine the population of Mexico and produce a history and geography of New Spain. While his work includes information related to his times, it also incorporates historical data taken from earlier writers such as Tello. Villa-Señor y Sánchez produced several volumes, including *Teatro*

Americano: Descripción General de los Reinos, y Provincias de la Nueva España, y sus jurisdicciones, first published in 1746 (see 1952).

Historical Geography of the Marismas Nacionales

As noted in the introduction to this chapter, as early as 1530 Nuño de Guzmán identified three sociopolitical divisions or provincias relevant to the Marismas study region (see Figure 12.1). The conquest-period settlements of Centispác were located on the coastal floodplain north of the Río Santiago, extending northward to the vicinity of the Río Santa Maria near Rosa Morado. Aztatlán's settlements of the same period were spread from the mountains near modern Acaponeta to the sea and extended from Rosa Morado to the Río Cañas, with the highest concentration of people residing along the Río Acaponeta. Chametla's settlements were between the Río Cañas and Río Piaxtla. The largest population concentration was located along the Río Baluarte, historically referred to as the Río Chametla, but some settlements were scattered in the foothills of the Sierra Madre.

It is of note that although the conquistadores and the earliest Spanish designated areas such as Centispác, Aztatlán, and Chametla as provincias, over time the Church and colonial entities grouped and regrouped areas of west Mexico and introduced other terms such as *district*, *independent territory*, and eventually *state* for these administrative units. This often causes great confusion when researching the contact and early historical periods of the region. With regard to the Marismas Nacionales area, one of the most important changes occurred in 1569 when Centispác, Aztatlán, Chametla, and parts of modern Durango became known as the Spanish District of Centispác.

Provincia de Centispác

When Nuño de Guzmán arrived with his armies in 1530, Centispác may have had 10,000 inhabitants (Tello 1974: 75), including a majority of Totorame and a scattering of Cora, Tepehuana, and Zayahueco Indians (Sauer 1934:10). The Totorame spoke Pinome, Pinult, and/or Pinonaquia (Tello 1974:71), and were lowland relatives of the nearby mountain Cora (see Coyle 1998; Grimes and Hinton 1969; Hinton 1972; Liffman and Coyle 2000; Neurath 2000; Sauer 1934:13; Weigand 1981, 1985; Weigand and Weigand 2000 for information as to the affinities of the various mountain and lowland Indians of this region).

Guzmán entered the coastal plains of Centispác accompanied by several hundred Spanish foot soldiers and thousands of allied and enslaved Indians from central Mexico, Jalisco, and Michoacán. The battle that ensued has been described as one of the four most difficult conflicts of the conquest of west Mexico (Brand 1971:652); it may be portrayed in Plate 64 of the Lienzo de Tlaxcala, a series of illustrations depicting the conquest that were drawn by indigenous allies of the Spanish (Figure 12.2) Although details of the battle are not well known, a portion of his army battled Indians near the modern settlement of Santiago Ixcuintla, after which Guzmán was welcomed to Centispác with offerings of gifts (López 1963:82). The tribute that followed the encounter set the tone for a rather peaceful interlude in the Guzmán Entrada, and both he and his forces remained in Centispác in relative peace (García del Pilar 1955:185).

Centispác was prosperous enough in 1530 to absorb all of Guzmán's army and Indian irregulars for at least 40 days. The economy was based on agriculture, fishing, and hunting. The regional food consisted of fish, oysters, fruits, corn, other vegetables, and fowl, including "chickens." These "chickens," sometimes called *gallinas de españa*, have been interpreted as domesticated indigenous fowl such as turkey and curassow (Sauer 1932:52). Since the Spanish made the distinction between *gallinas* and *gallinas de españa* in their accounts, some of the references could describe European domesticated fowl. Such varieties might have been traded into the region from central Mexico preceding Guzmán (de Fuentes 1963:244).

The conquest-period homes of Centispác were constructed of perishable materials. Tello (1974:72–73) later describes such houses as being made of palm mats and *tigre* (jaguar) skins. He also describes the homes as being well kept, decorated, and scented with copal, but on this point the original source leaves open the question of whether Tello is describing homes from earlier documents or the homes he saw in the seventeenth century. The conquistadores mention houses standing on *lomitas*. Remnants of such individual and clustered artificial mounds are visible throughout the region, which is often referred to as the *tierra caliente* (Kelly 1938:4; Sauer and Brand 1932:18). Regarding homes built in this environment Beals has the interesting comment:

> The abundance of palm in the tierra caliente affords a ready building material lacking in the highland. On the west coast its use is often furthered to the exclusion of heavier building materials by the circumstances that many of the villages and towns were in locations subject to overflow from the rivers. This would be destructive to adobe or stone and adobe structures. Consequently the presence or absence of the more permanent type of structures cannot be taken as an index of relative cultural levels (1932:107).

At the time of conquest, Centispác seems to have been politically integrated and influential on a regional

FIGURE 12.2. Plate 64, Lienzo de Tlaxcala, showing the 1530 battle between Indians of Centispác and the conquering Spanish.

level. Centispác's leader, Ocelotl, is said to have had the power to demand and receive gold, silver, honey, and cotton in tribute from an estimated 10,000 Cora, Tepehuane, and Zayahueco (Tello 1974:75). Although Tello uses the 10,000 count very freely in his accounts, this number, relative to figures he gives for other regions, indicates that the area was well populated and that Centispác had a vast number of supporting subjects. The conquistadores noted that much, if not all, of the political power was held within one family, with Ocelotl's sons and relatives holding positions of importance within Centispác and its subjected villages. At his personal service Ocelotl is said to have had 200 Indians to carry firewood and 100 women to grind corn (Tello 1974:72).

When he met Guzmán, Ocelotl wore a pointed coat and a jaguar skin cape and hat, all elaborately trimmed in feathers. He was adorned with shell and gold necklaces and carried a feather-decorated arrow, while an important relative carried a sword decorated in a similar fashion. Other adornments were made of shells, pearls, silver, gold, copper bells, and turquoise (Sauer and Brand 1932:54; Tello 1974:73).

Shells and feathers held a special cultural significance in Centispác. Shells were fashioned into horns for special occasions. Strings of snail shells are mentioned as symbols of rank and were given to Guzmán as a sign of peace and obedience. Among other gifts, Guzmán received food, a gold bracelet, and a bundle of many-colored feathers, along with the tasseled arrow and long decorated sword that Ocelotl and his relative had been carrying when they first met Guzmán. Beyond these gifts, Guzmán was promised women, land, and whatever else he desired. He is noted to have requested and received gold and silver (Sauer and Brand 1932:54; Tello 1974:73). Guzmán's allied Indians were given feathers, darts, and *macauitles*, which were long, flat, wooden clubs edged with a series of flat, rectangular blades (Tello 1974:72–75).

Three thousand warriors met Guzmán in Centispác. They were dressed in cotton, with many of them wearing only loincloths. Some were decorated with parrot and white night heron plumes, crested helmets, lip rings, and earrings. Weapons included spears, bows, arrows, and macauitles. Various warriors held round shields, displayed tasseled brazilwood darts, and brandished bows decorated with feathers (Tello 1974:72–73; Sauer and Brand 1932:54). At the time of Guzmán's appearance on the west coast, Centispác's warriors had been frequently warring with what they called cannibalistic Indians from the Sierras, who had been raiding the

TABLE 12.1. Changes in Governmental and Church Affiliations Related to the Marismas Nacionales.

North of the Río Cañas		South of the Río Cañas	
Church	Government	Church	Government
1530s Diocese of Michoacán	Viceroyalty of New Spain (Nueva Galicia)	1530s Diocese of Michoacán	Viceroyalty of New Spain (Nueva Galicia)
	1536 taken by Chametla		
1548 Diocese of Guadalajara		1548 Diocese of Guadalajara	
1563 Diocese of Durango (Guadiana)	1536–1564 Nueva Viscaya		
			1569 District of Centispác within Nueva Galicia
			1580 District of Acaponeta within Nueva Galicia
	1724 Province of Sinaloa		1786 Independencia of Guadalajara (Partido of Acaponeta)
	1820 Province of Arizpe (at times mapped as Independencia of Sonora)		1823 Province of Xalisco (Marismas area in the Cantado of Tepic)
	1830 Territory of Sinaloa		
	1835 State of Sinaloa		
			1867 The area of Tepic became part of the Distrito Militar of Nayarit
			1872 State of Jalisco
			1884 Federal Territory of Nayarit
			1917 State of Nayarit

Source: Data from Bancroft 1883:691; Cline 1972a; Gerhard 1972; Tello 1974.

coastal area for some time. In this fighting, Centispác had been allied with Aztatlán but had been unsuccessful in permanently stopping the raids. The Guzmán Entrada terminated most of this particular native warfare (Tello 1974:72–74), although other conflicts continued into the 1700s. In general, the conquering Spanish indicated that the lives and ceremonies within Centispác were similar to what they had seen in Jalisco (*Relación Anónima Segunda* 1963:313). Many Indians from central Mexico were relocated into the Centispác area shortly after conquest (Brand 1971:642), but missionaries to Centispác witnessed Indian cultural activities from the 1550s into the late 1600s similar to what the conquistadores had seen.

Centispác was the first region in which Guzmán established settlements along the west coast of Mexico. After 1531 many of the Indians within Centispác quickly became assimilated into the Hispanic culture established through the system of *encomiendas*, although thousands of natives died from imposed hardships or introduced epidemics. Many coastal settlements of Centispác, in particular, seem to have been depopulated by disease during an epidemic in 1576–1577. After the Spanish conquest, some of the original population of Centispác merged with introduced Aztec, Tarascan, Negro (black slaves brought by the Spanish), and other laborers. Others fled to the mountains to live with related groups and occasionally returned to visit, trade, or raid the Spanish frontier settlements.

As noted above, in 1569 the regions of Centispác, Aztatlán, and Chametla and parts of modern Durango became known as the Spanish District of Centispác (Tello 1945:108). Table 12.1 shows how complex the changes in governmental and church affiliations related to the Marismas Nacionales were. Some secondary sources do not take postconquest Spanish redistricting into account, and when discussing the conquest period they mistakenly link all of the area from the Río Santiago to the Río Baluarte together as one district. Also, some make the mistake of thinking that with the demise of the political entities, and the numerous deaths, there was a similar demise of the indigenous culture.

A convent was built at the settlement of Sentispac, and nearly 20 years later, in the mid- to late 1580s, Alonso Ponce, the Franciscan commissary general, visited the region (Figure 12.3). In his 1587 report Ponce notes that although Pinome was still used, the introduced Nahuatl

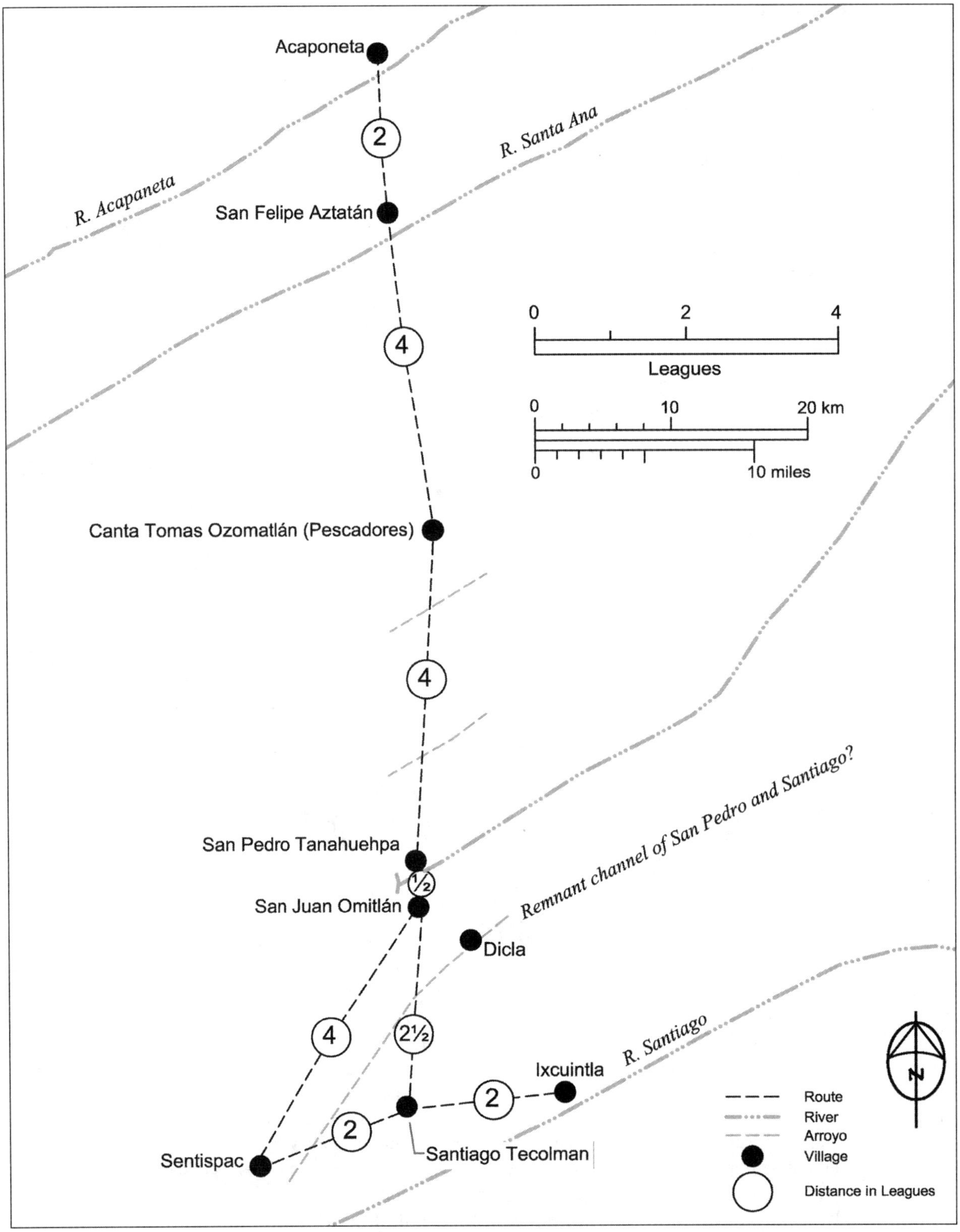

FIGURE 12.3. Route of Alonso Ponce's travels in Centispác and Aztatlán showing various river channels and arroyos Ponce mentions.

was spoken and understood throughout Centispác (see 1968:84). Cotton, corn, chile, fruits, and vegetables were cultivated, and chickens were abundant. Large quantities of oysters and fish were taken from rivers, estuaries, and the sea. Caravans of mules moved oysters from Centispác's estuaries to Mexico City. Ponce (1968:84) describes an oyster fishery and a fishing settlement near the banks of an estuary and at least two other Centispác settlements on islands within lagoons. Ponce also mentions that the mountain Vaynamota and Uzare Indians came to these coastal areas for salt and fish. "Uzare" is one of the oldest designations for the Huichol (Weigand 1985:147).

From Tello's mid-seventeenth-century account it is evident that Centispác was extremely hot and not an idyllic spot to inhabit. Various biting insects including those he (1945:146–148) calls *xexenes*, *rodadores*, *cinifis*, and *zancudos* forced the people of the estuaries within Centispác to stay cloistered in their homes during the worst part of the wet season. On his 1604 visit to Centispác, Mota y Escobar noted that this region provided fish, shrimp, and oysters to Spanish mining areas (see 1940:83). He described Centispác as having eight towns and a population of 330 married Indians of the Franciscan faith. Arregui described Centispác in the 1620s as a maritime province and noted that much of the indigenous population had been greatly reduced between 1610 and 1621. He (1946:98) also recorded that Centispác's fishermen were furnishing fish and shrimp to Zacatecas, Reamos, San Luis, and central Mexico.

Trade began with Peru from west Mexican ports in the mid-sixteenth century, and there was an active trade with China and the Philippines through Acapulco between 1591 and 1774 (Brand 1956:580). Brand (1956:579–581) notes that Matanchel, near present-day San Blas, was among the leading Mexican ports from the 1540s until San Blas was established in 1768. In fact, Brand notes that it was the most important port between Navidad and Mazatlán until the founding of San Blas. Mota y Escobar (1940:82) mentions Chinese ships in the ports of Matanchel (Matanchén) near San Blas in 1604. This port, now simply a beautiful bay, was one of the original ports for this Spanish trade (Brand 1956:579–583). Pearls, mother of pearl, and brazilwood were items Mexico furnished for trade with Asia, and all were readily available in or near Centispác. In the 1630s the District of Centispác became more highly populated as it became a shipbuilding center and an outlet for exporting brazilwood (Brand 1956:580).

Settlements of Centispác

In 1530 the Indian polity that Guzmán called Centispác consisted of more than 40 subject settlements (*Relación Anónima Tercera* 1955:137). The accounts of the conquistadores mention the settlements of Omitlán, Sentispac, Itzcuintlán, Tecomatlán, Sila, Dicla, and Cillan (Guzmán 1963:57; López 1963:82; *Relación Anónima Primera* 1963:296; Sámano 1963:139). These are the only settlements that can be recognized as part of Centispác at the time of conquest. There are, however, other settlements known to have existed in the early postconquest period. Missionary reports and secondary sources mention Acatlán, Acajala, Atotonilco, Mexcal, Mezcaltitlán, San Andrés, Santa Cruz, San Jose, San Juan, Santa Marie, San Migual, San Pedro Tanahuehpa, San Sebastian, Santiago, Tuxpan, and Zapotlán (Arregui 1946:98; Bordes 1944:145; Brotchie 1940:14–15; Navarro 1946:219; Ponce 1968:74; Tello 1945:108, 116; Villa-Señor y Sánchez 1952: 229). Settlements called Yago, Onache, and Tepehuacan are included within Centispác, but only by Brotchie and Bordes, who often mix early and late postconquest data. With less certainty, many of these locales may correspond to settlements that were subject to Centispác or otherwise had tributary relationships with the province. Some may have been subsequently renamed following the Spanish conquest. Most can be located on early postconquest or modern maps (Figure 12.4 shows approximate locations of various settlements as shown on pre-nineteenth-century maps).

1. Santiago Ixcuintla (Vitzcuintlan, Excuintla)

It was just south of the pueblo of Santiago Ixcuintla that Guzmán crossed the Santiago and entered Centispác (Sauer and Brand 1932:43). Near here fierce fighting erupted between the natives and Guzmán's forces. Exactly where the battle was fought is unclear. Toro (1952:23) says that the native settlement of Momote was destroyed in this fighting and later replaced by Santiago Ixcuintla. Santiago Tecolmán (Tecomatlán), a settlement near Ixcuintla, is also identified as the location of this battle.

Brotchie, Bordes, Navarro, Mota Padilla, Tello, and Ponce list Ixcuintla as part of Centispác. Ponce visited this settlement in 1587 and notes Pinome to be the native language (Sauer 1934:7). During his visit, Pinome-speaking Indians came from the mountains to welcome him, and he was given three bundles of decorated arrows as a sign of peace. Ponce mentions eating chickens, bananas, eggs, chile, honey, turtles, and fish during his stay in this settlement, which he locates at less than 2 leagues from Santiago Tecomatlán (see Figures 12.3 and 12.4). He also noticed that birds, especially gallinas, were an abundant source of food.

Mota y Escobar (1940:82) visited the area in 1604, long after the Spanish had created the District of Centispác, and treats it as a separate district composed of Ixcuintla and 10 subject villages and 50 married Indian residents of the Franciscan faith. In the 1620s Arregui (1946:97–98) also made a separate listing for Ixcuintla

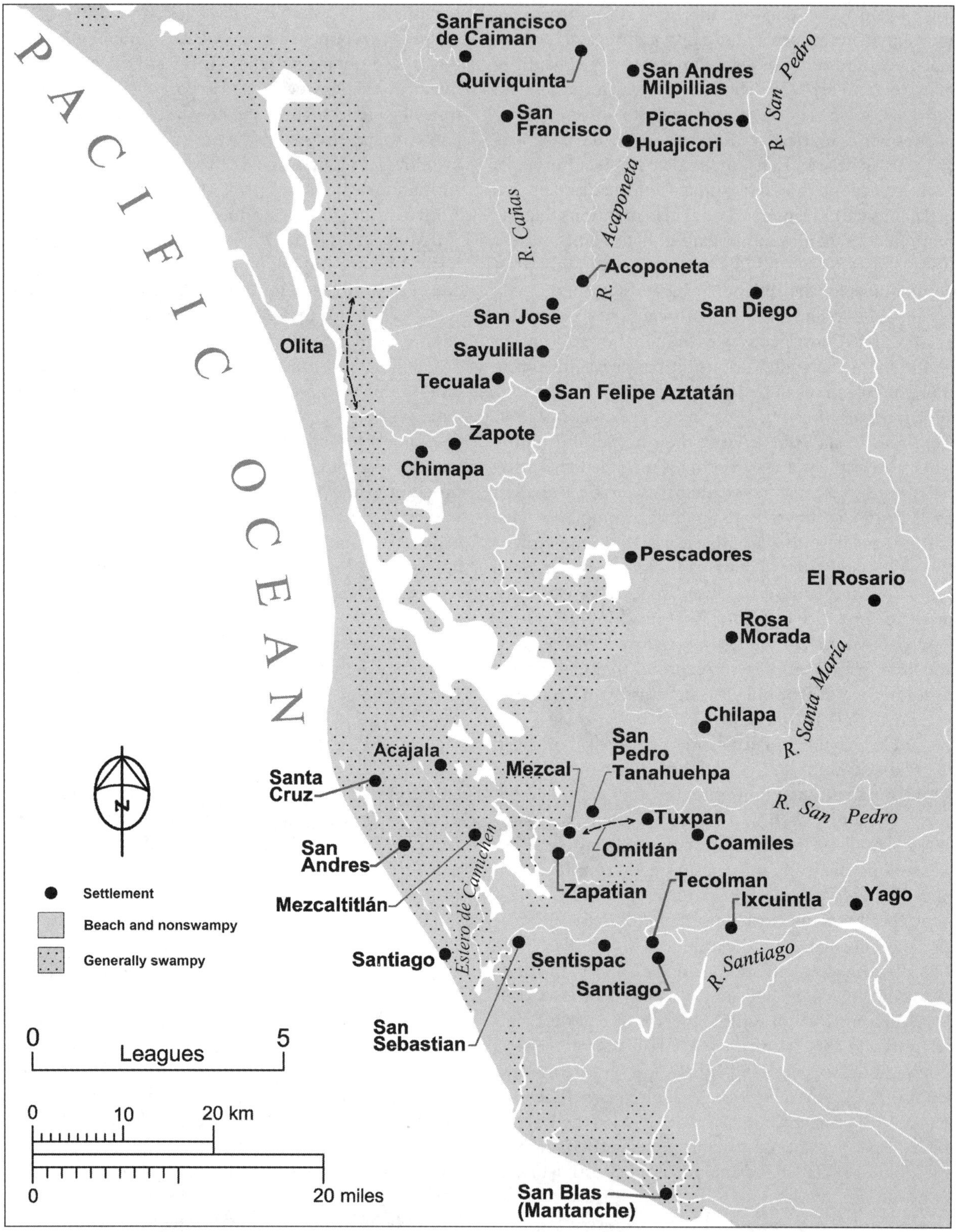

FIGURE 12.4. Approximate locations of sixteenth- and seventeenth-century settlements within Centispác and Aztatlán.

including the settlements of Izquintlán, Tlaxomulco, Mecatlán, Xaljocotlán, Autlán, San Adrés, and Macatlán. Early records often refer to Ixcuintla simply as Santiago, a name given to other towns in Centispác. In the early years of Spanish rule, Cora Indians were settled into the community as the Franciscans converted them. Ixcuintla is a viable community today and has underlying archaeological strata (Grosscup 1964:27).

2. Santiago Tecolmán (Tecomatlán, Atecomatlán)

The Santiago Tecolmán settlement of Centispác was probably the Atecomatlán said to be near Ixcuintla (López 1963:82). It is also a possible site for Guzmán's massive confrontation with the natives. Santiago Tecolmán could possibly be the destroyed Momote that Aranda del Toro (1952:23) mentions, if Momote is not in fact the site directly underlying Ixcuintla. Cook and Borah (1960:106) estimate that Santiago Tecolmán had 23 residents in 1568. Ponce (1968:74) visited the settlement in the 1580s, the location of which he described as 2 short leagues traveling northward from Ixcuintla. On his return route southward, Ponce indicates that Santiago Tecolmán was 2 leagues from Sentispac and 2.5 leagues from Omitlán (San Juan Omitlán on Figure 12.3).

In Ponce's time, the natives of Tecolmán were Pinome-speaking Indians. As in Ixcuintla, Indians from the mountains came to Santiago Tecolmán to see Ponce. These Indians carried painted shields. They threw fruit in mock battle demonstrations as they danced for the missionary. Ponce (1968:74) mentions eating sweet potatoes as well as bananas, ocean fish, and oysters in Santiago Tecolmán.

Arregui visited the settlement around 1620. In his journal he (1946:97–98) mentions that Santiago Tecolmán was within the jurisdiction of Centispác. Grosscup, in giving multiple hypotheses regarding the excavated site of Amapa, theorizes that Amapa may be the archaeological site of Santiago Tecolmán. Today there is no modern town of Tecomatlán or Tecolmán (Grosscup 1964:28); however, there is a Tecomán in the northern part of the city of Tepic, Nayarit.

3. Santiago

Besides the settlements of Santiago Tecolmán and Santiago Ixcuintla, there seems to have been another settlement simply called Santiago in the Centispác region. In his 1742 report Villa-Señor y Sánchez (see 1952) listed a Santiago as a coastal fishing town within Centispác, 6 leagues west of the settlement of Sentispac. In the mid-1740s there were 40 families there who fished for shrimp and fish (Villa-Señor y Sánchez 1952:229–230).

Whether this Santiago mentioned by Villa-Señor y Sánchez existed before the eighteenth century is unclear. Tello, however, lists a village of Santiago as one of many devastated by disease sometime between 1569, when the first convent was placed in Sentispac, and the time of his visit. It is possible that this Santiago was a fishing settlement depopulated in the 1500s and later repopulated permanently or seasonally. There is a modern fishing village of Santiago in the locale where Villa-Señor y Sanchez placed the eighteenth-century settlement (see Figure 12.4). Whether there is an archaeological site there or not is unknown.

4. Cillan (Cilla, Sila)

Guzmán mentions that he went to Sila after Tecomatlán, while López states that Guzmán went to Dicla, a settlement built upon an *arcabuco*, meaning a "mound" (Sauer and Brand 1932:44). The Third Anonymous Conquistador noted a settlement called Cillan 2 leagues beyond Atecomatlán on the way to Omitlán (*Relación Anónima Tercera* 1955). Grosscup (1964:24), as well as Sauer and Brand (1932:44), equated Cillan (or Sila) with Dicla, assuming that the conquistadores gave different names to the same settlement. Guzmán's forces, however, entered Centispác in separate parties, at times traveling slightly different routes. Dicla and Sila (Cilla or Cillan) could therefore be separate settlements close to each other. Accepting the distances given by the Anonymous Conquistador, the settlement or settlements would be located near Tecomatlán (see Figure 12.4). No available map shows a Dicla, Cilla, Cillan, or Sila in the vicinity of Centispác. However, among his multiple hypotheses, Grosscup (1964:31) believed that Amapa also might be the archaeological remnant of Cillan.

5. Yago

Only two sources placed the Yago settlement within Centispác (Bordes 1944:145; Brotchie 1940:14–15). The residents of Yago are said to be Cora by Bordes (1944: 145) and del Toro (1952:23). Perhaps this was one of the Cora settlements paying tribute to Centispác at the time of conquest or a village where Cora Indians were resettled when missionized. Yago appears on modern maps between Sentispac and Tuxpan and is an extant community, slightly northeast of Santiago Ixcuintla (see Figure 12.4).

6. Sentispac

Guzmán knew the settlement of Sentispac (Sauer and Brand 1932:43). Cook and Borah (1960:106) estimate that it had 1,409 residents in 1568. A Franciscan convent was built there in 1579, and Alonso de Badajoz was assigned as its first administrator (Kubler 1948:500). Alonso Ponce was in Sentispac in 1587. At that time the village had a thatch and adobe convent and 100 inhabitants (Kubler 1948:500). Ponce noted Sentispac to be 3 leagues from the sea, 4.5 leagues south of the Río San

Pedro, and 2 leagues northwest of Santiago Tecolmatlán (Tecolmán) (see Figure 12.3). Ponce (1968:83–84) also located Omitlán 4 leagues from Sentispac toward the Río San Pedro.

Ponce noted that during his visit to Sentispac the residents were obtaining drinking water from a big river 1 league away. He (1968:83–84) stated that previously the Río San Pedro flowed by the houses of Sentispac but that the river had shifted, leaving Sentispac without river water. Grosscup (1964:25–26) indicated that Ponce was mistaken in his observations about the Río San Pedro. Grosscup, however, noted the presence of an old river channel near Sentispac but said that there is no indication that water was flowing in this channel in historic times.

Geomorphological studies of this area published after Grosscup's Amapa study, however, show that Ponce may have been correct. Curray and his colleagues indicate that the region of Sentispac has been one of changing riverbeds. In particular, they feel that just prior to 500 BP the Río Santiago and the Río San Pedro entered the ocean together at about 21°48' latitude. They theorize that the mouths of the two rivers became separate as the Santiago shifted southward (Figure 12.5). After this change, a third channel remained in the Sentispac area between the Río San Pedro and the Río Santiago. Sometime thereafter the Río San Pedro shifted northward to its present position, and the third channel dried up (Curray et al. 1969:93–97).

Taking this information into account, Ponce's discussion of a shift in a river near Sentispac seems very plausible. Surely, if the dry channel had been permanently dry for any length of time before Ponce's visit, the people would have moved closer to a water source or dug wells. Ponce's account may well date the northward shift of the Río San Pedro to the 1580s. Ponce described very little of the daily life within Sentispac other than to say that the people were docile, domesticated, and dressed in a manner similar to styles prevalent in Jalisco and Acaponeta. The inhabitants welcomed him to Sentispac with great displays including shows of archers similar to those that welcomed Guzmán to Centispác (Ponce 1968:84).

Tello's account of Sentispac poses many questions. Again, as in many of his recordings for the Marismas region, it is hard to discern whether his references are to the conquest-period Sentispac or the settlement contemporary with his visit in the 1600s. It is also unclear whether he was actually talking about the settlement of Sentispac, because he only referred to this settlement as the "head town." Omitlán and even Ixcuintla also have been called the "head town" of Centispác. Tello, if he was indeed referring to Sentispac, located it at the time of conquest on the Río Santiago. He indicated that it was there that the leader or *cacique* of the Province of Centispác resided (Anguiano 1976:268–269). Tello described Sentispac as an agricultural community where the diet was dominated by fish, but again whether this was a historical or contemporary reference is unclear.

In the 1650s Sentispac was a major settlement within the District of Centispác. Tello (1945:108) reported that Franciscans from the Sentispac convent were visiting Santiago, San Andres, San Sebastian, Tzapotlán, Mezcaltitlán, Acaxala, Coatlán, and San Juan. These same missionaries ranged from Sentispac into areas of Aztatlán and Chametla. It is important to note Tello's (1945:116) comment that years before his visit, missionaries also served the communities of San Pedro, San Juan, San Miguel, Santa Cruz, Santiago, San Sebastian, and Atotonilco until these settlements were depopulated by disease. Sentispac has remained an active settlement to the present day. An extensive archaeological site is found near the modern village (Grosscup 1964:27).

7. Omitlán (Umitlán)

Guzmán and his forces stayed in Omitlán for 40 days in the summer of 1530. Considering the length of their stay, the conquistadores left little description of Omitlán other than that it was extremely hot and located on flat land (*Relación Anónima Primera* 1963:296). The conquistadores were fed beans, chickens, lots of fruits, fish, and maize, of which three crops were grown each year. Crocodiles were abundant in the rivers, and the settlement was full of scorpions (Guzmán 1963:57–58).

Guzmán depicts Omitlán to be the "head town" of the Provincia de Centispác. Toward the end of his stay, Guzmán built a church in Omitlán and erected two crosses. In September 1530 the rivers became swollen with rain, food sources were destroyed by flooding, and Guzmán was forced to move out of the region (López 1963:82). Before he left, many of his allied and enslaved Indians sickened and died (García del Pilar 1955:185). Even as Guzmán moved away from Omitlán, he lost more of his men and horses through drowning (López 1963:82).

Alonso Ponce visited Omitlán in January 1587. By his reckoning Omitlán was situated 4 leagues from Sentispac toward the Río San Pedro. The river was only a half league beyond Omitlán as he traveled northeast toward Acaponeta, a settlement 10 leagues beyond Omitlán. In another descriptive portion of his report, Ponce located Omitlán 2.5 leagues from Santiago Tecolmán. Between Tecolmatlán (Tecolmán) and Omitlán, Ponce noted an arroyo.

Ponce observed a church in Omitlán in 1587 and referred to the settlement as San Juan Omitlán. There were still Indians in Omitlán, and many more came from the mountains to greet him with displays resembling mock battles such as he had seen elsewhere in Centispác. Some

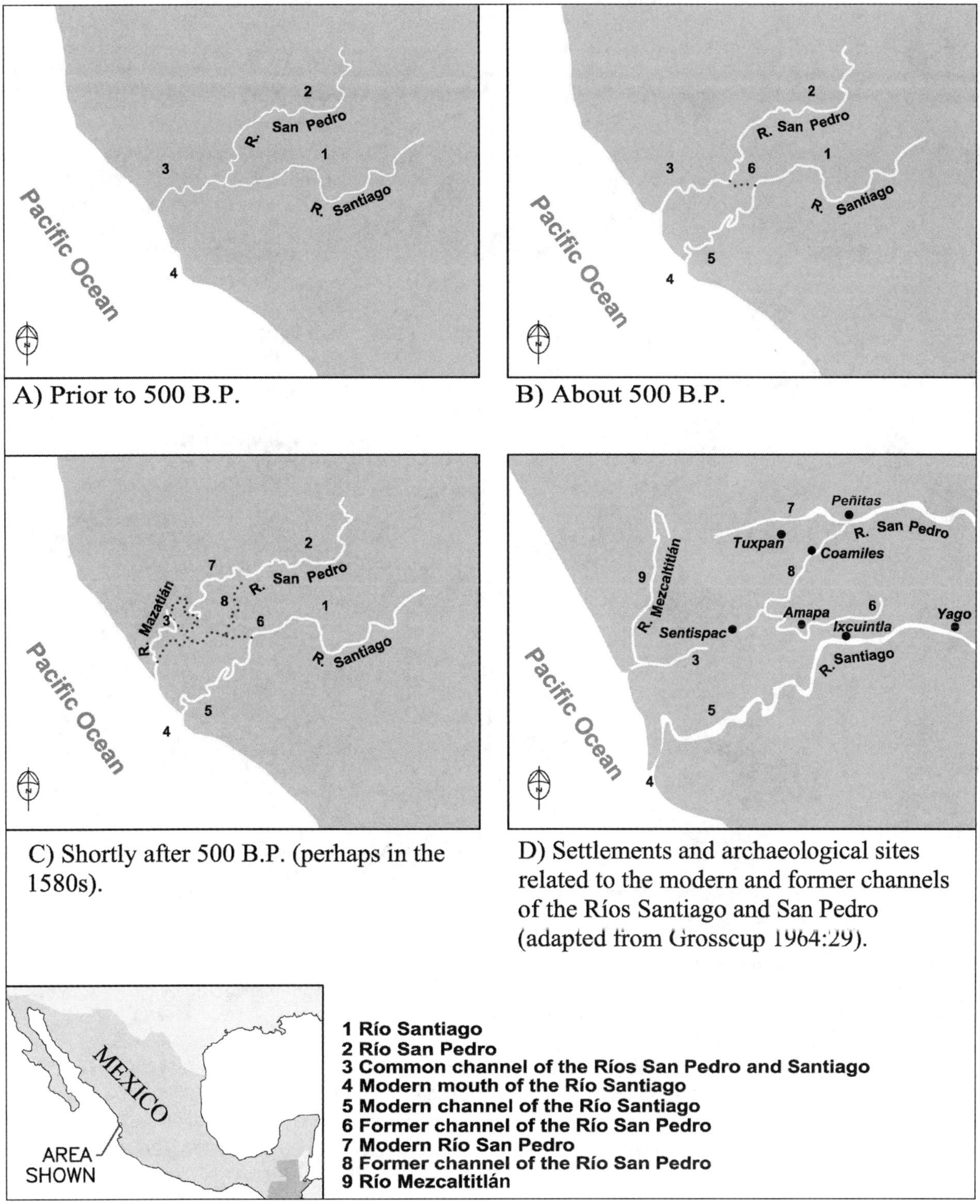

FIGURE 12.5. (*A–D*) Changing channels of the Ríos Santiago and San Pedro.

of the Indians were dressed in the style of the mountain Indians, while others remained naked. Those involved in the loud displays carried macauitles and round shields made from animal jawbone frames covered with crocodile skins (Ponce 1968:75). Thus, it is known that in 1587 the Indians of Omitlán maintained at least some elements of the native culture, which had been witnessed more than 50 years earlier by Guzmán.

Omitlán does not appear on modern maps, nor was it mapped on a 1541 map resulting from the chart of the

FIGURE 12.6. Area from Centispác to Chametla on Ortelius's *Hispaniae Novae Sive Magnae Recens Et Vera Descriptio*, published in 1579. The four islands in the Marismas Nacionales are labeled as (*1*) Copotitlán, (*2*) Amazcula, (*3*) Mezcacula, and (*4*) Tepuacan. Tepequacan on the peninsula may be the earlier site of Teacapán.

Gulf of California drawn by Domingo del Castillo. Yet its earlier importance is indicated by its appearance on early world maps, such as Diego Gutiérrez's 1562 *Americae sive qvartae orbis partis nova et exactissima descriptio*, Gerardus Mercator's 1569 *Orbis Terrae Compendiosa Descriptio*, André Thevet's 1575 *North America*, and Abraham Ortelius's 1570 [1564] map of the world published in his *Theatrum Orbis Terrarum* and *Hispaniae Novae Sive Magnae Recens Et Vera Descriptio* published in 1579 (Figure 12.6). All these maps were compiled from information learned from explorers, other cartographers, seamen, and smaller area maps and charts of the Sea of Cortés dating from the late 1530s, early 1540s, and later. For example, the Ortelius world map was adapted partially from Diego Gutiérrez's 1562 map (Cline 1962: 101–102). Each of these early world maps places Omitlán on the southern side of a large river, but all that can be said is that Omitlán must predate these maps. Maps detailing the local area, such as Juan Martínez's 1587 *Carta del Pacífico y de las costas de América* (see 1951 [1587a]), and others still put Omitlán in that location, but by the late sixteenth century Omitlán disappeared from maps other than from those known to be copies of older maps.

Ironically, Omitlán's location continues to be problematic in spite of its common occurrence on historical maps. In general its location was beyond Sentispac near the Río San Pedro (Grosscup 1964:22–23; Sauer and Brand 1932:43), but whether Omitlán was on the northern side or the southern side of the modern Río San Pedro is uncertain. The confusing factors are these: Guzmán (1963:57) locates Omitlán 10 leagues from the sea in a marshy area north of the Río Trinidad; López places it north of Dicla, "sobre una barranca de un rio grande" (1963:82); Ponce (1968:83) places it half a league south of the Río San Pedro; and finally López and Ponce locate Acaponeta 10 leagues beyond Omitlán in a northeastern direction.

Sauer and Brand (1932:43) have interpreted López's statement to mean that Omitlán was on the northern banks of the Río San Pedro. However, *un río grande* could simply mean "a big river." Thus, López may be placing Omitlán on an unspecified river rather than specifically the Río San Pedro. Sauer and Brand (1932:43) have also interpreted Guzmán's Río Trinidad to be the modern Río San Pedro. Perhaps the Río Trinidad was not the modern Río San Pedro but, rather, the remnant third channel of the Río Santiago and the Río San Pedro, which has been mentioned in the comments about Sentispac. If in 1530 what was called the Río Trinidad was the flowing remnant of the Río Santiago and the Río San Pedro, a possibility suggested by geographers (Curray et al. 1969:93–97), then the accounts of Guzmán, López, and Ponce are consistent. Omitlán would have existed south of the modern Río San Pedro, as Ponce says, and north of the old channel (the Río Trinidad), as Guzmán says. The arroyo shown on Grosscup's (1964:29) map may actually be the old channel of the Ríos Santiago and San Pedro. The arroyo Ponce (1968:83) mentions crossing after Tecolmán on his way to Omitlán may also be this same channel (Figure 12.3).

What happened to Omitlán? It is not an extant community. Ponce, who identified it as San Juan Omitlán in 1587, is the last to mention it as Omitlán. Arregui, who was in the area around 1620, mentioned a San Juan in the same locale. His (1946) reference to San Juan may be the shortened version of San Juan Omitlán (see Grosscup 1964:27), but on his detailed 1621 map he has Tuxpan and not Omitlán. Tello (1945:108) mentioned that missionaries visited a village known as San Juan from Sentispac. However, elsewhere in the same document, he (1945:116) noted that San Juan was a settlement no longer visited by missionaries. It is difficult, therefore, to follow Omitlán's history even if it is assumed to be San Juan.

Grosscup (1964:23, 28) suggests that there is a historical connection between Omitlán and modern Tuxpan. He notes that no one but Arregui (1946:98) mentions visiting both Omitlán and Tuxpan in the 1620s, but this can only be said if it is assumed that the San Juan that Arrequi mentions is also Omitlán. All other accounts subsequent to Arregui mention Tuxpan but not Omitlán or San Juan, and Tuxpan is named in sequence where Omitlán had been mentioned in earlier accounts.

Hypothetically, Omitlán may have lost its importance or ceased to exist as Tuxpan developed or expanded in the early seventeenth century. If this did occur, where is the archaeological site of Omitlán? Grosscup (1964:28) suggests that the site of Omitlán is most likely to be found west of Tuxpan. If indeed the Río San Pedro did shift to its modern bed near the time of Ponce's visit, then the old site of Omitlán would now be under alluvium near where the modern San Pedro enters marshlands, west of Tuxpan. This possible location of the old Omitlán places it as it seems to appear on the 1579 Ortelius map in the second additament of *Theatrum Orbis Terrarum* (Figure 12.6). If this is the case, a northward shift of the San Pedro would have forced the population of the original Omitlán to vacate the conquest-period site and move, presumably upriver toward Tuxpan sometime between the 1580s and the mid-1600s.

Grosscup (1964:28) makes the alternative suggestion that the site of Coamiles (see Figure 12.4) could be Omitlán, a location even swampier than that of modern Coamiles. If there were a third flowing channel between the Santiago and the modern San Pedro in the 1500s, Coamiles would have been situated in a wetter environment than it is today. If the San Pedro did shift its bed northward, the environment around Coamiles would have become drier, forcing resettlement near the new river channel. The area of modern Tuxpan represents a logical spot for such a relocation.

8. San Pedro Tanahuehpa

Ponce in the mid-1580s described the San Pedro Tanahuehpa settlement to be within Centispác across from Omitlán near the mouth of the Río San Pedro (see Figure 12.3). The settlement of Pescadores (Santo Tomás Ozomatlán) was 4 leagues northward beyond the river. One very interesting point in Ponce's account is his statement that as he traveled to San Pedro Tanahuehpa from Omitlán, he walked a half league on a flat road before crossing the mouth of the Río San Pedro. This clearly places both the settlement of San Pedro Tanahuehpa and Omitlán near the mouth of the San Pedro in Ponce's time.

In January 1587 Ponce witnessed Indians described as Chichimecas dancing and playing small guitars in San Pedro Tanahuehpa. The Indians carried clubs and shields, and one wore a breastplate of bells. Mountain Indians, who claimed to have been previously baptized, arrived on horseback during the festivities and requested to be taught the faith. They told Ponce that they had fallen away from the faith when they had retreated into the mountains to avoid constant Spanish mistreatment.

They claimed to have previously started a church on the coastal plain (Ponce 1968:75, 82).

9. Ayotuchpan (Ayotochpan, Ayutuzpa)

Neither Guzmán nor Ponce mentioned the Ayotuchpan settlement within the provincia of Centispác. However, the much later sources of Bordes (1944:145), Brotchie (1940:15, 24), and Navarro (1946:219) placed it within the district. Tello (1945:108) noted that it became part of the Spanish District of Centispác in 1569. Mota y Escobar (1940:83–84) visited Ayotuchpan between 1602 and 1605 and located it 6 leagues northward from Sentispac, 10 leagues before Acaponeta. This would place Ayotuchpan in the region of modern Tuxpan and near the projected site of Omitlán. Mota y Escobar wrote that Ayotuchpan was a village composed of resettled, newly converted Cora Indians.

Arregui (1946:89) listed Ayotuchpan around 1620 to be in the District of Acaponeta. His account concurs with Mota y Escobar's that the residents were converted Cora Indians. On Arregui's 1621 map this settlement is located between the Ríos San Pedro and Acaponeta (see 1946:13). This is a different and more northerly location than that attributed to Ayotuchpan by others. One could conclude that two different villages are being discussed or that the settlement moved. There is also the possibility that at least the settlement Mota y Escobar described may in fact be Tuxpan. Today there is no known Ayotuchpan in the Centispác area. From what little is known it seems that it may not have existed at the time of conquest. All the reports indicating that it was in Centispác were written after the Spanish District of Centispác was established.

10. Mezcaltitlán

Guzmán did not mention the Mezcaltitlán community by name. Toro (1952:3), however, said that it was subjected to Spanish rule in 1530 as it was within what became Nueva Galicia. This island settlement was listed as part of Centispác by Arregui (1946:97–98), Brotchie (1940:14–15), Navarro (1946:219), and Tello (1945:108). López (1963:84), in his 1530 exploratory travels around the Provincia de Centispác, viewed an island community from a hill as he traveled from Omitlán to Aztatlán, but it is unlikely that what he saw was Mezcaltitlán. It is far more likely that the community López saw was Pescadores (see Figure 12.4). The modern island community of Mezcaltitlán is situated in the midst of marshy lowlands. Even if the area was an open waterway at the time of conquest, there are no surrounding hills from which one could observe Mezcaltitlán. Furthermore, in order for López to see Mezcaltitlán between Omitlán and Aztatlán he would have had to travel in a southwesterly direction before going northeast toward Aztatlán. Nothing in his description suggests such a movement.

Around 1620 Arregui visited Mezcaltitlán and reached the island in a rush boat. He noted 25 to 30 houses there and mentions that the lagoon was full of caiman. The people fished but did not raise corn (Arregui 1946:98). Tello visited the island in the mid-seventeenth century and stated that it was just one of three island communities within the District of Centispác at that time. He indicated that Franciscans from Sentispac had missionized the island. He (1945:108), too, noted that corn was not raised on the island and that the people fished for robalos, shrimp, and small fish called *puyeque*. Regarding shrimp and fish it is interesting to note that the much later account of Antonio García Cubas (1890:25) stressed that in the 1800s there was an abundance of shrimp caught in the estuaries of Mezcaltitlán during October and November as well as robalos, *lesas* (lisas), and other fish from December to May.

Mezcaltitlán is situated in the southern portion of the Marismas, which is fed by the Ríos Cañas, Acaponeta, Rosa Morada, and San Pedro. Today the Río San Pedro channel goes through the lagoon and then flows to the sea at Camichín. As it meanders through the lagoons it leaves much sediment behind (Vann 1972:31). The channel where water flows from the lagoon to the sea is called the Río Mezcaltitlán but is in reality the mouth of the Río San Pedro. Mezcaltitlán is a surviving island community, isolated in the midst of estuaries. It is only possible to travel there from the north during times of high water. This may not have been the case before the conquest or even in early historic times. The Ortelius map (see Figure 12.6) shows what might be Mezcaltitlán as one of several islands located in a large body of water.

The village of Mezcaltitlán today covers virtually every square foot of the island, and its high-curbed streets are subject to annual flooding. There is no surface archaeological evidence for the early occupation of the island. The surrounding freshwater marshland, however, is a particularly rich archaeological zone. During the Marismas project inspection of scattered backshore ridges of beach deposits with earth mounds produced abundant quantities of Tuxpan Engraved and other "Amapa-type" ceramics (Stuart Scott, personal communication 2014).

11. Mezcal (Mexcal)

Neither Guzmán nor Ponce mentioned Mezcal. Arregui noted in his 1621 report that Mezcal was a maritime community within Centispác (see 1946:97–98). Tello (1945:108, 116) mentioned Mezcal as one of the towns within Centispác that the Franciscans visited prior to his arrival on the west coast and which was on his list of towns depopulated by disease. Villa-Señor y Sánchez located Mezcal 5 leagues west of Sentispac near the ocean. In the 1740s it had 60 Indian families who were sustained by fishing and shrimping (Villa-Señor y Sánchez 1952:229).

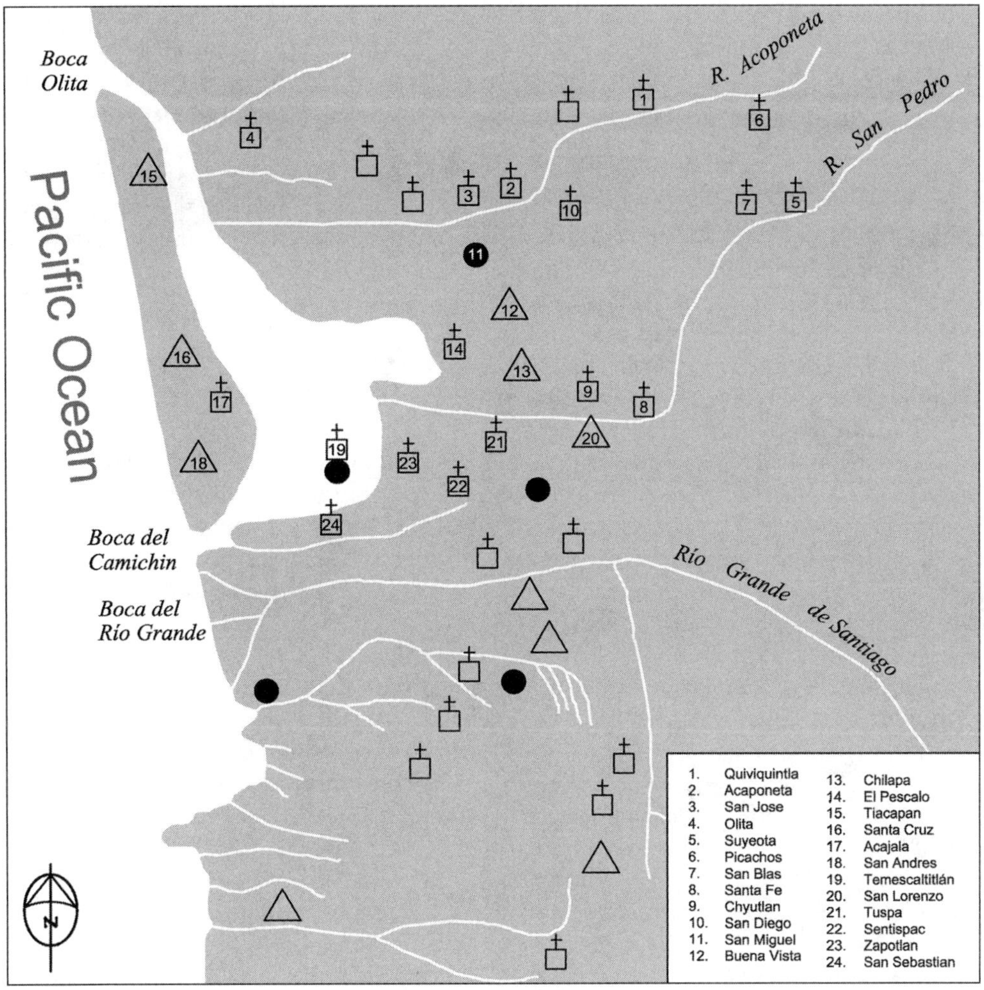

FIGURE 12.7. Settlements in and near the Marismas Nacionales adapted from Diego Joaquín Garavito's 1763 *Mapa particular de las costas del mar del Sur* (in Curray et al. 1969).

Grosscup (1964:28) indicates that Mezcal is only a shortened version of "Mezcal Tktlan," that is, Mezcaltitlán. However, Arregui (1946:98) clearly listed and discussed both Mezcal and Mezcaltitlán, and both appear together on some maps. Furthermore, Mezcal and Mezcaltitlán are both extant communities. Mezcal is located on the Estero el Mezcal at the mouth of the Río San Pedro east of Mezcaltitlán and just southwest of Tuxpan (see Figure 12.4). It is in an area exhibiting an abundance of archaeological remains and may be near where the archaeological site of Omitlán is located.

12. Atotonilco

Atotonilco is another settlement not mentioned by Guzmán or Ponce. Both Arregui (1946:59) and Tello (1945: 116) recorded it as part of Centispác. In his 1621 report, Arregui placed Atotonilco on the banks of the Mexcala lagoon near San Miguel, Ponzitlán, and Santa Cruz. Tello listed it as one of the many villages Franciscans had served prior to his visit. Atotonilco was ravaged by disease in the late 1500s (Tello 1945:108, 116). Today there is a pueblo Atotonilco Tecuala about 17 km southwest of Acaponeta.

13. San Miguel

Arregui (1946:59) located the San Miguel settlement near Santa Cruz, Mexcala, Atotonilco, and Ponzitlán. Tello (1945:108, 116) mentioned that it was ravaged by disease before he arrived but that previously Franciscans from Sentispac had missionized it. This settlement should not be mistaken for the San Miguel located between the Ríos San Pedro and Acaponeta on the Garavito 1763 map (see 1969 [1763]), which was settled by missionized Cora Indians from Nayarit in the 1650s (Figure 12.7).

14. Acaxala (Acajala)

Acaxala is not listed as a settlement in any of the accounts of the Guzmán Entrada. Acaxala was listed as part of the District of Centispác by Mota Padilla (1973:64), Arregui (1946:98), Tello (1945:108), Villa-Señor y Sánchez (1952:230), and Navarro (1946:219). Tello (1945:108) noted that it was missionized from Sentispac in the

1650s. Villa-Señor y Sánchez (1952:230) in his 1742 report described it as a settlement of 38 Indian families, located 7 leagues west of Sentispac. As Acajala, it appears on the Garavito map near Santa Cruz and San Andres (Figure 12.7). It appears on modern maps in the same area.

15. Santa Cruz

Arregui (1946:59) placed Santa Cruz in the margin of the Mezcal lagoon near Ponzitlán, San Miguel, and Atotonilco. Tello (1945:108) listed it as a once-missionized town within the District of Centispác, devastated by disease before his arrival. On the Garavito map it is seen in the estuary near the sea close to Acajala and San Andres (see Figure 12.7). According to Torres (1965:88), missionaries from Aztatlán founded this town near the sea. Although this seems to conflict with Tello's report that it was missionized from Centispác, it is possible that it was first missionized from Centispác before being devastated by disease and then reestablished by missionaries from Aztatlán. It is possible, of course, that Tello and Torres were discussing different communities. Today Santa Cruz is a small Marismas community located near Acajala and San Andres (see Figure 12.4).

16. San Andres

The first mention of the San Andres settlement is by Arregui (1946:97–98), who in his 1621 report listed it as a maritime community within the District of Centispác northwest of Sentispac. Tello (1945:108) said that it was missionized from Sentispac in his time. García Cubas (1890:25) helps locate San Andres by saying that it was 22 km west-northwest of Mezcaltitlán. It is a viable community in the same locale today (see Figure 12.4).

17. San Sebastian

Arregui (1946:98) listed San Sebastian as a maritime settlement within the jurisdiction of the District of Centispác. Tello (1945:116) listed it as one of the Indian villages missionized by the Franciscans before his visit but significantly depopulated by disease. In his 1742 report Villa-Señor y Sánchez also mentioned a San Sebastian in Centispác and located it 4 leagues west of Sentispac. When Villa-Señor y Sánchez (1952:230) visited the area, San Sebastian was a coastal town with a population of 25 families, all involved in the oyster commerce.

San Sebastian appears on the 1763 Garavito map of Mezcaltitlán north of the old river drainage of the Río San Pedro (see Figure 12.7). It is a small settlement today near the coast south of Mezcaltitlán on the southern banks of the Estero de Camichín (see Figure 12.4).

18. Zapotlán

The earliest mention of Zapotlán is by Arregui (1946:98). In his 1621 account, he described it as a maritime settlement of the District of Centispác. Tello (1945:108) listed it as one of the villages missionized from Sentispac in the mid-1600s. Navarro (1946:218) listed it within Centispác. As Tzapotlán, it appeared within Centispác's listings in Mota Padilla's *Historia* (1973:64).

Villa-Señor y Sánchez located Zapotlán in 1748 within Centispác, 20 leagues northwest of the settlement of Sentispac. It appears in this locale slightly southwest of Tuxpan on the 1763 Garavito map (see Figure 12.7). Zapotlán is located near a probable location of Omitlán (see Figure 12.4).

19. Acatlán

Acatlán was listed as part of the District of Centispác by Villa-Señor y Sánchez (1952:229) and Mota Padilla (1973:64) and the later historians Bordes (1944:145), Brotchie (1940:14–15), and Navarro (1946:219). Villa-Señor y Sánchez (1952) in his 1742 report located it 15 leagues northeast of Sentispac and stated that its population in the 1740s consisted of 42 Indian families, which Bordes listed as Tepehuana. Toro (1952:32), however, said that it was a Cora village that was first subjected to Spanish rule in 1723. Here again, the possibility exists that two towns having the same name are being described. It could also be that Acatlán was one of the Cora or Tepehuana settlements subject to the provincia of Centispác before conquest.

Provincia de Aztatlán

"The conquest period people of Aztatlán were lowlanders, tillers of the soil on a large scale, fishermen, and *salineros*, and they lived in warfare with the dwellers of the *barrancas*, who were cannibals" (Sauer and Brand 1932:7). The conquistadores reported that Aztatlán was populated from the mountains to the sea all along what is now called the Río Acaponeta (*Relación Anónima Cuarta* 1955:115). Nuño de Guzmán and his forces encountered the people of Aztatlán during a flood in late summer 1530 after leaving the provincia of Centispác. García del Pilar (1955:189), Gonzalo López (1963:83–90), Juan de Sámano (1963:140–142), and the *Relación Anónima Primera* (1963:226–228) provide details of the Spanish entrance into Aztatlán. The conquistadores' accounts seem to agree that there was a skirmish between native warriors and at least the first Spanish to arrive at the Río Aztatlán, the modern San Río Francisco (see Figure 12.4). This battle is probably the one depicted in the Lienzo de Tlaxcala, Plate 66 (Figure 12.8).

The majority of Guzmán's forces arrived in Aztatlán on the day following this battle, and they crossed the river and marshes with great difficulty. As the main body of the Spanish forces crossed the swollen river, the native warriors fled northward into a savanna area beyond another river. After entering Aztatlán, Guzmán met with a sizable group of feather-adorned Indians, who gathered peacefully. Tello (1945) states that there were 10,000

FIGURE 12.8. Plate 66, Lienzo de Tlaxcala, showing the 1530 battle between Indians of Aztatlán and the conquering Spanish.

warriors at this meeting, suggesting that Aztatlán's population was similar to that of Centispác.

The Spanish encamped between the Ríos Santa Ana and Acaponeta in a swampy, lowland region on the outskirts of the settlement of Aztlán, near modern San Felipe Aztatán (see Figure 12.4). The extensive archaeological site of La Cruz nearby may be Aztlán (Sauer and Brand 1932:44). From this settlement a contingency of the forces under the direction of Gonzalo López (Sámano 1963:142) explored the well-populated Río Acaponeta valley to the coast. These men must have passed through at least the southern edge of the area near the mouths of the Ríos Acaponeta and Cañas surveyed during the Marismas project.

While in Aztatlán, the conquistadores built numerous houses near modern San Felipe Aztatán and Tecuala (see Figure 12.4). They noted that there was an abundance of food, including enough fowl to feed their army for two years (López 1963:87; Sauer and Brand 1932:45). In the San Felipe area the Spanish were shown items supposedly belonging to a Christian trader who had died seven years before (*Relación Anónima Primera* 1963:297). If this account is true, the area was exposed to European culture as early as 1523.

Soon after arriving in Aztatlán some of the Spanish forces returned to Mexico City, while the majority of Guzmán's troops, Indian allies, and captives prepared to spend the winter in Aztatlán. In September 1530 all plans were abruptly altered by a sudden storm (García del Pilar 1955:185). First, winds demolished most of the native and Spanish houses. Then at midnight, when the winds stopped, the rivers began to swell from torrential rains in the mountains. By dawn the earth was a "sea" (López 1963:87–88). The land within 5 to 6 leagues in all directions was inundated (*Relación Anónima Primera* 1963:297).

Some of the natives of the region were killed as their homes were either flooded or destroyed by the wind. Many survived only by sensing the severity of the storm and quickly retreating into the mountains (López 1963:88; *Relación Anónima Segunda* 1963). More than a thousand of Guzmán's Indian allies drowned. Many of this number had been so weakened by sickness that they could not retreat to higher ground. The storm subsided two days later, and the waters returned to normal channels. By then, according to García del Pilar (1955:185–186), more than 8,000 Indians had taken sick, leaving only 200 Indians able to stand. Many of the natives accompanying Guzmán begged for permission to return to their homes in Jalisco and central Mexico for the season. Others tried to escape into the marshes. Guzmán, however, refused requests for even temporary leave. Moreover, to discourage dissension he hanged 50 natives (García del Pilar 1955:186).

After the storm, Guzmán stayed on in Aztatlán. Natives from outlying areas, especially the mountains, provided him with corn and birds (Tello 1974:77). García del Pilar and Samaniego were sent by Guzmán northward to pacify Chametla, the next province. López was sent southward for supplies and Indian reinforcements. While in the south, López carried out retaliatory raids against Indian dissenters in the previously conquered areas of Jalisco.

Samaniego soon returned from Chametla, bringing to Aztatlán 150 Indians loaded with gallinas. Among the natives was the son of Chametla's leader. Guzmán presented these Indians with blankets and cloth before letting them return home in peace. López was gone for several months, during which time García del Pilar journeyed southward to join him. When both finally returned to Aztatlán, they found that Guzmán and most of the remaining forces had moved to Chametla. López (1963) notes that Guzmán's ranks had been greatly diminished, as 3,000 of his soldiers, allies, and slaves had died in Aztatlán. Although the accounts of García del Pilar and López differ on the number that actually died, it is evident that illness and the flood at Aztatlán devastated Guzmán's forces. After traveling to Chametla and to regions farther north, Guzmán returned through all the areas he had previously conquered. Upon returning to these areas, some of his men established small settlements in the Río Acaponeta area, keeping with them enslaved Indians from central Mexico and Jalisco.

For a period Aztatlán was part of the Spanish District of Centispác; then in the course of time it evolved into the Partido of Acaponeta within the Independencia of Guadalajara. The Aztatlán region was incorporated into Nueva Galicia in 1548. At various times it was part of the Province of Xalisco, the Distrito Militar of Nayarit, the State of Jalisco, the Federal Territory of Nayarit, and finally the State of Nayarit (Table 12.1). Beyond the records of the conquistadores there is information regarding Aztatlán in missionary records and in the *Suma de Visitas* (see Paso y Troncoso 1905). The *Suma de Visitas* only records the following for the Aztatlán area: "The *estancias* on the level lands are in swamps. They could not be inspected for this reason and because they are in badly broken country. Moreover, the people are bestial and have no Christian training. They commit many abominations" (Paso y Troncoso 1905:SUMA LVI No. 289; my translation).

The Río Acaponeta area remained one of raids and conflict even as the Spanish established themselves there. Missionaries began to visit the Acaponeta valley area from Jalisco convents. Then, from 1569 to 1580, missionaries visited from convents established in Centispác (Kubler 1948:49). Juan de Padilla, a Franciscan missionary, mentioned that there were four Indian groups in the Aztatlán area: the Totorame, Tepehuane, Cora, and Sayahueco (Torres 1965:69). Alonso Ponce entered the region of Acaponeta after traveling 10 leagues from Omitlán. Within the area he noted that, although Nahuatl was used and understood, there were seven other languages or dialects spoken, including Pinutl (Pinome or Pinonuquia), Quachicanuquia, Guanuquia, Iruzanuquia, Naarinuguia, Newxinguia, and Quarinuguia (Sauer 1934:7).

Ponce (1968:79) recorded in the 1580s that the people of Aztatlán dressed like the natives of central Mexico and Jalisco with some slight variation. Silver was plentiful, and the Spanish were already mining the district. For three days before Ponce's arrival there were unseasonal rains in the mountains. Thus, when he arrived it was necessary for the Indians to help him cross the Río Acaponeta on a raft fabricated from layered gourds bound between an upper and lower surface of stick and reed matting. While Ponce was there in 1586, this flooding destroyed many crops, and the people suffered from hunger. He mentioned that maize, beans, melons, and cotton were the usual regional crops. Salt and fish such as robalos, bagres, mojarras, and puyeque were important products taken from the sea, estuaries, and rivers.

Ponce (1968:66–78) described coastal oyster fisheries located 6 leagues from Acaponeta and indicated that oysters from these extensive beds, as well as from Centispác, were taken in caravans to Mexico. He mentioned great amounts of oyster shells at the mouth of the Acaponeta, so great that there seemed to be mountains of these shells bordering the sea. He elaborated that the shells were used in making construction lime for Acaponeta. Ponce (1968:69) also revealed that the Vaznamota and Uzares, living in the mountains southeast of Acaponeta, came to the coast to obtain salt and fish. Another noteworthy comment in the accounts by Ponce is that numerous Indians who had previously populated Aztatlán had retreated to the uplands to escape Spanish mistreatment. He further noted that during his stay some previously missionized mountain Indians were relocated in the valley.

Alonso de la Mota y Escobar was in Aztatlán during his 1602–1605 visitations. He mentioned the abundance of fish in the river, sea, and estuaries and noted that 2,000 *hanegas* of white salt were produced here each year for the benefit of this region and many others. Since a hanega (fanega) equals about 1.5 bushels, the quantity of salt production was significant. There was also, according to Mota y Escobar (1940:83–84), an abundance of corn, cotton, vegetables, and both native and Spanish fruits. During the years of 1617 and 1618, while Francisco de Morga administered the convent of Acaponeta, some of the Cora and Tepehuane rebelled and burned Acaponeta. After the turmoil subsided, more Franciscans were sent to pacify the mountain Indians, whose villages skirted those of the Aztatlán coastal plain and provided refuge for dissidents. Numerous mountain Indian groups were relocated in new settlements along the Río Acaponeta. Fray Atanasio López details this missionary work in his *Misones o doctrinas en Jalisco en el siglo XVII* (1960; see also Warner 2001 related to later Jesuit missionaries to the Cora).

In his 1621 report Arregui (1946:98–102) emphasized the economic importance of fish and salt in the Acaponeta area. In particular he noted that fish, such as robalo and puyeque, as well as salt and fruit were taken from the Acaponeta area to Durango, Sombrerete, Zacatecas, Ramos, San Luis, and Mexico. Tello (1945) in his 1653 report discussed the richness of fish in the Río Acaponeta and in the estuaries located 8 leagues from the settlement of Acaponeta. He (1945:108, 118) also mentioned that shrimp were taken from the estuaries, and he noted an abundance of oyster fisheries in those estuaries. In reference to seventeenth-century salt production, Tello reported that 7,000 hanegas of salt were produced for Spain in this estuary region of Acaponeta and that from the two settlements of Olita and Ayocapa alone the Indians collected more than a thousand hanegas of a hard, white salt each year.

Antonio Arias de Saavedra's 1673 report (translated in McCarty and Matson 1975) included details of life near the Río Acaponeta. From this account we know that Tepehuanes, Cora, Xamucas, Caponetas, and Sayahuecos resided in the mountains near Acaponeta or near the mountains along the Río Acaponeta, while the Totorames lived in coastal areas and on islands (McCarty and Matson 1975:213). While visiting the coast Arias de Saavedra saw non-Christian Indians frequently traveling from the mountains to trade for salt, meat, and fish in exchange for mescal, fermented drinks, sugarcane, corn, beans, honey, and wine (McCarty and Matson 1975:196). Arias de Saavedra noted that many of the mountain Indians were descendants of people who had accepted conversion years before but left the faith. He also noted that converts were difficult to keep in the fold because if chastised, they simply fled into the mountains where they had friends and relatives.

Of all that Arias de Saavedra wrote, it is his descriptions of the Totorame, the religious aspects of the Nayarit Chora (Cora), and the settlement of Olita (Ollita) that bear most directly on the Marismas studies. Olita, according to Arias de Saavedra, was located on the Río Cañas near its mouth where both the Ríos Cañas and Acaponeta flow to the sea. This description of Olita matches what is shown on the Garavito and Narváez maps (see Figure 12.7, Figure 12.9). The outlet to the sea, today known as the Boca Teacapán, is even labeled "Boca Olita" on the Garavito map as well as others. Arias de Saavedra noted that the Totorame lived along the seashore as well as on little islands, and he described the Totorame as fishermen and salt gatherers. As Arias de Saavedra stated it: "Everyone calls them Themureles, which means 'frogs'" (in McCarty and Matson 1975:213). Their territory extended from Chametla to Autlán, which can be seen on the Ortelius map as Ayutlán (see Figure 12.6) and on the Narváez map as Aután (Figure 12.9) (based on ethnohistoric data, Spicer [1969:781] illustrates the approximate relative location of the Totorame). Essentially, the Totorame lived along the coast from Chametla to the area very near what is now San Blas and, based on these accounts, presumably in the Marismas area as late as the mid- to late 1600s.

In terms of population Arias de Saavedra (McCarty and Matson 1975:213) noted that the Totorame were second in size to the Cora. He described the Totorame as related to the Nayarit Chora, a general term along with *Nayarita* (*Nyalita*, *Náyari*), that at that time incorporated the various groups living in the mountains east of the coastal plain (Weigand 1985:134). According to Weigand (1981:9), the Totorame may have been related to the Tecual Huichol, who were considered Nayaritas by the early Spanish and closely related to the Mesa de Nayar and Guainamota Cora (see also Weigand and Weigand 2000:28–29 for Nayaritas in a regional context as it is understood today).

According to Tello, in the 1650s the Totorame worshipped a child-god, Teopiltzintli. This god was represented by a statue and was presented tribute such as cotton and shells. Tello (1974:72) mentions that the Totorame believed that this god had led their ancestors from Aztlán. Arias de Saavedra 10 to 20 years later defined the faith as a form of idolatry based on the four seasons controlled by the Piltzintli, Nicanori, Narama, and Uxuu. Arias de Saavedra further reports that

> they also believe that the Piltzintli has an influence on the others, and that when the Piltzintli Xucati (whom they call the son of God who is in the sun) comes on the 21st of March, he awakens the Nicanori so that he in turn can begin to create all the shellfish and fish and prepare the birds to lay eggs. When the sun has gone beyond the Nicanori, he touches the Narama with his light and arouses him to begin producing the salt and other fruits that are his assignment in the months of April, May, and June. At a certain point after the 21st of June, the sun Piltzintli comes and releases the mists in the form of water and sends lightning and thunder until the 21st of

FIGURE 12.9. Section of José María Narváez's 1824 *Carta corográfica de Jalisco, Zacatecas y Colima* (from the 1858 *Boletín de la Sociedad Mexicana de Geografía y Estadística de la México Tomo VI*).

> September. Then he again arouses the Nicanori, who begins from that day on to make the fish multiply and the birds lay. After the 21st of September, he awakens the Uxuu with his light so that she can prepare the earth with rains and mists for the production of fruits and seeds [in McCarty and Matson 1975:210].

There were specific places associated with each of the four deities. Piltzintli was thought to dwell on the mesa and was associated with the sun. Nicanori was present in the arm of the sea in the west, where the sun enters the first degree of the sign of Aries on March 21. Narama had once been associated with a hill called "Ichamet" but later lived in the sea, where the sun enters the first sign of Cancer on June 21. Arias de Saavedra translated Narama's former locations as "Horse Head" and added the meaning to be "the House of the Maguey and Mezcal." Uxuu was associated with a cliff on the sea called Matanche, where the sun enters the first degree of the sign of Capricorn on December 21 (McCarty and Matson 1975:210). In other words, in modern terms, these ritual activities and the locational "homes" of the four deities are associated with the spring and fall equinoxes as well as the summer and winter solstices.

Arias de Saavedra reported that during his 1656–1672 time in the Acaponeta region the natives had many other places related to positions of the sun and made idols that they placed near these sacred spots (McCarty and Matson 1975:207–211). What idols an individual made and which of the four personages was worshipped were related to occupation, but more than one deity could be worshipped. Salt gatherers, wine or pulque makers, and chile planters made figurines in the form of naked men to worship Narama. Fishermen and bird raisers made idols in the shape of upper human torsos, birds, eagles, or heads for Nicanori. Idols representing women pertained to Uxuu.

At least two devils were part of the belief system, which included aspects of magic and witchcraft. Arias de Saavedra explained that one could recognize caves and places where devils were worshipped by finding phallic symbols and serpents made of clay; irregularly shaped stones; offerings such as peyote, *tapat*, and *xihuitei*; wall paintings of female genitalia; and effigies of the devil. The devil Tzotonaric and the deity Uxuu were evoked for power over women, and men also used idols in the shape of women, phallic symbols, or irregularly shaped rocks for such power. Devils also had knowledge of adultery and the power to change shape, fly through the air, or make dead bodies talk (McCarty and Matson 1975:202–205).

These religious aspects were an integrated part of the social/political organization involving a hierarchy of elders who held special relationships with gods, ancestors, and sacred ceremonial places including mountain peaks, caves, lakes, wells, springs, and the ocean (Sea of Cortés). What Arias de Saavedra witnessed was also noted by Padre José de Ortega, the Jesuit who served the Cora for more than 25 years after the Spanish conquered the Mesa de Nayar in 1722 (Warner 2001:4). Even when the Cora on the Mesa de Nayar, which is east of the Marismas above Acaponeta, were put under Spanish rule, Ortega reported that the leaders accepted allegiance to the Spanish but did not accept the Spanish religion, because "they could not deprive the dignity of the sun, who they had always worshiped" (1754:84, in Coyle 1998: 517–518; see also Ortega 1887). The Spanish destroyed many idols and ceremonial shrines, including the most sacred center on the Mesa de Nayar, but the traditions continued. Aspects of the social, religious, and political systems that Arias de Saavedra and Ortega described are part of modern Cora and Huichol culture, even as the Catholic faith has been adopted by many (see Coyle 1998; Hinton 1972; Liffman and Coyle 2000; Neurath 2000; Warner 2001; Weigand and Weigand 2000).

Settlements of Aztatlán

Although the conquistadores described Aztatlán as populated from the mountains to the sea, they clearly identified only the settlement called Atzlán. López (1963: 80), however, seemed to describe the settlement of Pescadores (Santo Tomás Ozomatlán) when writing about viewing an island community after leaving Centispác. Because the conquistadores mentioned no other communities by name, determining which settlements might have been in Aztatlán at conquest is even more difficult than it was for Centispác.

Aztatlán as an area distinct from Chametla and Centispác is depicted on most of the early Mexican maps, but often its individual communities are not shown. One early exception to this is Diego Gutiérrez's 1562 map entitled *Americae sive qvartae orbis partis nova et exactissima descriptio*, which shows ASTATLAM in large letters and the settlement of Astatlam in small letters (see 1949 [1562]). Ortelius's 1570 [1564] map entitled *Americae Sive Novi Orbis, Nova Descriptio* shows almost the same, with ASTATLAN as a named area and Astatlán shown as a settlement. André Thevet's 1575 map also shows what appears to be the settlement of Astatlán. The Ortelius 1579 map (see Figure 12.6) shows various settlements between two rivers—probably the San Pedro and the Acaponeta—flowing into the Marismas and four settlements on islands within the Marismas. The settlements between these rivers probably correspond with what was known as the Province of Aztatlán. Saponita correlates with modern Acaponeta, and Haftatan, with Aztatlán. It has not been possible to relate the mapped island settlements seen on the Ortelius 1579 map to known communities other than to say that Tepeguacán, shown on a peninsula on the map, corresponds to an early known location for Teacapán, and Mezcacula may be Mezcaltitlán. Teacapán is now located on a peninsula, Isla del Palmito, on the eastern main estuarine channel. Currently it is under the political jurisdiction of Escuinapa, an area that probably would have been in the Province of Chametla at the time of conquest. It is conceivable, however, that when Teacapán was on the outer peninsula of the Marismas it might have been closely affiliated with settlements of the Río Acaponeta area, thus within what the Spanish saw as Aztatlán.

Within the missionary records available for study, the area of Aztatlán is most commonly called Acaponeta. Ponce (1968:76) in his report placed the settlement of Pescadores within Acaponeta. We know from Arregui that around 1620 the villages of Chilapa, Coyotlán, and Ayotuxpa were mountain villages of Acaponeta; Maya, Canamota, and San Francisco were villages near the settlement of Acaponeta; San Pedro, Ocumatlán, Ocelotepec, Chimapa, Yayoacan, Olita, and Astatlán were settlements of the district toward the coast; Picachos and Milpillas were on the road to Durango but within the District of Acaponeta; and Guaxicore, Namorita, and Quiviquinta were also part of Acaponeta. Arregui (1946:89), however, listed Quiviquinta and Milpillas as depopulated in 1617 following the rebellion.

Tello (1945:119) reported the following villages within the jurisdiction of the Acaponeta convent in the 1650s:

Atztalán, Ayacapan, Ayoacan, Acaponeta, Tequala, El Caiman, Colita, Chimapa, Zaulam, San Joseph, San Francisco, San Nicolan, Ytztapilla, San Diego, and San Antonia. Villa-Señor y Sánchez (1952:248–250) listed the following settlements in Acaponeta in his 1742 report: Milpillas, Cayman, Quiviquinta, Picachos, Pescadores, Aoyacan, Sallula, Tequala, Olita, San Joseph, San Diego, Guaxicoli, San Blas (the one north of Acaponeta, not the one in Centispác), San Cristoval of Guaynamota, and Santa Fe. Many of these settlements appear on the 1763 Garavito map (see Figure 12.8) as well as on the later Narváez map (see Figure 12.9), and the later histories of these settlements can be traced through references in missionary or governmental records. Some settlements that might have been extant communities within Aztatlán at the time of conquest are detailed below.

1. Pescadores (Santo Tomás Ozomatlán)

Ponce visited the Pescadores settlement in the 1580s, and it was probably the one the conquistador López noted while traveling from Omitlán to Aztatlán in 1530. López first viewed the island community from a nearby hill. Before reaching this settlement, he crossed several small rivers, along which he noted very small populations. Then, he entered the settlement after a skirmish with some warriors. He waded to the island through waist-high water and notes that the settlement was within a vast swamp. He saw hardly any corn but an abundance of fish and dogs, both of which were used as food. His Indian allies took as many of the dogs as they could (López 1963:84; Sauer and Brand 1932:52), which seems to suggest that the dogs were taken as a ready supply of what they considered a source of food. Later, on his way south, López chased some Indians who were attacking his forces from rafts.

Ponce (1968:76–82) visited the locale of Pescadores Ozomatlán twice sometime between 1584 and 1588, and he indicated that it was within the *guardiania* of Acaponeta. On January 22 he spent the evening and night in Ozomatlán, a ranch across from Pescadores. He noted the area to be very wet and that marshes and lagoons extended from there to the sea. There were many fish, especially in the lagoon. On his second visit, Ponce was transported across the smelly waters surrounding the island in a reed boat. He was welcomed to this small community by the Indians, stayed overnight, visited the church, and was fed eggs, chickens *de Castilla*, and special fish all cooked as a stew in oil and water. He noted that both Iruzanuquia and Quarinuquia were spoken in Pescadores.

Villa-Señor y Sánchez (1952:248–250) noted that Pescadores in the mid-1740s was situated on the banks of the lagoon at the mouth of the Río San Pedro. He located it 12 leagues southwest of Acaponeta and Sallula, which is 2.5 leagues southwest of Acaponeta in the same area as the Hacienda de Chilapa. Pescadores appears on modern maps on the eastern shores of a lagoon (see Figure 12.4). It appears as a peninsula and is very likely an island at times of high water.

2. Olita (Ollita)

Olita is not a settlement today. What is now known as the Boca de Teacapán is shown as the Boca de Olita on the Garavito map (see Figure 12.7), which shows both a settlement and a salina of Olita. In his 1673 report Arias de Saavedra noted that the water of the "river of reeds" (Cañas) dividing the kingdoms of Galicia and Viscaya "flows into the marshes and tidal pools of the pueblo of Ollita of the Province of Acaponeta" (in McCarty and Matson 1975:213). The Arias de Saavedra document provides this interesting description of a ceremony that he witnessed at Olita:

> It is the custom for most of the natives of this country to send the first fruits of all their crops as an offering to this temple of the Nayarit. In the pueblo of Olitta at the start of the rains, the natives and elders of the place all meet and choose the most worthy to be Captain Otzocuani who selects the site for the shrimp fishing. After the election, the old men go from door to door begging alms for the support of the Atzaquani. When the time comes, which they recognize by the position of the sun (who is the *Nicanori*, or in their native language, the Totorame yequi, which in the Mexican language is *Cuhiahuri*, "giver of rain and the creator of the waters and the fishes"), the Captain at once starts a five-day fast. During this time, he does not eat salt or chile, and he remains chaste during the whole season of the harvest. They believe that if he breaks the fast, he will die. After the fast, he offers a ball of cotton seeds wrapped in a disc of cotton with red heron feathers hanging from it. When he goes to bed at night, he is required to say these words: *Neamos, tamex yequi*, which means, "Lord, Son of God, the giver of rain and creator of the birds and the fishes, Give us shrimp!" He repeats these words many times, and they say that then the deity appears to the Captain in dreams and says to him: *Amin Moctamex Noxuu*, which means, "Friend, I will give you shrimp." In the morning, the Captain informs the Pueblo what happened. Then they make preparations for fishing. Along with some other things, they pour a wine that they call *Alassann* into the water. The first one

> to start fishing is the *Atzaquani*. As soon as he takes out the first shrimp, they prepare a drink for him in a jar that they call *paxnal*, and then he takes all the shrimp that he can hold in his hand and puts them in the paxnal. He drinks and gives what is left over to the dogs [in McCarty and Matson 1975:215].

The *Diccionario geográfico-histórico de las Indias Occidentales ó América* noted Olita to be the "pueblo de la cabezera de partido y alcaldia mayor de Acaponeta en el mismo reyno que los anteriores, situado a la orilla del río Cañas, 9 legunas al SO de su capital" (Alcedo 1788). Olita's location is east of Teacapán on a broad flat pan of land, which is flooded periodically. A large salina is noted near the town on the 1824 Narváez map (see 1858; Figure 12.9). Conditions there are perfect for salt production, and although salt is not produced there now, elderly local residents of the Teacapán area recall it being collected there within their lifetimes. As noted earlier, Olita was producing more than a thousand hanegas each year for the Spanish.

Olita at some time in the past was important not only for salt but apparently for other reasons. On the eastern margin of Olita stands a long, rather elliptical, but disturbed shell mound similar in composition to but much lower than the mound of El Calón, located farther to the north in Laguna Agua Grande. Although much smaller than El Calón, the shell mound at Olita is composed of mostly unarticulated *Anadara* shell and resembles the low mounds near El Calón more than anything else seen in the Marismas.

3. Aztatlán

Aztatlán was the settlement north of the Río San Francisco (Río Aztatlán) where Guzmán and his forces camped in the summer of 1530. Cook and Borah (1960: 120) estimate the population in 1532 at 138 and in 1568 at 212. Ponce called it San Felipe Aztatán and placed it 2 leagues south of Acaponeta between the Río Acaponeta and the Río Santa Ana (San Francisco). He noted that Pinome was the language spoken there during his visit, and while he was in Aztatlán sometime between 1584 and 1588, Indians in the clothes of Chichimecos danced for him. Ponce (1968:81) was served *pan de Castilla*, biscuits, bananas, sweet potatoes, chickens, and wines served from earthen jugs.

Arregui (1946:89) listed Aztatlán as a settlement in the District of Acaponeta in his 1621 report, as did Tello (1945:119) in the mid-1600s. The settlement appears on the Ortelius map as Haftatan (see Figure 12.6) and on later maps as Astatan, San Felipe, or San Felipe Astatan. The extensive sites in the area, particularly La Cruz, may be archaeological remnants of Aztatlán (Sauer and Brand 1932:44).

4. Acaponeta

Guzmán did not mention Acaponeta. As Saponita, it is seen on the Ortelius 1579 map (see Figure 12.6), and it appears on maps thereafter under its modern name. In 1580 a Franciscan convent was established in this settlement, with Fray Andres de Medina as its first guardian. On his visit between 1584 and 1588, Alonso Ponce saw a small adobe and straw church here, and he located Acaponeta 2 leagues north of the Río Santa Ana. He (1968:77) wrote that the houses of this moderately populated locale were built near the steep banks of the Río Acaponeta, which often flooded the nearby fields of corn, cotton, beans, and melons.

In 1617 Acaponeta was completely destroyed by the Cora and Tepehuane Indians (Arregui 1946:89). Tello, Arregui, and Arias de Saavedra all visited Acaponeta in the seventeenth century and noted a population of Spanish, mestizos, Cora, Tepehuane, and other Indians in the settlement. Throughout Arias de Saavedra's account of the Acaponeta area he lived in from 1653 to 1672 he noted specific locations of various indigenous groups (McCarty and Matson 1975). Just as in his day, today both Cora and Huichol Indians can be seen on the streets of Acaponeta, which is a thriving market center.

5. Tecuala

Modern Tecuala is located on the northern banks of the Río Acaponeta near San Felipe Aztatán (see Figure 12.4). Guzmán did not mention it. It is, however, assumed to be in the region of Guzmán's Aztatlán encampment and battle site (Sauer and Brand 1932:44). Ponce also made no mention of this settlement during his 1580s travels in the area. Mota Padilla (1973:76) mentioned it by name in his 1741 survey, and Villa-Señor y Sánchez (1952:248–250) placed it within Acaponeta in his 1742 commissioned report. Perhaps Tecuala was not a settlement at conquest or in the early historic period, unless it was one and the same settlement with San Felipe Aztatán. There is some evidence that this might have been true. There are archaeological sites west of the modern San Felipe Aztatán extending toward the Río Acaponeta (Sauer and Brand 1932:44; see Chapter 3).

6. Yayoacan (Ayoacan)

Although the Yayoacan settlement does not appear in the earliest accounts of the area, Arregui mentioned it in his 1621 report (see 1946:89). In Tello's 1653 writings he noted that it was a settlement that missionaries visited from Acaponeta. Furthermore, he (1945:117) described this as a major area for the production of salt. Villa-Señor y Sánchez (1952:248–250) located an Ayoacan 5 leagues south of Acaponeta, about 2.5 leagues south of Sallula.

7. Chilapa and Chimapa

Chilapa and Chimapa seem to be separate settlements, neither of which was mentioned by the conquistadores. Figure 12.4 shows a Chilapa just north of the Río Santa Maria and a Chimapa between the Ríos Santa Ana and Acaponeta. Arregui (1946:89) noted a Chilapa as a village of Acaponeta toward the mountains in his 1621 report. He also listed a settlement of Chimapa to be one of the villages of Acaponeta located toward the coast. Tello listed a settlement by the name of Chimapa within Acaponeta in his 1653 descriptions. Arias de Saavedra (McCarty and Matson 1975) said that it was attacked by mountain Indians around 1657. Chilapa appears on the 1763 Garavito map (see Figure 12.7). It is also located north of the Río San Pedro on the 1777 *Mapa de Acaponeta* (see 1961 [1777]). Today there is a pueblo called Chilapa about 22 km north of Santiago Ixcuintla that may or may not be related.

8. Tzapotzinco (Zapotzingo)

None of the primary sources listed a settlement by the name of Tzapotzinco as part of Acaponeta. However, Bordes (1944:108) noted a Tzapotzinco to be part of Acaponeta, while Brotchie (1940:14–15) described a Zapotzingo located within the same district. Navarro (1946:8) recorded Tzapotzinco as part of Aztatlán. Since these sources seem to have copied each other or a like source, this is not very revealing unless Tzapotzinco is the same as the hacienda of Zapote shown on the 1777 *Mapa de Acaponeta* south of the Río Acaponeta but north of the Río San Pedro.

9. Zaulam (Zaulan)

Tello (1945:119) indicated that the Franciscans visited a settlement within Acaponeta named Zaulam in the early seventeenth century. Navarro (1946:8) also listed it as part of Aztatlán.

Provincia de Chametla

From Aztatlán the conquistadores moved 40 leagues northward to Chametla in the months following the extensive flooding in Aztatlán (López 1963:89). The combined accounts of Carranza (1963), Guzmán (1955), García del Pilar (1955, 1963), López (1963), Sámano (1963), and the four anonymous conquistadores (*Relación Anónima Cuarta* 1955; *Relación Anónima Primera* 1963; *Relación Anónima Segunda* 1963; *Relación Anónima Tercera* 1955) provide descriptions of the initial Spanish interaction with Chametla. In November 1530 a conquistador named Samaniego traveled to Chametla from Aztatlán. Following an initial skirmish, he was accepted in peace and given many chickens. The battle may be that represented in Plate 67 of the Lienzo de Tlaxcala, which shows warriors at Ehiyametlan (Figure 12.10). Torres (1965:117) mentioned that the battle on the way to Chametla took place at the port of Penueles, where warriors resisted the Spaniards attempting to enter their territory.

From Penueles the Spanish entered what was called the "Plains of the Cows" before reaching the entrance to the District of Chametla and the Charco de los Caimanes located 2 leagues from the actual settlement. After a brief stay in Chametla, Samaniego returned to Aztatlán, delivering supplies of food that had been received as gifts. The natives of Chametla accompanying Samaniego then returned to their homeland laden with gifts from the Spanish. The groundwork for mutual acceptance was thus established but unfortunately not maintained.

When most of the troops who had been sent south of Aztatlán to procure food and additional Indian slaves returned to Aztatlán, Guzmán sent messengers to Chametla asking for carriers to help transport supplies northward. On route to Chametla the Spanish forced these Indians to carry heavy loads strung between poles supported over their shoulders in a manner to which the natives of Chametla were unaccustomed. Soon most of the carriers were injured, while at the same time, the Indians who had taken ill during the earlier flooding began to die on the road. The injured Indians rebelled. Guzmán's men hanged a few and attempted to catch others, who tried to escape into the marsh about a league from the settlement of Chametla.

When Guzmán arrived in Chametla, he found the natives agitated. Nevertheless, he was met with welcoming ceremonies. In recounting the conquest of Chametla, Tello noted that Guzmán's forces first encountered about 2,000 warriors highly decorated with feathers from birds such as parrots and herons. The natives' weapons included bows as well as brazilwood lances and darts. About a league from the settlement of Chametla, 5,000 warriors and the chief then received Guzmán. Upon entering the settlement 6,000 people welcomed Guzmán and his forces with displays of dancing. Chametla's leader had braided hair and red paint on his face and carried a feather-decorated bow along with an alligator skin shield, which was also adorned with colored feathers. At the ceremony there was a young male jaguar decorated with a collar of shells and a copper bell (Tello 1974:78).

What the conquistadores recorded leaves the impression that the Chametla area was hot, humid, and swampy. The *Relación Anónima Primera* (1963:300) even reported stands of what must be interpreted as mangrove trees extending to the sea. Food, especially fish,

FIGURE 12.10. Plate 67, Lienzo de Tlaxcala, showing the Spanish battle within Chametla.

was noted to be so plentiful that there were ample supplies for four armies. Tello listed oysters, birds, maize, bread, and honey as the foods of Chametla, but here he may have been referring to foods available to him when he visited the area in the mid-seventeenth century.

Some of Chametla's leaders resisted helping the conquistadores. One of Guzmán's men and many of his allied Indians were killed. Guzmán met this challenge by taking some leaders into custody and then hanging them in the marshes several days later. At Christmas the Spanish were still in Chametla. Gonzalo López arrived from the south with new slaves. Before his arrival, the Spanish cleared away the dead so that the reinforcements would not see them. Guzmán left Chametla for Culiacán in January 1531 following a road that the Indians cleared for him. He abandoned many of his embittered and debilitated Indians, both allied and enslaved, without concern for their care, and he burned what was left of the settlement of Chametla as well as a village at the mouth of the river (*Relación Anónima Primera* 1963:300).

The area north of Chametla was culturally different from what had been seen since the Spanish left Jalisco (*Relación Anónima Primera* 1963:301). After traveling on to discover and conquer the large Province of Culiacán, Guzmán returned to central Mexico through Chametla. Some of his forces established settlements within the district (Mecham 1968:25), just as they did in Aztatlán and Centispác. Cristobal de Barrios was assigned as the *alcalde mayor* in 1531, and from Chametla he founded Espiritu Santo, which later became Rosario (Figure 12.11).

In 1536–1537 the Indians of Chametla killed some of the early Spanish settlers and fled into the foothills of the Sierra Madre. The remaining Spaniards left the area, leaving Chametla totally controlled by the natives until 1565. Indians continuously raided Spanish parties attempting to pass through the district. When the Spanish tried to apprehend the raiding parties, they found that the Indians had vanished into the mountains to live in caves (Mecham 1968:144). Coronado had a difficult time passing through Chametla in 1540; however, he accidentally helped the natives by leaving some frail cattle

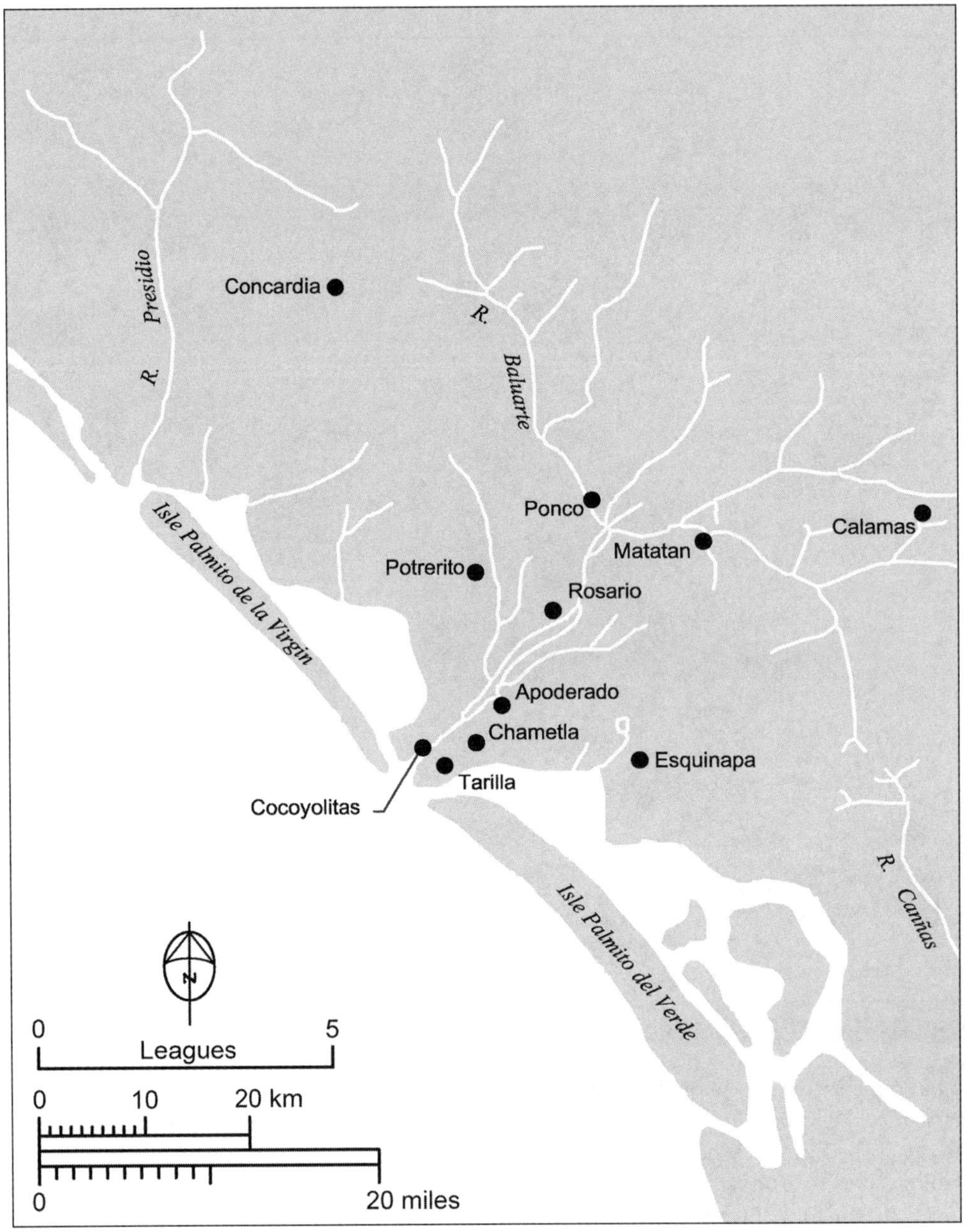

FIGURE 12.11. Map of the Chametla area (adapted from Kelly 1938:5).

behind (Mecham 1968:26). The native strength was such that the Spanish began to avoid the coastal route and traveled between Acaponeta and Culiacán via a difficult mountain route leading out from Durango.

As the years passed, pressure was put on the government to quiet these rebellious Indians and open the region to development. Thus, in 1564 Francisco de Ibarra was sent to pacify Chametla. He founded San Juan and occupied the area from there to Culiacán. He then entered Chametla, claiming the saltworks in the name of the crown. In reality it took Ibarra 11 months to conquer these Indians, who had maintained complete control of the area between Acaponeta and Culiacán for almost 30 years (Mecham 1968:144). Within a few years of Ibarra's victory, Chametla became a valued Spanish mining district.

Historically, the Jesuits are associated with the missionizing of the Chametla area; however, Bancroft (1883:118) reported that the Franciscan friars of Jalisco occasionally visited Chametla. Fray Juan de Tapia visited in 1558 from the mission in Acaponeta, but these early contacts did not seem to have the impact on the natives that was experienced in Centispác and Aztatlán, where many, at least nominally, accepted conversion. Much of the later missionary work in Chametla was from the Jesuit mission in Rosario. It is of interest to note here that because of unusual territorial changes, both Franciscans and Jesuits worked in Chametla.

Good descriptions of Chametla after it was conquered for the second time appear in the ca. 1604 writings of Mota y Escobar (1940, 1966). He noted that fishing played an important role in the native economy in the early years of the seventeenth century. Indians living at the mouth of the Río Chametla fished during two or three major periods of the year. At those times a lookout watched the pattern of fish swimming into the river. When large numbers entered the river during high tide, wicker sluice gates were lowered both at the mouth of the channel and a little farther upstream. As the water level receded, these gates imprisoned a variety of fish, such as robalos, pargos, lisas, *caballas*, picudas, pompano, and *salmonestes*. At the same time a variety of shellfish were collected. Caiman, sharks, and swordfish presented a constant danger as they were trapped in the enclosure, and at times these animals broke the gates, releasing themselves and others. After the fish were captured, useless varieties were quickly released, and edible fish were consumed or salted. Large catches were inevitable, and Mota y Escobar (1940:86–87) reported that the area of Chametla provided fish for Nueva Galicia as well as for many parts of Nueva España and Nueva Vizcaya.

In his 1673 report, Arias de Saavedra placed Chametla beyond the Cora area. He felt that the name meant House of Mezcal, and he says that it was also known as the House of Hell, Fire, or Purgatory (McCarty and Matson 1975:213). Arias de Saavedra described Chametla as being about 60 leagues wide, extending from the Province of Maloya to Autlán in Xalisco, indicating that he included the entire Marismas area in Chametla. Reconquered Chametla was incorporated into Spanish Nueva Vizcaya, after its creation in 1562, and then later into the Province of Sinaloa. For a short period it was within the Province of Arizpe and was sometimes even mapped as the Independencia of Sonora. In 1830 it became part of the Territory of Sinaloa, and then in 1835, the State of Sinaloa (see Table 12.1).

Settlements of Chametla

When the Spanish entered Chametla it had 22 subjected villages (*Relación Anónima Tercera* 1955:138), yet the conquistadores only mentioned the settlement of Chametla by name. Reconstructing the early post-conquest-period Chametla is therefore difficult, if not impossible, unless it can be done through governmental records regarding the reconquest. Early maps are not very helpful. The Río Chametla appears on the 1541 Castillo map, but unlike Omitlán and Aztatlán it does not appear on Gutiérrez's 1562 world map or Ortelius's 1570 world map. It is also not shown on Thevet's 1575 map, and Ortelius only charted the Río Chametla on his 1579 map. This lack of early mapped data is directly related to the period the Spanish were kept away from Chametla. Later, Chametla does appear as a settlement on Juan Martínez's 1587 stylized map entitled *Nueva España Occidental y Costa del Pacifico* (see 1951 [1587]). Even into the next two centuries, Chametla was often not clearly mapped. The following settlements are some that may have been in the area at the time of conquest or were established by the Spanish shortly thereafter.

1. Chametla (Chiametla, Ehiyametlan)

Chametla exists as a community today, but Isabel Kelly (1938:4) indicates that residents of the area consider Cocoyolitos (see Figure 12.11) to be the conquest-period Chametla. If this is true, the modern settlement of Chametla is probably one of those first visited by Guzmán on the outskirts of the "head" settlement. Today Chametla lies in a rich archaeological zone. Low mounds marking former house sites are plentiful, two taller mounds lacking surface sherds are still visible near Cocoyolitos, and other large mounds are located between Chametla and Apoderado (Kelly 1938:4). Any one of these taller mounds could be the one pictured in the Lienzo de Tlaxcala, Plate 67 (see Figure 12.10).

Mota y Escobar (1940:87–88) in 1604 located Chametla 1.5 leagues from the ocean on the banks of a river by the same name. At that time there were eight to 10 Spanish families and 40 Indian fishermen living in another settlement at the mouth of the river. The Spanish observed by Mota y Escobar during his 1602–1604 visits fished, collected oysters and shrimp, and obtained salt in a nearby estuary. The fish supplied areas of Nueva Galicia, part of Nueva Vizcaya, and Nueva España. Salt also received a wide commercial distribution. The area was known for abundant harvests of crops, which were used locally and exported to surrounding regions. At that time it was governed from San Sebastian. Brand (1956:579–581) reported Chametla to be an important port at various times, but in his discussion of the development of Pacific coast ports, he noted that "we know practically nothing about the salt merchants from Culiacan and Chametla" in relationship to trade between the Orient and Central and South America (1956:590).

2. Espiritu Santo

Espiritu Santo was a village on the Baluarte founded in 1532 by Captain Cristobal de Barrios, the *alcalde mayor* of Chametla. Its founders were 38 Spanish men from the maritime expedition of Diego Hurtado de Mendoza; however, they all left in 1536 when Indians demolished the village. It was here that the later settlement of Rosario was established (Pompa y Pompa 1960:93).

3. Esquinapa (Ichcuinapa, Isquinapa, Escuinapa)

Esquinapa (now spelled Escuinapa) is an extant and thriving municipal head of the Teacapán area (Figure 12.12). Its economy is based on fishing and shrimping. Mota y Escobar's (1966:86–87) 1604 description locates Esquinapa as a settlement of 20 *vecinos*, 6 leagues from Chametla. One source of the traditional history of the area is Dávila's *Diccionario Geográfico, Histórico, Biográfico y Estadístico del Estado de Sinaloa* (1959). Dávila believes that the original native inhabitants of Esquinapa were descendants of the Yuto-Azteca race who founded Auchén, a settlement located 6 km away from the present Escuinapa. Furthermore, he feels that the people from Auchén settled an area called "El Santo" on a hill 4 km from the present city before establishing Esquinapa. They moved to El Santo to get away from the mosquitoes and the increasing salinity of the land and water. It was in El Santo that they learned Spanish and constructed a church, renaming the settlement Pueblo Santo. They raised corn and made pottery. El Santo was burned three times, with the French burning it last in 1864. Dávila (1959:194–195) says that in the 1860s the population moved from El Santo to establish Esquinapa, yet both Auchén and Esquinapa (Isquinapa) appear on the same maps at least as far back as 1691 (Figure 12.12A).

4. Auchén

The exact location of Auchén, near modern Escuinapa, described above, is unknown. Auchén does appear on a crudely drawn 1691 map in Volume 8 of a publication entitled *Real Academia de la Historia* within the Colección Boturnini (*Descripción de esta parte de la America Septentrional, que se llama Nueva España* 1691; see Figure 12.12A). It also appears on a 1726 map by Don Francisco Álvarez Barreiro, dated 1770 (see Figure 12.12B). In both cases, Auchén and Esquinapa are charted on the same map, and the distance from Esquinapa to Auchén is shown to be about equal to the distance shown between Esquinapa and Chametla. Auchén is shown between the Ríos Cañas and Baluarte closer to the ocean than Esquinapa. Its placement on a modern map would therefore be near the shores of the Laguna Agua Grande, perhaps near what is now the settlement of Teilota. This is relatively close to large mounds located during the survey of the Marismas on the eastern edge of the Laguna Agua Grande.

5. Los Frisoles

Los Frisoles is the last settlement in which Sauer and Brand found Chametla pottery. They feel that this is where the native Province of Chametla ended. Los Frisoles may correspond to El Recodo (Sauer and Brand 1932:46).

6. Penuelas

The Penuelas port is only mentioned in a secondary accounting of the conquest. It was noted to be beyond Acaponeta on the way to Chametla (Torres 1965:19) and supposedly was a place where the Spanish stopped on their way to make initial contact with the people of that province. Its location and association with the modern settlement are unknown, although it is tempting to view Penuelas as a phonetic variation of Panales—the peninsular site exhibiting an abundance of archaeological material.

7. El Rosario

El Rosario was founded as Espiritu Santo in 1531 but destroyed by Indians in 1565. When the mining of silver began there, the settlement was repopulated. By 1763 Bishop Tamaron noted 2,400 inhabitants (Pompa y Pompa 1960:93). Both Franciscan and Jesuit missionaries served Rosario, but at different times. As Mazatlán became a major port on the west coast, Rosario became a trade center, and it is a thriving town today.

8. San Sebastian

San Sebastian was a settlement within Chametla. It was close to the mines founded by Ibarra following his conquest of the area (Bancroft 1883:110).

9. Teacapán

Teacapán, today, is a fishing settlement and growing tourist center, formerly located across the channel of the Boca de Teacapán. It is possible that the community of Tepequacan shown on the Ortelius map (see Figure 12.6) represents the earlier location of Teacapán. If this is true, then Teacapán can be traced back to at least the mid-1500s.

Summary and Observations

The information garnered from historians, historical geographers, ethnohistorians, and the primary literature of the Spanish colonial period led to the following observations related to the Marismas Nacionales. First, the records of Guzmán's 1530–1531 conquest indicate that the region was well populated and divided into three social-political entities the Spanish came to know as Centispác, Aztatlán, and Chametla. They were almost equal in population density, having possibly as many as 10,000 inhabitants each, with Centispác perhaps having a slightly larger population than the others, and Chametla, slightly lower. Centispác, where the conquistadores stayed for 40 days, seems to have been the strongest, but the Spanish perceived all three entities to have cohesive, indigenous governments and strong fighting forces.

At conquest the Spanish noted the cultures to be sim-

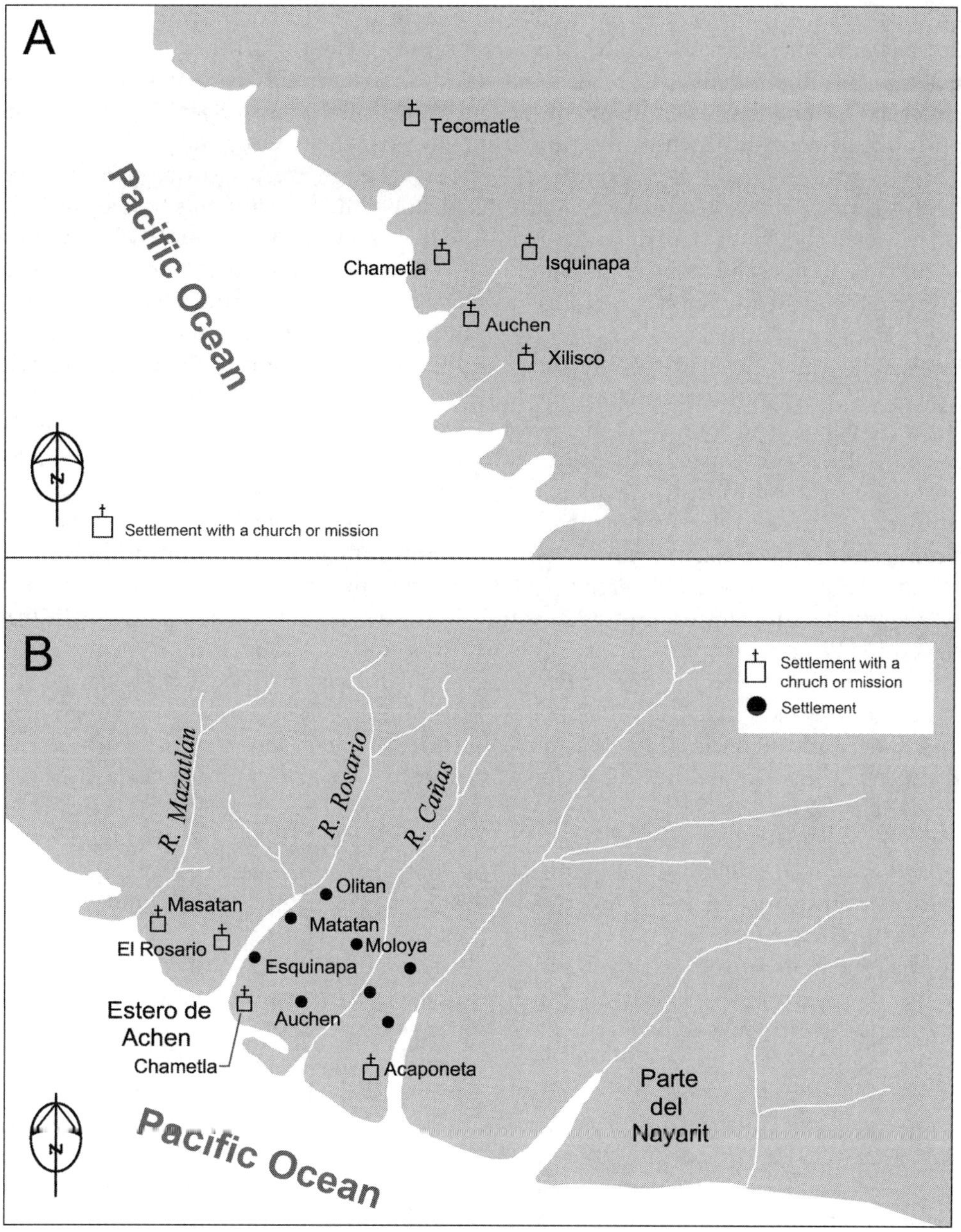

FIGURE 12.12. (*A*) Map of the Chametla area adapted from the 1691 *Descripción de esta parte de la America Septentrional, que se llama Nueva España* (in *Real Academia de la Historia*, Vol. 8). (*B*) Map of the Chametla area adapted from the Library of Congress copy of Don Francisco Álvarez Barreiro's 1726 map *Nueva Vizcaya*, originally published in 1770.

ilar to what they had seen in Jalisco (*Relación Anónima Segunda* 1963:313), but they noted that cultural aspects of Centispác and Aztatlán were most similar to each other, while some aspects of Chametla differed slightly. North of Chametla the conquistadores saw marked cultural differences. Thousands of warriors in Centispác, Aztatlán, and Chametla vigorously resisted the Spanish entrada, but each province in turn capitulated, welcomed the Spanish with great ceremonial flourishes, offered valued gifts, fed them, and assisted willingly and unwillingly in their movements even in severe flooded conditions. The Spaniards' reciprocations included cruelties, killings, and demands.

In earlier studies, Spicer (1969:787–789) expresses the opinion that areas such as these were each politically unified only when there was a need for defense. In peacetime their usual leadership came from an elder or group of elders. In times of peace there were no organized social units in this region larger than *ranchería* communities, at least not until after the area was missionized. Spicer further envisions the premissionized lowland river valley rancheria communities as groupings of 300 to 1,000 people, who held a mutual sense of territorial boundaries. Brand (1971:648–649) makes a similar assessment of west Mexican political geography. Meighan (1971:754–755), however, contends that perhaps the conquistadores' estimates of thousands of warriors should not be ignored or discounted, as they are indicative of large population sizes overall. He likens the political geography to a series of small city-states with tributary towns and villages. The literature reviewed here seems to support Meighan's point of view, particularly in regard to the military strength and coordinated response to the Nuño de Guzmán Entrada in Centispác, Aztatlán, and Chametla. This review seems to clearly show that the three provincias, as the Spanish called them, were independent, "with a recognizable social hierarchy, political structure, and institutions of formal leadership and power" (Weigand and Weigand 2000:16).

The conquistadores encountered inhabitants all around the Marismas area, and in Aztatlán people populated the banks of the Río Acaponeta all the way to the sea, where they were living on islands in what is today the Marismas Nacionales. The populations seemed well established, with many living in homes on top of *lomas* (mounds) in this area where flooding was frequent. The inhabitants were prosperous farmers, hunters, and fishermen. When discovered, they were not expending the majority of their energy exploiting their environment simply to survive. They had wealth, surpluses, and sufficient time and energy for ceremonial activities and trade. They were exploiting the sea for fish, shellfish, and salt and in some areas raising birds and dogs, both of which were used as food and in ceremonial ways. Cotton was used for clothing, and elaborate shields and other items were made from reeds and wood. They used wooden rafts and gourds to cross rivers and flooded areas. Weapons were elaborately decorated. They had formidable weapons for the time, for example, macauitles produced with numerous precisely formed obsidian blades embedded in rows on both sides of the long, flat wooden clubs. Bows and arrows, spears, and massive round stone and wood clubs were also used. Feathers, shells, copper bells, various animal skins, and dogs were cherished items with ceremonial significance.

As Guzmán's forces returned to central Mexico, some of his men and Indian allies settled in Chametla, Aztatlán, and Centispác. Some areas quickly became incorporated into the Spanish encomienda system. Spanish governmental policies replaced the native political entities of Centispác and Aztatlán, but some of the inhabitants fled to live with related groups in the mountains directly to the east of the Marismas. This was especially true of the Cora, who kept the Spanish out of the Mesa de Nayar of the Sierra Madre into the 1720s. The people of Chametla lived under Spanish rule for about five years, but in 1536 they rose up and pushed out the Spanish. The fact that they remained strong enough to keep the Spanish out of their territory for 30 years is further evidence of their conquest-era political cohesiveness.

Omitlán, Itzcuintlán, Tecomatlán, Sentispac, Sila, Dicla, Cillan, Aztlán, and Chametla are the only settlements definitively recorded by name in the area of the Marismas at the time of conquest, but it is known from conquistador accounts that as many as 66 settlements were in Centispác and Chametla alone. Some settlements were on islands along with Mezcaltitlán within what would be the Marismas Nacionales today. Of those settlements mentioned by the conquistadores, Omitlán is especially interesting in that it was so prominent at one time that it was noted as a substantial place on Ortelius's 1564 map of the world, published in *Theatrum Orbis Terrarum* in 1570, and on his 1579 map (see Figure 12.6). Omitlán was visited and described in detail by missionaries, yet in the late sixteenth century, this settlement, where Guzmán had stayed for 40 days, disappeared from the historic record. After a visit to Sentispac in 1587 Alonso Ponce (1968:83–84) concluded that the Río San Pedro had shifted. Did changing hydrology cause Omitlán to be relocated or abandoned (see Figure 12.5)? Is the archaeological site of Omitlán related to Tuxpan, as Grosscup (1964:23, 28) suggests it might be, or is it under the sediments between Tuxpan and the Marismas Nacionales? Wherever Omitlán is, it is an archaeological site worthy of discovery and future study.

The mystery of Omitlán's disappearance and the history of the Río San Pedro moving northward within historic times leads to speculations about what impact hydrologic changes might have had on the Marismas. Before the Río San Pedro shifted northward, only the Ríos Acaponeta and Cañas and their tributaries flowed into the Marismas. At that time the Marismas may have been a wider, more open body of water. Leaving aside other causal factors, such as earthquakes, subsidence, and any other land changes that were taking place, the water level in the Marismas may have changed after the Río San Pedro shifted northward to empty into it. Furthermore, over time sediments from the Río San Pedro would have been deposited into the Mezcaltitlán lagoon along with sediments from the Ríos Cañas and Acaponeta, resulting in enhanced conditions for the expansion of mangroves. With the passage of time, the southern extreme of the Marismas would have become a system of small lagoons connected by extensive estuaries, marshes, and expansive areas of mangroves, as it is today. Islands or portions of islands would have seemed to disappear into the mangroves, with possible remains from former island communities tangled in the roots.

A somewhat related question was raised in the historical literature. That is, throughout the literature the mention of varieties of fish within the Marismas region stays relatively the same, but it is of interest to note that the references to oyster and shrimp change with time. Oysters, but not shrimp, are mentioned most often in the earliest documents. Using Centispác as an example, oysters were noted there by the conquistadores in 1530, and Ponce reported oysters being transported by pack mules to Mexico City in the 1580s. Further, Mota y Escobar chronicled the shipment of fish, shrimp, and oysters to Spanish mining areas in the early 1600s. Accounts show that oysters not only were a valued commodity at the time of conquest but were also harvested during the early historic period in numbers far beyond those needed for local consumption. As well, a nineteenth-century source contends that the exploitation of this commodity continued well into the late 1800s (Ingersoll 1881:214). Oyster shells were extensively utilized for producing lime mortar, and Ponce noted that this was especially true in his time, with shells from what is now the Marismas being used for buildings in the Acaponeta area.

Oysters from this area may have also been traded farther than previously thought. As Brand (1956:579–584) writes, Mantache and Chametla, located at the southern and northern ends of the Marismas, were leading ports by 1540. Although Acapulco was the main port for the Philippine trade, Mota y Escobar noted Chinese ships in Mantache in 1604—a site that continued to be the most important port between Navidad and Mazatlán until San Blas was founded in 1768. Brand (1956:581) notes that the very last galleon arrived in 1815 and was unloaded in San Blas in 1816. Related to that, he also states:

> We know practically nothing about the salt merchants from Culiacan and Chametla, and the oyster and fish vendors from the coasts of modern Nayarit and Jalisco, who sometimes went directly by ship to Acapulco to trade for products of China, the Philippines, Chile and Peru [1956:590].

It is only reasonable to believe that some of the numerous shell mounds, especially the lineal mounds along the shoreline, not associated with prehistoric artifacts in the Marismas date to early postconquest and modern trade. Personal recollections of elderly residents of Teacapán confirm that the large-scale exploitation of oysters continued into the early twentieth century.

While there is an abundance of oyster reports, the lack of reports of shrimp in the earlier record, as mentioned above, may only reflect the small number of accounts available, seasonal differences when the accounts were written, or varying dietary preferences of the villagers or chroniclers. If, however, the suggested hydrologic changes at the southern end of the Marismas represent reality, the difference in historic references to shrimp and oysters in Centispác might be explained. Oysters need conditions of open-flowing water with little sedimentation, while shrimp thrive in the abundance of mangrove feeding and spawning grounds. Shrimp might naturally have replaced oysters in importance as a food source around the seventeenth century in many communities of Centispác if the southern Marismas did undergo the hydrologic changes theorized above. Although oysters would have survived in some areas, as they do now, the developing mangrove would have increased the abundance of shrimp, which are harvested in great quantities today.

There still remains another related question as to why the Marismas has been mapped as an open body of water with prominent islands. The Ortelius map (see Figure 12.6), produced in the mid-1500s partially from even earlier data and published in 1570, shows the Marismas as such an open body of water. It is considered stylistic, but it portrays all of Mexico in remarkably reliable detail considering the mapping technology of that time. It seems to indicate that in the mid-sixteenth century the Marismas was perceived as a large body of water comparable in size to Lake Chapala as mapped by Ortelius. On the same map there are three large islands and one other island clustered toward the southern and central portions of the Marismas, as well as six smaller islands located north of these.

Ponce mentions such island communities in his 1587 report. Tello recognizes two island settlements besides Mezcaltitlán in the mid-1600s, and Arias de Saavedra describes the Totorame living on islands within the estuary while he lived in Acaponeta from 1653 to 1672. Other simple maps, such as Garavito's 1763 map (see Figure 12.7), also show the Marismas as an open body of water. The 1824 Narváez map (see Figure 12.9); which was drawn in great detail by a renowned naval officer and even includes ocean soundings, also shows the Marismas as an open body of water with an arrangement of islands strikingly similar to the Ortelius map (Figure 12.13). All the reports and the maps indicate that the Marismas appeared, at least to some, from the 1570s into the mid-1800s as a wider and more open body of water with several prominent islands. Thus, the archaeological sites need to be considered not only in relationship to the current estuary and mangrove areas but also in terms of what might have been islands in the past.

Another question also remains. Where are the four island communities—Copotitlán, Amazcula, Mezcacula, and Tepuacan—shown on the Ortelius map (Figure 12.6)?

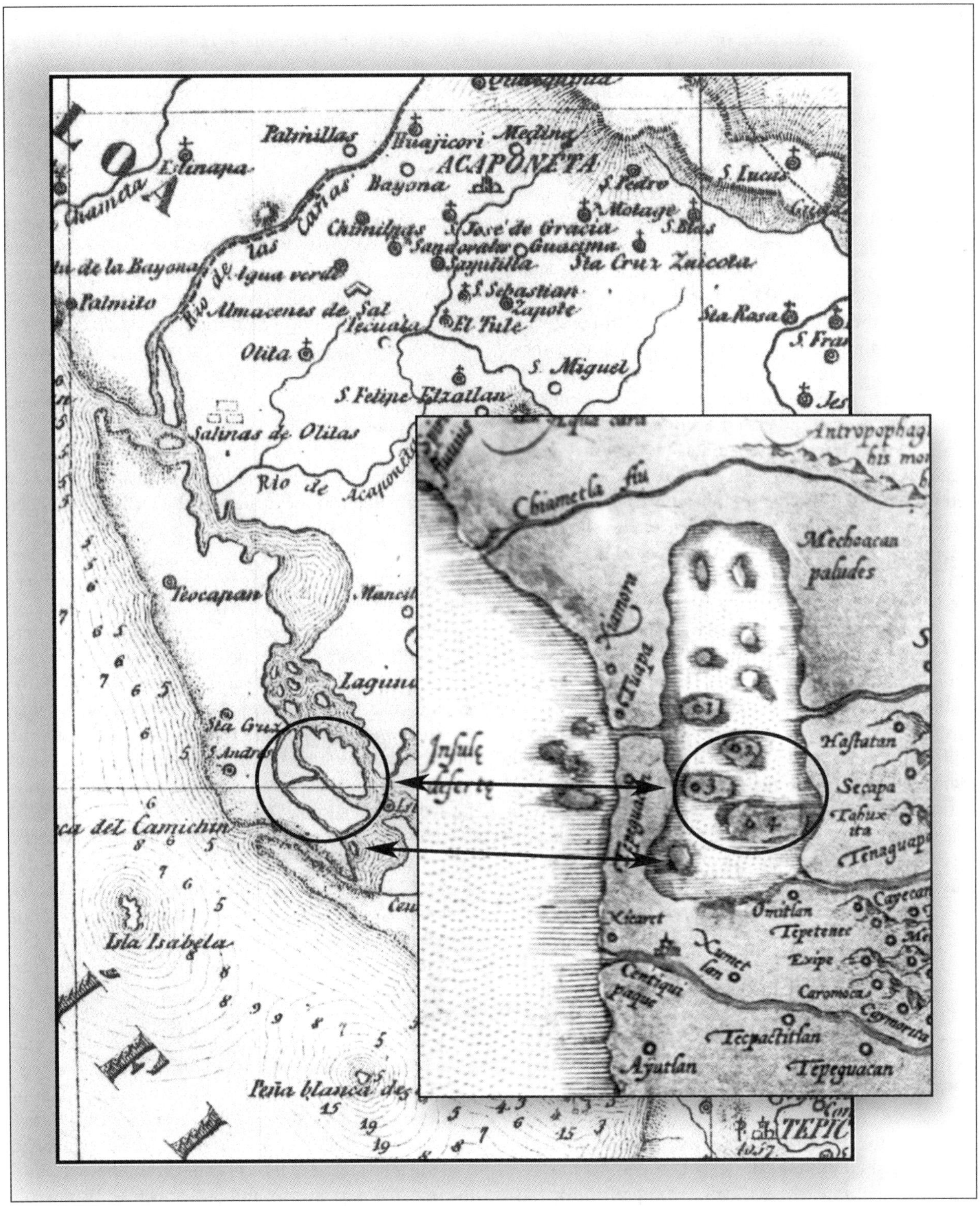

FIGURE 12.13. Possible island associations on the Ortelius 1579 map and the Narváez 1824 map of the Marismas Nacionales. The three large islands toward the south of the Marismas are also mapped in a similar way on Kelly's (1938:5) map of Chametla.

Are these sites underwater today within the estuary or on peninsulas? Or are they on patches of land in the midst of dense stands of mangrove? Are any of the Marismas sites investigated (see Figure 5.1) related to those mapped communities? Of the archaeological sites investigated, Chalpa, Arriñitas, Panales, Tecualilla, Cayetano, and sites near them appear to be the most likely candidates for sites that might have been on islands at one time, but so far no such associations have been made.

Of all that was revealed in the review of the historical records, the religious aspects of the contact-period inhabitants of the coastal area now within the Marismas Nacionales were the most interesting. In the Aztatlán section it was noted that Totorame were the primary inhabitants of the Marismas from Chametla to Rosamorada (Rosa Morada). The Totorame were most closely related to Indians now known as the Cora (Náyari) and Huichol (Wixárika) (see Coyle 1998; Hinton 1972; Liffman and Coyle 2000; Neurath 2000; Weigand 1981, 1985; Weigand and Weigand 2000). The early Spanish missionaries and chroniclers usually referred to these groups as the Nayarit Chora or the Nayaritas. In terms of religion, it is believed that the Totorame on the coast shared the faith of related Nayaritas, especially those living on the Mesa de Nayar.

The Nayaritas considered salt, mescal, and chile as sacred commodities, and their ceremonial life incorporated both sacred coastal locales and mountain peaks where the sun was seen to rise or set. Equinox and solstice sunrises and sunsets were especially noted. Harvested crops and idols, often in the form of figurines, were left in sacred places related to an elaborate, interwoven cosmology, ceremonialism, and sociopolitical organization involving ancestor worship. Ceremonial duties and attention to sacred sites were considered necessary to maintain the rains, crops, goodwill with controlling ancestors, and cultural survival (Coyle 1998). Many mountain peaks, caves, and water sources on the Mesa de Nayar and Cerro del Muerto to the east of the Marismas were considered sacred shrines or places in need of constant care and veneration. Likewise, an inlet of the Sea of Cortés to the west was considered sacred.

As noted in the earlier section about Aztatlán, Arias de Saavedra in 1673 records that there were four principal Nayarita deities. Each deity was related to a special sunrise. Beyond these there were other deities with special places where offerings were left in reverence and supplication (McCarty and Matson 1975). From Arias de Saavedra's report it is obvious that Olita, located at the mouth of the Río Cañas where it enters the estuaries of the Marismas, was a sacred ceremonial site for fishermen. There are similarities between a small, disturbed mound of *Anadara grandis* shell found at Olita and the much larger El Calón, composed almost solely of *A. grandis* shell. El Calón, where figurines have been found, has been interpreted as a ceremonial site also. The questions arise, Were there special sunrises related to Olita and El Calón, and is it possible that El Calón was also a sacred ceremonial site for the Nayaritas and their coastal Totorame relatives?

To address these questions, off-the-shelf software—Starry Night: Complete Space and Astronomy Pack by Imaginova—was used to calculate where people in the Marismas at Olita and El Calón, and where people on the sacred Mesa de Nayar where the sun resided, would have seen sunrises in the seventeenth century. By inputting the latitude and longitude of selected locations, the program allowed local azimuths for sunrises and sunsets to be calculated for 1673, the year of Arias de Saavedra's report (Table 12.2).[1] (In reality to the naked eye the same would appear no different today.) Also, azimuths were plotted in degrees on Google Maps of the Acaponeta area using the program's line drawing tool. Then, using a protractor azimuths were also plotted on a map entitled *Guadalajara* (published by ITMB Publishing Ltd., 530 West Broadway, Vancouver, Canada). On this map, with of a scale of 1 inch = 25 km, the locales on the Mesa de Nayar and within the Marismas were well labeled.

Everyone who has visited Teacapán, where the State University of New York Marismas field station was located on the Isla del Palmito, has seen spectacular sunsets, where the sun as a huge red ball appears to sink into the Sea of Cortés in a blaze of glory. And it is obvious that the sun rises over the Sierra Madre mountains east of the Marismas, where there are two prominent features the Nayaritas considered special. One is commonly called "The Sleeping Man" or "George Washington" by visitors, but it is more appropriately known as the sacred Cerro del Muerto (Figure 12.14). The other is the Mesa de Nayar, a prominent mesa above Acaponeta. Both features have multiple peaks and flat areas, where Nayaritas had hamlets, larger settlements, and ceremonial sites within the modern Cora (Náyari) and Huichol (Wixárika) homelands.

The astronomical calculations showed that the Totorame and their Nayaritas relatives living in the Marismas, near Acaponeta, or in the Sierra Madre east of the Marismas would have seen the sun set into the ocean's horizon during most of the year off what is now the Isla del Palmito near Teacapán (Table 12.3). At the summer solstice most of the Nayaritas would have seen sunsets farther north, off Laguna del Caimanero near Chametla. During the winter solstice most would have seen the sun set off the area of Playas Novillero, south of the Isla del Palmito. Therefore, for most of the Nayaritas the farthest north the sun would have seemed to move each year

Table 12.2. 1673 Azimuth Data.

Location	Spring Equinox	Fall Equinox	Summer Solstice	Winter Solstice
Acaponeta				
Sunrises	89 52.96 (90)	65 23.79 (65)	90 06.75 (90)	116 13.24 (116)
Sunsets	269 53.03 (270)	294 34.01 (295)	269 18.05 (270)	243 42.13 (244)
El Chico				
Sunrises	89 52.79 (90)	65 23.61 (65)	90 06.55 (90)	116 13.27 (116)
Sunsets	269 53.03 (270)	294 34.01 (295)	269 18.05 (270)	243 42.13 (244)
El Calón				
Sunrises	89 54.05 (90)	65 24.84 (65)	90 07.91 (90)	116 14.68 (116)
Sunsets	269 54.46 (270)	294 29.74 (295)	269 13.65 (270)	243 43.46 (244)
Huajicori				
Sunrises	89 54.43 (90)	65 25.04 (65)	90 08.19 (90)	116 09.01 (116)
Sunsets	269 49.88 (270)	294 30.92 (295)	269 14.92 (270)	243 44.62 (244)
Olita				
Sunrises	89 53.13 (90)	65 23.70 (65)	90 06.96 (90)	116 07.75 (116)
Sunsets	269 51.60 (270)	294 32.52 (295)	269 16.57	243 46.23 (244)
Escuinapa				
Sunrises	85 53.73 (90)	65 24.58 (65)	90 07.56 (90)	116 14.21 (116)
Sunsets	269 53.03 (270)	294 33.97 (295)	269 12.27	243 36.15 (244)

Note: The azimuth data from the Starry Night program are in degrees, minutes, and seconds from north, followed by the degrees rounded off. Degrees were rounded off, and azimuths were plotted using 90 and 270 degrees for equinox sun risings and settings, 25 degrees north of east for summer solstice sunrises, 26 degrees south of east for winter solstice sunrises, 25 degrees north of west for summer solstice sunsets, and 26 degrees south of west for winter solstice sunsets. In these cases north was 0 degrees, east was 90 degrees, south was 180 degrees, and west was 270 degrees.

Figure 12.14. Cerro del Muerto as seen from the Marismas Nacionales coastal plain, view to the east (photograph courtesy of Michael S. Foster).

TABLE 12.3. 1673 Sunset and Sunrise Locations in the Area of the Marismas Nacionales.

Viewing Location	Equinoxes	Summer Solstice	Winter Solstice
Sunset Locations Off the Shoreline of the Sea of Cortés			
Acaponeta	Over the Río Cañas and S of Teacapán	Laguna del Caimanero	S of Playas Novillero
Olita	Overhead into the Marismas S of Teacapán	La Tambora NW to Laguna del Caimanero	San Cayetano
El Calón	Las Cabras	Laguna del Caimanero	Between Las Cabras and Cristo
Escuinapa	Laguna Agua Grande	Laguna del Caimanero	Las Cabras
El Chico	Mouth of the Río Cañas	Laguna del Caimanero	S of Playas Novillero
Huajicori	N of Cristo	Laguna del Caimanero	San Cayetano
Santa Teresa	Tip of the Isla del Palmito	Laguna del Caimanero	S of Puerto del Palapares
Yánhke	S of Palmas de Cuatla	S of Las Cabras NW to Laguna del Caimanero	N of Boca de Camichín
El Jesús María	N of Palmas de Cuatla	Just S of San Cayetano	S of Las Coloradas
Sunrise Locations Over the Mountains East of the Marismas			
Acaponeta	Mesa de Nayar just N of Santa Teresa	Cerro del Muerto, southern part	Mesa de Nayar in line with Santa Teresa
Olita	Mesa de Nayar N in line with Santa Teresa	Cerro del Muerto just S of Picachos	Mesa de Nayar S of Yánhke
El Calón	Cerro del Muerto, over highest peak	Cerro del Muerto, northern part	Mesa de Nayar between Yánhke and Big Jesús María
Escuinapa	Cerro del Muerto in line with Mesa de Ricos	Cerro del Muerto N of Barbacoa	Mesa de Nayar in line with Small Jesús María
El Chico	Mesa de Nayar N of Santa Teresa	Cerro del Muerto S of the highest peak	Mesa de Nayar just N of Santa Teresa
Huajicori	Cerro del Muerto well S of Picaso	Cerro del Muerto just S of Picaso	Mesa de Nayar in line with Small Jesús María
Santa Teresa	E into the Sierra Madre	NE into the Sierra Madre	Mesa de Nayar S of San Juan Peyotán
Yánhke	E into the Sierra Madre, S of Small Jesús María	NE toward Jesús María	SE into the Sierra Madre
El Jesús María	E into the Sierra Madre in line with Small Jesús María	NE into the Sierra Madre	SE into the Sierra Madre

was off Laguna del Caimanero near Chametla, and the farthest south the sun would have seemed to move was near Playas Novillero north of Laguna Agua Brava.

A few Nayaritas living slightly north from the general area of Acaponeta would have viewed the summer solstice sunset beyond the Laguna del Caimanero in Chametla toward Mazatlán. A few living at the southern extreme of their settlements would have seen the winter solstice off Laguna Agua Brava and even farther south toward San Blas. Even from these more northern and southern locations most of the year's sunsets would have appeared off the Isla del Palmito or areas just to the north and south of it. It is interesting to note that the extremes of the southern sunsets seen at San Blas and the northern sunsets at Chametla encompass the range of coastal land associated with the Totorame and Nayaritas territory that Arias de Saavedra describes. These extremes also encompass the mountain territories of many of the modern Cora (Náyari) and Huichol (Wixárika), who revere sacred coastal areas.

The general conclusions about sunsets were not surprising, but what is surprising is that when Olita and El Calón were studied it was discovered that the spring and fall equinoxes were the most special in terms of sunrises. Most of the time from Olita looking east the sun would appear to rise over one area of the Mesa de Nayar or another, but on the equinoxes the sun would appear over the highest part of the sacred Mesa de Nayar, and on the summer solstice it would rise over the sacred Cerro del Muerto. Likewise, on the equinoxes the sun as seen from the Mesa de Nayar would appear to slide into the Sea of Cortés in a line with Olita. From El Calón a similar situation is noted. That is, most of the time from there the sun would rise to the north or south of the highest peak of Cerro del Muerto, but at the equinoxes the sun would rise over the highest, most prominent feature of that section of the Sierra Madre. Likewise, people living in Cerro del Muerto settlements, who could look down over the Marismas and see El Calón, would have seen the sun set over that high mound on the equinoxes.

When researchers consider astronomical aspects related to cultural cosmologies, they often consider the summer solstice as the most important. Staff and crewmembers of the State University of New York Marismas research team spent June 21, 1975, on top of El Calón from before sunrise to after sunset. They saw nothing spectacular in the sunrise but did note that the sunset was "right smack dab in the middle of a saddle formed by the low offshore hills" (Gordon Connally to Stuart Scott, personal communication June 30, 1975). It is interesting to note that from the villages on the Mesa de Nayar the sun sets in this same place on the summer solstice. The saddle is between two prominent lomas near Chametla and the Laguna del Caimanero, representing the farthest north the sun seems to travel from the Marismas area. Was this, then, perhaps an important sunset for the Totorame and Nayarita, marking the end of their territory?

The sunrise data showed the possible importance of the Mesa de Nayar and Cerro del Muerto to sun-worshipping people living along the coast. McCarty and Matson have interpreted Arias de Saavedra's account as follows:

> The Indians believe that the mesa is the representation of the *Pilzintli*, whose seat of power is the sun, that an arm of the sea is the image of Nicanori, that a mountain peak which they call "Horse Head" is the *Narama* (it is also called Ychamet, "house of Maguey and mescal"), and that a cliff in the sea which they call Matanche, meaning "tick or louse" is a representation of the *Uxuu* [1975:210].

The Arias de Saavedra report and the findings about sunrises and sunsets lead to speculations that Pilzintli resides on the Mesa de Nayar, the sacred mesa of the Nayaritas and modern Cora; the Marismas Nacionales south from the area of San Blas and north to Chametla, with its spectacular sunsets, is perhaps the sacred arm of the sea associated with Nicanori; perhaps Narama is Cerro del Muerto; and as is already known, Uxuu is the cliff at Matanche, near modern San Blas, still revered by Cora and Huichol. Arias de Saavedra, however, is not consistent in describing these deities, and other interpretations could be made.

One cannot say that El Calón was the place of residence of one of the principal Nayaritas deities. But the relationship of equinox sunrises from there appearing over the highest part of Cerro del Muerto indicates that it certainly would be of importance to any culture, such as the Nayaritas, that has a strong cosmology related to sunrises. El Calón is obviously due west of that sacred mountain. Is it a coincidence, or did the builders, whoever they were, choose the location because it was in line with the prominent shape of the Cerro del Muerto? If so, did they know that equinox suns would rise due east over the prominent mountain peak? In terms of the passage of time, the sunsets and sunrises would be no different to the eye today from what the Nayaritas would have seen in 1673. Furthermore, based on the information generated with the Starry Night software, there has been little visible change going back several thousand years. Considering the long tradition of sun worship and astronomical observations in Mesoamerica, it seems possible that if El Calón was built thousands of years ago, as possibly suggested by the radiocarbon assays from the shell used in its construction, the builders may have had such knowledge. Certainly, if El Calón was a Postclassic era construction, as Luis Alfonso Grave Tirado (2010:20–35) suggests, the builders would have had such knowledge. And certainly, if the Nayaritas *knew* of El Calón, and its relations to Cerro del Muerto in terms of equinox sunrises and sunsets, it would have been at least one of their many sacred places.

In conclusion, this historical research was initiated with the hope of finding specific associations of conquest communities and early historic-period sites with the archaeological sites investigated during the Marismas project. Although this specific objective was not met, the research did point to areas for future research, and it has demonstrated that the area of the Marismas Nacionales on the West Mexican Coastal Plain was a zone of highly developed precolumbian cultures at the time of Spanish contact. The Spanish chroniclers and subsequent researchers have documented and elucidated a variety of lifeways and worldviews that provide a framework with which to better understand the peoples in and around the Marismas Nacionales whom the conquistadores and missionaries encountered. The documents show that while many of the coastal inhabitants in and near the Marismas were killed, died of disease, or were temporarily or permanently displaced, cultural aspects witnessed by conquistadores were also observed by missionaries in the sixteenth and seventeenth centuries; furthermore this cultural continuum can be observed today with the Cora (Náyari) and Huichol (Wixárika).

Note

1. As noted above, missionaries to the Marismas area in the mid-1600s noted that solar equinoxes and solstices were important to the Totorame, who resided in what is now the Marismas, and their related Nayarit Chora (Cora), who lived to the east on and near the Mesa de Nayar. Collectively the Totorame and Nayarit Chora, often referred to as the Nayaritas, held certain peaks, caves, and water sources and a western arm of the sea as sacred places related to sunrises and sunsets. In 1673 Arias de Saavedra witnessed ceremonies related to the solar progression at Olita, where the Río Cañas enters the Marismas. Some people even today associate El Calón with the summer solstice sunrise, but the State University of New York at Buffalo researchers saw no such association with that sunrise. It therefore became interesting to see where the equinox and solstice sunrises and sunsets were in 1673 and today related to Olita, El Calón, and the Marismas. Astronomical calculations of the sun's path in 1673 were calculated using the off-the-shelf computer program Starry Night: Complete Space and Astronomy Pack, by Imaginova. The first important finding was that to the naked eye there would be no difference in where the sun would seem to set or rise in 1673 or today.

   Among the locations studied were Escuinapa, near the research area; Olita; Acaponeta, where Arias de Saavedra was headquartered; El Chico and Huajicori in the mountains near Acaponeta; and Santa Teresa, Yánhke, and Jesús María on the Mesa de Nayar—the sacred site of the Nayarit Chora. Table 12.2 gives the azimuth data in degrees, and Table 12.3 describes where the sun would rise or set in relationship to specific locations or geologic features.

CHAPTER 13

Summary and Conclusion

Stuart D. Scott and Michael S. Foster

The preceding chapters have addressed a wide range of issues regarding prehistoric human occupation of and adaptation to the Marismas Nacionales of coastal west Mexico. The multidisciplinary project undertaken by the State University of New York at Buffalo brought archaeologists, bioanthropologists, and natural scientists together in order to identify and reconstruct the geomorphology, paleoenvironment, human cultural and physical adaptation, and ethnohistory of the vast estuary–lagoon–mangrove forest system on the coastal frontier of northwestern Mesoamerica. Despite the intervening 40 years since the fieldwork was initiated and completed, the region remains poorly understood archaeologically. While the Marismas archaeological investigations focused on the recovery of material culture, identification of features, and recovery of human remains to reconstruct aspects of synchronic and diachronic human adaptation to a seemingly inhospitable mangrove swamp and estuary system, the geomorphologists and other natural scientists set forth to reconstruct the history of the development and growth of the coastal plain in which the Marismas is situated. Thus, the following considers the contribution of this research to our greater understanding of the natural setting of the Marismas, the cultural sequence and culture history area, intra- and interregional site relationships and settlement, the relationship of the Marismas to the Aztatlán tradition, and the nature of the Marismas during the sixteenth century. Finally, the overall importance of this multidisciplinary approach to broadening our understanding of human behavior and adaptation in ecologically diverse and highly sensitive environments such as marine estuaries is emphasized.

The Marismas Cultural Sequence

The Marismas Nacionales provides one example of the regional variability within the extensive archaeological record of west Mexico. Viewed in that wider context, the peripheral northwest Mexican Pacific cultures might be seen as simply a collective entity of marginal, coastal-dwelling people, drawn to the predictably rich yield of aquatic resources from the sea, estuary, and lagoon. Although such a picture may be generally true for the earliest occupiers, that is, Archaic and early Preclassic period people, archaeological evidence from the past 70 years shows that the evolution of these coastal cultures was more diverse and complex.

Regions of Mexico's west coast were utilized and occupied by Archaic period hunters and gatherers and early agricultural groups exploiting shellfish and other maritime resources, and from the coastal zones of Chiapas and Guatemala a relatively detailed picture of Archaic and Preclassic Mesoamerican lifeways has emerged (e.g., Blake et al. 1992; Blake et al. 1995; Coe 1961; Flannery 1976; Pye and Demarest 1991; Shook and Hatch 1979; Voorhies 1976, 1989; Voorhies and Kennett 1995; Voorhies et al. 2002). For coastal west Mexico, some of the best information on this period comes from the area of Matanchén Bay, south of San Blas (Mountjoy 1970a, 1970b, 1974, 2000). Recently, however, excavations in coastal Sinaloa north of the Marismas have resulted in the identification of Middle Archaic period remains. The discovery of multiple projectile points at the stratified site of Flor del Océano, 50 km north of Mazatlán, has led Instituto Nacional de Antropología e Historia archaeologists to infer the presence of temporary camps and possibly a lithic manufacturing workshop. Although the results of the excavations are yet to be synthesized, these materials may represent the oldest human presence yet found in Sinaloa (Santos Ramírez and Vicente López 2012). To date, however, there has been no evidence for the widespread utilization of the Marismas in the Archaic period.

As for the west Mexican Preclassic period, it is best seen in the Capacha (Kelly 1980; Mountjoy 2000) and Shaft Tomb traditions of the southern west coast and west Mexican highlands (Beltrán Medina 2001; Foster 2001; Mountjoy 2000; Townsend 1998). Although some evidence of Shaft Tomb complex material culture has been associated with the early ceramic cultures of the region, the use of tombs did not spread northward into northern coastal Nayarit and Sinaloa. Nevertheless, the recovery of Chinesco complex (Shaft Tomb) ceramics (Chinesco Polychrome and Chinesco Black-on-cream) in northern Nayarit and southern Sinaloa led archaeologists working in the area to add a Chinesco phase to local sequences where these materials have been identified.

The Marismas Nacionales itself may have witnessed the first human settlement as early as the second millennium BC, when immigrants arrived and colonized on the relatively new soils of an emergent coastal plain. Early on, Moriarty, based on existing archaeological evidence correlated with climatological and oceanographic reconstruction, offered a few cautious inferences regarding the presence of early human use of the Costa de Nayarit that explain the lack of remains but yet acknowledge the possibility of early remains in the area:

> It is possible to venture a prediction that radiocarbon dates of some antiquity (6,000 to 8,000 BP) will be forth coming in the future. As the age of the strand plain is relatively recent in the geologic sense, occupation could not have occurred previous to 5,000 BP. Carbon dates taken from sites on the strand plain will demonstrate a progression of very recent dates on or near the coast. The middens will increase in age in relation to their distance from the coast with the oldest being found on the eastern edge where the beginning of the deltaic deposits border the older landward formations [1965:70].

The traces of early humans in the area, here termed Costanero, may be seen only at El Calón, a shell mound unique for its height and configuration, for the builders' use of articulated shell as a construction medium, and for its age. The discovery of El Calón (Shenkel 1971a, 1971b) deep within a well-developed mangrove forest and its apparent age generated considerable interest regarding its potential role in regional prehistory. If, as the dating of marine shell from the mound suggests, El Calón was constructed between 2600 and 1200 BC, it is, then, central to the question of the early presence of humans in the Marismas.

El Calón was situated at or near the margin of a gradually prograding coastal plain. To infer from the volume of articulated ribbed clam (*Anadara grandis*) and other shell used in its construction, it is presumed that those who built El Calón had an impressive array of edible foodstuffs available. If the feature was purposely built as a platform mound, as initially proposed in the early stages of the Marismas project, it reflects a remarkable energy requirement for such a massive accumulation of surplus shellfish to produce a sort of "public works" structure. In seeking a simple descriptor, there is risk of overinterpreting this structure with few clues to its origin. "Temple mound" and even "pyramid" are generally understood as structures sanctified as religious architectural components of a complex society. Until further research can verify such an assertion of form and function, it is probably best to describe El Calón as an immense platform mound, built perhaps in the Late Archaic or Early Preclassic period and possibly intended for some unspecified ceremonial function. Despite some recent work at the mound and the recovery of a limited number of artifacts, other than the mound itself, the site offers little in the way of material culture, associated features, or structures that would allow further assessment and interpretation of the mound.

In all probability, geomorphic processes have obscured the earliest archaeological record of El Calón; these too account for the subsequent period, based on the lack of archaeological remains, when humans abandoned this coastal sector during a lengthy hiatus. In characterizing this deltaic coastal study area, Curray and his associates point to the distinctive coalescence of deltas from the Río Grande de Santiago, Río San Pedro, Río Acaponeta, and Río Baluarte, which is complicated further by the changing conditions of sea level, climate, direction of longshore currents, and abundance of the sand supply. From their study of the sediments, morphology, structure, and history of the coastal plain, Curray and his associates conclude that around 3600 BP, the direction of longshore transport of sand along the north part of the area reversed from north-flowing to south-flowing: "This is interpreted as possible regional climatic cooling which caused a change of wind and wave regime and longshore drift" (1969:63). Of interest is the fact that El Calón has been dated to the same approximate time as Curray's postulated "significant local climatic changes" (Curray et al. 1969:99).

Mountjoy's (1970a, 1970b, 1974, 2000) work at Matanchén Bay near San Blas remains the dominant study of Archaic period use of the northwest Mexican coast; however, few other Archaic sites have been identified. At Matanchén, marine resources represent an important proportion of the diet, yet coastal, estuarine, and marine resources appear not to have been widely exploited at that time. Mountjoy offers a possible explanation:

> The lack of Archaic shellfishing sites in the regions is probably due in great part to the rarity of localities that, like Matanchén Bay, are rich enough in molluscan resources to have supported humans dependent on exploiting shellfish for survival [2000:83].

Citing the instability of the coastline, Mountjoy (2000:84) sees erosion and the submergence of deposits as negative factors in finding, let alone interpreting, Archaic coastal occupations. Lacking any known associations between Matanchén and El Calón beyond the use of shellfish at a roughly comparable early date, we are left with sparse and inconclusive evidence for the early presence of humans on the coast.

A critical aspect of the overall study is the physiological context of the Marismas. During the two or more millennia since the arrival of the first humans, in Moriarty's terms, at the border of older landward formations, and perhaps colonizing on an emergent soil, the subsequent seaward growth of the coastal plain provided a landform on which later Classic and Postclassic settlement could occur. In the simplest terms, the geology of the strand plain coast is a history of the deposition of sediments from various rivers, complicated by changing conditions of sea level, climate, beach ridge formation, episodic flooding, and so on (see Connally 1977; Curray et al. 1969). The normal human ecological relationships, then, may have been particularly dynamic there, evoking broader questions: What is the suitability of the designation "Marismas Nacionales" from a cultural as well as a geographic and ecological perspective? What sets the area apart from neighboring regions?

For reconstructing the human occupational history of the Marismas, the apparent abandonment of the coast during high sea level occurs in tandem with or follows on the heels of climatic and topographic change. Following the long interval of postulated rising sea level that ultimately submerged, or partially submerged, the coastal plain, the initiation of the Marismas cultural sequence encompassing the Classic (Estero) and Postclassic (Conchera) is coincident with the emergence of Medial Beach deposits. Here, at the edge of land and sea, the low-lying region eventually became the habitat for a relatively dense population of indigenous people, growing the traditional assemblage of maize, beans, chiles, and squash and otherwise living off of an apparent bounty of abundant marine and estuarine resources, especially shellfish.

In addition to the prodigious quantities of oysters harvested for export within the historic era, millions of cubic meters of archaeological mollusks are found in the form of spatially segregated mounds. In the Marismas, the start of the Early Estero period around AD 300 is marked by the particular presence of shell mounds on high, well-developed beach ridges (Curray et al. 1969:91). Ceramic and radiocarbon dates indicate that the earliest Estero occupants focused on harvesting the small smooth clam *Tivela byronensis*. One example, the distribution of *Tivela* mounds on a portion of Panales, as can be seen from Figure 5.18, may provide us with some semblance of a land-use pattern, with a series of shell middens aligned on a series of Medial Beach ridges. These appear to represent shellfish processing loci or perhaps habitation sites that may equate to a habitation zone. From a simple adaptational perspective, these Early Estero colonizers selected their habitat on the seaward side of the estuary for its proximity to the Pacific beach. In his description of these sampled middens, Shenkel reports that the sandy beach-dwelling molluscan component *Tivela* presents "over 90% of the numbers of individuals, the remaining 10% consisting of other beach species, with only an occasional brackish water estuarine oyster or clam" (1974:58).

Sometime following the appearance of those who deposited the *Tivela* middens, the sketchy outline of Marismas prehistory expands as the Estero period advances. The archaeological record now reflects a more complex scenario involving, among other factors, a broader range of settlement locations, which here is ascribed to a presumed population expansion and the availability of more land.

Throughout the entire middle of the study area, at shoreline and nearshore locations, are oyster middens, varying in size from small surface scatters a few meters in diameter to large symmetrical mounds reaching 60 m in diameter and up to 6 m in height. Less conspicuous but possibly even more numerous are artificial earth mounds. The Marismas mounds, both earth and shell, each kind numbering in the hundreds, produce predictable assemblages of Classic and Postclassic period artifacts, including elaborate engraved and polychrome pottery, manos and metates, obsidian, spindle whorls, pipes, copper, and faunal bone. The abundance of refuse recovered demonstrates the use of some of these mounds as platforms for houses, which, as observed by the Spanish, stood on *lomitas* or low-rounded artificial mounds (see Chapter 12).

It was from the earthen mounds that the large sample of human remains was recovered. Of all the Marismas sites excavated, none failed to produce human remains. It generally appears that these burial mounds grew by accretion as more and more individuals were interred; however, the lack of stratigraphy within these mounds hinders any observations or conclusions regarding the sequence of their construction and growth. The earlier burials, it is surmised, were in graves dug into undisturbed native soil, laid on the surface, or buried just below the surface. As burials were added, new soil was excavated from adjacent areas to cover the bodies, and as the soil was of the same composition as that used in the earlier interments, no textural, color, or compositional distinctions were observable. The internment of additional individuals also churned the mound fill and mixed mound deposits. This inevitably resulted in the disturbance of earlier burials, which may have been unmarked or whose location was gradually forgotten. Although the Spanish reported Indian houses on earthen mounds, only a few discrete living floors were identified. Because of the lack of culturally stratified deposits and well-defined contexts, the culture-chronological goals of Marismas archaeology became dependent on the relative dating of ceramics and other artifacts as well as correlations with neighboring archaeological sequences.

Relationships to Other Sites and the Later Cultural Sequence

Of special interest is the location of the Marismas Nacionales between Chametla (Kelly 1938) and Amapa (Meighan, ed. 1976). Unlike the problematic nature of the Costanero period, the later Estero and Conchera periods are more easily contextualized in reference to the long occupational histories and especially the ceramic variability and sequences at those two sites. In totality, the Marismas sites span the Classic and Postclassic period occupations as represented at both Chametla and Amapa. A number of features excavated provided evidence of sequential occupations at sites with earlier earthen mounds capped by later shell midden deposits. These sites provide clear evidence of a break in the occupational sequence of the area, which is also noted at Chametla and Amapa but for the Marismas may be related to the subsidence and inundation of the coastal plain between ca. AD 750 and 900.

The Marismas period assignments to Estero and Conchera rest on the contents of several multicomponent sites of earth and shell, which demonstrate clear ties with pottery types at Chametla and Amapa as well as, occasionally, more tenuous connections to Peñitas and Culiacán. Specifically, there was a prevalence of diagnostic ceramic types—Black-banded Engraved, Middle Chametla Polychrome, and so forth—found in the Marismas Late Estero period mounds. These Classic period types correlate to the Baluarte phase at Chametla. In contrast, the shell middens capping many of the earlier mounds produced Conchera period Postclassic ceramics noted at both Chametla (El Taste, Black-banded Ware, etc.) and Amapa (Sentispac Buff, Tuxpan Red-on-orange, and other more decorative Postclassic pottery types).

Let us consider, then, the central position of the Marismas subregion between Chametla and Amapa, with attention to other similarities and differences in terms of environmental potentials and cultural contingencies. The site of Amapa lies near a river's edge on fertile soils that would have been well suited to maize agriculture. In fact, along with beans and squash, the Amapa diet included chiles, eggplants, guavas, plums, and other fruits, with the addition of domesticated fowl and bees (Bell 1960:26). Agriculture seems to have been based on dry-farming techniques, although Beals (1932:100, 158), drawing on Tello's account, notes the existence in historic times of irrigation in the coastal lowlands.

The location of this important coastal site near the sea raises the issue of the role of marine resources in the subsistence economy. Meighan notes that in spite of Amapa's proximity to the ocean, "marine shells seem not to have been of particular consequence to the inhabitants of the site, either as food or ornament" (1976e:122). Sanburg concurs: "Of primary importance in this [faunal] collection is the almost total absence of material that can be associated with the sea only a few kilometers distant" (1976:201). Allowing for the possibility that mollusks may be underrepresented in the archaeological record at Amapa, Meighan notes that some shellfish could have had the meat extracted from their shells at or near where they were harvested. He cites the example of a few *Donax* species—very small mollusks—found at Amapa, "which could be readily transported from the beach and kept alive by leaving them in the shell for a few hours until they could be cooked" (1976e:122). In general, though, in affirming the primacy of agriculture at Amapa, Meighan concludes:

> It is only the southern part of this delta plain, in the immediate area of Amapa, that is highly favorable for agriculture today, for the northern 80 percent of this delta region, extending up into the State of Sinaloa, is largely occupied by the extensive swamplands of the Marismas Nacionales. Indeed it is only the alluvial deposition of the San Pedro and Santiago Rivers that prevents the Amapa area from being part of the Marismas Nacionales [ed. 1976:5].

It seems clear that Amapa may well owe its importance as an "urban" or regional civic center, at least partially, to its location on fertile, river delta soils. In other words, Amapa's architecture of authority might never have developed in what is characterized as "the extensive swamplands of the Marismas Nacionales." Yet the case for agricultural vs. aquatic (marine/estuarine) food resources is more complicated than Meighan suggests. In archaeological literature that focuses on human subsistence strategies, there is a tendency to emphasize certain differences in terrestrial vs. marine food; see, for example, Schalk's (1977:239–240) discussion of the energy base of fishing adaptations as opposed to agrarian adaptations. The oceans are generally less productive than are the continental landmasses. Marine vs. terrestrial species and agricultural produce are often seen as "low return resources due to a need for labor intensification, in the case of shellfish and small food package-sized marine organisms, and due to their low protein content" (Osborn 1977b:177). Zeitlin has documented cases from the Isthmus of Tehuantepec contrasting groups having an agricultural economy supplemented by fishing and hunting with the other coastal-estuarine adaptation:

> Confined to the agriculturally marginal barrier beaches and estuarine shores of the Pacific lagoons, egalitarian Huave villages depended on fish, shrimp, mollusks, and turtle eggs collected from the lagoons for both food and trade, as they have to the present day [1989:34].

A third adaptive alternative is neither isolated nor divided by an ocean-land boundary and instead relies on the complementarity manifested by each environmental zone. The Marismas is located between two adjacent ecological communities or ecosystems, with the *edge effect* at their ecotonal boundaries producing a greater-than-normal diversity of species (Odum 1971:157–158). The Marismas inhabitants exploited both their marine and estuarine resource base and the fertile deltaic soils for agriculture, with evidence for the latter found archaeologically in both indirect (manos, metates) and direct (charred maize) forms. Allowing for episodes of flooding and gradual geomorphological changes over time, we might suppose that arable land was available to a greater or lesser degree throughout the Teacapán estuarine system during the Classic and Postclassic eras. Thus, characterizations of the northern coastal delta extending into Sinaloa as swampland not only peripheralize or diminish the importance of aquatic resources and food cultivation; they ignore the significant role of environmental and adaptive verticality, which unites lowland and upland environments in the overall sustainability of human populations.

Today the Marismas Nacionales of northern Nayarit and southern Sinaloa is a 1,100-k^{m2} wetland of international importance, and the portion of the Nayarit coastal plain south of the Río Cañas marine terrace that is under cultivation or used for livestock grazing represents roughly twice the volume area of mangrove, marsh, salt flat, and open water (Bojórquez-Tapia et al. 1997:19). Coastal habitats along the vast shorelines represent 20 percent of Mexico's mangroves and support 44 percent of Mexico's avifauna. Similarly, the inshore fisheries are the rearing grounds for commercially important species of shrimp and, in former times, oysters and fish (McGoodwin 1979:52–57). In the precontact era before significant environmental problems, including agrochemical contamination, domestic sewage, habitat destruction due to expanding shrimp farms, and seawater incursion through man-made canals, the numerous wetlands of the Marismas must have been a treasure of biodiversity. Sauer offered the general view that "fertile flood plains supplied abundant crops, which were supplemented by a wealth of fish and shellfish from estuaries and streams" (1932:7). Furthermore, as Spanish conquest and missionary records assert, fish, shrimp, and especially oysters were harvested and exported in great quantities from the estuaries and lagoons. With respect to economics, then, the Marismas can be spoken of as a sphere of resource prodigality. But questions remain about when the export or exchange of food commodities might have begun and where.

Given the acknowledged natural resource potential of wetland ecosystems, why, then, do the archaeological records at Chametla and Amapa reflect a more exclusive dependence on cultivation and hunting? Undoubtedly, much of this biased view has to do with the preservation and lack of emphasis on the archaeological recovery of subsistence remains—faunal, pollen, and macrobotanical. Meighan (1976e:122) suggests that at the time that Amapa was flourishing, trade in marine and estuarine foodstuffs had not yet developed. Nevertheless, it seems reasonable to suppose that, in pre-Spanish times, valued inshore marine resources would have been exported to surrounding regions, even before a demand was created by the colonial market system. This may have been particularly true during periods of low agricultural production due to flood, drought, pest, or disease, when the need for marine food supplements intensified.

Meighan (ed. 1976) indicates that the seacoast, being just a few kilometers away, was the source for the apparently small, fractional part that marine resources played in the Amapa economy. Yet, within that same distance, the Río San Pedro offered a navigable connection to the heart of the Marismas ecosystem via the Laguna Agua Brava. Travel and transport within the coastal plain habitats, therefore, do not appear to have been a serious obstacle (Mathiowetz 2011:618–620; Nelson 2004; Publ 1990:226–232), and therefore, agriculturalists and aquatic foragers had ready access to abundant and nutritious resources that were exploited as export commodities. It appears that some pottery and other nonlocal items such as copper and obsidian were brought into the Marismas sites, perhaps in exchange for marine and estuarine edibles and salt. Interpretations of the human ecosystem and particularly its practitioners as providers and consumers, however, remain incomplete and speculative.

To whatever extent the Marismas is thought of as unique in terms of its environment or resources, it is surely less separable with respect to the sociocultural sphere. The geographic area encompassed by Amapa, Chametla, and the intermediate Marismas is small enough to expect a certain amount of cultural overlap. In fact, by the measure of shared or exchanged ceramics during the Classic (Estero) and Postclassic (Conchera) periods, the Marismas and neighboring sites may have constituted a single cultural entity. Yet ceramic cross-ties alone may not indicate uniformity. Were the shared ceramic types fabricated in each area, or was there a regional site or sites from which all the shared types originated? What was the distribution and consumption of pottery in relation to possible related trade to and from the Marismas? These questions remain.

In this region of the Pacific coastal lowlands, settlement is another factor in the nature of preconquest Marismas society and culture. As detailed elsewhere in this volume, the Marismas exhibits a large congregation of small and broadly dispersed household mound sites, apparently not that distinct from or contrastive with each other. This echoes a conclusion drawn earlier by Meighan that

> in delta, estuary, and floodplain regions the sites sometimes appear to go on for miles with no apparent major centers. These remains represent many of the "pueblos" adjacent to one another and show a dense but rather evenly distributed population without major urban centers [1971:755].

Other than shell vs. earth mound construction, we found little site type variability in the Marismas, confirming Meighan's interpretation. This stands in contrast to the site of Amapa, with its cut-stone mound and plaza construction and other features suggestive of a concentrated regional center. Unfortunately, little can be said at this time about the social organizational factors that might have determined how these coastal sites stood in relationship to each other or to centers such as Amapa.

The Wider Setting—Aztatlán Intrusion?

Ekholm's discovery of a distinctive funerary assemblage at Guasave some 70 years ago was a pivotal event in the course of west Mexican research. The finding there of an elaborate and distinctive class of iconographic polychrome pottery and related artifacts collectively termed Aztatlán was, in many respects, a major stimulus for much of the research on prehispanic west Mexico in the years that followed. Pottery, it has been said, is only a thin gloss to understanding the total scheme of things (Woosley and McIntyre 1996:1), yet it often evokes, as it did at Guasave, major cultural and interpretive questions. For north and west Mexican prehistory, one such question revolves around Aztatlán itself—a confusingly common name for a native region (Sauer and Brand 1932:5), an artistic horizon style (Ekholm 1942), and a trade or mercantile system (Kelley 1971, 1986, 2000). Beyond these general definitions, the Mesoamerican Aztatlán tradition—its content, its time span, its associations, its distribution, and even its usefulness as a cultural concept—has been an enduring topic of discussion. Many new sources have emerged over the last two decades reflecting an evolving outlook that takes into account, for example, the time depth and nature of the sociopolitical complexity of culture within the Aztatlán region. A few among the numerous published sources are Beekman 2010; Bell 1971; Bojórquez Diego 2009; Carpenter 1996; Di Peso et al. 1974; Ekholm 1942; Foster 1999; Kelley 1971, 1986, 2000; Kelley and Winters 1960; Kelly 1945; Lister 1955; Mathiowetz 2011; Meighan, ed. 1976; Mountjoy 1990, 2000; Pendergast 1962a; Publ 1986, 1990; and Sauer and Brand 1932. Of related interest, too, is the work of Vidal Aldana (2011) in evaluating Aztatlán-era interaction between the coastal lowlands and the altiplano. Using several lines of evidence centering on petrography and style, Vidal identifies the coast between the Ríos Presidio and Santiago as the source of intrusive pottery found in the cultural assemblages of the Guadiana Valley and other sites of the Durango highlands.

Vidal's sourcing studies, although limited, carry implications for interregional exchange and long-distance trade. More specifically, our concern here is Aztatlán from the perspective of the Marismas Nacionales. This wetlands region lies at the heart of Ekholm's (1942) "coastal corridor" and coincides directly with what some investigators have termed the core coastal zone of the Aztatlán tradition or the coastal plain of Nayarit and southern Sinaloa (Mathiowetz 2011:565; Publ 1985:1, 71). No one should find it surprising, then, that the Aztatlán assemblage recovered in the Marismas mirrors descriptions given for other northwest Mexican sites: generally a cultural suite of the engraved and polychrome pottery types reflecting the purported Mixteca-Puebla iconography, accompanied by the traditional trait complex of Mazapan figurines, spindle whorls, beads, copper objects, and stemmed smoking pipes. Other diagnostics, however, among them *molcajetes*, plumbate pottery, and cylindrical stamps, were not found as part of the Marismas material inventory.

What do our findings in the Marismas add to the larger and more variable research universe of the general Aztatlán expression? Based on biological as well as archaeological evidence, there is a strong indication that the standing population of the Marismas was transformed by the arrival of a new ethnic group into this portion of the Pacific lowlands during the Postclassic. To consider this further let us briefly review several important aspects of the scope and origin of Aztatlán.

A basic theme of archaeological interpretation is the identification, often through inference, of population movement from cultural remains. To account for something apparently "new" in the material inventory, some interpretations emphasize the transfer of styles or concepts through intermediary cultural agencies, while alternative perspectives emphasize either internal changes or the actual movement of migrants. A familiar example from elsewhere in Mesoamerica is the recently renewed debate over Olmec origins in the Gulf of Mexico lowlands. Based on their work at San Lorenzo, Coe and Diehl have argued that changes in the ceramic assemblage mark the arrival of a new group "who were to become the San Lorenzo Olmec" (1980:150). In response, Arnold (2011:95) sees the appearance of a vessel form or decorative style as too easily transposed into the immigration of a particular "ethnic" group. In this case, he suggests that observed changes "appear to be internal and related to the context of production—the gradual transitions in settlement organization and subsistence activities though time" (2011:95). There is no need, therefore, to invoke an influx of new residents.

The same conceptual issue is relevant to the widespread appearance of Aztatlán throughout the west coast and possible mechanisms to explain it. Mountjoy's inclusive vision of the Aztatlán tradition is that of a "widespread and relatively uniform cultural expansion over a considerable part of the west Mexican coast and up into the northwestern part of the central Mexican plateau, at least into the Durango area" (2000:95). For Ekholm, there were insufficient data at the time to explain confidently how various Mixteca-Puebla traits were dispersed as far as northern Sinaloa, but he nevertheless proposed that the Aztatlán complex culture of Guasave appeared there "through the actual migration of a small group of people, including a number of religious devotees" (1942:128). In the intervening years a broad spectrum of anthropological research has continued to contribute similarly divergent interpretations of the Aztatlán phenomenon.

Following Ekholm, various researchers have suggested the migration of Mixteca-Puebla people as the principal mechanism for the appearance of Aztatlán at Guasave. Meighan notes:

> The West Mexican Postclassic also shows another wave of influences which extended the cultures of Central Mexico to Amapa and far to the north into Sinaloa. In this period, also, the changes that appear beginning about AD 900 are so extensive and striking that they cannot be explained as due to passive acceptance of traits from another area; it is necessary to propose a transfer of considerable numbers of people to bring about the observed influx of Postclassic features [1976g:161].

From his review of the striking similarities and correlations between Aztatlán symbolic motifs, artistic conventions, and deities and highland central Mexican cosmology, Mathiowetz comments:

> This topic is expansive, but it is my preliminary estimation that either some Aztatlán individuals travelled to distant regions to obtain knowledge of highland Central Mexican or Oaxacan motifs along with a complex of religious practices and forms of social organization, or individuals or groups from highland Central Mexico or Oaxaca came to the Aztatlán region with this cultural knowledge [2011:532].

Grosscup's view, that "the interpretation that the Late Period marks the invasion of the area [Nayarit] by a new people seems inescapable" (1961:404), may be typical, although we note as well a more recent and conflicting view:

> Large-scale migration of populations from central Mexico such as that proposed by Ekholm is not supported by the archaeological evidence. Subsequent work has both demonstrated that Mixteca-Puebla traits are widespread throughout Postclassic Mexico and failed to produce tangible evidence for significant population movement in the intervening areas [Carpenter 1996:206].

Thus, Carpenter, while acknowledging the Mesoamerican origin of the "material and ideological component," rejects the role of migration, emphasizing instead a picture of responses by indigenous peoples to "external stimuli" (1996:361). For more complete discussions of the migration vs. nonmigration argument, see Carpenter 1997; Kelley 2000:154; Mathiowetz 2011:530–531; and Scott and Foster 2000:112–122.

A second explanatory model offers, both literally and figuratively, a more far-reaching theoretical consideration of Aztatlán. It takes the logical but unsubstantiated view that market exchange via itinerant merchants brought Postclassic expansion from nuclear Mesoamerica to north and west Mexico, with trade links between towns extending even as far as the American Southwest (e.g., Di Peso et al. 1974; Kelley 1986, 1993, 2000). And yet another variation on the migrationary hypothesis is the notion that the Mixteca-Puebla concept was a derivative of an earlier (Early Postclassic) "Religious Style," which preexisted "through extensive trade and communication networks within the context of an increasingly standardized pan-Mesoamerican religion" (Smith and Heath-Smith 1980:38; see also Foster and

Weigand 1985:2). Smith and Heath-Smith (1980:38) also question a strategy that seems biased toward highland central Mexico as the sociocultural center of gravity and therefore the source of innovation and influence.

Does the Aztatlán assemblage represent an incremental spread of the commercial or symbolic constituents of ancient interregional trade, commerce, or religion or a combination of these? Or as Arnold raised the question earlier with regard to Olmec origins, should a change in ceramic assemblage (Aztatlán) be explained in terms of a new immigrant population? Or both? The continuing debate has been argued almost entirely in terms of archaeologist-defined comparisons (see, for example, Carpenter 1996; Mathiowetz 2011; Smith and Heath-Smith 1980). In the Marismas, however, there is other significant evidence in support of migration, specifically in the presence of two precolumbian Marismas populations. As Gill and Case (see Chapter 11) have framed their argument, heritable osteological and odontological traits are indicative of a preexisting, genetically stable population, described as "a homogeneous and possibly endogamous people representing a very early population element of central and western Mexico" (this volume). This occupation spans the Late Classic and Early Postclassic and is best seen in skeletal material from Chalpa, Venadillo, and Tecualilla. Subsequently, in Late Postclassic contexts beginning ca. AD 1200, other skeletal traits indicate the presence of an immigrant group with apparent affinities to Aztec skeletal morphology. An important point to stress here is that the skeletal data indicate genetic continuity through time for the earlier population, in sustained coexistence with the newer immigrant population. As Gill and Case explain, "The continuation of some of these rather unusual physical traits... appears explained best by migration followed by settlement and intermarriage, not replacement" (this volume). From the perspective of population genetics, then, the evidence seems reliable enough to postulate a composite population of "ancestral" and "foreign" occupants of the Marismas during the late prehistoric period.

The notion that some population movement occurred no longer seems as speculative. The case rests heavily on skeletal evidence, but certain cultural associations would seem to substantiate that principal conclusion. For example, excavators recognized a late Postclassic change in mortuary practices—something that might reasonably be taken as a change in the social landscape in the base society or standing population. Also, a large and consistent percentage of Aztatlán Ware was found widely dispersed in most, if not all, of the Late Postclassic contexts. A logical inference might be that Postclassic colonists, once having relocated in the Marismas, tended to acculturate, following the social-integrative model of settlement and intermarriage.

If, as we believe, there was a late influx of new residents into the Marismas, were they the agents for the introduction of new ceramic types and an array of other material culture? Was material culture of the Aztatlán complex produced locally or imported as commodities, as Kelley (1986) and others have suggested? Trace analysis using heavy minerals has shown that the mineral suite from local sand samples is the same as that found in sherds from the Marismas, Chametla, and Culiacán (Scott and Cazeau 1976:214). One key problem, of course, is the need to more narrowly define the mineralogical provinces of west Mexico or, alternatively, to apply other methods of geochemical sourcing. Nevertheless, heavy mineral sourcing indicates local manufacture vs. the importing of foreign-made vessels, a realization consistent with a picture of movement of peoples and resettlement but quite inconsistent with an idea of trade by exchange and communication network (see Gill and Case, Chapter 11, regarding evidence for the movement of people into the area).

The new ceramic styles and assemblages could be indicative of both continuing trade contacts and products imported by new settlers in the context of increasing evidence for market exchange, both in marketplaces and via itinerant merchants, during the Postclassic (Pool and Bey 2007:36). In our view, however, itinerant merchants would leave only a minor biological imprint, whereas the data from the Marismas skeletal series seem to point reliably to both established resident and later migrant populations. Because the arrival of a new genetic population in the Marismas was so readily identifiable in the skeletal material, it is likely that the initial arrival of the migrants was relatively sudden; otherwise the identifiable skeletal traits may have been masked or at least less obvious if a gradual and sporadic influx of migrants had occurred.

In addition to the migrants, Aztatlán cultural practice warrants some consideration as well. At Guasave, Ekholm recognized the particular religious or ceremonial importance of the Aztatlán culture complex in connection with the use of pottery as mortuary offerings. It was impossible to know, he added, how this elaborate complex of traits, having achieved ceremonial significance "by being put into graves," would compare with artifacts "centered around everyday existence" (1942:123). In the Marismas, the Aztatlán component, evident in the pottery, pipes, and other artifact types interred with the dead (see Chapters 7 and 9), also appeared to be more pervasively distributed in mound fill and random locations. At Arriñitas, for example, a meter-thick unit of flood-deposited silt overlies an oyster shell midden containing a suite of Aztatlán-related ceramic sherd types.

Pieces of copper and Postclassic period ceramic types, namely, Aztatlán types, found at random in earth mound fill were perhaps dispersed from disturbed burials.

The occurrence of Aztatlán ceramics in the upper, Postclassic shell components of compound earth and shell mounds in the Marismas has already been noted, although some earlier and presumably local pottery types continue to appear during the Postclassic era of shell deposition. It would be a mistake, however, to equate the Postclassic newcomers with a shellfishing economy. The same inventory of Late Postclassic Aztatlán Ware occurs in other sites on the Teacapán Peninsula where there is little or no evidence of shellfish resource utilization.

For the Marismas Nacionales, the evidence suggests that the principal ceramic manifestation of the west Mexican Postclassic was the Aztatlán complex, a term that in Bell's (1971:699) opinion has almost taken on the status of a culture by force of common usage. Its broad characteristics apply across much of northern and western Mexico, but in the Marismas there is now both archaeologically and biologically demonstrable support for the long-posited theory of migration, which in turn may account for the appearance of Aztatlán as a nonlocal cultural complex.

Mortuary Behavior and Bioanthropology

Three chapters of this volume, one on mortuary practices and two on the physical remains themselves, are dedicated to the 248 individuals whose skeletal remains were recovered during the Marismas project. The physical remains ranged from well-preserved complete skeletons to barely intact splinters of bone. Many of these individuals were accompanied by funerary offerings, and many exhibited cultural modifications to their skulls and dentition. Additionally, the skeletons conveyed a wealth of information regarding the health and genetic composition of those who had inhabited the Marismas prior to the arrival of the Spanish.

The goal of the mortuary study was to seek evidence of vertical and horizontal ranking within the families and small communities that occupied the Marismas and to gain some insight into their worldview as manifested in the mortuary customs represented. The small prehispanic villages and hamlets that dotted the Marismas were undoubtedly tied to larger towns that functioned as centers of ceremonial and economic activities. Such sites would include Amapa and San Felipe Aztatán; these would have been sites overseen by highly ranked elites who managed both the ceremonial and economic components of their domains. However, the communities within the Marismas do not appear to have been defined by the presence of highly ranked societies. Rather, the evidence, based on the analysis of the presence or absence of funerary offerings, the treatment of individuals based on age and sex, and the presence or absence of cranial and dental modifications, suggests that social differentiation within the communities was likely based on prestige. No evidence of centralized leadership or of inherited rank was apparent. The patterns seen appear to imply that the social standing of an individual within the family and perhaps the community was based on age, and an individual's accomplishments within the family and perhaps within the Marismas communities had a greater influence on material expressions of grief at the time of burial. Funerary offerings generally consisted of ceramic vessels, personal adornment, tools, perhaps some ceremonial items (pipes), items of personal use, gifts or tokens of bereavement, and perhaps items of membership, prestige, and authority. There are no clear qualitative differences in funerary offerings or mortuary treatments to suggest sharp sociopolitical differences within the Marismas communities investigated.

Perhaps the greatest insight into worldview comes from the orientation of the Marismas burials. As Carr (1995) notes, based on his cross-cultural study, the primary determinant of body orientation was based in the philosophical-religious realm and linked to the perceived direction of the afterworld. Burial orientation for the Marismas burials was highly variable. Nevertheless, west was the direction in which bodies were most commonly oriented, followed by east. Together, west- and east-oriented burials account for nearly half of the 122 individuals for which orientation could be determined. Males were more frequently oriented to the east, while females were more commonly oriented to the west. The east was seen as important because of the rising sun (the sun being reborn) and its association with fertility, birth, rebirth of the soul, creation, light, rain, corn, and spiritual power, while the west was the place where the sun began its daily journey through the underworld.

It is also clear that about AD 1200, with the apparent arrival of a new genetic population from the south, a shift in the local worldview occurred. This change is manifested in the shift from predominately extended burials to predominately flexed burials. Although the influx of a new genetic population, the size of which is undetermined, affected the physical makeup of the local population, it apparently also marked a change in burial practices as well as how the socioreligious component of their world was viewed (e.g., Mathiowetz 2011). There is little doubt that the precolumbian peoples of the Marismas actively participated in a ceremonial calendar that was centered on the agricultural cycle as well as fishing.

The identification of disease, pathology, trauma, and other anomalies in the skeletal Marismas remains, as well as age-at-death distributions, has resulted in considerable insights into the health and lifeways of the precolumbian people who occupied the area. In addition to evidence of arthritis, periostitis, osteomyelitis, and cranial hyperostosis, a few cases of tuberculosis (Pott's disease) were clearly identified in some of the Postclassic period remains. The differences in frequencies of dental caries from site to site suggest that there may have been a greater reliance on agriculture at some sites, while at others, such as Tecualilla, Chalpa, and Venadillo, the agricultural diet was more heavily supplemented with animal protein. A high death rate among young males is taken by Gill to represent evidence of warfare or some other form of interpersonal conflict. However, no archaeological evidence for this, such as an abundance of projectile points or other weapons such as mace heads or axes, was recovered.

Also of importance is the identification of cultural modifications, both purposeful and accidental, to the human skeleton. These include a variety of artificial cranial and dental modifications to both males and females. Evidence of accidental injury to the skeleton was seen in swollen tibiae, and it is thought that the inflammation seen was caused by wearing ligatures around the upper lower leg.

Additionally, analyses of skeletal traits yielded information regarding biological relationships, migration, and population change within the Marismas over time. Classic and early Postclassic period populations were a homogeneous and possibly endogamous population not dissimilar to the earliest inhabitants of central Mexico. However, around AD 1200 some new genetic traits begin to appear in the Marismas populations, and such traits increase in later populations. Despite the introduction of these traits, derived from the migration of new peoples into the area, there is nevertheless a significant amount of genetic continuity within the Marismas population as a whole. The identification of gene flow via the migration of people, likely from the south or from the central part of Mexico, is perhaps one of the most significant discoveries of the bioanthropological studies, and its important archaeological implications cannot be overstated.

The Marismas in the Sixteenth Century

Sauer and Brand (1932:41–58) and others have reviewed Spanish accounts of the sixteenth-century conquest and subjugation of the northwestern native populations (see Anguiano 1976). The use of the direct historical approach assumes some cultural continuity between documented and ancestral prehistoric cultures. In that connection, it is interesting to note that in Kelly's opinion, at Chametla "the series [cultural sequence] may not continue to Conquest days" (1938:2). Similarly, Meighan adds:

> It is also clear from the archaeology of Amapa that the major occupation there took place between AD 600–1300, with very little occupation, if any, for the past century-and-a-half prior to the arrival of the first Spanish in the region [ed. 1976:8].

Pendergast (1962a:370) argues for an even earlier date for the abandonment of Amapa and the surrounding area while allowing for the possibility that occupation of the site may have continued into the contact period.

If, indeed, Amapa and Chametla were abandoned prior to contact, the *tierra caliente* itself was clearly not depopulated at the time of conquest. The Spanish in 1530–1531 found a heavily populated region of agriculturalists who also exploited an array of coastal and estuarine resources, divided into three *provincias* or sociopolitical territories—Centispác, Aztatlán, and Chametla. These three sociopolitical entities shared overlapping tribal units, common languages, many cultural likenesses, and common enemies. Their leaders were noted to possess, display, and wear status objects, especially shells. Each of these *provincias* was said to have jurisdiction over numerous settlements, and at least in the case of Centispác, tribute was collected from subject communities (Tello 1974:75).

Within the three entities, the people of Centispác and Aztatlán were allied and most similar at the time of conquest, and the conquest narratives note that this had been the case for some time. In her synthesis of the ethnohistoric period (see Chapter 12), P. Scott notes that the conquistadores found the people of Aztatlán populating the Río Acaponeta to the sea, and many in Centispác were also residing in areas now within the Marismas Nacionales. Owing to the same difficulties of rectifying variable Spanish accounts and other problems cited by Grosscup (1964:15–30) in his attempt to link Amapa with a historic village or town, we cannot identify any specific Marismas archaeological sites with any contact-period Indian settlement named by the Spanish conquistadores. We can assume, however, that with further excavations some sites may later be associated with settlements that appear on early Spanish maps, especially the island communities labeled on the Ortelius 1579 map (Figure 12.6).

Marismas settlements in the archaeological record appear to consist of simple, local nuclear or extended family households that were typically situated on small *lomitas* or earth mounds. From the archaeological record a minimal local political organization has been assumed. As discussed in Chapter 9, mortuary practices and grave offerings indicate that social standing was based on age and perhaps economic importance to the community rather than a highly ranked society.

Spicer (1962:789) describes the coastal lowland conquest settlement pattern to consist of scattered *rancherías*. These clusters of houses, he says, were autonomous but shared a sense of boundaries, defended only in wartime by an alliance of rancherias. Meighan (1971), however, likens the political geography to a series of city-states with tributary towns and villages. As discussed in Chapter 12, the provinces of Centispác, Aztatlán, and

Chametla had recognizable social hierarchies, political structures, and institutions of formal leadership and power (see also Weigand and Weigand 2000:16). Even if the Marismas populations at the time of conquest lacked local, corporate organization in peacetime, as Spicer suggests, the ability to form temporary and reliable wartime alliances implies a longer and more durable period of cultural union. Publ observes in reference to these independent coastal political units: "In essence, the Aztatlán region reflects a notable cultural complexity in West Mexico at contact time and bespeaks an extended period of development long before the Spanish contact" (1990:206).

Mid-seventeenth-century missionaries note that the people living in what is now the Marismas were known as the Totorame—now considered lowland relatives of the nearby Cora (Sauer 1934:13) and the Tecual Huichol (Phil Weigand, personal communication). As Arias de Saavedra (1975 [1899]) noted in the mid-1600s, the Totorame lived along the seashore as well as on little islands and were fishermen and salt gatherers. Everyone called them Themureles, meaning "Frogs" (McCarty and Matson 1975:208), because they tended to retreat to the mountains in times of high water or other stresses. What remains unclear is how long the Totorame had been inhabiting and utilizing the coastal resources. Were they the coastal inhabitants Guzmán's forces saw in the coastal areas of Aztatlán and Centispác in 1530? How long had they and their close mountain relatives been utilizing what is now the Marismas Nacionales—an area that at least in the seventeenth century they seem to have considered sacred?

The Spanish chroniclers clearly define three social-political entities with territorial boundaries, leaders, and a tribute system, showing a greater societal complexity than is apparent in the excavated Marismas sites. The question remains as to who the conquest-period Marismas inhabitants owed allegiance and possible tribute to, if anyone, and how long these social-political alliances had existed. What ancestral relationship, if any, did the conquest-period inhabitants have to the people represented in the Marismas excavations? Whatever the case, while disease and mistreatment annihilated many of the sixteenth-century coastal inhabitants, the Marismas and nearby areas remained populated and in some cases in power. Chametla was able to hold off Spanish intrusion for several decades. The Cora not only maintained control of the Mesa de Nayar to the east of the Marismas into the eighteenth century but have maintained their cultural identity to the present.

Final Observations

The State University of New York's field studies were undertaken with the following goals: to expand our knowledge of human cultural and physical adaptations to an ecologically diverse coastal and estuarine landscape, to better understanding the origin and timing of precolumbian cultural developments within the Marismas area and interpret them within a regional context, to identify and elucidate subsistence practices and local human adaptations and examine them within our understanding of regional and panregional economies, and to understand the developmental history of a large portion of the West Mexican Coastal Plain. The chapters in this volume have addressed the resulting archaeological, bioanthropological, historical, and geologic findings, reinforcing and expanding our knowledge of this subregion of northwest coastal Mexico as first delineated by Sauer and Brand some 70 years ago. These new data, namely, evidence of a combination of environmental, climactic, cultural, and population changes, offer a more complete and, therefore, better (if not more beneficial) understanding of this portion of the Pacific lowlands and its role in the wider and evolving panorama of west Mexican and Mesoamerican prehistory.

APPENDIX A

A Green Glaze Anomaly

Stuart D. Scott

Of all the artifacts, ecofacts, or tangible material residue found that might reflect the use activities of aboriginal Marismas Nacionales populations, it is an anomalous glaze, or slag-like material occurring in moderate abundance, that presents a continuing archaeological puzzle (Figure A.1). Two key components for discussion are the circumstances of discovery and how the glaze was produced.

Discovery

The subject material can perhaps best be described as fragments of glaze-coated rock, most of which seem to have been randomly discarded on the surface or as refuse in mound fill. Like debitage, the majority are dissimilar pieces, in this case porous basalt and other less porous base rock fragments, on which a green to brown vitrified surface has formed. As a generality, without the glaze, the core fragments might have gone unnoticed or would have been simply noted as detritus. In succeeding seasons, however, field notes reflect a growing awareness of glazed rock samples found in more specific archaeological contexts. At Venadillo, for example, their occurrence with oyster shells in a kitchen area suggests their use in food preparation, perhaps being heated to cook shellfish (Figure A.2).

More significantly, however, glazed baseball-size spheres were found in apparent association with human burials. Although they were not found in intimate association with skeletal remains, as were primary mortuary furnishings such as vessels, pipes, spindle whorls, and so on, these glazed balls nevertheless were discovered consistently close enough to the deceased, usually within a meter or two, to support the notion that these may have had a role in whatever cosmology dictated the burial practices. Parenthetically, Gill (1969a) also notes the finding of similar-sized reddish balls of burned clay with burials at Chalpa. Additionally, one small potsherd exhibits the same apparent green glaze seen elsewhere but now applied to the body of a vessel, rather evenly on both sides of a raised medial(?) ridge or flange (Figure A.3).

Just as the cumulative sample of glazed pieces is small, so too is the known literature on this regional anomaly. Beyond the Marismas, glaze/slag examples have been reported from various sites and contexts. In Sinaloa, for example, during her 1930s reconnaissance to excavate and classify various types of precolonial pottery, Isabel Kelly made note of scattered fragments of "slag," seemingly similar if not identical to the Marismas examples. In her letter to Carl Sauer from the field, she writes:

> About the slag—personally, I'm more confused than ever. McLeod wrote immediately, but I did not get the letter until my return. He made a microscopic examination, from which he was virtually certain that it was volcanic stone; but he went further and had the stuff analyzed: Silica 63.0, Alumina 17.1, Iron 4.6, Lime 3.4, and Magnesia 2.3. Remainder: partly oxygen and some alkalies, together with a few traces of odds and ends. He says that the melting points of silica and alumina are so high that he doubts they could be fused today in a reverbatory at the smelter: "Therefore I am quite certain that this was never any kind of slag. Its microscopic appearance did not give any indication of such a possibility."... To return to the slag, so-called. It occurs, in our experience, almost exclusively at sites of prehistoric occupation, and generally associated with a great litter of shell. But this association is not consistent. Just downstream from La Carbonera, at Las Calaveras, is a low peninsula jutting from the bank into the bed of the estero. We cut through this, as it was "slag" covered, and found again absolutely clean-clay beneath. Still further downstream, at Llacuato, is a Late site, with several heaps of "slag." One was pretty sharply conical, and we trenched it. It seems definitely to have been an oven—more or less pot-shaped, with layers of charcoal, ash, and several pieces of blackened clam shell. Another heap at this site proved again

FIGURE A.1. Examples of glazed balls (scale approximate).

FIGURE A.2. Detail of hearth area with green-glazed balls, Test Pit A, Venadillo.

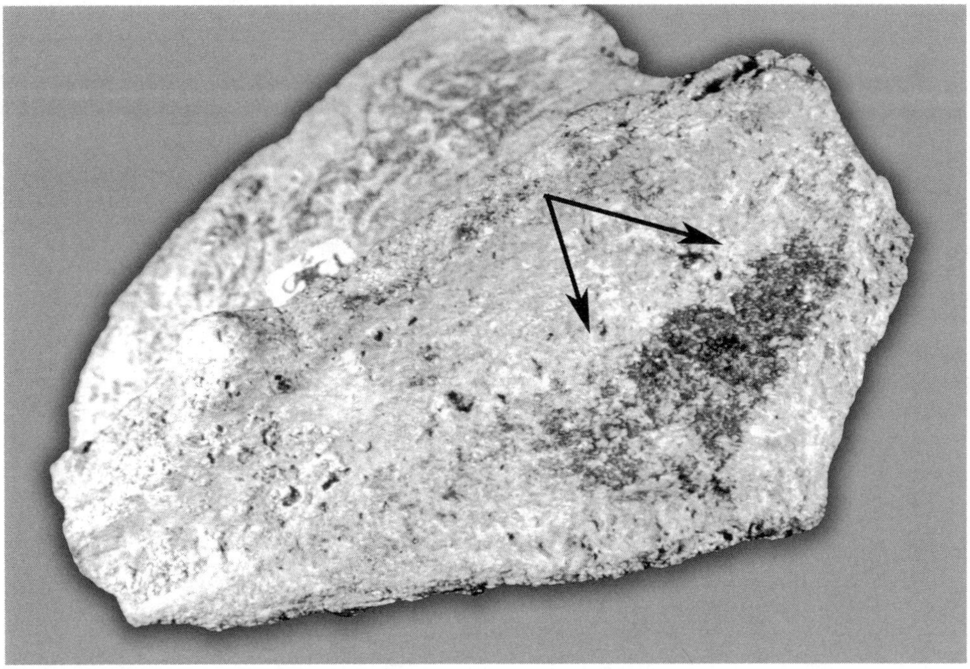

FIGURE A.3. Sherd with green glaze like that found on the green-glazed balls.

to have clean clay beneath [Kelly to Sauer, Culiacán, February 3, 1939, in González 1989:44–45].

Weigand (1975a) offered an overview of glazing in western Mexico, to the extent that the scarcity of data would permit. From lowland Nayarit he reports the finding of a "probable glazed jar" (1975a:5) in association with copper artifacts from Crabtree's 1961 excavations at Santa Cruz. He notes, as well, that Mountjoy (1970b:279–281) discovered small quantities of decoratively glazed sherds in the San Blas/Santa Cruz sector of the Nayarit coast. Weigand comments: "The glaze is 'fugitive' in the sense that it is light and runny. Some pocking occurs, and the glaze is lead based" (1975a:5). He also points to other material evidence, that is to say, small collections of glazed ceramics from the Ameca Valley of highland Jalisco near the Nayarit border, with similarities to examples described by Glassow (1967) from Huistla, near Etzatlán, Jalisco. In his survey of these somewhat rare occurrences of ceramic glazing in west Mexico, Weigand suggests that it seems to be related to the production of copper artifacts in that region. In his (1975a:6) view, the glazing was conceivably concomitant with the development of the west Mexican metallurgical traditions and ancestral to the appearance of glazes in the Southwest.

From his excavations at Amapa, Meighan (1976f) arrived at the same tentative conclusion based on the finding of seven slag fragments in pit excavations, each bearing traces of copper and dating to the Cerritos and Ixcuintla phase deposits. The Amapa report included a compositional analysis by W. C. Root (Meighan 1976f: 118, Table XVII), which included the results for slag seen in Table A.1.

Finally, another puzzling example comes from Pueblo Grande, the large Hohokam platform mound in the Phoenix basin. There, a similar green vitrified glaze-like cover was found on basalt rock fragments in large subterranean agave-roasting pits (Mitchell and Merewether 1994).

How Was the Glaze Produced?

Several attempts at analysis at different laboratories have been conducted, and the results are reported here. Soon after we recognized the glaze anomaly as a phenomenon with possible cultural associations, Patricia St. George (1971) undertook experiments in replicating the formation of the glaze on stone. St. George's data and findings are repeated here as well.

Duplicate samples of the core and glaze were examined by neutron activation analysis at the Western New York Nuclear Research Center, Buffalo (Table A.2). Although present, the amounts of U, Rb, Zn, Cr, Co, and Ni were extremely low (Table A.3). No quantification of Ca and Si was done, although these elements are known to be present in samples in significant amounts.

As a mineral, clay or kaolinite has the following formula: $Al_2O_3 \bullet 2SiO_2 \bullet 2H_2O$. In this formula the relative amounts of the oxides present are stated as a molecular ratio rather than by percentage weight. In clay, one molecule of Al_2O_3 (alumina) is associated with two molecules of SiO_2 (silica) and two molecules of H_2O (water). This formula, which is typical, overlooks the complex array of

TABLE A.1. Constituent Analysis of Copper Slag from Amapa, Nayarit.

Catalog Number	Element								
	Gold	Silver	Copper	Tin	Lead	Arsenic	Stibium	Bismuth	Indium
2150	—	—	T	—	—	—	—	—	—
1437	—	*	T	—	—	—	—	—	—
1147	—	*	T	—	—	—	—	—	—
2213	—	*	*	—	—	—	—	—	—
415	—	—	*	—	—	—	—	—	—
265	—	—	*	—	—	—	—	—	—
1171	—	—	*	—	—	—	—	—	—

Note: * 0.001%; T = 0.01%–0.1%.
Source: Meighan 1976f:118.

TABLE A.2. Results of Neutron Activation of Samples of Core and Glaze Material.

Sample	Aluminum (%)	Iron (%)	Manganese (ppm)	Lanthanum (ppm)	Scandium (ppm)
Core 1	10.31	8.20	275	32	13.7
Core 2	9.46	5.29	783	11	11.7
Glaze 1	8.70	4.03	154	16	9.9
Glaze 2	10.14	3.75	213	16	10.5

"impurities" always present in actual samplings. That no quantification of calcium and silica was done in testing is unfortunate since silica is the principal glass-forming agent and the amount and type of flux determines the melting temperature. Feldspars, naturally occurring minerals that supply fluxes to help melt a glaze, contain alumina (Al_2O_3) and silica (SiO_2), combined with one or more oxides, generally of an alkaline or calcium nature. The formulas for commonly occurring feldspars are illustrated in Table A.4.

These are typical formulas. As in the case of actual rock specimens, it would be probable that two or more alkalis would be present in any feldspar and that there would be some impurities such as iron. When feldspar is disintegrated by geologic weathering, the alkali part, the soda potash, or lime, which is relatively soluble, is carried off by water. This leaves the alumina and silica. This kaolinization of feldspar tells nothing of the physical properties of clay but accounts only for the chemical composition. The melting point, or temperature at which clay fuses, tends to be high. Even common surface clays fuse at temperatures above 1,000°C (Rhodes 1957:6).

Patricia St. George's testing of composition and firing temperatures concluded with the examination of a large fragment of metate from the Marismas excavations, seemingly fused with globs of this same material "stuck" to it but apparently not fired to the glaze maturing point. Duane Anderson, a ceramist at the State University College at Buffalo, New York, fired several pieces of this

TABLE A.3. Other Elements Identified in the Core and Glaze Material.

Element	Amount (ppm)
Uranium	0.05
Rubidium	5.00
Zinc	1.00
Chromium	1.00
Cobalt	1.00
Nickel	5.00

TABLE A.4. Common Feldspar Formulas.

Type of Feldspar	Formula
Orthoclase	$K_2O \cdot Al_2O_3 \cdot 6SiO_2$
Albite	$Na_2O \cdot Al_2O_3 \cdot 6SiO_2$
Anorthite	$CaO \cdot Al_2O_3 \cdot 2SiO_2$

material with the following results: At cone 4 (approximately 1,050°C) in an oxidation atmosphere, a definite color change was noticed, with the sample changing to a reddish brown, with tiny specks of glaze appearing at the surface. A second sample of the same piece was then fired at cone 9 (approximately 1,250°C), in which the sample "puddled." The smoky reduction firing, even though at a much higher temperature, is more closely associated with the atmosphere of primitive kiln firings than the oxidation. This material, St. George (1971:14)

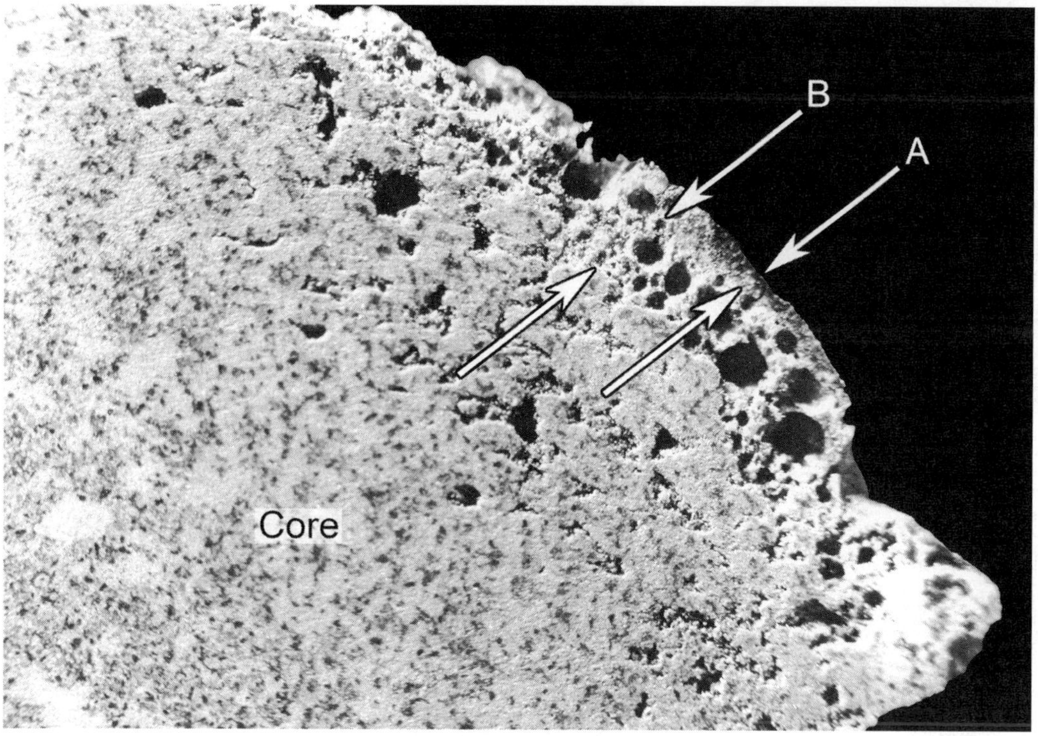

FIGURE A.4. Cross section of a piece of a green-glazed ball. The enlarged cross section shows that Zone A is a glassy glaze and Zone B is a fine clay layer (approximately 1 mm). The core below the A and B layers is a coarse-textured clay with a high calcite content.

concludes, has a maturing point between cones 4 and 6 (approximately 1,165°C–1,190°C).

Given the intrinsic cultural significance of glazed artifacts vs. "slag," a sample of the balls and the sherd were sent for analysis by the Brookhaven National Laboratory. The analysis was as follows:

> The large irregular lumps and the sherd were examined with a normal optical microscope. The large lumps seem to have been roughly hand formed from coarse textured clay with high calcite content. These balls were probably sun dried and then coated with a thin irregular (approximately 1mm) layer of fine smooth clay; most probably to seal and produce a smooth surface for glazing. This fine layer is quite porous, possibly caused by steam or gas evolved upon heating of the composite ball before glaze firing. The glaze forms a layer, of the order of the thickness as the fine undercoating that is now crazed but fairly free of bubbles or gas holes.
>
> X-ray fluorescence studies indicate that the major elements present in the glaze are Si, Ca, Ti, K, Fe, with measurable quantities of Mn, Cu, Zn, As, Br, and Sr. Pb, a component of some green glazes was checked for but not found.
>
> A comparison of composition of the glaze on the large lumps and the sherd indicates that they are the same in composition and color [Figure A.4]. Melting point tests were performed on the glaze and results indicate that the glaze melts at a relatively low temperature: 850 ± 40°C [Carriveau 1976].

Carriveau's analysis confirms his earlier impressions of the glaze conveyed via Weigand to Scott:

> The "balls" (perhaps weights or volume displacers?) are made of two layers of clay—inner is coarser and an outer near slip, which is very thin and interacted with the inner silicate glaze. Firing temperature ca. 750°C which is enough to vitrify the glaze and form the union between the glaze and the slip clay layer. No lead is in the glaze. The glaze is purposeful and for its own sake, i.e., probably decorative. It does not seem to be a function of another metallurgical process [Weigand 1975b].

As stated, the Southwest's Pueblo Grande produced another component of the general inquiry, in this case, glazed basalt rock fragments found in agave-roasting pits. The late Phil Weigand conducted a preliminary analysis of the glaze constituents and original firing temperatures, which he discussed in a preliminary, informal report submitted to Soil Systems during the Pueblo Grande Project. The following is taken from Weigand's 1990 report:

> The base rock from which the vitrified, glaze-like material was generated is semicompact basalt, with some phenocrysts embedded in the groundmass but certainly not enough for the basalt to be classified as porphyritic or vesicular. The vitrified, glaze-like materials were determined to be iron silicate ($FeSiO_3$). The basalt is probably mafic, given the abundance of iron silicate on its surface. Mafic basalt commonly has 44–55 percent silicon (SiO_2) and, just as important, more than 9 percent iron (Fe_2O_3 at 3.2 percent and FeO at 6 percent). Iron's melting point (MP) is 1,536°C, though it will begin to fuse or frit together within the temperature range of 1,310°C–1,480°C. Iron's boiling point (BP) is 2,750°C. Silicon's MP is 1,410°C, and its BP is 2,355°C.
>
> The first experiment in the kiln involved a sample that weighed 2.85 g and was a combination of the groundmass and the vitrified, glaze-like material. The purpose was to determine the temperatures at which the material would begin to liquefy and then boil (or more descriptively, bubble). The kiln could only be calibrated at 25°C intervals, and even this is probably a too-optimistic characterization of its accuracy. At 1,450°C, most parts of the glaze-like surface had begun to reliquefy. At this point, it was clear that the glaze base was silicon. At 1,550°C, the glaze underwent a slight color shift, indicating that iron was rejoining in the melt. Since the surface of the original sample was pocked, indicating that at least one of these two elements had boiled, the kiln's temperature was raised by 50°C increments until bubbling began. Some minor bubbling began around 2,000°C, but the surface was completely reconfigured only above 2,100°C. The kiln's temperature certainly did not reach much above this temperature, which appears to be the natural capacity that it has. Hence, it is most likely that it was only the silicon that reached its BP and produced the pocks on the sample surface (both on the original surface and on the reconfigured, revitrified surface). Aside from the kiln's temperature, the reason for this statement is that the edges of the pocks are very slightly more greenish than elsewhere on the sample's original surface. This probably means that iron did not reach its BP.
>
> A second sample, more closely chosen to represent only the glazed material and weighing 0.81 g, was brought up to the MPs of both elements but not beyond, due to the very high heat buildup in the kiln's lead wiring. It is interesting to note that the MPs for both elements obtained by this experiment were within 25°C on the kiln's thermostat of the first set of readings.
>
> One conclusion: it is certainly not the lack of oven temperatures that can be used as the explanator for the lack of metallurgy or highly fired ceramics at Pueblo Grande. Temperatures of ca. 2,100°C are truly high. A pit at Chavez Pass, Arizona, also shows that these temperatures could be reached in significantly smaller ovens than Feature 586 at Pueblo Grande. The presence of iron silicate glazes elsewhere in Mesoamerica and the Southwest strongly suggests that high temperatures per se were not the technological problem—indeed metallic glazes are a form of metallurgy. Rough glazes can be obtained without reaching for BP temperatures, of course, and indeed are smoother if only MP temperatures are used. The temperatures obtained in Pueblo Grande Feature 586 are far higher than what was required for agave cooking. Pit design and other cultural factors, rather than temperatures, should be used to explain the lack of metallurgy and vitrified ceramics at the site. Given the rather low temperatures apparent for the ceramics at Pueblo Grande, it appears that these truly great temperatures were not directly employed for any purpose within the realm of pyrotechnology.

Wood Ash?

In trying to describe the origin and character of the glaze, whether an iron silicate or aluminosilicate, the working assumption by St. George and Weigand is that it is a product of the heating of the parent material. But what of the possibility of reproducing the coating or vitreous layer by the deliberate addition of ash? Cornwall writes:

> Flints or other siliceous stones heated to a high temperature in contact with wood-ash, which, as we have seen, contains potash and lime, are sometimes found bearing a fine glaze. These are generally not "crackled," having cooled gradually as the fire died down. The glaze is a true glass, due to fusion of the surface as a bright red heat in combination with the alkali and the lime. No doubt, the observation of some such accidental glazing was the foundation of the idea of deliberate application of a glaze to pottery. Crushed flint

> or fine sand, mixed with wood-ash in some kind of organic excipient or clay-slip would yield an application fusible to a glaze on firing at a temperature considerably below that necessary for the biscuit-firing of the pot. Any iron present in the minerals—almost certain to be there—would give an elegant bottle-green colour to the glaze [1958:67–68].

Today wood ash is an important feature of pottery glaze-making, a process thought to have been produced originally by accident when white-hot wood ash was carried with the draft of a wood-fueled fire onto pots where the searing heat melted it to a glass (Rogers 2003). Using several native wood species from the Marismas coastal forests, ceramist Karl Sandmeyer of the Carborundum Corporation attempted to reproduce the glaze from ash with kiln firing at various temperatures, without success.

Implications

Weigand (1975a) hypothesizes a relationship between west Mexico's areas of copper metallurgy and its glazed ceramics and in doing so, raises the issue of the possible spread of glaze paint technology across the macro-area of west/northwestern Mexico and the Southwest. There is, then, an implied link among the random findings of slag, the production of copper, and concomitantly, glazed pottery. Glaze-painted pottery was known in the Southwest and particularly among the Late Prehispanic Pueblo populations of the Rio Grande (Huntley et al. 2012). Snow (1982:239–243) suggests a tentative model of Mesoamerica as the origin of a copper-based glaze paint, seen in a similar form at Paquimé, and by extension, part of the Postclassic resurgence of ceremonial art and stylistic traditions. Phillips (2012:41–42) offers a similar version of the same idea, namely, that, hypothetically, Mesoamerica's northwest frontier may have been the ultimate source of a constellation of traits, including glaze paint–decorated pottery.

The use of glazing as a means of ceramic decoration is an issue that spans the contiguous but distant regions of west Mexico and the American Southwest, yet the evidence for the spread of glaze technology on the scale of such long-distance interaction is patchy at best, if not totally speculative. There is no direct evidence associating the residual slag glazes found in west Mexico with copper production. At Amapa the finding of seven isolated and scattered pieces of slag containing traces of copper was taken to be evidence of the working of metal; however, no metalworking area was found at Amapa. Interestingly, although Kelly reported slag from prehistoric coastal Sinaloan contexts, no copper artifacts were reported from Chametla. It may indeed be shown in the future that the slag is the product of local west Mexican copper smelting, as Meighan (ed. 1976) and Mountjoy (2000) have suggested.

More important, however, for the purposes of this summary volume is to focus on the smaller universe of random glazing, a material phenomenon found widely in the Marismas and reported by Kelly in coastal Sinaloa as the product of apparent intentional human action. Researchers studying the cultural importance of glaze ware technology agree that more compositional studies are needed, a prescription that applies expressly to the Marismas case study of glazes. The use, accidental or otherwise, of ash or other organic ingredients to produce the glaze is not proved. But neither does the assumption that these rocks are of a self-glazing nature fully answer the question of origin. Unlike the random, scattered distribution of detrital and glazed-core rock fragments, that is, basalt erratics, the aforementioned near-perfect spheres found with burials bear the finger imprints from having been hand-molded into a ball shape before the glaze was formed. Logically, the parent material in that case must have been of a sufficiently clayey and malleable material to achieve a desired shape and the hypothetical status as an artifact. Also, among the thousands of sherds recovered from Marismas excavations, one small potsherd exhibits the same apparent green glaze seen elsewhere. During the period of the State University of New York Marismas Nacionales project fieldwork, the finding of a single glazed sherd was provocative but clearly insufficient to suggest that we were seeing an experimental decorative technique. Besides the dispersed, irregular, and coated basalt fragments, in the case of the sherd and mortuary balls, the glaze does not appear to be random splashings from cooking, metallurgical, or other fire pit heating activities but, instead, suggests a deliberate attempt to create the colorful glassy coating. Much more experimental testing to distinguish how and why the glaze was produced will be needed to resolve this geoarchaeological puzzle.

APPENDIX B

Discrete and Continuous Nonmetric Traits

George W. Gill

Within this appendix are illustrations of most of the 31 discrete traits (Figure B.1) utilized within this study, as well as one continuous nonmetric trait—orbital form (Figure B.2). All 31 discrete traits and all continuous traits used in this study have been previously defined and illustrated in Gill (1971), and more recently Rogers (2008) defines most of the same traits, but with some additional traits added and without the two postcranial traits and mandibular traits included here.

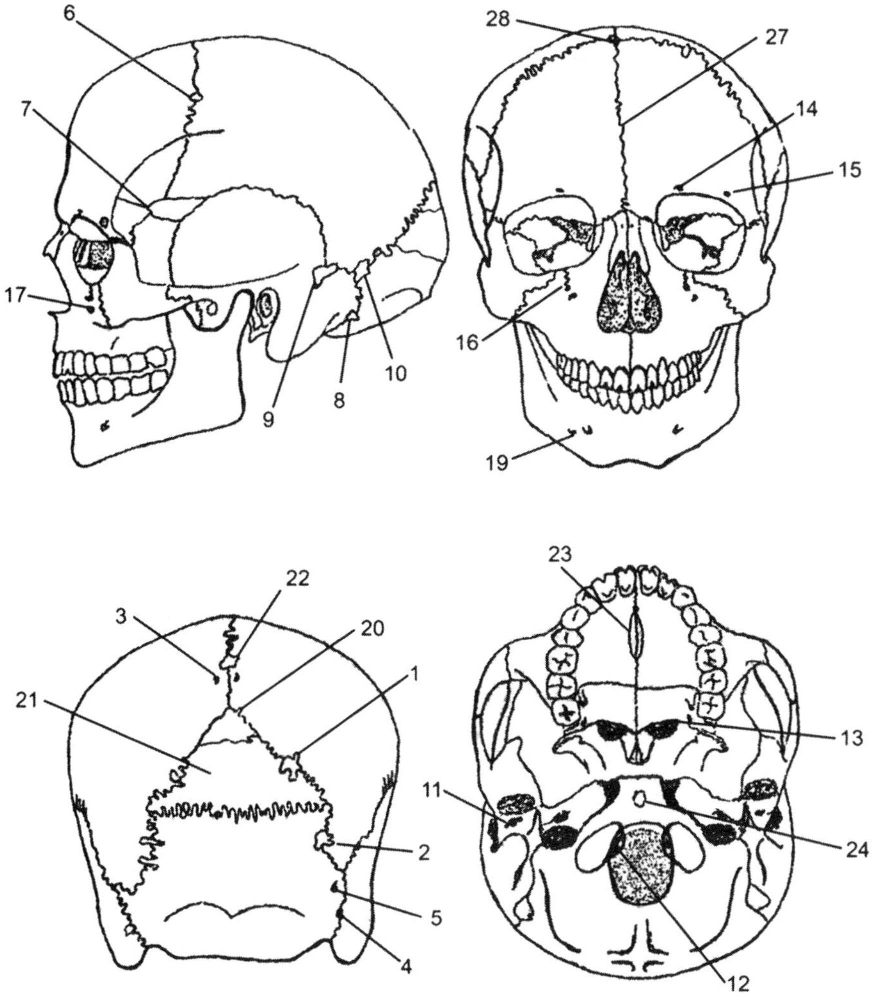

FIGURE B.1. Discrete nonmetric traits of the external cranium.
1 = lambdoid ossicle—medial;
2 = lambdoid ossicle—lateral;
3 = parietal foramen;
4 = mastoid foramen;
5 = mastoid foramen exsutural;
6 = coronal ossicle;
7 = epiteric bone;
8 = mastoid sutural ossicle;
9 = parietal notch bone;
10 = asterion ossicle;
11 = tympanic dehiscence;
12 = double anterior condylar canal;
13 = accessory lesser palatine foramen;
14 = supraorbital foramen;
15 = frontal foramen;
16 = suture into the infraorbital foramen;
17 = accessory infraorbital foramen;
18 = mylohyoid bridge (not shown);
19 = accessory mental foramina;
20 = epactal bone (ossicle at lambda);
21 = Inca bone;
22 = sagittal ossicle;
23 = palatine torus;
24 = pharyngeal fossa;
25 = superior sagittal sinus (not shown);
26 = frontotemporal articulation (not shown);
27 = metopic suture;
28 = bregmatic bone;
29 = mandibular torus (not shown);
30 = sternal foramen (not shown; postcranial);
31 = olecranon foramen (not shown; postcranial).

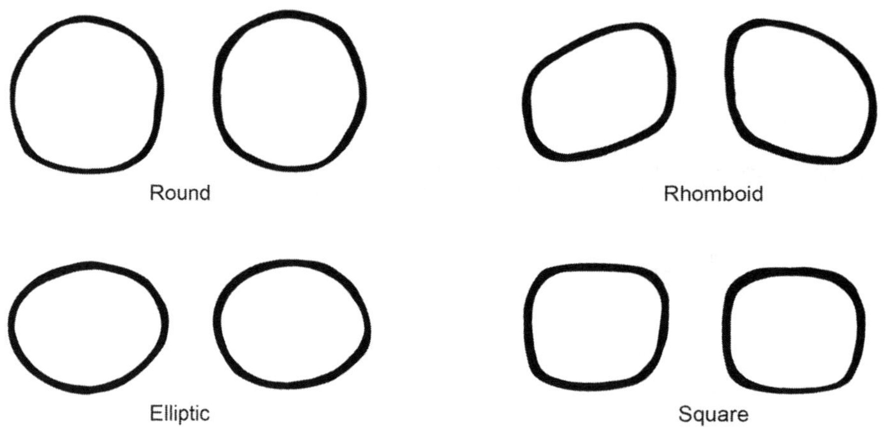

FIGURE B.2. Orbital forms defined by Olivier (1969) and utilized in this study.

APPENDIX C

Craniometric Data—Classic Period

George W. Gill

Within this appendix are two tables, C.1 and C.2, that contain cranial measurements and indices from the Classic period site of Venadillo Mound C, the only Classic period site within the Marismas Nacionales for which craniometric data exist. These have been reproduced from Gill's (1984) Tables II and III.

TABLE C.1. Craniometry of Venadillo Mound C Males and Females (in Millimeters).

	Males			Females			
Measurement	*N*	Mean	SD	*N*	Mean	SD	Range
Cranial							
Maximum frontal breadth	—	—	—	1	130.00	—	—
Minimum frontal breadth	—	—	—	1	93.00	—	—
Facial							
Nasion-prosthion	—	—	—	2	67.00	0.71	67–68
Nasal height	—	—	—	2	53.00	4.24	50–56
Nasal breadth	—	—	—	2	25.00	1.41	24–26
Left orbital height	—	—	—	2	33.50	0.71	33–34
Left orbital breadth	—	—	—	2	40.50	0.71	40–41
Biorbital breadth	—	—	—	1	98.00	—	—
Interorbital breadth	—	—	—	2	20.00	—	—
Cheek height	—	—	—	2	22.00	1.41	21–23
Palatal length	—	—	—	1	52.00	—	—
Palatal breadth	—	—	—	1	63.00	—	—
Mandibular							
Symphyseal height	—	—	—	3	31.00	3.61	27–34
Bigonial diameter	1	87.00	—	2	102.00	7.07	97–107
Bicondylar diameter	1	112.00	—	1	118.00	—	—
Ascending ramus breadth	1	34.00	—	3	32.33	2.89	29–34
Ascending ramus height	1	56.00	—	3	58.33	7.09	52–66
Corpal length	1	77.00	—	3	77.00	3.00	74–80
Gonial angle	1	31.00°	—	3	29.33°	9.24	24°–40°

Source: Gill 1984:Table II.

TABLE C.2. Craniofacial Indices of Venadillo Mound C Females.

Index	*N*	Mean	SD	Range
Nasal index	2	47.43	6.46	42.86–52.00
Orbital index	2	82.75	3.20	80.49–85.00

Source: Gill 1984:Table III.

APPENDIX D

Craniometric Data—Postclassic Period

George W. Gill

Within this appendix are Tables D.1–D.7, which contain means, standard deviations, and ranges for each of the key craniofacial measurements obtained in this study. Males and females are tabulated separately by site. All measurable adult specimens from the following Postclassic sites have been included: Tecualilla, Chalpa, Venadillo Mound B, Rincon de Panal, and Cristo Rey.

TABLE D.1. Craniometry of Tecualilla Males (in Millimeters).

	Unmodified				Modified			
Measurement	*N*	Mean	SD	Range	*N*	Mean	SD	Range
Cranial								
Cranial length	5	178.40	3.05	175–182	10	164.80	7.81	150–177
Cranial breadth	5	144.00	5.10	137–150	10	158.80	7.27	148–169
Basion-bregma	5	134.20	5.45	129–141	9	133.78	5.78	124–143
Basion-nasion	5	101.40	3.85	97–107	8	99.50	3.66	94–104
Basion-prosthion	4	103.00	3.64	99–108	8	98.38	5.50	91–105
Basion-gnathion	3	115.67	2.87	112–119	7	107.14	6.31	97–115
Maximum frontal breadth	5	115.60	3.65	113–122	9	126.89	6.29	119–136
Minimum frontal breadth	5	96.40	3.65	92–100	11	98.90	4.23	93–107
Biauricular breadth	4	127.00	2.74	123–130	8	134.50	5.34	127–145
Auricular height	3	122.00	0.82	121–123	7	121.00	4.31	114–128
Facial								
Nasion-prosthion	4	74.75	1.92	73–78	7	70.57	3.61	67–76
Nasion-gnathion	3	123.67	4.50	120–130	7	117.71	6.23	109–130
Bizygomatic breadth	4	137.75	1.92	135–140	8	144.50	6.52	135–156
Nasal height	5	52.80	0.40	52–53	6	51.83	2.04	49–55
Nasal breadth	5	26.60	2.94	23–32	7	25.86	1.73	24–28
Left orbital height	4	34.25	1.09	33–36	6	34.33	1.11	33–36
Left orbital breadth	4	43.50	2.29	41–47	6	43.33	1.80	41–46
Biorbital breadth	4	98.50	2.18	96–102	7	100.71	3.37	96–107
Interorbital breadth	4	17.50	0.87	17–19	7	18.71	0.88	18–20
Cheek height	5	27.00	1.67	25–30	10	25.10	2.43	23–29
Palatal length	5	57.60	2.58	53–61	9	56.22	3 33	50–60
Palatal breadth	5	69.00	2.90	65–72	9	70.11	3.21	64–75
Palatal depth	5	17.20	0.74	16–18	8	16.00	2.06	13–20
Porion-nasion	3	95.67	2.36	94–99	7	93.29	2.49	90–98
Porion-subnasale	3	98.67	1.70	97–101	7	96.14	3.90	91–102
Porion-prosthion	3	106.67	3.40	102–110	7	102.29	4.40	96–109
Porion-gnathion	3	123.33	2.50	120–126	7	117.86	7.53	105–128

Table D.1. (cont'd.) Craniometry of Tecualilla Males (in Millimeters).

	Unmodified				Modified			
Measurement	*N*	Mean	SD	Range	*N*	Mean	SD	Range
Mandibular								
Symphyseal height	5	38.00	3.16	33–40	10	36.30	3.16	31–41
Bigonial diameter	5	105.60	4.80	99–111	9	101.67	7.79	93–113
Bicondylar diameter	5	125.00	6.07	117–133	9	129.44	3.17	125–135
Ascending ramus breadth	5	35.60	3.93	30–42	10	34.00	2.86	29–38
Ascending ramus height	5	66.80	2.32	64–69	9	62.89	4.41	54–69
Corpal length	5	81.80	1.17	80–83	10	78.00	5.22	72–87
Gonial angle	5	26.80°	4.45	20°–33°	9	27.89°	3.78	23°–37°

Table D.2. Craniometry of Chalpa Males (in Millimeters).

Measurement	*N*	Mean	SD	Range
Cranial				
Cranial length	6	160.17	5.70	148–167
Cranial breadth	7	157.86	6.75	140–168
Basion-bregma	4	132.50	5.72	125–139
Basion-nasion	3	99.67	0.95	94–101
Basion-prosthion	3	98.33	4.11	93–103
Basion-gnathion	3	107.67	2.63	104–110
Maximum frontal breadth	5	124.60	5.95	115–132
Minimum frontal breadth	5	95.80	4.12	82–100
Biauricular breadth	4	131.25	3.83	127–136
Auricular height	3	117.67	4.19	112–122
Facial				
Nasion-prosthion	6	72.83	3.98	66–78
Nasion-gnathion	5	119.00	6.16	108–127
Bizygomatic breadth	2	134.00	2.00	132–136
Nasal height	8	53.13	2.03	49–56
Nasal breadth	9	25.22	0.63	24–26
Left orbital height	5	36.40	2.73	33–40
Left orbital breadth	5	41.80	1.72	40–45
Biorbital breadth	4	98.00	3.61	93–103
Interorbital breadth	5	19.20	1.47	18–22
Cheek height	9	24.44	1.34	23–27
Palatal length	6	55.33	2.69	52–59
Palatal breadth	6	65.00	3.46	58–68
Palatal depth	7	15.57	2.92	13–21
Porion-nasion	4	92.75	2.28	90–95
Porion-subnasale	3	94.67	0.48	94–95
Porion-prosthion	3	100.67	2.06	98–103
Porion-gnathion	3	118.33	4.64	112–123
Mandibular				
Symphyseal height	9	35.78	4.49	31–47
Bigonial diameter	7	96.86	5.03	92–105
Bicondylar diameter	5	126.80	4.26	122–132
Ascending ramus breadth	10	33.20	2.04	31–37
Ascending ramus height	8	58.63	4.27	49–63
Corpal length	7	77.86	3.48	73–85
Gonial angle	7	31.43°	8.31	20°–42°

Note: All male crania within this sample show evidence of artificial cranial deformation.

TABLE D.3. Craniometry of Tecualilla and Chalpa Females (in Millimeters).

	Tecualilla				Chalpa			
Measurement	*N*	Mean	SD	Range	*N*	Mean	SD	Range
Cranial								
Cranial length	6	158.50	7.50	145–173	14	158.50	8.37	143–171
Cranial breadth	6	156.33	10.37	144–172	13	156.31	8.95	142–179
Basion-bregma	5	130.00	8.79	117–140	13	130.85	5.14	118–138
Basion-nasion	4	93.50	1.80	91–96	13	96.23	3.38	92–103
Basion-prosthion	4	95.75	5.76	90–105	7	97.14	3.00	92–100
Basion-gnathion	4	102.25	5.49	95–110	7	106.86	3.18	105–111
Maximum frontal breadth	6	122.17	7.24	112–132	13	122.08	5.50	115–132
Minimum frontal breadth	6	96.50	3.45	91–102	14	93.79	5.05	83–104
Biauricular breadth	6	129.17	5.34	121–138	14	126.93	4.47	120–138
Auricular height	5	119.80	8.03	112–133	9	120.44	5.78	109–131
Facial								
Nasion-prosthion	6	67.50	2.99	63–71	9	67.22	3.16	62–72
Nasion-gnathion	6	109.33	5.74	101–116	8	112.25	2.28	108–115
Bizygomatic breadth	5	135.20	8.82	125–147	10	134.10	3.96	129–140
Nasal height	6	49.50	1.89	46–52	9	50.78	2.20	48–55
Nasal breadth	7	25.57	1.40	23–28	9	24.56	1.83	21–27
Left orbital height	7	33.71	1.83	31–37	10	35.10	1.30	33–37
Left orbital breadth	7	42.29	1.83	40–46	10	41.80	2.44	38–45
Biorbital breadth	6	98.00	1.15	96–99	9	98.33	4.08	91–105
Interorbital breadth	7	19.14	1.81	16–21	11	19.36	2.90	15–24
Cheek height	7	23.29	1.83	20–25	13	22.46	1.65	20–27
Palatal length	7	54.00	3.02	48–57	9	51.89	2.89	46–56
Palatal breadth	7	65.43	3.62	61–72	8	64.50	4.64	54–70
Palatal depth	5	12.80	1.47	11–15	7	13.86	1.55	12–17
Porion-nasion	5	90.00	2.76	87–95	12	88.58	3.55	82–96
Porion-subnasale	5	91.20	4.75	83–95	7	91.29	3.28	84–95
Porion-prosthion	5	97.40	4.13	91–103	7	97.71	4.43	88–103
Porion-gnathion	5	108.00	3.85	102–114	7	113.00	2.93	108–117
Mandibular								
Symphyseal height	8	31.25	2.33	29–35	13	31.23	3.80	23–36
Bigonial diameter	7	90.86	4.91	82–97	13	92.54	7.07	85–106
Bicondylar diameter	7	122.71	6.14	111–131	11	121.91	5.15	115–132
Ascending ramus breadth	8	33.88	1.76	31–36	14	30.93	1.79	26–34
Ascending ramus height	7	54.43	5.21	44–63	14	56.14	3.54	52–62
Corpal length	8	75.25	2.44	71–80	14	75.93	3.24	69–81
Gonial angle	8	29.75°	5.43	20°–37°	14	32.14°	5.68	20°–41°

Note: All crania within the two samples presented here show at least slight amounts of cranial deformation.

TABLE D.4. Craniometry of Venadillo Mound B Males (in Millimeters).

Measurement	N	Mean	SD	Range
Cranial				
Cranial length	10	167.40	9.16	150–185
Cranial breadth	10	158.70	7.32	145–167
Basion-bregma	8	129.00	4.84	121–136
Basion-nasion	7	97.29	2.93	94–101
Basion-prosthion	2	100.00	0.00	—
Basion-gnathion	4	113.50	3.70	110–118
Maximum frontal breadth	10	125.60	6.04	110–130
Minimum frontal breadth	10	95.20	5.90	88–104
Biauricular breadth	9	135.00	3.57	130–140
Auricular height	5	113.80	9.28	98–120
Facial				
Nasion-prosthion	4	76.75	2.50	74–80
Nasion-gnathion	4	124.25	5.25	121–132
Bizygomatic breadth	3	145.33	2.08	143–147
Nasal height	6	55.16	1.17	54–57
Nasal breadth	5	25.20	1.30	24–26
Left orbital height	6	38.66	1.57	37–40
Left orbital breadth	6	45.83	1.60	43–47
Biorbital breadth	5	104.60	2.41	102–108
Interorbital breadth	6	18.33	3.67	14–24
Cheek height	10	24.30	2.31	21–30
Palatal length	3	55.67	2.89	54–59
Palatal breadth	6	66.33	3.93	61–71
Palatal depth	5	13.60	2.07	11–16
Porion-nasion	10	92.30	3.80	85–97
Porion-subnasale	5	93.80	4.15	89–97
Porion-prosthion	3	101.67	2.52	99–102
Porion-gnathion	4	116.50	1.29	115–118
Mandibular				
Symphyseal height	10	37.10	3.96	31–42
Bigonial diameter	7	101.57	7.25	92–114
Bicondylar diameter	4	125.25	8.81	118–137
Ascending ramus breadth	11	34.00	3.29	30–41
Ascending ramus height	11	64.36	3.56	59–70
Corpal length	10	79.60	4.77	70–86
Gonial angle	11	27.27°	4.43	21°–35°

Note: One cranium within this series is without evidence of artificial cranial deformation. All others show signs of deformation or else could not be assessed.

Table D.5. Craniometry of Venadillo Mound B Females (in Millimeters).

Measurement	*N*	Mean	SD	Range
Cranial				
Cranial length	9	157.44	6.33	149–168
Cranial breadth	12	152.75	9.65	136–167
Basion-bregma	4	132.00	2.83	130–136
Basion-nasion	4	95.00	5.10	88–100
Basion-prosthion	3	100.67	5.13	95–105
Basion-gnathion	3	107.00	3.46	103–109
Maximum frontal breadth	10	122.60	8.26	113–135
Minimum frontal breadth	8	96.50	4.28	88–102
Biauricular breadth	9	128.78	8.30	113–137
Auricular height	6	113.83	4.45	109–119
Facial				
Nasion-prosthion	5	67.60	4.10	69–71
Nasion-gnathion	3	115.33	5.03	110–120
Bizygomatic breadth	3	134.00	8.54	126–143
Nasal height	5	45.80	2.17	43–49
Nasal breadth	4	25.50	1.00	24–26
Left orbital height	6	33.83	1.33	33–36
Left orbital breadth	5	41.60	0.55	41–42
Biorbital breadth	5	98.80	3.11	94–102
Interorbital breadth	5	18.40	1.67	17–21
Cheek height	8	22.63	2.20	18–25
Palatal length	3	56.67	4.04	52–59
Palatal breadth	4	64.50	6.56	55–70
Palatal depth	4	15.25	0.96	14–16
Porion-nasion	9	85.56	4.07	78–91
Porion-subnasale	3	91.00	3.46	87–93
Porion-prosthion	3	100.33	4.73	95–104
Porion-gnathion	3	110.33	5.69	104–115
Mandibular				
Symphyseal height	7	33.86	4.26	28–39
Bigonial diameter	10	99.40	5.56	91–110
Bicondylar diameter	6	126.17	7.31	116–137
Ascending ramus breadth	10	35.10	1.73	33–38
Ascending ramus height	10	58.58	2.37	54–61
Corpal length	11	76.73	3.07	71–80
Gonial angle	11	31.00°	6.62	19°–43°

Note: One skull within this sample shows no apparent signs of deformation, but all others are either modified or so poorly preserved as to prohibit definite assessment.

TABLE D.6. Craniometry of Rincon de Panal and Cristo Rey Males (in Millimeters).

	Rincon de Panal				Cristo Rey			
Measurement	*N*	Mean	SD	Range	*N*	Mean	SD	Range
Cranial								
Cranial length	4	162.75	4.03	157–166	5	164.20	5.93	157–173
Cranial breadth	3	159.33	6.66	155–167	5	160.40	7.92	153–172
Basion-bregma	3	132.00	7.55	124–139	5	135.60	4.77	128–141
Basion-nasion	3	96.00	1.00	95–97	4	96.50	5.69	93–105
Basion-prosthion	2	99.50	0.71	99–100	2	92.00	6.66	88–96
Basion-gnathion	2	117.00	2.83	115–119	—	—	—	—
Maximum frontal breadth	3	129.33	3.06	126–132	4	128.00	6.22	123–137
Minimum frontal breadth	4	95.00	1.63	93–97	5	94.60	3.51	89–98
Biauricular breadth	1	121.00	—	—	1	133.00	—	—
Auricular height	4	120.25	6.55	112–128	2	114.50	—	111–118
Facial								
Nasion-prosthion	3	72.00	2.65	70–75	4	65.50	1.73	64–68
Nasion-gnathion	3	121.67	5.51	118–128	1	110.00	—	—
Bizygomatic breadth	2	139.00	2.83	137–141	1	126.00	—	—
Nasal height	4	51.50	1.29	50–53	3	51.33	2.31	50–54
Nasal breadth	4	26.25	1.26	25–28	3	25.67	3.21	22–28
Left orbital height	4	35.75	1.50	34–37	3	35.33	1.53	34–37
Left orbital breadth	4	42.50	2.33	40–45	3	43.33	1.53	42–45
Biorbital breadth	4	100.25	2.36	97–102	3	96.33	3.51	93–100
Interorbital breadth	4	19.00	0.82	18–20	3	16.00	1.00	15–17
Cheek height	4	23.50	1.91	22–26	4	25.00	2.45	22–28
Palatal length	4	54.50	3.11	52–59	4	52.00	2.16	50–55
Palatal breadth	4	66.25	4.57	61–71	4	60.50	3.79	55–63
Palatal depth	2	16.00	0.00	—	—	—	—	—
Porion-nasion	4	86.75	2.36	85–90	3	85.00	1.73	84–87
Porion-subnasale	2	89.00	4.24	86–92	1	94.00	—	—
Porion-prosthion	3	97.33	2.31	96–100	2	92.50	4.95	89–96
Porion-gnathion	3	119.67	7.51	112–127	—	—	—	—
Mandibular								
Symphyseal height	3	36.33	2.08	34–38	9	34.33	3.28	30–41
Bigonial diameter	2	94.00	15.56	83–105	3	101.33	1.15	100–102
Bicondylar diameter	1	122.00	—	—	3	129.33	6.81	124–137
Ascending ramus breadth	3	34.67	3.06	32–38	8	35.00	2.33	33–39
Ascending ramus height	2	65.50	2.12	64–67	5	59.00	4.74	51–63
Corpal length	2	82.00	7.07	77–87	6	89.33	8.52	74–97
Gonial angle	2	29.00°	2.83	27°–31°	1	34.00°	—	—

Note: All crania in the samples presented here reveal at least some degree of cultural modification.

Table D.7. Craniometry of Rincon de Panal and Cristo Rey Females (in Millimeters).

	Rincon de Panal				Cristo Rey			
Measurement	*N*	Mean	SD	Range	*N*	Mean	SD	Range
Cranial								
Cranial length	2	157.00	7.07	152–162	2	172.50	7.50	165–180
Cranial breadth	2	156.00	5.66	152–160	2	146.00	0.00	—
Basion-bregma	1	126.00	—	—	1	132.00	—	—
Basion-prosthion	1	93.00	—	—	1	99.00	—	—
Maximum frontal breadth	1	126.00	—	—	1	16.00	—	—
Minimum frontal breadth	2	96.00	1.41	95–97	1	95.00	—	—
Biauricular breadth	1	127.00	—	—	—	—	—	—
Auricular height	1	115.00	—	—	—	—	—	—
Facial								
Nasion-prosthion	1	64.00	—	—	—	—	—	—
Nasion-gnathion	1	108.00	—	—	—	—	—	—
Bizygomatic breadth	1	134.00	—	—	—	—	—	—
Nasal height	1	45.00	—	—	—	—	—	—
Nasal breadth	1	24.00	—	—	—	—	—	—
Left orbital height	1	34.00	—	—	—	—	—	—
Left orbital breadth	1	41.00	—	—	—	—	—	—
Biorbital breadth	1	96.00	—	—	—	—	—	—
Interorbital breadth	1	19.00	—	—	—	—	—	—
Cheek height	1	23.00	—	—	1	22.50	—	—
Palatal length	1	52.00	—	—	1	55.00	—	—
Palatal breadth	1	62.00	—	—	1	65.00	—	—
Porion-nasion	1	88.00	—	—	—	—	—	—
Porion-subnasale	1	85.00	—	—	—	—	—	—
Porion-prosthion	1	92.00	—	—	—	—	—	—
Porion-gnathion	1	105.00	—	—	—	—	—	—
Mandibular								
Symphyseal height	2	33.00	4.24	30–26	3	31.00	2.82	30–32
Bigonial diameter	2	93.50	4.95	90–97	2	91.50	2.12	90–93
Bicondylar diameter	1	120.00	—	—	1	127.00	—	—
Ascending ramus breadth	2	32.50	2.13	31–34	4	34.00	1.53	32–35
Ascending ramus height	2	59.00	2.83	57–61	2	57.00	7.07	52–62
Corpal length	2	73.00	4.24	70–76	2	91.50	6.35	87–96
Gonial angle	2	30.50°	12.0	21°–40°	1	25.00°	—	—

Note: One of the two Cristo Rey skulls in the samples treated here appears to be unmodified. All other female crania from these sites show at least some degree of artificial modification.

APPENDIX E

Discrete Nonmetric Data—Classic Period

George W. Gill

Within this appendix (Table E.1) are the few nonmetric discrete traits that could be tabulated from a Classic period site within the Marismas Nacionales. The skeletal material from this site, Venadillo Mound C, has been reported earlier (Gill 1984).

TABLE E.1. Discrete Nonmetric Traits at Mound C, Venadillo.

	Males (n = 1)		Females (n = 4)	
Trait	Number	Percent	Number	Percent
Cranial				
Lambdoid ossicle—medial	—	—	1/1	100.0
Tympanic dehiscence	0/2	0.0	2/4	50.0
Double anterior condylar canal	0/1	0.0	0/4	0.0
Accessory lesser palatine foramen	—	—	1/3	33.3
Supraorbital foramen	—	—	1/4	25.0
Frontal foramen	1/1	100.0	0/4	0.0
Suture to infraorbital foramen	—	—	1/4	25.0
Accessory infraorbital foramen	—	—	1/3	33.3
Mylohyoid bridge	0/2	0.0	0/4	0.0
Accessory mental foramen	0/1	0.0	0/5	0.0
Epactal bone	—	—	0/1	0.0
Inca bone	—	—	0/1	0.0
Palatal torus	—	—	0/1	0.0
Pharyngeal fossa	—	—	0/1	0.0
Superior sagittal sinus—left	—	—	2/3	66.7
Metopic suture	0/1	0.0	0/3	0.0
Mandibular torus	0/2	0.0	0/6	0.0
Postcranial				
Perforated olecranon fossa	—	—	2/3	66.7

APPENDIX F

Discrete Nonmetric Data—Postclassic Period

George W. Gill

Tables F.1–F.4 within this appendix are tabulations of the discrete nonmetric traits found on adult and adolescent specimens from the sites of Tecualilla, Chalpa, Venadillo Mound B, Rincon de Panal, and Cristo Rey. The number of traits assessed may exceed the total number of individuals in those cases where paired traits are involved. That is, both the right and left sides have been included for each individual when possible.

Table F.1. Discrete Nonmetric Traits Among Males at Tecualilla, Chalpa, and Venadillo Mound B.

	Tecualilla (*n* = 19)		Chalpa (*n* = 15)		Venadillo Mound B (*n* = 13)	
Trait	Number	Percent	Number	Percent	Number	Percent
Cranial						
Lambdoid ossicle—medial	19/22	86.4	13/15	86.7	14/16	87.5
Lambdoid ossicle—lateral	20/26	76.9	12/12	100.0	14/16	87.5
Parietal foramen	10/26	38.5	11/14	78.6	4/17	23.5
Mastoid foramen	21/30	70.0	8/11	72.7	15/19	78.9
Mastoid foramen exsutural	16/30	53.3	6/11	54.5	10/19	52.6
Coronal ossicle	6/16	37.5	8/11	72.7	10/14	71.4
Epiteric bone	2/21	9.5	0/4	0.0	1/7	14.3
Mastoid sutural ossicle	7/24	29.2	4/8	50.0	3/10	30.0
Parietal notch bone	1/31	3.2	2/11	18.2	1/12	08.3
Asterion ossicle	7/31	22.6	3/9	33.3	0/13	00.0
Tympanic dehiscence	3/29	10.3	4/18	22.2	2/24	08.3
Double anterior condylar canal	7/29	24.1	2/11	18.2	3/17	17.6
Accessory lesser palatine foramen	17/23	73.9	2/7	28.6	4/8	50.0
Supraorbital foramen	23/31	74.2	15/22	55.6	10/23	43.5
Frontal foramen	9/33	27.3	1/20	5.0	1/19	05.3
Suture into infraorbital foramen	10/23	43.5	8/16	50.0	0/14	00.0
Accessory infraorbital foramen	3/21	14.3	0/15	0.0	0/8	00.0
Mylohyoid bridge	3/29	10.3	4/14	28.6	3/19	15.8
Accessory mental foramen	4/30	13.3	2/19	10.5	0/19	00.0
Epactal bone	5/12	41.7	4/8	50.0	2/8	25.0
Inca bone	2/17	11.8	1/8	12.5	3/9	33.3
Sagittal ossicle	4/11	36.4	3/4	75.0	5/8	62.5
Palatal torus	8/14	57.1	4/6	66.7	7/8	87.5
Pharyngeal fossa	3/14	21.4	2/5	40.0	1/10	10.0
Superior sagittal sinus—left	7/15	46.7	4/9	44.4	3/11	27.3

TABLE F.1. (cont'd.) Discrete Nonmetric Traits Among Males at Tecualilla, Chalpa, and Venadillo Mound B.

	Tecualilla (n = 19)		Chalpa (n = 15)		Venadillo Mound B (n = 13)	
Trait	Number	Percent	Number	Percent	Number	Percent
Frontotemporal articulation	1/21	4.8	0/3	0.0	0/16	00.0
Metopic suture	1/18	5.6	0/12	0.0	0/11	00.0
Bregmatic bone	0/13	0.0	0/8	0.0	0/10	00.0
Mandibular torus	0/30	0.0	3/20	15.0	0/22	00.0
Postcranial						
Perforated olecranon fossa	5/26	19.2	4/14	28.6	0/16	00.0
Perforated sternal blade	4/11	36.4	1/4	25.0	0/2	00.0

TABLE F.2. Discrete Nonmetric Traits Among Females at Tecualilla, Chalpa, and Venadillo Mound B.

	Tecualilla (n = 9)		Chalpa (n = 19)		Venadillo Mound B (n = 15)	
Trait	Number	Percent	Number	Percent	Number	Percent
Cranial						
Lambdoid ossicle—medial	12/13	92.3	9/24	79.2	15/18	83.3
Lambdoid ossicle—lateral	9/13	69.2	19/24	79.2	8/14	57.1
Parietal foramen	8/12	66.7	16/26	61.5	9/16	56.3
Mastoid foramen	5/9	55.6	12/26	46.2	10/15	66.7
Mastoid foramen exsutural	4/9	44.4	7/26	26.9	7/15	38.5
Coronal ossicle	3/13	23.1	7/23	21.9	5/13	38.5
Epiteric bone	0/7	0.0	1/20	5.0	0/17	0.0
Mastoid sutural ossicle	1/7	14.3	7/25	28.0	5/14	35.7
Parietal notch bone	0/12	0.0	0/25	0.0	2/19	10.5
Asterion ossicle	3/11	27.3	4/22	18.2	3/19	15.8
Tympanic dehiscence	1/15	6.7	15/33	45.5	0/23	0.0
Double anterior condylar canal	3/1	27.3	4/24	16.7	1/8	12.5
Accessory lesser palatine foramen	4/7	57.1	8/15	53.3	4/9	44.4
Supraorbital foramen	2/13	15.4	14/28	50.0	6/19	31.6
Frontal foramen	2/13	15.4	1/30	3.3	1/21	4.8
Suture to infraorbital foramen	4/10	40.0	7/23	30.4	7/12	58.3
Accessory infraorbital foramen	1/10	10.0	0/17	0.0	2/11	18.2
Mylohyoid bridge	0/15	0.0	1/31	3.2	2/21	9.5
Accessory mental foramen	0/15	0.0	2/29	6.9	0/23	0.0
Epactal bone	3/6	50.0	2/12	16.7	5/11	45.5
Inca bone	0/7	0.0	0/13	0.0	0/10	0.0
Sagittal ossicle	4/6	66.7	6/11	54.5	3/9	33.3
Palatal torus	5/7	71.4	6/11	54.5	2/5	40.0
Pharyngeal fossa	4/7	57.1	1/12	8.3	0/5	0.0
Superior sagittal sinus—left	4/6	66.7	3/16	18.8	2/11	18.2
Frontotemporal articulation	0/7	0.0	0/22	0.0	0/15	0.0
Metopic suture	1/7	14.3	0/16	0.0	0/10	0.0
Bregmatic bone	0/8	0.0	0/13	0.0	0/10	0.0
Mandibular torus	0/17	0.0	0/31	0.0	0/26	0.0
Postcranial						
Perforated olecranon fossa	13/17	76.5	24/32	75.0	6/13	46.6
Perforated sternal blade	1/4	25.0	0/4	0.0	0/3	0.0

TABLE F.3. Discrete Nonmetric Traits Among Males at Rincon de Panal and Cristo Rey.

	Rincon de Panal (n = 4)		Cristo Rey (n = 9)	
Trait	Number	Percent	Number	Percent
Cranial				
Lambdoid ossicle—medial	8/8	100.0	14/14	100.0
Lambdoid ossicle—lateral	6/6	100.0	9/9	100.0
Parietal foramen	2/4	50.0	8/11	72.7
Mastoid foramen	6/6	100.0	8/12	66.7
Mastoid foramen exsutural	5/6	83.3	8/8	100.0
Coronal ossicle	6/6	100.0	5/5	100.0
Epiteric bone	1/4	25.0	0/4	0.0
Mastoid sutural ossicle	2/4	50.0	4/8	50.0
Parietal notch bone	1/3	33.3	1/9	11.1
Asterion ossicle	1/3	33.3	3/9	33.3
Tympanic dehiscence	5/8	62.5	5/14	35.7
Double anterior condylar canal	1/6	16.7	1/10	10.0
Accessory lesser palatine foramen	3/6	50.0	6/8	75.0
Supraorbital foramen	5/7	71.4	6/11	54.5
Frontal foramen	4/8	50.0	2/10	20.0
Suture to infraorbital foramen	3/8	37.5	3/11	27.3
Accessory infraorbital foramen	2/8	25.0	4/9	44.4
Mylohyoid bridge	1/5	20.0	2/14	14.3
Accessory mental foramen	0/6	0.0	0/18	0.0
Epactal bone	1/4	25.0	5/6	83.3
Inca bone	0/4	0.0	0/6	0.0
Sagittal ossicle	2/2	100.0	2/4	50.0
Palatal torus	1/4	25.0	2/7	28.6
Pharyngeal fossa	3/3	100.0	1/5	20.0
Superior sagittal sinus—left	2/4	50.0	1/3	33.3
Frontotemporal articulation	0/5	0.0	0/8	0.0
Metopic suture	0/4	0.0	0/7	0.0
Bregmatic bone	0/4	0.0	0/4	0.0
Mandibular torus	0/6	0.0	0/18	0.0
Postcranial				
Perforated olecranon fossa	0/6	0.0	1/10	10.0
Perforated sternal blade	0/1	0.0	0/1	0.0

TABLE F.4. Discrete Nonmetric Traits Among Females at Rincon de Panal and Cristo Rey.

	Rincon de Panal (n = 2)		Cristo Rey (n = 5)	
Trait	Number	Percent	Number	Percent
Cranial				
Lambdoid ossicle—medial	3/4	75.0	1/1	100.0
Lambdoid ossicle—lateral	3/3	100.0	2/2	100.0
Parietal foramen	4/4	100.0	1/1	100.0
Mastoid foramen	1/1	100.0	3/3	100.0
Mastoid foramen exsutural	1/1	100.0	1/3	33.3
Coronal ossicle	3/3	100.0	2/2	100.0
Epiteric bone	1/2	50.0	0/2	0.0
Mastoid sutural ossicle	1/3	33.3	—	—
Parietal notch bone	2/3	66.7	—	—
Asterion ossicle	3/3	100.0	—	0.0
Tympanic dehiscence	0/3	0.0	0/8	0.0
Double anterior condylar canal	0/2	0.0	0/5	0.0
Accessory lesser palatine foramen	0/2	0.0	3/4	75.0
Supraorbital foramen	1/4	25.0	6/6	100.0
Frontal foramen	1/4	25.0	5/7	71.4
Suture to infraorbital foramen	0/2	0.0	2/3	66.7
Accessory infraorbital foramen	0/2	0.0	0/3	0.0
Mylohyoid bridge	3/4	75.0	0/7	0.0
Accessory mental foramen	0/4	0.0	1/8	12.5
Epactal bone	0/2	0.0	—	—
Inca bone	0/2	0.0	—	—
Sagittal ossicle	2/2	100.0	—	—
Palatal torus	0/1	0.0	0/3	0.0
Pharyngeal fossa	0/1	0.0	0/3	0.0
Superior sagittal sinus—left	1/2	50.0	0/3	0.0
Frontotemporal articulation	0/3	0.0	0/2	0.0
Metopic suture	0/2	0.0	0/3	0.0
Bregmatic bone	0/2	0.0	0/1	0.0
Mandibular torus	0/4	0.0	0/9	0.0
Postcranial				
Perforated olecranon fossa	1/3	33.3	3/6	50.0
Perforated sternal blade	—	—	0/2	0.0

APPENDIX G

Dental Metrics

Henry W. Case

The four tables, G.1–G.4, within this appendix reflect the means for mesiodistal and buccolingual crown diameters for all adult dentitions within the Marismas Nacionales. Metrics are listed by individual site group—Panales, Tecualilla, Chalpa, Venadillo Mound B, Rincon de Panal, and Cristo Rey. Sexes are combined for each tooth category for all Marismas sites.

TABLE G.1. Mesiodistal Crown Diameter of Maxillary Teeth (in Millimeters) by Site (Sexes Combined).

Tooth and Site	*N*	Mean	SD	Range	Tooth and Site	*N*	Mean	SD	Range
I1					P2				
PAN	2	8.05	—	7.1–9.0	PAN	4	7.28	—	6.8–7.6
TEC	8	8.54	0.90	7.3–10.0	TEC	7	6.91	0.49	6.2–7.6
CHA	9	8.96	0.54	8.1–10.0	CHA	16	6.94	0.30	6.4–7.5
VNB	7	9.14	0.42	8.6–9.7	VNB	11	7.04	0.46	6.3–7.7
RNC	2	8.45	—	8.0–8.9	RNC	5	6.76	0.46	6.2–7.2
CSR	2	8.25	—	8.1–8.4	CSR	6	6.58	0.59	5.9–7.6
I2					M1				
PAN	2	6.95	—	6.5–7.4	PAN	3	10.20	—	9.9–10.8
TEC	5	6.60	0.50	6.2–7.4	TEC	11	10.81	0.69	9.7–11.9
CHA	11	7.24	0.75	6.4–8.8	CHA	18	10.74	0.62	9.9–12.1
VNB	8	7.40	0.47	6.8–8.2	VNB	13	10.73	0.46	10.0–11.3
RNC	2	7.75	—	7.7–7.8	RNC	5	10.00	0.35	9.7–10.5
CSR	4	7.00	—	6.1–7.4	CSR	4	10.10	—	9.7–10.8
C					M2				
PAN	3	8.33	—	8.1–8.6	PAN	3	10.20	—	10.0–10.4
TEC	10	8.05	0.51	7.5–9.2	TEC	11	10.16	0.50	9.2–10.9
CHA	16	8.15	0.36	7.5–8.9	CHA	20	10.17	0.60	9.3–11.1
VNB	9	7.99	0.27	7.6–8.5	VNB	14	10.31	0.48	9.4–11.3
RNC	3	7.97	—	7.3–8.6	RNC	5	9.56	0.65	8.7–10.2
CSR	4	7.80	—	7.4–8.5	CSR	6	9.63	0.64	8.7–10.4
P1					M3				
PAN	4	7.20	—	6.8–7.5	PAN	2	9.75	—	9.7–9.8
TEC	9	7.18	0.40	6.6–7.8	TEC	7	9.50	0.76	8.3–10.5
CHA	13	7.19	0.40	6.4–7.8	CHA	12	9.56	0.97	8.2–11.1
VNB	10	7.16	0.38	6.4–7.5	VNB	12	9.01	0.79	7.0–9.9
RNC	4	7.10	—	6.7–7.7	RNC	4	8.78	—	7.2–10.8
CSR	4	7.22	—	6.5–8.2	CSR	5	9.94	0.20	9.8–10.2

Note: PAN = Panales; TEC = Tecualilla; CHA = Chalpa; VNB = Venadillo Mound B; RNC = Rincon de Panal; CSR = Cristo Rey.

TABLE G.2. Mesiodistal Crown Diameter of Mandibular Teeth (in Millimeters) by Site (Sexes Combined).

Tooth and Site	*N*	Mean	SD	Range	Tooth and Site	*N*	Mean	SD	Range
I1					P2				
PAN	1	5.10	—	—	PAN	4	7.20	—	6.9–7.6
TEC	6	4.97	0.54	4.5–5.7	TEC	7	6.89	0.60	6.4–8.2
CHA	13	5.80	0.92	4.6–6.4	CHA	20	7.02	0.60	6.0–8.3
VNB	6	5.50	0.37	5.0–5.9	VNB	8	6.84	0.24	6.4–7.3
RNC	2	5.40	—	5.3–5.5	RNC	3	6.57	—	6.4–6.8
CSR	4	5.32	—	5.1–5.8	CSR	9	6.78	0.44	6.2–7.7
I2					M1				
PAN	1	5.70	—	—	PAN	4	11.65	—	10.7–12.5
TEC	8	6.11	0.39	5.6–6.8	TEC	11	11.18	0.69	10.3–12.7
CHA	16	6.29	0.55	5.5–7.4	CHA	22	11.50	0.62	10.6–13.1
VNB	8	6.34	0.41	5.7–6.8	VNB	8	11.52	0.41	10.7–12.0
RNC	3	5.73	—	5.2–6.1	RNC	2	11.25	—	11.1–11.4
CSR	5	5.62	0.29	5.2–6.0	CSR	7	10.97	0.49	10.2–11.6
C					M2				
PAN	4	7.12	—	6.3–7.6	PAN	4	11.28	—	10.5–12.3
TEC	11	7.07	0.58	6.3–8.0	TEC	12	10.63	0.64	9.7–11.9
CHA	19	7.11	0.40	6.4–7.7	CHA	22	10.66	0.73	9.3–11.8
VNB	10	7.05	0.28	6.6–7.0	VNB	12	10.85	0.61	9.9–11.8
RNC	3	7.07	—	6.8–7.3	RNC	3	9.97	—	9.7–10.4
CSR	5	6.62	0.39	6.0–7.0	CSR	10	10.79	0.7	9.7–11.7
P1					M3				
PAN	4	6.92	—	6.5–7.3	PAN	1	11.90	—	—
TEC	10	6.83	0.51	6.1–7.8	TEC	10	10.56	1.08	8.1–12.1
CHA	17	6.92	0.40	6.1–7.5	CHA	14	11.04	0.98	9.0–13.2
VNB	6	6.75	0.55	6.1–7.3	VNB	13	10.73	0.88	8.7–11.8
RNC	3	6.63	—	6.4–6.8	RNC	3	10.40	—	9.4–11.0
CSR	6	6.70	0.24	6.4–7.1	CSR	8	11.21	0.62	10.5–12.3

Note: PAN = Panales; TEC = Tecualilla; CHA = Chalpa; VNB = Venadillo Mound B; RNC = Rincon de Panal; CSR = Cristo Rey.

TABLE G.3. Buccolingual Crown Diameter of Maxillary Teeth (in Millimeters) by Site (Sexes Combined).

Tooth and Site	*N*	Mean	SD	Range
I1				
PAN	2	7.85	—	7.6–8.1
TEC	8	7.35	0.42	6.6–7.8
CHA	11	7.24	0.53	6.7–8.0
VNB	7	7.11	0.24	6.9–7.6
RNC	2	7.10	—	6.8–7.4
CSR	2	7.30	—	7.2–7.4
I2				
PAN	2	6.90	—	6.4–7.4
TEC	5	6.22	0.75	5.5–7.2
CHA	12	6.43	0.42	5.8–7.1
VNB	8	6.35	0.34	6.0–7.1
RNC	2	6.45	—	5.8–7.1
CSR	4	6.58	—	6.0–7.0
C				
PAN	3	8.33	—	8.1–9.6
TEC	10	8.05	0.51	7.8–9.2
CHA	15	8.55	0.37	8.3–9.2
VNB	11	7.99	0.27	7.8–9.1
RNC	3	7.97	—	8.1–9.6
CSR	3	7.78	—	8.3–9.2
P1				
PAN	4	9.38	—	9.0–10.2
TEC	9	9.27	0.44	8.6–10.0
CHA	15	9.31	0.46	8.2–10.0
VNB	10	9.05	0.53	8.3–10.1
RNC	4	9.35	—	8.5–10.2
CSR	4	9.25	—	8.8–9.8

Tooth and Site	*N*	Mean	SD	Range
P2				
PAN	3	9.47	—	9.2–9.7
TEC	8	9.19	0.62	8.4–10.5
CHA	16	9.20	0.37	8.5–9.7
VNB	11	9.14	0.47	8.1–9.8
RNC	5	9.42	0.69	8.5–10.1
CSR	6	8.77	0.48	8.3–9.3
M1				
PAN	3	11.27	—	11.0–11.7
TEC	11	11.54	0.45	10.7–12.3
CHA	19	11.48	0.60	10.4–12.3
VNB	13	11.32	0.31	10.6–11.8
RNC	5	11.32	0.76	10.1–12.0
CSR	4	11.22	—	11.1–11.4
M2				
PAN	3	10.47	—	9.5–11.4
TEC	11	11.13	0.58	10.2–12.4
CHA	20	10.84	0.83	9.0–12.3
VNB	15	10.47	0.60	9.3–11.5
RNC	5	11.08	1.18	10.1–13.0
CSR	5	10.94	0.43	10.5–11.5
M3				
PAN	2	10.05	—	9.6–10.5
TEC	7	10.33	0.89	9.1–11.5
CHA	12	10.17	1.06	8.1–11.6
VNB	13	10.07	0.84	8.8–11.5
RNC	4	10.08	—	7.2–12.5
CSR	5	10.38	0.94	9.1–11.6

Note: PAN = Panales; TEC = Tecualilla; CHA = Chalpa; VNB = Venadillo Mound B; RNC = Rincon de Panal; CSR = Cristo Rey.

TABLE G.4. Buccolingual Crown Diameter of Mandibular Teeth (in Millimeters) by Site (Sexes Combined).

Tooth and Site	N	Mean	SD	Range	Tooth and Site	N	Mean	SD	Range
I1					P2				
PAN	1	6.20	—	—	PAN	4	8.20	—	8.0–8.6
TEC	6	5.65	0.32	5.3–6.1	TEC	8	8.30	0.45	7.3–8.7
CHA	14	5.96	0.63	5.3–7.3	CHA	20	8.05	0.41	7.4–8.8
VNB	5	5.96	0.11	5.8–6.1	VNB	8	7.96	0.29	7.5–8.3
RNC	2	5.25	—	5.2–5.3	RNC	3	7.87	—	7.4–8.1
CSR	4	5.50	—	5.1–5.9	CSR	9	7.71	0.68	6.8–8.7
I2					M1				
PAN	1	6.60	—	—	PAN	4	10.88	—	10.5–11.2
TEC	8	6.23	0.41	5.6–6.7	TEC	11	10.83	0.55	10.2–11.8
CHA	17	6.22	0.35	5.7–7.1	CHA	23	10.81	0.57	9.7–12.1
VNB	9	6.02	0.39	5.4–6.5	VNB	8	10.72	0.40	10.4–11.6
RNC	3	6.03	—	5.7–6.2	RNC	3	10.30	—	9.6–10.7
CSR	5	5.86	0.36	5.3–6.2	CSR	7	10.11	0.30	9.7–10.5
C					M2				
PAN	3	8.20	—	8.2–8.2	PAN	4	10.30	—	9.7–11.1
TEC	10	8.01	0.43	7.5–8.9	TEC	13	10.26	0.48	9.2–11.0
CHA	20	7.68	0.20	7.3–8.2	CHA	22	10.30	0.57	8.9–11.3
VNB	11	7.53	0.67	6.0–8.3	VNB	12	10.35	0.44	9.7–11.3
RNC	3	7.83	—	7.3–8.8	RNC	3	10.03	—	9.5–10.5
CSR	5	7.74	0.38	7.3–8.2	CSR	10	10.00	0.63	9.2–11.4
P1					M3				
PAN	4	8.12	—	7.7–8.8	PAN	1	10.30	—	—
TEC	11	7.89	0.33	7.2–8.3	TEC	10	10.23	0.65	9.0–11.2
CHA	18	7.71	0.38	7.0–8.3	CHA	14	10.39	0.62	9.3–11.2
VNB	7	7.67	0.38	7.1–8.1	VNB	11	10.17	0.26	9.6–10.5
RNC	3	7.20	—	7.1–7.4	RNC	3	9.93	—	9.7–10.3
CSR	6	7.27	0.34	6.8–7.6	CSR	8	10.25	0.53	9.7–11.2

Note: PAN = Panales; TEC = Tecualilla; CHA = Chalpa; VNB = Venadillo Mound B; RNC = Rincon de Panal; CSR = Cristo Rey.

APPENDIX H

Penrose Comparisons—Postclassic Period Craniometrics and Odontometrics

Henry W. Case

Within this appendix are two tables, H.1 and H.2, that present Penrose's distance (CH^2), shape (CZ^2), and size (CQ^2) comparisons for the various Postclassic populations based upon both craniometrics (15 variables) and odontometrics (16 variables).

TABLE H.1. Penrose's Distance (CH^2), Shape (CZ^2), and Size (CQ^2) Comparisons for Postclassic Males and Females.

Sex and Site Pair	CH^2	CZ^2	CQ^2
Males			
Chalpa-Tecualilla	0.65183	0.49270	0.19198
Chalpa–Rincon de Panal	0.69850	0.68999	0.05451
Tecualilla–Venadillo Mound B	0.70332	0.70543	0.04445
Chalpa–Venadillo Mound B	0.79665	0.52101	0.31037
Chalpa–Cristo Rey	0.81971	0.70641	0.33612
Tecualilla–Rincon de Panal	0.98093	0.69087	0.33612
Cristo Rey–Rincon de Panal	1.02954	1.07319	0.02790
Tecualilla–Cristo Rey	1.29044	0.78509	0.55769
Rincon de Panal–Venadillo Mound B	1.33978	0.76581	0.62502
Cristo Rey–Venadillo Mound B	1.87711	1.02867	0.91702
Females			
Tecualilla–Venadillo Mound B	0.26636	0.27159	0.01288
Chalpa-Tecualilla	0.282557	0.26084	0.04212
Chalpa–Venadillo Mound B	0.57269	0.60458	0.00341
Chalpa–Rincon de Panal	1.0730	0.93462	0.20063
Tecualilla–Rincon de Panal	1.2103	0.83966	0.42661
Rincon de Panal–Venadillo Mound B	1.2211	0.9963	0.29122

Note: Comparisons based upon 15 craniometric variables.

Table H.2. Penrose's Distance (CH^2), Shape (CZ^2), and Size (CQ^2) Comparisons for Marismas Nacionales Sites.

	$CZ^2 \times m - 1$		$CQ^2 \times 100$	
Site Pair	CH^2	m	CQ^2	CH^2
Chalpa-Tecualilla	0.05784	0.05773	0.00012	0.20
Chalpa-Venadillo	0.07426	0.05888	0.01548	20.82
Tecualilla-Venadillo	0.15420	0.13594	0.01826	11.84
Chalpa-Panales	0.19369	0.16566	0.02802	14.47
Panales-Venadillo	0.22069	0.13552	0.08516	38.59
Cristo Rey–Rincon de Panal	0.23009	0.22832	0.00178	0.77
Tecualilla-Panales	0.27090	0.25455	0.02455	8.80
Cristo Rey–Venadillo	0.48694	0.30829	0.17866	36.69
Chalpa–Rincon de Panal	0.49036	0.23537	0.25499	52.00
Chalpa–Cristo Rey	0.49280	0.19348	0.29931	60.74
Tecualilla–Rincon de Panal	0.53407	0.26813	0.26594	49.79
Rincon de Panal–Venadillo	0.54210	0.39729	0.14481	26.71
Tecualilla–Cristo Rey	0.54769	0.23652	0.31117	56.81
Cristo Rey–Panales	0.88306	0.37254	0.51051	57.81
Panales–Rincon de Panal	0.91777	0.46568	0.45208	49.26

Note: Male and female dimensions are combined. Sixteen odontometric variables are used in the calculations: the mesiodistal and buccolingual crown diameters of upper and lower P1, P2, M1, and M2.

APPENDIX I

Principal Components Results

Henry W. Case

This appendix presents principal component scores for individual skeletons from the Marismas Nacionales. Figure I.1 shows values for males, and Figure I.2, for females.

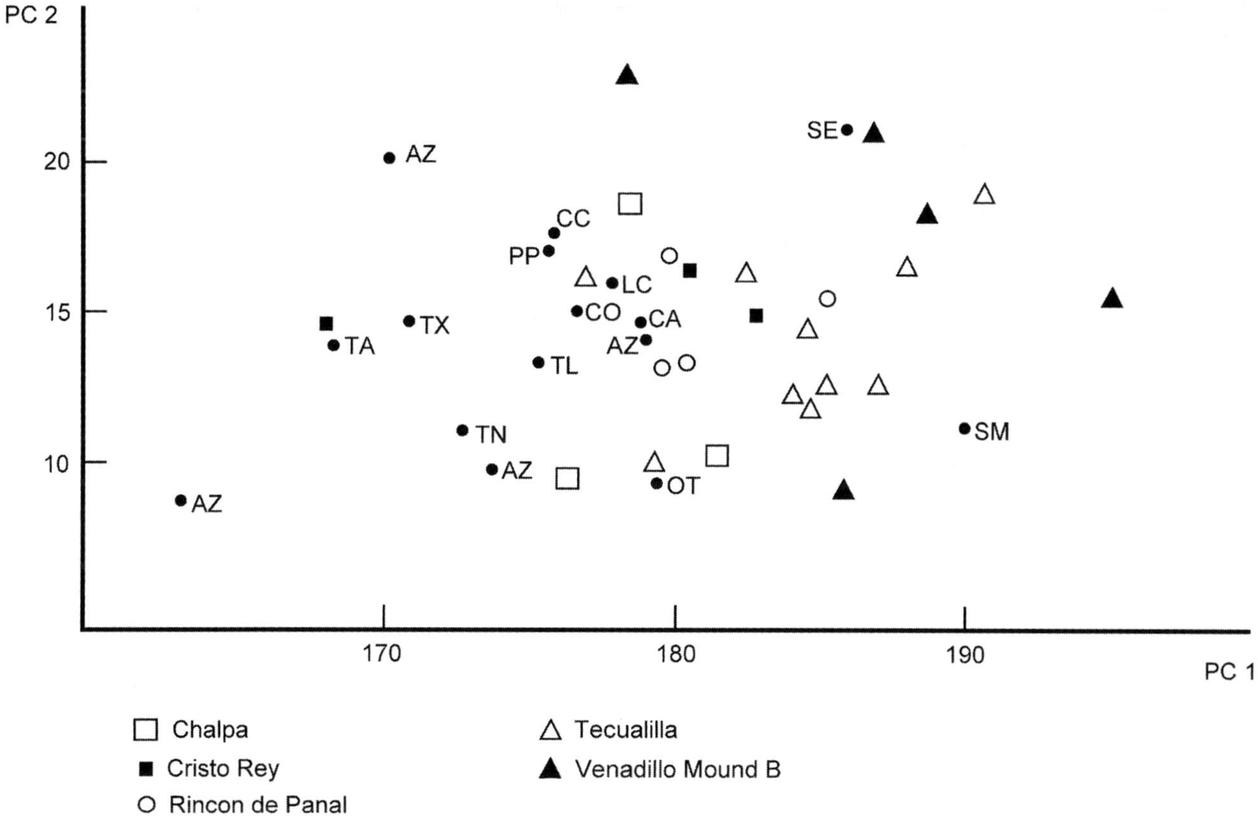

FIGURE I.1. First two principal component (PC) scores for males. AZ = Aztec; CA = Campeche; CC = Coahuila Caves; CO = Coahuila; LC = La Candelair; OT = Otomi; PP = Pecos Pueblo (Hooton); SE = Seri; SM = Santa Maria A. (Romano); TA = Tarahumara; TL = Tlatelolco; TN = Teparec; TX = Tlaxaltec.

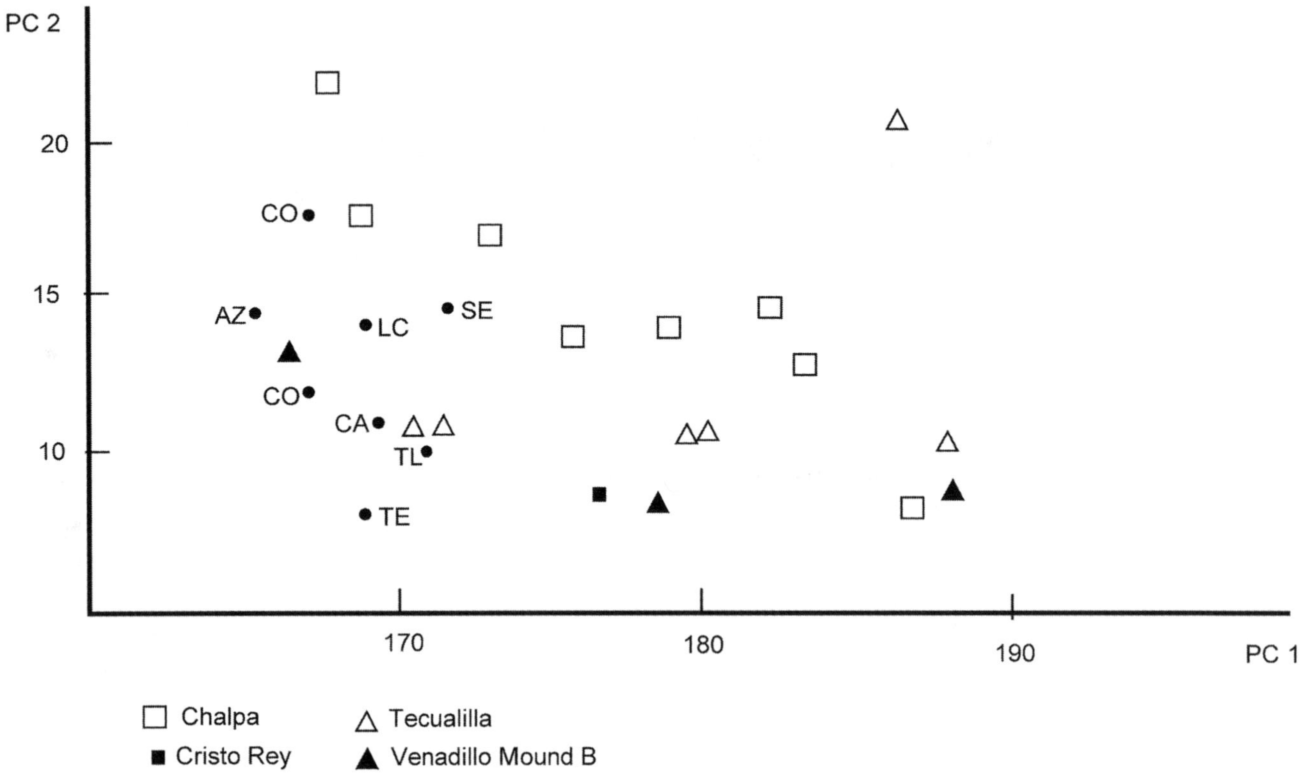

FIGURE I.2. First two principal component (PC) scores for females. AZ = Aztec; CA = Campeche; CO = Coahuila; LC = La Candelair; SE = Seri; TE = Tepanecas; TL = Tlatelolco.

References Cited

Abbott, Isabella A., and Eleanor Horswill Williamson

1974 *Límu: An Ethnobotanical Study of Some Edible Hawaiian Seaweeds*. Pacific Tropical Botanical Garden, Kauai.

Acosta, Jorge R.

1940 Exploraciónes en Tula, Hidalgo, 1940. *Revista Mexicana de Estudios Antropológicos* 4:172–194. Mexico City.

Alcedo, Antonio de

1788 *Diccionario geográfico-histórico de las Indias Occidentales ó América: Es á saber: De los reynos del Perú, Nueva España, Tierra Firme, Chile, y Nuevo reyno de Granada*, Vol. 4. Madrid.

Alducin Hidalgo y Terán, Rafael

1995 *Informe técnico del rescate de urnas funerarias en Chametla, Estado de Sinaloa*. Photocopy, technical report. Coordinación Nacional de Arqueología del Instituto Nacional de Antropología e Historia, Mazatlán.

1997 Las Urnas funerarias. Una tradición prehispánica del occidente de México. *Boletín Informativa del Centro INAH Sinaloa*, July: 6–7.

Allen, Glover M.

1920 Dogs of the American Aborigines. *Bulletin of the Museum of Comparative Zoology* 63:431–517. Harvard University, Cambridge.

Anderies, John M., Ben A. Nelson, and Ann P. Kinzig

2008 Analyzing the Impact of Agave Cultivation on Famine Risk in Arid Prehispanic Northern Mexico. *Human Ecology* 36:409–422.

Anderson, Arthur J. O., and Charles E. Dibble (editors and translators)

1952 *Florentine Codex: General History of the Things of New Spain, Bk. 3: The Origin of the Gods*, by Fray Bernardino de Sahagún. Monograph of the School of American Research, No. 14, Pt. 4. Santa Fe, New Mexico.

Anguiano, Marina

1976 Nayarit: Costa y Altiplanicie en el Momento del Contacto, Ecologia y Fuentes. Unpublished Master's thesis, Universidad Nacional Autónoma de México, Mexico City.

1980 La Costa y la Altiplanicie en el Momento del Contacto. In *Rutas de Intercambio en Mesoamerica y el Norte de México, XVI Reunion de Mesa Redonda, Saltillo, 1979*, Vol. II:233–240. Sociedad Mexicana de Antropología, Mexico City.

Arias de Saavedra, Antonio

1975 [1899] *Información acerca del estado de Nayarit*. In *Nayarit: Colección de Documentos Inéditos, Históricos y Etnográficos, Acerca de la Sierra de ese Nombre*, compiled by Alberto Santoscoy. Guadalajara, Mexico.

Armstrong, David M., and J. Knox Jones, Jr.

1971 Mammals from the Mexican State of Sinaloa. I. Marsupialia, Insectivora, Edentata, Lagomorpha. *Journal of Mammalogy* 52(4):747–757.

Armstrong, David M., J. Knox Jones, Jr., and Elmer C. Birney

1972 Mammals from the Mexican State of Sinaloa. III. Carnivora and Artiodactyla. *Journal of Mammalogy* 53(1):48–61.

Arnold, Dean

2011 Ceramic Theory and Cultural Process. *Ethnoarchaeology* 3(1):63–98.

Arregui, Domingo Lázaro de

1621 *Descripción de la Nueva Galicia*. Manuscript produced for Presidente de la Audiencia, Obispado de Durango.

1946 *Descripción de la Nueva Galicia*. Edited by François Chevalíer. Vol. XXIV, Serie 3a, No. 3. Escuela de Estudios Hispano-Americanos de la Universidad de Sevilla, Seville.

Arriaza, Bernardo T., Wilmar L. Salo, Arthur C. Aufderheide, and Todd A. Holcomb

1995 Pre-Columbian Tuberculosis in Northern Chile: Molecular and Skeletal Evidence. *American Journal of Physical Anthropology* 98:37–45.

Baker, Howard, John Cammarata, Jean Clark, and Edward Weil

1972 Excavations at Cristo Rey. In *Archaeological Reconnaissance and Excavations in the Marismas Nacionales, Sinaloa and Nayarit, Mexico. West Mexican Prehistory*, edited by Stuart D. Scott, Pt. 6:64–76. Mimeograph, Department of Anthropology, State University of New York at Buffalo.

Bancroft, Hubert H.

1883 *The Native Races, Wild Tribes*, Vol. I. A. L. Bancroft and Company, San Francisco.

Bass, William M.

2005 *Human Osteology: A Laboratory Manual*. 5th ed. Missouri Archaeological Society, Columbia.

Bass, William M., and Jon C. Barlow

1964 A Human Skeleton from the Pryor Creek Burial, 24YL404, Yellowstone County, Montana. *Plains Anthropologist* 9(23):29–36.

Bass, William M., III, and Donald C. Lacy

1963 Three Human Skeletons from the PK Burial, Sheridan County, Wyoming. *Plains Anthropologist* 8(21):142–157.

Beals, Ralph L.
1932 *The Comparative Ethnology of Northern Mexico Before 1750*. Ibero-Americana 2. University of California Press, Berkeley.
1945 *The Contemporary Culture of the Cahita Indians*. Smithsonian Institution, Bureau of American Ethnology, Bulletin 142. Washington, D.C.

Beekman, Christopher S.
2010 Recent Research in Western Mexican Archaeology. *Journal of Archaeological Research* 18:41–109.

Bell, Betty B.
1958 *The Decorated Pottery of Peñitas*. Manuscript on file, Department of Anthropology and Sociology, University of California, Los Angeles.
1960 Analysis of Ceramic Style: A West Mexican Collection. Unpublished Ph.D. dissertation, Department of Anthropology, University of California at Los Angeles.
1971 Archaeology of Nayarit, Jalisco, and Colima. In *Archaeology of Northern Mesoamerica, Pt. 2*, edited by Gordon F. Ekholm and Ignacio Bernal, pp. 694–753. Handbook of Middle American Indians, Vol. 11, Robert Wauchope, general editor, University of Texas Press, Austin.

Bell, Betty B. (editor)
1974 *The Archaeology of West Mexico*. West Mexican Society for Advanced Study, Ajijic, Mexico.

Beltrán Medina, Jose Carlos
2001 *La explotación de la costa en el occidente de Mesoamérica y los contactoscon Sudamérica y con otras regiones culturales*. Universidad Autónoma de Nayarit, Tepic.

Benfer, Robert A.
1968 An Analysis of a Prehistoric Skeletal Population, Casas Grandes, Chihuahua, Mexico. Unpublished Ph.D. dissertation, Department of Anthropology, University of Texas, Austin.

Bennett, C. F.
1959 The Procurement and Utilization of Native Animals by Three Different Culture Groups in the Republic of Panama: A Study of an Aspect of Tropical Biogeography. Unpublished Ph.D. dissertation, Department of Geology, University of California, Los Angeles.

Bernor, Raymond L.
1976 Appendix B. Human Skeletal Remains from Amapa. In *The Archaeology of Amapa, Nayarit*, edited by Clement W. Meighan, pp. 182–186. Monumenta Archaeologica 2. Institute of Archaeology, University of California, Los Angeles.

Binford, Lewis R.
1971 Mortuary Practices: Their Study and Their Potential. In *Approaches to the Social Dimensions of Mortuary Practices*, edited by James A. Brown, pp. 6–29. Memoirs 25. Society for American Archaeology, Washington, D.C.

Bird, Douglas W., and Rebecca B. Bird
2000 The Ethnoarchaeology of Juvenile Foragers: Shellfishing Strategies Among Meriam Children. *Journal of Anthropological Archaeology* 19:461–476.

Birkby, Walter H., and William M. Bass
1963 Possible Shoshonean Skeletal Material from the Turk Burial Site, 48WA301. *Plains Anthropologist* 8(20):103–114.

Blake, Michael, John E. Clark, Brian S. Chisholm, and Karen Mudar
1992 Non-Agricultural Staples and Agricultural Supplements: Early Formative Subsistence in the Soconusco Region, Mexico. In *Transitions to Agriculture in Prehistory*, edited by Anne B. Gebauer and T. D. Price, pp. 133–151. Monographs in World Prehistory, No. 4. Prehistory Press, Madison, Wisconsin.

Blake, Michael, John E. Clark, Barbara Voorhies, George Michaels, Michael W. Love, Mary E. Pye, Arthur A. Demarest, and Barbara Arroyo
1995 Radiocarbon Chronology for the Late Archaic and Formative Periods on the Pacific Coast of Southeastern Mesoamerica. *Ancient Mesoamerica* 6:161–183.

Blanco y Correa, José Manuel (editor)
2011 *Diagnóstico Funcional de Marismas Nacisonales: Informe final de los convenios de coordinación entre le Universidad Autónoma de Nayarit y la Comisión Nacional Forestal, con el patricinio del Gobierno del Reino Unido*. Universidad Autónoma de Nayarit, Tepic.

Blanton, Richard E., Jill Appel, Laura Finsten, Steve Kowalewski, Gary Feinman, and Eva Fisch
1979 Regional Evolution in the Valley of Oaxaca. *Journal of Field Archaeology* 6(4):369–390.

Block, Terry
1968 Juana Gomez: Gully Squares. In *Archaeological Reconnaissance and Excavations in the Marismas Nacionales, Sinaloa and Nayarit, Mexico. West Mexican Prehistory*, edited by Stuart D. Scott, Pt. 2:22–23. Mimeograph, Department of Anthropology, State University of New York at Buffalo.

Boicourt, William C.
1993 Where the River Meets the Sea. *Oceanus* 36(2):29–37.

Bojórquez Diego, V. L.
2009 El Complejo Ritual Aztatlán del Occidente: Interacción e Innovación Durante el Posclásico Temprano. Unpublished Master's thesis, Archaeology, Escuela Nacional de Antropología e Historia, Mexico City.

Bojórquez-Tapia, Luis, S. Díaz-Mondragón, and R. Saunier
1997 *Ordenamiento Ecológico de la Costa Norte de Nayarit*. Organización de los Estados Americanosz–Instituto de Ecología de la Universidad Nacional Autónoma de México, Mexico City.

Bordaz, Jacques
1964 Precolumbian Ceramic Kilns at Peñitas, a Postclassic Site in Coastal Nayarit, Mexico. Unpublished Ph.D. dissertation, Department of Anthropology, Columbia University, New York.

Bordes, Luis Topete
1944 *Jalisco Precortesiano: Estudio Histórico y Etnogénico*. El Sobre Azul, Mexico City.

Borgstrom, Georg A. (editor)
1962 *Fish as Food, Vol. IV: Processing*, Pt. 2. Academic Press Inc., New York.

Bowdler, Sandra
1976 Hook, Line, and Dilly Bag: An Interpretation of an Australian Coastal Shell Midden. *Mankind* 10(4): 248–258.

Bowen, Thomas
1976 *Seri Prehistory: The Archaeology of the Central Coast of Sonora, Mexico*. University of Arizona Press, Tucson.

Bradley, Ronna Jane
1993 Marine Shell Exchange in Northwest Mexico and the Southwest. In *The American Southwest and Mesoamerica: Systems of Prehistoric Exchange*, edited by Jonathon E. Ericson and Timothy G. Baugh, pp. 121–151. Plenum Press, New York.

Brand, Donald D.
1956 *The Development of Pacific Coast Ports During the Spanish Colonial Period in Mexico*. Estudios Antropológicos. Sociedad Mexicana de Antropología, Universidad Nacional Autónoma de México, Mexico City.
1971 Ethnohistoric Synthesis of Western Mexico. In *Archaeology of Northern Mesoamerica*, edited by Gordon F. Ekholm and Ignacio Bernal, pp. 632–656. Handbook of Middle American Indians, Vol. 11, Robert Wauchope, general editor, University of Texas Press, Austin.

Braun, David P.
1979 Illinois Hopewell Burial Practices and Social Organization: A Reexamination of the Klunk-Gibson Mound Group. In *Hopewell Archaeology: The Chillicothe Conference*, edited by David S. Brose and N'omi Greber, pp. 66–79. Kent State University Press, Kent, Ohio.

Breitburg, Emanuel
1993 The Evolution of Turkey Domestication in the Greater Southwest and Mesoamerica. In *Culture and Contact: Charles C. Di Peso's Gran Chichimeca*, edited by Anne I. Woosley and John C. Ravesloot, pp. 153–172. Amerind Foundation Publication, Dragoon, Arizona; and University of New Mexico Press, Albuquerque.

Brenner, Anita
1931 *The Influence of Technique on the Decorative Style in the Domestic Pottery of Culhuacán*. Columbia University Press, New York.

Broom, M. J.
1985 The Biology and Culture of Marine Bivalve Molluscs of the Genus *Anadara*. *ICLARM Contribution* 263:1–37.

Brotchie, Luis Páez
1940 *Jalisco, Historia Minimo*, Vol. 1. Edited by Ricardo Delgado. Guadalajara, Mexico.

Brown, James A. (editor)
1971 *Approaches to the Social Dimensions of Mortuary Practices*. Society for American Archaeology, Memoirs 25. Washington, D.C.

Brues, Alice M.
1990 *People and Races*. Waveland Press, Inc., Long Grove, Illinois.

Brunson, Judy L.
1989 The Social Organization of the Las Muertos Hohokam: A Reanalysis of Cushing's Hemenway Expedition Data. Unpublished Ph.D. dissertation, Department of Anthropology, Arizona State University, Tempe.

Buikstra, Jane E.
1981 Introduction. In *Prehistoric Tuberculosis in the Americas*, edited by Jane E. Buikstra, pp. 13–23. Northwestern University Archaeological Program, Evanston.

Buikstra, Jane E., and D. C. Cook
1981 Pre-Columbian Tuberculosis in West-Central Illinois: Prehistoric Disease in Biocultural Perspective. In *Prehistoric Tuberculosis in the Americas*, edited by Jane E. Buikstra, pp. 115–139. Northwestern University Archaeological Program, Evanston.

Byers, Steven N.
2011 *Introduction to Forensic Anthropology*. 4th ed. Pearson/Prentice Hall, Boston.

Cabrero G., Maria Teresa
1989 *Civilización en el norte de México; Arqueología de la Cañada del Río Bolaños*. Serie Antropología 103. Instituto de Investigaciones Antropológicas, Universidad Nacional Autónoma de México, Mexico City.
1995 *La Muerte en el Occidente del México Prehispánico*. Instituto de Investigaciones Antropológicas, Mexico City.

Calkin, Parker E., and Dan Cottrell
1973 Contemporary Beach Forms and Processes, and Origin of the Regressive Sand Ridge System of the Teacapán Peninsula, Southern Sinaloa. In *The Marismas Nacionales of Mexico: Report on Continuing Investigations of the Archaeology and Related Natural Science Studies. West Mexican Prehistory*, edited by Stuart D. Scott, Pt. 7:80–87. Mimeograph, Department of Anthropology, State University of New York at Buffalo.

Carpenter, John Philip
1996 *El ombligo en la Labor*: Differentiation, Interaction, and Integration in Prehispanic Sinaloa, Mexico. Unpublished Ph.D. dissertation, Department of Anthropology, University of Arizona, Tucson.
1997 Passing Through the Netherworld: New Insights from the American Museum of Natural History's Sonora-Sinaloa Archaeological Project (1937–1940). In *Prehistory of the Borderlands: Recent Research in the Archaeology of Northern Mexico and the Southern Southwest*, edited by John P. Carpenter and G. Sánchez, pp. 113–127. Arizona State Museum, University of Arizona, Tucson.

Carr, Christopher
1995 Mortuary Practices: Their Social, Philosophical-Religious, Circumstantial, and Physical Determinants. *Journal of Archaeological Method and Theory* 2(2):105–201.

Carranza, Pedro de
1963 Relación sobre la Jornada que Hizo Nuño de Guzmán, de la Entrada y Sucesos de la Nueva Galicia, Hecha por Pedro de Carranza. In *Crónicas de la Conquista del Reino de Nueva Galicia en Territorio de la Nueva España*, pp. 153–180. H. Ayuntamiento de la Ciudad de Guadalajara e Instituto Jalisciense de Antropología e Historia, Guadalajara, Mexico.

Carriveau, Gary
1976 *Green Glazed Objects from West Mexico. Preliminary Report, Department of Chemistry, Brookhaven National Laboratory, Upton, Long Island, New York.* Report on file, Department of Anthropology, State University of New York at Buffalo.

Carson, E. Ann
2006 Maximum Likelihood Estimation of Human Craniometric Heritabilities. *American Journal of Physical Anthropology* 131(2):169–180.

Case, Henry W.
1976 Odontometrics and Taxonomy of Prehistoric Inhabitants of the Marismas Nacionales, Sinaloa and Nayarit, Mexico. Unpublished Master's thesis, Department of Anthropology, University of Wyoming, Laramie.

Casserino, Christopher M.
2009 Bioarchaeology of Violence and Site Abandonment at Casas Grandes, Chihuahua, Mexico. Unpublished Ph.D. dissertation, Department of Anthropology, University of Oregon, Eugene.

Castetter, Edward F., and Willis H. Bell
1942 *Pima and Papago Indian Agriculture.* University of New Mexico Press, Albuquerque.

Cazeau, Charles J.
1972 Of Ceramics and Sediments: A Geologist Speaks. In *Archaeological Reconnaissance and Excavations in the Marismas Nacionales, Sinaloa and Nayarit, Mexico. West Mexican Prehistory*, edited by Stuart D. Scott, Pt. 6:83–93. Mimeograph, Department of Anthropology, State University of New York at Buffalo.

Chadwick, Robert E.
1971 Archaeological Synthesis of Michoacán and Adjacent Regions. In *Archaeology of Northern Mesoamerica*, edited by Gordon F. Ekholm and Ignacio Bernal, pp. 657–693. Handbook of Middle American Indians, Vol. 11, Pt. 2, Robert Wauchope, general editor, University of Texas Press, Austin.

Chapman, Patrick, and George W. Gill
1993 Use of the Palate and Palatine Suture in Race Identification. Paper presented at the 45th Annual Meeting of the American Academy of Forensic Sciences, Boston.

Claassen, Cheryl
1986 Shellfishing Seasons in the Prehistoric Southeastern United States. *American Antiquity* 51(1):21–37.
1998 *Shells.* Cambridge Manuals in Archaeology. Cambridge University Press, Cambridge.

Cline, Howard F.
1962 The Ortelius Maps of New Spain, 1579, and Related Contemporary Materials, 1560–1610. *Imago Mundi: The International Journal for the History of Cartography* 16(1):98–115.
1972a Introductory Notes on Territorial Division of Middle America. In *Guide to Ethnohistoric Resources, Pt. I*, edited by Howard F. Cline, pp. 17–63. Handbook of Middle American Indians, Vol. 12, Robert Wauchope, general editor, University of Texas Press, Austin.
1972b Viceroyalty to Republics, 1786–1952: Historical Notes on the Evolution of Middle American Political Units. In *Guide to Ethnohistoric Resources, Pt. I*, edited by Howard F. Cline, pp. 138–165. Handbook of Middle American Indians, Vol. 12, Robert Wauchope, general editor, University of Texas Press, Austin.

Coe, Michael D.
1961 *La Victoria: An Early Site on the Pacific Coast of Guatemala.* Peabody Museum Archaeological and Ethnological Papers 53. Harvard University, Cambridge, Massachusetts.

Coe, Michael D., and Richard A. Diehl
1980 *In the Land of the Olmec.* University of Texas Press, Austin.

Coe, Michael D., and Kent V. Flannery
1967 *Early Cultures and Human Ecology in South Coastal Guatemala.* Contributions to Anthropology 3. Smithsonian Press, Washington, D.C.

Comas, Juan
1960 *Manual of Physical Anthropology.* Charles C. Thomas, Springfield, Illinois.

Comisión Nacional para el Conocimiento y Uso de la Biodiversidad
2008 *Manglares de México.* Comisión Nacional para el Conocimiento y Uso de la Biodiversidad, Mexico City.

Connally, G. Gordon
1974a Soil Stratigraphy in the Rio Acaponeta Valley and the Tectonic History of the West Mexican Coastal Plain. In *The Marismas Nacionales of Mexico: Report on Continuing Investigations of the Archaeology and Related Natural Science Studies. West Mexican Prehistory*, edited by G. Gordon Connally, Pt. 8:1–4. Mimeograph, Department of Anthropology, State University of New York at Buffalo.
1974b A Soil Stratigraphic Model for the Age and Origin of the West Mexican Coastal Plain. In *The Marismas Nacionales of Mexico: Report on Continuing Investigations of the Archaeology and Related Natural Science Studies. West Mexican Prehistory*, edited by G. Gordon Connally, Pt. 8:5–20. Mimeograph, Department of Anthropology, State University of New York at Buffalo.
1975 Coastal Plain Development and Inferred Tectonic Events. In *The Marismas Nacionales of Mexico: Report on Continuing Investigations of the Archaeology and Related Natural Science Studies. West Mexican Prehistory*, edited by G. Gordon Connally, Pt. 9:12–17. Mimeograph, Department of Anthropology, State University of New York at Buffalo.
1977 The West Mexican Coastal Plain: Physiographic Definition. Abstract. *Geological Society of America, Abstracts with Programs* 5. Boulder.
1984 Soil Stratigraphy and Inferred Tectonic History of the West Mexican Coastal Plain. In *Neotectonics and Sea Level Variations in the Gulf of California Area, a Symposium*, edited by V. Malpica-Cruz, S. Celis-Gutierrez, J. Guerrero-Garcia, and L. Ortlieb, pp. 55–73. Universidad Nacional Autónoma de México, Instituto de Geología, Mexico City.

Connally, G. Gordon (editor)
1974 *The Marismas Nacionales of Mexico: Report on Con-*

tinuing Investigations of the Archaeology and Related Natural Science Studies. West Mexican Prehistory, Pt. 8. Mimeograph, Department of Anthropology, State University of New York at Buffalo.

1975 *The Marismas Nacionales of Mexico: Report on Continuing Investigations of the Archaeology and Related Natural Science Studies. West Mexican Prehistory*, Pt. 9. Mimeograph, Department of Anthropology, State University of New York at Buffalo.

Connally, G. Gordon, and J. T. Sullivan

1984 Inferred Tectonic History of the West Mexican Coastal Plain: Mazatlán to San Blas. Paper presented at the symposium "Neotectonics and Sea Level Variations in the Gulf of California Area," Hermosillo, April 21–23.

Cook, Sherburne F., and Woodrow Borah

1960 *The Indian Population of Central Mexico, 1531–1610*. Ibero-Americana 44. University of California Press, Berkeley.

Cornwall, I. W.

1958 *Soils for the Archaeologist*. Phoenix House, London.

Cottrell, Daniel J.

1972 Some Geomorphological Aspects of the Marismas Nacionales. In *Archaeological Reconnaissance and Excavations in the Marismas Nacionales, Sinaloa, and Nayarit, Mexico. West Mexican Prehistory* Pt. 6, edited by Stuart D. Scott, Pt. 6:95–107. Mimeograph, Department of Anthropology, State University of New York at Buffalo.

1973 The Sedimentary Petrology and Some Geomorphological Aspects of the Marismas Nacionales, West Mexico. Unpublished Master's thesis, Department of Geology, State University of New York at Buffalo.

Coyle, Philip E.

1998 The Customs of Our Ancestors: Cora Religious Conversion and Millennialism, 2000–1722. *Ethnohistory* 45(3):509–542.

Crabtree, Robert

1961 *Two Urn Burials from Santa Cruz, Nayarit*. Manuscript on file, Department of Anthropology, University of California, Los Angeles.

Cronin, L. E.

1967 The Role of Man in Estuarine Processes. In *Estuaries*, edited by George H. Lauff, pp. 667–689. American Association for the Advancement of Science, Publication 83. Washington, D.C.

Cucina, Andrea, Cristina P. Cantillo, Thelma Sierra Sosa, and Vera Tiesler

2011 Carious Lesions and Maize Consumption Among the Prehispanic Maya: An Analysis of a Coastal Community in Northern Yucatan. *American Journal of Physical Anthropology* 145(4):560–567.

Cumbaa, Stephen L.

1973 Subsistence Strategy at Venadillo, Sinaloa, Mexico. In *Archaeological Reconnaissance and Excavations in the Marismas Nacionales, Sinaloa and Nayarit, Mexico. West Mexican Prehistory*, edited by Stuart D. Scott, Pt. 7:66–79. Mimeograph, Department of Anthropology, State University of New York at Buffalo.

Curray, Joseph R., F. J. Emmel, and Perry J. Crampton

1969 Holocene History of a Strand Plain, Lagoonal Coast, Nayarit, México. In *Costeras, Lagunas, a Symposium*, edited by A. A. Castanares and F. B. Phleger, pp. 62–100. Universidad Nacional Autónoma de México, Mexico City.

Curray, Joseph R., and David G. Moore

1964 Holocene Regressive Littoral Sand, Costa de Nayarit, Mexico. In *Deltaic and Shallow Marine Deposits*, edited by I. M. J. U. Martin Van Straaten, pp. 76–82. Elsevier Press, Amsterdam.

Dahlberg, Albert A.

1963 Analysis of the American Indian Dentition. In *Dental Anthropology*, edited by Don R. Brothwell, pp. 149–177. Pergamon Press, Oxford.

Darling, J. Andrew

1993 Notes on Obsidian Sources of the Southern Sierra Madre Occidental. *Ancient Mesoamerica* 4:245–253.

1998 Obsidian Distribution and Exchange in the North-Central Frontier of Mesoamerica. Unpublished Ph.D. dissertation, Department of Anthropology, University of Michigan, Ann Arbor.

Darling, J. Andrew, and Michael D. Glascock

1998 Acquisition and Distribution of Obsidian in the North-Central Frontier of Mesoamerica. In *Rutas de Intercambio en Mesoamérica*, edited by Evelyn Childs Rattry, pp. 345–364. Universidad Nacional Autónoma de México, Instituto de Investigaciones Antropológicas, Mexico City.

Darling, J. Andrew, and Frances M. Hayashida

1995 Compositional Analysis of the Huitzila and La Lobera Obsidian Sources in the Southern Sierra Madre Occidental, México. *Journal of Radioanalytical and Nuclear Chemistry* 196(2):245–254.

Dávila, Amado Gonzalez

1959 *Diccionario Geográfico, Histórico, Biográfico y Estadístico del Estado de Sinaloa*. State of Sinaloa Government, Culiacán.

Davis, Betty

1969 Ceramics. In *Archaeological Reconnaissance and Excavations in the Marismas Nacionales, Sinaloa and Nayarit, Mexico. West Mexican Prehistory*, edited by Stuart D. Scott, Pt. 3:88–95. Mimeograph, Department of Anthropology, State University of New York at Buffalo.

de Fuentes, Patricia

1963 *The Conquistadors*. Orion Press, New York.

Degerbøl, Magnus

1961 On a Find of a Preboreal Domestic Dog (*Canis familiaris*) from Star Carr, Yorkshire, with Remarks on Other Mesolithic Dogs. *Proceedings of the Prehistoric Society* (N.S.) 27:35–55.

Dembo, Adolfo, and Jose Imbelloni

1938 *Deformaciónes Intencionales del Cuerpo Humano de Carácter Étnico*. Editori Nova. José Anesi, Buenos Aires.

Díaz-Romo, Patricia, and Samuel Salinas Alvarez

2000 The Huichol Indians, Tobacco, and Pesticides. In *Tobacco Use by Native Americans: Sacred Smoke and*

Silent Killer, edited by Joseph C. Winter, pp. 342–352. University of Oklahoma Press, Norman.

Di Peso, Charles C., John B. Renaldo, and Gloria J. Fenner
1974 *Casas Grandes: A Fallen Trading Center of the Gran Chichimeca, Vol. 8: Bone-Economy-Burials*. Amerind Foundation Publications, Dragoon, Arizona; and Northland Press, Flagstaff.

Dobzhansky, Theodosius
1962 *Mankind Evolving*. Yale University Press, New Haven.

Drögemüller, Cord, Elinor K. Karlsson, Marjo K. Hytönen, Michele Perloski, Gaudenz Dolf, Kirsi Sainio, Hannes Lohi, Kerstin Lindblad-Toh, and Tosso Leeb
2008 The Mutation in Hairless Dogs Implicates FOXI3 in Ectodermal Development. *Science* 321(5895):14–62.

Duverger, Christian
1998 Coamiles, Nayarit: Hacia una Periodización. *Antropología e Historia del Occidente de México* 1:609–628. Memorias de la XXIV Mesa Redonda de la Sociedad Mexicana de Antropología. Universidad Nacional Autónoma de México, Mexico City.

Duverger, Christian, and Daniel Levine
1993 *Informe Relativo a la Exploración Arqueológica del Sitio de Coamiles, Municipio de Tuxpan, Estado de Nayarit*. Technical report, Instituto Nacional de Antropología e Historia, Nayarit, Tepic.

Dwight, Thomas
1905 The Size of the Articular Surfaces of the Long Bones as Characteristics of Sex: An Anthropological Study. *American Journal of Anatomy* 4:19–32.

Eerkens, Jelmer W., Kevin J. Vaughn, Tim R. Carpenter, Christina A. Conlees, Moises Linares Grados, and Katharina Schreiber
2008 Obsidian Hydration Dating on the South Coast of Peru. *Journal of Archaeological Science* 35(8):2231–2239.

Ekholm, Gordon F.
1939a Recent Archaeological Work in Sonora and Northern Sinaloa. *Actas del XXVII Congreso Internacional de Americanistas* 2:64–73.
1939b Results of an Archaeological Survey of Sonora and Northern Sinaloa. *Revista Mexicana de Antropología* 3(1):7–11.
1942 Excavations at Guasave, Sinaloa, Mexico. *Anthropological Papers of the American Museum of Natural History* 38(2):23–139. New York.

El-Najjar, Mahmoud Y., and Kenneth R. McWilliams
1978 *Forensic Anthropology*. Charles C. Thomas, Springfield, Illinois.

Emery, Kitty
2001 Fauna. In *Archaeology of Ancient Mexico and Central America*, edited by Susan Toby Evans and David L. Webster, pp. 255–265. Garland Publishing, New York.

Ericson, J. E., and J. Kimberlin
1977 Obsidian Sources, Chemical Characterization and Hydration Rates in West Mexico. *Archaeometry* 19(2):157–166.

Erlandson, Jon M.
1988 The Role of Shellfish in Prehistoric Economies: A Protein Perspective. *American Antiquity* 53(1): 102–109.

Fastlicht, Samuel, and Javier Romero
1951 *El Arte de las Mutilaciónes Dentarias*. Enciclopedia Mexicana de Arte, 14. Ediciones Mexicana, S.A., Mexico City.

Fay, George E.
1959 *Handbook of Pottery Types of Nayarit, Mexico*. Instituto Interamericano, Miscellaneous Papers, Archaeological Series, No. 1. Mimeograph, Department of Sociology and Anthropology, Southern State College, Magnolia, Arkansas.

Feinman, Garry M., Stephen A. Kowalewski, Laura Finsten, Richard E. Blanton, and Linda Nicholas
1985 Long-Term Demographic Change: A Perspective from the Valley of Oaxaca, Mexico. *Journal of Field Archaeology* 12(3):333–362.

Feldman, Lawrence H.
1972 A Note on Fish Found in Sonoran, Sinaloan and North Nayaritan Archaeological Sites and Ethnohistoric Records. In *Archaeological Reconnaissance and Excavations in the Marismas Nacionales, Sinaloa and Nayarit, Mexico. West Mexican Prehistory*, edited by Stuart D. Scott, Pt. 6:25–29. Mimeograph, Department of Anthropology, State University of New York at Buffalo.
1976 Greater Nayarit Archaeological Molluscs and Archaeo-Molluscan Units in Northern Mesoamerica. In *The Archaeology of Amapa, Nayarit*, edited by Clement W. Meighan, pp. 163–181. Monumenta Archaeologica 2, Institute of Archaeology, University of California, Los Angeles.

Felger, Richard Stephen, and Mary Beck Moser
1985 *People of the Desert and Sea: Ethnobotany of the Seri Indians*. University of Arizona Press, Tucson.

Fikes, Jay C.
2010 *Unknown Huichol: Shamans and Immortals, Allies Against Chaos*. AltaMira Press, Lanham, Maryland.

Fine, Bonnie, Judd Pilossof, and Ronald Yablon
1973 Excavations at Venadillo, 1973. In *Archaeological Reconnaissance and Excavations in the Marismas Nacionales, Sinaloa and Nayarit, Mexico. West Mexican Prehistory*, edited by Stuart D. Scott, Pt. 7:8–16. Mimeograph, Department of Anthropology, State University of New York at Buffalo.

Flannery, Kent V.
1976 Empirical Determination of Site Catchments in Oaxaca and Tehuacán. In *The Early Mesoamerican Village*, edited by Kent V. Flannery, pp. 103–117. Academic Press, New York.

Flannery, Kent V., and Joyce Marcus
1983 The Origins of the State in Oaxaca. In *The Cloud People*, edited by Kent Flannery and Joyce Marcus, pp. 79–83. Academic Press, New York.

Flores Verdugo, Francisco Javier
1986 *Ecología de los manglares y perfil de comunidades en dos sistemas lagunares de Agua Brava y Marismas Nacionales, Nayarit*. Technical report, Consejo Nacional de Ciencia y Technología. Clave PCECBNA-022068. Mexico City.

Foster, Michael S.
1974 Excavations in Mound C, Venadillo, Sinaloa, Mexico.

Unpublished Master's thesis, Department of Anthropology, University of Colorado, Boulder.

1978 Loma San Gabriel: A Prehistoric Culture of Northwest Mexico. Unpublished Ph.D. dissertation, Department of Anthropology, University of Colorado, Boulder.

1984 Archaeological Investigations in the Marismas Nacionales: Mound C, Venadillo, Sinaloa, Mexico. *Publicaciones Antropológicos del Occidente* 4:1–60. Universidad Autónoma de Guadalajara, Guadalajara, Mexico.

1995 The Chalchihuites Chronological Sequences: A View from the West Coast of Mexico. In *Arqueología del orte y del ccidente de México: Homenaje al Doctor J. Charles Kelley*, edited by Barbro Dahlgren and Ma. Dolores Soto de Arechavaleta, pp. 67–92. Universidad Nacional Autónoma de México, Mexico City.

1999 The Aztatlán Tradition of West and Northwest Mexico and Casas Grandes: Speculations on the Medio Period Florescence. In *The Casas Grandes World*, edited by Curtis F. Schaafsma and Carroll L. Riley, pp. 149–163. University of Utah Press, Salt Lake City.

2000 The Archaeology of Durango. In *Greater Mesoamerica: The Archaeology of West and Northwest Mexico*, edited by Michael S. Foster and Shirley Gorenstein, pp. 197–219. University of Utah Press, Salt Lake City.

2001 Shaft Tombs. In *Archaeology of Ancient Mexico and Central America*, edited by Susan Toby Evans and David L. Webster, pp. 661–662. Garland Publishing, New York.

Foster, Michael S., and George W. Gill

1977 *Prehistoric Human Burials from the Marismas Nacionales of West Mexico: Cultural Practices and Social Implications*. Manuscript on file with the authors.

Foster, Michael S., and Shirley Gorenstein (editors)

2000 *Greater Mesoamerica: The Archaeology of West and Northwest Mexico*. University of Utah Press, Salt Lake City.

Foster, Michael S., Douglas R. Mitchell, Gary Huckleberry, and David Dettman

2008 Observations on the Archaeology, Paleoenvironment, and Geomorphology of the Puerto Peñasco Area of Northern Sonora, Mexico. *Kiva* 73:263–290.

Foster, Michael S., and Phil C. Weigand (editors)

1985 *The Archaeology of West and Northwest Mesoamerica*. Westview Press, Boulder.

Foy, Constance Anne

1970 Ceramics. In *Archaeological Reconnaissance and Excavations in the Marismas Nacionales, Sinaloa and Nayarit, Mexico. West Mexican Prehistory*, edited by Stuart D. Scott, Pt. 4:1–32. Mimeograph, Department of Anthropology, State University of New York at Buffalo.

Freter, Ann Corinne

1993 Obsidian-Hydration Dating: Its Past, Present, and Future Application in Mesoamerica. *Ancient Mesoamerica* 4(2):285–303.

Friedman, Irving, and John Obradovich

1981 Obsidian Hydration Dating of Volcanic Events. *Quaternary Research* 16(1):37–47.

Friedman, Irving, Fred W. Trembour, and Richard E. Hughes

1997 Obsidian Hydration Dating. In *Chronometric Dating in Archaeology*, edited by R. E. Taylor and M. J. Aitken, pp. 297–321. Plenum Press, New York.

Friedman, Irving, Fred W. Trembour, Franklin L. Smith, and George I. Smith

1994 Is Obsidian Hydration Dating Affected by Relative Humidity? *Quaternary Research* 41(2):185–190.

Ganot Rodríguez, Jaime, and A. Peschard Fernández

1997 *Aztatlán: Apuntes para la Historia y Arqueologia de Durango*. Secretaria de Educación, Cultura y Deporte, Gobierno del Estado de Durango, Durango.

García Cubas, Antonio

1890 *Diccionario Geográfico, Histórico y Biográfico de los Estados Unidos Mexicanos*. Imprenta Murguía y Sría. de Fomento, Mexico City.

García del Pilar

1955 Relación de la Entrada de Nuño de Guzmán que dió García del Pilar, su intérprete. In *Memoria de los Servicios que habia hecho Nuño de Guzmán, desde que fue nombrado Gobernador de Pánuco en 1525*, notes by Manuel Carrera Stampa, pp. 179–193. José Porrúa e Hijos Sucs., Mexico City.

1963 Relación de la Entrada de Nuño de Guzmán a Nueva Galicia que dio García del Pilar su Intérprete en la Jornada. In *Crónicas de la Conquista del Reinode Nueva Galicia en Territorio de la Nueva España*, pp. 215–238. H. Ayuntamiento de la Ciudad de Guadalajara e Instituto Jalisciense de Antropología e Historia, Guadalajara, Mexico.

Garduño, Mauricio

2008 *Investigaciónes Arqueológicas en el Cerro de Coamiles, Nayarit. Reporte Técnico Final Temporada de Campo 2007*. Technical report, Instituto Nacional de Antropología e Historia, Nayarit, Tepic.

2009 Descubren Observatorio Astronómico en Coamiles. *Arqueología Mexicana* 98(July–August):12. Editorial Raices. Instituto Nacional de Antropología e Historia, Mexico City.

2012 Cerro de Coamiles, Nayarit: Un Sitio Emblemático Aztatlán del Septentrión Costero Mesoamericano. *Diario de Campo*, July–September, 24–32.

2013 *Analisis de Materiales de Excavación del Proyecto Arqueológico Coamiles*. Technical report, Instituto Nacional de Antropología e Historia, Nayarit, Tepic.

Garduño, Mauricio, and Lorena Gámez Eternod

2005 *Programa Emergente de Rescate Arqueológico en San Felipe Aztatán, Municipio de Tecuala (Nayarit). Informe Técnico Final Trabajos de Sondeo Arqueológico*. Technical report, Instituto Nacional de Antropología e Historia, Nayarit, Tepic.

Garduño, Mauricio, and Miguel Angel Vásquez del Mercado

2005 Investigaciones Arqueológicas en el Cerro de Coamiles, Nayarit. *Boletin Diario de Campo* 81:65–75. Coordinación Nacional de Antropológia, Instituto Nacional de Antropología e Historia, Mexico City.

Gavan, James A.

1949 The Skeletal Material from Tuxcacuesco. In *The Archaeology of the Autlán-Tuxcacuesco Area of Jalisco, II: The Tuxcacuesco-Zapotitlán Zone*, by

Isabel Kelly, pp. 213–224. Ibero-Americana 27. University of California Press, Berkeley.

Genovés, Santiago T.

1967 Proportionality of the Long Bones and Their Relation to Stature Among Mesoamericans. *American Journal of Physical Anthropology* 26(1):67–78.

1970 Anthropometry of Late Prehistoric Human Remains. In *Physical Anthropology*, edited by T. Dale Stewart, pp. 35–49. Handbook of Middle American Indians, Vol. 9, Robert Wauchope, general editor, University of Texas Press, Austin.

Gentry, Howard Scott

1942 *Rio Mayo Plants*. Carnegie Institute of Washington Publication 527. Washington, D.C.

Gerhard, Peter

1972 Colonial New Spain 1519–1786: Historical Notes on the Evolution of Minor Political Jurisdictions. In *Guide to Ethnohistorical Sources, Pt. I*, edited by Howard F. Cline, pp. 63–137. Handbook of Middle American Indians, Vol. 12, Robert Wauchope, general editor, University of Texas Press, Austin.

1993a *A Guide to the Historical Geography of New Spain*. Rev. ed. University of Oklahoma Press, Norman.

1993b *The Northern Frontier of New Spain*. Rev. ed. University of Oklahoma Press, Norman.

1993c *The Southeast Frontier of New Spain*. University of Oklahoma Press, Norman.

Giles, Eugene, and H. K. Bleibtreu

1961 Cranial Evidence in Archaeological Reconstruction: A Trial of Multivariate Techniques for the Southwest. *American Anthropologist* 63(1):48–61.

Giles, Eugene, and Orville Elliot

1962 Race Identification from Cranial Measurements. *Journal of Forensic Sciences* 7:147–157.

Gill, George W.

1968 Human Skeletal Remains. In *Archaeological Reconnaissance and Excavations in the Marismas Nacionales, Sinaloa and Nayarit, Mexico. West Mexican Prehistory*, edited by Stuart D. Scott, Pt. 2:153–173. Mimeograph, Department of Anthropology, State University of New York at Buffalo.

1969a Excavation of Burial Mounds Within the Marismas Nacionales, Chalpa and Tecualilla Sites. In *Archaeological Reconnaissance and Excavations in the Marismas Nacionales, Sinaloa and Nayarit, Mexico. West Mexican Prehistory*, edited by Stuart D. Scott, Pt. 3:48–65. Mimeograph, Department of Anthropology, State University of New York at Buffalo.

1969b Human Skeletal Remains: Chalpa and Tecualilla Sites. In *Archaeological Reconnaissance and Excavations in the Marismas Nacionales, Sinaloa and Nayarit, Mexico. West Mexican Prehistory*, edited by Stuart D. Scott, Pt. 3:112–132. Mimeograph, Department of Anthropology, State University of New York at Buffalo.

1971 The Prehistoric Inhabitants of Northern Coastal Nayarit: Skeletal Analysis and Description of Burials. Unpublished Ph.D. dissertation, Department of Anthropology, University of Kansas, Lawrence.

1974 Toltec-Period Burial Customs Within the Marismas Nacionales of Western Mexico. In *The Archaeology of West Mexico*, edited by Betty B. Bell, pp. 83–105. Sociedad de Estudios Avanzados del Occidente de México, Ajijic, Mexico.

1976 Appendix B. Human Skeletal Remains from Amapa: Morphology. In *The Archaeology of Amapa, Nayarit*, edited by Clement W. Meighan, pp. 187–200. Monumenta Archaeologica. Institute of Archaeology, University of California, Los Angeles.

1977 Manifestations of a Conceptual Artistic Ideal Within a Prehistoric Culture. *American Antiquity* 42(1): 101–110.

1978 The Sand Creek Skeleton. Appendix I in William E. Scoggin, The Sand Creek Burial from Central Wyoming. *Wyoming Archaeologist* 21(3):21–30.

1984 Classic Period Human Skeletal Remains from Venadillo, Mound C. In *Archaeological Investigations in the Marismas Nacionales: Mount C, Venadillo, Sinaloa, Mexico*, by Michael S. Foster, pp. 47–57. Publicaciones Antropológicos del Occidente 7. Universidad Autónoma de Guadalajara, Guadalajara.

1985 Cultural Implications of Artificially Modified Human Remains from Northwestern Mexico. In *The Archaeology of West and Northwest Mesoamerica*, edited by Michael S. Foster and Phil C. Weigand, pp. 193–215. Westview Press, Boulder.

1998 Craniofacial Criteria in Skeletal Race Attribution. In *Forensic Osteology*, edited by Kathleen Reichs, pp. 293–317. Charles C. Thomas, Springfield, Illinois.

Gill, George W., and Henry W. Case

1983 Migration and Population Change in the Prehistoric Marismas Nacionales of Northwestern Mexico. Paper presented at the 18th Mesa Redonda, Sociedad Mexicana de Antropología, Taxco.

Gill, George W., and Cresta V. Deeds

2008 Temporal Changes in Nasal Sill Development Within the Northwestern Plains Region. In *Skeletal Biology and Bioarchaeology of the Northwestern Plains*, edited by George W. Gill and Rick L. Weathermon, pp. 263–270. University of Utah Press, Salt Lake City.

Gill, George W., and Susan Prince

1970 Figurines and Skulls from Coastal Nayarit. In *Archaeological Reconnaissance and Excavations in the Marismas Nacionales, Sinaloa and Nayarit, Mexico. West Mexican Prehistory*, edited by Stuart D. Scott, Pt. 4:85–93. Mimeograph, Department of Anthropology, State University of New York at Buffalo.

Glascock, Judith

1980 Craniometrics, Odontometrics, and Anthroposcopic Observations of a Prehistoric Population at Guasave, Sinaloa, Mexico. Unpublished Master's thesis, Department of Anthropology, University of Wyoming, Laramie.

Glascock, Michael D.

2007 *X-Ray Fluorescence Analysis of Obsidian Artifacts from Chacalilla, Nayarit*. Report prepared for Dr. Michael Ohnersorgen, University of Missouri, Saint Louis.

Glascock, Michael D., R. S. Herrera, and H. Neff
1997 *Hydration and Source Analysis of Obsidian from Excavated Archaeological Contexts in the Municipality of Puerto Vallarta, Mexico*. Report from the Research Reactor Center, University of Missouri, Columbia.

Glascock, Michael D., Phil C. Weigand, Rodrigo Esparza Lopez, Michael A. Ohnersorgen, Mauricio Garduno Ambriz, Joseph B. Mountjoy, and J. Andrew Darling
2010 Geochemical Characterization of Obsidian Sources in Western Mexico: The Sources of Jalisco, Nayarit, and Zacatecas. In *Crossing the Straights: Prehistoric Obsidian Source Exploitation Along the Pacific Rim*, edited by Yaroslav Kuzmin and Michael D. Glascock, pp. 201–217. BAR International Series S2152. Archaeopress, Oxford.

Glassow, Michael A.
1967 The Ceramics of Huistla. *American Antiquity* 32(1): 64–83.

Goldstein, Lynne G.
1976 Spatial Structure and Social Organization: Regional Manifestations of Mississippian Society. Unpublished Ph.D. dissertation, Department of Anthropology, Northwestern University, Evanston.

González, Yólotl (editor)
1989 *Homenaje a Isabel Kelly*. Serie Arqueología, Instituto Nacional de Antropología e Historia, Mexico City.

González Barajas, Lourdes, and Jose Carlos Beltrán Medina
2007 Arqueología de la Bahía de Banderas. In *El Occidente de México Perspectivas Multidisciplinaria. Memorias del V Coloquio Internacional de Occidentalistas*, edited by Rosa Yañez, pp. 312–324. Universidad de Guadalajara, Guadalajara, Mexico.

Gorenstein, Shirley, and Michael S. Foster
2000 West and Northwest Mexico: The Ins and Outs of Mesoamerica. In *Greater Mesoamerica: The Archaeology of West and Northwest Mexico*, edited by Michael S. Foster and Shirley Gorenstein, pp. 3–19. University of Utah Press, Salt Lake City.

Grave Tirado, Luis Alfonso
2003a *Informe Projecto Arqueológico El Calón. Primera Temporada de Campo*. Technical report, Coordinación Nacional de Arqueología del Instituto de Antropología e Historia, Mexico City.
2003b La region fundada en la tradición. El norte de Nayarit y el sur de Sinaloa, una region a lo largo del tiempo. Unpublished Master's thesis in Mesoamerican studies, Universidad Nacional Autónoma de México, Mexico City.
2003c Patrón de asentimiento prehispánico en la region Totorame (el norte de Nayarit y el sur de Sinaloa). *Arqueología* 30:5–26. Segunda Época Mayo–Agosto, Mexico City.
2006 *Informe de la segunda temporada de campo del Proyecto Arqueológico El Calón*. Technical report, Instituto de Antropología e Historia, Sinaloa, Mazatlán.
2007 *Una Historia Prehispánica de Escuinapa*. El Nayarit Editorial, Tepic.
2010 El Calón, un Espacio Sagrado en las Marismas del Sur de Sinaloa. *Estudios Mesoamericanos* 8:19–39. Universidad Nacional Autónoma de México, Mexico City.
2011 *Informe Final de la Primera Temporada de Campo del Proyecto Arqueológico Marismas del Sur de Sinaloa*. Instituto Nacional de Antropología e Historia, Centro Instituto de Antropología e Historia, Mazatlán.
2012 *Informe de la Segunda Temporada de Campo del Proyecto Arqueológico Marismas del Sur de Sinaloa*. Instituto Nacional de Antropología e Historia, Centro Instituto de Antropología e Historia, Mazatlán.

Grave Tirado, Luis Alfonso, and Angélica C. Nava Burgueño
2010 *Informe final de la Primera Temporada de Campo del Projecto Arqueológico Rio Baluarte*. Technical report, Coordinación Nacional de Arqueologia, Instituto de Antropología e Historia, Mexico City.
2012 *Informe final de la segunda temporada de Campo del Proyecto Arqueológico Rio Baluarte, Sinaloa, México*. Instituto Nacional de Antropología e Historia, Centro Instituto de Antropología e Historia, Mazatlán.

Grave Tirado, Luis Alfonso, and S. Pulido Méndez
2005 Proyecto arqueológico El Calón. Investigaciónes en las marismas del sur de Sinaloa. *Anales de Arqueología* 1:191–197. Instituto Nacional de Antropología e Historia, Mexico City.

Grimes, Joseph E., and Thomas B. Hinton
1969 The Huichol and Cora. In *Ethnology Pt. II*, edited by Evon Z. Vogt, pp. 792–813. Handbook of Middle American Indians, Vol. 8, Robert Wauchope, general editor, University of Texas Press, Austin.

Grosscup, Gordon L.
1961 A Sequence of Figurines from West Mexico. *American Antiquity* 26(3):390–406.
1964 The Ceramics of West Mexico. Unpublished Ph.D. dissertation, Department of Anthropology, University of California at Los Angeles.
1976 The Ceramic Sequence at Amapa. In *The Archaeology of Amapa, Nayarit*, edited by Clement W. Meighan, pp. 206–272. Monumenta Archaeologica 2. Institute of Archaeology, University of California, Los Angeles.

Guzmán, Nuño Beltrán de
1955 *Memoria de los Servicios que habia hecho Nuño de Guzmán, desde que fue nombrado Gobernador de Pánuco en 1525*. Notes by Manuel Carrera Stampa. José Porrúa e Hijos Sucs., Mexico City.
1963 Carta a S.M. del Presidente de la Audiencia de Mejico, Nuño Beltrán de Guzmán en que Refiere la Jornada que Hizo a Mechuacan, a Conquistar la Provincia de los Tebles Chichimecas que Confina con Nueva España, 1530. In *Crónicas de la Conquista del Reino de Nueva Galicia en Territorio de la Nueva España*. H. Ayuntamiento de la Ciudad de Guadalajara e Instituto Jalisciense de Antropología e Historia, Guadalajara, Mexico.

Hard, Robert J.
1986 Ecological Relationships Affecting the Rise of Farming Economies: A Test from the American Southwest. Unpublished Ph.D. dissertation, Department of Anthropology, University of New Mexico, Albuquerque.

Hard, Robert J., and John R. Roney
1998 A Massive Terraced Village Complex in Chihuahua, Mexico, 3000 Years Before Present. *Science* 279(5357): 1661–1664.

Hardy, Laurence M., and Roy W. McDiarmid
1969 The Amphibians and Reptiles of Sinaloa, Mexico. *Museum of Natural History* 18(3):40–252. University of Kansas, Lawrence.

Hartney, Patrick C.
1981 Tuberculous Lesions in a Prehistoric Population Sample from Southern Ontario. In *Prehistoric Tuberculosis in the Americas*, edited by Jane E. Buikstra, pp. 141–160. Northwestern University Archaeological Program, Evanston.

Healy, Paul F.
1980 *Archaeology of the Rivas Region, Nicaragua*. Wilfred Laurier University Press, Waterloo, Canada.

Hertz, Robert
1907 Contribution a une Etude sure la Representation Collective de la Mort. *Année Sociologique* 10:48–137.

Hewitt, Marilyn, Daniel Cottrell, and Charles Cazeau
1971 Geological Investigations in Sinaloa and Nayarit, Mexico. In *Archaeological Reconnaissance and Excavations in the Marismas Nacionales. West Mexican Prehistory*, edited by Stuart D. Scott, Pt. 5:1–10. Mimeograph, Department of Anthropology, State University of New York at Buffalo.

Hinton, Thomas B.
1972 *Coras, Huicholes y Tepehuanes*. Instituto Nacional Indigenista, Mexico City.

Hoffman, Stephen A.
1974 Provenance, Depositional Environments, and Origin of the Units of the West Mexican Coastal Plain. In *The Marismas Nacionales of Mexico: Report on Continuing Investigations of the Archaeology and Related Natural Science Studies. West Mexican Prehistory*, edited by G. Gordon Connally, Pt. 8:15–21. Mimeograph, Department of Anthropology, State University of New York at Buffalo.
1975a Re-Examination of the Barrier Bar–Lagoon Complex of the West Mexican Coastal Plain by Scanning Electron Microscopy. In *The Marismas Nacionales of Mexico: Report on Continuing Investigations of the Archaeology and Related Natural Science Studies. West Mexican Prehistory*, edited by G. Gordon Connally, Pt. 9:1–11. Mimeograph, Department of Anthropology, State University of New York at Buffalo.
1975b Provenance, Depositional Environment and Progressive Diagenetic Changes of Sands from the West Mexican Coastal Plain. Unpublished Master's thesis, Department of Geology, State University of New York at Buffalo.

Hooton, Earnest A.
1930 *The Indians of Pecos Pueblo*. Yale University Press, New Haven.

Hosler, Dorothy
1988 Ancient West Mexican Metallurgy: South and Central American Origins and West Mexican Transformations. *American Anthropologist* 90(4):832–855.
1992 The Huastec Region: A Second Locus for the Production of Bronze Alloys in Ancient Mesoamerica. *Science* 257:1215–1220.
1994 *The Sounds and Colors of Power: The Sacred Metallurgical Technology of Ancient West Mexico*. Massachusetts Institute of Technology Press, Cambridge.

Howells, W. W.
1973 *Cranial Variation of Man*. Papers of the Peabody Museum of Archaeology and Ethnology, Vol. 67. Harvard University, Cambridge.

Hrdlička, Aleš
1909 Tuberculosis Among Certain Indian Tribes. *Bureau of American Ethnology Bulletin* 42:1–48. Washington, D.C.

Hubbs, Carl L., and Gunnar I. Roden
1964 Oceanography and Marine Life Along the Pacific Coast. In *Natural Environment and Early Cultures*, edited by Robert C. West, pp. 143–186. Handbook of Middle American Indians, Vol. 1, Robert Wauchope, general editor, University of Texas Press, Austin.

Hughen, Konrad A., Michael G. L. Baillie, Edouard Bard, A. Bayliss, J. Warren Beck, Chanda Bertrand, Paul G. Blackwell, Caitlin E. Buck, George S. Burr, Kirsten B. Cutler, Paul E. Damon, Richard L. Edwards, Richard G. Fairbanks, Michael Friedrich, Thomas P. Guilderson, Bernard Kromer, F. Gerry McCormac, Sturt Manning, Christopher B. Ramsey, Paula J. Reimer, Ron W. Reimer, Sabine Remmele, John R. Southon, Minze Stuiver, Sahra Talamo, F. W. Taylor, Johannes van der Plicht, and Constanza E. Weyhenmeyer
2004 Marine04 Marine Radiocarbon Age Calibration, 26–0 cal kyr. *Radiocarbon* 46(3):1059–1086.

Hull, Kathleen L.
2001 Reasserting the Utility of Obsidian Hydration Dating: A Temperature-Dependent Empirical Approach to Practical Temporal Resolution with Archaeological Obsidians. *Journal of Archaeological Science* 28(10):1025–1040.

Hulse, Frederick S.
1945 Appendix III: Skeletal Material. In *Excavations at Culiacán, Sinaloa*, by Isabel Kelly, pp. 187–198. Ibero-Americana 25. University of California, Berkeley.

Huntley, Deborah L., T. Fenn, Judith A. Habicht-Mauche, and Barbara J. Mills
2012 Embedded Networks? Pigments and Long-Distance Procurement Strategies in the Late Prehispanic Southwest. In *Potters and Communities of Practice: Glaze Paint and Polychrome Pottery in the American Southwest, AD 1250–1700*, edited by Linda S. Cordell and Judith A. Habicht-Mauche, pp. 8–18. Anthropological Papers of the University of Arizona, No. 75. Tucson.

Hurtado, Eusebio Dávalos
1970 Pre-Hispanic Osteopathology. In *Physical Anthropology*, edited by T. Dale Stewart, pp. 68–81. Handbook of Middle American Indians, Vol. 9, Robert Wauchope, general editor, University of Texas Press, Austin.

Ingersoll, Ernest
1881 *The History and Present Condition of the Fishing Industry: The Oyster Industry*. Government Printing Office, Washington, D.C.

Ingle, J. C., Jr.
1984 Neogene Stratigraphy and Tectonic History of the Gulf of California. Paper presented at the symposium "Neotectonics and Sea Level Variations in the Gulf of California Area," Hermosillo, April 21–23.

Jacobs, Bonnie Fine
1982 Modern Pollen Spectra from Surface Soil Samples, Northern Nayarit, Southern Sinaloa, Mexico. *Journal of the Arizona-Nevada Academy of Science* 17:1–14.

Jantz, Richard L.
1970 Change and Variation in Skeletal Populations of Arikara Indians. Unpublished Ph.D. dissertation, Department of Anthropology, University of Kansas, Lawrence.

Jimenez, Publio, Arturo Martinez-Insua, Jaime Franco-Vasquez, Xose Luis Otero-Cepeda, and Urbano Santana
2012 Maxillary Changes and Occlusal Traits in Crania with Artificial Fronto-Occipital Deformation. *American Journal of Physical Anthropology* 147(1): 40–51.

Johannsdottir, Berglind, F. Thorarinsson, A. Thordarson, and T. E. Magnusson
2005 Heritability of Craniofacial Characteristics Between Parents and Offspring Estimated from Lateral Cephalograms. *American Journal of Orthodontics and Dentofacial Orthopedics* 127(2):200–207.

Johnston, F. E.
1968 *Report on Tikal, Burial 166. March 13, 1968.* University Museum, University of Pennsylvania, Philadelphia.

Kan, Michael, Clement W. Meighan, and H. B. Nicholson
1989 *Sculpture of Ancient West Mexico: Nayarit, Jalisco, Colima: A Catalogue of the Proctor Stafford Collection at the Los Angeles County Museum of Art.* Los Angeles County Museum of Art, Los Angeles.

Keen, A. Myra
1971 *Sea Shells of Tropical West America.* 2nd ed. Stanford University Press, Stanford.

Kelley, J. Charles
1971 Archaeology of the Northern Frontier: Zacatecas and Durango. In *Archaeology of Northern Mesoamerica*, Pt. 2, edited by Gordon F. Ekholm and Ignacio Bernal, pp. 768–801. Handbook of Middle American Indians, Vol. 11, Robert Wauchope, general editor, University of Texas Press, Austin.
1985 The Chronology of the Chalchihuites Culture. In *The Archaeology of West and Northwest Mesoamerica*, edited by Michael S. Foster and Phil C. Weigand, pp. 269–287. Westview Press, Boulder.
1986 The Mobile Merchants of Molino. In *Ripples in the Chichimec Sea: New Considerations of Southwestern-Mesoamerican Interactions*, edited by Frances J. Mathien and Randall H. McGuire, pp. 81–105. Southern Illinois University Press, Carbondale.
1990 The Early Post-Classic in Northern Zacatecas and Durango: IX to XII Centuries. In *Mesoamérica y norte de México, siglos IX–XII*, edited by F. Sodi Miranda, pp. 487–519. Sociedad Mexicana de Antropología, Instituto Nacional de Antropología e Historia, Mexico City.
1993 Zenith Passage: The View from Chalchihuites. In *Culture and Contact: Charles C. Di Peso's Gran Chichimeca*, edited by Anne I. Woosley and John C. Ravesloot, pp. 227–250. Amerind Foundation Publication, Dragoon, Arizona; and University of New Mexico Press, Albuquerque.
1995 Trade Goods, Traders, and Status in Northwestern Greater Mesoamerica. In *The Gran Chichimeca: Essays on the Archaeology and Ethnohistory of Northern Mesoamerica*, edited by J. E. Reyman, pp. 102–145. Avebury Press, Brookfield, Vermont.
2000 The Aztatlán Mercantile System: Mobile Traders and the Northwestward Expansion of Mesoamerican Civilization. In *Greater Mesoamerica: The Archaeology of West and Northwest Mexico*, edited by Michael S. Foster and Shirley Gorenstein, pp. 137–154. University of Utah Press, Salt Lake City.

Kelley, J. Charles, and Ellen Abbott Kelley
1971 *An Introduction to the Ceramics of the Chalchihuites Culture of Zacatecas and Durango, Mexico, Pt. I: The Decorated Wares.* Mesoamerican Studies, No. 5. Research Records of the University Museum, Southern Illinois University, Carbondale.

Kelley, J. Charles, and Howard D. Winters
1960 A Revision of the Archaeological Sequence in Sinaloa, Mexico. *American Antiquity* 25(4):547–561.

Kelly, Isabel T.
1938 *Excavations at Chametla, Sinaloa.* Ibero-Americana 14. University of California Press, Berkeley.
1945 *Excavations at Culiacán, Sinaloa.* Ibero-Americana 25. University of California Press, Berkeley.
1947 *Excavations at Apatzingán, Michoacán.* Viking Fund Publications in Anthropology 7. New York.
1949 *The Archaeology of the Autlán-Tuxcacuesco Area of Jalisco, II: The Tuxcacuesco-Zapotitlán Zone.* Ibero-Americana 27. University of California Press, Berkeley.
1974 Stirrup Pots from Colima: Some Implications. In *The Archaeology of West Mexico*, edited by Betty B. Bell, pp. 206–211. Sociedad de Estudios Avanzados del Occidente de México, Ajijic, Mexico.
1980 *Ceramic Sequence in Colima: Capacha, an Early Phase.* Anthropological Papers of the University of Arizona, No. 37. Tucson.

Kelly, William H.
1977 *Cocopa Ethnography.* Anthropology Papers of the University of Arizona, No. 29. Tucson.

Khonsari, Roman H., M. Friess, Johan Nysjö, G. Ordi, Filip Malmberg, Ingela Nyström, E. Messo, Jan M. Hirsch, E. A. M. Cabanis, K. H. Kunzelmann, J. M. Salagnac, P. Corre, A. Ohazama, Paul T. Sharpe, Philippe Charlier, and Raphael Olszewski
2013 Shape and Volume of the Craniofacial Cavities in Intentional Skull Deformation. *American Journal of Physical Anthropology* 151(1):110–119.

Kreger, Meggan M. B.
2010 Urban Population Dynamics in a Preindustrial New World City: Morbidity, Mortality, and Immigration in Postclassic Cholula. Unpublished Ph.D. dissertation, Department of Anthropology, Pennsylvania State University, State College.

Krogman, Wilton M.
1962 *The Human Skeleton in Forensic Medicine*. Charles C. Thomas, Springfield, Illinois.
Krogman, Wilton M., and M. Yaysar Iscan
1986 *The Human Skeleton in Forensic Medicine*. 2nd ed. Charles C. Thomas, Springfield, Illinois.
Kubler, George
1948 *Mexican Architecture of the 16th Century*. Yale University Press, New Haven.
LeBlanc, Steven A.
1999 *Prehistoric Warfare in the American Southwest*. University of Utah Press, Salt Lake City.
Leopold, A. Starker
1950 Vegetation Zones of Mexico. *Ecology* 31(4):507–518.
1959 *Wildlife of Mexico: The Game Birds and Mammals*. University of California Press, Berkeley.
Lewis, Oscar
1951 *Life in a Mexican Village: Tepotzlán Restudied*. University of Illinois Press, Urbana.
Liffman, Paul M., and Philip E. Coyle
2000 Introduction: Ritual and Historical Territoriality of the Cora (Náyari) and Huichol (Wixárika) Peoples. *Journal of the Southwest* 42(1):1–11.
Linne, Sigvald
2003 *Archaeological Researches at Teotihuacan, Mexico*. Reprint. University of Alabama Press, Tuscaloosa. Originally published 1934, V. Petterson, Stockholm.
Lister, Robert H.
1949 *Excavations at Cojumatlán, Michoacán, Mexico*. University of New Mexico Publications in Anthropology. University of New Mexico, Albuquerque.
1955 *The Present Status of the Archaeology of Western Mexico: A Distributional Study*. University of Colorado Studies, Series in Anthropology, No. 5.
Long, Stanley V., and Marcia V. V. Wire
1966 *Excavations at Barra de Navidad, Jalisco*. Antropológica No. 18. Instituto Caribe de Antropología y Sociología, Fundación La Salle de Ciencias Naturales, Caracas.
López, Atanasio
1960 *Misiones o doctrinas en Jalisco en el siglo XVII*. Estudios Históricos, Vol. II, No. 1. Librería Font, S.A., Guadalajara, Mexico.
López, Gonzalo
1963 Relación del Descubrimiento y Conquista que se Hizo por el Gobernador Nuño de Guzmán y su Ejército en las Provincias de la Nueva Galicia. In *Crónicas de la Conquista del Reino de Nueva Galicia en Territorio de la Nueva España*, pp. 61–113. H. Ayuntamiento de la Ciudad de Guadalajara e Instituto Jalisciense de Antropología e Historia, Guadalajara, Mexico.
López Portillo y Weber, José
1935 *La Conquista de Nueva Galicia*. Secretaría de Educacíon Publica, Talls. Grafs. De la Nación, Mexico.
Lugo, Ariel E., and Samuel C. Snedaker
1974 The Ecology of Mangroves. *Annual Review of Ecology and Systematics* 5:39–64.
Lumholtz, Carl
1902 *Unknown Mexico*. 2 vols. Charles Scribner's Sons, New York.
Macias, Jorge A.
1996 *Viñetas Escuinapenses*. Colección de Documentos para la Historia de Sinaloa 15. Centro de Estudios Históricos del Noroeste, A.C. Campus, Culiacán.
MacKenzie, Clyde L., Jr.
2001 The Fisheries for Mangrove Cockles, *Anadara* spp., from Mexico to Peru, with Descriptions of Their Habitats and Biology, the Fishermen's Lives, and the Effects of Shrimp Farming. *Marine Fisheries Review* 63(1):1–39.
Manley, Frederick H.
1975a Clay Mineralogy of the Soils of the West Mexican Coastal Plain. In *The Marismas Nacionales of Mexico: Report on Continuing Investigations of the Archaeology and Related Natural Science Studies. West Mexican Prehistory*, edited by G. Gordon Connally, Pt. 9: 32–41. Mimeograph, Department of Anthropology, State University of New York at Buffalo.
1975b Clay Petrology of the Geomorphic Surfaces on the Rio Acaponeta Valley. In *The Marismas Nacionales of Mexico: Report on Continuing Investigations of the Archaeology and Related Natural Science Studies. West Mexican Prehistory*, edited by G. Gordon Connally, Pt. 9:18–31. Mimeograph, Department of Anthropology, State University of New York at Buffalo.
Mann, Robert W., and David R. Hunt
2005 *Photographic Regional Atlas of Bone Disease: A Guide to Pathologic and Normal Variation in the Human Skeleton*. 3rd ed. Charles C. Thomas, Springfield, Illinois.
Manrique, Jimena
2004 *Informe Arqueozoológico de los Materiales de CECYTEN (Plantel 06) y San Felipe Aztatlán (Huachotita), Municipio de Tecuala, Nayarit*. Technical report, Instituto Nacional de Antropología e Historia, Nayarit, Tepic.
Manzanilla López, R., and Jorge A. Talavera Gonzáles
1988 *Informe de los trabajos de salvamento e investigación arqueológica el la población de Mochicahui, municipio de Fuerte, estado de Sinaloa*. Report on file, Dirección de Antropología Física, Departmento de Salvamento Arqueología, Instituto Nacional de Antropología e Historia, Mexico City.
Martin, Paul S., David A. Yetman, Mark Fishbein, Phil Jenkins, Thomas R. Van Devender, and Rebeca K. Wilson (editors)
1998 *Gentry's Rio Mayo Plants: The Tropical Deciduous Forest and Environs of Northwest Mexico*. University of Arizona Press, Tucson.
Mathiowetz, Michael D.
2011 The Diurnal Path of the Sun: Ideology and Interregional Interaction in Ancient Northwest Mesoamerica and the American Southwest. Unpublished Ph.D. dissertation, University of California, Riverside.
McAnany, Patricia A.
1987 Structure and Dynamics of Intercommunity Exchange. In *Maya Stone Tools: Selected Papers from the Second Maya Lithics Conference*, edited by Thomas R.

Hester and Harry J. Shafer, pp. 271–293. Monographs in World Archaeology No. 1. Prehistory Press, Madison, Wisconsin.

McCaa, Robert

2002 Paleodemography of the Americas: From Ancient Times to Colonialism and Beyond. In *The Backbone of History: Health and Nutrition in the Western Hemisphere*, edited by Richard H. Steckel and Jerome C. Rose, pp. 94–124. Cambridge University Press, Cambridge.

McCafferty, Geoffrey G.

2008 Domestic Practice in Postclassic Santa Isabel, Nicaragua. *Latin American Antiquity* 19(1):64–82.

McCarty, Karen, and Dan S. Matson

1975 Franciscan Report on the Indians of Nayarit, 1673. *Ethnohistory* 22(3):192–221.

McDowell, Fred W., and Richard P. Keizer

1977 Timing of Mid-Tertiary Volcanism in the Sierra Madre Occidental Between Durango City and Mazatlán, Mexico. *Bulletin of the Geological Society of America* 88:1479–1487.

McGoodwin, James R.

1973 Economy and Work on the Northwest Mexican Littoral: An Analysis of Labor Recruitment Among the Shark Fishermen of Teacapán, Sinaloa. Unpublished Ph.D. dissertation, Department of Anthropology, University of Texas, Austin.

1976 Society, Economy, and Shark-Fishing Crews in Rural Northwest Mexico. *Ethnology* 15:377–391.

1979 The Decline of Mexico's Pacific Inshore Fisheries. *Oceanus* 22(2):51–59.

McGuire, Randall H.

1987 *Death, Society, and Ideology in a Hohokam Community: Colonial and Sedentary Period Burials from La Ciudad.* Arizona State University Office of Cultural Resource Management Report No. 68. Tempe.

1992 *Death, Society, and Ideology in a Hohokam Community*. Westview Press, Boulder.

2001 Ideologies of Death and Power in the Hohokam Community of La Ciudad. In *Ancient Burial Practices in the American Southwest: Archaeology, Physical Anthropology, and Native American Perspectives*, edited by Douglas R. Mitchell and Judy L. Brunson-Hadley, pp. 27–44. University of New Mexico Press, Albuquerque.

Mecham, J. Lloyd

1968 *Francisco de Ibarra and Nueva Vizcaya*. Greenwood Press, New York.

Meehan, Betty

1977 The Role of Seafood in the Economy of a Contemporary Aboriginal Society in Coastal Arnhem Land. In *Joint Select Committee on Land Rights in the Northern Territory*, pp. 1085–1112. Hansard Report. Government Printer, Canberra.

1982 *Shell Bed to Shell Midden*. Australian Institute of Aboriginal Studies, New Series 37. Canberra.

Meighan, Clement W.

1969 Cultural Similarities Between Western Mexico and Andean Regions. In *Precolumbian Contact Within Nuclear America*, edited by J. Charles Kelley and Carroll L. Riley, pp. 11–25. Mesoamerican Studies, Series '69M4A. Research Records of the University Museum, Southern Illinois University, Carbondale.

1970 Molluscs as Food Remains in Archaeological Sites. In *Science in Archaeology: A Survey of Progress and Research*, edited by Don Brothwell and Eric Higgs, pp. 415–422. Prager Publishers, New York.

1971 Archaeology of Sinaloa. In *Archaeology of Northern Mesoamerica*, Pt. 2, edited by Gordon F. Ekholm and Ignacio Bernal, pp. 754–767. Handbook of Middle American Indians, Vol. 11, Robert Wauchope, general editor, University of Texas Press, Austin.

1972 *Archaeology of the Morett Site, Colima*. University of California Publications in Anthropology, Vol. 7. University of California Press, Berkeley.

1974 Prehistory of West Mexico. *Science* 184(4143):1254–1261.

1976a Pottery Artifacts Other than Vessels. In *The Archaeology of Amapa, Nayarit*, edited by Clement W. Meighan, pp. 66–95. Monumenta Archaeologica 2. Institute of Archaeology, University of California, Los Angeles.

1976b Chronology. In *The Archaeology of Amapa, Nayarit*, edited by Clement W. Meighan, pp. 51–59. Monumenta Archaeologica 2. Institute of Archaeology, University of California, Los Angeles.

1976c Ceramics. In *The Archaeology of Amapa, Nayarit*, edited by Clement W. Meighan, pp. 127–152. Institute of Archaeology, University of California, Los Angeles.

1976d Chipped Stone Artifacts. In *The Archaeology of Amapa, Nayarit*, edited by Clement W. Meighan, pp. 105–111. Institute of Archaeology, University of California, Los Angeles.

1976e Shell Artifacts. In *The Archaeology of Amapa, Nayarit*, edited by Clement W. Meighan, pp. 122–123. Monumenta Archaeologica 2. Institute of Archaeology, University of California, Los Angeles.

1976f Metal Artifacts. In *The Archaeology of Amapa, Nayarit*, edited by Clement W. Meighan, pp. 112–121. Institute of Archaeology, University of California, Los Angeles.

1976g Summary and Conclusion. In *The Archaeology of Amapa, Nayarit*, edited by Clement W. Meighan, pp. 153–162. Institute of Archaeology, University of California, Los Angeles.

Meighan, Clement W. (editor)

1976 *The Archaeology of Amapa, Nayarit*. Monumenta Archaeologica 2. Institute of Archaeology, University of California, Los Angeles.

Meighan, Clement W., and Leonard J. Foote

1968 *Excavations at Tizapan El Alto, Jalisco*. Latin American Studies, Vol. 11. University of California, Los Angeles.

Meighan, Clement W., Leonard J. Foote, and Paul V. Aiello

1968 Obsidian Dating in West Mexican Archaeology. *Science* 160(3832):211–214.

Michels, Joseph W.

1973 *Dating Methods in Archaeology*. Seminar Press, New York.

Michels, Joseph W., and Ignatius S. T. Tsong
1980 Obsidian Hydration Dating: A Coming of Age. In *Advances in Archaeological Method and Theory* 3:405–444. Academic Press, New York.

Miller, Mark E., and George W. Gill
1980 A Late Prehistoric Bundle Burial from Southern Wyoming. *Plains Anthropologist* 25:235–245.

Milner, George R., James W. Wood, and Jesper L. Boldsen
2008 Advances in Paleodemography. In *Biological Anthropology of the Human Skeleton*, edited by M. Anne Katzenberg and Shelley R. Saunders, pp. 561–600. John Wiley and Sons, Hoboken.

Mitchell, Douglas R.
1994a Summary and Conclusions. In *The Pueblo Grande Project, Vol. 7: An Analysis of Classic Period Mortuary Patterns*, edited by Douglas R. Mitchell, pp. 207–215. Soil Systems Publications in Archaeology, No. 20. Soil Systems, Phoenix.
1994b The Pueblo Grande Burial Artifact Analysis: A Search for Wealth, Ranking, and Prestige. In *The Pueblo Grande Project, Vol. 7: An Analysis of Classic Period Mortuary Patterns*, edited by Douglas R. Mitchell, pp. 129–180. Soil Systems Publications in Archaeology, No. 20. Soil Systems, Phoenix.

Mitchell, Douglas R., and Judy L. Brunson-Hadley
2001 An Evaluation of Classic Period Hohokam Burials and Society: Chiefs, Priests, or Acephalous Complexity? In *Ancient Burial Practices in the American Southwest: Archaeology, Physical Anthropology, and Native American Perspectives*, edited by Douglas R. Mitchell and Judy L. Brunson-Hadley, pp. 45–67. University of New Mexico Press, Albuquerque.

Mitchell, Douglas R., and J. A. Merewether
1994 Pit Descriptions. In *The Pueblo Grande Project, Vol. 2: Feature Descriptions, Chronology, and Site Structure*, edited by Douglas R. Mitchell, pp. 85–105. Soil Systems Publications in Archaeology, No. 20. Soil Systems, Phoenix.

Moore, David G., and Edwin C. Buffington
1968 Transform Faulting and Growth of the Gulf of California Since the Late Pliocene. *Science* 161(3847): 1238–1241.

Moorrees, Coenraad F. A.
1957 *The Aleut Dentition: A Correlative Study of Dental Characteristics in an Eskimoid People*. Harvard University Press, Cambridge.

Morganstein, Maury E., Carolyn L. Wicket, and Aaron Barkatt
1999 Considerations of Hydration-Rind Dating of Glass Artefacts: Alteration Morphologies and Experimental Evidence of Hydrogeochemical Soil-Zone Pore Water Control. *Journal of Archaeological Science* 26(9):1193–1210.

Moriarty, James M.
1964 The Influence of Strand Plain Morphology on the Development of Primitive Industries Along the Costa de Nayarit. Pt. I. *America Indigena* 24:365–379.
1965 The Influence of Strand Plain Morphology on the Development of Primitive Industries Along the Costa de Nayarit. Pt. II. *America Indigena* 25:65–77.

Mota Padilla, Matías Ángel de la
1742 *Conquista del Reino de la Nueva Galicia*. Manuscript on file, Biblioteca Digital Hispánica, Universidad de Guadalajara, Guadalajara, Mexico.
1920 *Historia de la conquista del reino de la Nueva Galicia*. Talleres gráficos de Gallardo y Alvarez del Castillo, Guadalajara, Mexico.
1973 *Historia del Reino de Nueva Galicia en la América Septentrional*. Colección Histórica de Obras Facsimilares 3. H. Ayuntamiento de la Ciudad de Guadalajara e Instituto Jaliciense de Antropología e Historia, Guadalajara, Mexico.

Mota y Escobar, Alonso de la
ca. 1604 *Descripción Geográfica de los Reinos de Nueva Galicia, Nueva Vizcaya y Nuevo León*. Guadalajara, Mexico.
1940 *Descripción Geográfica de los Reinos de Nueva Galicia, Nueva Vizcaya y Nuevo León*. Edited by Pedro Robredo. Mexico City.
1966 *Descripción Geográfica de los Reinos de Nueva Galicia, Nueva Vizcaya y Nuevo León*. Colección Histórica de Obras Facsimilares. H. Ayuntamiento de la Ciudad de Guadalajara e Instituto Jalisciense de Antropología e Historia, Guadalajara, Mexico.

Mountjoy, Joseph B.
1969 On the Origin of West Mexican Metallurgy. In *Precolumbian Contact Within Nuclear America*, edited by J. Charles Kelley and Carroll L. Riley, pp. 26–42. Mesoamerican Studies, '69M4A. Research Records of the University Museum, Southern Illinois University, Carbondale.
1970a La sucesión cultural en San Blas. *Boletín del Instituto Nacional de Antropología e Historia* 39:41–48. Mexico City.
1970b Prehispanic Culture History and Cultural Contact on the Southern Coast of Nayarit, Mexico. Unpublished Ph.D. dissertation, Department of Anthropology, Southern Illinois University, Carbondale.
1974 San Blas Complex Ecology. In *The Archaeology of West Mexico*, edited by Betty B. Bell, pp. 106–119. Sociedad de Estudios Avanzados del Occidente de México, Ajijic, Mexico.
1990 El Desarrollo de la Cultura Aztatlán en el Occidente de México Visto Desde Su Frontera Sur-Oeste. In *Mesoamerica y Norte de Mexico, Siglo IX–XII*, Vol. 2, edited by Federica Sodi Miranda, pp. 541–564. Instituto Nacional de Antropología e Historia, Mexico City.
2000 Prehispanic Cultural Development Along the Southern Coast of West Mexico. In *Greater Mesoamerica: The Archaeology of West and Northwest Mexico*, edited by Michael S. Foster and Shirley Gorenstein, pp. 81–106. University of Utah Press, Salt Lake City.
2002 El desarrollo de cultura indígena en la costa de Jalisco, Municipio de Puerto Vallarta. *Mexicoa* 3:25–37.

2012 *El Pantano y otros Sitios del Formativo Medio en el Valle de Mascota, Jalisco*. Secretaría de Cultura, Gobierno de Jalisco, Guadalajara, Mexico.

Mountjoy, Joseph B., Sergio Aburto, Luis Barba, and Salvador Gutierrez

1982 Late Postclassic Commerce in the Tomatlán River Valley, Jalisco: Clay Mining and Analysis of Fine Paste Pottery. *Anthropology* 6(1–2):128–198.

Mountjoy, Joseph B., and Cheryl P. Claassen

2005 Middle Formative Diet and Seasonality on the Central Coast of Nayarit, Mexico. In *Archaeology Without Limits: Papers in Honor of Clement W. Meighan*, edited by Brian D. Dillon and Matthew A. Boxt, pp. 267–282. Labyrinthos Press, Lancaster, California.

Mountjoy, Joseph B., R. E. Taylor, and Lawrence H. Feldman

1972 Matanchén Complex: New Radiocarbon Dates on Early Coastal Adaptation in West Mexico. *Science* 175(4027):1242–1243.

Mountjoy, Joseph B., and Luis Torres M.

1985 The Production and Use of Prehispanic Metal Artifacts in the Central Coastal Area of Jalisco, Mexico. In *The Archaeology of West and Northwest Mesoamerica*, edited by Michael S. Foster and Phil C. Weigand, pp. 133–152. Westview Press, Boulder.

Muriá, José Maria

2000 Una Descripción de la Nueva Galicia en 1621: La de Domingo Lázaro de Arregui y a la Espera de una Nueva Edición. *Relaciones* 21(84):253–271. El Colegio de Michoacán, Zamora, Mexico.

Navarro, Everardo Peña

1946 *Estudio Histórico del Estado de la Conquista a la Independencia*. Tepic.

Nelson, Ben A.

2004 Current and Future Directions in Northwest Mexican Archaeology. In *Surveying the Archaeology of Northwest Mexico*, edited by Gillian E. Newell and Emiliano Gallaga, pp. 289–296. University of Utah Press, Salt Lake City.

Nelson, Ben A., J. Andrew Darling, and David A. Kice

1992 Mortuary Practices and the Social Order at La Quemada, Zacatecas. *Latin American Antiquity* 3(4):298–315.

Neurath, Johannes

2000 Tukipa Ceremonial Centers in the Community of Tuapurie (Santa Catarina Cuexcomatitlán): Cargo Systems, Landscape, and Cosmovision. *Journal of the Southwest* 42(1):81–110.

Nicholson, Henry B., and Clement W. Meighan

1974 The University of California at Los Angeles Department of Anthropology Program in West Mexican Archaeology-Ethnology, 1965–1970. In *The Archaeology of West Mexico*, edited by Betty B. Bell, pp. 6–18. Sociedad de Estudios Avanzados del Occidente de México, Ajijic, Mexico.

Nietschmann, Bernard

1973 *Between Land and Water: The Subsistence Ecology of the Miskito Indians, Eastern Nicaragua*. Seminar Press, New York.

Noguera, Eduardo

1942 Exploraciones en El Opeño, Michoacán. *XXVII Congreso Internacional de Americanistas* 1:574–586. Mexico City.

Noli, Dieter, and Graham Avery

1988 Protein Poisoning and Coastal Subsistence. *Journal of Archaeological Science* 15(4):395–401.

O'Brien, Patricia A.

1975 Foraminiferal Assemblages of El Calón II. In *The Marismas Nacionales of Mexico: Report on Continuing Investigations of the Archaeology and Related Natural Science Studies. West Mexican Prehistory*, edited by G. Gordon Connally, Pt. 9:57–63. Mimeograph, Department of Anthropology, State University of New York at Buffalo.

Odum, Eugene P.

1971 *Fundamentals of Ecology*. 3rd ed. W. B. Saunders Company, Philadelphia.

Ohnersorgen, Michael A.

2004 *Informe Tecnico Final de Proyecto "Investigaciones Arqueológica Preliminar de Chacalilla, Nayarit, México, Temporada 2003."* Instituto Nacional de Antropología e Historia, Nayarit, Tepic.

2007 *La Organización Socio-economica y la Interacción Regional de un Centro Aztatlán: Investigaciones Arqueológicos en Chacalilla, Nayarit. Informe Tecnico Parcial, Temporada de 2005*. Instituto Nacional de Antropología e Historia, Nayarit, Tepic.

Ohnersorgen, Michael A., Michael D. Glascock, Joseph B. Mountjoy, and Mauricio Garduño A.

2012 *Chemical Analysis of Obsidian Artifacts from the West Coast of Mexico: Implications for Regional and Interregional Economic Organization*. Manuscript on file with the authors.

Oliveros, José Arturo

1974 Nuevas Exploraciones en El Opeño, Michoacán. In *The Archaeology of West Mexico*, edited by Betty B. Bell, pp. 182–201. Sociedad de Estudios Avanzados del Occidente de México, Ajijic, Mexico.

Olivier, Georges

1969 *Practical Anthropology*. Charles C. Thomas, Springfield, Illinois.

Ortega, José de

1754 *Marvillosa conquista y reducción de la provincial de San Joseph del Gran Nayar, Nuevo Reino de Toledo*. Mexico City.

1887 *Historia del Nayarit, Sonora, Sinaloa y Ambas Californias*. Mexico City.

Ortner, Donald J.

2003 *Identification of Pathological Conditions in Human Skeletal Remains*. Smithsonian Institution Press, Washington, D.C.

Osborn, Alan J.

1977a Aboriginal Exploitation of Marine Food Resources. Unpublished Ph.D. dissertation, Department of Anthropology, University of New Mexico, Albuquerque.

1977b Strandloopers, Mermaids, and Other Fairy Tales: Ecological Determinants of Marine Resource

Utilization—The Peruvian Case. In *For Theory Building in Archaeology: Essays on Faunal Remains, Aquatic Resources, Spatial Analysis, and Systemic Modeling*, edited by Lewis R. Binford, pp. 157–205. Academic Press, New York.

O'Shea, James

1981 Social Configurations and the Archaeological Study of Mortuary Practices: A Case Study. In *The Archaeology of Death*, edited by Robert Chapman, Ian Kinnes, and Klavs Randsborg, pp. 39–52. Cambridge University Press, Cambridge.

Owsley, Douglas W., Ann M. Mires, and George W. Gill

1985 Carious Lesions in Permanent Dentitions of Protohistoric Easter Islanders. *Journal of the Polynesian Society* 94(4):415–422.

Parsons, Jeffery R., and Mary Hrones Parsons

1990 *Maguey Utilization in Highland Central Mexico: An Archaeological Ethnography*. University of Michigan, Anthropology Papers No. 82. Museum of Anthropology, Ann Arbor.

Parsons, Mary H.

1972 Spindle Whorls from the Teotihuacán Valley, Mexico. In *Miscellaneous Studies in Mexican Prehistory*, by Jeffery R. Parsons, Michael W. Spence, and Mary Hrones Parsons, pp. 45–80. Anthropology Papers No. 45. Museum of Anthropology, University of Michigan, Ann Arbor.

Paso y Troncoso, Francisco del

1905 *Suma de Visitas de pueblos de la Nueva España 1548–1550*. Edited by René Garcia Castro. Madrid.

Pearson, Emerson L.

1968 Geocultural Survey of the West Mexican Marismas Nacionales. Unpublished Master's thesis, Department of Anthropology, State University of New York at Buffalo.

Pendergast, David M.

1962a Metal Artifacts from Amapa, Nayarit, Mexico. *American Antiquity* 27(3):370–379.

1962b Metal Artifacts in Prehispanic Mesoamerica. *American Antiquity* 27(4):520–545.

Pennington, Campbell W.

1963 *The Tarahumara of Mexico*. University of Utah Press, Salt Lake City.

1969 *The Tepehuan of Chihuahua*. University of Utah Press, Salt Lake City.

1980 *The Pima Bajo of Central Sonora, Mexico*, Vol. I. University of Utah Press, Salt Lake City.

Penrose, Lionel S.

1954 Distance, Size, and Shape. *Annals of Eugenics* 18: 337–343.

Perez, Manuel, Lorena Gámez, and Mauricio Garduño

2000 *Proyecto de Salvamento Arqueológico "Autopista Entronque San Blas-Mazatlán, tramo Nayarit." Informe Técnico, Trabajos de Reconocimiento de Superficie y Excavación (1998)*. Dirección de Salvamento Arqueológico Centro Instituto National de Antropología e Historia, Nayarit, Tepic.

Phillips, David A., Jr.

2012 The Northwest Mexican Polychrome Traditions. In *Potters and Communities of Practice: Glaze Paint and Polychrome Pottery in the American Southwest, AD 1250–1700*, edited by Linda S. Cordell and Judith A. Habicht-Mauche, pp. 34–44. Anthropological Papers of the University of Arizona, No. 75. Tucson.

Pierce, Daniel E.

2012 Obsidian Source Frequencies as a Social Attribute at San Felipe Aztatán, Mexico. Unpublished Master's thesis, Department of Anthropology, University of Missouri, Columbia.

Pompa y Pompa, Antonio

1960 *Memorias y Revista del Congreso Mexicano de Historia*. Mexico City.

Ponce, Alonso

1872 *Tratado curioso y docto de las grandezas de la Nueva Espana: Relación breve y verdadera de algunas cosas de las muchas que sucedieron al fray Alonso Ponce en las provincias de la Nueva España, siendo comisario general de aquellas partes*. Madrid.

1968 *Tratado curioso y docto de las grandezas de la Nueva Espana: Relación breve y verdadera de algunas cosas de las muchas que sucedieron al fray Alonso Ponce en las provincias de la Nueva España, siendo comisario general de aquellas partes*. In *Viajes de Fray Alonso Ponce al Occidente de México*, edited by José Ramírez Flores, pp. 68–86. Correponsalia del Seminario de Cultura Méxicana, Guadalajara, Mexico.

Pool, Christopher, and George J. Bey

2007 *Pottery Economies in Mesoamerica*. University of Arizona Press, Tucson.

Pool, Douglas J., Samuel C. Snedaker, and Ariel E. Lugo

1977 Structure of Mangrove Forests in Florida, Puerto Rico, Mexico, and Costa Rica. *Biotropica* 9(3):195–212.

Porter, Muriel Noe

1956 Excavations at Chupícuaro, Guanajuato, Mexico. *Transactions of the American Philosophical Society* 46:515–637. Philadelphia.

Prat, Jordi Gómez i, and Sheila M. F. Mendonça de Souza

2003 Prehistoric Tuberculosis in America: Adding Comments to a Literature Review. *Memórias do Instituto Oswaldo Cruz* 98(Suppl. 1):151–159. Institute of Oswaldo Cruz, Rio de Janeiro.

Pritchard, D. W.

1967 What Is an Estuary: Physical Viewpoint. In *Estuaries*, edited by George H. Lauff, pp. 3–5. American Association for the Advancement of Science, Publication 83. Washington, D.C.

Publ, Helmut

1985 Prehispanic Exchange Networks and the Development of Social Complexity in Western Mexico: The Aztatlán Interaction Sphere. Unpublished Ph.D. dissertation, Department of Anthropology, Southern Illinois University.

1986 *Prehispanic Exchange Networks and the Development of Social Complexity in Western Mexico: The Aztatlán Interaction Sphere*. Southern Illinois University Press, Carbondale.

1990 Interaction Spheres, Merchants, and Trade in Prehispanic West Mexico. In *Research in Economic Anthropology: A Research Annual*, edited by Barry L. Isaac, Vol. 12:201–242. JAI Press, Greenwich, Connecticut.

Pye, Mary E., and Arthur A. Demarest
1991 The Evolution of Complex Societies in Southeastern Mesoamerica: New Evidence from El Mesak, Guatemala. In *The Formation of Complex Society in Southeastern Mesoamerica*, edited by William R. Fowler, Jr., pp. 77–100. CRC Press, Boca Raton.

Ramírez de Swartz, Susana, C. Liot, J. Reveles, O. Schöndube, C. Cérdenas, F. Mata, C. Melgarejo, L. Santoyo, and V. B. Diego
2005 La Peña: Un Sitio Transición Entre el Epiclasico y el Postclásico Temprano en La Cuenca de Sayula, Jalisco. In *El Antiguo Occidente de México: Nuevas Perspectivas sobre el Pasado Prehispánico*, edited by Eduardo Williams, Phil C. Weigand, Lorenza López Mestas, and David C. Grove, pp. 305–329. El Colegio de Michoacán, Zamora, Mexico.

Rao, Dinesh
2013 Muscular Changes. *Dr. Dinseh Rao's Forensic Pathology*. Electronic document, http://www.forensicpathologyonline.com/e-book/post-mortem-changes/muscular-changes.

Ravesloot, John C.
1988 *Mortuary Practices and Social Differentiation at Casas Grandes, Chihuahua, Mexico*. Anthropological Papers of the University of Arizona, No. 49. Tucson.

Rawlings, Kristen J.
2002 Racial Variation in the Palate Shape and Form of the Transverse Palatine Suture. Unpublished Master's thesis, Department of Anthropology, University of Wyoming, Laramie.

Reimer, Paula J., Michael G. L. Baille, Edouard Bard, Alex Bayliss, J. Warren Beck, Chanda Bertrand, Paul G. Blackwell, Caitlin E. Buck, George Burr, Kirsten B. Cutler, Paul E. Damon, R. Lawrence Edwards, Richard G. Fairbanks, Michael Friedrich, Thomas P. Guilderson, Alan G. Hogg, Konrad A. Hughen, Bernd Kromer, F. Gerry McCormac, Sturt Manning, Christopher B. Ramsey, Ron W. Reimer, Sabine Remmele, John R. Southon, Minze Stuiver, Sahra Talamo, F. W. Taylor, Johannes van der Plicht, and Costanze E. Weyhenmeyer
2004 INTCAL04 Terrestrial Radiocarbon Age Calibration, 0–26 cal kyr BP. *Radiocarbon* 46(3):1029–1058.

Reitz, Elizabeth J., and Elizabeth S. Wing
2008 *Zooarchaeology*. 2nd ed. Cambridge Manuals in Archaeology. Cambridge University Press, Cambridge.

Relación Anónima Cuarta
1955 In *Memoria de los Servicios que habia hecho Nuño de Guzmán, desde que fue nombrado Gobernador de Pánuco en 1525*. Notes by Manuel Carrera Stampa. José Porrúa e. Hijos Sucs., Mexico City.

Relación Anónima Primera
1963 In *Crónicas de la Conquista del Reino de Nueva Galicia en Territorio de la Nueva España*. H. Ayuntamiento de la Ciudad de Guadalajara e Instituto Jalisciense de Antropología e Historia, Guadalajara, Mexico.

Relación Anónima Segunda
1963 In *Crónicas de la Conquista del Reino de Nueva Galicia en Territorio de la Nueva España*. H. Ayuntamiento de la Ciudad de Guadalajara e Instituto Jalisciense de Antropología e Historia, Guadalajara, Mexico.

Relación Anónima Tercera
1955 In *Memoria de los Servicios que habia hecho Nuño de Guzmán, desde que fue nombrado Gobernador de Pánuco en 1525*. Notes by Manuel Carrera Stampa. José Porrúa e Hijos Sucs., Mexico City.

Rhine, Stanley
1990 Non-Metric Skull Racing. In *Skeletal Attribution of Race: Methods for Forensic Anthropology*, edited by George W. Gill and Stanley Rhine, pp. 9–20. Maxwell Museum of Anthropology, Anthropological Papers No. 4. Albuquerque.

Rhodes, Daniel
1957 *Clay and Glazes for the Potter*. Chilton Book Company, Philadelphia.

Ridings, Rosanna
1996 Where in the World Does Obsidian Hydration Dating Work? *American Antiquity* 61(1):136–148.

Riley, Carroll L.
1963 Color-Directional Symbolism: An Example of Mexican-Southwestern Contacts. *América Indigena* 23(1):49–60.
1987 *The Frontier People. The Greater Southwest in the Protohistoric Period*. University of New Mexico Press, Albuquerque.
2005 *Becoming Aztlán: Mesoamerican Influence in the Greater Southwest, AD 1200–1500*. University of Utah Press, Salt Lake City.

Rogers, Martha
2008 An Analysis of Discrete Cranial Traits of Northwestern Plains Indians. In *Skeletal Biology and Bioarchaeology of the Northwestern Plains*, edited by George W. Gill and Rick L. Weathermon, pp. 177–199. University of Utah Press, Salt Lake City.

Rogers, Phil
2003 *Ash Glazes*. University of Pennsylvania Press, Philadelphia.

Romano, Arturo
1965 *Estudio Morfologico de la Deformación Craneana en Tamuín, S.L.P. y en la Isla del Idolo, Ver*. Serie Investigaciones 10. Instituto Nacional de Antropología e Historia, Mexico City.
1970 Preceramic Human Remains. In *Physical Anthropology*, edited by T. Dale Stewart, pp. 22–34. Handbook of Middle American Indians, Vol. 9, Robert Wauchope, general editor, University of Texas Press, Austin.

Romero, Javier
1958 *Mutilaciónes Dentarias Prehispánicas de México y América en general*. Serie Investigaciones 3. Instituto Nacional de Antropología e Historia, Mexico City.
1970 Dental Mutilation, Trephination, and Cranial Deformation. In *Physical Anthropology*, edited by T. Dale Stewart, pp. 50–67. Handbook of Middle American Indians, Vol. 9, Robert Wauchope, general editor, University of Texas Press, Austin.

Rossman, David L.
1976 A Site Catchment Analysis of San Lorenzo, Veracruz. In *The Early Mesoamerican Village*, edited by Kent V. Flannery, pp. 95–103. Academic Press, New York.

Rucker, Mark
1969 Excavations at La Tarjea. In *Archaeological Reconnaissance and Excavations in the Marismas Nacionales, Sinaloa and Nayarit, Mexico. West Mexican Prehistory*, edited by Stuart D. Scott, Pt. 3:1–23. Mimeograph, Department of Anthropology, State University of New York at Buffalo.

Russell, Frank
1908 *The Pima Indians.* 26th Annual Report of the Bureau of American Ethnology. Washington, D.C.

Rydout, Gary B.
1975 El Calón: A New Proposal. In *The Marismas Nacionales of Mexico: Report on Continuing Investigations of the Archaeology and Related Natural Science Studies. West Mexican Prehistory*, edited by G. Gordon Connally, Pt. 9:64–70. Mimeograph, Department of Anthropology, State University of New York at Buffalo.

Salo, Wilmar L., Arthur C. Aufderheide, Jane E. Buikstra, and Todd A. Holcolm
1994 Identification of *Mycobacterium tuberculosis* DNA in a Pre-Columbian Peruvian Mummy. *Proceedings of the National Academy of Sciences* 91:2091–2094.

Sámano, Juan de
1963 Relación de la Conquista de los Teules Chichimecas que Dió el Capitán de Emergencia Juan de Sámano. In *Crónicas de la Conquista del Reino de Nueva Galicia en Territorio de la Nueva España*, pp. 115–152. H. Ayuntamiento de la Cidudad de Guadlajara e Instituto Jalisciense de Antropología e Historia, Guadalajara, Mexico.

Sanburg, Delmer E.
1976 Appendix D: Faunal Remains from Amapa. In *The Archaeology of Amapa, Nayarit*, edited by Clement W. Meighan, pp. 201–203. Monumenta Archaeologica 2. Institute of Archaeology, University of California, Los Angeles.

Sanders, William T., and Deborah L. Nichols
1988 Ecological Theory and Cultural Evolution in the Valley of Oaxaca. *Current Anthropology* 29(1):33–80.

Santos Ramírez, Víctor Joel
2002 *Informe del Rescate Arqueológico Realizado en el Sitio "Tierra del Padre," Chametla, Sinaloa.* Technical report, Coordinación Nacional de Arqueología del Instituto Nacional de Antropología e Historia, Mexico City.

2008a *Análisis de los Materiales Arqueológicos del Rescate Arqueológico Realizado en el Sitio Loma del Panteón, Chametla, Sinaloa.* Technical report, Coordinación Nacional de Arqueología del Instituto Nacional de Antropología e Historia, Mexico City.

2008b *Informe Técnico del Rescate Arqueológico Realizado en el Sitio Loma del Panteón, Chametla, Sinaloa.* Technical report, Coordinación Nacional de Arqueología del Instituto Nacional de Antropología e Historia, Mexico City.

2012a Excavaciónes en Tierra del Padre Chametla, Sinaloa. In *Trópico de Cáncer: Estudios de Historia y Arqueología sobre el Sur de Sinaloa*, edited by Alfonso Grave Tirado, Víctor Joel Santos Ramírez, and Gilberto López Castillo, pp. 73–101. Centro Instituto Nacional de Antropología e Historia, Sinaloa, Culiacán.

2012b Las Figurillas Prehispánicas de Chametla. In *Trópico de Cáncer: Estudios de Historia y Arqueología sobre el Sur de Sinaloa*, edited by Alfonso Grave Tirado, Víctor Joel Santos Ramírez, and Gilberto López Castillo, pp. 103–116. Centro Instituto Nacional de Antropología e Historia, Sinaloa, Culiacán.

Santos Ramírez, Víctor Joel, and Julio César Vicente López
2012 Las Labradas y las Culturas Costeras del Trópico de Cáncer. In *Trópico de Cáncer: Estudios de Historia y Arqueología sobre el Sur de Sinaloa*, edited by Alfonso Grave Tirado, Víctor Joel Santos Ramírez, and Gilberto López Castillo, pp. 17–32. Centro Instituto Nacional de Antropología e Historia, Sinaloa, Culiacán.

Sattenspiel, Lisa, and Henry Harpending
1983 Stable Populations and Skeletal Age. *American Antiquity* 48(3):489–498.

Sauer, Carl O.
1932 *The Road to Cíbola.* Ibero-Americana 3. University of California Press, Berkeley.

1934 *The Distribution of Aboriginal Tribes and Languages in Northwestern México.* Ibero-Americana 5. University of California Press, Berkeley.

1935 *Aboriginal Populations of Northwestern México.* Ibero-Americana 10. University of California Press, Berkeley.

Sauer, Carl O., and Donald D. Brand
1932 *Aztatlán: Prehistoric Mexican Frontier on the Pacific Coast.* Ibero-Americana 1. University of California Press, Berkeley.

Saul, Frank P.
1972 *The Human Skeletal Remains of Altar de Sacrificios: An Osteobiographic Analysis.* Papers of the Peabody Museum of Archaeology and Ethnology, 63(2). Peabody Museum, Cambridge, Massachusetts.

Schalk, Randall
1977 The Structure of an Anadromous Fish Resource. In *For Theory Building in Archaeology: Essays on Faunal Remains, Aquatic Resources, Spatial Analysis, and Systemic Modeling*, edited by Lewis R. Binford, pp. 207–249. Academic Press, New York.

Scheiber, Laura L.
2006 Skeletal Biology: Plains. In *Environment, Origins, and Population*, edited by Douglas Ubelaker, pp. 595–609. Handbook of North American Indians, Vol. 3, William C. Sturtevant, general editor, Smithsonian Institution Press, Washington, D.C.

2008 Life and Death on the Northwestern Plains: Mortuary Practices and Cultural Transformations. In *Skeletal Biology and Bioarchaeology of the Northwestern Plains*, edited by George W. Gill and Rick L. Weathermon, pp. 22–41. University of Utah Press, Salt Lake City.

Scheiber, Laura L., and George W. Gill
1996 Bioarchaeology of the Northwestern Plains. In *Archaeological and Bioarchaeological Resources of the*

Northern Plains, edited by George C. Frison and Robert C. Mainfort, pp. 91–119. Arkansas Archaeological Survey Research Series No. 47. Fayetteville.

Schlesinger, Roger, and Arthur P. Stabler
1986 *André Thevet's North America: A Sixteenth-Century View*. McGill-Queen's University Press, Kingston, Canada.

Schöndube, Otto
1998 Natural Resources and Human Settlements in Ancient West Mexico. In *Ancient West Mexico: Art and Archaeology of the Unknown Past*, edited by Richard F. Townsend, pp. 205–215. Art Institute of Chicago, Chicago and Thames and Hudson, London.

Schutte, James E.
1973 Prehistoric Inhabitants of Central West Mexico: Cristo Rey and Rincon de Panal. Unpublished Master's thesis, Department of Anthropology, University of Wyoming, Laramie.

Schutte, James E., and George W. Gill
1972 Preliminary Analysis of Human Burials from the Marismas Nacionales: Recent Findings. In *Archaeological Reconnaissance and Excavations in the Marismas Nacionales, Sinaloa and Nayarit, Mexico. West Mexican Prehistory*, edited by Stuart D. Scott, Pt. 6:1–24. Mimeograph, Department of Anthropology, State University of New York at Buffalo.

Schutte, James E., and Marla Speas
1973 Burial Practices at Venadillo Site. In *Archaeological Reconnaissance and Excavations in the Marismas Nacionales, Sinaloa and Nayarit, Mexico. West Mexican Prehistory*, edited by Stuart D. Scott, Pt. 7:17–22. Mimeograph, Department of Anthropology, State University of New York at Buffalo.

Scott, Patricia Kay
1975 An Ethnohistoric Note Related to Geomorphological Aspects of the Marismas Nacionales, Mexico. In *The Marismas Nacionales of Mexico: Report on Continuing Investigations of the Archaeology and Related Natural Science Studies. West Mexican Prehistory*, edited by G. Gordon Connally, Pt. 9:42–45. Mimeograph, Department of Anthropology, State University of New York at Buffalo.

Scott, Stuart D.
1985 Core Versus Marginal Mesoamerica: A Coastal West Mexican Perspective. In *The Archaeology of West and Northwest Mesoamerica*, edited by Michael S. Foster and Phil C. Weigand, pp. 181–192. Westview Press, Boulder.
1999 The Marismas Nacionales Project, Sinaloa and Nayarit, Mexico. In *Pacific Latin America in Prehistory: The Evolution of Archaic and Formative Cultures*, edited by Michael Blake, pp. 13–24. Washington State University Press, Pullman.

Scott, Stuart D. (editor)
1967 *Archaeological Reconnaissance and Excavations in the Marismas Nacionales, Sinaloa, and Nayarit, Mexico. West Mexican Prehistory*, Pt. 1. Mimeograph, Department of Anthropology, State University of New York at Buffalo.
1968 *Archaeological Reconnaissance and Excavations in the Marismas Nacionales, Sinaloa, and Nayarit, Mexico. West Mexican Prehistory*, Pt. 2. Mimeograph, Department of Anthropology, State University of New York at Buffalo.
1969 *Archaeological Reconnaissance and Excavations in the Marismas Nacionales, Sinaloa, and Nayarit, Mexico. West Mexican Prehistory*, Pt. 3. Mimeograph, Department of Anthropology, State University of New York at Buffalo.
1970 *Archaeological Reconnaissance and Excavations in the Marismas Nacionales, Sinaloa, and Nayarit, Mexico. West Mexican Prehistory*, Pt. 4. Mimeograph, Department of Anthropology, State University of New York at Buffalo.
1971 *Archaeological Reconnaissance and Excavations in the Marismas Nacionales, Sinaloa, and Nayarit, Mexico. West Mexican Prehistory*, Pt. 5. Mimeograph, Department of Anthropology, State University of New York at Buffalo.
1972 *Archaeological Reconnaissance and Excavations in the Marismas Nacionales, Sinaloa, and Nayarit, Mexico. West Mexican Prehistory*, Pt. 6. Mimeograph, Department of Anthropology, State University of New York at Buffalo.
1973 *Archaeological Reconnaissance and Excavations in the Marismas Nacionales, Sinaloa, and Nayarit, Mexico. West Mexican Prehistory*, Pt. 7. Mimeograph, Department of Anthropology, State University of New York at Buffalo.

Scott, Stuart D., and Charles J. Cazeau
1976 Interdisciplinary Approach to Ceramic History: Cultural and Mineralogic Sherd Analysis. *American Ceramic Society Bulletin* 55(2):213–214.

Scott, Stuart D., and Michael S. Foster
2000 The Prehistory of Mexico's Northwest Coast: A View from the Marismas Nacionales of Sinaloa and Nayarit. In *Greater Mesoamerica: The Archaeology of West and Northwest Mexico*, edited by Michael S. Foster and Shirley Gorenstein, pp. 107–135. University of Utah Press, Salt Lake City.

Sharer, Robert J., and Wendy Ashmore
1979 *Fundamentals of Archaeology*. Benjamin/Cummings, Menlo Park.

Shenkel, J. Richard
1968 The Marismas Nacionales. In *Archaeological Reconnaissance and Excavations in the Marismas Nacionales, Sinaloa and Nayarit, Mexico. West Mexican Prehistory*, edited by Stuart D. Scott, Pt. 2:24–42. Mimeograph, Department of Anthropology, State University of New York at Buffalo.
1969a Shell Mound Archaeology in the Marismas Nacionales. In *Archaeological Reconnaissance and Excavations in the Marismas Nacionales, Sinaloa and Nayarit, Mexico. West Mexican Prehistory*, edited by Stuart D. Scott, Pt. 3:24–46. Mimeograph, Department of Anthropology, State University of New York at Buffalo.
1969b Interpretational Comment on Tecualilla. In *Archaeological Reconnaissance and Excavations in the*

Marismas Nacionales, Sinaloa and Nayarit, Mexico. West Mexican Prehistory, edited by Stuart D. Scott, Pt. 3:80–87. Mimeograph, Department of Anthropology, State University of New York at Buffalo.

1971a Cultural Adaptation to the Mollusk: A Methodological Survey of Shellmound Archaeology and a Consideration of the Shellmounds of the Marismas Nacionales, West Mexico. Unpublished Ph.D. dissertation, Department of Anthropology, State University of New York at Buffalo.

1971b El Calón: Revisited. In *Archaeological Reconnaissance and Excavations in the Marismas Nacionales, Sinaloa and Nayarit, Mexico. West Mexican Prehistory*, edited by Stuart D. Scott, Pt. 5:19–24. Mimeograph, Department of Anthropology, State University of New York at Buffalo.

1972 A Zoomorphic Shell Pendent from Chalpa. In *Archaeological Reconnaissance and Excavations in the Marismas Nacionales, Sinaloa and Nayarit, Mexico. West Mexican Prehistory*, edited by Stuart D. Scott, Pt. 6:77–82. Mimeograph, Department of Anthropology, State University of New York at Buffalo.

1974 Quantitative Analysis and Population Estimates of the Shell Mounds of the Marismas Nacionales, Sinaloa and Nayarit, Mexico. In *The Archaeology of West Mexico*, edited by Betty B. Bell, pp. 57–67. Sociedad de Estudios Avanzados del Occidente de México, Ajijic, Mexico.

Shook, Edwin M., and M. P. Hatch

1979 Early Preclassic Sequence in the Ocós-Salinas La Blanca Area, South Coast of Guatemala. *Contributions of the University of California Archaeological Research Facility* 4:143–195. Berkeley.

Shreve, Forrest

1934 Vegetation of the Northwestern Coast of Mexico. *Bulletin of the Torrey Botanical Club* 61(7):373–380.

1937 Lowland Vegetation of Sinaloa. *Bulletin of the Torrey Botanical Club* 64:605–613.

Sirkin, Leslie A.

1974 A Palynologic Model for Reconstructing Vegetation and Environments in the Marismas Nacionales, Sinaloa, Mexico. In *The Marismas Nacionales of Mexico: Report on Continuing Investigations of the Archaeology and Related Natural Science Studies. West Mexican Prehistory*, edited by G. Gordon Connally, Pt. 8:22–31. Mimeograph, Department of Anthropology, State University of New York at Buffalo.

1975 Palynological Research in the West Mexican Project: An Overview. In *The Marismas Nacionales of Mexico: Report on the Continuing Investigations of the Archaeology and Related Natural Science Studies. West Mexican Prehistory*, edited by G. Gordon Connally, Pt. 9:46–50. Mimeograph, Department of Anthropology, State University of New York at Buffalo.

1984 Late Pleistocene Stratigraphy and Environments of the West Mexican Coastal Plain. In *Neotectonics and Sea Level Variations in the Gulf of California, a Symposium*, edited by V. Malpica-Cruz, S. Celis-Gutiérrez, J. Guerrero-Garcia, and L. Ortlieb, pp. 309–318. Hermosillo.

1985 Late Quaternary Stratigraphy and Environments of the West Mexican Coastal Plain. *Palynology* 9:3–25.

Sirkin, Leslie A., and G. Gordon Connally

1977 Quaternary Geomorphology and Environments of the West Mexican Coastal Plain. *Geological Society of America, Abstracts with Programs* 9:1178–1179.

Sirkin, Leslie A., G. Gordon Connally, and Robert Stuckenrath

1981 The West Mexican Coastal Plain: Stratigraphy and Geologic History. *Geological Society of America, Abstracts with Programs* 13(2):106.

Sirkin, Leslie A., and Deborah Gilbert

1975 Palynology and Sea Level Change in the Marismas Nacionales: Further Evidence from El Calón. In *The Marismas Nacionales of Mexico: Report on Continuing Investigations of the Archaeology and Related Natural Science Studies. West Mexican Prehistory*, edited by G. Gordon Connally, Pt. 9:51–56. Mimeograph, Department of Anthropology, State University of New York at Buffalo.

Skinner, Craig E., and Kimberly J. Tremaine

1993 *Obsidian: An Interdisciplinary Bibliography*. International Association for Obsidian Studies Occasional Paper No. 1. San Jose.

Smith, A. Ledyard

1972 *Excavations at Altar de Sacrificios: Architecture, Settlement, Burials and Caches.* Papers of the Peabody Museum of Archaeology and Ethnology, Harvard University, 62(2). Peabody Museum, Cambridge, Massachusetts.

Smith, Michael E., and Cynthia M. Heath-Smith

1980 Waves of Influence in Postclassic Mesoamerica? A Critique of the Mixteca-Puebla Concept. *Anthropology* 4(2):15–48. Department of Anthropology, State University of New York at Stony Brook.

Snedaker, Samuel C.

1971 The Calón Shell Mound: An Ecological Anachronism. In *Archaeological Reconnaissance and Excavations in the Marismas Nacionales, Sinaloa and Nayarit, Mexico. West Mexican Prehistory*, edited by Stuart D. Scott, Pt. 5:15–18. Mimeograph, Department of Anthropology, State University of New York at Buffalo.

Snow, David H.

1982 The Rio Grande Glaze, Matte-Paint, and Plainware Tradition. In *Southwestern Ceramics: A Comparative Review*, edited by Albert H. Schroeder, pp. 235–278. *Arizona Archaeologist* 15. Arizona Archaeological Society, Phoenix.

Soustelle, Jacques

1980 *Los Aztecas*. S.A. Editions, Oikos-Tau, Barcelona, Spain.

Spence, Michael W., Phil C. Weigand, and Dolores Soto de Arechavaleta

1980 Obsidian Exchange in West Mexico. In *Rutas de Intercambio en Mesoamerica y Norte de México*, Vol. I:357–361. XVI Reunión de Mesa Redonda, Saltillo, 1979. Sociedad Mexicana de Antropología, Mexico City.

Spicer, Edward H.

1962 *Cycles of Conquest: The Impact of Spain, Mexico, and*

the United States on the Indians of the Southwest, 1533–1960. University of Arizona Press, Tucson.
1969 Northwest Mexico: Introduction. In *Ethnology Pt. II*, edited by Evon Z. Vogt, pp. 777–791. Handbook of Middle American Indians, Vol. 8, Robert Wauchope, general editor, University of Texas Press, Austin.

Spielmann, Katherine A.
1990 Diet and Subsistence Studies. In *A Design for Salado Research*, edited by G. E. Rice, pp. 123–127. Roosevelt Monograph Series 1, Anthropological Field Studies 22. Office of Cultural Resource Management, Department of Anthropology, Arizona State University, Tempe.

Stark, Barbara L.
1977 *Prehistoric Ecology at Patarata 52, Veracruz, Mexico: Adaptation to the Mangrove Swamp*. Publications in Anthropology 18. Vanderbilt University, Nashville.
2001 Figurines and Other Artifacts. In *Classic Period Mixtequilla, Veracruz, Mexico: Diachronic Inferences from Residential Investigations*, edited by Barbara L. Stark, pp. 179–226. Institute for Mesoamerican Studies 12. University at Albany, Albany.

Stark, Barbara L. (editor)
2001 *Classic Period Mixtequilla, Veracruz, Mexico: Diachronic Inferences from Residential Investigations*. Institute for Mesoamerican Studies 12. University at Albany, Albany.

Stark, Barbara L., and Barbara Voorhies
1978 Future Research Directions. In *Prehistoric Coastal Adaptations: The Economy and Ecology of Maritime Middle America*, edited by Barbara L. Stark and Barbara Voorhies, pp. 275–304. Academic Press, New York.

Stark, Barbara L., and Barbara Voorhies (editors)
1978 *Prehistoric Coastal Adaptations: The Economy and Ecology of Maritime Middle America*. Academic Press, New York.

Steele, D. Gentry, and Claud A. Bramblett
1988 *The Anatomy and Biology of the Human Skeleton*. Texas A&M University Press, College Station.

Stevenson, Christopher M., Mike Gottesman, and Michael Macko
2000 Redefining the Working Assumptions of Obsidian Hydration Dating. *Journal of California and Great Basin Anthropology* 22:223–236. Malki Museum, Banning.

Stevenson, Christopher M., Elizabeth Knaus, James J. Mazer, and John K. Bates
1993 Homogeneity of Water Content in Obsidian from the Coso Volcanic Field: Implications for Obsidian Hydration Dating. *Geoarchaeology* 8(5):371–384.

Stevenson, Christopher M., James J. Mazer, and Barry E. Scheetz
1998 Laboratory Obsidian Hydration Rates. In *Archaeological Obsidian Studies: Method and Theory*, edited by M. Steven Shackley, pp. 181–204. Advances in Archaeological and Museum Science Series. Plenum Publishing Co., New York.

Stewart, Thomas D.
1941 New Examples of Tooth Mutilation from Middle America. *American Journal of Physical Anthropology* 28(1):117–124.
1978 The Decoration of Teeth in Pre-Columbian Societies. *Dental Dimensions* 2(1):36.
1979 *Essentials of Forensic Anthropology*. Charles C. Thomas, Springfield, Illinois.

St. George, Patricia
1971 A Glazed Rock Anomaly from Western Mexico. In *Archaeological Reconnaissance and Excavations in the Marismas Nacionales, Sinaloa and Nayarit, Mexico. West Mexican Prehistory*, edited by Stuart D. Scott, Pt. 5:11–14. Mimeograph, Department of Anthropology, State University of New York at Buffalo.

Stocker, Terry
1983 Clay Figurines from Tula, Hidalgo, Mexico. Unpublished Ph.D. dissertation, Department of Anthropology, University of Illinois, Urbana.

Stone, Tammy, and Michael S. Foster
1994 Miscellaneous Artifacts. In *The Pueblo Grande Project, Vol. 4: Material Culture*, edited by Michael S. Foster, pp. 203–262. Soil Systems Publications in Archaeology No. 20. Soil Systems, Phoenix.

Stuiver, Minze, Gordon W. Pearson, and Thomas F. Braziunas
1986 Radiocarbon Age Calibration of Marine Samples Back to 9000 cal yr BP. *Radiocarbon* 28(2B):980–1021.

Stuiver, Minze, Paula J. Reimer, and Ron Reimer
2005 *CALIB Radiocarbon Calibration. Marine Reservoir Correction Database*. Electronic document, http://calib.org/calib/.

Sullivan, J. Timothy
1980 A Geophysical Study of the Continental Margin, Southeastern Gulf of California, Mexico. Unpublished Master's thesis, Department of Geology, State University of New York at Buffalo.

Sweet, A. Porter S., Michael G. Buonocore, and Irwin F. Buck
1963 Pre-Hispanic Indian Dentistry. *Dental Radiography and Photography* 36(1):3–22.

Sweetman, Rosemary
1968 Pottery Types: Sequence and Geographic Connections. In *Archaeological Reconnaissance and Excavations in the Marismas Nacionales, Sinaloa and Nayarit, Mexico. West Mexican Prehistory*, edited by Stuart D. Scott, Pt. 2:43–149. Mimeograph, Department of Anthropology, State University of New York at Buffalo.
1970 Analysis of Pottery from Excavations in Southern Sinaloa and Northern Nayarit, Mexico. Unpublished Master's thesis, Department of Anthropology, State University of New York at Buffalo.
1974 Prehistoric Pottery from Coastal Sinaloa and Nayarit. In *The Archaeology of West Mexico*, edited by Betty B. Bell, pp. 68–82. Sociedad de Estudios Avanzados del Occidente de México, Ajijic, Mexico.

Szuter, Christine R.
1991 Hunting by Hohokam Desert Farmers. *Kiva* 56: 227–291.

Tainter, Joseph A.
1978 Mortuary Practices and the Study of Prehistoric Social Systems. *Advances in Archaeological Method and Theory* 1:105–141.

Talavera, Arturo, and Nancy Gelover
2008 Algunas Características Bioarqueológicas de los Hacedores del Complejo Aztatlán de la Planicie Costera del Norte de Nayarit y Sur de Sinaloa. Paper presented at Primer Seminario-Taller Regional de la Arqueología Aztatlán, Museo Regional de Guadalajara, Guadalajara, Mexico, September 11–13.

Talavera González, Jorge Arturo
1997 Resumen del informe bioarqueológico de una de cinco urnas funerarias, recuperadas en la localidad de Chametla, Rosario, Sinaloa. *Boletín Informativo*. Centro Instituto Nacional de Antropología e Historia, Sinaloa, Culiacán.

Taylor, Steve, and Doug Rautenkranz
1971 Excavations on Panales, 1971. In *Archaeological Reconnaissance and Excavations in the Marismas Nacionales, Sinaloa and Nayarit, Mexico. West Mexican Prehistory*, edited by Stuart D. Scott, Pt. 5:25–36. Mimeograph, Department of Anthropology, State University of New York at Buffalo.

Teague, Lynn S.
1998 *Textiles in Southwestern Prehistory*. University of New Mexico Press, Albuquerque.

Tello, Antonio
1891 *Crónica Miscelánea en que se trata de la Conquista Espiritual y Temporal de la Sancta Provincia de Xalisco en el Nuevo Reino de Galicia y Nueva Vizcaya y Descubrimiento del Nuevo México*. Partial publication of Tello's 1653 report. La República Literaria de C. L. De Guevaray Cía., Guadalajara, Mexico.
1945 *Crónica Miscelánea de la Sancta Provincia de Xalisco, Libro IV*. Edited by J. C. Franco. Editorial Font, Guadalajara, Mexico.
1974 *Crónica Miscelánea de la Sancta Provincia de Xalisco*. In *El Prehispanico en Occidente Segun (Recopilación de textos Los Historia de res neo gallegos)*, compiled by José Maria Muriá. Departmento de Historia del Centro Regional de Occidente, Guadalajara, Mexico.

Thomas, David H.
1971 On Distinguishing Natural from Cultural Bone in Archaeological Sites. *American Antiquity* 36(3): 366–371.

Toro, Luis Aranda del
1952 *Nayarit: Aportación para Algunos Capítulos de la Historia*. Mexico City.

Torres, Francisco Mariano de
1965 *Crónica de la Sancta Provincia de Xalisco Escrita por Fray Francisco Mariano de Torres 1755*. Edited by Luis del Refugio de Palacio. H. Ayuntamiento de la Ciudad de Guadalajara e Instituto Jalisciense de Antropología e Historia, Serie de Historia, 7. Guadalajara, Mexico.

Tower, Donald B.
1945 *The Use of Marine Mollusca and Their Value in Reconstructing Prehistoric Trade Routes in the American Southwest*. Papers of the Excavators Club 2(3). Cambridge.

Townsend, Richard F. (editor)
1998 *Ancient West Mexico: Art and Archaeology of the Unknown Past*. Art Institute of Chicago, Chicago; and Thames and Hudson, London.

Trubitt, Mary Beth D.
2003 The Production and Exchange of Marine Shell Prestige Goods. *Journal of Archaeological Research* 11(3):243–277.

Turner, Christy G., II
1979 Dental Anthropological Indications of Agriculture Among the Jomon People of Central Japan. *American Journal of Physical Anthropology* 51(4):619–635.

Uphof, Johannes C. T.
1968 *Dictionary of Economic Plants*. J. Cramer, Lehre, Germany.

Vaillant, George C.
1935 Excavations at El Arbolillo: Pt. II. *Anthropological Papers of the American Museum of Natural History* 35:137–279.
1940 Patterns in Middle American Archaeology. In *The Maya and Their Neighbors*, edited by Clarence L. Hay, Ralph L. Linton, Samuel K. Lothrop, Harry L. Shapiro, and George C. Vaillant, pp. 295–305. Appleton-Century, New York.

Van Horne, John
1936 Fray Antonio Tello, Historian. *Hispania* 19(2):191–200.

Vann, John H.
1972 *Physical Geography of the Pacific Coastal Lowland of Sinaloa and Nayarit, Mexico*. Final report for the Office of Naval Research, Project NR 388-028. Office of Naval Research, Washington, D.C.

Verna, Emeline, M. Piercecchi-Marti, Kathia Chaumoitre, Christophe Bartoli, Georges Leonetti, and Pascal Adalian
2013 Discrete Traits of the Sternum and Ribs: A Useful Contribution to Identification in Forensic Anthropology and Medicine. *Journal of Forensic Sciences* 58(3):571–577.

Vidal Aldana, C. I.
2011 El Intercambio en el Noroccidente Prehispánico. La Relación Entre la Rama Guadiana de la Tradición Arqueológica Chalchihuites y la Tradición Aztatlán, Entre El 600–1300 D.C. Unpublished Ph.D. dissertation, Escuela Nacional de Antropología e Historia, Mexico City.

Villa-Señor y Sánchez, Joseph Antonio
1746 *Teatro Americano: Descripción General de los Reinos, y Provincias de la Nueva España, y sus jurisdicciones*. Joseph Bernardo de Hogal, Mexico City. Electronic document, http://www.cervantesvirtual.com/obra/teatro-americano-descripcion-general-de-los-reinos-y-provincias-de-la-nueva-espana-y-sus-jurisdicciones-su-author-d-joseph-antonio-de-villasenor-y-sanchez--0/.
1952 *Teatro Americano: Descripción General de los Reinos y Provincias de la Nueva España y sus Jurisdicciónes, II*. Editorial Nacional, S.A., Mexico City.

Vivó Escoto, Jorge A.
1964 Weather and Climate of Mexico and Central America. In *Natural Environment and Early Cultures*, edited by Robert C. West, pp. 187–215. Handbook of

Middle American Indians, Vol. 1, Robert Wauchope, general editor, University of Texas Press, Austin.

Volkmann, John, and George Gill
1970 Continued Excavations at Tecualilla, 1970. In *Archaeological Reconnaissance and Excavations in the Marismas Nacionales, Sinaloa and Nayarit, Mexico. West Mexican Prehistory*, edited by Stuart D. Scott, Pt. 4:45–74. Mimeograph, Department of Anthropology, State University of New York at Buffalo.

Von Winning, Herbert
1956 Offerings from a Burial Mound in Coastal Nayarit. *Masterkey* 30:157–170.

Voorhies, Barbara
1976 *The Chantuto People: An Archaic Period Society of the Chiapas Littoral, Mexico*. Papers of the New World Archaeological Foundation, No. 41. Brigham Young University, Provo.

Voorhies, Barbara (editor)
1989 *Ancient Trade and Tribute: Economies of the Soconusco Region of Mesoamerica*. University of Utah Press, Salt Lake City.
2000 *Reconstructing Mobility Patterns of Late Hunter-Gatherers in Coastal Chiapas, Mexico: The View from the Shellmounds. Anais do IX Congresso da Sociedade de Arqueologia Brasleira. Rio de Janerio, August 2000*. CD-ROM.

Voorhies, Barbara, and Douglas J. Kennett
1995 Burial Sites on the Soconusco Coastal Plain, Chiapas, Mexico. *Journal of Field Archaeology* 22(1):65–79.

Voorhies, Barbara, Douglas J. Kennett, John G. Jones, and Thomas A. Wake
2002 A Middle Archaic Archaeological Site on the West Coast of Mexico. *Latin American Antiquity* 13(2): 179–200.

Wagner, Philip L.
1964 Natural Vegetation of Middle America. In *Natural Environment and Early Cultures*, edited by Robert C. West, pp. 216–264. Handbook of Middle American Indians, Vol. 1, Robert Wauchope, general editor, University of Texas Press, Austin.

Walker, Phillip L., Rhonda R. Bathurst, Rebecca Richman, Thor Gjerdrum, and Valerie A. Andrushko
2009 The Causes of Porotic Hyperostosis and Cribra Orbitalia: A Reappraisal of the Iron-Deficiency-Anemia Hypothesis. *American Journal of Physical Anthropology* 139(2):109–125.

Ward, Sue Palmer
1967 Report on Mesoamerican Pottery Analysis. Mimeograph.

Warner, Rick
2001 Jesuits and Coras in Colonial Nayarit: Negotiating Evangelization. Paper presented at the Latin American Studies Association, Washington, D.C., September 6–8.

Waselkov, Gregory A.
1987 Shellfish Gathering and Shell Midden Archaeology. In *Advances in Archaeological Method and Theory* 10:93–210. Academic Press, San Diego.

Webb, William S., and Charles E. Snow
1981 *The Adena People*. University of Tennessee Press, Knoxville.

Weigand, Phil C.
1975a Aboriginal West Mexican Glazes and Their Possible Relationship to the Southwest. *Pottery Southwest* 2:5–6.
1975b Correspondence from Weigand to Scott. On file, Marismas Nacionales Project Files, Center for Archaeology and Society Repository, School of Human Evolution and Social Change, Arizona State University, Tempe.
1981 Differential Acculturation Among the Huichol Indians. In *Themes of Indigenous Acculturation in Northwest Mexico*, edited by Phil C. Weigand and Thomas B. Hinton, pp. 9–21. Anthropological Papers of the University of Arizona, No. 38. Tucson.
1985 Considerations on the Archaeology and Ethnohistory of the Mexicaneros, Tequales, Coras, Huicholes, and Caxcanes of Nayarit, Jalisco, and Zacatecas. In *Contributions to the Archaeology and Ethnohistory of Greater Mesoamerica*, edited by William J. Folan, pp. 126–187. Southern Illinois University Press, Carbondale.
1990 *Preliminary Observations on an Iron Silicate Sample from Feature 586, Pueblo Grande*. Report submitted to Soil Systems, Phoenix.

Weigand, Phil C., and Michael W. Spence
1982 The Obsidian Mining Complex at La Joya, Jalisco. In *Mining and Mining Techniques in Ancient Mesoamerica*, edited by Phil C. Weigand and G. Gwynne, pp. 175–188. Special Issue of *Anthropology* 6(1–2). State University of New York, Stony Brook.

Weigand, Phil C., and Acelia G. de Weigand
1991 Death and Mourning Among the Huicholes of Western Mexico. In *Coping with the Final Tragedy: Cultural Variations in Dying and Grieving*, edited by David Counts and Dorothy Counts, pp. 53–69. Baywood Publishing, Amityville, New York.
2000 Huichol Society Before the Arrival of the Spanish. *Journal of the Southwest* 42(1):13–36.

Weil, Edward B.
1974 Site Location and Local Ceramic Chronology in the Marismas Nacionales, Sinaloa and Nayarit, Mexico. Unpublished Master's thesis, Department of Anthropology, State University of New York at Buffalo.

West, Robert C.
1964 The Natural Regions of Middle America. In *Natural Environment and Early Cultures*, edited by Robert C. West, pp. 363–383. Handbook of Middle American Indians, Vol. 1, Robert Wauchope, general editor, University of Texas Press, Austin.

White, Timothy D., Michael T. Black, and Pieter A. Folkens
2012 *Human Osteology*. 3rd ed. Academic Press, Boston.

Williams, J. A., and J. Signe Snortland-Coles
1986 Pre-Contact Tuberculosis in a Plains Woodland Mortuary. *Plains Anthropologist* 31(113):249–252.

Wing, Elizabeth S.
1968 Preliminary Note on the Faunal Remains Excavated from Several Sites in Sinaloa and Nayarit, Mexico. In *Archaeological Reconnaissance and Excavations in the Marismas Nacionales, Sinaloa and Nayarit, Mexico. West Mexican Prehistory*, edited by Stuart D. Scott, Pt. 2:150–152. Mimeograph, Department of Anthropology, State University of New York at Buffalo.
1969 Preliminary Analysis of the Subsistence Economy of Two Small Communities in the Marismas Nacionales of Sinaloa and Nayarit, Mexico. In *Archaeological Reconnaissance and Excavations in the Marismas Nacionales, Sinaloa and Nayarit, Mexico. West Mexican Prehistory*, edited by Stuart D. Scott, Pt. 3:96–109. Mimeograph, Department of Anthropology, State University of New York at Buffalo.
1978 Use of Dogs for Food: An Adaptation to the Coastal Environment. In *Prehistoric Coastal Adaptations: The Economy and Ecology of Maritime Middle America*, edited by Barbara L. Stark and Barbara Voorhies, pp. 29–41. Academic Press, New York.
2013 Dog Remains from the Marismas Nacionales. In *The Archaeology of Mesoamerican Animals*, edited by Christopher Götz and Kitty Emery, pp. 23–47. Lockwood Press, Atlanta.

Winter, Joseph C.
2000a Botanical Descriptions of North American Tobacco Species. In *Tobacco Use by Native Americans: Sacred Smoke and Silent Killer*, edited by Joseph C. Winter, pp. 87–127. University of Oklahoma Press, Norman.
2000b Traditional Uses of Tobacco by Native Americans. In *Tobacco Use by Native Americans: Sacred Smoke and Silent Killer*, edited by Joseph C. Winter, pp. 9–58. University of Oklahoma Press, Norman.

Wolynec, Renata B.
1970a Obsidian Hydration Dating: A Practical Application. Senior thesis project, Department of Anthropology, State University of New York at Buffalo.
1970b Arriñitas Site Report. In *Archaeological Reconnaissance and Excavations in the Marismas Nacionales, Sinaloa and Nayarit, Mexico. West Mexican Prehistory*, edited by Stuart D. Scott, Pt. 4:33–44. Mimeograph, Department of Anthropology, State University of New York at Buffalo.

Wolynec, Renata B., Terry Block, Andrew Anderson, and Betty Davis
1968 Juana Gomez. In *Archaeological Reconnaissance and Excavations in the Marismas Nacionales, Sinaloa and Nayarit, Mexico. West Mexican Prehistory*, edited by Stuart D. Scott, Pt. 2:12–21. Mimeograph, Department of Anthropology, State University of New York at Buffalo.

Woosley, Anne I., and Allan J. McIntyre
1996 *Mimbres Mogollon Archaeology: Charles C. Di Peso's Excavations at Wind Mountain*. University of New Mexico Press, Albuquerque.

Wright, Norman P.
1960 *El Enigma del Xoloitzcuintli*. Instituto Nacional de Antropología e Historia, Mexico City.

Yorston, R. M., Christopher F. Gaffney, and Vincent L. Gaffney
1987 A Manifesto for Pragmatic Archaeology. In *Pragmatic Archaeology: Theory in Crisis?* edited by Christopher F. Gaffney and Vincent L. Gaffney, pp. 107–114. British Archaeological Reports 167. Oxford.

Zarky, Alan
1976 Statistical Analysis of Site Catchments at Ocós, Guatemala. In *The Early Mesoamerican Village*, edited by Kent V. Flannery, pp. 117–128. Academic Press, New York.

Zeitlin, Judith
1989 Ranchers and Indians on the Southern Isthmus of Tehuantepec: Economic Change and Indigenous Survival in Colonial Mexico. *Hispanic American Historical Review* 69:23–60.

Zitt, Jennifer O.
2008 Prehistoric and Early Historic Subsistence Patterns and Dental Pathology of the Northern Great Plains. In *Skeletal Biology and Bioarchaeology of the Northwestern Plains*, edited by George W. Gill and R. L. Weathermon, pp. 226–232. University of Utah Press, Salt Lake City.

Maps

Álvarez Barreiro, Francisco
1770 [1726] *Nueva Vizcaya*. Copy from the U.S. Library of Congress, Washington, D.C. (Original in the Cultural Deporte Archivo General de Indies M.P., 121, Ministerio de Educación, Madrid).

Castillo, Domingo del
1541 *The Gulf of California Chart*. Map produced during the 1540–1542 exploration of Hernando de Alarcón. *Descripción de esta parte de la America Septentrional, que se llama Nueva España*
1691 In *Real Academia de la Historia*, Vol. 8. Colección Boturnini. Madrid.

Garavito, Diego Joaquín
1969 [1763] *Mapa particular de las costas del mar del Sur*. In "Holocene History of a Strand Plain, Lagoonal Coast, Nayarit, México," by Joseph R. Curray, F. J. Emmel, and Perry J. Crampton. In *Costeras, Lagunas, a Symposium*, edited by A. A. Castanares and F. B. Phleger, pp. 62–100. Universidad Nacional Autónoma de México, Mexico City. (Original in the British Museum, London.)

Gutiérrez, Diego
1949 [1562] *Americae sive qvartae orbis partis nova et exactissima descriptio*. Rosenwald Collection, Library of

Congress, No. 1303. *Quarterly Journal of the Library of Congress* 6(3), May: 18–20.

Mapa de Acaponeta

1961 [1777] In *Cartografía de la Nueva Galicia*. Ediciónes del Banco Industrial de Jalisco, Guadalajara, Mexico.

Martínez, Juan

1951 [1587a] *Carta del Pacífico y de las costas de América*. In *Mapas Españoles de America Siglos XV–XVII*. Madrid.

1951 [1587b] *Nueva España Occidental y Costa del Pacico*. In *Mapas Españoles de America Siglos XV–XVII*. Madrid.

Narváez, José María

1858 [1824] *Carta corográfica de Jalisco, Zacatecas y Colima. Boletín de la Sociedad Mexicana de Geografía y Estadística de la México*6. Mexico City. (Original at http://bdmx.mx/detalle_documento/?id_cod=49#.KZnyhBBlc.)

Ortelius, Abraham

1570 [1564] *America Sive Novi Orbis, Nova Descriptio*. In *Theatrum Orbis Terrarum*. Apud Gillis Coppens van Diest, Antwerp.

1579 *Hispaniae Novae Sive Magnae Recens Et Vera Descriptio*. In *Theatrum Orbis Terrarum*, 2nd additament. Antwerp.

Contributors

Michael S. Foster earned his Ph.D. from the University of Colorado, Boulder, in 1978. He has participated in and directed numerous projects in west and northwest Mexico, Central America, the American Southwest, and the mountain west of the United States. He retired in 2013 and resides in Phoenix, Arizona.

Stuart D. Scott earned a Ph.D. from the University of Arizona in 1963 and in 1997 retired from the State University of New York, where he is an associate professor emeritus. He initiated and directed the Marismas Nacionales project and has published extensively on west Mexico; on Old Fort Niagara, New York; on Americans and Canadians transported to Tasmania in the 1840s; and on the modern pottery of Mata Ortiz, Chihuahua, Mexico. He lives in Tucson, Arizona.

George W. Gill is a Distinguished Professor Emeritus with the Department of Anthropology, University of Wyoming, where he taught bioanthropology for 40 years. He studied prehistoric human skeletal remains from west Mexico, Easter Island, and the Great Plains of North America, as well as analyzing human skeletal remains for various law enforcement agencies and the U.S. military. He resides in Laramie, Wyoming.

Patricia Kay Scott graduated from the University of Buffalo in 1967 with a B.A. in anthropology. Her Marismas Nacionales research was conducted in the late 1960s and early 1970s in archives and libraries in Mexico City and Guadalajara. In the following years she devoted her research and writing to the history and archaeology of Old Fort Niagara in Youngstown, New York. She lives in Tucson, Arizona.

Henry W. Case earned his M.A. in anthropology at the University of Wyoming in 1976, with additional graduate studies at the University of Tennessee. He is now retired, after more than 40 years of bioanthropological research with several research firms and the Department of the Army. He resides in Framingham, Massachusetts.

Index

Numbers in *italics* indicate figures and tables